Springer Series on Biofilms

Volume 11

Springer Series on Biofilms presents carefully refereed volumes on selected topics on this field of research. All volumes reflect the latest findings and developments. Once anchored to a surface, biofilm microorganisms carry out a variety of detrimental (like the plaque on teeth) or beneficial (bioremediating hazardous waste sites) reactions. International experts will write, or edit, volumes on specific aspects of biofilms, like research methods, or the roles of biofilms in diseases. While based in microbiology, biofilms are of intense interest to many other scientists, because they affect whole ecosystems, and lie at the root of chronic bacterial infections.

Katharina Richter • Kasper Nørskov Kragh
Editors

Antibiofilm Strategies

Current and Future Applications to Prevent, Control and Eradicate Biofilms

Springer

Editors
Katharina Richter
Basil Hetzel Institute for Translational Health Research and The Queen Elizabeth Hospital
The University of Adelaide
Adelaide, SA, Australia

Kasper Nørskov Kragh
Costerton Biofilm Center
University of Copenhagen
Copenhagen, Denmark

ISSN 1863-9607 ISSN 1863-9615 (electronic)
Springer Series on Biofilms
ISBN 978-3-031-10991-1 ISBN 978-3-031-10992-8 (eBook)
https://doi.org/10.1007/978-3-031-10992-8

This Springer imprint is published by the registered company Springer Nature Switzerland AG
The registered company address is: Gewerbestrasse 11, 6330 Cham, Switzerland

Preface

Bacteria and fungi have the ability to aggregate together or on surfaces in densely packed microcolonies, facilitated by extracellular polymeric substances protecting cells within and holding the cells together. These biofilms have proven extremely hard to eradicate and remove when first established. In chronic infections, this condition can result in a high degree of morbidity and mortality as regular antibiotic treatments are ineffective against biofilms. In industrial facilities, the formation of biofilms can ruin production and result in enormous financial losses.

The last few decades have seen comprehensive research into new strategies and treatments to prevent, control, and eradicate biofilms. In this book, we have assembled experts from around the world to present the current state of antibiofilm research. Novel cutting-edge techniques and new optimized strategies based on more established methods will be presented in chapters focused on biofilm prevention, treatment, and control for application in clinical, industrial, and veterinary settings. Antibiofilm strategies such as chemical and enzymatic treatments, surface modification and coatings, quorum sensing inhibition and dispersal induction, phage therapy, cold plasma treatment, as well as hyperbaric oxygen treatment, and metal-based nanomedicine will be covered among many others.

Adelaide, Australia — Katharina Richter
Copenhagen, Denmark — Kasper Nørskov Kragh

Keywords

Biofilms, therapeutics, infectious diseases, antibiofilm, biofilm treatment, chronic infections, antibiotic tolerance, antimicrobial resistance, bacteriophages, biofouling

Acknowledgments

We kindly acknowledge the following people who assisted in the peer review of all book chapters.

Adrian Abdo, University of Adelaide, Australia
Aurélie Crabbé, Ghent University, Belgium
Bastiaan Krom, Academic Centre for Dentistry Amsterdam (ACTA), The Netherlands
Birthe Kjellerup, University of Maryland, USA
Darla Goeres, Montana State University, USA
David Cameron, Bern University Hospital, Switzerland
Eva Armengol Rivero, University of Barcelona, Spain
Gordon Ramage, University of Glasgow, UK
Jason Brown, University of Glasgow, UK
Jeffrey Carey, University of Washington, USA
Karin Sauer, Binghamton University, USA
Karishma Kaushik, Savitribai Phule Pune University, India
Kasper Nørskov Kragh, University of Copenhagen, Denmark
Katharina Richter, University of Adelaide, Australia
Kendra Rumbaugh, Texas Tech University, USA
Lasse Kvich, University of Copenhagen, Denmark
Livia Tenuta, University of Michigan, USA
Lucas Pedraz, University of British Columbia, Canada
Melanie Dostert, University of British Columbia, Canada
Melissa Brown, Flinders University, Australia
Mette Kolpen, Rigshospitalet, Copenhagen, Denmark
Michel Hoogenkamp, Academic Centre for Dentistry Amsterdam (ACTA), The Netherlands
Mirja Nolff, University of Zurich, Switzerland
Oana Ciofu, University of Copenhagen, Denmark
Peter Østrup Jensen, University of Copenhagen, Denmark

Rebekah Dedrick, University of Pittsburgh, USA
Sepideh Zununi Vahed, Tabriz University of Medical Sciences, Iran
Simona Gatej, University of Adelaide, Australia
Sonia Troeira Henriques, Queensland University of Technology, Australia
Stefania Vitale, University of Strasbourg, France
Tagbo Niepa, University of Pittsburgh, USA
Tim Holm Jakobsen, University of Copenhagen, Denmark
Tim Tolker-Nielsen, University of Copenhagen, Denmark
Tom Coenye, Ghent University, Belgium
Trine Rolighed Thomsen, Aalborg University, Denmark
Vaughn Cooper, University of Pittsburgh, USA
Yok-Ai Que, Bern University Hospital, Switzerland
Yue Qu, Monash University, Australia

Contents

About the Editors

Katharina Richter is an enthusiastic biomedical scientist, dedicated to improving therapies against antibiotic-resistant bacteria and biofilms. She has global work and research stints in Germany, Switzerland, Denmark, Belgium, New Zealand, and Australia. A trained pharmacist she completed her PhD in medicine and applied microbiology in 2017 and started her own research group in 2019 at the University of Adelaide, Australia. Her group has two priorities: (i) developing new treatments against biofilm infections and bringing them from bench to bedside and (ii) improving the health literacy of society by effective science communication through public speaking, science outreach activities, and STEM workshops at schools. Katharina collaborates with scientists, clinicians, and industry partners to ensure a real-life impact of her work.

Kasper Nørskov Kragh obtained a Master's in Science and a PhD in Health and Medical Sciences from the University of Copenhagen, Denmark. He is now an assistant professor at the Costerton Biofilm Center, University of Copenhagen. During his scientific career, Kasper has focused his attention on the use of microscopy in biofilm research and how the phenotype of non-attached biofilm aggregates is involved and influences both pathogenesis and *in vitro* systems. The study of the mechanisms behind the chronic infectious state of biofilms, in *in vivo*-like *in vitro* models and direct infections, has been a red thread throughout his career.

Part I
Introduction

Introduction: Biofilms 101

Kasper Nørskov Kragh and Katharina Richter

Abstract Biofilm is the ancient way of life for bacteria. A biofilm is a group of bacteria which have clumped together in a dense colony surrounded by a self-produced matrix. This form of life has proven to be particularly advantageous for bacteria by enabling them to live in a favorable and stable environment while protecting them from external threats. The biofilm mode of growth protects bacteria against anti-microbial agents, such as antibiotics, by increasing their tolerance. Additionally, a biofilm will enable pathogenic bacteria to repel the host's immune system by resisting phagocytosis and attacks with anti-microbial substances, such as reactive oxygen and nitrogen species. This ability to withstand the immune response and antibiotic treatment means that infections of bacteria living in biofilms are complicated to overcome. As clinicians, biologists and engineers become more aware of the biofilm mode of growth for bacteria, the amount of research dedicated to eradicating them has increased significantly.

1 History of Biofilms

Microorganisms attaching in large quantities to a surface were already described by Antonie van Leeuwenhoek (1632–1723) in 1684 in a report to the Royal Society on dental plaques. Since then, the work of great microbiologists, such as Robert Koch (1843–1910), Louis Pasteur (1822–1895), and others have expanded our knowledge of the correlation between bacteria, sanitation, and infections. The first century of experimental microbiology has primarily used pure cultures for investigations, with

K. N. Kragh
Costerton Biofilm Center, University of Copenhagen, Copenhagen, Denmark
e-mail: kct@sund.ku.dk

K. Richter (✉)
Basil Hetzel Institute for Translational Health Research and The Queen Elizabeth Hospital, The University of Adelaide, Adelaide, Australia
e-mail: Katharina.Richter@adelaide.edu.au

K. Richter, K. Nørskov Kragh (eds.), *Antibiofilm Strategies*, Springer Series on Biofilms 11, https://doi.org/10.1007/978-3-031-10992-8_1

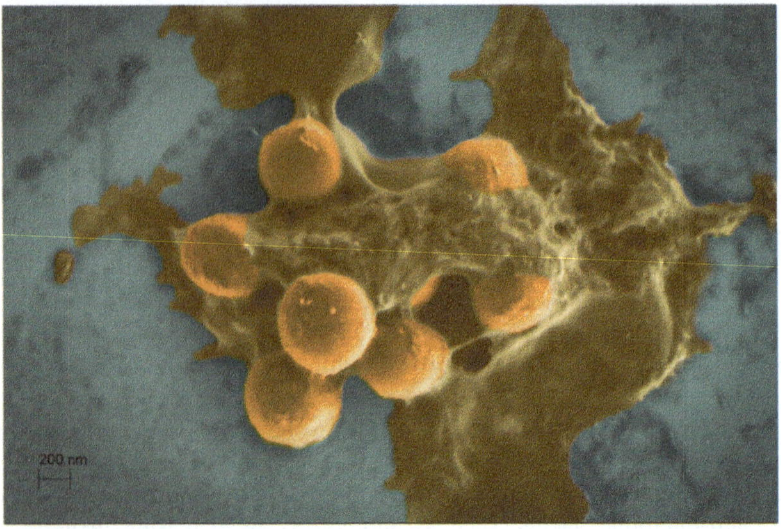

Fig. 1 Scanning electron microscopy image of *Staphylococcus aureus* biofilm. Courtesy of Dr. Katharina Richter

the implied understanding that these cultures consisted of single-celled planktonic organisms (Center for Biofilm Engineering n.d.; Costerton 1999a).

The Golden Age of research into bacteria aggregating together, so-called biofilms (Fig. 1), really started with Bill Costerton's (1934–2012) defining paper in 1978, *How Bacteria Stick* (Costerton et al. 1978). However, the term "biofilm" was previously used in 1975 to describe microbial film development on water filters (Mack et al. 1975). In 1977, Niels Høiby described "heaps of bacteria" in the chronically infected lungs of patients with cystic fibrosis (CF) and, for the first time, linked aggregated bacteria to chronic infections (Hoiby et al. 1977). The last decades have shown an explosion in the number of publications dealing with biofilm. With the increased knowledge of biofilms, it has become apparent how biofilms play a very central role in most microbial life and have a key involvement in chronic infections and biofouling (Alhede and Alhede 2014; Costerton et al. 1987).

1.1 Biofilms Prevalence

Biofilms are ubiquitous, they can be found in various environments, from stones in rivers to implants in patients (Patiniott et al. 2022). Biofilm formation is a basic survival mechanism in the environment, offering protection against grazing from protozoa, UV, and providing a stable environment for the embedded microbes (Espeland and Wetzel 2001; Le Magrex-Debar et al. 2000). Biofilms may represent an ancient way of protecting a sessile population of bacteria in inhabitable environments or a way of ensuring the position of a population in a favorable environment (Hall-Stoodley et al. 2004; Westall et al. 2001).

In infections, biofilm formation has been proposed to protect bacterial cells within and enable them to stick in advantageous niches of the host, as well as provide a stable chemical environment, besides allowing the cells to survive a substantial amount of antibiotics and the host immune response (Costerton 1999a, 1999b; Costerton et al. 1995; Donlan and Costerton 2002). The National Institutes of Health estimate that 80% of the 17 million chronic infections reported in the USA each year are biofilm-related (Wolcott and Dowd 2011). However, the actual percentage may very well be closer to 100%, as biofilm may be difficult to diagnose through traditional diagnostic techniques (Høiby et al. 2015).

Both bacteria and yeast populations have a natural drive toward aggregation and biofilm formation, both under natural and artificial settings. For example, in a process called flocculation, every brewer appreciates how yeast aggregate together, forming free-floating biofilm aggregates during the fermentation of beer and wine. This yeast biofilm formation has been known for centuries and has been used to remove yeast after the end of fermentation as an integrated part of the purification process of beer or wine (Jin and Alex Speers 1998). Conversely, it is a known problem that unwanted bacteria can form biofilms in food and biotech production systems, and thereby contaminate and spoil the production.

Even under strictly controlled laboratory conditions, bacteria will prefer to form biofilms (Kragh et al. 2018). Many traditional methods of culturing bacteria have been developed to promote planktonic phase bacteria in liquid culture (Høiby 2014; Isenberg 1990). In most liquid cultures, biofilms will appear over time, most often seen as a surface-attached biofilm inside the flask or test tube, or at the culture–air interphase. *Pseudomonas aeruginosa* has been shown to aggregate in liquid batch cultures long before surfaces have been properly colonized (Schleheck et al. 2009). Even without any selective pressure for non-attached aggregates, *P. aeruginosa* will aggregate eventually. These aggregates are primarily being produced by accumulation and, to a lesser extent, by clonal growth (Alhede et al. 2011; Kragh et al. 2018). The advantage of aggregating in the liquid phase may be the stable chemical environment the biofilm colony provides floating in a liquid, as also seen in activated sludge (Bossier and Verstraete 1996).

1.2 How to Define a Biofilm

A bacterial biofilm has traditionally been defined as "a structured community of bacterial cells enclosed in a self-produced polymeric matrix and adherent to an inert or living surface". The biofilm phenotype is considered the primary mode of growth for bacteria both in natural environments and in the colonization of a host (Costerton et al. 1987).

The first three decades of biofilm research have almost exclusively been focused on surface-attached biofilms (Barken et al. 2008; Chiang et al. 2013; Costerton et al.

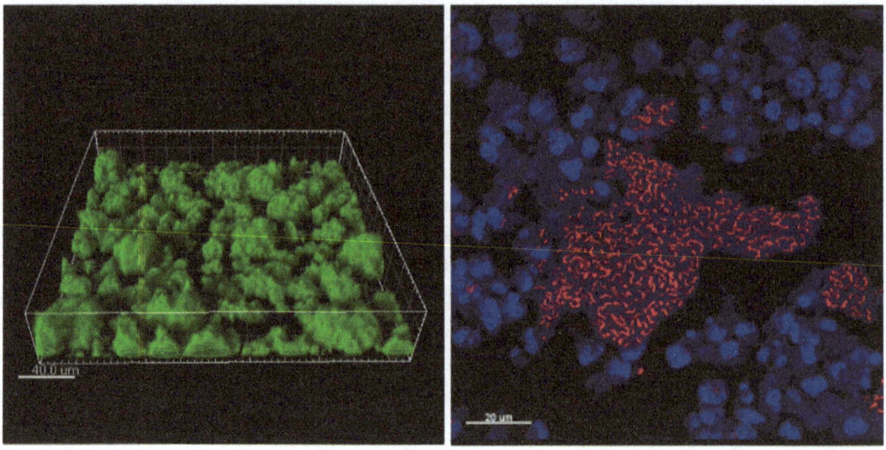

Fig. 2 Biofilms in vitro and in vivo. Left) Classically structured biofilm formed by *Pseudomonas aeruginosa* tagged with green fluorescence protein. The cells were grown in a flow cell with a constant flow of media over a surface. Right) A biofilm was imaged in ex vivo lung tissue from a patient with cystic fibrosis and chronic pneumonia. *P. aeruginosa* cells were stained with PNA-FISH probes (red). Tissue and host immune cells were counterstained with DAPI (blue). Courtesy of Dr. Kasper Kragh

1995; Flemming and Wingender 2010; Klausen et al. 2003a, b; Pamp et al. 2008; Stoodley et al. 2002). Surface attached biofilms can be observed as well-structured colonies with towering structures and water channels, as seen in Fig. 2 (Barken et al. 2008; de Beer et al. 1997; Klausen et al. 2003a; Walters et al. 2003). However, a discussion has arisen in recent years as to whether the definition of biofilm should be altered, as a biofilm consortium does not necessarily have to adhere to a surface and as mushroom-like structures have not been identified in in vivo biofilms.

2 Physiology of a Biofilm

2.1 Matrix

One of the main components of a biofilm (at least in vitro) is the matrix. The self-produced polymeric matrix provides the biofilm with most of its properties. In an in vitro biofilm, the matrix can account for over 90% of the biomass, the rest being cells, in addition to a large volume of water (Flemming and Wingender 2010). The matrix can consist of a varying blend of biopolymers, including polysaccharides, lipids, extracellular DNA (eDNA), and proteins (including extracellular enzymes and structural proteins) (Flemming et al. 2007; Flemming and Wingender 2010; Mayer et al. 1999). Besides these, secreted biopolymer cell-bound structures, such as fimbriae, flagellae, and pilii may form part of the matrix (Zogaj et al. 2001).

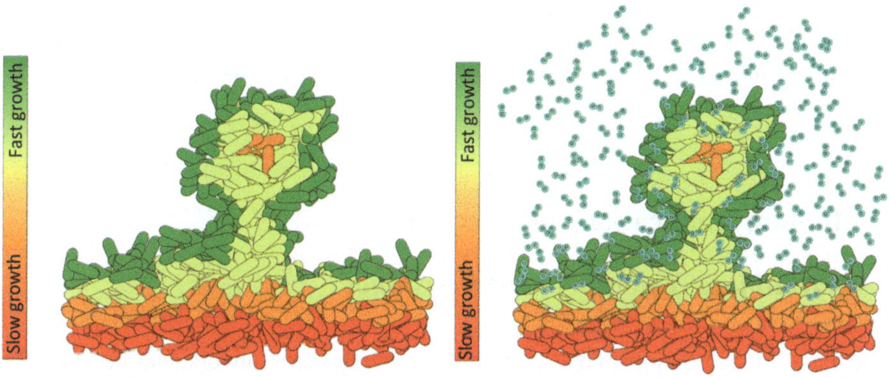

Fig. 3 Schematic drawing of chemical gradients (e.g., of nutrients, oxygen, signal molecules, and waste products) influencing growth and metabolic activity in biofilms. Left) Diffraction of growth rates from top to bottom of an in vitro biofilm. Central and deeper regions of the biofilm will have slow (orange) to non-growth (red). Right) In most in vitro biofilms, the limiting factor for growth is slow and inhibited diffusion of molecules through the biofilm. Especially, oxygen (blue circles) plays a central role for biofilm growth. The high oxygen consumption of the outer layers of cells restricts the amount of oxygen reaching the central and deeper layer of the biofilm. Courtesy of Dr. Kasper Kragh

Mechanically, the matrix enables a microcolony of cells to form a stable, cohesive structure.

2.1.1 Chemical Gradients in the Biofilm

The densely packed community of cells and matrix components can lead to a very heterogenic state of growth and metabolic activity (Fig. 3). This heterogeneity is thought to be the result of differences in the ecological and physical conditions on the micro-scale (Stewart and Franklin 2008). Conditions change because of steep chemical gradients of nutrients, oxygen, signal molecules, and waste products through the biofilm resulting from outcompeting of supply by consumption, i.e., the high consumption of nutrients and oxygen. In addition, the accumulation of waste may result from the highly dense cell community combined with the restricted diffusion through the biofilm (Folsom et al. 2010; Stewart 2013; Stewart and Franklin 2008). In particular, oxygen concentration has been shown to drop to anoxic levels within 40–50 μm from the surface of a biofilm (Fig. 3, de Beer et al. 1994; Ramsing et al. 1993; Wessel et al. 2014). These considerable differences in the local environment will lead to cells with a high degree of diversity in phenotype within the same biofilm (Folsom et al. 2010). Cells will exhibit decreased growth if they have limited access to either an electron donor (e.g., carbon source) or an electron acceptor (e.g., oxygen or nitrate) (Zumft 1997). Interestingly, these chemical gradients have only been demonstrated in in vitro biofilms. The majority of biofilms found in infections are thought to be completely anoxic (Aanaes et al. 2011;

Kolenbrander 2000). Indeed, endobronchial secretions from infected CF patients harbor immobile sources of microbial activity in the anoxic zones, suggesting the presence of biofilm with anaerobic metabolism (Cowley et al. 2015; Kolpen et al. 2014a, b).

3 Biofilm Implications in Infections

3.1 Antibiotic Tolerance and Resistance

The main driver for biofilm research has been the desire to understand the tolerance toward antimicrobial agents that biofilm cells exhibit. The three pathogenic bacteria, *E. coli*, *P. aeruginosa,* and *S. aureus,* have been shown in in vitro studies to survive >1000 times the minimal inhibitory concentration (MIC) of a wide range of antibiotics when cells are embedded in a biofilm (Ceri et al. 1999; Mah and O'Toole 2001). These tolerances are attributed to several physical and chemical mechanisms of the biofilm.

A heterogenic growth pattern in a biofilm (Fig. 3) protects the population within against attacks by antimicrobial agents. For example, an inactive sub-population will be able to survive treatment with the majority of the most common type of antibiotics (e.g., ciprofloxacin, tobramycin, etc.), which have reduced effectiveness on dormant or slow-growing cells (Evans et al. 1991; Pamp et al. 2008; Walters et al. 2003). On the contrary, other types of antibiotics, such as colistin, which targets the membrane, will achieve higher efficiency on the slow-growing sub-population than the metabolic active cells (Haagensen et al. 2007; Pamp et al. 2008). This heterogeneity in growth may result from the distinct chemical gradients within the biofilm (Folsom et al. 2010; Walters et al. 2003).

The matrix itself may act as a physical barrier that will ultimately hinder access to some types of antibiotics or considerably reduce the diffusion rate of antimicrobial agents through the biofilm (Chiang et al. 2013; de Beer et al. 1997; Gordon et al. 1988; Hatch and Schiller 1998; Suci et al. 1994). An upregulation of β-lactamase and efflux pumps can, on its own and combined with a reduced diffusion rate, facilitate resistance for cells in a biofilm (Dibdin et al. 1996; Drenkard 2003; Drenkard and Ausubel 2002). Other possible factors increasing the tolerance of a population embedding itself in a biofilm include the presence of so-called persister cells and the immobilization of antimicrobial molecules (Keren et al. 2004; Lewis 2010; Purdy Drew et al. 2009).

A detailed description of antibiotic tolerance and resistance mechanisms in bacterial biofilms can be found in the following book chapters (Chapter "Antimicrobial and innate immune tolerance mechanisms in biofilms" by Elizabeth L. Sandvik, Timothy R. Borgogna & Philip S. Stewart; Chapter "Antibiotic resistance development in bacterial biofilms" by María D. Macià & Antonio Oliver).

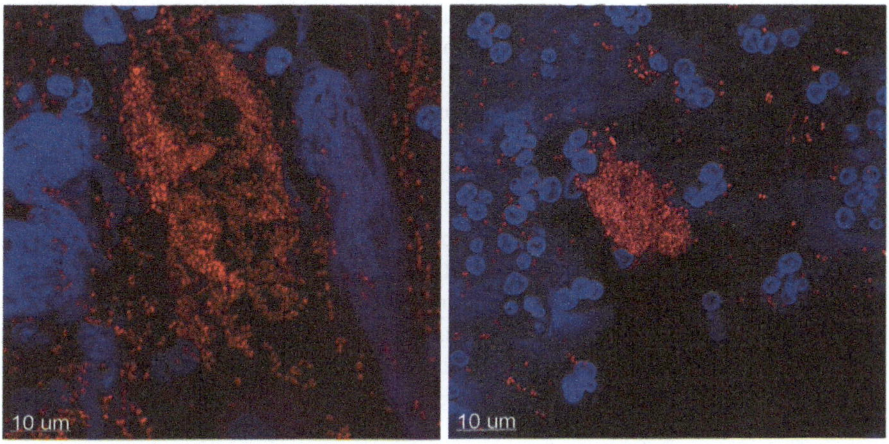

Fig. 4 Examples of biofilms in pneumonic expectorate from cystic fibrosis patients. Specimens from the patients were fixated and stained with bacterial-specific PNA-FISH probes (red), as well as counterstained with DAPI (blue). The red cells form distinct biofilms surrounded by host immune cells, primarily PMNs. Blue shadows mucus can be seen around the aggregates. Courtesy of Dr. Kasper Kragh

3.1.1 Immune Evasion/Defense

The increased tolerance toward antibiotics does not explain why the host's immune defense does not clear chronic biofilm infections. A biofilm infection will activate a severe inflammatory response, primarily from the innate immune system, with the recruitment of both macrophages and neutrophils (Wagner and Roth 2000). Polymorphonuclear leukocytes (PMNs) mainly can be observed in abundance at the site of many biofilm infections surrounding the biofilms and embedding them (Fig. 4, Alhede et al. 2009; Bjarnsholt et al. 2009; Kirketerp-Møller et al. 2008; Kragh et al. 2014; Van Gennip et al. 2009). PMNs have the ability to phagocytose pathogenic microorganisms and produce reactive oxygen species for antimicrobial use (Segal 2005). The recruitment and effect of PMNs in infected mucosa in the lungs of patients with CF levied much focus, as this response is responsible for the majority of the tissue damage (Bals et al. 1999). With the use of microscopy, the PMNs surrounding the bacterial biofilm in the mucus can be visualized (Fig. 4, Bjarnsholt et al. 2009; Kragh et al. 2014). The PMNs appear unable to penetrate the matrix, which may hamper the phagocytosis of the cells within the biofilm. This resilience against PMNs has been shown in *P. aeruginosa* to be linked to the active killing of PMNs by secreting rhamnolipids, which can act as a shield against the actions of PMNs (Jensen et al. 2007; Van Gennip et al. 2009).

It may be a bit surprising that bacterial cells can establish a biofilm in the first place. One would expect colonization of pathogenic single bacterial cells in a host to be cleared rapidly upon intrusion. Yet, bacteria are able to establish themselves in biofilms before clearing. A deviation from the normal health environment of a host can give the bacteria the opening, allowing the establishment of biofilms. Such a deviation could be the implantation of a foreign body, like an implant (Patiniott et al. 2022), immunodeficiency, compromised clearing of mucus in patients with CF, or impaired blood circulation, such as diabetes (Bjarnsholt et al. 2013; Jensen et al. 2010).

Biofilms play a central role when an infection becomes chronic. The list of chronic infections linked to the biofilm mode of growth is long and continues to expand as more and more infections are examined (Bjarnsholt et al. 2013; Costerton 1999a, 1999b). Biofilms have been identified, e.g., in chronic wounds, soft tissue fillers, chronic otitis media, on catheters and stents, pacemakers, implants, endodontic infections, and chronic pneumonia in patients with CF (Kaul et al. 2021; Patiniott et al. 2022; Jacombs et al. 2020; Bjarnsholt et al. 2009; Chávez de Paz et al. 2010; Ciofu et al. 2013; Costerton et al. 2011; Høiby et al. 2015; Homøe et al. 2009; Kirketerp-Møller et al. 2008; Marrie and Costerton 1984a, 1984b; Parsa et al. 2010). Biofilms appear very difficult to clear in clinical infections due to high resilience toward the immune response, and high tolerance and resistance toward antibiotic treatment.

The last decades have seen comprehensive research into new strategies and treatments to prevent, control, and eradicate biofilms. In this book, we have assembled experts from around the world to present the current state of antibiofilm research. Novel, cutting-edge techniques and new optimized strategies based on more established methods will be presented in chapters focused on biofilm prevention, treatment, and control for the application in clinical, industrial, and veterinary settings. Antibiofilm strategies, such as chemical and enzymatic treatments, surface modification and coatings, quorum sensing inhibition and dispersal induction, phage therapy, cold plasma treatment, as well as hyperbaric oxygen treatment, and metal-based nanomedicine will be covered among many others.

References

Aanaes K, Rickelt LF, Johansen HK, von Buchwald C, Pressler T, Høiby N, Jensen PØ (2011) Decreased mucosal oxygen tension in the maxillary sinuses in patients with cystic fibrosis. J Cyst Fibros: Official Journal of the European Cystic Fibrosis Society 10(2):114–120. https://doi.org/10.1016/j.jcf.2010.12.002

Alhede M, Alhede M (2014) The biofilm challenge. EWMA J 14(1):54–58

Alhede M, Bjarnsholt T, Jensen PØ, Phipps RK, Moser C, Christophersen L, Christensen LD, van Gennip M, Parsek M, Høiby N, Rasmussen TB, Givskov M (2009) Pseudomonas aeruginosa recognizes and responds aggressively to the presence of polymorphonuclear leukocytes. Microbiology 155(Pt 11):3500–3508. https://doi.org/10.1099/mic.0.031443-0

Alhede M, Kragh KN, Qvortrup K, Allesen-Holm M, van Gennip M, Christensen LD, Jensen PØ, Nielsen AK, Parsek M, Wozniak D, Molin S, Tolker-Nielsen T, Høiby N, Givskov M, Bjarnsholt T (2011) Phenotypes of non-attached Pseudomonas aeruginosa aggregates resemble surface attached biofilm. PLoS One 6(11):e27943. https://doi.org/10.1371/journal.pone.0027943

Bals R, Weiner DJ, Wilson JM (1999) The innate immune system in cystic fibrosis lung disease. J Clin Invest 103(3):303–307. https://doi.org/10.1172/JCI6277

Barken KB, Pamp SJ, Yang L, Gjermansen M, Bertrand JJ, Klausen M, Givskov M, Whitchurch CB, Engel JN, Tolker-Nielsen T (2008) Roles of type IV pili, flagellum-mediated motility and extracellular DNA in the formation of mature multicellular structures in Pseudomonas aeruginosa biofilms. Environ Microbiol 10(9):2331–2343. https://doi.org/10.1111/j.1462-2920.2008.01658.x

Bjarnsholt T, Tolker-Nielsen T, Givskov M, Janssen M, Christensen LH (2009) Detection of bacteria by fluorescence in situ hybridization in culture-negative soft tissue filler lesions. Dermatol Surg: Official Publication for American Society for Dermatologic Surgery [et al.] 35(Suppl 2):1620–1624. https://doi.org/10.1111/j.1524-4725.2009.01313.x

Bjarnsholt T, Alhede M, Alhede M, Eickhardt-Sørensen SR, Moser C, Kühl M, Jensen PØ, Høiby N (2013) The in vivo biofilm. Trends Microbiol 21(9):466–474. https://doi.org/10.1016/j.tim.2013.06.002

Bossier P, Verstraete W (1996) Triggers for microbial aggregation in activated sludge? Appl Microbiol Biotechnol. http://link.springer.com/article/10.1007/s002530050640

Center for Biofilm Engineering, U. of M (n.d.) Brief history of "biofilm" | Center for Biofilm Engineering. Retrieved May 7, 2015, from https://www.biofilm.montana.edu/node/2930

Ceri H, Olson ME, Stremick C, Read RR, Morck D, Buret A (1999) The Calgary biofilm device: new technology for rapid determination of antibiotic susceptibilities of bacterial biofilms. J Clin Microbiol 37(6):1771–1776. http://www.pubmedcentral.nih.gov/articlerender.fcgi?artid=84946&tool=pmcentrez&rendertype=abstract

Chávez de Paz LE, Bergenholtz G, Svensäter G (2010) The effects of antimicrobials on endodontic biofilm bacteria. J Endod 36(1):70–77. https://doi.org/10.1016/j.joen.2009.09.017

Chiang W-C, Nilsson M, Jensen PØ, Høiby N, Nielsen TE, Givskov M, Tolker-Nielsen T (2013) Extracellular DNA shields against aminoglycosides in Pseudomonas aeruginosa biofilms. Antimicrob Agents Chemother 57(5):2352–2361. https://doi.org/10.1128/AAC.00001-13

Ciofu O, Hansen CR, Høiby N (2013) Respiratory bacterial infections in cystic fibrosis. Curr Opin Pulm Med 19(3):251–258. https://doi.org/10.1097/MCP.0b013e32835f1afc

Costerton JW (1999a) Bacterial biofilms: a common cause of persistent infections. Science 284(5418):1318–1322. https://doi.org/10.1126/science.284.5418.1318

Costerton JW (1999b) Introduction to biofilm. Int J Antimicrob Agents 11(3–4):217–221. discussion 237-9. http://www.ncbi.nlm.nih.gov/pubmed/10394973

Costerton JW, Geesey GG, Cheng KJ (1978) How bacteria stick. Sci Am 238(1):86–95. http://www.ncbi.nlm.nih.gov/pubmed/635520

Costerton JW, Cheng KJ, Geesey GG, Ladd TI, Nickel JC, Dasgupta M, Marrie TJ (1987) Bacterial biofilms in nature and disease. Annu Rev Microbiol 41:435–464. https://doi.org/10.1146/annurev.mi.41.100187.002251

Costerton JW, Lewandowski Z, Caldwell DE, Korber DR, Lappin-Scott HM (1995) Microbial biofilms. Annu Rev Microbiol 49(1):711–745. https://doi.org/10.1146/annurev.mi.49.100195.003431

Costerton JW, Post JC, Ehrlich GD, Hu FZ, Kreft R, Nistico L, Kathju S, Stoodley P, Hall-Stoodley L, Maale G, James G, Sotereanos N, DeMeo P (2011) New methods for the detection of orthopedic and other biofilm infections. FEMS Immunol Med Microbiol 61(2):133–140. https://doi.org/10.1111/j.1574-695X.2010.00766.x

Cowley ES, Kopf SH, LaRiviere A, Ziebis W, Newman DK (2015) Pediatric cystic fibrosis sputum can be chemically dynamic, anoxic, and extremely reduced due to hydrogen sulfide formation. MBio 6(4):e00767–e00715. https://doi.org/10.1128/mBio.00767-15

de Beer D, Stoodley P, Roe F, Lewandowski Z (1994) Effects of biofilm structures on oxygen distribution and mass transport. Biotechnol Bioeng 43(11):1131–1138. https://doi.org/10.1002/bit.260431118

de Beer D, Stoodley P, Lewandowski Z (1997) Measurement of local diffusion coefficients in biofilms by microinjection and confocal microscopy. Biotechnol Bioeng 53(2):151–158. https://doi.org/10.1002/(SICI)1097-0290(19970120)53:2<151::AID-BIT4>3.0.CO;2-N

Dibdin GH, Assinder SJ, Nichols WW, Lambert PA (1996) Mathematical model of beta-lactam penetration into a biofilm of Pseudomonas aeruginosa while undergoing simultaneous inactivation by released beta-lactamases. J Antimicrob Chemother 38(5):757–769. http://www.ncbi.nlm.nih.gov/pubmed/8961045

Donlan RM, Costerton JW (2002) Biofilms: survival mechanisms of clinically relevant microorganisms. Clin Microbiol Rev 15(2):167–193. http://www.pubmedcentral.nih.gov/articlerender.fcgi?artid=118068&tool=pmcentrez&rendertype=abstract

Drenkard E (2003) Antimicrobial resistance of Pseudomonas aeruginosa biofilms. Microbes and Infection / Institut Pasteur 5(13):1213–1219. http://www.ncbi.nlm.nih.gov/pubmed/14623017

Drenkard E, Ausubel FM (2002) Pseudomonas biofilm formation and antibiotic resistance are linked to phenotypic variation. Nature 416(6882):740–743. https://doi.org/10.1038/416740a

Espeland EM, Wetzel RG (2001) Complexation, stabilization, and UV photolysis of extracellular and surface-bound glucosidase and alkaline phosphatase: implications for biofilm microbiota. Microb Ecol 42(4):572–585. https://doi.org/10.1007/s00248-001-1023-7

Evans DJ, Allison DG, Brown MR, Gilbert P (1991) Susceptibility of Pseudomonas aeruginosa and Escherichia coli biofilms towards ciprofloxacin: effect of specific growth rate. J Antimicrob Chemother 27(2):177–184. http://www.ncbi.nlm.nih.gov/pubmed/1905285

Flemming H-C, Wingender J (2010) The biofilm matrix. Nat Rev Microbiol 8(9):623–633. https://doi.org/10.1038/nrmicro2415

Flemming H-C, Neu TR, Wozniak DJ (2007) The EPS matrix: the "house of biofilm cells". J Bacteriol 189(22):7945–7947. https://doi.org/10.1128/JB.00858-07

Folsom JP, Richards L, Pitts B, Roe F, Ehrlich GD, Parker A, Mazurie A, Stewart PS (2010) Physiology of Pseudomonas aeruginosa in biofilms as revealed by transcriptome analysis. BMC Microbiol 10:294. https://doi.org/10.1186/1471-2180-10-294

Gordon CA, Hodges NA, Marriott C (1988) Antibiotic interaction and diffusion through alginate and exopolysaccharide of cystic fibrosis-derived Pseudomonas aeruginosa. J Antimicrob Chemother 22(5):667–674. http://www.ncbi.nlm.nih.gov/pubmed/3145268

Haagensen JAJ, Klausen M, Ernst RK, Miller SI, Folkesson A, Tolker-Nielsen T, Molin S (2007) Differentiation and distribution of colistin- and sodium dodecyl sulfate-tolerant cells in Pseudomonas aeruginosa biofilms. J Bacteriol 189(1):28–37. https://doi.org/10.1128/JB.00720-06

Hall-Stoodley L, Costerton JW, Stoodley P (2004) Bacterial biofilms: from the natural environment to infectious diseases. Nat Rev Microbiol 2(2):95–108. https://doi.org/10.1038/nrmicro821

Hatch RA, Schiller NL (1998) Alginate lyase promotes diffusion of aminoglycosides through the extracellular polysaccharide of mucoid Pseudomonas aeruginosa. Antimicrob Agents Chemother 42(4):974–977. http://www.pubmedcentral.nih.gov/articlerender.fcgi?artid=105585&tool=pmcentrez&rendertype=abstract

Høiby N (2014) A personal history of research on microbial biofilms and biofilm infections. Pathog Dis 70(3):205–211. https://doi.org/10.1111/2049-632X.12165

Hoiby N, Flensborg EW, Beck B, Friis B, Jacobsen SV, Jacobsen L (1977) Pseudomonas aeruginosa infection in cystic fibrosis. Diagnostic and prognostic significance of Pseudomonas aeruginosa precipitins determined by means of crossed immunoelectrophoresis. Scand J Respir Dis 58(2):65–79. http://www.ncbi.nlm.nih.gov/pubmed/404701

Høiby N, Bjarnsholt T, Moser C, Bassi GL, Coenye T, Donelli G, Hall-Stoodley L, Holá V, Imbert C, Kirketerp-Møller K, Lebeaux D, Oliver A, Ullmann AJ, Williams C (2015) ESCMID guideline for the diagnosis and treatment of biofilm infections 2014. Clin Microbiol Infect: The Official Publication of the European Society of Clinical Microbiology and Infectious Diseases 21(Suppl 1):S1–S25. https://doi.org/10.1016/j.cmi.2014.10.024

Homøe P, Bjarnsholt T, Wessman M, Sørensen HCF, Johansen HK (2009) Morphological evidence of biofilm formation in Greenlanders with chronic suppurative otitis media. Eur Arch Oto-Rhino-Laryngol: Official Journal of the European Federation of Oto-Rhino-Laryngological

Societies (EUFOS) : Affiliated with the German Society for Oto-Rhino-Laryngology - Head and Neck Surgery 266(10):1533–1538. https://doi.org/10.1007/s00405-009-0940-9

Isenberg HD (1990) Where we've been and where we may be going. Adv Exp Med Biol 263:1–9. http://www.ncbi.nlm.nih.gov/pubmed/2191542

Jacombs ASW, Karatassas A, Klosterhalfen B, Richter K, Patiniott P, Hensman C (2020) Biofilms and effective porosity of hernia mesh: are they silent assassins? Hernia 24(1):197–204. https://doi.org/10.1007/s10029-019-02063-y

Jensen PØ, Bjarnsholt T, Phipps R, Rasmussen TB, Calum H, Christoffersen L, Moser C, Williams P, Pressler T, Givskov M, Høiby N (2007) Rapid necrotic killing of polymorphonuclear leukocytes is caused by quorum-sensing-controlled production of rhamnolipid by Pseudomonas aeruginosa. Microbiology 153(Pt 5):1329–1338. https://doi.org/10.1099/mic.0.2006/003863-0

Jensen PØ, Givskov M, Bjarnsholt T, Moser C (2010) The immune system vs. Pseudomonas aeruginosa biofilms. FEMS Immunol Med Microbiol 59(3):292–305. https://doi.org/10.1111/j.1574-695X.2010.00706.x

Jin Y-L, Alex Speers R (1998) Flocculation of Saccharomyces cerevisiae. Food Res Int 31(6–7):421–440. https://doi.org/10.1016/S0963-9969(99)00021-6

Kaul L, Süss R, Zannettino A, Richter K (2021) The revival of dithiocarbamates: from pesticides to innovative medical treatments. IScience 24(2):102092. https://doi.org/10.1016/j.isci.2021.102092

Keren I, Kaldalu N, Spoering A, Wang Y, Lewis K (2004) Persister cells and tolerance to antimicrobials. FEMS Microbiol Lett 230(1):13–18. http://www.ncbi.nlm.nih.gov/pubmed/14734160

Kirketerp-Møller K, Jensen PØ, Fazli M, Madsen KG, Pedersen J, Moser C, Tolker-Nielsen T, Høiby N, Givskov M, Bjarnsholt T (2008) Distribution, organization, and ecology of bacteria in chronic wounds. J Clin Microbiol 46(8):2717–2722. https://doi.org/10.1128/JCM.00501-08

Klausen M, Aaes-Jørgensen A, Molin S, Tolker-Nielsen T (2003a) Involvement of bacterial migration in the development of complex multicellular structures in Pseudomonas aeruginosa biofilms. Mol Microbiol 50(1):61–68. http://www.ncbi.nlm.nih.gov/pubmed/14507363

Klausen M, Heydorn A, Ragas P, Lambertsen L, Aaes-Jørgensen A, Molin S, Tolker-Nielsen T (2003b) Biofilm formation by Pseudomonas aeruginosa wild type, flagella and type IV pili mutants. Mol Microbiol 48(6):1511–1524. http://www.ncbi.nlm.nih.gov/pubmed/12791135

Kolenbrander PE (2000) Oral microbial communities: biofilms, interactions, and genetic systems 1. Annu Rev Microbiol 54(1):413–437. https://doi.org/10.1146/annurev.micro.54.1.413

Kolpen M, Bjarnsholt T, Moser C, Hansen CR, Rickelt LF, Kühl M, Hempel C, Pressler T, Høiby N, Jensen PØ (2014a) Nitric oxide production by polymorphonuclear leucocytes in infected cystic fibrosis sputum consumes oxygen. Clin Exp Immunol 177(1):310–319. https://doi.org/10.1111/cei.12318

Kolpen M, Kühl M, Bjarnsholt T, Moser C, Hansen CR, Liengaard L, Kharazmi A, Pressler T, Høiby N, Jensen PØ (2014b) Nitrous oxide production in sputum from cystic fibrosis patients with chronic Pseudomonas aeruginosa lung infection. PLoS One 9(1):e84353. https://doi.org/10.1371/journal.pone.0084353

Kragh KN, Alhede M, Jensen PØ, Moser C, Scheike T, Jacobsen CS, Poulsen SS, Eickhardt-Sørensen SR, Trøstrup H, Christoffersen L, Hougen H-P, Rickelt LF, Kühl M, Høiby N, Bjarnsholt T (2014) Polymorphonuclear leukocytes restrict growth of Pseudomonas aeruginosa in the lungs of cystic fibrosis patients. Infect Immun 82(11). https://doi.org/10.1128/IAI.01969-14

Kragh KN, Alhede M, Rybtke M, Stavnsberg C, Jensen PØ, Tolker-Nielsen T, Whiteley M, Bjarnsholt T (2018) The inoculation method could impact the outcome of microbiological experiments. Appl Environ Microbiol 84(5):e02264-17. https://doi.org/10.1128/AEM.02264-17

Le Magrex-Debar E, Lemoine J, Gellé MP, Jacquelin LF, Choisy C (2000) Evaluation of biohazards in dehydrated biofilms on foodstuff packaging. Int J Food Microbiol 55(1–3):239–243. http://www.ncbi.nlm.nih.gov/pubmed/10791750

Lewis K (2010) Persister cells. Annu Rev Microbiol 64:357–372. https://doi.org/10.1146/annurev.micro.112408.134306

Mack WN, Mack JP, Ackerson AO (1975) Microbial film development in a trickling filter. Microb Ecol 2(3):215–226. https://doi.org/10.1007/BF02010441

Mah TF, O'Toole GA (2001) Mechanisms of biofilm resistance to antimicrobial agents. Trends Microbiol 9(1):34–39. http://www.ncbi.nlm.nih.gov/pubmed/11166241

Marrie TJ, Costerton JW (1984a) Scanning and transmission electron microscopy of in situ bacterial colonization of intravenous and intraarterial catheters. J Clin Microbiol 19(5):687–693. http://www.pubmedcentral.nih.gov/articlerender.fcgi?artid=271156&tool=pmcentrez&rendertype=abstract

Marrie TJ, Costerton JW (1984b) Morphology of bacterial attachment to cardiac pacemaker leads and power packs. J Clin Microbiol 19(6):911–914. http://www.pubmedcentral.nih.gov/articlerender.fcgi?artid=271211&tool=pmcentrez&rendertype=abstract

Mayer C, Moritz R, Kirschner C, Borchard W, Maibaum R, Wingender J, Flemming HC (1999) The role of intermolecular interactions: studies on model systems for bacterial biofilms. Int J Biol Macromol 26(1):3–16. http://www.ncbi.nlm.nih.gov/pubmed/10520951

Pamp SJ, Gjermansen M, Johansen HK, Tolker-Nielsen T (2008) Tolerance to the antimicrobial peptide colistin in Pseudomonas aeruginosa biofilms is linked to metabolically active cells, and depends on the pmr and mexAB-oprM genes. Mol Microbiol 68(1):223–240. https://doi.org/10.1111/j.1365-2958.2008.06152.x

Parsa K, Schaudinn C, Gorur A, Sedghizadeh PP, Johnson T, Tse DT, Costerton JW (2010) Demonstration of bacterial biofilms in culture-negative silicone stent and jones tube. Ophthal Plast Reconstr Surg 26(6):426–430. https://doi.org/10.1097/IOP.0b013e3181cff61f

Patiniott P, Jacombs A, Kaul L, Hu H, Warner M, Klosterhalfen B, Karatassas A, Maddern G, Richter K (2022) Are late hernia mesh complications linked to Staphylococci biofilms? Hernia:1–7. https://doi.org/10.1007/s10029-022-02583-0

Purdy Drew KR, Sanders LK, Culumber ZW, Zribi O, Wong GCL (2009) Cationic amphiphiles increase activity of aminoglycoside antibiotic tobramycin in the presence of airway polyelectrolytes. J Am Chem Soc 131(2):486–493. https://doi.org/10.1021/ja803925n

Ramsing NB, Kühl M, Jørgensen BB (1993) Distribution of sulfate-reducing bacteria, O2, and H2S in photosynthetic biofilms determined by oligonucleotide probes and microelectrodes. Appl Environ Microbiol 59(11):3840–3849. http://www.pubmedcentral.nih.gov/articlerender.fcgi?artid=182539&tool=pmcentrez&rendertype=abstract

Schleheck D, Barraud N, Klebensberger J, Webb JS, McDougald D, Rice SA, Kjelleberg S (2009) Pseudomonas aeruginosa PAO1 preferentially grows as aggregates in liquid batch cultures and disperses upon starvation. PLoS One 4(5):e5513. https://doi.org/10.1371/journal.pone.0005513

Segal AW (2005) How neutrophils kill microbes. Annu Rev Immunol 23:197–223. https://doi.org/10.1146/annurev.immunol.23.021704.115653

Stewart PS (2013) Biophysics of biofilm infection. Pathogens and Disease 70(3):212–218. https://doi.org/10.1111/2049-632X.12118

Stewart PS, Franklin MJ (2008) Physiological heterogeneity in biofilms. Nat Rev Microbiol 6(3):199–210. https://doi.org/10.1038/nrmicro1838

Stoodley P, Sauer K, Davies DG, Costerton JW (2002) Biofilms as complex differentiated communities. Annu Rev Microbiol 56:187–209. https://doi.org/10.1146/annurev.micro.56.012302.160705

Suci PA, Mittelman MW, Yu FP, Geesey GG (1994) Investigation of ciprofloxacin penetration into Pseudomonas aeruginosa biofilms. Antimicrob Agents Chemother 38(9):2125–2133. http://www.pubmedcentral.nih.gov/articlerender.fcgi?artid=284696&tool=pmcentrez&rendertype=abstract

Van Gennip M, Christensen LD, Alhede M, Phipps R, Jensen PØ, Christophersen L, Pamp SJ, Moser C, Mikkelsen PJ, Koh AY, Tolker-Nielsen T, Pier GB, Høiby N, Givskov M, Bjarnsholt T (2009) Inactivation of the rhlA gene in Pseudomonas aeruginosa prevents rhamnolipid production, disabling the protection against polymorphonuclear leukocytes. APMIS: Acta Pathologica, Microbiologica, et Immunologica Scandinavica 117(7):537–546. https://doi.org/10.1111/j.1600-0463.2009.02466.x

Wagner JG, Roth RA (2000) Neutrophil migration mechanisms, with an emphasis on the pulmonary vasculature. Pharmacol Rev 52(3):349–374. http://www.ncbi.nlm.nih.gov/pubmed/10977867

Walters MC, Roe F, Bugnicourt A, Franklin MJ, Stewart PS (2003) Contributions of antibiotic penetration, oxygen limitation, and low metabolic activity to tolerance of Pseudomonas aeruginosa biofilms to ciprofloxacin and tobramycin. Antimicrob Agents Chemother 47(1): 317–323. https://doi.org/10.1128/AAC.47.1.317-323.2003

Wessel AK, Arshad TA, Fitzpatrick M, Connell JL, Bonnecaze RT, Shear JB, Whiteley M (2014) Oxygen limitation within a bacterial aggregate. MBio 5(2):e00992. https://doi.org/10.1128/mBio.00992-14

Westall F, de Wit MJ, Dann J, van der Gaast S, de Ronde CE, Gerneke D (2001) Early Archean fossil bacteria and biofilms in hydrothermally-influenced sediments from the Barberton greenstone belt, South Africa. Precambrian Res 106(1–2):93–116. https://doi.org/10.1016/S0301-9268(00)00127-3

Wolcott R, Dowd S (2011) The role of biofilms: are we hitting the right target? Plast Reconstr Surg 127(Suppl):28S–35S. https://doi.org/10.1097/PRS.0b013e3181fca244

Zogaj X, Nimtz M, Rohde M, Bokranz W, Römling U (2001) The multicellular morphotypes of salmonella typhimurium and Escherichia coli produce cellulose as the second component of the extracellular matrix. Mol Microbiol 39(6):1452–1463. http://www.ncbi.nlm.nih.gov/pubmed/11260463

Zumft WG (1997) Cell biology and molecular basis of denitrification. Microbiol Mol Biol Rev 61(4):533–616. http://www.ncbi.nlm.nih.gov/pubmed/9409151

Antimicrobial and Innate Immune Tolerance Mechanisms in Biofilms

Elizabeth L. Sandvik, Timothy R. Borgogna, and Philip S. Stewart

Abstract It is well documented that biofilms have inherent tolerance to killing and removal by biocides, antibiotics, and the innate immune system. While the matrix may impede antimicrobial transport, many antimicrobials reach the full depth of the biofilm but are inactive against the cells they reach. Clearance of biofilm infections can be hindered by delayed recruitment of immune responses. Nutrient depletion in the biofilm leads to slow or dormant growth but the transition to these dormant states can also trigger active metabolic adaptations and stress responses lending the cells in the biofilm to be unaffected. Transition to these tolerant states and changes in gene expression happen early in biofilm formation and even small biofilm clusters can challenge clearance. Biofilm growth as a multicellular community has a diversity of survival strategies and incomplete clearance leads to rapid recolonization. In this chapter, we will discuss how antimicrobial penetration, the distribution of physiological states, immune system recruitment, and bacterial responses to immune attack contribute to biofilm tolerance.

1 Introduction

Biofilms are communities of microorganisms embedded in a matrix of cell-produced extracellular polymeric substances. The matrix affords a protective and multicellular mode of growth providing a nutrient reservoir, mechanical protection, a spectrum of microenvironmental niches, ability to harbor diversity, and proximity between cells facilitating the exchange of metabolites and communication molecules. The variety of species and cell states that coexist within the biofilm suggests that microorganisms

E. L. Sandvik · T. R. Borgogna
Center for Biofilm Engineering, Montana State University, Bozeman, MT, USA

P. S. Stewart (✉)
Center for Biofilm Engineering, Montana State University, Bozeman, MT, USA

Department of Chemical and Biological Engineering, Montana State University, Bozeman, MT, USA
e-mail: phil_s@montana.edu

K. Richter, K. Nørskov Kragh (eds.), *Antibiofilm Strategies*, Springer Series on Biofilms 11, https://doi.org/10.1007/978-3-031-10992-8_2

are no longer acting as single-celled organisms but as a community, a strategy that has been compared to cell differentiation seen in multicellular organisms (Watnick and Kolter 2000; Nikolaev and Plakunov 2007; Shapiro 1998; Penesyan et al. 2021).

A hallmark feature of biofilm is tolerance to physical and chemical challenges. Biofilm tolerance to antimicrobials such as biocides or antibiotics is well established. Antimicrobial tolerance is distinct from antibiotic resistance in which mutations or acquisition of genetic material confer heritable protection at the single cell level. In contrast, microbial cells in biofilms exhibit tolerance that can be reversed if the microorganisms are dispersed from the biofilm. In addition to antimicrobials, biofilms are also protected against immune system attack. These tolerances come from the organisms collectively expressing mechanical resilience along with a spectrum of cellular states and responses that afford protection at the community level. Even though a subset of the biofilm population may succumb to antimicrobials or immune system attack, the surviving cells are typically sufficient to permit rapid recolonization once the antibiofilm pressure is no longer present. This recalcitrance and resilience explain why the biofilm mode of growth is so successful and continues to challenge mitigation strategies in applications from industrial water systems to chronic infections in medicine.

Prior reviews have delved into biofilm tolerance (Ciofu and Tolker-Nielsen 2019; Yan and Bassler 2019; Stewart 2015; Hall and Mah 2017; Harrison et al. 2007; Ciofu et al. 2022). The transport of nutrients, metabolic bi-products, and antimicrobials through the physical structure of the matrix and reaction or utilization by cells within it set up microenvironments that dictate a wide diversity of cell states. The purpose of this chapter is to sketch out major mechanisms of biofilm tolerance, to both antimicrobials and the innate immune system, as a backdrop to the detailed chapters that follow. Proposed mechanisms of biofilm tolerance due to the physical structure of the biofilm matrix and the diversity of cell states within it are summarized in Fig. 1 and will be discussed in more detail in the sections that follow.

1.1 Antimicrobial Penetration

The most obvious explanation for biofilm tolerance to soluble antimicrobials is that these agents fail to fully penetrate throughout the biofilm. Penetration is governed by the balance of solute reaction and diffusion. Direct measurements of the penetration of highly reactive biocides like chlorine (Debeer et al. 1994; Davison et al. 2010), which is neutralized by reactions with matrix material as well as cellular biomass, demonstrate its poor penetration. Likewise, hydrogen peroxide is actively degraded by catalases faster than it can diffuse into the depths of a biofilm (Stewart et al. 2000). In contrast, some antimicrobial agents that are much larger than these reactive oxidants can penetrate readily. This is because these agents are not being rapidly reacted or modified as they move in the biofilm. Examples include the antimicrobial peptide nisin (Davison et al. 2010), and the cyclic lipopeptide antibiotic daptomycin (Stewart et al. 2009). In the absence of high reactivity, the matrix can still retard penetration by binding the antimicrobial. For example, direct measurements of antibiotic penetration in the same biofilm model revealed rapid and complete penetration of four quinolones and two β-lactams but reduced penetration of the positively charged aminoglycosides amikacin and gentamycin (Shigeta et al. 1997). This theme recurred in the microscopic visualization of incomplete penetration of fluorescently-tagged tobramycin into *Pseudomonas* biofilms. Delayed penetration of positively-charged antibiotics is presumably due to their complexing with negatively charged matrix polymers or cell surface moieties (Walters et al. 2003). However, even when agents penetrate, they may not be effective at killing. Additional studies of fluorescently tagged antibiotics or direct antibiotic measurement show no diffusion limitation, with vancomycin (Jefferson et al. 2005), daptomycin (Stewart et al. 2009), ciprofloxacin (Tseng et al. 2013; Walters et al. 2003; Vrany et al. 1997), and levofloxacin (Vrany et al. 1997) penetrating the full depth of the biofilm but lacking efficacy against the cells they reach.

2 Reduced Envelope Permeability by Matrix Polysaccharides

A recurrent theme emerging from studies of mutant strains deficient in the synthesis of extracellular polysaccharides is that these polymers afford protection against antibiotics in the biofilm state (Colvin et al. 2011; Billings et al. 2013; Hentzer et al. 2001; Goltermann and Tolker-Nielsen 2017; Stewart et al. 2019) While some of this may be an effect of retarded penetration through the biofilm as a whole, we offer an alternative conjecture. We hypothesize that extracellular polymers, and possibly periplasmic glucans (Mah et al. 2003), when sufficiently locally concentrated, provide a shell of reduced permeability that restricts access of antimicrobials into the cell interior (Fig. 1).

2.1 Cell Physiological States

Biofilms harbor heterogeneous cell states including dormant or slow-growing cells, cells with metabolic adaptations, and cells activating stress responses (Fig. 1). Many of these responses are linked to environmental triggers and divergent strategies for community survival. We will discuss how these cell states can impact biocide and antibiotic activity before discussing immune system tolerance.

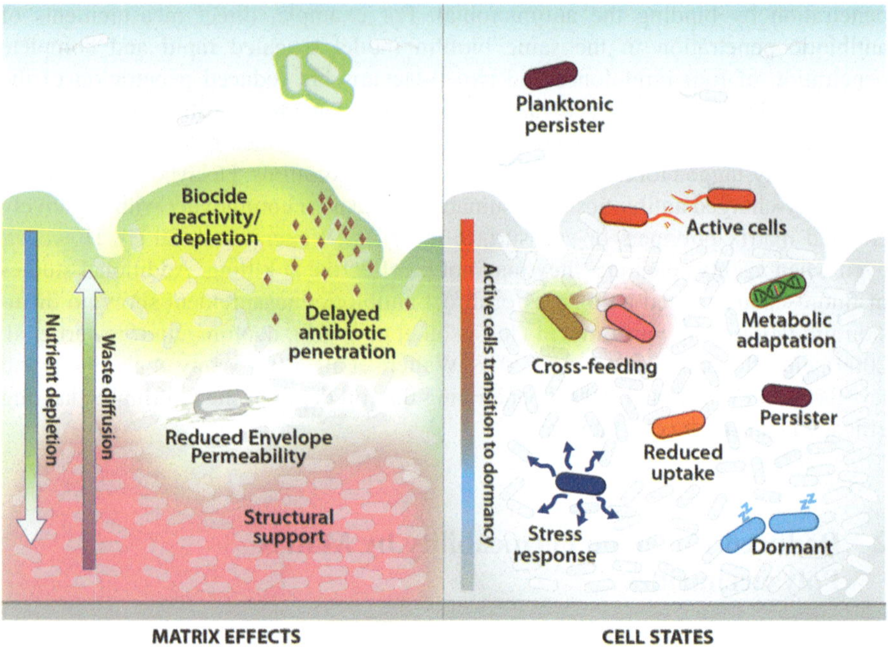

Fig. 1 Summary of proposed mechanisms contributing to biofilm tolerance. Microenvironments exist throughout the biofilm due to microbial consumption of nutrients diffusing into the biofilm and reverse diffusion of waste products. This leads to slow and dormant growth while other cells adapt metabolically or initiate stress responses to maintain viability through a variety of strategies. Mechanisms of biofilm tolerance due to matrix and cell states are discussed in more detail in the text

2.1.1 Dormancy and Slow Growth

Biofilm tolerance is often attributed to slow growth and dormancy. In planktonic systems, decreases in growth rates and the transition to stationary phase are triggered by nutrient depletion or waste product buildup. These lead to a progressive shutdown of multiple metabolic processes, increased DNA/protein ratio, and cell shrinkage (Nystrom 2004). While these cells are no longer growing and are considered growth arrested, they maintain the capacity for protein synthesis long after the transition to stationary phase (Gefen et al. 2014). Analogous signals and nutrient depletion occur in different regions of the biofilm and similar growth arrests and evidence of active transition to stationary phase may allow cells in the biofilm to transition to a non-growing state such that cells maintain viability and are prepared to resuscitate when environmental conditions change.

One way in which bacteria decrease growth while maintaining viability is through ribosome hibernation aided by small ribosome-associated proteins. These proteins include hibernation-promoting factor (HPF), the HPF homolog YfiA, and ribosome modulation factor (RMF). All prokaryotes possess one or more genes such as these that produce proteins that reversibly bind to ribosomes and turn off protein synthesis while maintaining intact ribosomes during starvation transitions when many other proteins are being degraded (Polikanov et al. 2012). These small ribosome-associated genes are some of the only genes expressed at higher levels in the base of the biofilm versus active regions at the top of biofilm (Williamson et al. 2012) and deletions of these genes in a multitude of organisms cause reduced starvation survival and ribosome depletion. A striking example of the importance of this function is the flexibility of Gram(−) *Pseudomonas aeruginosa* *Δhpf* mutant to utilize mutated copies of native HPF and even distantly-related long-HPF of the Gram (+) organisms *Staphylococcus aureus* and *Clostridium difficile* to restore starvation survival and rRNA preservation to wild-type levels (Franklin et al. 2020). Ribosome hibernation has been proposed to be activated by the stringent response during nutrient depletion via the alarmone ppGpp (Song and Wood 2020). Single deletion mutants of these small ribosomal-associated proteins have been demonstrated to have 12- to 600-fold decreases in cells surviving ampicillin treatment (Song and Wood 2020). Resuscitation studies demonstrate persister cells wake following antibiotic treatment based on ribosomal content (Kim 2020) suggesting ribosome hibernation is not only important for survival during nutrient depletion but also in biofilm recolonization following an antibiotic challenge. Overall, this reversible protection of intact ribosomes allows for rapid cellular responses upon nutrient restoration and represents a mechanism of maintaining viability in nutrient-limited regions of the biofilm.

A variety of metabolic markers have been used to study the distribution of cellular activity in biofilms, often associated with parallel measurements of nutrients and antibiotic susceptibility. Metabolic markers of protein synthesis (Rani et al. 2007), DNA replication (Rani et al. 2005), expression of alkaline phosphatase activity (Huang et al. 1998), and the response to IPTG induction (Werner et al. 2004; Borriello et al. 2004; Walters et al. 2003) have all been used to study stratified growth through the depth of biofilms. All of these studies use cross-sections of the biofilm to show beautiful stratification with often narrow bands of cellular activity at interfaces adjoining nutrient and energy sources (bulk fluid interfaces, agar surfaces, or air interfaces), whereas the vast majority of cells have very low to no metabolic activity. Slow or dormant growth can contribute to biofilm tolerance as antibiotics that target active metabolic pathways are ineffective against cells not actively metabolizing.

A subpopulation associated with antimicrobial tolerance is persister cells. Persister cells are non-dividing cells that tolerate antibiotic exposure and quickly exit the persister state to regrow when the antibiotic challenge is removed. These populations typically constitute a small subset of slow-growing cells, often making up less than 1% of the population in mature biofilms (Keren et al. 2004). While often associated with biofilm, persister cells also form in planktonic cultures during active growth, increasing significantly during the transition to stationary phase. Depletion of primary nutrients, transition to a secondary carbon source, and other metabolic cues can trigger persister cell formation (reviewed in Amato et al. 2014). Again, the tolerance of the persister population is not resistance as the tolerance is not heritable.

Cultures grown from the surviving cells repeat the same kill curves, antibiotic susceptibility, and survival fractions as exhibited with ampicillin persistence in recurrently transferred and treated planktonic *Escherichia coli* cultures (Keren et al. 2004). Persister cells can present challenges to study for this reason as only during the time of non-division are they considered persister cells, exiting the persister state to recolonize the population upon removal of the antibiotic challenge.

2.1.2 Metabolic Adaptation

While slow growth and dormancy are a consequence of regions of the biofilm lacking nutrients, intermediate regions may have nutrients available while others are lacking leading to metabolic adaptations that may contribute to biofilm tolerance (Crabbe et al. 2019). For example, hypoxia may lead to the induction of denitrification or fermentation pathways (Goodwine et al. 2019) and reduced antimicrobial susceptibility (Borriello et al. 2004; Kolpen et al. 2017; Field et al. 2005; Gupta et al. 2016; Schaible et al. 2012; King et al. 2010). Recent spatiotemporal mapping of metabolic adaptations reveals extensive microscale heterogeneity between even adjacent cells within biofilm (Dar et al. 2021). Some of these changes can be explained by rapid nutrient depletion and microenvironment changes. In glucose-fed experiments of monolayer bacteria, reporters of glucose uptake versus acetate utilization have demonstrated cells further from the glucose source cross-feed on acetate produced by glucose metabolism of closer cells with the highest expression rates of glucose versus acetate utilization occurring over very short distances of only several cell lengths (Dal Co et al. 2019). Alternatively, cooperative strategies may allow for the diversification of metabolic states. In similar cross-feeding studies, synthetic communities of two auxotrophs requiring metabolic products from the other to grow can exchange metabolic products over multi-cell lengths (Dal Co et al. 2020). Cells diversifying metabolic states, some of which are associated with different stress responses, while benefiting from alternative strategies of their neighbors indicate cooperative survival of the biofilm as a community.

In the context of biofilm tolerance, dormant or slowed growth as well as active but altered metabolic states may dictate antimicrobial efficacy. Antibiotics targeting metabolic processes would naturally have less activity against cells not actively metabolizing (Nguyen et al. 2011). Even in planktonic studies, stationary phase cells have decreased susceptibility compared to cells treated during exponential growth and biofilm tolerance has often been compared to stationary phase tolerances of the predominantly inactive cells in mature biofilms (Spoering and Lewis 2001; Anderl et al. 2003; Marques and Nelson 2019; Folsom et al. 2010; Meylan et al. 2017; Levin and Rozen 2006). Modified metabolisms and microenvironments can also contribute to decreased antimicrobial efficacy. Multiple antibiotic tolerances have been associated with the transition from aerobic to anaerobic respiration in oxygen-depleted niches within biofilm (Hill et al. 2005; Hassett et al. 2002; Sonderholm et al. 2017) and biofilms often have decreased efficacy in minimal media compared to rich media environments. While many studies have demonstrated antibiotic effects of metabolically active biofilm regions, this is not always the case. Imaging of flow-cell biofilms show ciprofloxacin or tetracycline kills metabolically active cells on the outer edge of biofilm clusters while colistin treatment has the inverse effect, killing dormant inner cells while the metabolically active population adapt to exposure through antibiotic efflux and survive (Pamp et al. 2008), an example of an active stress response discussed in the next section.

One strategy to combat antibiotic tolerance exhibited by slow-growing, dormant, and persister cells is through metabolic stimulation. Antibiotic uptake can decrease due to cellular changes of dormant or slow-growing cells and antibiotics targeting metabolic activities may not impact dormant cells. In anaerobic *Pseudomonas aeruginosa* biofilms, ciprofloxacin and tobramycin sensitivity can be increased by stimulating metabolism by the addition of an electron acceptor (nitrate) to stimulate denitrification or addition of fermentable arginine (Borriello et al. 2006). In the case of aminoglycosides, antibiotic uptake is an active process facilitated by a base level of proton-motive force and can contribute to reduced antibiotic uptake in non-growing cells (Taber et al. 1987). Studies of aminoglycoside uptake have been enhanced while cells remain non-dividing by co-dosing antibiotics with a carbon source that increases the PMF and subsequent killing of dormant and antibiotic tolerant *E. coli* or *S. aureus* (Allison et al. 2011) and *P. aeruginosa* persister cells (Meylan et al. 2017). Sensitization of anoxic biofilm regions with hyperbaric oxygen enhances ciprofloxacin lethality (Kolpen et al. 2017).

2.1.3 Stress Responses

Many stress responses have been associated with biofilm tolerance and a genetic basis for genes or active responses beyond growth arrest or cellular inactivity has been studied extensively. These stress responses are often expressed simultaneously within the biofilm offering complementary or alternative strategies contributing to biofilm tolerance (Stewart et al. 2015).

Stress responses are often coupled with or triggered by metabolic adaptation or nutrient starvation. The stringent response and production of the alarmone (p)ppGpp by RelA/SpoT is activated by starvation of carbon, amino acid, or iron and while considered a universal stress regulator of growth rate, phase transitions, and biofilm formation, activation of the stringent response also induces many virulence factors contributing to tolerance (Hobbs and Boraston 2019). Activation of the stringent response during diauxic transitions from glucose to a secondary fumarate carbon source stimulates the formation of both ampicillin and ofloxacin persisters through the same metabolic trigger but distinct pathways, both requiring ppGpp and regulation of RNA polymerase activity through DksA (Amato and Brynildsen 2015). The stringent response also plays a role in oxidative defenses to reactive oxygen species (ROS) (Nguyen et al. 2011). The accumulation of reactive oxygen species (ROS) in biofilm can be associated with signaling at low levels but high accumulation leads to oxidative stress. In biofilms, genes that code for peroxidase, catalase, or antioxidant activity (e.g., glutathione) contribute to protection against certain antimicrobials

(Nilsson et al. 2019; Elkins et al. 1999; Hahn et al. 2021; van Acker et al. 2013). The glyoxylate shunt has been implicated in the biofilm response to antibiotic-induced oxidative stress, but not in a consistent fashion (van Acker et al. 2013; Ahn et al. 2016). The stationary phase sigma-factor is associated with energy-limited growth, upregulated in biofilm, particularly in less active regions of the biofilm and deletion of *rpoS* in *P. aeruginosa* leads to increased ciprofloxacin susceptibility (Stewart et al. 2019). While coupled with starvation in aging biofilms, activation of the SOS-response in response to DNA damage triggered by oxidative stress or antimicrobials has been tied to increased ofloxacin resistance (Bernier et al. 2013). Cellular communication via quorum sensing is associated with increased tolerance (Bjarnsholt et al. 2005; Stewart et al. 2019; Stewart 2015). Toxin–antitoxin modules contribute to biofilm tolerance (Harrison et al. 2009; Ma et al. 2019; van Acker et al. 2014), perhaps by arresting macromolecular synthetic processes.

These are just some of the suite of stress responses exhibited by organisms within biofilm and while mutants deficient in these responses often have decreased tolerance, mutation rarely completely eliminates biofilm tolerance.

2.2 *Innate Immune Response*

Host-associated biofilms develop and persist despite the presence of an active immune response. Specific immune tolerance mechanisms vary with biofilm composition and location; however, the strategies microorganisms employ to combat immune clearance can be generalized into three categories: (1) growth and physical enlargement (2) immune modulation, and (3) virulence factor production. As the biofilm community matures these strategies synergize resulting in enhanced protection from immune defenses. Growth of nascent biofilms into aggregates reduces leukocyte phagocytic efficacy and increases the consumption of local nutrients. Leukocytes engaged with the biofilm respond to locally diminished nutrient availability by expressing an array of anti-inflammatory cytokines and markers that deescalate immune recruitment, killing, and priming. The reduced immune burden enables the biofilm to continue to survive and expand. Increases in biomass are accompanied by the accumulation of virulence factors and secreted toxins within the matrix and at the biofilm surface. These virulence factors can directly facilitate leukocyte killing and promote biofilm dispersal. Cells dispersed from the biofilm community disseminate to new locations and the cycle towards establishing immune tolerance repeats.

2.2.1 Early Growth and Establishment

Decades of research aimed at understanding biofilm dynamics and eradication strategies have led to one overwhelming consensus: a mature biofilm demonstrates remarkable recalcitrance to antimicrobial therapies and immune clearance (Gunther et al. 2009; Verderosa et al. 2019; Yamada and Kielian 2019). Biofilms are well characterized to efficiently withstand clearance from host immune interactions compared to planktonic counterparts of the same species. These observations imply that biofilms express innate characteristics that impede host immune defenses (Fig. 2a–c) (Leid 2009; Gonzalez et al. 2018; Yamada and Kielian 2019). Given the productivity of innate immune defenses against non-biofilm microbial contaminants, the transition into a biofilm community can be a major determinant of microbial survival. Thus, the early interactions between the innate immune system and nascent biofilms are of extreme interest as they may direct the trajectory and outcome of infection.

Polymorphonuclear cells (PMNs) and macrophages are largely recognized as the dominate leukocytes immediately recruited to a site of damage or infection. These cell types are primarily responsible for the immediate control of microbial

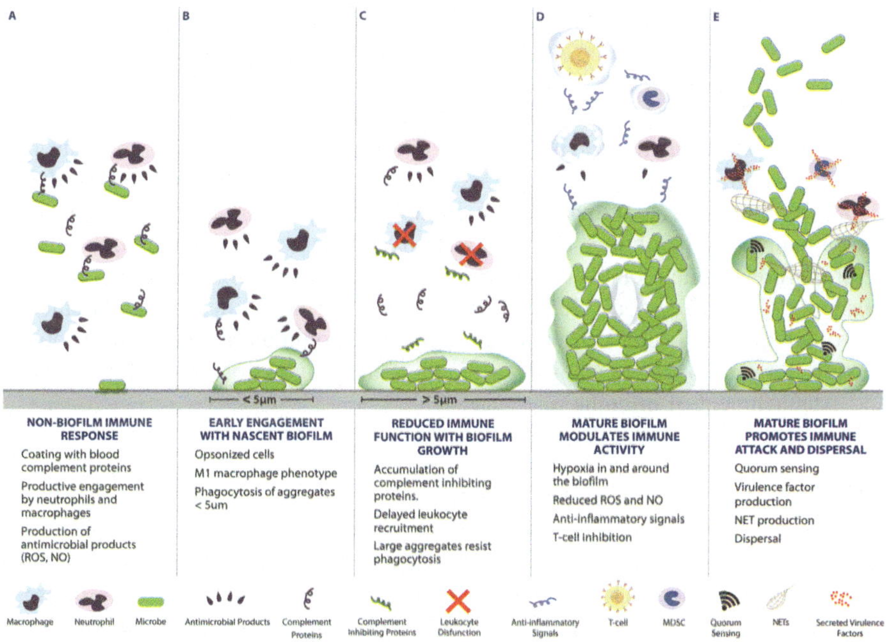

Fig. 2 Biofilm tolerance and modulation of the host immune response. (**a**) Detection and clearance of non-biofilm-associated microorganisms by innate defenses through complement deposition, phagocyte recruitment, and the production of antimicrobial products. (**b**) During nascent biofilm growth inflammatory signals recruit M1 macrophages and neutrophils that readily clear microbial aggregates <5 μm in diameter. (**c**) As the biofilm enlarges, aggregate size mechanically resists phagocytosis. The accumulation of microbial products inhibits innate immune function and promotes the transition to anti-inflammatory environment. (**d**) Maturation of the biofilm modulates the host immune response. Hypoxia in and around the biofilm reduces leukocyte killing potential. M2 macrophage and MDSCs prevalence increases leading to reduced T-cell and adaptive immune activation. (**e**) Quorum sensing in the mature biofilm enhances virulence factor production leading to toxin-mediated leukocyte killing, net production, and biofilm dispersal

contamination through phagocytosis, production of antimicrobial products, and early priming of adaptive immunity. The efficacy of this initial response is paramount to containing the initial wave of an infection and enabling further recruitment of leukocytes. During the early stages of biofilm establishment, the ability of PMNs to engage and phagocytose contaminating microbes has been demonstrated to be dependent on aggregate size and multiplicity of infection (Ghimire et al. 2019; Alhede et al. 2020a). One study reported PMN phagocytosis of *Pseudomonas aeruginosa*, *Staphylococcus aureus*, *Escherichia coli*, or *Staphylococcus epidermidis* was only consistent in aggregates less than 5 μm in diameter (Fig. 2b and c) (Alhede et al. 2020a). Similarly, others have demonstrated *S. aureus* aggregates grown for 3 hours prior to neutrophil engagement are able to evade neutrophil defenses (Ghimire et al. 2019). These data imply that only a short window exists wherein microbial aggregates remain small enough to be sufficiently cleared by first-line PMN defenses and emphasize the necessity of rapid leukocyte recruitment to the infection.

Device-associated biofilms may have the competitive advantage of delayed leukocyte recruitment due to devices frequently being implanted in poorly vascularized tissues. PMNs must be able to extravasate from adjacent vessels in order to survey and interact with infectious materials. A recent study by Pettygrove et al. used confocal and intravital imaging to assess the consequences of neutrophil recruitment time to *S. aureus* aggregates. Using a murine subcutaneous model of *S. aureus* implant contamination, the authors observed no significant neutrophil recruitment within 3 hours of implantation and demonstrated that the delayed neutrophil recruitment enabled *S. aureus* biofilm aggregate formation and immune evasion (Pettygrove et al. 2021). Taken together, the ability of contaminating microbial species to quickly grow into aggregate communities prior to encounter by patrolling leukocytes is a key determinant of the progression of biofilm formation and establishment of immune tolerance.

2.2.2 Immune Evasion and Modulation

Biofilms are notoriously resistant to host defenses and interfere with multiple stages of immune activation. The cellular component of the innate immune response is often referred to as the host's first line of defense; however, before pathogens are detected by patrolling leukocytes, it is the humoral component of the innate response that engages and sounds the alarm on invading organisms. To combat this first wave of immune attack, nearly all dominant species implicated in host-associated biofilms express strategies to evade humoral immune detection and skew the host response to an anti-inflammatory phenotype. This is accomplished by inhibiting complement activation, reducing neutrophil and macrophage function, and ultimately impairing the priming of adaptive immunity.

Upon pathogen detection, complement circulating in blood immediately initiates an enzymatic cascade that promotes pathogen killing through opsonin deposition, inflammation, and lysis (Dunkelberger and Song 2010). In response, biofilms produce proteases and proteins that degrade complement components, block opsonization, or interfere with complement receptors on leukocytes (Dunkelberger and Song 2010; Hovingh et al. 2016; Sultan et al. 2018). For example, *P. aeruginosa* production of *Pseudomonas* elastase (PaE) degrades C1q and C3, *S. aureus* Staphylococcal complement inhibitor (SCIN) interferes with C3 convertase and C5a receptors, and *Candida* spp. secreted aspartyl proteases (Sap) 1–3 proteolytically degrade C3b, C4b, and C5 (Rooijakkers et al. 2006; Hovingh et al. 2016; Harpf et al. 2020). The redundant complement inhibition genes within and among common biofilm-associated species strongly imply complement inhibition is an important mechanism for delaying innate immune recruitment and enabling biofilm establishment (Fig. 2c).

As biofilms mature, detection by the immune system is inevitable. Thus, tolerance mechanisms shift from evasion to modulation. Increases in cellular respiration and oxygen consumption due to biofilm growth cause hypoxia in the local microenvironment. This is exacerbated by O_2 consumption by PMNs at the leukocyte biofilm interface and directly impairs PMN and macrophage killing by reducing reactive oxygen species (ROS) and nitric oxide (NO) production (Jesaitis et al. 2003; Kolpen et al. 2010; Wu et al. 2018). Furthermore, hypoxia reduces macrophage glycolysis rates and promotes polarization to the anti-inflammatory M2 phenotype (Yamada and Kielian 2019). Of note, increased oxygen consumption and hypoxia are most often a function of cell density not identity; therefore, this likely represents an additional phenotype common to biofilms that aids in persistence. M2 macrophages are described for their role in inhibiting T-cell proliferation and are characterized by the expression of arginase-1 (Arg-1) and the anti-inflammatory cytokine IL-10 (Heim et al. 2014; Scherr et al. 2014; Heim et al. 2018). Production of Arg-1 inhibits inducible nitric oxide synthase (iNOS) promoting a positive feedback loop that suppresses NO production (Yang and Ming 2014). Similar to M2 macrophages, Arg-1 and IL-10 are produced by myeloid-derived suppressor cells (MDSCs) at the site of biofilm infection. In a *S. aureus* orthopedic biofilm model, MDSCs were reported to constitute >75% of all leukocytes recruited to infection (Heim et al. 2014). Importantly, MDCS have described roles promoting M2 macrophage polarization, inhibiting NK cell cytotoxicity, and blocking T-cell activation. Taken together, nutrient depletion mediated MDCS recruitment and expansion may provide a primary explanation for the poor innate and adaptive immune response observed in chronic biofilm infections (Fig. 2d).

2.2.3 Virulence Factor Production

Concurrent with the immunosuppressive effects described above, biofilm maturation is accompanied by transcriptional changes and virulence production. As cell densities increase with growth and maturation, many pathogenic biofilm species sense and respond to the changing environment through quorum sensing (QS) mechanisms. Often, QS leads to the upregulation of an array of species-specific virulence factors. The local accumulation of specific virulence factors within and around the biofilm contributes to immune tolerance by directly combating leukocyte viability and function and promoting biofilm dispersal.

The QS systems Las and Rhl have been demonstrated to sense PMNs and upregulate the production of rhamnolipids in *P. aeruginosa* biofilms (Christensen et al. 2007; Jensen et al. 2007; Dusane et al. 2010). Rhamnolipids are a class of biological surfactants capable of inducing PMN necrosis. In addition to killing the PMNs, these necrotic events expel neutrophil DNA into the environment that can subsequently be incorporated into the biofilm matrix (Alhede et al. 2020b). Furthermore, the accumulation of rhamnolipids at the biofilm surface can shield the underlying community from host bactericidal products (Roilides et al. 2015). Interestingly, Parks et al. reported *P. aeruginosa* biofilm formation increased up to 3.5-fold in the presence of PMNs, suggesting that QS detection of PMNs may be an active stratagem utilized by *P. aeruginosa* biofilms to enhance matrix structure and community growth (Parks et al. 2009).

The role of the accessory gene regulator (agr) QS system has been extensively studied in *Staphylococcal* species (Queck et al. 2008; Khan et al. 2015; Tan et al. 2018; Rapun-Araiz et al. 2020). During early biofilm growth agr activity is relatively low; however, as cell density increases *Staphylococcal* spp. upregulate *agr* in response to increases in auto-inducing peptide concentrations (Podbielski and Kreikemeyer 2004; Queck et al. 2008; Wang and Muir 2016). The *agr* system along with the *S. aureus* exoprotein expression via the two-component regulatory system *saeRS*, contribute to the regulation of bicomponent pore-forming leukocidins such as HlgAB, LukAB, and LukF-PVL (Voyich et al. 2009; Nygaard et al. 2010; Alonzo and Torres 2014). Bicomponent toxins generally function by forming of pores in the lipid membranes of host cells. Recently, biofilm populations expressing LukAB and LukF-PVL have been demonstrated to induce contact-independent NETosis in infiltrating neutrophils. NETosis is a noninflammatory cell death process that is broadly characterized by the controlled release of neutrophil chromatin, granules, and antimicrobial products (Bhattacharya et al. 2018). Toxin-induced NETosis inhibits the release of proinflammatory cytokines associated with similar necrotic events. Thus, this strategy not only directly removes neutrophils engaging with the biofilm but also quiets chemotactic signals that would otherwise enhance extravasation and recruitment. Furthermore, NET formation may promote a physical barrier precluding leukocyte access to the biofilm interface.

In addition to pore-forming toxins, *agr* directly regulates a class of toxins knowns as phenol soluble modulins (PSMs). Similar to rhamnolipids, PSMs are biological surfactants that have been demonstrated to have leukocidal activity but also play a vital role in biofilm dispersal (Peschel and Otto 2013). Observations using isogenic gene deletion mutants of *agr* reported that strains lacking *agr* had significant defects in biofilm dispersal compared to wildtype (Guggenberger et al. 2012). Notably, recent in vivo models using isogenic gene deletion mutants of PSMs have recapitulated the biofilm dispersal defect associated with *agr* mutants (Le et al. 2019). Taken together, these data imply Staphylococcal biofilm dispersal is highly dependent on *agr*-regulated PSM production. Moreover, QS enables biofilm communities to sense and respond to the surrounding environment allowing reciprocations to host defenses that enhance biofilm survival (Fig. 2e).

3 Conclusion

Whether in an industrial pipe or on an implanted medical device, biofilms manifest tolerance to antimicrobial agents and host defenses. They do this mostly through multicellular mechanisms. Elaboration of extracellular polymers and cross-linking proteins deploys collective resilience to mechanical challenges including chemical removal agents and phagocytosis. The concerted metabolism of the biofilm community establishes graded chemical microenvironments that can harbor not only a multitude of species but a variety of physiological states for each organism. Arrested growth, dormancy, and persister cells are among the cell states that may afford protection. The collective expression of active stress responses or toxins that target innate immune cells is naturally strongest in an aggregate of microorganisms. While vast diversity in biofilm composition, structure, and metabolism and growth are present, the mechanisms described here represent recurrent themes in the persistence and resilience of biofilm communities against antimicrobial and host immune attack.

References

Ahn S, Jung J, Jang IA, Madsen EL, Park W (2016) Role of Glyoxylate shunt in oxidative stress response. J Biol Chem 291:11928–11938

Alhede M, Lorenz M, Fritz BG, Jensen PO, Ring HC, Bay L, Bjarnsholt T (2020a) Bacterial aggregate size determines phagocytosis efficiency of polymorphonuclear leukocytes. Med Microbiol Immunol 209:669–680

Alhede M, Qvortrup K, Kragh KN, Jensen PO, Stewart PS, Bjarnsholt T (2020b) The origin of extracellular DNA in bacterial biofilm infections *in vivo*. Pathogens and Disease 78(2):ftaa018

Allison KR, Brynildsen MP, Collins JJ (2011) Metabolite-enabled eradication of bacterial persisters by aminoglycosides. Nature 473:216-+

Alonzo F, Torres VJ (2014) The Bicomponent pore-forming leucocidins of *Staphylococcus aureus*. Microbiol Mol Biol Rev 78:199–230

Amato SM, Brynildsen MP (2015) Persister heterogeneity arising from a single metabolic stress. Curr Biol 25:2090–2098

Amato SM, Fazen CH, Henry TC, Mok WWK, Orman MA, Sandvik EL, Volzing KG, Brynildsen MP (2014) The role of metabolism in bacterial persistence. Front Microbiol 5:70

Anderl JN, Zahller J, Roe F, Stewart PS (2003) Role of nutrient limitation and stationary-phase existence in *Klebsiella pneumoniae* biofilm resistance to ampicillin and ciprofloxacin. Antimicrob Agents Chemother 47:1251–1256

Bernier SP, Lebeaux D, Defrancesco AS, Valomon A, Soubigou G, Coppee JY, Ghigo JM, Beloin C (2013) Starvation, together with the SOS response, mediates high biofilm-specific tolerance to the fluoroquinolone Ofloxacin. PLoS Genet 9(1):e1003144

Bhattacharya M, Berends ETM, Chan R, Schwab E, Roy S, Sen CK, Torres VJ, Wozniak DJ (2018) *Staphylococcus aureus* biofilms release leukocidins to elicit extracellular trap formation and evade neutrophil-mediated killing. Proc Natl Acad Sci U S A 115:7416–7421

Billings N, Millan MR, Caldara M, Rusconi R, Tarasova Y, Stocker R, Ribbeck K (2013) The extracellular matrix component Psl provides fast-acting antibiotic defense in *Pseudomonas aeruginosa* biofilms. PLoS Pathog 9(8):e1003526

Bjarnsholt T, Jensen PO, Burmolle M, Hentzer M, Haagensen JAJ, Hougen HP, Calum H, Madsen KG, Moser C, Molin S, Hoiby N, Givskov M (2005) *Pseudomonas aeruginosa* tolerance to tobramycin, hydrogen peroxide and polymorphonuclear leukocytes is quorum-sensing dependent. Microbiology-SGM 151:373–383

Borriello G, Werner E, Roe F, Kim AM, Ehrlich GD, Stewart PS (2004) Oxygen limitation contributes to antibiotic tolerance of *Pseudomonas aeruginosa* in biofilms. Antimicrob Agents Chemother 48:2659–2664

Borriello G, Richards L, Ehrlich GD, Stewart PS (2006) Arginine or nitrate enhances antibiotic susceptibility of *Pseudomonas aeruginosa* in biofilms. Antimicrob Agents Chemother 50:382–384

Christensen LD, Moser C, Jensen PO, Rasmussen TB, Christophersen L, Kielleberg S, Kumar N, Hoiby N, Givskov M, Bjarnsholt T (2007) Impact of *Pseudomonas aeruginosa* quorum sensing on biofilm persistence in an *in vivo* intraperitoneal foreign-body infection model. Microbiology-SGM 153:2312–2320

Ciofu O, Tolker-Nielsen T (2019) Tolerance and resistance of *Pseudomonas aeruginosa* biofilms to antimicrobial agents-how *P. aeruginosa* can escape antibiotics. Front Microbiol 10:913

Ciofu O, Moser C, Jensen PO, Hoiby N (2022) Tolerance and resistance of microbial biofilms. Nat Rev Microbiol. https://doi.org/10.1038/s41579-022-00682-4

Colvin KM, Gordon VD, Murakami K, Borlee BR, Wozniak DJ, Wong GCL, Parsek MR (2011) The Pel polysaccharide can serve a structural and protective role in the biofilm matrix of *Pseudomonas aeruginosa*. PLoS Pathog 7(1):e1001264

Crabbe A, Jensen PO, Bjarnsholt T, Coenye T (2019) Antimicrobial tolerance and metabolic adaptations in microbial biofilms. Trends Microbiol 27:850–863

Dal Co A, van Vliet S, Ackermann M (2019) Emergent microscale gradients give rise to metabolic cross-feeding and antibiotic tolerance in clonal bacterial populations. Philos Trans R Soc Lond B Biol Sci 374(1786):20190080

Dal Co A, van Vliet S, Kiviet DJ, Schlegel S, Ackermann M (2020) Short-range interactions govern the dynamics and functions of microbial communities. Nature Ecology & Evolution 4:366-+

Dar D, Dar N, Cai L, Newman DK (2021) Spatial transcriptomics of planktonic and sessile bacterial populations at single-cell resolution. Science 373:758-+

Davison WM, Pitts B, Stewart PS (2010) Spatial and temporal patterns of biocide action against *Staphylococcus epidermidis* biofilms. Antimicrob Agents Chemother 54:2920–2927

Debeer D, Srinivasan R, Stewart PS (1994) Direct measurement of chlorine penetration into biofilms during disinfections. Appl Environ Microbiol 60:4339–4344

Dunkelberger JR, Song WC (2010) Complement and its role in innate and adaptive immune responses. Cell Res 20:34–50

Dusane DH, Zinjarde SS, Venugopalan VP, Mclean RJC, Weber MM, Rahman P (2010) Quorum sensing: implications on Rhamnolipid biosurfactant production. Biotechnol Genet Eng Rev 27(27):159–184

Elkins JG, Hassett DJ, Stewart PS, Schweizer HP, Mcdermott TR (1999) Protective role of catalase in *Pseudomonas aeruginosa* biofilm resistance to hydrogen peroxide. Appl Environ Microbiol 65:4594–4600

Field TR, White A, Elborn JS, Tunney MM (2005) Effect of oxygen limitation on the *in vitro* antimicrobial susceptibility of clinical isolates of *Pseudomonas aeruginosa* grown planktonically and as biofilms. Eur J Clin Microbiol Infect Dis 24:677–687

Folsom JP, Richards L, Pitts B, Roe F, Ehrlich GD, Parker A, Mazurie A, Stewart PS (2010) Physiology of *Pseudomonas aeruginosa* in biofilms as revealed by transcriptome analysis. BMC Microbiol 10:294

Franklin MJ, Sandvik E, Yanardag S, Williamson KS (2020) Functional characterization of the *Pseudomonas aeruginosa* ribosome hibernation-promoting factor. J Bacteriol 202(19): e00280-20

Gefen O, Fridman O, Ronin I, Balaban NQ (2014) Direct observation of single stationary-phase bacteria reveals a surprisingly long period of constant protein production activity. Proc Natl Acad Sci 111:556–561

Ghimire N, Pettygrove BA, Pallister KB, Stangeland J, Stanhope S, Klapper I, Voyich JM, Stewart PS (2019) Direct microscopic observation of human neutrophil-*Staphylococcus aureus* interaction *in vitro* suggests a potential mechanism for initiation of biofilm infection on an implanted medical device. Infect Immun 87(12):e00745-19

Goltermann L, Tolker-Nielsen T (2017) Importance of the exopolysaccharide matrix in antimicrobial tolerance of *Pseudomonas aeruginosa* aggregates. Antimicrob Agents Chemother 61(4): e02696-16

Gonzalez JF, Hahn MM, Gunn JS (2018) Chronic biofilm-based infections: skewing of the immune response. Pathogens and Disease 76(3):fty023

Goodwine J, Gil J, Doiron A, Valdes J, Solis M, Higa A, Davis S, Sauer K (2019) Pyruvate-depleting conditions induce biofilm dispersion and enhance the efficacy of antibiotics in killing biofilms *in vitro* and *in vivo*. Sci Rep 9(1):3763

Guggenberger C, Wolz C, Morrissey JA, Heesemann J (2012) Two distinct coagulase-dependent barriers protect *Staphylococcus aureus* from neutrophils in a three dimensional *in vitro* infection model. PLoS Pathog 8(1):e1002434

Gunther F, Wabnitz GH, Stroh P, Prior B, Obst U, Samstag Y, Wagner C, Hansch GM (2009) Host defence against *Staphylococcus aureus* biofilms infection: phagocytosis of biofilms by polymorphonuclear neutrophils (PMN). Mol Immunol 46:1805–1813

Gupta S, Laskar N, Kadouri DE (2016) Evaluating the effect of oxygen concentrations on antibiotic sensitivity, growth, and biofilm formation of human pathogens. Microbiology insights 9:37–46

Hahn MM, Gonzalez JF, Gunn JS (2021) *Salmonella* biofilms tolerate hydrogen peroxide by a combination of extracellular polymeric substance barrier function and catalase enzymes. Front Cell Infect Microbiol 11:683081

Hall CW, Mah TF (2017) Molecular mechanisms of biofilm-based antibiotic resistance and tolerance in pathogenic bacteria. FEMS Microbiol Rev 41:276–301

Harpf V, Rambach G, Wurzner R, Lass-Florl C, Speth C (2020) *Candida* and Complement: new aspects in an old Battle. Front Immunol 11:1471

Harrison JJ, Ceri H, Turner RJ (2007) Multimetal resistance and tolerance in microbial biofilms. Nat Rev Microbiol 5:928–938

Harrison JJ, Wade WD, Akierman S, Vacchi-Suzzi C, Stremick CA, Turner RJ, Ceri H (2009) The chromosomal toxin gene *yafQ* is a determinant of multidrug tolerance for *Escherichia coli* growing in a biofilm. Antimicrob Agents Chemother 53:2253–2258

Hassett DJ, Cuppoletti J, Trapnell B, Lymar SV, Rowe JJ, Yoon SS, Hilliard GM, Parvatiyar K, Kamani MC, Wozniak DJ, Hwang SH, Mcdermott TR, Ochsner UA (2002) Anaerobic metabolism and quorum sensing by *Pseudomonas aeruginosa* biofilms in chronically infected cystic fibrosis airways: rethinking antibiotic treatment strategies and drug targets. Adv Drug Deliv Rev 54:1425–1443

Heim CE, Vidlak D, Scherr TD, Kozel JA, Holzapfel M, Muirhead DE, Kielian T (2014) Myeloid-derived suppressor cells contribute to *Staphylococcus aureus* orthopedic biofilm infection. J Immunol 192:3778–3792

Heim CE, Vidlak D, Odvody J, Hartman CW, Garvin KL, Kielian T (2018) Human prosthetic joint infections are associated with myeloid-derived suppressor cells (MDSCs): implications for infection persistence. J Orthop Res 36:1605–1613

Hentzer M, Teitzel GM, Balzer GJ, Heydorn A, Molin S, Givskov M, Parsek MR (2001) Alginate overproduction affects *Pseudomonas aeruginosa* biofilm structure and function. J Bacteriol 183:5395–5401

Hill D, Rose B, Pajkos A, Robinson M, Bye P, Bell S, Elkins M, Thompson B, Macleod C, Aaron SD, Harbour C (2005) Antibiotic susceptibilities of *Pseudomonas aeruginosa* isolates derived from patients with cystic fibrosis under aerobic, anaerobic, and biofilm conditions. J Clin Microbiol 43:5085–5090

Hobbs JK, Boraston AB (2019) (p)ppGpp and the stringent response: an emerging threat to antibiotic therapy. ACS Infectious Diseases 5:1505–1517

Hovingh ES, van den Broek B, Jongerius I (2016) Hijacking complement regulatory proteins for bacterial immune evasion. Frontiers in Microbiology, p 7

Huang CT, Xu KD, Mcfeters GA, Stewart PS (1998) Spatial patterns of alkaline phosphatase expression within bacterial colonies and biofilms in response to phosphate starvation. Appl Environ Microbiol 64:1526–1531

Jefferson KK, Goldmann DA, Pier GB (2005) Use of confocal microscopy to analyze the rate of vancomycin penetration through *Staphylococcus aureus* biofilms. Antimicrob Agents Chemother 49:2467–2473

Jensen PO, Bjarnsholt T, Phipps R, Rasmussen TB, Calum H, Christoffersen L, Moser C, Williams P, Pressler T, Givskov M, Hoiby N (2007) Rapid necrotic killing of polymorphonuclear leukocytes is caused by quorum-sensing-controlled production of rhamnolipid by *Pseudomonas aeruginosa*. Microbiology-SGM 153:1329–1338

Jesaitis AJ, Franklin MJ, Berglund D, Sasaki M, Lord CI, Bleazard JB, Duffy JE, Beyenal H, Lewandowski Z (2003) Compromised host defense on *Pseudomonas aeruginosa* biofilms: characterization of neutrophil and biofilm interactions. J Immunol 171:4329–4339

Keren I, Kaldalu N, Spoering A, Wang YP, Lewis K (2004) Persister cells and tolerance to antimicrobials. FEMS Microbiol Lett 230:13–18

Khan BA, Yeh AJ, Cheung GYC, Otto M (2015) Investigational therapies targeting quorum-sensing for the treatment of *Staphylococcus aureus* infections. Expert Opin Investig Drugs 24:689–704

King P, Citron DM, Griffith DC, Lomovskaya O, Dudley MN (2010) Effect of oxygen limitation on the *in vitro* activity of levofloxacin and other antibiotics administered by the aerosol route against *Pseudomonas aeruginosa* from cystic fibrosis patients. Diagn Microbiol Infect Dis 66: 181–186

Kolpen M, Hansen CR, Bjarnsholt T, Moser C, Christensen LD, van Gennip M, Ciofu O, Mandsberg L, Kharazmi A, Doring G, Givskov M, Hoiby N, Jensen PO (2010) Polymorphonuclear leucocytes consume oxygen in sputum from chronic *Pseudomonas aeruginosa* pneumonia in cystic fibrosis. Thorax 65:57–62

Kolpen M, Lerche CJ, Kragh KN, Sams T, Koren K, Jensen AS, Line L, Bjarnsholt T, Ciofu O, Moser C, Kuhl M, Hoiby N, Jensen PO (2017) Hyperbaric oxygen sensitizes anoxic *Pseudomonas aeruginosa* biofilm to ciprofloxacin. Antimicrob Agents Chemother 61(11):e01024-17

Le KY, Villaruz AE, Zheng Y, He L, Fisher EL, Nguyen TH, Ho TV, Yeh AJ, Joo HS, Cheung GYC, Otto M (2019) Role of phenol-soluble Modulins in *Staphylococcus epidermidis* biofilm formation and infection of indwelling medical devices. J Mol Biol 431:3015–3027

Leid, J. G. 2009. Bacterial biofilms resist key host defenses

Levin BR, Rozen DE (2006) Opinion - non-inherited antibiotic resistance. Nat Rev Microbiol 4: 556–562

Ma DZ, Mandell JB, Donegan NP, Cheung AL, Ma WY, Rothenberger S, Shanks RMQ, Richardson AR, Urish KL (2019) The toxin-antitoxin MazEF drives *Staphylococcus aureus* biofilm formation, antibiotic tolerance, and chronic infection. mBio 10(6):e01658-19

Mah TF, Pitts B, Pellock B, Walker GC, Stewart PS, O'Toole GA (2003) A genetic basis for *Pseudomonas aeruginosa* biofilm antibiotic resistance. Nature 426:306–310

Marques CNH, Nelson SM (2019) Pharmacodynamics of ciprofloxacin against *Pseudomonas aeruginosa* planktonic and biofilm-derived cells. Lett Appl Microbiol 68:350–359

Meylan S, Porter CBM, Yang JH, Belenky P, Gutierrez A, Lobritz MA, Park J, Kim SH, Moskowitz SM, Collins JJ (2017) Carbon sources tune antibiotic susceptibility in *Pseudomonas aeruginosa* via tricarboxylic acid cycle control. Cell Chemical Biology 24:195–206

Nguyen D, Joshi-Datar A, Lepine F, Bauerle E, Olakanmi O, Beer K, Mckay G, Siehnel R, Schafhauser J, Wang Y, Britigan BE, Singh PK (2011) Active starvation responses mediate antibiotic tolerance in biofilms and nutrient-limited bacteria. Science 334:982–986

Nikolaev YA, Plakunov VK (2007) Biofilm – "City of microbes" or an analogue of multicellular organisms? Microbiology 76:125–138

Nilsson M, Jakobsen TH, Givskov M, Twetman S, Tolker-Nielsen T (2019) Oxidative stress response plays a role in antibiotic tolerance of *Streptococcus mutans* biofilms. Microbiology-SGM 165:334–342

Nygaard TK, Pallister KB, Ruzevich P, Griffith S, Vuong C, Voyich JM (2010) SaeR binds a consensus sequence within virulence gene promoters to advance USA300 pathogenesis. J Infect Dis 201:241–254

Nystrom T (2004) Stationary-phase physiology. Annu Rev Microbiol 58:161–181

Pamp SJ, Gjermansen M, Johansen HK, Tolker-Nielsen T (2008) Tolerance to the antimicrobial peptide colistin in *Pseudomonas aeruginosa* biofilms is linked to metabolically active cells, and depends on the *pmr* and *mexAB-oprM* genes. Mol Microbiol 68:223–240

Parks QM, Young RL, Poch KR, Malcolm KC, Vasil ML, Nick JA (2009) Neutrophil enhancement of *Pseudomonas aeruginosa* biofilm development: human F-actin and DNA as targets for therapy. J Med Microbiol 58:492–502

Penesyan A, Paulsen IT, Kjelleberg S, Gillings MR (2021) Three faces of biofilms: a microbial lifestyle, a nascent multicellular organism, and an incubator for diversity. NPJ Biofilms and Microbiomes 7(1):80

Peschel A, Otto M (2013) Phenol-soluble modulins and staphylococcal infection (vol 11, pg 667, 2013). Nat Rev Microbiol 11(10):667–673

Pettygrove BA, Kratofil RM, Alhede M, Jensen PO, Newton M, Qvortrup K, Pallister KB, Bjarnsholt T, Kubes P, Voyich JM, Stewart PS (2021) Delayed neutrophil recruitment allows nascent *Staphylococcus aureus* biofilm formation and immune evasion. Biomaterials 275: 120775

Podbielski A, Kreikemeyer B (2004) Cell density - dependent regulation: basic principles and effects on the virulence of gram-positive cocci. Int J Infect Dis 8:81–95

Polikanov YS, Blaha GM, Steitz TA (2012) How hibernation factors RMF, HPF, and YfiA turn off protein synthesis. Science 336:915–918

Queck SY, Jameson-Lee M, Villaruz AE, Bach THL, Khan BA, Sturdevant DE, Ricklefs SM, Li M, Otto M (2008) RNAIII-independent target gene control by the *agr* quorum-sensing system: insight into the evolution of virulence regulation in *Staphylococcus aureus*. Mol Cell 32:150–158

Rani SA, Pitts B, Stewart PS (2005) Rapid diffusion of fluorescent tracers into *Staphylococcus epidermidis* biofilms visualized by time lapse microscopy. Antimicrob Agents Chemother 49: 728–732

Rani SA, Pitts B, Beyenal H, Veluchamy RA, Lewandowski Z, Davison WM, Buckingham-Meyer K, Stewart PS (2007) Spatial patterns of DNA replication, protein synthesis, and oxygen concentration within bacterial biofilms reveal diverse physiological states. J Bacteriol 189: 4223–4233

Rapun-Araiz B, Haag AF, de Cesare V, Gil C, Dorado-Morales P, Penades JR, Lasa I (2020) Systematic reconstruction of the complete two-component sensorial network in *Staphylococcus aureus*. mSystems 5(4):e00511-20

Roilides E, Simitsopoulou M, Katragkou A, Walsh TJ (2015) How biofilms evade host defenses. Microbiol Spectr 3(3). https://doi.org/10.1128/microbiolspec.MB-0012-2014

Rooijakkers SHM, Ruyken M, van Roon J, van Kessel KPM, van Strijp JAG, van Wamel WJB (2006) Early expression of SCIN and CHIPS drives instant immune evasion by *Staphylococcus aureus*. Cell Microbiol 8:1282–1293

Schaible B, Taylor CT, Schaffer K (2012) Hypoxia increases antibiotic resistance in *Pseudomonas aeruginosa* through altering the composition of multidrug efflux pumps. Antimicrob Agents Chemother 56:2114–2118

Scherr TD, Heim CE, Morrison JM, Kielian T (2014) Hiding in plain sight: interplay between staphylococcal biofilms and host immunity. Front Immunol 5:37

Shapiro JA (1998) Thinking about bacterial populations as multicellular organisms. Annu Rev Microbiol 52:81–104

Shigeta M, Tanaka G, Komatsuzawa H, Sugai M, Suginaka H, Usui T (1997) Permeation of antimicrobial agents through *Pseudomonas aeruginosa* biofilms: A simple method. Chemotherapy 43:340–345

Sonderholm M, Bjarnsholt T, Alhede M, Kolpen M, Jensen PO, Kuhl M, Kragh KN (2017) The consequences of being in an infectious biofilm: microenvironmental conditions governing antibiotic tolerance. Int J Mol Sci 18(12):2688

Song S, Wood TK (2020) ppGpp ribosome dimerization model for bacterial persister formation and resuscitation. Biochem Biophys Res Commun 523:281–286

Spoering AL, Lewis K (2001) Biofilms and planktonic cells of *Pseudomonas aeruginosa* have similar resistance to killing by antimicrobials. J Bacteriol 183:6746–6751

Stewart PS (2015) Antimicrobial tolerance in biofilms. Microbiol Spectr 3(3). https://doi.org/10.1128/microbiolspec.MB-0010-2014

Stewart PS, Roe F, Rayner J, Elkins JG, Lewandowski Z, Ochsner UA, Hassett DJ (2000) Effect of catalase on hydrogen peroxide penetration into *Pseudomonas aeruginosa* biofilms. Appl Environ Microbiol 66:836–838

Stewart PS, Davison WM, Steenbergen JN (2009) Daptomycin rapidly penetrates a *Staphylococcus epidermidis* biofilm. Antimicrob Agents Chemother 53:3505–3507

Stewart PS, Franklin MJ, Williamson KS, Folsom JP, Boegli L, James GA (2015) Contribution of stress responses to antibiotic tolerance in *Pseudomonas aeruginosa* biofilms. Antimicrob Agents Chemother 59:3838–3847

Stewart PS, White B, Boegli L, Hamerly T, Williamson KS, Franklin MJ, Bothner B, James GA, Fisher S, Vital-Lopez FG, Wallqvist A (2019) Conceptual model of biofilm antibiotic tolerance that integrates phenomena of diffusion, metabolism, gene expression, and physiology. J Bacteriol 201(22):e00307-19

Sultan AR, Swierstra JW, Lemmens-Den Toom NA, Snijders SV, Manaskova SH, Verbon A, van Wamel WJB (2018) Production of staphylococcal complement inhibitor (SCIN) and other immune modulators during the early stages of *Staphylococcus aureus* biofilm formation in a mammalian cell culture medium. Infect Immun 86(8):e00352-18

Taber HW, Mueller JP, Miller PF, Arrow AS (1987) Bacterial uptake of aminoglycoside antibiotics. Microbiol Rev 51:439–457

Tan L, Li SR, Jiang B, Hu XM, Li S (2018) Therapeutic targeting of the *Staphylococcus aureus* accessory gene regulator (*agr*) System. Front Microbiol 9:55

Tseng BS, Zhang W, Harrison JJ, Quach TP, Song JL, Penterman J, Singh PK, Chopp DL, Packman AI, Parsek MR (2013) The extracellular matrix protects *Pseudomonas aeruginosa* biofilms by limiting the penetration of tobramycin. Environ Microbiol 15:2865–2878

van Acker H, Sass A, Bazzini S, de Roy K, Udine C, Messiaen T, Riccardi G, Boon N, Nelis HJ, Mahenthiralingam E, Coenye T (2013) Biofilm-grown *Burkholderia cepacia* complex cells survive antibiotic treatment by avoiding production of reactive oxygen species. PLoS One 8(3): e58943

van Acker H, Sass A, Dhondt I, Nelis HJ, Coenye T (2014) Involvement of toxin-antitoxin modules in *Burkholderia cenocepacia* biofilm persistence. Pathogens and Disease 71:326–335

Verderosa AD, Totsika M, Fairfull-Smith KE (2019) Bacterial biofilm eradication agents: A current review. Frontiers in Chemistry, p 7

Voyich JM, Vuong C, Dewald M, Nygaard TK, Kocianova S, Griffith S, Jones J, Iverson C, Sturdevant DE, Braughton KR, Whitney AR, Otto M, Deleo FR (2009) The SaeR/S gene regulatory system is essential for innate immune evasion by *Staphylococcus aureus*. J Infect Dis 199:1698–1706

Vrany JD, Stewart PS, Suci PA (1997) Comparison of recalcitrance to ciprofloxacin and levofloxacin exhibited by *Pseudomonas aeruginosa* bofilms displaying rapid-transport characteristics. Antimicrob Agents Chemother 41:1352–1358

Walters MC, Roe F, Bugnicourt A, Franklin MJ, Stewart PS (2003) Contributions of antibiotic penetration, oxygen limitation, and low metabolic activity to tolerance of *Pseudomonas aeruginosa* biofilms to ciprofloxacin and tobramycin. Antimicrob Agents Chemother 47:317–323

Wang BY, Muir TW (2016) Regulation of virulence in *Staphylococcus aureus*: molecular mechanisms and remaining puzzles. Cell Chemical Biology 23:214–224

Watnick P, Kolter R (2000) Biofilm, city of microbes. J Bacteriol 182:2675–2679

Werner E, Roe F, Bugnicourt A, Franklin MJ, Heydorn A, Molin S, Pitts B, Stewart PS (2004) Stratified growth in *Pseudomonas aeruginosa* biofilms. Appl Environ Microbiol 70:6188–6196

Williamson KS, Richards LA, Perez-Osorio AC, Pitts B, Mcinnerney K, Stewart PS, Franklin MJ (2012) Heterogeneity in *Pseudomonas aeruginosa* biofilms includes expression of ribosome hibernation factors in the antibiotic-tolerant subpopulation and hypoxia-induced stress response in the metabolically active population. J Bacteriol 194:2062–2073

Wu YL, Klapper I, Stewart PS (2018) Hypoxia arising from concerted oxygen consumption by neutrophils and microorganisms in biofilms. Pathogens and Disease 76(4):fty043

Yamada KJ, Kielian T (2019) Biofilm-leukocyte cross-talk: impact on immune polarization and Immunometabolism. J Innate Immun 11:280–288

Yan J, Bassler BL (2019) Surviving as a community: antibiotic tolerance and persistence in bacterial biofilms. Cell Host Microbe 26:15–21

Yang ZH, Ming XF (2014) Functions of arginase isoforms in macrophage inflammatry responses: impact on cardiovascular diseases and metabolic disorders. Front Immunol 5:533

Antibiotic Resistance Development in Bacterial Biofilms

María D. Macià and Antonio Oliver

Abstract Biofilms are organized bacterial communities embedded in an extracellular polymeric matrix attached to living or abiotic surfaces or forming aggregates. This mode of growth is characteristically associated with chronic infections, marked by the persistence of the bacteria regardless of proper antimicrobial therapies. Biofilms are up to 100–1000 times more resistant to antibiotics than planktonic cells. The basis for resistance to antimicrobials in biofilms is complex and involves both mechanisms of tolerance and classical resistance. Physical, physiological, adaptive and in vivo tolerance together with the *persister* phenomenon and the expression of specific genes lay at the origin of the recalcitrant (or tolerant) nature of the biofilm to antibiotics. On the other hand, mutational resistance and horizontal gene transfer determine the genetic changes leading to genuine resistance to antimicrobials. In this chapter, using the bacterium *Pseudomonas aeruginosa* as an archetypal microorganism, we explore the different mechanisms of antibiotic resistance development in biofilms.

1 Introduction

Biofilms are classically defined as organized bacterial communities embedded in an extracellular polymeric matrix attached to living or abiotic surfaces (Stoodley et al. 2002). This social bacterial structure is highly prevalent in nature (Petrova and Sauer 2012) and arises as an adaptation strategy for survival in hostile environments, including the human host (Fux et al. 2005). From a clinical point of view, the transition from a planktonic to a biofilm mode of growth frequently determines the establishment of a chronic infection (Smith et al. 2006; Oliver et al. 2007). Consequently, biofilms are representatively implicated in chronic infections, in contrast to

M. D. Macià (✉) · A. Oliver
Microbiology Service, Hospital Son Espases, Instituto de Investigación Sanitaria Illes Balears (IdISBa), Palma de Mallorca and CIBER Enfermedades Infecciosas. Instituto de Salud Carlos III, Madrid, Spain
e-mail: mariad.macia@ssib.es

K. Richter, K. Nørskov Kragh (eds.), *Antibiofilm Strategies*, Springer Series on Biofilms 11, https://doi.org/10.1007/978-3-031-10992-8_3

planktonic bacteria that are involved in acute processes (Costerton et al. 1999). Biofilm-related chronic infections are characterized by the persistence of the aetiological agent, despite (in principle) adequate antibiotic therapy and host immune responses. In general, up to 65–80% of all infections are associated with biofilm formation (Jamal et al. 2018). These infections encompass a wide range, from those related to indwelling medical devices (e.g., catheters, prosthetic joints, and surgical implants) to tissue infections, such as otitis media, osteomyelitis, rhinosinusitis, wound infections or chronic respiratory infections in cystic fibrosis (CF) patients (Høiby et al. 2015a).

In device-associated chronic infections, replacing the implant to remove biofilms is an effective intervention strategy. However, for tissue or respiratory chronic biofilm infections, the only available therapy to date is antibiotic treatment. Long-term therapies with antibiotics, despite in vitro susceptibility, frequently lead to a longstanding inflammatory response, together with a high risk of antibiotic resistance development, thus triggering an unavoidable chronic infection extremely difficult to eradicate.

One of the most important characteristics of biofilms is their increased tolerance to antimicrobial agents (Wimpenny et al. 2000). Compared to planktonic cells biofilms are up to 100–1000 times less susceptible to antibiotics and disinfectants (Costerton et al. 1987, 1999; Bjarnsholt et al. 2005).

The mechanisms of resistance and tolerance differ depending on the specific antimicrobial agent, the bacterial strain and species, the age and developmental stage of the biofilm, and the biofilm growth conditions (Ito et al. 2009; Alhede et al. 2011; Bowler et al. 2012; Haaber et al. 2012; Stewart et al. 2015).

Commonly for biofilm infections, the terms tolerance and resistance are used synonymous. Moreover, the term tolerance recurrently encompasses, or even replaces, that of resistance. However, it is necessary to distinguish their different conceptual meaning.

In general, tolerance of biofilms to antibiotics is multifactorial and includes physical, physiological, environmental, and inducible genetic factors, whereas antibiotic resistance needs the involvement of acquired or heritable processes that alter the genetic background of bacteria and do not revert in the absence of antibiotics. In other words, tolerance can be defined as the survival in the presence of the antibiotic due to the biofilm itself. In this case, single dispersed cells from biofilms are susceptible to antibiotics. The development of conventional resistance requires stable genetic changes like point mutations or horizontal gene transfer (Lewis 2008). In all likelihood, the tolerance phenomenon precedes and promotes antibiotic resistance development in biofilms. Though ultimately, both tolerance and resistance mechanisms act synergistically, thus seriously reducing the options of an effective antimicrobial treatment for chronic infections.

Understanding the mechanisms underlying biofilm-specific antibiotic resistance and tolerance will help guide the selection of efficient treatments to fight chronic infections.

In this chapter, using the bacterium *Pseudomonas aeruginosa* as a model, we examine the different mechanisms of resistance development in biofilms.

2 Mechanisms of Tolerance to Antibiotics in Biofilms

2.1 *Physical Tolerance*

Traditionally, it was considered that diffusion of antibiotics was restricted by the extracellular biofilm matrix. While biofilm matrices do not inhibit the diffusion of antibiotic molecules, restricted penetration through biofilms may occur when antibiotics bind to components of the matrix or the bacterial membranes (Walters et al. 2003; Tseng et al. 2013). As an example, using fluorescently labeled antibiotics it has been demonstrated that positively charged peptides and aminoglycosides, like tobramycin, are sequestered in the negatively charged extracellular matrix (Walters et al. 2003; Chiang et al. 2013), whereas the neutral ciprofloxacin readily penetrates into the biofilms (Tseng et al. 2013). The presence of anionic extracellular DNA (eDNA) and alginate in the extracellular matrix are responsible for the capture and consequently the reduced activity observed for positively charged antibiotic classes, rather than an inability to diffuse across the biofilm mass (Cao et al. 2016; Mulcahy et al. 2008). This retardation in the entrance of the antibiotic, which ultimately exposes the biofilm to lower antibiotic concentrations, gives time and helps the bacteria to adapt to a more tolerant state (Bagge et al. 2004a; Pamp et al. 2008a; Kindrachuk et al. 2011; Ciofu and Tolker-Nielsen 2019).

2.2 *Physiological Tolerance*

Another striking characteristic of biofilms is their compartmentalization in different metabolic states. This heterogeneity is due to and at the same time a consequence of the decreasing concentrations of oxygen and nutrients along the biofilm that established gradients of metabolic stages. Thus, the consumption of oxygen and nutrients by the metabolically active bacteria located in the periphery of the biofilm leave scarce oxygen and nutrients to the bacteria placed inside the biofilm (Stewart et al. 2016). Consequently, subpopulations at the periphery of the biofilm show an important physiological activity, whereas subpopulations positioned at the inner parts of the biofilm show a significantly reduced physiological activity or even a total lack of activity with no evidence of growth (Pamp et al. 2008a; Ciofu and Tolker-Nielsen 2019). Numerous antibiotics target processes that take place in metabolically active bacteria (e.g. replication, transcription, translation, and cell wall synthesis), thus biofilm areas with low metabolic activity may display increased antimicrobial tolerance with an impact on the general biofilm susceptibility. Likewise, it has been shown that some antibiotics that work under aerobic conditions (e.g., tobramycin, ciprofloxacin and tetracycline) are only effective on biofilm cells located at the air interface and not at other areas of the microcolony (Walters et al. 2003; Pamp et al. 2008a; Stewart et al. 2016). In addition, it has been recently shown that reactive oxygen species (ROS) are thought to play a role in antibiotic-induced cell death, so hypoxia may additionally promote biofilm cell survival upon antibiotic exposure due to the fact that ROS production requires molecular oxygen (Van Acker et al. 2013; Jensen et al. 2014; Van Acker and Coenye 2016). Thus, other than the gradient of nutrients, low oxygen concentrations along the biofilm impair the efficacy of ROS-dependent bactericidal antibiotics and decrease bacterial metabolism (Jensen et al. 2014; Van Acker and Coenye 2016; Van Acker et al. 2016).

Related to this, it is well known that hypoxia is associated with increasing antibiotic tolerance through the upregulation of efflux pumps, which is an induced, non-specific antibiotic tolerance mechanism. Concerning biofilms, it is noteworthy that various studies have revealed differential expression of several conventional and biofilm-specific antibiotic resistance genes in biofilms (Whiteley et al. 2001). It has been shown that in oxygen-starved planktonic cells, MexEF-OprN efflux pump genes are upregulated and contribute to antibiotic tolerance (Schaible et al. 2010) like MexEF-OprN or MexCD-OprJ. Moreover, the MexCD-OprJ efflux pump locus was shown to be transcriptionally upregulated in *P. aeruginosa* biofilms grown under low oxygen conditions (Tata et al. 2016), establishing a close connection between physiological tolerance and adaptive resistance mechanisms.

In addition to the reduction of metabolic activity, impairment of the efficacy of ROS-dependent antibiotics, and upregulation of efflux pumps; the compartmentalization of the biofilm microenvironment due to the declining gradients of oxygen and nutrients could lead to areas of starved cells. It has been shown that responses to nutrient and iron starvation, like the stringent response, also play an important role in the tolerance of biofilms to antibiotics (Nguyen et al. 2011; Bernier et al. 2013).

2.3 Persisters

The most extreme stage regarding metabolic inactivity could be represented by the characteristic dormant or persistent cells of biofilms; popularly called *persisters*. This is a spore-like cell state, dormant and incapable of growth, that protects bacteria from the action of antibiotics (Lewis 2008, 2010). However, as a tolerance mechanism, the refractoriness to antibiotics is not heritable and when *persister* cells are segregated from the biofilm and grown individually, they display the same pattern of antibiotic susceptibility as the original population (Brauner et al. 2016). Importantly, after treatment they can revert to the vegetative state, repopulating the biofilm, and thus reconstituting the infection, leading to prolonged treatment duration (Lewis 2008, 2010; Barraud et al. 2013). *Persisters* represent less than 0.1% of the biofilm population and have been documented in a number of bacterial species, including *Staphylococcus aureus*, *Mycobacterium tuberculosis*, *Escherichia coli*, *Salmonella entérica* subsp. *enterica* serovar Typhimurium and *P. aeruginosa*. Several mechanisms, such as reduced levels of cellular energy (low intracellular ATP levels), and adaptive responses have been associated in *persister* cell formation (Wilmaerts et al. 2019).

2.4 Expression of Specific Genes

A number of genes have been identified as being specifically involved in the tolerance of biofilms to antibiotics. In general, these genes are characteristically overexpressed in biofilm growth and this increase in their expression is related to tolerance to antibiotics. This is a different concept from adaptive resistance which requires the presence of the antibiotic to induce the resistance mechanism.

For instance, the biofilm resistance locus regulator, BrlR, was found to be a Mer-like transcriptional activator of biofilm-specific antibiotic tolerance in *P. aeruginosa* by its interaction with a variety of different potential targets (Poudyal and Sauer 2018). For example; activating the expression of the MexAB-oprM and MexEF-oprN efflux pumps (Liao et al. 2013), inducing the production of the ABC transporters, like the ABC transporter PA1874–1877, which has been found to be 10 times more expressed in *P. aeruginosa* biofilms compared to planktonic cells (Poudyal and Sauer 2018; Zhang and Mah 2008), or altering the expression of genes encoding modification of LPS, membrane protein composition, or metabolism and energy generation.

As mentioned before, multidrug efflux pumps efficiently eject antibiotics out of the cell and their induction through different ways is an important factor for antibiotic tolerance in biofilms. Different forms of stresses can induce efflux pumps in paradigmatic biofilms of *P. aeruginosa*, such as oxidative stress, nitrosative stress and membrane-damaging agents can activate MexXY-OprM, MexEF-OprN and MexCD-OprJ, respectively (Poole 2014).

2.5 Adaptive Tolerance

We refer to adaptive tolerance as an inducible mechanism that depends on the antibiotic presence and does not require genetic changes, consequently; it is considered a tolerance mechanism. Adaptive tolerance is mediated by regulatory networks that lead to its switch off or reversion when the antibiotic molecules disappear. Thus, similarly to *persisters*, an on/off model with long-term consequences can be evidenced; if cells are not completely eradicated, once the treatment is stopped resettlement of susceptible biofilms can occur.

Examples of adaptive tolerance in *P. aeruginosa* are the induction of AmpC β-lactamase, LPS alteration and the upregulation of efflux pumps.

The induction of AmpC β-lactamase through the exposure to β-lactam antibiotics, as imipenem or ceftazidime, is likely the main adaptive tolerance mechanisms in *P. aeruginosa* biofilms (Giwercman et al. 1991, 1992). Bagge et al. demonstrated that the expression of this enzyme followed a special structural distribution being characteristically concentrated at the periphery of the biofilms (Bagge et al. 2004b).

Similarly, exposure of *P. aeruginosa* biofilms to colistin was shown to upregulate the *pmr* and *arn* genes in the metabolically active peripheral subpopulations (Pamp et al. 2008a). The *arn* gene products alter the negative charge of LPS that leads to a decrease in the uptake of polymyxins and consequently, an increase in the biofilm tolerance to the cationic peptides (Lewenza 2013). Upregulation of MexAB-OprM efflux pump is also involved in colistin biofilm tolerance (Pamp et al. 2008a). Interestingly, tolerance to colistin is linked to metabolically active cells, thus establishing a pattern for combination of antibiotics with different mechanisms of action (Pamp et al. 2008a).

2.6 In Vivo Tolerance

What has been exposed so far refers to intrinsic factors of the biofilm special atmosphere that leads to the physiological tolerance but, as well as biofilm micro-environment inherent conditions, there is also the in vivo contribution of the host immune system by, for example, the involvement of polymorphonuclear leukocytes (PMNs). These cells are known to consume oxygen, thus, further impacting the mechanisms of tolerance described previously, and to release eDNA which, also contributes to the antibiotic tolerance through the sequestration of cationic antibiotics (Kolpen et al. 2010; Kragh et al. 2014; Worlitzsch et al. 2002).

Apart from the PMNs action, several circumstances are implicated in the in vivo tolerance of biofilms, for example, the different in vitro/in vivo sizes of biofilms (Høiby et al. 2015b), their placement in clusters (Bjarnsholt et al. 2013) or the polymicrobial nature of in vivo biofilms (Burmolle et al. 2010). Regarding the accessibility of antibiotics, biofilm has been considered as a third compartment after tissue (second) and blood (first) (Cao et al. 2015).

3 Antibiotic Resistance Development in Biofilms

3.1 Mutational Resistance

The described tolerance mechanisms have been extensively studied and clearly related to the recalcitrance of biofilms to antibiotics; however, there is still an important gap of knowledge regarding the effect and relevance of conventional resistance mutational mechanisms in the reduced antibiotic susceptibility of biofilms.

According to current knowledge, it would be reasonable to summarize that tolerance mechanisms provide a fertile ground for the emergence of antibiotic-resistant mutants (Levin-Reisman et al. 2017).

Traditionally, the balance between the evolution, driven by mutation and selection and determining the adaptability to new environments or to adverse conditions, and stability of genetic information, related with the population fitness and the perpetuation of species, shapes the mutation rate (Levin-Reisman et al. 2017).

A priori, exceeding limits of the normal mutation rate could lead to an evolutionary death by no adaptability (hypomutation) or to short-time death by the accumulation of deleterious mutations (hypermutation), (Radman et al. 1999). However, in some situations, e.g., when confronting new, fluctuating or stressful environments, mutators can be amplified by being co-selected with adaptive mutations (Sniegowski et al. 1997; Taddei et al. 1997; Mao et al. 1997) and thus hypermutation could be advantageous.

Hypermutators have a spontaneous mutation rate increased from 100- to 1000-fold due to defects in DNA repair or errors in the avoidance systems, frequently affecting the key genes of the methyl-directed mismatch repair system (MMSR) *mutS*, *mutL* or *uvrD* (*mutU*), (Miller 1996). Moreover, mutator phenotype can arise from mutations affecting the 8-oxo-dG (or GO) repair system genes (*mutM*, *mutY* and *mutT*), as well as genes involved in the prevention of oxidative damage produced by reactive oxygen species (ROS) (*oxyR*, *sodA* and *pfpI*), (Miller 1996; Rodríguez-Rojas and Blázquez 2009).

It has been reported that hypermutator strains were extremely high prevalent in the context of chronic respiratory infections in CF patients (Oliver et al. 2000). Mutators, mainly due to deficiencies in the DNA mismatch repair (MMR) system (frequently in the *mutS* gene), represented 20% of the total number of *P. aeruginosa* isolates and were found in 37% of patients (Oliver et al. 2000, 2002), which is in contrast to the low prevalence (<1%) of this feature in acute infections (Gutiérrez et al. 2004). Likewise, in other respiratory infections such as bronchiectasis, or chronic obstructive pulmonary disease (COPD) (Maciá et al. 2005), a very high prevalence of hypermutable strains was also documented and again, hypermutation was mainly due to errors in MMRS, especially in *mutS* (Oliver et al. 2002; Maciá et al. 2005). Similarly, an association between hypermutation and chronic infections was also established in other microorganisms (Watson et al. 2004; Labat et al. 2005; Prunier et al. 2003).

The proportion of hypermutable isolates can vary from less than 1% at acute infections, going through 6% at environmental settings, or 10% at the onset of chronic infection, to up to 65% at chronical phases (Oliver et al. 2000; Gutiérrez et al. 2004; Maciá et al. 2005; Martínez-Solano et al. 2008; Montanari et al. 2007; Kenna et al. 2007; Mena et al. 2008).

Nevertheless, one of the characteristics of hypermutator strains with the greatest clinical impact is their association with high antibiotic resistance rates (Oliver et al. 2000; Maciá et al. 2005). Indeed, it was shown by in vitro and in vivo experiments that hypermutation dramatically favors resistance development during antibiotic exposure (Oliver et al. 2000; Maciá et al. 2005; Woodford and Ellington 2007; Blázquez et al. 2002).

In recent years, studies focussing on the effects of hypermutation on biofilms development showed that mutagenesis is intrinsically increased in biofilms and that hypermutation plays an important role in development, adaptation and diversification processes (Conibear et al. 2009; Webb et al. 2003; Boles and Singh 2008; Driffield et al. 2008).

In fact, it was demonstrated that MMRS deficient mutants were associated with enhanced micro-colony development and increased rates of phenotypic diversification, which conferred improved competitiveness (Luján et al. 2011).

Recently, studies have been conducted to assess the effect of hypermutation on the resistance to antibiotics in biofilms. For example, the antibiotic azithromycin showed promising results in the treatment of chronic respiratory infections by *P. aeruginosa*. Although macrolides are not active against *P. aeruginosa* in standard in vitro susceptibility tests, these antibiotics showed bactericidal activity on stationary phase cells (Imamura et al. 2005; Saiman et al. 2003). In a biofilm model using the Calgary device it was shown that azithromycin was active against *P. aeruginosa* biofilms, however, resistant mutants were readily selected, especially in hypermutable strains (MMRS deficient mutants) (Mulet et al. 2009). The main resistance mechanism was found to be the hyperexpression of MexCD-OprJ, due to *nfxB* mutations, which collaterally showed cross-resistance to other unrelated antipseudomonal agents, such as ciprofloxacin or cefepime, and hypersusceptibility to other drugs, such as imipenem or tobramycin (Mulet et al. 2009). This work demonstrated the impact of hypermutation on antibiotic resistance in biofilms and provides guidance in the selection of appropriate antipseudomonal therapies in CF patients under azithromycin maintenance treatment (Mulet et al. 2009).

Similarly, in a model of theoretically optimized pharmacokinetic/pharmacodynamic (PK/PD) parameters of ciprofloxacin to treat flow cell biofilms formed by hypermutable strains, deficient in *mutS*, this antibiotic, which was one of the most active antibiotics on biofilms according to previous works (Moskowitz et al. 2004), led to the selection and amplification of resistant mutants (Macià et al. 2011). Particularly relevant was the demonstration that mutational mechanisms played a major role in biofilm antibiotic resistance and that theoretically optimized PK/PD parameters failed to suppress resistance development. The results from this work supported the hypothesis that the increased antibiotic tolerance (physical, physiological and adaptive), driven by the special biofilm microenvironment, may raise the effective mutant preventive concentration (MPC) and favors gradual mutational resistance development, especially in mutator strains. Accordingly, overexpression of MexCD-OprJ or MexEF-OprN was the primary cause of ciprofloxacin resistance development in biofilms. Importantly, it was also demonstrated for the first time that mutator populations were dramatically amplified under antibiotic treatment by co-selection with resistance mutations in *P. aeruginosa* biofilms. In addition, the results highlighted the need to adapt the classical PK/PD target values to the biofilm mode of growth, in order to optimize the therapeutic management of chronic infections (Macià et al. 2011).

More recently, with the same aim of optimizing PK/PD strategies, the impact of ciprofloxacin and meropenem as monotherapy and in combination was evaluated on hypermutable strains using the dynamic in vitro CDC biofilm reactor (CBR) (Bilal et al. 2020). Concentration-time profiles attainable in epithelial lining fluid (ELF) following FDA-approved doses were simulated in the CBR. Overall, all monotherapies failed and combination treatments demonstrated synergistic bacterial killing and resistance suppression in hypermutable strains (Bilal et al. 2020).

In an analogous study, different dosage regimens of meropenem (standard and in continuous infusion) and tobramycin, as monotherapies and in combination against hypermutable carbapenem-resistant *P. aeruginosa* were studied (Bilal et al. 2019). In accordance with the work by (Bilal et al. 2020), monotherapy of meropenem and tobramycin were ineffective against planktonic and biofilm bacteria, while the combination treatment in continuous infusion exhibited enhanced bacterial killing and resistance suppression of carbapenem-resistant hypermutable *P. aeruginosa*. The results from these works emphasize the importance of using PK/PD optimized combinations of antibiotics when treating hypermutable *P. aeruginosa* biofilm-related infections (Bilal et al. 2020, 2019).

The role of mutational resistance has been revealed as a key mechanism for antibiotic resistance in biofilms. In a recent study, *P. aeruginosa* mutants showing diverse resistance mechanisms, such as β-lactamase or efflux pumps hyperexpression, were studied in biofilms (Rojo-Molinero et al. 2019). As expected, biofilms formed by AmpC (*dacB*) and MexAB-OprM (*mexR*) hyperproducing mutants showed enhanced β-lactam (cefepime) resistance (Rojo-Molinero et al. 2019). Resistant mutants (*dacB*) were selected and amplified in the competition with the wild type under cefepime treatment (Rojo-Molinero et al. 2019). Under the pressure of antibiotics, resistant mutants were also selected and amplified in biofilms. Interestingly, in mixed biofilms, resistant mutants shielded susceptible populations from antibiotic treatment (Rojo-Molinero et al. 2019). This finding could help to decipher the role of the frequent phenomenon of multi-strain/species in biofilms (Ciofu et al. 2012) and to determine why susceptible variants are not effortlessly eradicated from CF patients despite long-term intensive antimicrobial treatments (Rojo-Molinero et al. 2019).

The impact of the resistance mechanisms on biofilm community was also studied in preceding studies. In a work by Mulet et al. (2011) it was demonstrated that protective resistance mechanisms are different in biofilm vs planktonic growth. As it is well known, the most relevant *P. aeruginosa* resistance mechanisms include the overexpression of the chromosomal β-lactamase AmpC, the inactivation of the carbapenem porin OprD or the overexpression of different efflux pumps encoded in its genome (Livermore 2002; Poole 2004). Unfortunately, the combinations of these resistance pathways are often cumulative or synergistic, frequently driving to multidrug resistance. However, one of these mechanisms, the overexpression of the MexCD-OprJ efflux pump (caused by inactivation of its negative regulator NfxB), collaterally yields a notable hypersusceptibility to other relevant antipseudomonal agents, including most β-lactams and aminoglycosides (Gotoh et al. 1998). The key to this is that overexpression of the MexCD-OprJ efflux pump causes major changes in the cell envelope physiology, impairing the backbone of *P. aeruginosa* intrinsic resistance, including the major constitutive (MexAB-OprM) and inducible (MexXY-OprM) efflux pumps and the inducible AmpC β-lactamase. In addition, it also impaired the constitutive AmpC overexpression through a dramatic decrease in periplasmic-lactamase activity, produced by an abnormal permeation of AmpC out of the cell. Interestingly, although non-protective on planktonic cells, this permeation of AmpC into the matrix resulted in a protective mechanism for biofilm communities against β-lactams. Therefore, in accordance with other works previously mentioned, hyperexpression of MexCD-OprJ efflux pump by *nfxB* mutants is a resistance mechanism that has been found to be beneficial and protective in biofilms (Mulet et al. 2009, 2011; Macià et al. 2011).

Similar findings have been confirmed in other bacterial species like *Acinetobacter baumanii* where, by using experimental evolution experiments, it was demonstrated that antibiotic resistance mechanisms were dependent upon environmental structure and bacterial lifestyle (Santos-Lopez et al. 2019). Similarly, in evolution experiments with *A. baumanii* and *P. aeruginosa* exposed to increasing concentrations of tobramycin in planktonic and biofilm environments, both species acquired antibiotic resistance-associated mutations that were more prevalent in the biofilm lifestyle than in planktonic cells (Scribner et al. 2020).

The development of mutational resistance has been demonstrated for almost all antibiotic classes in the chronic respiratory infection context. For example, in addition to the selection of mutants hyperexpressing MexXY-OprM, high-level aminoglycoside resistance driven by mutations in *fusA1*/*fusA2* has been documented in the CF setting (López-Causapé et al. 2017, 2018). Similarly, the PhoPQ regulatory system, PmrB, and ParS/ParR have been identified as hot spots of mutation that promote polymyxin resistance of *P. aeruginosa* isolated from colistin-treated CF patients (Gutu et al. 2013; Moskowitz et al. 2012; Greipel et al. 2016).

In summary, mutation-driven resistance has been shown to be a key factor in the development of resistance to antibiotics in biofilms. As a whole, the tolerance processes, which precede and support the development of classical resistance, as well as the biofilm microenvironment, which enhances mutagenesis and genomic and phenotypic diversification, are convergent forces that promote the development of mutational resistance. Figure 1 schematically illustrates this sequential process.

3.2 *Horizontal Gene Transfer*

In general, horizontal gene transfer, either by conjugation, transformation, or phage transduction mechanisms, is a principal source of evolution.

Within a biofilm, horizontal transfer of resistance-conferring genes between bacterial cells has been reported as being 700 times more efficient than among free-living, planktonic bacterial cells (Flemming et al. 2016).

Biofilms are ideally suited to the exchange of genetic material of various origins since conditions, such as high cell density and proximity, as well as increased genetic competence and accumulation of genetic elements are provided in their microenvironment (Flemming et al. 2016; Molin and Tolker-Nielsen 2003). Indeed, the conjugation mechanism itself may stimulate biofilm development (Madsen et al. 2012). Likewise, transformation seems to be part of a biofilm-related life cycle and both of these gene-transfer mechanisms may be auto-catalytically promoted in biofilms (Madsen et al. 2012).

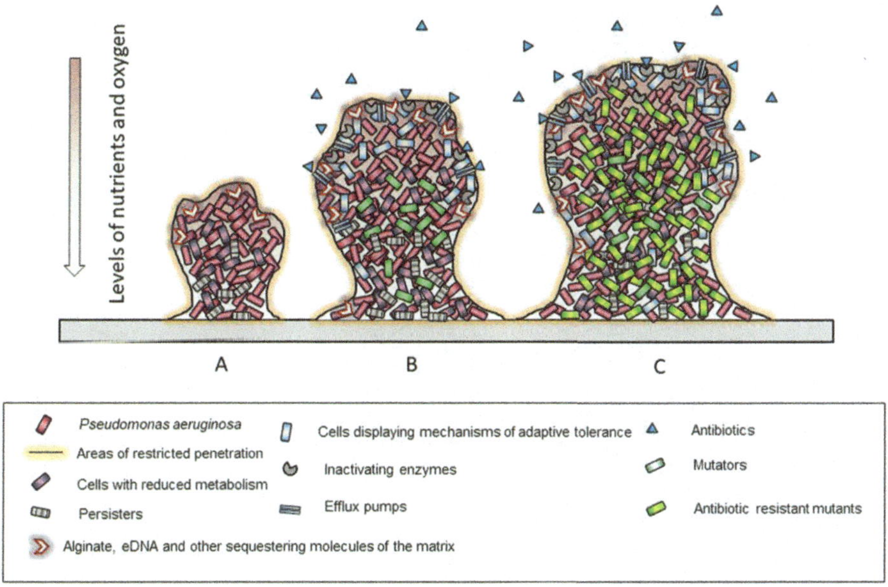

Fig. 1 Graphic representation of the sequential process of antibiotic resistance development in biofilm. (**a**) Innate tolerance of biofilms represented by physical (restricted penetration, sequestering molecules), physiological (gradients of oxygen and nutrients, reduced metabolism) and *persisters* phenomena. (**b**) Addition of adaptive tolerance represented by cells displaying mechanisms such as hyperexpression of inactivating enzymes and/or efflux pumps. Presence of mutators. (**c**) Addition of antibiotic conventional resistance mechanisms represented by the selection and amplification of antibiotic-resistant mutants

Moreover, it has been shown that biofilms facilitate horizontal gene transfer and plasmids induce biofilm formation (Ghigo 2001). Plasmid transfer has been proven to occur at a higher frequency in biofilm cultures of different bacterial species (Kouzel et al. 2015; Savage et al. 2013; Hennequin et al. 2012; Tanner et al. 2017).

Table 1 summarizes the mechanisms of antibiotic resistance described in biofilms.

3.3 Conclusion and Future Perspectives

Biofilms are hotspots for the development of antibiotic resistance due to the intrinsic phenotypic tolerance linked to the mutagenic microenvironment that helps emergence, selection and amplification of mutational resistance, together with the proficiency in transferring resistance genes.

In recent years, many studies have investigated the different mechanisms underlying antibiotic resistance in biofilms; however, there are still important knowledge

Table 1 Mechanisms of biofilm antibiotic resistance

	Tolerance			Acquired resistance	
Antibiotics	Physical tolerance	Physiological tolerance	Adaptive tolerance	Mutational resistance	Horizontal gene transfer
β-lactams	• Restricted penetration	• Low metabolic activity (Walters et al. 2003) • *Persisters* (Lewis 2008; Lewis 2010) • Non-specific upregulation of efflux pumps (MexAB-oprM) (Liao et al. 2013; Poole 2014)	• Induction of AmpC β-lactamase (Giwercman et al. 1991; Giwercman et al. 1992; Bagge et al. 2004b) • Repression of OprD (Livermore 2002; Poole 2004)	• Mutants overexpressing AmpC (*dacB*) (Rojo-Molinero et al. 2019) • Mutants overexpressing MexAB-OprM (*mexR*) (Rojo-Molinero et al. 2019) • Mutations in OprD (Bilal et al. 2019, 2020)	• bla_{NDM-1} plasmids (Tanner et al. 2017) • conjugative plasmid carring bla(CTX-M-15), bla(OXA-1) and aac (6′)-Ib-cr genes (Hennequin et al. 2012)
Fluoroquinolones	• No effect	• Low metabolic activity (Walters et al. 2003; Bernier et al. 2013) • Anaerobic conditions (Jensen et al. 2014) • Non-specific upregulation of efflux pumps (MexCD-OprJ, MexEF-OprN) (Whiteley et al. 2001; Liao et al. 2013; Poole 2014) • *Persisters* (Lewis 2008; Lewis 2010)	• Upregulation of efflux pumps (MexCD-OprJ, MexEF-OprN) (Poole 2014)	• Mutants overexpressing MexEF-OprN/MexCD-OprJ (Macià et al. 2011) • Mutations in *GyrA*/*GyrB* (Macià et al. 2011)	
Aminoglycosides	• Restricted penetration (binding alginate, eDNA) (Tseng et al. 2013; Chiang et al. 2013; Cao et al. 2016; Mulcahy et al. 2008)	• Low metabolic activity (Walters et al. 2003) • Anaerobic conditions (Walters et al. 2003) • *Persisters* (Lewis 2008, 2010)	• Upregulation of efflux pumps (MexXY-OprM) (Poole 2014)	• Mutants overexpressing • MexXY-OprM, *fusA1*/*fusA2* mutants (López-Causapé et al. 2017, 2018)	

		• Non-specific upregulation of efflux pumps (MexXY-OprM) (Liao et al. 2013; Zhang and Mah 2008; Poole 2014)			
Colistin	• Restricted penetration (binding alginate, eDNA) (Pamp et al. 2008a)		• Upregulation of efflux pumps (MexAB-OprM) (Pamp et al. 2008a) • Upregulation of *pmr*/*arn* genes and LPS modification (Mulcahy et al. 2008)	• PhoPQ (Gutu et al. 2013), PmrAB (Moskowitz et al. 2012), and ParRS mutants (Greipel et al. 2016)	
Azithromcycin			• Upregulation of efflux pumps (MexCD-OprJ)	• Mutants overexpressing MexCD-OprJ (Mulet et al. 2009)	

gaps. A better understanding of these molecular mechanisms could lead to improvements in the effective treatment of biofilm-related infections.

Considering the multifactorial nature of antibiotic resistance in biofilms, strategies adopted to overcome resistance should take into account the different systems acting together and efficiently combine different strategies to treat biofilm-related infections. For instance, the use of quorum sensing inhibitors to prevent and/or disrupt biofilms (Jakobsen et al. 2013), the use of SDS, EDTA, chlorhexidine or antibiotics like colistin to activate metabolically inactive cells (Pamp et al. 2008b), or restocking lacking metabolites and electron acceptors to revert tolerance to susceptibility (Crabbé et al. 2019), hyperbaric oxygen treatment (HBOT) to enhance antibiotic activity (Gade et al. 2018; Kolpen et al. 2016; Lerche et al. 2017), treatment with eDNAses (Yu et al. 2015) and/or oligo G to attack the matrix (Hengzhuang et al. 2016), use of efflux pumps inhibitors (EPIs) directed to the adaptive resistance (Kvist et al. 2008), use of antioxidants like N-Acetylcysteine to target hypermutation, SOS, and oxidative stress (Skov et al. 2015), or improving therapeutic antibiotic strategies, as well as PK/PD parameters, to overwhelm mutation selection like combination of antibiotics (Bilal et al. 2019, 2020; Jakobsen et al. 2013; Maciá et al. 2006; Plasencia et al. 2007), innovative sequential regimens (Rojo-Molinero et al. 2016), or topical administration (Krajewski et al. 2014; Heijerman et al. 2009). Progress in new alternative therapeutic approaches like formulation of antibiotics in nanoparticles (Pelgrift and Friedman 2013), novel anti-biofilm compounds, bacteriophage therapy (Pires et al. 2017), CRISPR gene editing technologies and photodynamic therapies is necessary. In addition, an important effort on transferring all these treatment options into real and practical therapeutic guidelines for the treatment of biofilm-related infections is urgently needed.

Acknowledgments María D. Macià and Antonio Oliver are supported by the Spanish Network for Research in Infectious Diseases (REIPI RD16/0016).

References

Alhede M, Kragh KN, Qvortrup K, Allesen-Holm M, van Gennip M, Christensen LD, Jensen PØ, Nielsen AK, Parsek M, Wozniak D, Molin S, Tolker-Nielsen T, Høiby N, Givskov M, Bjarnsholt T (2011) Phenotypes of non-attached *Pseudomonas aeruginosa* aggregates resemble surface attached biofilm. PLoS One 6(11):e27943. https://doi.org/10.1371/journal.pone.0027943. Epub 2011 Nov 21. PMID: 22132176; PMCID: PMC3221681

Bagge N, Schuster M, Hentzer M, Ciofu O, Givskov M, Greenberg EP et al (2004a) *Pseudomonas aeruginosa* biofilms exposed to imipenem exhibit changes in global gene expression and beta-lactamase and alginate production. Antimicrob Agents Chemother 48:1175–1187. https://doi.org/10.1128/AAC.48.4.1175-1187.2004

Bagge N, Hentzer M, Andersen JB, Ciofu O, Givskov M, Høiby N (2004b) Dynamics and spatial distribution of beta-lactamase expression in *Pseudomonas aeruginosa* biofilms. Antimicrob Agents Chemother 48:1168–1174. https://doi.org/10.1128/AAC.48.4.1168-1174

Barraud N, Buson A, Jarolimek W, Rice SA (2013) Mannitol enhances antibiotic sensitivity of persister bacteria in *Pseudomonas aeruginosa* biofilms. PLoS One 8:e84220. https://doi.org/10.1371/journal.pone.0084220

Bernier SP, Lebeaux D, DeFrancesco AS, Valomon A, Soubigou G, Coppee JY et al (2013) Starvation, together with the SOS response, mediates high biofilm-specific tolerance to the fluoroquinolone ofloxacin. PLoS Genet 9:e1003144. https://doi.org/10.1371/journal.pgen.1003144

Bilal H, Bergen PJ, Kim TH, Chung SE, Peleg AY, Oliver A, Nation RL, Landersdorfer CB (2019 Oct 22) Synergistic meropenem-tobramycin combination dosage regimens against clinical hypermutable *Pseudomonas aeruginosa* at simulated epithelial lining fluid concentrations in a dynamic biofilm model. Antimicrob Agents Chemother 63(11):e01293–19. https://doi.org/10.1128/AAC.01293-19. PMID: 31427301; PMCID: PMC6811397

Bilal H, Bergen PJ, Tait JR, Wallis SC, Peleg AY, Roberts JA, Oliver A, Nation RL, Landersdorfer CB (2020 Jun 23) Clinically relevant epithelial lining fluid concentrations of meropenem with ciprofloxacin provide synergistic killing and resistance suppression of hypermutable *pseudomonas aeruginosa* in a dynamic biofilm model. Antimicrob Agents Chemother 64(7): e00469–20. https://doi.org/10.1128/AAC.00469-20. PMID: 32366710; PMCID: PMC7318014

Bjarnsholt T, Jensen PO, Burmolle M et al (2005) *Pseudomonas aeruginosa* tolerance to tobramycin, hydrogen peroxide and polymorphonuclear leukocytes is quorum-sensing dependent. Microbiology 151:373–383

Bjarnsholt T, Alhede M, Alhede M, Eickhardt-Sorensen SR, Moser C, Kuhl M et al (2013) The in vivo biofilm. Trends Microbiol 21:466–474. https://doi.org/10.1016/j.tim.2013.06.002

Blázquez J, Oliver A, Gómez-Gómez JM (2002 Aug) Mutation and evolution of antibiotic resistance: antibiotics as promoters of antibiotic resistance? Curr Drug Targets 3(4):345–349. https://doi.org/10.2174/1389450023347579

Boles BR, Singh PK (2008 Aug 26) Endogenous oxidative stress produces diversity and adaptability in biofilm communities. Proc Natl Acad Sci U S A 105(34):12503–8. https://doi.org/10.1073/pnas.0801499105. Epub 2008 Aug 21. PMID: 18719125; PMCID: PMC2527941

Bowler PG, Welsby S, Towers V, Booth R, Hogarth A, Rowlands V, Joseph A, Jones SA (2012 Aug) Multidrug-resistant organisms, wounds and topical antimicrobial protection. Int Wound J 9(4):387–96. https://doi.org/10.1111/j.1742-481X.2012.00991.x. Epub 2012 May 29. PMID: 22640181; PMCID: PMC7950580

Brauner A, Fridman O, Gefen O, Balaban NQ (2016) Distinguishing between resistance, tolerance and persistence to antibiotic treatment. Nat Rev Microbiol 14:320–330

Burmolle M, Thomsen TR, Fazli M, Dige I, Christensen L, Homoe P et al (2010) Biofilms in chronic infections–a matter of opportunity– monospecies biofilms in multispecies infections. FEMS Immunol Med Microbiol 59:324–336. https://doi.org/10.1111/j.1574-695X.2010.00714.x

Cao B, Christophersen L, Thomsen K, Sønderholm M, Bjarnsholt T, Jensen PØ, Høiby N, Moser C (2015 Jul) Antibiotic penetration and bacterial killing in a *Pseudomonas aeruginosa* biofilm model. J Antimicrob Chemother 70(7):2057–2063. https://doi.org/10.1093/jac/dkv058. Epub 2015 Mar 18

Cao B, Christophersen L, Kolpen M, Jensen PØ, Sneppen K, Høiby N, Moser C, Sams T (2016 Apr 21) Diffusion retardation by binding of tobramycin in an alginate biofilm model. PLoS One 11(4):e0153616. https://doi.org/10.1371/journal.pone.0153616. PMID: 27100887; PMCID: PMC4839563

Chiang WC, Nilsson M, Jensen PO, Hoiby N, Nielsen TE, Givskov M et al (2013) Extracellular DNA shields against aminoglycosides in *Pseudomonas aeruginosa* biofilms. Antimicrob Agents Chemother 57:2352–2361. https://doi.org/10.1128/AAC.00001-13

Ciofu O, Tolker-Nielsen T (2019 May 3) Tolerance and resistance of *Pseudomonas aeruginosa* biofilms to antimicrobial agents-how *P. aeruginosa* can escape antibiotics. Front Microbiol 10: 913. https://doi.org/10.3389/fmicb.2019.00913. PMID: 31130925; PMCID: PMC6509751

Ciofu O, Mandsberg LF, Wang H, Høiby N (2012 Jul) Phenotypes selected during chronic lung infection in cystic fibrosis patients: implications for the treatment of *Pseudomonas aeruginosa* biofilm infections. FEMS Immunol Med Microbiol 65(2):215–25. https://doi.org/10.1111/j.1574-695X.2012.00983.x. Epub 2012 Jun 1. Erratum in: FEMS Immunol Med Microbiol. 2012 Oct;66(1):120

Conibear TC, Collins SL, Webb JS (2009) Role of mutation in *Pseudomonas aeruginosa* biofilm development. PLoS One 4:e6289

Costerton JW, Cheng KJ, Geesey GG et al (1987) Bacterial biofilms in nature and disease. Annu Rev Microbiol 41:435–464

Costerton JW, Stewart PS, Greenberg EP (1999) Bacterial biofilms: a common cause of persistent infections. Science 284:1318–1322

Crabbé A, Jensen PØ, Bjarnsholt T, Coenye T (2019 Oct) Antimicrobial tolerance and metabolic adaptations in microbial biofilms. Trends Microbiol. 27(10):850–863. https://doi.org/10.1016/j.tim.2019.05.003. Epub 2019 Jun 6

Driffield K, Miller K, Bostock JM, O'Neill AJ, Chopra I (2008 May) Increased mutability of *Pseudomonas aeruginosa* in biofilms. J Antimicrob Chemother 61(5):1053–1056. https://doi.org/10.1093/jac/dkn044. Epub 2008 Feb 6

Flemming H-C, Wingender J, Szewwzyk U, Steinberg P, Rice SA, Kjelleberg S (2016) Biofilms: an emergent form of bacterial life. Nat Rev Microbiol 14(9):563–575

Fux CA, Costerton JW, Stewart PS, Stoodley P (2005 Jan) Survival strategies of infectious biofilms. Trends Microbiol 13(1):34–40. https://doi.org/10.1016/j.tim.2004.11.010

Gade PAV, Olsen TB, Jensen PØ, Kolpen M, Høiby N, Henneberg KÅ, Sams T (2018 Jun 14) Modelling of ciprofloxacin killing enhanced by hyperbaric oxygen treatment in *Pseudomonas aeruginosa* PAO1 biofilms. PLoS One 13(6):e0198909. https://doi.org/10.1371/journal.pone.0198909. PMID: 29902223; PMCID: PMC6002088

Ghigo JM (2001 Jul 26) Natural conjugative plasmids induce bacterial biofilm development. Nature 412(6845):442–445. https://doi.org/10.1038/35086581

Giwercman B, Jensen ET, Høiby N, Kharazmi A, Costerton JW (1991 May) Induction of beta-lactamase production in *Pseudomonas aeruginosa* biofilm. Antimicrob Agents Chemother 35(5):1008–10. https://doi.org/10.1128/AAC.35.5.1008. PMID: 1906694; PMCID: PMC245148

Giwercman B, Meyer C, Lambert PA, Reinert C, Høiby N (1992 Jan) High-level beta-lactamase activity in sputum samples from cystic fibrosis patients during antipseudomonal treatment. Antimicrob Agents Chemother 36(1):71–6. https://doi.org/10.1128/AAC.36.1.71. PMID: 1590704; PMCID: PMC189229

Gotoh N et al (1998) Characterization of the MexC-MexD-OprJ multidrug efflux system in ΔmexA-mexB-oprM mutants of *Pseudomonas aeruginosa*. Antimicrob Agents Chemother 42:1938–1943

Greipel L, Fischer S, Klockgether J, et al. (2016) Molecular epidemiology of mutations in antimicrobial resistance loci of *pseudomonas aeruginosa* isolates from airways of cystic fibrosis patients. Antimicrob Agents Chemother 60(11):6726–6734. Published 2016 Oct 21. https://doi.org/10.1128/AAC.00724-16

Gutiérrez O, Juan C, Pe'rez JL, Oliver A (2004) Lack of association between hypermutation and antibiotic resistance development in *Pseudomonas aeruginosa* isolates from intensive care unit patients. Antimicrob Agents Chemother 48:3573–3575

Gutu AD, Sgambati N, Strasbourger P, Brannon MK, Jacobs MA, Haugen E, Kaul RK, Johansen HK, Høiby N, Moskowitz SM (2013 May) Polymyxin resistance of *Pseudomonas aeruginosa* phoQ mutants is dependent on additional two-component regulatory systems. Antimicrob Agents Chemother 57(5):2204–15. https://doi.org/10.1128/AAC.02353-12. Epub 2013 Mar 4. PMID: 23459479; PMCID: PMC3632916

Haaber J, Cohn MT, Frees D, Andersen TJ, Ingmer H (2012) Planktonic aggregates of *Staphylococcus aureus* protect against common antibiotics. PLoS One 7(7):e41075. https://doi.org/10.1371/journal.pone.0041075. Epub 2012 Jul 18. Erratum in: PLoS One. 2012;7(10). https://doi.

org/10.1371/annotation/08d0f2a8-0c40-4a0c-b546-0025648e73f0. PMID: 22815921; PMCID: PMC3399816

Heijerman H, Westerman E, Conway S, Touw D, Döring G (2009) Inhaled medication and inhalation devices for lung disease in patients with cystic fibrosis: a European consensus. J Cyst Fibros 8:295–315

Hengzhuang W, Song Z, Ciofu O, Onsøyen E, Rye PD, Høiby N (2016 Apr 22) OligoG CF-5/20 disruption of mucoid *pseudomonas aeruginosa* biofilm in a murine lung infection model. Antimicrob Agents Chemother 60(5):2620–6. https://doi.org/10.1128/AAC.01721-15. PMID: 26833153; PMCID: PMC4862494

Hennequin C, Aumeran C, Robin F, Traore O, Forestier C (2012 Sep) Antibiotic resistance and plasmid transfer capacity in biofilm formed with a CTX-M-15-producing Klebsiella pneumoniae isolate. J Antimicrob Chemother 67(9):2123–2130. https://doi.org/10.1093/jac/dks169. Epub 2012 May 10

Høiby N, Bjarnsholt T, Moser C, Bassi GL, Coenye T, Donelli G et al (2015a) ESCMID guideline for the diagnosis and treatment of biofilm infections 2014. Clin Microbiol Infect 21(Suppl. 1): S1–S25. https://doi.org/10.1016/j.cmi.2014.10.024)

Høiby N, Bjarnsholt T, Moser C, Bassi GL, Coenye T, Donelli G et al (2015b) ESCMID guideline for the diagnosis and treatment of biofilm infections 2014. Clin Microbiol Infect 21(Suppl. 1): S1–S25. https://doi.org/10.1016/j.cmi.2014.10.024

Imamura Y, Higashiyama Y, Tomono K, Izumikawa K, Yanagihara K, Ohno H, Miyazaki Y, Hirakata Y, Mizuta Y, Kadota J, Iglewski BH, Kohno S (2005 Apr) Azithromycin exhibits bactericidal effects on *Pseudomonas aeruginosa* through interaction with the outer membrane. Antimicrob Agents Chemother 49(4):1377–80. https://doi.org/10.1128/AAC.49.4.1377-1380. 2005. PMID: 15793115; PMCID: PMC1068619

Ito A, Taniuchi A, May T, Kawata K, Okabe S (2009 Jun) Increased antibiotic resistance of *Escherichia coli* in mature biofilms. Appl Environ Microbiol 75(12):4093–100. https://doi.org/10.1128/AEM.02949-08. Epub 2009 Apr 17. PMID: 19376922; PMCID: PMC2698376

Jakobsen TH, Bjarnsholt T, Jensen PØ, Givskov M, Høiby N (2013 Jul) Targeting quorum sensing in *Pseudomonas aeruginosa* biofilms: current and emerging inhibitors. Future Microbiol 8(7): 901–921. https://doi.org/10.2217/fmb.13.57

Jamal M, Ahmad W, Andleeb S, Jalil F, Imran M, Nawaz MA, Hussain T, Ali M, Rafiq M, Kamil MA (2018) Bacterial biofilm and associated infections. J Chin Med Assoc 81(1):7–11. https://doi.org/10.1016/j.jcma.2017.07.012

Jensen PO, Briales A, Brochmann RP, Wang H, Kragh KN, Kolpen M et al (2014) Formation of hydroxyl radicals contributes to the bactericidal activity of ciprofloxacin against *Pseudomonas aeruginosa* biofilms. Pathog Dis 70:440–443. https://doi.org/10.1111/2049-632X.12120

Kenna DT, Doherty CJ, Foweraker J, Macaskill L, Barcus VA, Govan JRW (2007 Jun) Hypermutability in environmental *Pseudomonas aeruginosa* and in populations causing pulmonary infection in individuals with cystic fibrosis. Microbiology (Reading). 153(Pt 6): 1852–1859. https://doi.org/10.1099/mic.0.2006/005082-0

Kindrachuk KN, Fernandez L, Bains M, Hancock RE (2011) Involvement of an ATP-dependent protease, PA0779/AsrA, in inducing heat shock in response to tobramycin in *Pseudomonas aeruginosa*. Antimicrob Agents Chemother 55:1874–1882. https://doi.org/10.1128/AAC.00935-10

Kolpen M, Hansen CR, Bjarnsholt T, Moser C, Christensen LD, van Gennip M et al (2010) Polymorphonuclear leucocytes consume oxygen in sputum from chronic *Pseudomonas aeruginosa* pneumonia in cystic fibrosis. Thorax 65:57–62. https://doi.org/10.1136/thx.2009.114512

Kolpen M, Appeldorff CF, Brandt S, Mousavi N, Kragh KN, Aydogan S, Uppal HA, Bjarnsholt T, Ciofu O, Høiby N, Jensen PØ (2016 Feb) Increased bactericidal activity of colistin on *Pseudomonas aeruginosa* biofilms in anaerobic conditions. Pathog Dis 74(1):ftv086. https://doi.org/10.1093/femspd/ftv086. Epub 2015 Oct 12. PMID: 26458402; PMCID: PMC4655427

Kouzel N, Oldewurtel ER, Maier B (2015 Jul) Gene transfer efficiency in gonococcal biofilms: role of biofilm age, architecture, and pilin antigenic variation. J Bacteriol 197(14):2422–31. https://doi.org/10.1128/JB.00171-15. Epub 2015 May 11. PMID: 25962915; PMCID: PMC4524189

Kragh KN, Alhede M, Jensen PØ, Moser C, Scheike T, Jacobsen CS, Seier Poulsen S, Eickhardt-Sørensen SR, Trøstrup H, Christoffersen L, Hougen HP, Rickelt LF, Kühl M, Høiby N, Bjarnsholt T (2014 Nov) Polymorphonuclear leukocytes restrict growth of *Pseudomonas aeruginosa* in the lungs of cystic fibrosis patients. Infect Immun 82(11):4477–86. https://doi.org/10.1128/IAI.01969-14. Epub 2014 Aug 11. PMID: 25114118; PMCID: PMC4249348

Krajewski J, Bode-Böger SM, Tröger U, Martens-Lobenhoffer J, Mulrooney T, Mittelstädt H, Russlies M, Kirchner R, Knobloch JK (2014 Oct) Successful treatment of extensively drug-resistant *Pseudomonas aeruginosa* osteomyelitis using a colistin- and tobramycin-impregnated PMMA spacer. Int J Antimicrob Agents 44(4):363–366. https://doi.org/10.1016/j.ijantimicag.2014.05.023. Epub 2014 Aug 13

Kvist M, Hancock V, Klemm P (2008 Dec) Inactivation of efflux pumps abolishes bacterial biofilm formation. Appl Environ Microbiol 74(23):7376–82. 10.1128/AEM.01310-08. Epub 2008 Oct 3. PMID: 18836028; PMCID: PMC2592912

Labat F, Pradillon O, Garry L, Peuchmaur M, Fantin B, Denamur E (2005 Jun 1) Mutator phenotype confers advantage in *Escherichia coli* chronic urinary tract infection pathogenesis. FEMS Immunol Med Microbiol 44(3):317–321. https://doi.org/10.1016/j.femsim.2005.01.003

Lerche CJ, Christophersen LJ, Kolpen M, Nielsen PR, Trøstrup H, Thomsen K, Hyldegaard O, Bundgaard H, Jensen PØ, Høiby N, Moser C (2017 Sep) Hyperbaric oxygen therapy augments tobramycin efficacy in experimental *Staphylococcus aureus* endocarditis. Int J Antimicrob Agents 50(3):406–412. https://doi.org/10.1016/j.ijantimicag.2017.04.025. Epub 2017 Jun 29

Levin-Reisman I, Ronin I, Gefen O, Braniss I, Shoresh N, Balaban NQ (2017) Antibiotic tolerance facilitates the evolution of resistance. Science 355:826–830. https://doi.org/10.1126/science.aaj2191

Lewenza S (2013) Extracellular DNA-induced antimicrobial peptide resistance mechanisms in *Pseudomonas aeruginosa*. Front Microbiol 4:21. https://doi.org/10.3389/fmicb.2013.00021

Lewis K (2008) Multidrug tolerance of biofilms and persister cells. Curr Top Microbiol Immunol 322:107–131. https://doi.org/10.1007/978-3-540-75418-3_6

Lewis K (2010) Persister cells. Annu Rev Microbiol 64:357–372. https://doi.org/10.1146/annurev.micro.112408.134306

Liao J, Schurr MJ, Sauer K (2013) The MerR-like regulator BrlR confers biofilm tolerance by activating multidrug efflux pumps in biofilms. J Bacteriol 195:3352–3363. https://doi.org/10.1128/JB.00318-13

Livermore DM (2002) Multiple mechanisms of antimicrobial resistance in *Pseudomonas aeruginosa*: our worst nightmare? Clin Infect Dis 34:634–640

López-Causapé C, Sommer LM, Cabot G, Rubio R, Ocampo-Sosa AA, Johansen HK, Figuerola J, Cantón R, Kidd TJ, Molin S, Oliver A (2017 Jul 17) Evolution of the *Pseudomonas aeruginosa* mutational resistome in an international Cystic Fibrosis clone. Sci Rep 7(1):5555. https://doi.org/10.1038/s41598-017-05621-5. PMID: 28717172; PMCID: PMC5514035

López-Causapé C, Rubio R, Cabot G, Oliver A (2018) Evolution of the *pseudomonas aeruginosa* aminoglycoside mutational resistome in vitro and in the cystic fibrosis setting. Antimicrob Agents Chemother 62(4):e02583–17. Published 2018 Mar 27. https://doi.org/10.1128/AAC.02583-17

Luján AM, Maciá MD, Yang L, Molin S, Oliver A, Smania AM (2011) Evolution and adaptation in *Pseudomonas aeruginosa* biofilms driven by mismatch repair system-deficient mutators. PLoS One 6(11):e27842. https://doi.org/10.1371/journal.pone.0027842. Epub 2011 Nov 17. PMID: 22114708; PMCID: PMC3219696

Maciá MD, Blanquer D, Togores B, Sauleda J, Pérez JL, Oliver A (2005) Hypermutation is a key factor in development of multiple-antimicrobial resistance in *Pseudomonas aeruginosa* strains causing chronic lung infections. Antimicrob Agents Chemother 49(8):3382–3386. https://doi.org/10.1128/AAC.49.8.3382-3386.2005

Maciá MD, Borrell N, Segura M, Gómez C, Pérez JL, Oliver A (2006 Mar) Efficacy and potential for resistance selection of antipseudomonal treatments in a mouse model of lung infection by hypermutable *Pseudomonas aeruginosa*. Antimicrob Agents Chemother 50(3):975–83. https://doi.org/10.1128/AAC.50.3.975-983.2006. PMID: 16495260; PMCID: PMC1426455

Macià MD, Pérez JL, Molin S, Oliver A (2011 Nov) Dynamics of mutator and antibiotic-resistant populations in a pharmacokinetic/pharmacodynamic model of *Pseudomonas aeruginosa* biofilm treatment. Antimicrob Agents Chemother 55(11):5230–7. https://doi.org/10.1128/AAC.00617-11. Epub 2011 Aug 22. PMID: 21859941; PMCID: PMC3195006

Madsen JS, Burmølle M, Hansen LH, Sørensen SJ (2012 Jul) The interconnection between biofilm formation and horizontal gene transfer. FEMS Immunol Med Microbiol 65(2):183–195. https://doi.org/10.1111/j.1574-695X.2012.00960.x. Epub 2012 Apr 23

Mao EF, Lane L, Lee J, Miller JH (1997 Jan) Proliferation of mutators in A cell population. J Bacteriol 179(2):417–22. https://doi.org/10.1128/jb.179.2.417-422.1997. PMID: 8990293; PMCID: PMC178711

Martínez-Solano L, Macia MD, Fajardo A, Oliver A, Martinez JL (2008 Dec 15) Chronic *Pseudomonas aeruginosa* infection in chronic obstructive pulmonary disease. Clin Infect Dis 47(12): 1526–1533. https://doi.org/10.1086/593186

Mena A, Smith EE, Burns JL, Speert DP, Moskowitz SM, Perez JL, Oliver A (2008 Dec) Genetic adaptation of *Pseudomonas aeruginosa* to the airways of cystic fibrosis patients is catalyzed by hypermutation. J Bacteriol 190(24):7910–7. https://doi.org/10.1128/JB.01147-08. Epub 2008 Oct 10. PMID: 18849421; PMCID: PMC2593214

Miller JH (1996) Spontaneous mutators in bacteria: insights into pathways of mutagenesis and repair. Annu Rev Microbiol 50:625–643. https://doi.org/10.1146/annurev.micro.50.1.625

Molin S, Tolker-Nielsen T (2003 Jun) Gene transfer occurs with enhanced efficiency in biofilms and induces enhanced stabilisation of the biofilm structure. Curr Opin Biotechnol 14(3): 255–261. https://doi.org/10.1016/s0958-1669(03)00036-3

Montanari S, Oliver A, Salerno P, Mena A, Bertoni G, Tümmler B, Cariani L, Conese M, Döring G, Bragonzi A (2007 May) Biological cost of hypermutation in *Pseudomonas aeruginosa* strains from patients with cystic fibrosis. Microbiology (Reading). 153(Pt 5):1445–1454. https://doi.org/10.1099/mic.0.2006/003400-0

Moskowitz SM, Foster JM, Emerson J, Burns JL (2004 May) Clinically feasible biofilm susceptibility assay for isolates of *Pseudomonas aeruginosa* from patients with cystic fibrosis. J Clin Microbiol 42(5):1915–22. https://doi.org/10.1128/JCM.42.5.1915-1922.2004. PMID: 15131149; PMCID: PMC404629

Moskowitz SM, Brannon MK, Dasgupta N, Pier M, Sgambati N, Miller AK, Selgrade SE, Miller SI, Denton M, Conway SP, Johansen HK, Høiby N (2012 Feb) PmrB mutations promote polymyxin resistance of *Pseudomonas aeruginosa* isolated from colistin-treated cystic fibrosis patients. Antimicrob Agents Chemother 56(2):1019–30. https://doi.org/10.1128/AAC.05829-11. Epub 2011 Nov 21. PMID: 22106224; PMCID: PMC3264203

Mulcahy H, Charron-Mazenod L, Lewenza S (2008) Extracellular DNA chelates cations and induces antibiotic resistance in *Pseudomonas aeruginosa* biofilms. PLoS Pathog 4:e1000213. https://doi.org/10.1371/journal.ppat.1000213

Mulet X, Maciá MD, Mena A, Juan C, Pérez JL, Oliver A (2009 Apr) Azithromycin in *Pseudomonas aeruginosa* biofilms: bactericidal activity and selection of nfxB mutants. Antimicrob Agents Chemother 53(4):1552–60. https://doi.org/10.1128/AAC.01264-08. Epub 2009 Feb 2. PMID: 19188376; PMCID: PMC2663112

Mulet X, Moyá B, Juan C, Macià MD, Pérez JL, Blázquez J, Oliver A (2011 Oct) Antagonistic interactions of *Pseudomonas aeruginosa* antibiotic resistance mechanisms in planktonic but not biofilm growth. Antimicrob Agents Chemother 55(10):4560–8. https://doi.org/10.1128/AAC.00519-11. Epub 2011 Aug 1. PMID: 21807976; PMCID: PMC3186965

Nguyen D, Joshi-Datar A, Lepine F, Bauerle E, Olakanmi O, Beer K et al (2011) Active starvation responses mediate antibiotic tolerance in biofilms and nutrient-limited bacteria. Science 334: 982–986. https://doi.org/10.1126/science.1211037

Oliver A, Cantón R, Campo P, Baquero F, Blázquez J (2000) High frequency of hypermutable *Pseudomonas aeruginosa* in cystic fibrosis lung infection. Science 288:1251–1253

Oliver A, Baquero F, Blázquez J (2002) The mismatch repair system (mutS, mutL, and uvrD genes) in *Pseudomonas aeruginosa*: molecular characterization of naturally occurring mutants. Mol Microbiol 43:1641–1650

Oliver A, Mena A, Macià MD (2007) Evolution of *Pseudomonas aeruginosa* pathogenicity: from acute to chronic infections. In: Baquero F et al (eds) Evolutionary biology of bacterial and fungal pathogens. American Society for Microbiology, Washington, DC, pp 433–434

Pamp SJ, Gjermansen M, Johansen HK, Tolker-Nielsen T (2008a) Tolerance to the antimicrobial peptide colistin in *Pseudomonas aeruginosa* biofilms is linked to metabolically active cells, and depends on the pmr and mexAB-oprM genes. Mol Microbiol 68:223–240. https://doi.org/10.1111/j.1365-2958.2008.06152.x

Pamp SJ, Gjermansen M, Johansen HK, Tolker-Nielsen T (2008b Apr) Tolerance to the antimicrobial peptide colistin in *Pseudomonas aeruginosa* biofilms is linked to metabolically active cells, and depends on the pmr and mexAB-oprM genes. Mol Microbiol 68(1):223–240. https://doi.org/10.1111/j.1365-2958.2008.06152.x. Epub 2008 Feb 28

Pelgrift RY, Friedman AJ (2013) Nanotechnology as a therapeutic tool to combat microbial resistance. Adv Drug Deliv Rev 65:1803–1815

Petrova OE, Sauer K (2012 May) Sticky situations: key components that control bacterial surface attachment. J Bacteriol 194(10):2413–25. https://doi.org/10.1128/JB.00003-12. Epub 2012 Mar 2. PMID: 22389478; PMCID: PMC3347170

Pires DP, Melo L, Vilas Boas D, Sillankorva S, Azeredo J. Phage therapy as an alternative or complementary strategy to prevent and control biofilm-related infections. Curr Opin Microbiol 2017 Oct;39:48–56. https://doi.org/10.1016/j.mib.2017.09.004. Epub 2017 Sep 28

Plasencia V, Borrell N, Maciá MD, Moya B, Pérez JL, Oliver A (2007 Jul) Influence of high mutation rates on the mechanisms and dynamics of in vitro and in vivo resistance development to single or combined antipseudomonal agents. Antimicrob Agents Chemother 51(7):2574–81. https://doi.org/10.1128/AAC.00174-07. Epub 2007 Apr 30. PMID: 17470655; PMCID: PMC1913281

Poole K (2004) Efflux mediated multiresistance in gram-negative bacteria. Clin Microbiol Infect 10:12–26

Poole K (2014) Stress responses as determinants of antimicrobial resistance in *Pseudomonas aeruginosa*: multidrug efflux and more. Can J Microbiol 60:783–791. https://doi.org/10.1139/cjm-2014-0666

Poudyal B, Sauer K (2018) The ABC of biofilm drug tolerance: the MerR-like regulator BrlR is an activator of ABC transport systems, with PA1874-77 contributing to the tolerance of *Pseudomonas aeruginosa* biofilms to tobramycin. Antimicrob Agents Chemother 62:e01981–e01917. https://doi.org/10.1128/AAC.01981-17

Prunier AL, Malbruny B, Laurans M, Brouard J, Duhamel JF, Leclercq R (2003 Jun 1) High rate of macrolide resistance in *Staphylococcus aureus* strains from patients with cystic fibrosis reveals high proportions of hypermutable strains. J Infect Dis 187(11):1709–1716. https://doi.org/10.1086/374937. Epub 2003 May 15

Radman M, Matic I, Taddei F (1999 May 18) Evolution of evolvability. Ann N Y Acad Sci 870: 146–155. https://doi.org/10.1111/j.1749-6632.1999.tb08874.x

Rodríguez-Rojas A, Blázquez J (2009 Feb) The *Pseudomonas aeruginosa* pfpI gene plays an antimutator role and provides general stress protection. J Bacteriol 191(3):844–50. https://doi.org/10.1128/JB.01081-08. Epub 2008 Nov 21. PMID: 19028889; PMCID: PMC2632063

Rojo-Molinero E, Macià MD, Rubio R, Moyà B, Cabot G, López-Causapé C, Pérez JL, Cantón R, Oliver A (2016 Apr 22) Sequential treatment of biofilms with aztreonam and tobramycin is a novel strategy for combating *Pseudomonas aeruginosa* chronic respiratory infections. Antimicrob Agents Chemother 60(5):2912–22. https://doi.org/10.1128/AAC.00196-16. PMID: 26926631; PMCID: PMC4862541

Rojo-Molinero E, Macià MD, Oliver A (2019 Mar 22) Social behavior of antibiotic resistant mutants within *Pseudomonas aeruginosa* biofilm communities. Front Microbiol 10:570. https://doi.org/10.3389/fmicb.2019.00570. PMID: 30967851; PMCID: PMC6438888

Saiman L, Marshall BC, Mayer-Hamblett N, Burns JL, Quittner AL, Cibene DA, Coquillette S, Fieberg AY, Accurso FJ, Campbell PW (2003 Oct 1) 3rd; macrolide study group. Azithromycin in patients with cystic fibrosis chronically infected with *Pseudomonas aeruginosa*: a randomized controlled trial. JAMA 290(13):1749–1756. https://doi.org/10.1001/jama.290.13.1749

Santos-Lopez A, Marshall CW, Scribner MR, Snyder DJ, Cooper VS (2019 Sep 13) Evolutionary pathways to antibiotic resistance are dependent upon environmental structure and bacterial lifestyle. Elife 8:e47612. https://doi.org/10.7554/eLife.47612. PMID: 31516122; PMCID: PMC6814407

Savage VJ, Chopra I, O'Neill AJ (2013 Apr) *Staphylococcus aureus* biofilms promote horizontal transfer of antibiotic resistance. Antimicrob Agents Chemother 57(4):1968–70. https://doi.org/10.1128/AAC.02008-12. Epub 2013 Jan 28. PMID: 23357771; PMCID: PMC3623343

Schaible B, Schaffer K, Taylor CT (2010 Dec 31) Hypoxia, innate immunity and infection in the lung. Respir Physiol Neurobiol 174(3):235–243. https://doi.org/10.1016/j.resp.2010.08.006. Epub 2010 Aug 13

Scribner MR, Santos-Lopez A, Marshall CW, Deitrick C, Cooper VS (2020) Parallel evolution of tobramycin resistance across species and environments. mBio 11(3):e00932–20. Published 2020 May 26. https://doi.org/10.1128/mBio.00932-20

Skov M, Pressler T, Lykkesfeldt J, Poulsen HE, Jensen PØ, Johansen HK, Qvist T, Kræmer D, Høiby N, Ciofu O (2015 Mar) The effect of short-term, high-dose oral N-acetylcysteine treatment on oxidative stress markers in cystic fibrosis patients with chronic *P. aeruginosa* infection -- a pilot study. J Cyst Fibros 14(2):211–218. https://doi.org/10.1016/j.jcf.2014.09.015. Epub 2014 Oct 23

Smith EE et al (2006) Genetic adaptation by *Pseudomonas aeruginosa* to the airways of cystic fibrosis patients. Proc Natl Acad Sci U S A 103:8487–8492

Sniegowski PD, Gerrish PJ, Lenski RE (1997 Jun 12) Evolution of high mutation rates in experimental populations of E. coli. Nature 387(6634):703–705. https://doi.org/10.1038/42701

Stewart EJ, Ganesan M, Younger JG, Solomon MJ (2015 Aug 14) Artificial biofilms establish the role of matrix interactions in staphylococcal biofilm assembly and disassembly. Sci Rep 5: 13081. https://doi.org/10.1038/srep13081. PMID: 26272750; PMCID: PMC4536489

Stewart PS, Zhang T, Xu R, Pitts B, Walters MC, Roe F et al (2016) Reaction-diffusion theory explains hypoxia and heterogeneous growth within microbial biofilms associated with chronic infections. NPJ Biofilms Microbiomes 2:16012. https://doi.org/10.1038/npjbiofilms.2016.12

Stoodley P, Sauer K, Davies DG, Costerton JW (2002) Biofilms as complex differentiated communities. Annu Rev Microbiol 56:187–209. https://doi.org/10.1146/annurev.micro.56.012302.160705. Epub 2002 Jan 30

Taddei F, Radman M, Maynard-Smith J, Toupance B, Gouyon PH, Godelle B (1997 Jun 12) Role of mutator alleles in adaptive evolution. Nature 387(6634):700–702. https://doi.org/10.1038/42696

Tanner WD, Atkinson RM, Goel RK, Toleman MA, Benson LS, Porucznik CA, VanDerslice JA (2017 Apr 1) Horizontal transfer of the blaNDM-1 gene to *Pseudomonas aeruginosa* and Acinetobacter baumannii in biofilms. FEMS Microbiol Lett 364(8). https://doi.org/10.1093/femsle/fnx048

Tata M, Wolfinger MT, Amman F, Roschanski N, Dötsch A et al (2016) RNASeq based transcriptional profiling of *Pseudomonas aeruginosa* PA14 after short- and long-term anoxic cultivation in synthetic cystic fibrosis sputum medium. PLoS One 11(1):e0147811. https://doi.org/10.1371/journal.pone.0147811

Tseng BS, Zhang W, Harrison JJ, Quach TP, Song JL, Penterman J, Singh PK, Chopp DL, Packman AI, Parsek MR (2013 Oct) The extracellular matrix protects *Pseudomonas aeruginosa* biofilms by limiting the penetration of tobramycin. Environ Microbiol 15(10):2865–78. https://

doi.org/10.1111/1462-2920.12155. Epub 2013 Jun 10. PMID: 23751003; PMCID: PMC4045617

Van Acker H, Coenye T (2016 Jun 10) The role of efflux and physiological adaptation in biofilm tolerance and resistance. J Biol Chem 291(24):12565–12572. https://doi.org/10.1074/jbc.R115.707257. Epub 2016 Apr 21. PMID: 27129224; PMCID: PMC4933474

Van Acker H, Sass A, Bazzini S, De RK, Udine C, Messiaen T et al (2013) Biofilm-grown Burkholderia cepacia complex cells survive antibiotic treatment by avoiding production of reactive oxygen species. PLoS One 8:e58943. https://doi.org/10.1371/journal.pone.0058943

Van Acker H, Gielis J, Acke M, Cools F, Cos P, Coenye T (2016 Jul 20) The role of reactive oxygen species in antibiotic-induced cell death in Burkholderia cepacia Complex Bacteria. PLoS One 11(7):e0159837. https://doi.org/10.1371/journal.pone.0159837. PMID: 27438061; PMCID: PMC4954720

Walters MC III, Roe F, Bugnicourt A, Franklin MJ, Stewart PS (2003) Contributions of antibiotic penetration, oxygen limitation, and low metabolic activity to tolerance of *Pseudomonas aeruginosa* biofilms to ciprofloxacin and tobramycin. Antimicrob Agents Chemother 47:317–323. https://doi.org/10.1128/AAC.47.1.317-323.2003

Watson ME, Burns JL, Smith AL (2004 Sep) Hypermutable Haemophilus influenzae with mutations in mutS are found in cystic fibrosis sputum. Microbiology (Reading) 150(Pt 9): 2947–2958. https://doi.org/10.1099/mic.0.27230-0

Webb JS, Thompson LS, James S, Charlton T, Tolker-Nielsen T, Koch B, Givskov M, Kjelleberg S (2003 Aug) Cell death in *Pseudomonas aeruginosa* biofilm development. J Bacteriol 185(15): 4585–92. https://doi.org/10.1128/JB.185.15.4585-4592.2003. PMID: 12867469; PMCID: PMC165772

Whiteley M, Bangera MG, Bumgarner RE, Parsek MR, Teitzel GM, Lory S, Greenberg EP (2001 Oct 25) Gene expression in *Pseudomonas aeruginosa* biofilms. Nature 413(6858):860–864. https://doi.org/10.1038/35101627

Wilmaerts D, Windels EM, Verstraeten N, Michiels J (2019) General mechanisms leading to persister formation and awakening. Trends Genet 35:401–411

Wimpenny J, Manz W, Szewzyk U (2000) Heterogeneity in biofilms. FEMS Microbiol Rev 24: 661–671

Woodford N, Ellington MJ (2007 Jan) The emergence of antibiotic resistance by mutation. Clin Microbiol Infect 13(1):5–18. https://doi.org/10.1111/j.1469-0691.2006.01492.x

Worlitzsch D, Tarran R, Ulrich M, Schwab U, Cekici A, Meyer KC et al (2002) Effects of reduced mucus oxygen concentration in airway pseudomonas infections of cystic fibrosis patients. J Clin Invest 109:317–325. https://doi.org/10.1172/JCI0213870

Yu S, Su T, Wu H, Liu S, Wang D, Zhao T, Jin Z, Du W, Zhu MJ, Chua SL, Yang L, Zhu D, Gu L, Ma LZ (2015 Dec) PslG, a self-produced glycosyl hydrolase, triggers biofilm disassembly by disrupting exopolysaccharide matrix. Cell Res 25(12):1352–67. https://doi.org/10.1038/cr.2015.129. Epub 2015 Nov 27. PMID: 26611635; PMCID: PMC4670989

Zhang L, Mah TF (2008) Involvement of a novel efflux system in biofilm-specific resistance to antibiotics. J Bacteriol 190:4447–4452. https://doi.org/10.1128/JB.01655-07

Part II
Biofilm Prevention & Treatment

Prevention of Biofilms in Catheter-Associated Urinary Tract Infections (CAUTIs): A Review

Dana Ronin, Ryan B. Felix, Candace M. Williams, Stacey Audrey Mannuel, Darla Goeres, Jennifer Summers, John E. LaFleur, and Birthe V. Kjellerup

Abstract Acquisition of catheter-associated urinary tract infections (CAUTIs) is of significant detriment to patients' health and comfort since such infections increase hospital stay durations, health care costs, and patient morbidity. In addition, CAUTIs can perpetuate antibiotic-resistant bacterial presence, when acquired in the hospital or healthcare setting, thus causing problems after the CAUTI has been treated. The pathogenesis of CAUTIs stems from biofilm growth on an implanted urinary catheter surface and the altered bacterial phenotype. Biofilms are microorganisms that are attached to an inert and/or biotic surface and embedded in an extracellular polymeric substance (EPS). Several techniques have been developed to produce catheters that aim to delay biofilm formation and subsequently reduce biofilm growth, thus delaying the onset of CAUTI, but these have shown varying degrees of success. In this review, we discuss the etiology of CAUTIs, existing laboratory models of pathogenesis, approaches to disease prevention, and the interconnectivity among these topics.

Dana Ronin, Ryan B. Felix and Candace M. Williams contributed equally with all other contributors.

D. Ronin · R. B. Felix · C. M. Williams · B. V. Kjellerup (✉)
College Park, MD, USA
e-mail: bvk@umd.edu

S. A. Mannuel
College Park, MD, USA

University of Illinois College of Medicine, Chicago, IL, USA

D. Goeres · J. Summers
Center for Biofilm Engineering, Montana State University, Bozeman, MT, USA

J. E. LaFleur
Department of Emergency Medicine, George Washington University, Washington, DC, USA

K. Richter, K. Nørskov Kragh (eds.), *Antibiofilm Strategies*, Springer Series on Biofilms 11, https://doi.org/10.1007/978-3-031-10992-8_4

1 Introduction

Intermittent catheterization of the urinary bladder has been utilized as a necessary treatment for urinary retention as early as 1500 B.C. (Hanafy et al. 1974). Prior to the nineteenth century and the onset of what is colloquially referenced as "modern medicine," catheters consisted of rigid, hollow tubes (Feneley et al. 2015). The invention of the conventional and flexible Foley catheter was initially designed for hemostatic regulation following prostatic resection; however, it saw countless other uses as it allowed for facile indwelling catheterization, a practice defined by leaving a catheter in place for greater than 30 days (Foley 1937; Feneley et al. 2015). Unfortunately, the medical community was slow to react to problems that arose from patient catheterization. For example, open drainage of catheters was a significant cause of infection initially commented on in a 1928 article, yet the usage of closed drainage was not widespread until 1966 (Dukes 1928; Kunin and McCormack 1966).

Catheterization is applied if a patient is suffering from urinary retention or urinary incontinence with an inability to use incontinence pads as a means of bladder irrigation or chemotherapy delivery (Badalato et al. 2014), and as a method of measuring urinary output, among other reasons (Otero et al. 2010). In the late 1960s it was realized that indwelling catheterization was a primary cause of infection in hospitalized patients, leading to the development of intermittent catheterization and intermittent self-catheterization, techniques that are still in use today (Guttmann and Frankel 1966; Lapides et al. 1972; Seth et al. 2014). The detriment of indwelling catheterization stems from the increasing probability of Catheter-associated Urinary Tract Infections (CAUTI) contraction the longer the catheter is in place, a risk that increases by approximately 5% each day (Garibaldi et al. 1974). In fact, patients who wear an indwelling catheter for the majority of a year have an associated three-fold increase in mortality (Platt et al. 1982; Kunin et al. 1992). Other negative effects are an increase in patient health care costs duration of hospitalization and reduced quality of life (Saint 2000; Hooton et al. 2010).

The risk of CAUTI has led to the research and development of antimicrobial catheters, some of which have gone to market, such as those coated with silver or nitrofurazone (Lam et al. 2014). Though catheters of new materials are being developed and marketed, the clinical benefit of these devices remains elusive and widely debated (Riley et al. 1995; Schierholz et al. 1998; Johnson et al. 2006, p. 200; Drekonja et al. 2008; Al-Qahtani et al. 2019). Studies looking at the efficacy of antimicrobial urinary catheters have found confounding results for short-term versus long-term catheter use. While antimicrobial coated catheters show some promise for short-term catheterization, there was little antimicrobial effect in long-term catheterization (Vaitkus et al. 2021). From the current literature, it is unclear where the inconsistency arises between developmental research and clinical evaluation. A potential cause of the discrepancy may be the unforeseeable effects of the presence of biofilm in clinical pathogenesis. Discussed in this review is the intersection of CAUTI etiology with biofilm character, and a commentary on how these topics pertain to the diagnosis and treatment of both CAUTIs and asymptomatic bacteriuria, as well as the efficacy of current preventative technologies and the development of clinically relevant future technologies.

2 Etiology of Catheter-Associated Urinary Tract Infection

2.1 Means of Infection

A healthy urinary tract resists infection primarily through the body's innate immune defenses (Abraham and Miao 2015). An infection may occur when uropathogens colonize the urinary tract system and form unhealthy microbiota. Depending on the patient's underlying condition, UTIs are categorized into complicated and uncomplicated infections. In an uncomplicated urinary tract infection, microorganisms are free-floating in urine in their planktonic state (Trautner and Darouiche 2004). CAUTIs are classified as complicated infections due to the introduction of the foreign catheter which compromises' the patient's urinary tract system (Wei Tan and Chlebicki 2016). When a urinary catheter is inserted, it gives the bacteria a surface to attach to and the ability to develop into a biofilm (Cortese et al. 2018). On the surface of the bladder, the binding of bacteria to the bladder mucosa triggers an inflammatory response, resulting in the removal of the bacteria-bound cells (Trautner and Darouiche 2004). The surface of a catheter, however, lacks this inflammatory response, hence allowing the buildup of bacterial cells on the surface. Patient CAUTI contraction is a consequence of biofilm colonization of the urinary catheter, which occurs through the migration of bacteria on the catheter surfaces and into the bladder (Nickel et al. 1992; Trautner and Darouiche 2004). Catheter insertion directly bypasses the skin's protective barrier, sidestepping the body's first defense mechanism, and serves as a substrate for biofilm formation (Danese 2002; Delcaru et al. 2016). CAUTI contraction is thought to follow two primary routes of bacterial contamination and subsequent migration: the intraluminal route, which is along the surface of the catheter lumen, and the extraluminal route which is between the catheter exterior and the patient urethra (Fünfstück 1998; Tambyah et al. 1999; Barford and Coates 2009; Mathur 2018). With the onset of intraluminal contamination, uropathogenic microorganisms can gain entrance to the lumen through a break in the closed drainage system via tubing disruptions, contamination of the collection bag, or the catheter-collection bag junction. To avoid intraluminal contamination, nurses and physicians follow the maintenance guidelines of the Centers for Disease Control (CDC) which emphasize aseptic technique and avoidance of unnecessary and/or prolonged catheterization (Gould et al. 2010). In the case of extraluminal contamination, uropathogenic microorganisms arise on the surface through contact contamination by a physician or nurse during insertion, descension from the patient's perineum by capillary action via the thin mucosal film on the catheter surface, or ascension of periurethral skin colonies through the patient meatus during catheter insertion (Schaeffer and Chmiel 1983; Tambyah et al. 1999; Barford et al. 2008;

Mathur 2018). It has been suggested that extraluminal, rather than intraluminal, contamination of the catheter is the most common route of infection; about two-thirds of infections are extraluminal (Tambyah et al. 1999). This discrepancy may be due to the adoption of closed drainage systems and the decreased likelihood of drainage system breakage compared to contact contamination of the catheter exterior, a result that has been corroborated by an experimental animal model (Kunin and McCormack 1966; Nickel et al. 1985). With the widespread adoption of closed drainage systems, it has been suggested that extraluminal migration of the biofilm precedes intraluminal colonization (Garibaldi et al. 1980; Barford et al. 2008). Although research suggests that the primary contamination route is extraluminal, more studies are necessary to further understand the pathogenic process of bacterial introduction to urinary catheters.

Following extraluminal or intraluminal contamination of the surface, the biofilm appears to undergo several phenotypically distinct stages of development: reversible attachment, irreversible attachment, maturation, and dispersion (Sauer et al. 2002). These linear phenotypic differences are characterized by a vast array of genetic variations (Sauer et al. 2002; Prüß et al. 2006). The uropathogen first adheres to the surface by establishing a conditioning layer (Denstedt et al. 1998; Trautner and Darouiche 2004). The conditioning layer may either be derived from the in vivo fluid, in vitro media, or bacterial secretion of adhesins (Gristina 1987; Denstedt et al. 1998). This conditioning layer allows for both transient and irreversible adhesion of the biofilm to the surface. Cell motility mechanisms (such as flagella or type IV pili) play a significant role in adhesion to abiotic surfaces (O'Toole and Kolter 1998a; Pratt and Kolter 1998). It has been suggested that flagella mediate an initial reversible attachment to the surface by allowing cells to come into close proximity by overcoming repulsive forces at the boundary, while pili and fimbriae play a more critical role in irreversible adhesion and mature biofilm structure (Korber et al. 1989). However, it seems that for any given bacterial strain more than one adhesion mechanism is possible and that these means of attachment and propagation vary considerably among bacterial genera, species, strains, and with differing environmental conditions and surface energies (O'Toole and Kolter 1998b; Pratt and Kolter 1998; Flores-Mireles et al. 2015).

Once adhered to the surface, the bacteria reproduce and secrete a hygroscopic matrix of polysaccharides, nucleic acids, proteins, and lipids termed extracellular polymeric substances (EPS) (Lawrence et al. 1987; Flemming and Wingender 2010). Beyond the initial attachment phase, the diverse bacterial EPS provides additional permanence to the anchoring of biofilm to the surface (Hoyle and Costerton 1991; Costerton et al. 1995; Flemming and Wingender 2010). The biofilm EPS secretion also acts as a protective environment for the film, with each constituent of the EPS serving various roles (Flemming et al. 2007). Compared to the rapid-growing, planktonic (free-floating) bacteria, the slower-growing biofilm is phenotypically distinct and works as a complex community (Davies et al. 1993, 1998; Costerton et al. 1995). The combined effect of the EPS matrix contributes to a hydrated environment that is protective from external forces, porous to mitigate nutrient transfer, contains an extracellular means for horizontal gene transfer and recycling of intracellular components, and confers resistance to an array of biocides and antibiotics (Hoyle and Costerton 1991, p. 199; de Beer and Stoodley 1995; Høiby et al. 2010; Flemming and Wingender 2010). While attached to the surface, the biofilm matures until it enters the dispersion stage of development, where motile bacteria migrate out of the film into the planktonic environment and subsequently colonize uninhabited surfaces, including the urethra wall (tissue) (Lawrence et al. 1987, p. 19; Sauer et al. 2002). In intraluminal contamination, fluid flow from the contaminated intravascular catheter also provides bacteria easy access to the vasculature (Danese 2002), which is also true of urinary catheters. Virulence of bacteria has been linked—in part—to migration means such as twitching, swimming, and swarming motility (Jones et al. 2004; Lane et al. 2005; Wright et al. 2005; Hola et al. 2012). This is likely due to the role of motility mechanisms in surface attachment and the spread of the film into the bladder, which act as contributing factors to uropathogenicity in CAUTIs.

The modulation of these phenotypically diverse stages in biofilm development is a complex cellular-signaling process. Quorum sensing (QS) is the cell-density regulation of bacterial gene expression by secretion of small, diffusible autoinducer (AI) molecules which is involved in the innate development of biofilms (Fuqua et al. 1994; McLean et al. 1997; Davies et al. 1998). In Gram-negative biofilms, QS is primarily mediated by acyl-homoserine lactone (AHL)-type molecules that accumulate in the extracellular environment until reaching sufficient concentration to bind transcriptional regulators (Ivanova et al. 2015). In Gram-positive biofilms, autoinducing peptides (AIPs) are the primary signaling units and require transport across the membrane and generate signals via membrane-associated receptors (Ivanova et al. 2015). The involvement of QS in the maturation of biofilms has implicated quorum-quenching (QQ) in future prevention strategies (discussed later). Despite the significant role that QS plays in biofilm development and uropathogenic virulence, it is not the sole means of essential endogenous sensory communication (Sauer and Camper 2001; Senturk et al. 2012).

Some people are never cured once diagnosed with CAUTI, having to deal with flare-ups for the rest of their lives. The danger of CAUTI is that it can lead to the development of serious, and potentially lethal, diseases if bacteria begin disseminating to other organs. When bacteria attach to the urethral epithelial cells via a urinary catheter and travel past the bladder, the kidneys may also become infected, as shown in Fig. 1. The inflammation resulting from this infection is pyelonephritis. Symptoms include fever, nausea, increased urgency and frequency for urination, and back pain (Ashraf et al. 2020). If severe, pyelonephritis may lead to acute renal failure and require patient hospitalization or even renal transplantation. The spread of bacteria from an untreated CAUTI can also lead to other systemic infections including bacteremia, sepsis, multiple organ system dysfunction, and shock. Approximately 13,000 patients die from CAUTI per year in the US (Podkovik et al. 2019).

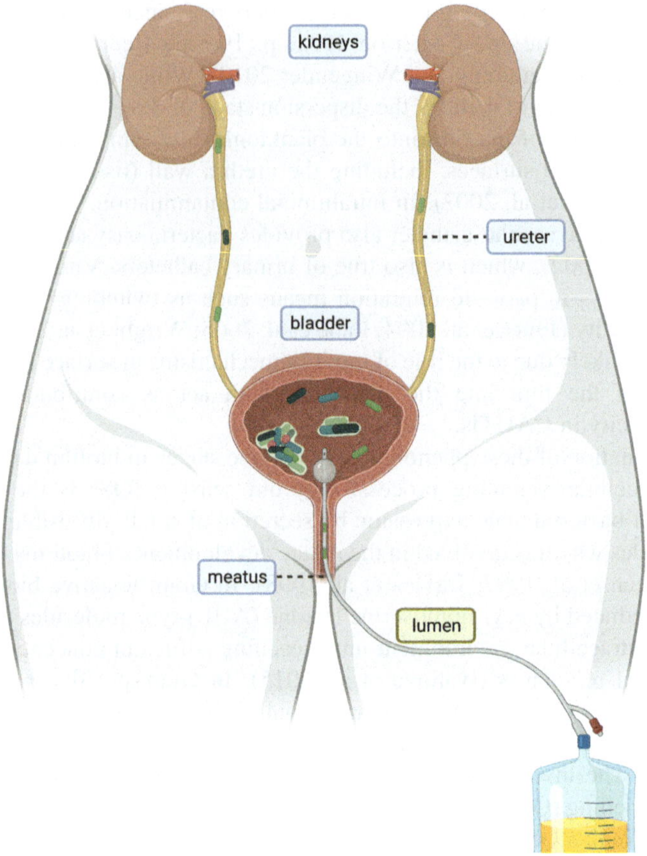

Fig. 1 Clinical Map of Catheter-Associated Urinary Tract Infection. In this diagram of a CAUTI in a female bladder, bacteriuria travels intraluminally into the bladder and attaches to the surface. This leads to the formation of a biofilm. The bacteria can also travel up the ureters and into the kidneys, leading to pyelonephritis. Freely after Belyayeva and Jeong (2021)

2.2 *Organisms of Infection*

Contraction of catheter-associated urinary tract infection (CAUTI) during short-term catheterization is believed to be primarily monomicrobial (Tambyah and Maki 2000). However, patients subjected to a longer duration of indwelling catheterization present with CAUTIs that are polymicrobial in nature, with a clinical study reporting that only approximately 13% of such CAUTIs are monomicrobial (Nicolle 2005; Hola et al. 2012, p. 201). The organisms of infection consist of a diverse array of uropathogenic bacteria colonizing the inserted catheter, bladder, and stagnant urine within the bladder (Jacobsen et al. 2008). The most common organism of infection is *Escherichia coli*, sometimes broadly referred to as uropathogenic *E. coli* (UPEC) (Nicolle 2005). Other commonly implicated bacteria are *Klebsiella spp.*, *Serattia spp.*, *Citrobacter spp.*, *Pseudomonas aeruginosa*, and the urease-producing *Proteus mirabilis*, *Morganella morganii*, and *Providencia stuartii* (Nicolle 2005; Hola et al. 2012). Yeasts are also implicated, though typically only *Candida spp.* (Hola et al. 2012). The results of sampling CAUTI microbiota are largely the same whether culture-dependent or culture-independent techniques are utilized (Xu et al. 2012).

3 Diagnosis of CAUTI

3.1 Clinical Guidelines and Symptoms

Clinical guidelines for the classification, surveillance, and diagnosis of CAUTIs have been established by the Infectious Disease Society of America (IDSA) and the National Healthcare Safety Network (NHSN), respectively (Nicolle 2005; Hooton et al. 2010). Adult patients who had an indwelling urethral, suprapubic, or condom catheter for 48 hours, even if they were no longer catheterized, may be diagnosed with either symptomatic or asymptomatic CAUTIs (Nicolle 2005; Hooton et al. 2010). Typically, the guideline for a symptomatic patient's urine sample must contain more than 10^3 CFU/mL of at least one bacterial species (Maki and Tambyah 2001, p. 200; Nicolle 2005; Hooton et al. 2010). For diagnosis of asymptomatic CAUTIs, the guideline standard is 10^5 CFU/mL of urine (Hooton et al. 2010). False-positive urine cultures for CAUTI diagnostic tests have been reported to occur with equal frequency in patients with or without infections, illustrating the necessity for a more accurate diagnostic tool (Frontera et al. 2020). False positives and false negatives for UTIs typically result from an increase or decrease in bacterial cell count, respectively. Bacteriuria research suggests that low bacterial cell counts can result from brisk diuresis or higher urine dilution due to increased fluid intake, antimicrobial drugs, or slow-growing or fastidious microorganisms (Kunin et al. 1992). In contrast, high bacterial cell counts for false-positive bacteriuria in women have been linked to vaginal contamination (Chen et al. 2013).

For symptomatic patients, CAUTI-associated symptoms may include: dysuria, fever, altered mental state, urgent or frequent urination, lethargy, and suprapubic pain (Hooton et al. 2010). Physicians are cautioned not to rely on symptoms as the sole basis for diagnosis (Tambyah and Maki 2000, p. 200). Instead, symptoms can be used to provide additional diagnostic evidence for symptomatic CAUTIs when there is no other identifiable source of infection. Thus, the broad-spectrum signs and symptoms represent a significant challenge for clinicians and infectious disease (ID) consultants in diagnosing CAUTI patients (Al-Qas Hanna et al. 2013). Additionally, diagnosis discrepancies as variations in adherence to the NHSN definition of CAUTIs have been reported among clinicians and ID consultants (Al-Qas Hanna et al. 2013). The IDSA's guidelines focus on clinical management, while NHSN's guidelines are for epidemiological surveillance purposes. Therefore, the thresholds are lower for IDSA's bacterial counts as it relies more on the clinician's assessment based on patient symptoms. Further research should be conducted to determine if

Table 1 Comparison of diagnostic guidelines

CAUTI diagnostic	IDSA guideline	NHSN guideline
Nomenclature	Symptomatic (CA-UTI) Asymptomatic (CA-ASB)	Symptomatic (SUTI) Asymptomatic (ABUTI)
Duration of Catheterization	48 hrs. (2 days)	48 hrs. (2 days)
Minimum Urine Sample Cell Count	Symptomatic ($\geqslant 10^3$ CFU/mL) Asymptomatic ($\geqslant 10^5$ CFU/mL)	Symptomatic ($\geqslant 10^5$ CFU/mL) Asymptomatic ($\geqslant 10^5$ CFU/mL)
Adult Symptoms for CA-UTI	• fever • suprapubic pain or tenderness • costovertebral angle pain or tenderness • urgent or frequent urination • dysuria • altered mental status • malaise • lethargy • flank pain • acute hematuria • pelvic discomfort	• fever (>38.0°C) • suprapubic tenderness • costovertebral angle tenderness • urinary urgent or frequent urination • dysuria

Comparative summary of diagnostic guidelines for asymptomatic and symptomatic catheter-associated infections by the Infectious Disease Society of America (IDSA) and the National Healthcare Safety Network (NHSN)

differences between IDSA and NHSN guidelines for CAUTI classification and diagnosis also contribute to alternative diagnosis patterns. These differences are summarized in Table 1.

For asymptomatic patients, the catheter itself may segregate inflammatory cells from the urethra and decompress the bladder (Maki and Tambyah 2001). This segregation may prevent symptoms of suprapubic pain and inflammation and decompression may prevent bladder distillation and reflux symptoms (Maki and Tambyah 2001). More recently in 2009, the definition of CAUTIs was modified to exclude asymptomatic bacteriuria (Press and Metlay 2013). Treatment of asymptomatic bacteriuria has been found to not affect patient outcomes and may instead increase the risk of antibiotic-resistant bacteria. Changes in the definitions of CAUTI reflect the ongoing movement to avoid the spread of antibiotic resistance, adverse reactions to antibiotics, and increased health care costs (Khawcharoenporn et al. 2011; Press and Metlay 2013).

3.2 *Role of Biofilms in CAUTI Diagnosis*

Biofilm species in urine samples do not always appropriately reflect the bacterial species that have colonized the bladder epithelia and may even be absent in the urine sample (Reid et al. 1992; Elvy and Colville 2009). Thus, researchers have theorized that microorganisms in urine may be more representative of biofilm formation within or on the catheter device rather than the bladder itself (Elvy and Colville 2009). This theory could explain why one research study found that about 16% (8 of 49) of spinal cord injury patients' bladders were colonized but their urine samples were negative for bacteriuria (Reid et al. 1992). Additionally, urine samples for bacteriuria laboratory identification had limited accuracy in one ureteral stent study in which 68% of patients were colonized but only 30% were positive for bacteriuria, emphasizing the prospect for more sophisticated diagnostic tests for bacteriuria (Farsi et al. 1995; Choong and Whitfield 2000). Finally, one research study found that in vitro biofilm growth may be diagnostic of recurring UTIs (Soto et al. 2006). Identification of these in vitro biofilm organisms may help identify the most ideal patients for therapeutic treatment (Soto et al. 2006).

The main issue is that biofilms are not considered in the definition of CAUTI. This leads to the possibility that the biofilm is not detected while it is still developing. Additionally, when the catheter is removed, there is a high chance that pieces of the biofilm slough off, further leading to the underestimation of bacterial presence. A study compared the colony counts of catheterized, untreated patients to non-catheterized, symptomatic UTI patients. Symptomatic UTI patients had colony counts as low as 10^2 CFU/mL. Initially in catheterized patients, the colony counts were approximately $\geq 10^2$ CFU/mL. After 24–48 hours, the colony counts for catheterized patients increased to $>10^5$ CFU/mL (Hooton et al. 2010). This demonstrates the rapid development of bacteriuria into a biofilm.

4 Laboratory-Based Biofilm Quantification Methods

Biofilm growth on medical devices has a significant, detrimental, impact on patient health and hospital care. Biofilms are responsible for 65% of nosocomial infections and can grow on all implanted medical devices such as urinary catheters, central venous catheters (CVCs), pacemakers, mechanical heart valves, and even contact lenses (Mah and O'Toole 2001; Donlan 2001a). For most patients, the threat of biofilm infection can only be eliminated by the removal of the contaminated device (Donlan 2001a, b).

The proper estimation of biofilm growth on catheters plays a significant role in CAUTI diagnosis and estimating patient risk of bacteriuria. CAUTI is mostly from bacteria accessing the bladder via the extra lumen. Thus, by the time bacteria are detected in the urine and the patient has symptoms, it is likely that both the urethra and bladder are fully colonized. The result of inefficient diagnostics sets up the patient for recurring infections. Based on IDSA guidelines, the diagnosis of symptomatic CAUTIs is more dependent on patient symptoms than laboratory quantification (Hooton et al. 2010). The estimation of biofilm growth plays a significant role in the diagnosis of asymptomatic CAUTIs by allowing earlier detection and management. Assessing the biofilm will give clinicians a clear idea of what—and how much—bacteria are present, allowing them to develop a better treatment plan. The real-time patient data will also allow for further innovation on how to prevent CAUTIs. Unfortunately, there is no gold standard for biofilm quantification in CAUTIs (Mandakhalikar et al. 2018). This section offers a brief introduction to alternative bacteriuria quantification strategies other than urine sample analysis and provides commentary on how biofilm quantification techniques are specifically relevant to CAUTI research.

4.1 Culture

Culture is a conventional technique utilized for biofilm infection diagnostics. Some benefits that come with this semi-quantitative method include low cost, simple methodology, and no need for intricate laboratory equipment (Ronin et al. 2021). However, culture analysis requires the removal of the catheter, has a high false-negative rate, and misses viable but non-culturable organisms (Sutrave et al. 2019). The Roll Plate method, developed in 1977, is considered the reference diagnostic when working with catheters; with this technique, the catheter tip is rolled several times on an agar plate, effectively culturing the bacterial species present on the external surface of the catheter (Maki et al. 1977; Slobbe et al. 2009; Riboli et al. 2014). Bonkat et al. showed that the Roll Plate method was significantly more effective at detecting bacterial colonization of urinary catheters compared to urine culture (Bonkat et al. 2013). Addressing a limitation of this method, Cleri et al. developed a modified culture method to also culture bacterial species present on the interior of the catheter lumen (Cleri et al. 1980). Culture—as a method for biofilm quantification—while effective to a degree, results in the loss of critical infection information such as the spatial organization of the biofilm on the catheter. The low sensitivity and specificity of this method also contribute to the need of the development of more effective biofilm quantification methods in CAUTI (Portillo et al. 2013).

4.2 Sonication

Sonication is a technique that uses ultrasonic energy to effectively dislodge a biofilm from its surface while maintaining the viability of the bacteria (Bjerkan et al. 2009). With the strong adherence of biofilms to their surfaces, sonication could isolate more of the biofilm from a catheter surface (Ronin et al. 2021). When compared to urine culture, sonication showed great promise in its ability to detect colonization. Interestingly, while less sensitive than the Roll Plate method for colonization detection, sonication was more effective at identifying multispecies biofilms (Bonkat et al. 2013). The incorporation of vortexing prior to sonication has also improved the detachment of the biofilm from its surface. The microbubbles created by the vortexing step create microbubbles on the surface, increasing interfacial tension, and thereby priming the biofilm for more effective dislodgement from the surface (Peel 2019). However, sonication does not have a standardized protocol, making it difficult to control variation between different laboratories (Dibartola et al. 2017). Additionally, the extensive handling required with this method leaves room for contamination.

4.3 Microscopy for Visualization and Quantification

Urologic microscopy and Gram staining have been utilized as a diagnostic tool for CAUTIs (Wiwanitkit et al. 2005; Wilson and Gaido 2004). Noteworthy benefits of this technique were identified to be relatively low cost (Wiwanitkit et al. 2005; Blodgett 2009) and rapid availability of results (Blodgett 2009; Wilson and Gaido 2004). Microscopy can provide critical information regarding the spatial organization of the biofilm on the catheter (Ronin et al. 2021). Yet, this method is labor-intensive (Wilson and Gaido 2004), subject to technician interpretation (Blodgett 2009), and ultimately, has been associated with poor sensitivity (Pezzlo 1988; Wilson and Gaido 2004). Also, any microscopic method assumes that the biofilm was not altered when the catheter was removed. Unless CFU counts are high, it is challenging to find the bacteria and or clumps of biofilm on the catheter surface. Even if clumps of biofilm are identified, the heterogenous nature of biofilm growth on surfaces makes it difficult to extrapolate microscopy results and yield a holistic infection diagnostic. With a minimum sensitivity of 10^5 CFU/mL, this technique cannot detect infections of 10^2-10^3 CFU/mL (Bedrossian et al. 2017; Wilson and Gaido 2004) and its utilization is recommended only for additional evidence of bacteriuria (Pezzlo 1988). Additionally, microscopy techniques have a relatively small field of view relative to the surface area of the catheter. The question remains: how many fields of view are required for a researcher to have confidence that what they are seeing under the scope is typical of the catheter surface? A common unofficial, yet accepted, standard in microscopy is to select 10 random fields of view for qualitative and quantitative assessment. However, this value may be too low. For visualizing individual bacteria on a urinary catheter, the field of view should be on the order of 100 μm^2. Urinary catheters are usually on the order of cm^2, meaning that one million fields of view would be required to visualize an individual bacterium (Hall-Stoodley et al. 2012). There is a lack in standardization in the number of fields of view used for microscopy. The curvature of the catheter only adds another challenge of visualizing the biofilm (Mandakhalikar et al. 2018). Due to the limitations of diagnostic tests like the Gram-stain urine test for identifying bacteriuria, various microscopic techniques, and 3-D imaging software for quantification have also been employed.

4.3.1 Scanning Electron Microscope

With respect to CAUTI research, scanning electron microscopy (SEM) has been employed in many crystalline biofilm encrustation studies (Cox and Hukins 1989; Morris and Stickler 1998; Jones et al. 2005), biofilm interaction studies (Macleod and Stickler 2007) and studies regarding formation on potential catheter surfaces (Reddy et al. 2011). This technique applies a beam of accelerated electrons to generate a 3-D image of the biological specimen. Pre-processing steps include fixation, dehydration (or freeze-drying), and coating (Hannig et al. 2010). Dehydration typically involves acetone or ethanol washes with the gradual displacement of fluid carbon dioxide for artifact reduction (Holling et al. 2014). Next, a metallic coating such as gold or plutonium is typically applied to account for the lack of conductivity in biologic specimens (Hannig et al. 2010; Holling et al. 2014). Since SEM requires pre-processing steps (such as conductive coatings) artifacts could be introduced due to biofilm disruption (Hannig et al. 2010). Thus, environmental scanning electron microscopy (ESEM), which permits specimen observation in its natural environment, has been suggested to be the more superior microscopy method (Hannig et al., 2010; Holl). Yet, a significant disadvantage of both microscopy practices is that neither permits the differentiation of bacterial species (Hannig et al. 2010). Since CAUTIs are often the result of polymicrobial biofilms, organisms associated with encrustation and more severe symptoms of CAUTIs cannot be distinguished from the other species within the biofilm with SEM or ESEM alone.

4.3.2 Atomic Force Microscopy

For the past few decades, Atomic Force Microscopy (AFM) has been utilized for biotechnological and biochemical research studies (Ducker et al. 1991; Hoh et al. 1991; Lee et al. 1994; Schabert et al. 1995). These studies involve ligand-receptor interactions (Butt et al. 1990a; Dufrêne et al. 1997; Wong et al. 1998), cell surface structure and morphology (Kasas and Ikai 1995; Dufrêne et al. 1999; Camesano et al. 2000; Dufrêne 2000; Shiga et al. 2000), and bacterial adhesion and aggregation on different surfaces like glass, silicone, and mica (Dufrêne 2002; Hinterdorfer and Dufrêne 2006). The AFM has four main parts: the cantilever probe, the cantilever probe tip, the sample stage, and the optical deflection system (Hinterdorfer and Dufrêne 2006). Unlike other microscopy techniques, AFM is unique in that it does not require a beam for image development (Dufrêne 2002). Instead, AFM records the slight force changes between a sharp, cantilever probe tip and the sample surface (Dufrêne 2002; Hinterdorfer and Dufrêne 2006). These force changes are determined by the deflection of the spring-like cantilever probe (Butt et al. 1990b; Camesano et al. 2000; Dufrêne 2002; Hinterdorfer and Dufrêne 2006). Since SEM requires a vacuum environment, researchers cannot observe biological specimens in real time (Dufrêne 2002). In contrast, AFM has been regarded as the superior microscopic technique for real-time, high-resolution observation of biological samples (Radmacher et al. 1992; Dufrêne 2002). Yet, the high resolution associated with AFM requires the adhesion of biologic specimens such as bacterial biofilms to its substrate to be stronger than the force that is being applied by the cantilever probe tip (Dufrêne 2002; Hinterdorfer and Dufrêne 2006). This is because weak adhesion will result in cell detachment from the substrate, producing poor resolution (Razatos et al. 1998; Dufrêne 2002; Hinterdorfer and Dufrêne 2006). Common strategies to prevent detachment include air drying, covalent bonding of biofilm to a substrate, or other novel immobilization approaches (Dufrêne 2002). Unfortunately, air drying and covalent bonding can lead to protein denaturation and potentially alter the surface of the cell or biofilm (Camesano et al. 2000; Dufrêne 2002; Hinterdorfer and Dufrêne 2006). Additionally, one author's novel immobilization approach was successful in preventing detachment of only spherical cells rather than rod-shaped cells (Dufrêne 2002). Thus, further research is necessary to address detachment concerns. Nevertheless, the AFM allows CAUTI researchers to monitor and record biofilm formation on different catheter surfaces in real time and at high resolution.

4.3.3 Confocal Fluorescence Microscopy

The emergence of confocal fluorescence microscopy or confocal laser scanning microscopy (CLSM) revolutionized the way fluorescence is imaged (Dokland et al. 2006). In epifluorescence microscopy, the focal light receives interference from light outside of the focus plane, resulting in hazy images with poor contrast and resolution (Semwogerere and Weeks 2005; Dokland et al. 2006). In 1955, Marvin Minsky invented confocal microscopy and resolved this issue with the addition of a pinhole aperture (Semwogerere and Weeks 2005). The confocal pinhole limited light to the focal plane, making sectioning and 3-D imaging of the biological specimen possible (Pawley 2010). Yet, confocal fluorescence microscopy has several disadvantages. Appropriate pinhole size is critical for ensuring that enough photons can pass through (Semwogerere and Weeks 2005). Increasing the intensity to adjust for a small pinhole size can result in photobleaching, or the loss of the ability of the fluorophores to fluoresce (Cox and Sheppard 2004; Semwogerere and Weeks 2005). Additionally, the slow movement of a mechanical stage limits the scan time to 10 seconds, preventing the viewer from observing the specimen directly (White et al. 1987). Nevertheless, fluorescent labeling of bacterial cells or cell components for quantification represents a significant advantage to CAUTI researchers. Several authors have utilized confocal microscopy with fluorescent staining techniques to identify distinct species within a bacterial biofilm (Frank et al. 2009), establish the efficacy of antimicrobial catheter coatings (Dave et al. 2011), monitor the ability of catheter surfaces to inhibit bacterial adhesion (Guiton et al. 2010), and observe the progression of CAUTIs in murine bladders (Jacobsen and Shirtliff 2011).

4.3.4 Image Software for Biofilm Quantification

Imaging software is essential for producing and processing large, complex data sets of biological images (Schindelin et al. 2012). Within the past few decades, image analysis software has been utilized in several biofilm quantification studies (Yang et al. 2000; Heydorn et al. 2000; Xavier et al. 2003; Staudt et al. 2004; Truong et al. 2010; Zhang and Fang 2001). Useful features associated with the image analysis of bacterial biofilms include registration, segmentation, and tracking (Schindelin et al. 2012).

Imaging software may be an additional feature supplied by the manufacturer of laboratory microscopes like CLSM (Zhang and Fang 2001). Yet, commercial imaging software programs such as CellProfiler (Carpenter et al. 2006; Lamprecht et al. 2007) or open-source imaging software such as ImageJ and FIJI (Schindelin et al. 2012; Schneider et al. 2012) COMSTAT (Heydorn et al. 2000), and ISA (Yang et al. 2000) are also available. Due to the proprietary nature of most commercial programs, open-source software has become increasingly popular among researchers (Lamprecht et al. 2007). One reason for this popularity is that the researcher is free to edit the software to suit his or her specific queries of interest (Heydorn et al. 2000; Selinummi et al. 2005; Schindelin et al. 2012; Schneider et al. 2012). However, successful adaptation of these programs requires extensive programming knowledge more commonly associated with computer scientists rather than biologists (Lamprecht et al. 2007; Schindelin et al. 2012). Yet, many authors have successfully utilized imaging software in various biofilm quantification studies to analyze bacterial adhesion and formation on potential catheter surface materials (Reddy et al. 2011), to assess the growth of specific uropathogens (Vejborg and Klemm 2008; Walker et al. 2017), and to establish the efficacy of potential antimicrobial drugs (Rahmani-Badi et al. 2014) and polymers (Nowatzki et al. 2012; Bovis et al. 2015).

5 Current and Future Prevention Strategies

Although glass catheters are available, the most common materials comprising modern catheter tubing are vinyl, latex, and silicone (Mandakhalikar et al. 2016). Unfortunately, all catheter materials remain susceptible to biofilm contamination and infection (Nicolle 2005). The goal is to prevent CAUTIs, rather than treat them. By taking steps to minimize biofilm formation on the catheter in the first place, the chances of avoiding infection are much higher. Manufacturers and researchers have begun to produce anti-biofilm coatings to reduce bacterial colonization, hence reducing the risk of CAUTIs. The antimicrobial coatings discussed in this section are summarized in Table 2 and have been divided into five main categories: antimicrobial, antibiotic, enzymatic, surface energy, and iron scavenging coatings. However, it is important to note that not all these coatings have been clinically tested or have been approved for manufacturing.

Table 2 Coating prevention strategies, effects, and commercial availability

Coating prevention strategy	Coating agent	Biofilm effect	Commercially available
Antimicrobial Coatings (Metals/ Metal Oxides)	1. Silver 2. Zinc 3. Copper	Ionizes DNA and proteins, inhibiting reproduction and biofilm formation	Yes; Controversial clinical results
Hydrogel Coatings	Hydrogel	Establishes anti-fouling by formation of hydration layer on surface, inhibiting nonspecific protein adsorption	Yes
Antibiotic Coatings	1. Nitrofurazone 2. Rifampicin Minocycline	1. Inhibits bacterial DNA replication 2. Inhibits protein synthesis	Yes
Enzymatic Coatings	1. Acidase 2. Oxidase 3. Cellobiose Dehydrogenase	Direct killing of bacterial cells; disrupt quorum signaling	No; Not clinically tested
Surface Energy Coatings	1. Trimethylsilane 2. Plasma nanocoatings 3. Poly(ethylene oxide) 4. Poly(L-lysine)-grafted-poly(ethylene glycol) 5. Diamond-Like Carbon	Disrupts bacterial binding to conditioning layer via intramolecular forces, inhibiting the aggregation of bacteria required to form biofilm	No
Iron-Scavenging Coatings	1. Lactoferrin 2. Deferoxamine	Iron chelating agents that compete against bacteria for the iron nutrient, preventing bacterial clustering	No

This table represents the products or prototypes of six different catheter coatings and their biofilm prevention method

5.1 *Antimicrobial Coatings (Metals/Metal Oxides)*

The use of metal catheters—such as silver—originated in 1000–1500 BC and only began to evolve into more flexible, soft catheters during the Renaissance (Beiko et al. 2004; Feneley et al. 2015). Studies have shown that silver ions can disrupt microbial reproduction and cellular membranes through the ionization of DNA and protein binding (Dunnill et al. 2011; Lo et al. 2014). With cells that are no longer able to reproduce, biofilm formation is stalled. Silver coatings are the most popular, clinically used anti-biofilm strategy for CAUTI prevention (Andersen and Flores-Mireles 2020). While they have been extensively tested in in vitro, animal models, and clinical studies, the outcomes from these experiments yield varied results, leading to skepticism of their efficacy. Some research studies had success in showing that silver is an effective antiseptic capable of reducing bacteriuria (Jahn et al. 2012, p. 2). In a

1996–2005 systematic review of 12 trials of 13,392 participants, patients with short-term (less than 30 days) of in-dwelling, silver alloy-coated catheters were associated with a reduced risk of CAUTIs (Johnson et al. 2006). While silver-coated catheters can reduce bacterial colonization when a patient is catheterized for less than a week, they have no significant anti-biofilm effect for catheters inserted for longer periods of time (Wang et al. 2019). One reason for this short-term protection is that even the most effective silver-based coatings, such as silver-latex, are still susceptible to the crystallized blockages formed by *Proteus mirabilis*. This susceptibility results because *P. mirabilis* can create several layers of crystallization between itself and the inert silver atoms coating the interior walls of the catheter. Thus, these blockages can lead to bacteriuria presence, septicemia, and endotoxic shock (Ha and Cho 2006; Stickler 2014). This mechanism is true for many antimicrobial surfaces. Urine, other bodily fluids, or bacteria can coat the catheter surface, rendering the antimicrobial coating ineffective. An additional drawback of silver-based coatings is the toxicity of silver to human cells (Vaterrodt et al. 2016). Silver, in high concentrations or prolonged exposure, is not only capable of ionizing microbial cells but also human cells (Noimark et al. 2009; Vaterrodt et al. 2016). As a result, the accumulation of silver correlates to gray or gray-purple skin pigmentation, a condition known as argyria, or commonly called "Blue Man's Syndrome," which in rare cases has led to vision problems (Wadhera and Fung 2005; Noimark et al. 2009). Silver-coated catheters—along with other antimicrobial-coated catheters—are less cost-effective compared to non-coated catheters and are considered clinically irrelevant (Milo et al. 2016). Yet, silver coatings remain controversial due to the few studies showcasing their efficacy in high-risk patients (Schierholz et al. 1998, p. 19).

Like silver, other metal-based coatings (alloys and oxides) such as zinc and copper can have similar detrimental effects on human cells (Noimark et al. 2009). Zinc, a biologically safe material, has been used in combination with silver as an antimicrobial coating on catheters (Vaitkus et al. 2021). One study showed that the zinc-silver combination, used in an antimicrobial wound fabric, produced hydrogen peroxide and successfully inhibited biofilm formation on agar plates (Banerjee et al. 2015). Hence, the combination of zinc and silver in a coating could potentially be a successful method of inhibiting biofilm formation in urinary catheters. Using copper-based coatings is another promising antimicrobial strategy for urinary catheters. Copper acts as an antimicrobial by binding to the DNA-phosphate site and degrading DNA, thus leading to cell damage (Vaitkus et al. 2021). However, understanding the toxicity levels of copper coatings is critical prior to their clinical use (Andersen and Flores-Mireles 2020).

5.2 Hydrogel Coatings

Some coating types, rather than aiming to kill the bacteria colonizing on the surface, repel them from attaching in the first place. Hydrogel-coated catheters are the most popular of these anti-fouling coatings. The mechanism of function for these hydrogel-coated catheters is that due to their hydrophilic and insoluble nature, they form a barrier on the catheter surface that prevents nonspecific protein adsorption (Andersen and Flores-Mireles 2020). While clinical trials have yielded controversial results, hydrogel catheters are commercially approved (Lam et al. 2014). Hydrogel-coated catheters have also been embedded with silver; this combination of catheter coating is commonly used in hospital settings (Andersen and Flores-Mireles 2020). One study assessing the efficacy of silver-hydrogel-coated catheters found a 47% reduction in the rate of CAUTI in such coatings compared to standard, non-coated, catheters (Lederer et al. 2014). However, a separate study found that hydrogel-coated catheters increased the rate of catheter blockage by bacterial growth when compared to silicone-coated catheters (Kazmierska et al. 2010). While hydrogel-coated catheters are widely used in the clinic and show great promise, their inconsistent anti-biofilm efficacy is a source of concern.

5.3 Antibiotic Coatings

Antibiotic coatings tend to be more expensive and have a limited impact on biofilm formation (Maki and Tambyah 2001; Lam et al. 2014). Of the many antibiotic coatings, the two that are cost-effective and of clinical relevance are nitrofurazone (NF) and Rifampicin-minocycline (Pickard et al. 2012; Lam et al. 2014). Unfortunately, only NF is commercially available (Lam et al. 2014). Although studies have shown that NF reduces intraluminal biofilm formation, NF is linked to increased patient discomfort and is ineffective against infections caused by some microbes such as *P. aeruginosa* and *P. mirabilis* (Stickler 2014; Lam et al. 2014; Andersen and Flores-Mireles 2020). Additionally, studies suggest that the efficacy of NF catheter coatings does not surpass more than 2–3 weeks (Maki and Tambyah 2001). The limited efficacy of NF is likely due to its natural depletion from the intraluminal coating during this short time period (Stickler 2014). In 2002, the U.S. Food and Drug Administration (FDA) banned the use of NF as a topical antiseptic due to its carcinogenic properties. While this ban only applies to topical use—thus not affecting the use of NF-coated catheters—further research of NF-coated catheters has stalled (Andersen and Flores-Mireles 2020).

Complications can result from antibiotic coatings (Flores-Mireles et al. 2015). First, antibiotics often exacerbate the problem by increasing the risk of multidrug resistance (Subramanian et al. 2012; Sievert et al. 2013; Flores-Mireles et al. 2015). Several studies reveal growing resistance to antibiotic treatments among ESBL producing Enterobacteriaceae (Garau 2008) and in Gram-positive (Khadri and Alzohairy 2010) or many other Gram-negative bacteria (Sievert et al. 2013). Second, biofilms are intrinsically antibiotic tolerant. Studies have shown that antibiotic tolerance increases with increasing biofilm thickness because the diffusion of antibiotics to the bottom layer of biofilm cells is either slowed or blocked due to the extracellular polymeric matrix or glycocalyx (Costerton et al. 1999; Mah and O'Toole 2001).

Additionally, researchers have discovered that biofilm cells can exist in a state of "starvation" in which their metabolism is slowed due to insufficient nutrient access; such cells are called persister cells. This state appears to be directly linked to antibiotic tolerance. Persister cells create a gradient of nutrients and lead to increased heterogeneity (Costerton et al. 1999; Mah and O'Toole 2001; Lewis 2005). Heterogeneity in this context refers to the placement of cells in the biofilm with respect to their physiological state. Biofilm cells vary in rates of metabolism, respiration, and protein synthesis. Studies have shown that cells in a specific variation of one or more physiological functions reside in specific layers of the biofilm (Costerton et al. 1999; Mah and O'Toole 2001). Furthermore, there are genes responsible for antibiotic resistance. Since a single plasmid may carry antibiotic resistance and biofilms have increased mutation rates, an increased rate of horizontal gene transfer is believed to be responsible for multi-antibiotic resistant biofilms (Høiby et al. 2010; Dale et al. 2015). The genes transferred may encode for various efflux pumps or enzymes that degrade specific antibiotics (Høiby et al. 2010). While antibiotic coatings may be an immediate solution to CAUTI, they could lead to the long-term problem of antibiotic resistance in catheterized patients.

5.4 *Enzyme Coatings*

Enzyme-coated catheters are not commercially available and there have not been any significant clinical research trials. However, research studies showed that enzymes such as acylase and oxidase—which produce the antimicrobial, biocidal hydroxide—have successfully reduced or inhibited biofilm formation in the short term. Another enzyme, cellobiose dehydrogenase (CDH), also has an anti-biofilm effect in the short term. CDH forms both hydroxide (^{-}OH) and hydroxyl radicals (•OH) by the Fenton reaction, and thus, the hydroxyl radical (•OH) is the main antimicrobial product of CDH activity, not the hydroxide ion (^{-}OH). This is attributed to their ability to disrupt QS, if not directly kill, the bacterial cells comprising the biofilm (Ivanova et al. 2015; Zumstein et al. 2017). In several of these studies, placing enzymes such as acylase or CDH layer by layer successfully reduced biofilm growth, suggesting that a multilayer enzyme coating may become clinically relevant in the future (Djordjevic et al. 2002; Chutipongtanate and Thongboonkerd 2010; O'Toole 2011). Successful regulation of enzymatic activity is limited by storage, layer depth, and number of zwitterion structures present (Vaterrodt et al. 2016). Storage in a buffer solution rather than freeze-drying enzymes can decrease enzymatic activity. The enzymatic activity also decreases with increasing layer depth and a large concentration of zwitterion structures (Vaterrodt et al. 2016). For example, the diffusion of hydrogen peroxide produced by CDH was either stalled or hindered by the multiple layers (Ivanova et al. 2015; Vaterrodt et al. 2016). Consequently, clinical trials are necessary to determine the efficacy of multilayer enzymatic coatings in various environments.

5.5 Surface Energy Coatings

Bacterial adhesion is the first and most important step in implant infection (Schierholz and Beuth 2001). There are three central theories regarding the physico-chemical properties responsible for bacterial adhesion to catheter surfaces. These theories are the Derjaguin-Landau Verwey-Overbeek (DLVO) theory, the thermodynamic theory, and the extended DLVO theory (XDLVO) (Habash and Reid 1999; Katsikogianni and Missirlis 2004).

The DLVO theory primarily attributes repulsion and attraction between a bacterium and the catheter substratum to electrostatic charges (Katsikogianni and Missirlis 2004). These electrostatic charge interactions are called Van der Waals forces (Habash and Reid 1999; Katsikogianni and Missirlis 2004). However, the DLVO theory does not account for the conditioning layer or other cell surface structures influencing distance (Katsikogianni and Missirlis 2004).

The thermodynamic theory considers the effects of the dipole-dipole interactions and the polarity (the hydrophilic/hydrophobic nature of the bacterium and the substratum) in addition to the Van der Waals forces (Habash and Reid 1999; Katsikogianni and Missirlis 2004). The adhesion surface free energy described in this theory can be thought of as a function of these intramolecular forces (Habash and Reid 1999; Katsikogianni and Missirlis 2004). However, studies have shown that adhesion occurs at positive or negative values for this function, indicating that the adhesion surface free energy values cannot reliably predict bacterial adhesion to the substratum (Bayoudh et al. 2006).

Thirdly, the XLDVO theory considers all the intramolecular forces involved in the DLVO and thermodynamic theory, as well as the acid–base interactions which account for the effects of hydrogen bonding (Habash and Reid 1999; Katsikogianni and Missirlis 2004).

To briefly summarize the process of bacterial adhesion, the attractive Van der Waals forces predispose the bacterium to the catheter substratum at a long-range distance >50 nm (Habash and Reid 1999; Schierholz and Beuth 2001; Katsikogianni and Missirlis 2004). At about 10-20 nm, electrostatic repulsion occurs and bacterial adhesion is still reversible (Habash and Reid 1999). Finally, the hydrophobic interactions remove water molecules between the cell and the catheter surface, bringing them close enough to bind at less than 1.5 nm (Habash and Reid 1999). Binding is facilitated by adhesins, slime production, fimbriae, atomic structures, electronegative state, dipole-dipole forces, oxidation states, and other physico-chemical properties (Schierholz and Beuth 2001; Katsikogianni and Missirlis 2004; Campoccia et al. 2013).

During the adhesion process, bacteria are attracted to the nutrient-rich conditioning film composed of the host molecules and bodily fluids that covers the implanted portions of the catheter (Schierholz and Beuth 2001). Anti-fouling strategies attempt to disrupt this conditioning film by repelling bacteria, bacterial proteins, or bacterial lipids with hydrophobic, hydrophilic, or amphiphilic coatings (Legeay et al. 2006; Banerjee et al. 2011; Drelich et al. 2011). Materials associated with anti-adhesion coatings are trimethylsilane (TMS) plasma nanocoatings (Chen et al. 2013), polymeric brush coatings like poly(ethylene oxide) (PEO) (Drelich et al. 2011; Chen et al. 2013), diamond-like carbon (DLC) (Roy and Lee 2007; Page et al. 2009), and Poly(L-lysine)-grafted-poly(ethylene glycol) (PLL-g-PEG) (Page et al. 2009; Banerjee et al. 2011). In vitro studies have been successful at reducing pathogenic bacterial adhesion from catheter surfaces. For example, the trimethylsilane plasma coating was found to inhibit *Staphylococcus epidermidis* biofilm (Ma et al. 2012). Another in vitro study found that Poly(L-lysine)-grafted-poly(ethylene glycol) (PLL-g-PEG) coating decreased *Staphylococcus aureus* adhesion (Harris et al. 2004). Although in vitro results have been successful in significantly reducing bacterial adhesion, the in vivo efficacy is still unknown for many of the coatings (Chen et al. 2013). Tran et al. used mice models to test the efficacy of oranoselenium as an anti-adhesive coating; it was able to inhibit the development of *S. aureus* biofilms in vivo for 3 days (Tran et al. 2012). Another study assessed the antimicrobial efficacy of an anti-adhesive antimicrobial peptide coating using a mouse model. This coating showed excellent antimicrobial activity toward *P. aeruginosa, S. aureus,* and *S. saprophyticus* (Yu et al. 2017). While there are some in vivo studies looking at anti-adhesive coatings on catheters, they are sparse, making it difficult to draw conclusions regarding the effectiveness and clinical potential of these anti-adhesive coatings. Thus, further research should be conducted to identify the source (s) of this discrepancy.

5.6 Iron Scavenging Coatings

Another attractive prevention strategy is to create catheter coatings that scavenge or "compete" against bacteria for nutrients, such as iron, that are essential for biofilm development (Stewart 2003; Trautner and Darouiche 2004). Multiple bacterial species secrete siderophores for iron acquisition from host transferrin proteins (Modun et al. 1998; Orsi 2004, p. 200; Kim et al. 2006). Additionally, other iron-collection proteins, like hemolysin and enterochelin, are utilized by *P. mirabolis* (Stickler 2014) and multiple strains of *K. pneumoniae* and *K. oxytoca* (Tarkkanen et al. 1992), respectively. Furthermore, individual studies have shown that the opportunistic uropathogenic *E. coli* utilizes yersiniabactin (Klemm et al. 2007) or both enterochelin and aerobactin (Tarkkanen et al. 1992) for iron acquisition. Iron uptake can also be stimulated by catecholamine inotropes in *S. epidermidis* (Lyte et al. 2003). While *S. epidermidis* is more clinically relevant for intravascular catheterization rather than CAUTI, the concept can be applied in urinary catheters in the future. However, there is a lack of studies about the use of biomaterials in urinary catheters which scavenge iron (Jacobsen et al. 2008; O'May et al. 2008). Disrupting these bacterial scavenging systems for iron inhibits the formation of bacterial clusters from forming on catheter surfaces (Götz 2002; Singh et al. 2002; Orsi 2004), ultimately inhibiting biofilm formation. Individual studies have shown that iron chelating lactoferrin (Singh et al. 2002; Orsi 2004) and deferoxamine

(Singh et al. 2002) successfully reduced biofilm formation by decreasing the concentration of available iron. Conversely, high iron concentrations can reduce biofilm formation by suppressing iron scavenging gene products (Klemm et al. 2007). Yet, one significant disadvantage of iron scavenging catheter coatings is that host-adapted bacteria have developed siderophores that can remove iron directly from lactoferrin (Orsi 2004) and potentially from other iron-chelating molecules. Further research should be conducted to determine the limitations represented by host adaptation in iron scavenging catheter coatings before clinical testing.

5.7 *Methods to Test Antimicrobial Surfaces*

The goal of in vitro efficacy test methods is for the laboratory data to predict how the antimicrobial will perform in vivo. There are two basic strategies researchers use when developing a predictive model. The first strategy is to engineer a biofilm to have specific characteristics that emulate the biofilm in vivo, matching for example, the architecture, thickness, and strength of attachment. The second strategy uses a model that incorporates the most important physiochemical and biological characteristics. Summers conducted a review of methods often used to assess antimicrobial catheters and the pros and cons of their use (Summers and Goeres 2019). ASTM Method E3321, known as the Intraluminal Catheter Model (ICM), was designed based on the physiochemical properties present in a catheter lumen. The ICM was subjected to a rigorous statistical evaluation of its ruggedness, specifically, the amount the biofilm log density on untreated catheters and reduction in biofilm log density on treated catheters changed with small adjustments to key operational factors. The results of the analysis highlighted that biofilm growth is most sensitive to changes in pH, which indicates that the procedure to make the artificial urine must be tightly controlled. The analysis also demonstrated the importance of the technique used to harvest biofilm from the catheter surface. The test method exhibited acceptable repeatability for the controls (Sect. 12.2 of ASTM Method E3321), and was able to distinguish between catheters of varying degrees of efficacy (Fig. 2). Paired with data from a method that evaluates the extra lumen of a catheter, the in vitro methods should provide data to support moving forward with an animal model and/or clinical trial.

6 Treatment Strategies

While approaching CAUTIs with a preventative mindset is a more effective method at reducing CAUTIs, many prevention strategies yield mixed results when tested in clinical trials. Concrete treatment guidelines are difficult to establish due to this lack of definite results (Trautner 2010). Many of treatment guidelines emphasize the need for aseptic handling during catheter insertion and management. Also, while a

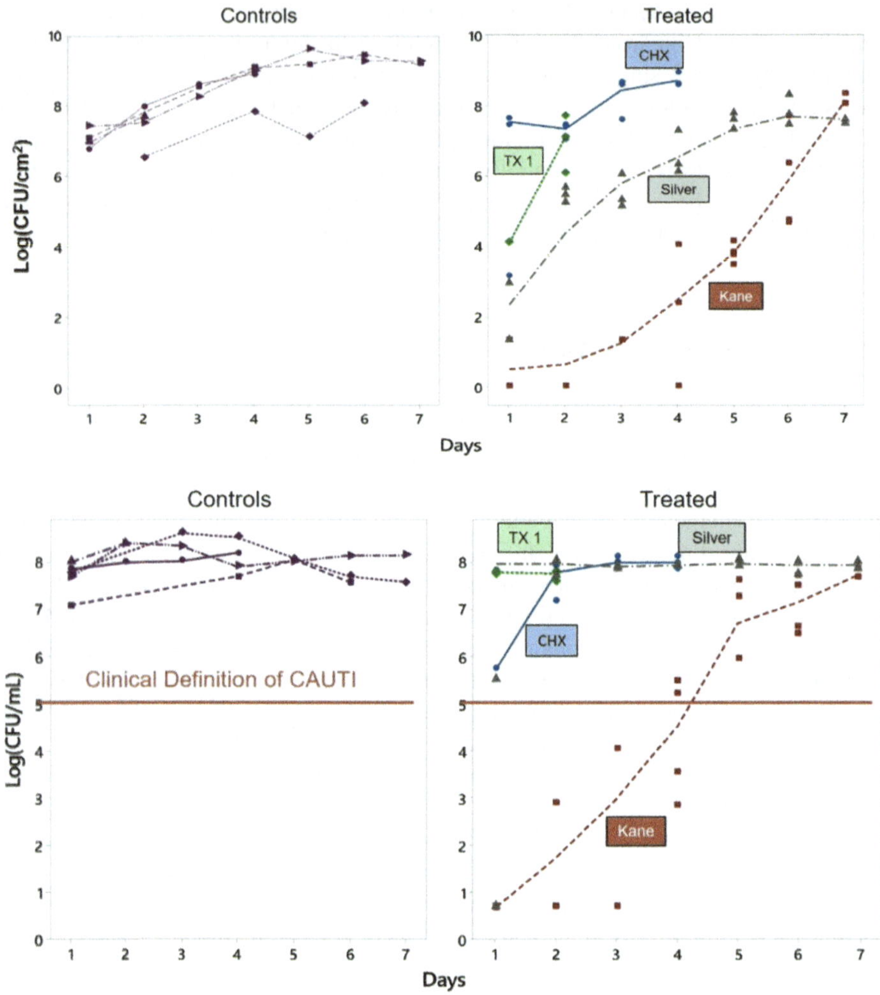

Fig. 2 The ICM demonstrated responsiveness for both biofilm samples, top two panels, and planktonic samples, bottom two panels. The control data for biofilm and the planktonic counts are on the left. Using the clinical definition of CAUTI, the Kane catheters delayed infection for 4 days (Summers 2019)

common treatment for CAUTI are antibiotics, the possibility for drug resistance increases greatly among uropathogens with this solution (Flores-Mireles et al. 2019). Conventional treatment methods, and their drawbacks, are outlined below.

6.1 Proper Catheter Use and Monitoring

In recent years, CAUTI prevention efforts have focused on limiting catheter use (Fakih et al. 2014). The inappropriate use of urinary catheters is a big problem in healthcare settings (Wald et al. 2008; Hooton et al. 2010). Wald et al. conducted a multicenter, retrospective cohort study that looked at Medicare patients in the USA over the age of 65 who were admitted to a nursing home postoperatively. With over 170,000 participants, Wald et al. found that 23% of these patients were discharged to a nursing home with a urinary catheter. Many of these patients did not have a reason for catheterization (Wald et al. 2008). Adherence to specific monitoring protocols for inserting urinary catheters is critical to serve as a preventative tool for CAUTI.

When catheter use is unavoidable, aseptic insertion and proper monitoring of the catheter decreases the chance of CAUTI development (Fakih et al. 2014). Knowledge of proper catheter use and prevention of CAUTI is lacking for both nurses and doctors. In a study looking at 719 U.S. hospitals, less than 25% of the hospitals utilized a monitoring system for patients with urinary catheters (Trautner 2010). A potential reason to explain the lack of regulatory compliance of catheterized patients is that CAUTI is not viewed as a serious problem (Jain et al. 2015). Healthcare workers around the world are not emphasizing the significance of bacteriuria. While the presence of bacteriuria is often used as an indicator for CAUTI, it is not always an accurate indicator for infection. Since the antibiotic treatment of asymptomatic bacteriuria is not recommended, healthcare workers are encouraged to hesitate to emphasize the presence of bacteriuria. Nevertheless, bacteriuria increases at a rate of 3–10% in a CAUTI patient. Without the presence of symptoms, the infection often goes unrecognized. Even in literature, there is confusion on the proper next steps in the presence of bacteriuria (Cope et al. 2009). Also, catheters have short-term benefits as they are a convenient option for nurses to manage patients with urinary incontinence. At times, family members or even the actual patient request a catheter (Krein et al. 2013). In these situations, it is important for healthcare workers to inform the patient of the risks associated with catheterization. As recommended by the CDC, limiting the duration of urinary catheters is a very effective and sustainable technique to prevent CAUTI due to the low risk and cost associated (Flores-Mireles et al. 2019). The proper monitoring of patients with urinary catheters would allow proper and prompt removal when appropriate, thus reducing further infectious and noninfectious complications (Fakih et al. 2014). Another strategy to reduce infections include placing patients on intermittent catheterization instead of long-term indwelling catheterization (Han et al. 2017). Ensuring proper clinical protocol—such as hand hygiene, aseptic catheter insertions, and education of hospital workers—can help decrease CAUTI occurrences (Flores-Mireles et al. 2019). When nurses are given a sterile space to prepare for procedures, infections are reduced. However, many hospitals today are overcrowded, forcing nurses to work on surfaces such as patient beds or the small area near sinks. It should be a priority to ensure that nurses have the proper space to perform aseptic procedures.

6.2 Antibiotic Treatment

Once a CAUTI diagnosis is confirmed, a commonly recommended treatment is systemic antibiotics. However, the use of prophylaxis with systemic antibiotics is not recommended without the presence of clinical indicators, such as bacteriuria presence and fever (Table 1) (Flores-Mireles et al. 2019). Prior to proceeding with antimicrobial therapy, the catheter should be removed and/or replaced (Flores-Mireles et al. 2019). It is also critical to distinguish between CAUTI and catheter-associated asymptomatic bacteriuria (CA-ASB); many times, CA-ASB are incorrectly categorized as CAUTI. Various studies have demonstrated that asymptomatic bacteriuria has no clinical significance and requires no treatment (Trautner and Darouiche 2004; Cope et al. 2009; Hooton et al. 2010). In fact, treating CA-ASB with antibiotics is ineffective and can be harmful by allowing antibiotic-resistant pathogens to develop (Trautner 2010). Improving the process of distinguishing between CAUTI and CA-ASB would reduce the use of unnecessary antibiotics, and therefore, reduce the number of catheter-associated infections (Cope et al. 2009).

An analysis done by Pereira et al. (2014) demonstrated that the prophylactic use of antibiotics after catheter removal reduced CAUTI development in greater than 50% of surgical patients. However, this study only focused on surgical patients. Therefore, further research is needed to investigate the effect of prophylactic antibiotics use after catheter removal on non-surgical patients. In general, prophylactic antibiotic treatment remains unrecommended as it promotes antibacterial resistance.

If a patient demonstrates CAUTI symptoms, antibiotic treatment may be appropriate. The type of antibiotic selected varies, among other factors such as bacterial culture, susceptibility, and drug interactions. One study suggested a 5-day levofloxacin treatment for CAUTI patients who are not severely ill (Hooton et al., 2010). There are differing recommendations based on the severity of patient morbidity and setting, (outpatient/inpatient) as well as the extent of community multidrug resistance (Pereira et al. 2014). Ultimately, the clinical decision for antibiotic treatment should be individualized based on patient status and setting. Nevertheless, it is critical to distinguish between the planktonic bacteria used to make the diagnosis and the different bacteria forming the biofilm. Antibiotic treatments, which may be effective for planktonic bacteria, may not be effective against the biofilm. By submitting the colonized catheter for enumeration along with the urine sample, clinicians could ensure that the treatment utilized covers all bacteria present.

Regarding the length of treatment, different studies recommend different treatment durations. Pereira et al. demonstrated that a 3-day antibiotic treatment—prior to the onset of symptoms—was sufficient to reduce CAUTI risk after catheter removal by a statistically significant amount; the rate of CAUTI was reduced by 5.8% in the group treated with antibiotics. A separate study states that if a patient demonstrates CAUTI symptoms, antibiotic treatment is recommended over the course of 7 days; the treatment time is extended to between 10 and 14 days if the patient has a delayed response to the treatment (Hooton et al. 2010). Deciding on the optimal treatment duration remains a challenge (Pereira et al. 2014). Overall, there is conflicting information regarding treating CAUTI with antibiotics. The treatment strategies for CAUTI are lacking in standardization and significant clinical results.

7 Conclusions

It is critical to consider biofilms in CAUTI diagnosis. The biofilm is often the root cause of the infection; not addressing it can lead to infection recurrence and the need for future interventions. Regarding treatment, the solution offered by various researchers is abstaining from catheter insertion in the first place. While this solution feels primitive, it is the conclusion many professionals draw due to the improper catheter etiquette widely practiced. Reliance on urinary catheters—even when they are not necessary—is one of the main issues causing such high incidences of CAUTIs. Deciding whether to catheterize a patient is nuanced. While it comes with clear benefits—such as convenience—the treatment and quality of life of the patient are also affected. One way to discourage unnecessary catheterization in hospitals has been with financial incentives; some hospitals face a potential reduction in funding for having high rates of CAUTI patients (Davis 2019). This would ideally encourage healthcare workers to take greater precautions in both catheterizing patients and diagnosing CAUTI prior to the start of potentially unnecessary antibiotic treatment. While there are a great number of studies with recommendations for CAUTI preventative strategies and managements, their results are highly variable (Kranz et al. 2020). The lack of consistency in the suggested guidelines is oftentimes the reason behind the mismatch in clinical management decisions. Revising the official definition of CAUTI could allow for a more nuanced approach to choosing whether or not to proceed with catheterization.

For all the preventative technologies discussed, there is a dependence on time. This 'time-to-failure' is the drawback of most *in vitro* studies, which look at one-time point. Therefore, it is necessary to look at the data in a new way in order to assess when and for how long a technology will work, and when it will fail. One idea to assess the data in a new way is by incorporating machine learning in the image analysis of these studies. Collecting time evolution data regarding the effectiveness of the CAUTI preventative technologies is tedious. However, by inputting the image data into a machine learning model, it would be possible to train this model to predict the success rates of future preventative technologies being tested. This would also address the issue of only analyzing one time point in these studies. The key with machine learning is that while to create a predictive model, a lot of data is required; when this model is built though, much less data is needed to make accurate predictions. If time evolution image data of a preventative technology was used to make this predictive tool, it would be possible to input an image of a different technology and make a hypothesis on its potential success. Streamlining the process of data collection would allow for improved data analysis of preventative CAUTI technologies.

Acknowledgments The authors wish to thank The McNair's Scholars Program at the University of Maryland for summer research funding for Ryan Felix in addition to the Department of Civil and Environmental that provided startup funding for Dr. Birthe V. Kjellerup.

Conflict of Interest The authors all declare that they do not have any conflicts of interest.

References

Abraham SN, Miao Y (2015) The nature of immune responses to urinary tract infections. Nat Rev Immunol 15:655–663. https://doi.org/10.1038/nri3887

Al-Qahtani M, Safan A, Jassim G, Abadla S (2019) Efficacy of anti-microbial catheters in preventing catheter associated urinary tract infections in hospitalized patients: a review on recent updates. J Infect Public Health 12:760–766. https://doi.org/10.1016/j.jiph.2019.09.009

Al-Qas Hanna F, Sambirska O, Iyer S et al (2013) Clinician practice and the National Healthcare Safety Network definition for the diagnosis of catheter-associated urinary tract infection. Am J Infect Control 41:1173–1177. https://doi.org/10.1016/j.ajic.2013.05.024

Andersen MJ, Flores-Mireles AL (2020) Urinary catheter coating modifications: the race against catheter-associated infections. Coatings 10:23. https://doi.org/10.3390/coatings10010023

Ashraf MS, Gaur S, Bushen OY et al (2020) Diagnosis, treatment, and prevention of urinary tract infections in post-acute and long-term care settings: a consensus statement from AMDA's infection advisory subcommittee. J Am Med Dir Assoc 21:12–24.e2. https://doi.org/10.1016/j.jamda.2019.11.004

Badalato G, Korets R, Matulay J, Gupta M (2014) Pd13-07 delivery of upper tract chemotherapy to patients with urinary diversion: a novel approach. J Urol 191:e354–e354. https://doi.org/10.1016/j.juro.2014.02.992

Banerjee I, Pangule RC, Kane RS (2011) Antifouling coatings: recent developments in the design of surfaces that prevent fouling by proteins, bacteria, and marine organisms. Adv Mater Deerfield Beach Fla 23:690–718. https://doi.org/10.1002/adma.201001215

Banerjee J, Ghatak PD, Roy S et al (2015) Silver-zinc redox-coupled electroceutical wound dressing disrupts bacterial biofilm. PLOS One 10:e0119531. https://doi.org/10.1371/journal.pone.0119531

Barford J, Coates A (2009) The pathogenesis of catheter-associated urinary tract infection. J Infect Prev 10:50–56. https://doi.org/10.1177/1757177408098265

Barford JMT, Anson K, Hu Y, Coates ARM (2008) A model of catheter-associated urinary tract infection initiated by bacterial contamination of the catheter tip. BJU Int 102:67–74. https://doi.org/10.1111/j.1464-410X.2008.07465.x

Bayoudh S, Othmane A, Bettaieb F et al (2006) Quantification of the adhesion free energy between bacteria and hydrophobic and hydrophilic substrata. Mater Sci Eng C 26:300–305. https://doi.org/10.1016/j.msec.2005.10.045

Bedrossian M, Barr C, Lindensmith CA et al (2017) Quantifying microorganisms at low concentrations using Digital Holographic Microscopy (DHM). J Vis Exp JoVE (129):56343. https://doi.org/10.3791/56343

Beiko DT, Knudsen BE, Watterson JD et al (2004) Urinary tract biomaterials. J Urol 171:2438–2444. https://doi.org/10.1097/01.ju.0000125001.56045.6c

Belyayeva M, Jeong JM (2021) Acute pyelonephritis. In: StatPearls. StatPearls Publishing, Treasure Island (FL)

Bjerkan G, Witsø E, Bergh K (2009) Sonication is superior to scraping for retrieval of bacteria in biofilm on titanium and steel surfaces in vitro. Acta Orthop 80:245–250. https://doi.org/10.3109/17453670902947457

Blodgett TJ (2009) Reminder systems to reduce the duration of indwelling urinary catheters: a narrative review. Urol Nurs 29(369–378):quiz 379

Bonkat G, Braissant O, Rieken M et al (2013) Comparison of the roll-plate and sonication techniques in the diagnosis of microbial ureteral stent colonisation: results of the first prospective randomised study. World J Urol 31:579–584. https://doi.org/10.1007/s00345-012-0963-5
Bovis M, Noimark S, Woodhams J et al (2015) Photosensitisation studies of silicone polymer doped with methylene blue and nanogold for antimicrobial applications †. RSC Adv 5:54830. https://doi.org/10.1039/c5ra09045h
Butt HJ, Downing KH, Hansma PK (1990a) Imaging the membrane protein bacteriorhodopsin with the atomic force microscope. Biophys J 58:1473–1480. https://doi.org/10.1016/S0006-3495(90)82492-9
Butt H-J, Wolff EK, Gould SAC et al (1990b) Imaging cells with the atomic force microscope. J Struct Biol 105:54–61. https://doi.org/10.1016/1047-8477(90)90098-W
Camesano TA, Natan MJ, Logan BE (2000) Observation of changes in bacterial cell morphology using tapping mode atomic force microscopy. Langmuir 16:4563–4572. https://doi.org/10.1021/la990805o
Campoccia D, Montanaro L, Arciola CR (2013) A review of the biomaterials technologies for infection-resistant surfaces. Biomaterials 34:8533–8554. https://doi.org/10.1016/j.biomaterials.2013.07.089
Carpenter AE, Jones TR, Lamprecht MR et al (2006) CellProfiler: image analysis software for identifying and quantifying cell phenotypes. Genome Biol 7:R100. https://doi.org/10.1186/gb-2006-7-10-r100
Chen M, Yu Q, Sun H (2013) Novel strategies for the prevention and treatment of biofilm related infections. Int J Mol Sci 14:18488–18501. https://doi.org/10.3390/ijms140918488
Choong S, Whitfield H (2000) Biofilms and their role in infections in urology. BJU Int 86:935–941. https://doi.org/10.1046/j.1464-410x.2000.00949.x
Chutipongtanate S, Thongboonkerd V (2010) Systematic comparisons of artificial urine formulas for in vitro cellular study. Anal Biochem 402:110–112. https://doi.org/10.1016/j.ab.2010.03.031
Cleri DJ, Corrado ML, Seligman SJ (1980) Quantitative culture of intravenous catheters and other intravascular inserts. J Infect Dis 141:781–786. https://doi.org/10.1093/infdis/141.6.781
Cope M, Cevallos ME, Cadle RM et al (2009) Inappropriate treatment of catheter-associated asymptomatic bacteriuria in a tertiary care hospital. Clin Infect Dis Off Publ Infect Dis Soc Am 48:1182–1188. https://doi.org/10.1086/597403
Cortese YJ, Wagner VE, Tierney M et al (2018) Review of catheter-associated urinary tract infections and in vitro urinary tract models. J Healthc Eng 2018:2986742. https://doi.org/10.1155/2018/2986742
Costerton JW, Lewandowski Z, Caldwell DE et al (1995) Microbial biofilms. Annu Rev Microbiol 49:711–745. https://doi.org/10.1146/annurev.mi.49.100195.003431
Costerton JW, Stewart PS, Greenberg EP (1999) Bacterial biofilms: a common cause of persistent infections. Science 284:1318–1322. https://doi.org/10.1126/science.284.5418.1318
Cox AJ, Hukins DW (1989) Morphology of mineral deposits on encrusted urinary catheters investigated by scanning electron microscopy. J Urol 142:1347–1350. https://doi.org/10.1016/s0022-5347(17)39095-x
Cox G, Sheppard CJR (2004) Practical limits of resolution in confocal and non-linear microscopy. Microsc Res Tech 63:18–22. https://doi.org/10.1002/jemt.10423
Dale JL, Cagnazzo J, Phan CQ et al (2015) Multiple roles for Enterococcus faecalis glycosyltransferases in biofilm-associated antibiotic resistance, cell envelope integrity, and conjugative transfer. Antimicrob Agents Chemother 59:4094–4105. https://doi.org/10.1128/AAC.00344-15
Danese PN (2002) Antibiofilm approaches: prevention of catheter colonization. Chem Biol 9:873–880. https://doi.org/10.1016/S1074-5521(02)00192-8
Dave RN, Joshi HM, Venugopalan VP (2011) Novel Biocatalytic polymer-based antimicrobial coatings as potential ureteral biomaterial: preparation and in vitro performance evaluation. Antimicrob Agents Chemother 55:845–853. https://doi.org/10.1128/AAC.00477-10

Davies DG, Chakrabarty AM, Geesey GG (1993) Exopolysaccharide production in biofilms: substratum activation of alginate gene expression by Pseudomonas aeruginosa. Appl Environ Microbiol 59:1181–1186

Davies DG, Parsek MR, Pearson JP et al (1998) The involvement of cell-to-cell signals in the development of a bacterial biofilm. Science 280:295–298. https://doi.org/10.1126/science.280.5361.295

Davis M (2019) When guidelines conflict: patient safety, quality of life, and CAUTI reduction in patients with spinal cord injury. Spinal Cord Ser Cases 5:1–3. https://doi.org/10.1038/s41394-019-0198-4

de Beer D, Stoodley P (1995) Relation between the structure of an aerobic biofilm and transport phenomena. Water Sci Technol 32:11–18. https://doi.org/10.1016/0273-1223(96)00002-9

Delcaru C, Alexandru I, Podgoreanu P et al (2016) Microbial biofilms in urinary tract infections and prostatitis: etiology, pathogenicity, and combating strategies. Pathogens 5:65. https://doi.org/10.3390/pathogens5040065

Denstedt JD, Wollin TA, Reid G (1998) Biomaterials used in urology: current issues of biocompatibility, infection, and encrustation. J Endourol 12:493–500. https://doi.org/10.1089/end.1998.12.493

Dibartola AC, Swearingen MC, Granger JF et al (2017) Biofilms in orthopedic infections: a review of laboratory methods. APMIS 125:418–428. https://doi.org/10.1111/apm.12671

Djordjevic D, Wiedmann M, McLandsborough LA (2002) Microtiter plate assay for assessment of Listeria monocytogenes biofilm formation. Appl Environ Microbiol 68:2950–2958. https://doi.org/10.1128/AEM.68.6.2950-2958.2002

Dokland T, Hutmacher DW, Ng MM-L, Schantz J-T (2006) Techniques in microscopy for biomedical applications. World Scientific, Singapore

Donlan RM (2001a) Biofilm formation: a clinically relevant microbiological process. Clin Infect Dis Off Publ Infect Dis Soc Am 33:1387–1392. https://doi.org/10.1086/322972

Donlan RM (2001b) Biofilms and device-associated infections. Emerg Infect Dis 7:277–281

Drekonja DM, Kuskowski MA, Wilt TJ, Johnson JR (2008) Antimicrobial urinary catheters: a systematic review. Expert Rev Med Devices 5:495–506. https://doi.org/10.1586/17434440.5.4.495

Drelich J, Chibowski E, Meng D, Terpilowski K (2011) Hydrophilic and superhydrophilic surfaces and materials. Soft Matter 7:9804–9828. https://doi.org/10.1039/C1SM05849E

Ducker WA, Senden TJ, Pashley RM (1991) Direct measurement of colloidal forces using an atomic force microscope. Nature 353:239–241. https://doi.org/10.1038/353239a0

Dufrêne YF (2000) Direct characterization of the physicochemical properties of fungal spores using functionalized AFM probes. Biophys J 78:3286–3291

Dufrêne YF (2002) Atomic force microscopy, a powerful tool in microbiology. J Bacteriol 184:5205–5213. https://doi.org/10.1128/JB.184.19.5205-5213.2002

Dufrêne YF, Barger WR, Green J-BD, Lee GU (1997) Nanometer-scale surface properties of mixed phospholipid monolayers and bilayers. Langmuir 13:4779–4784. https://doi.org/10.1021/la970221r

Dufrêne YF, Boonaert CJP, Gerin PA et al (1999) Direct probing of the surface ultrastructure and molecular interactions of dormant and germinating spores of Phanerochaete chrysosporium. J Bacteriol 181:5350–5354

Dukes C (1928) Urinary infections after excision of the rectum: their cause and prevention. Proc R Soc Med 22:259–269. https://doi.org/10.1177/003591572802200267

Dunnill CW, Page K, Aiken ZA et al (2011) Nanoparticulate silver coated-titania thin films—Photo-oxidative destruction of stearic acid under different light sources and antimicrobial effects under hospital lighting conditions. J Photochem Photobiol Chem 220:113–123. https://doi.org/10.1016/j.jphotochem.2011.04.001

Elvy J, Colville A (2009) Catheter associated urinary tract infection: what is it, what causes it and how can we prevent it? J Infect Prev 10:36–41. https://doi.org/10.1177/1757177408094852

Fakih MG, Krein SL, Edson B et al (2014) Engaging health care workers to prevent catheter-associated urinary tract infection and avert patient harm. Am J Infect Control 42:S223–S229. https://doi.org/10.1016/j.ajic.2014.03.355

Farsi HM, Mosli HA, Al-Zemaity MF et al (1995) Bacteriuria and colonization of double-pigtail ureteral stents: long-term experience with 237 patients. J Endourol 9:469–472. https://doi.org/10.1089/end.1995.9.469

Feneley RCL, Hopley IB, Wells PNT (2015) Urinary catheters: history, current status, adverse events and research agenda. J Med Eng Technol 39:459–470. https://doi.org/10.3109/03091902.2015.1085600

Flemming H-C, Wingender J (2010) The biofilm matrix. Nat Rev Microbiol 8:623–633. https://doi.org/10.1038/nrmicro2415

Flemming H-C, Neu TR, Wozniak DJ (2007) The EPS matrix: the "house of biofilm cells.". J Bacteriol 189:7945–7947. https://doi.org/10.1128/JB.00858-07

Flores-Mireles AL, Walker JN, Caparon M, Hultgren SJ (2015) Urinary tract infections: epidemiology, mechanisms of infection and treatment options. Nat Rev Microbiol 13:269–284. https://doi.org/10.1038/nrmicro3432

Flores-Mireles A, Hreha TN, Hunstad DA (2019) Pathophysiology, treatment, and prevention of catheter-associated urinary tract infection. Top Spinal Cord Inj Rehabil 25:228–240. https://doi.org/10.1310/sci2503-228

Foley FEB (1937) A hemostatic bag catheter: a one piece latex rubber structure for control of bleeding and constant drainage following prostatic resection. J Urol 38:134–139. https://doi.org/10.1016/S0022-5347(17)71935-0

Frank DN, Wilson SS, St Amand AL, Pace NR (2009) Culture-independent microbiological analysis of foley urinary catheter biofilms. PloS One 4:e7811. https://doi.org/10.1371/journal.pone.0007811

Frontera JA, Wang E, Phillips M et al (2020) Protocolized urine sampling is associated with reduced catheter-associated urinary tract infections: a pre- and postintervention study. Clin Infect Dis. https://doi.org/10.1093/cid/ciaa1152

Fünfstück R (1998) Urinary tract infections: detection, prevention, and management. Infection 26: 183–183. https://doi.org/10.1007/BF02771849

Fuqua WC, Winans SC, Greenberg EP (1994) Quorum sensing in bacteria: the LuxR-LuxI family of cell density-responsive transcriptional regulators. J Bacteriol 176:269–275

Garau J (2008) Other antimicrobials of interest in the era of extended-spectrum beta-lactamases: fosfomycin, nitrofurantoin and tigecycline. Clin Microbiol Infect Off Publ Eur Soc Clin Microbiol Infect Dis 14(Suppl 1):198–202. https://doi.org/10.1111/j.1469-0691.2007.01852.x

Garibaldi RA, Burke JP, Dickman ML, Smith CB (1974) Factors predisposing to bacteriuria during indwelling urethral catheterization. N Engl J Med 291:215–219. https://doi.org/10.1056/NEJM197408012910501

Garibaldi RA, Burke JP, Britt MR et al (1980) Meatal colonization and catheter-associated bacteriuria. N Engl J Med 303:316–318. https://doi.org/10.1056/NEJM198008073030605

Götz F (2002) Staphylococcus and biofilms. Mol Microbiol 43:1367–1378. https://doi.org/10.1046/j.1365-2958.2002.02827.x

Gould CV, Umscheid CA, Agarwal RK et al (2010) Guideline for prevention of catheter-associated urinary tract infections 2009. Infect Control Hosp Epidemiol 31:319–326. https://doi.org/10.1086/651091

Gristina AG (1987) Biomaterial-centered infection: microbial adhesion versus tissue integration. Science 237:1588–1595. https://doi.org/10.1126/science.3629258

Guiton PS, Hung CS, Hancock LE et al (2010) Enterococcal biofilm formation and virulence in an optimized murine model of foreign body-associated urinary tract infections. Infect Immun 78: 4166–4175. https://doi.org/10.1128/IAI.00711-10

Guttmann L, Frankel H (1966) The value of intermittent catheterisation in the early management of traumatic paraplegia and tetraplegia. Paraplegia 4:63–84. https://doi.org/10.1038/sc.1966.7

Ha U-S, Cho Y-H (2006) Catheter-associated urinary tract infections: new aspects of novel urinary catheters. Int J Antimicrob Agents 28:485–490. https://doi.org/10.1016/j.ijantimicag.2006.08.020

Habash M, Reid G (1999) Microbial biofilms: their development and significance for medical device-related infections. J Clin Pharmacol 39:887–898. https://doi.org/10.1177/00912709922008506

Hall-Stoodley L, Stoodley P, Kathju S et al (2012) Towards diagnostic guidelines for biofilm-associated infections. FEMS Immunol Med Microbiol 65:127–145. https://doi.org/10.1111/j.1574-695X.2012.00968.x

Han CS, Kim S, Radadia KD et al (2017) Comparison of urinary tract infection rates associated with transurethral catheterization, suprapubic tube and clean intermittent catheterization in the postoperative setting: a network meta-analysis. J Urol 198:1353–1358. https://doi.org/10.1016/j.juro.2017.07.069

Hanafy HM, Saad SM, Al-Ghorab MM (1974) Ancient Egyptian medicine: contribution to urology. Urology 4:114–120. https://doi.org/10.1016/0090-4295(74)90124-1

Hannig C, Follo M, Hellwig E, Al-Ahmad A (2010) Visualization of adherent micro-organisms using different techniques. J Med Microbiol 59:1–7. https://doi.org/10.1099/jmm.0.015420-0

Harris LG, Tosatti S, Wieland M et al (2004) Staphylococcus aureus adhesion to titanium oxide surfaces coated with non-functionalized and peptide-functionalized poly(l-lysine)-grafted-poly (ethylene glycol) copolymers. Biomaterials 25:4135–4148. https://doi.org/10.1016/j.biomaterials.2003.11.033

Heydorn A, Nielsen AT, Hentzer M et al (2000) Quantification of biofilm structures by the novel computer program COMSTAT. Microbiol Read Engl 146(Pt 10):2395–2407. https://doi.org/10.1099/00221287-146-10-2395

Hinterdorfer P, Dufrêne YF (2006) Detection and localization of single molecular recognition events using atomic force microscopy. Nat Methods 3:347–355. https://doi.org/10.1038/nmeth871

Hoh JH, Lal R, John SA et al (1991) Atomic force microscopy and dissection of gap junctions. Science 253:1405–1408. https://doi.org/10.1126/science.1910206

Høiby N, Bjarnsholt T, Givskov M et al (2010) Antibiotic resistance of bacterial biofilms. Int J Antimicrob Agents 35:322–332. https://doi.org/10.1016/j.ijantimicag.2009.12.011

Hola V, Peroutkova T, Ruzicka F (2012) Virulence factors in *Proteus* bacteria from biofilm communities of catheter-associated urinary tract infections. FEMS Immunol Med Microbiol 65:343–349. https://doi.org/10.1111/j.1574-695X.2012.00976.x

Holling N, Dedi C, Jones CE et al (2014) Evaluation of environmental scanning electron microscopy for analysis of Proteus mirabilis crystalline biofilms in situ on urinary catheters. FEMS Microbiol Lett 355:20–27. https://doi.org/10.1111/1574-6968.12451

Hooton TM, Bradley SF, Cardenas DD et al (2010) Diagnosis, prevention, and treatment of catheter-associated urinary tract infection in adults: 2009 international clinical practice guidelines from the infectious diseases society of America. Clin Infect Dis 50:625–663. https://doi.org/10.1086/650482

Hoyle BD, Costerton JW (1991) Bacterial resistance to antibiotics: the role of biofilms. Prog Drug Res Fortschritte Arzneimittelforschung Progres Rech Pharm 37:91–105. https://doi.org/10.1007/978-3-0348-7139-6_2

Ivanova K, Fernandes MM, Mendoza E, Tzanov T (2015) Enzyme multilayer coatings inhibit Pseudomonas aeruginosa biofilm formation on urinary catheters. Appl Microbiol Biotechnol 99: 4373–4385. https://doi.org/10.1007/s00253-015-6378-7

Jacobsen SM, Shirtliff ME (2011) Proteus mirabilis biofilms and catheter-associated urinary tract infections. Virulence 2:460–465. https://doi.org/10.4161/viru.2.5.17783

Jacobsen SM, Stickler DJ, Mobley HLT, Shirtliff ME (2008) Complicated catheter-associated urinary tract infections due to Escherichia coli and Proteus mirabilis. Clin Microbiol Rev 21: 26–59. https://doi.org/10.1128/CMR.00019-07

Jahn P, Beutner K, Langer G (2012) Types of indwelling urinary catheters for long-term bladder drainage in adults. Cochrane Database Syst Rev 10:CD004997. https://doi.org/10.1002/14651858.CD004997.pub3

Jain M, Dogra V, Mishra B et al (2015) Knowledge and attitude of doctors and nurses regarding indication for catheterization and prevention of catheter-associated urinary tract infection in a tertiary care hospital. Indian J Crit Care Med Peer-Rev Off Publ Indian Soc Crit Care Med 19: 76–81. https://doi.org/10.4103/0972-5229.151014

Johnson JR, Kuskowski MA, Wilt TJ (2006) Systematic review: antimicrobial urinary catheters to prevent catheter-associated urinary tract infection in hospitalized patients. Ann Intern Med 144: 116–126. https://doi.org/10.7326/0003-4819-144-2-200601170-00009

Jones BV, Young R, Mahenthiralingam E, Stickler DJ (2004) Ultrastructure of Proteus mirabilis swarmer cell rafts and role of swarming in catheter-associated urinary tract infection. Infect Immun 72:3941–3950. https://doi.org/10.1128/IAI.72.7.3941-3950.2004

Jones BV, Mahenthiralingam E, Sabbuba NA, Stickler DJ (2005) Role of swarming in the formation of crystalline Proteus mirabilis biofilms on urinary catheters. J Med Microbiol 54: 807–813. https://doi.org/10.1099/jmm.0.46123-0

Kasas S, Ikai A (1995) A method for anchoring round shaped cells for atomic force microscope imaging. Biophys J 68:1678–1680

Katsikogianni M, Missirlis YF (2004) Concise review of mechanisms of bacterial adhesion to biomaterials and of techniques used in estimating bacteria-material interactions. Eur Cell Mater 8:37–57. https://doi.org/10.22203/ecm.v008a05

Kazmierska KA, Thompson R, Morris N et al (2010) In vitro multicompartmental bladder model for assessing blockage of urinary catheters: effect of hydrogel coating on dynamics of Proteus mirabilis growth. Urology 76:515.e15–515.e20. https://doi.org/10.1016/j.urology.2010.04.039

Khadri H, Alzohairy M (2010) High prevalence of Multi-Drug-Resistance (MDR) and Extended Spectrum β-Lactamases (ESBL) producing bacteria among community-acquired urinary tract .... J Bacteriol Res 1:105–110

Khawcharoenporn T, Vasoo S, Ward E, Singh K (2011) Abnormal urinalysis finding triggered antibiotic prescription for asymptomatic bacteriuria in the ED. Am J Emerg Med 29:828–830. https://doi.org/10.1016/j.ajem.2011.04.007

Kim S-J, Park R-Y, Kang S-M et al (2006) Pseudomonas aeruginosa alkaline protease can facilitate siderophore-mediated iron-uptake via the proteolytic cleavage of transferrins. Biol Pharm Bull 29:2295–2300. https://doi.org/10.1248/bpb.29.2295

Klemm P, Hancock V, Kvist M, Schembri MA (2007) Candidate targets for new antivirulence drugs: selected cases of bacterial adhesion and biofilm formation. Future Microbiol 2:643–653. https://doi.org/10.2217/17460913.2.6.643

Korber DR, Lawrence JR, Sutton B, Caldwell DE (1989) Effect of laminar flow velocity on the kinetics of surface recolonization by Mot(+) and Mot (-) Pseudomonas fluorescens. Microb Ecol 18:1–19. https://doi.org/10.1007/BF02011692

Kranz J, Schmidt S, Wagenlehner F, Schneidewind L (2020) Catheter-associated urinary tract infections in adult patients. Dtsch Ärztebl Int 117:83–88. https://doi.org/10.3238/arztebl.2020.0083

Krein SL, Kowalski CP, Harrod M et al (2013) Barriers to reducing urinary catheter use: a qualitative assessment of a statewide initiative. JAMA Intern Med 173:881–886. https://doi.org/10.1001/jamainternmed.2013.105

Kunin CM, McCormack RC (1966) Prevention of catheter-induced urinary-tract infections by sterile closed drainage. N Engl J Med 274:1155–1161. https://doi.org/10.1056/NEJM196605262742101

Kunin CM, Douthitt S, Dancing J et al (1992) The association between the use of urinary catheters and morbidity and mortality among elderly patients in nursing homes. Am J Epidemiol 135: 291–301. https://doi.org/10.1093/oxfordjournals.aje.a116283

Lam TBL, Omar MI, Fisher E et al (2014) Types of indwelling urethral catheters for short-term catheterisation in hospitalised adults. Cochrane Database Syst Rev (9):CD004013. https://doi.org/10.1002/14651858.CD004013.pub4

Lamprecht MR, Sabatini DM, Carpenter AE (2007) CellProfiler: free, versatile software for automated biological image analysis. BioTechniques 42:71–75. https://doi.org/10.2144/000112257

Lane MC, Lockatell V, Monterosso G et al (2005) Role of motility in the colonization of Uropathogenic Escherichia coli in the urinary tract. Infect Immun 73:7644–7656. https://doi.org/10.1128/IAI.73.11.7644-7656.2005

Lapides J, Diokno AC, Silber SJ, Lowe BS (1972) Clean, intermittent self-catheterization in the treatment of urinary tract disease. J Urol 107:458–461. https://doi.org/10.1016/s0022-5347(17)61055-3

Lawrence JR, Delaquis PJ, Korber DR, Caldwell DE (1987) Behavior ofPseudomonas fluorescens within the hydrodynamic boundary layers of surface microenvironments. Microb Ecol 14:1–14. https://doi.org/10.1007/BF02011566

Lederer JW, Jarvis WR, Thomas L, Ritter J (2014) Multicenter cohort study to assess the impact of a silver-alloy and hydrogel-coated urinary catheter on symptomatic catheter-associated urinary tract infections. J Wound Ostomy Continence Nurs 41:473–480. https://doi.org/10.1097/WON.0000000000000056

Lee G, Chrisey L, Colton R (1994) Direct measurement of the forces between complementary strands of DNA. Science 266:771–773. https://doi.org/10.1126/science.7973628

Legeay G, Poncin-Epaillard F, Arciola CR (2006) New surfaces with hydrophilic/hydrophobic characteristics in relation to (no)bioadhesion. Int J Artif Organs 29:453–461. https://doi.org/10.1177/039139880602900416

Lewis K (2005) Persister cells and the riddle of biofilm survival. Biochem Biokhimiia 70:267–274. https://doi.org/10.1007/s10541-005-0111-6

Lo J, Lange D, Chew BH (2014) Ureteral stents and Foley catheters-associated urinary tract infections: the role of coatings and materials in infection prevention. Antibiotics 3:87–97. https://doi.org/10.3390/antibiotics3010087

Lyte M, Freestone PPE, Neal CP et al (2003) Stimulation of Staphylococcus epidermidis growth and biofilm formation by catecholamine inotropes. Lancet Lond Engl 361:130–135. https://doi.org/10.1016/S0140-6736(03)12231-3

Ma Y, Chen M, Jones JE et al (2012) Inhibition of Staphylococcus epidermidis biofilm by trimethylsilane plasma coating. Antimicrob Agents Chemother 56:5923–5937. https://doi.org/10.1128/AAC.01739-12

Macleod SM, Stickler DJ (2007) Species interactions in mixed-community crystalline biofilms on urinary catheters. J Med Microbiol 56:1549–1557. https://doi.org/10.1099/jmm.0.47395-0

Mah TF, O'Toole GA (2001) Mechanisms of biofilm resistance to antimicrobial agents. Trends Microbiol 9:34–39. https://doi.org/10.1016/s0966-842x(00)01913-2

Maki DG, Tambyah PA (2001) Engineering out the risk for infection with urinary catheters. Emerg Infect Dis 7:342–347

Maki DG, Weise CE, Sarafin HW (1977) A semiquantitative culture method for identifying intravenous-catheter-related infection. N Engl J Med 296:1305–1309. https://doi.org/10.1056/NEJM197706092962301

Mandakhalikar KD, Chua RR, Tambyah PA (2016) New technologies for prevention of catheter associated urinary tract infection. Curr Treat Options Infect Dis 8:24–41. https://doi.org/10.1007/s40506-016-0069-5

Mandakhalikar KD, Rahmat JN, Chiong E et al (2018) Extraction and quantification of biofilm bacteria: method optimized for urinary catheters. Sci Rep 8:8069. https://doi.org/10.1038/s41598-018-26342-3

Mathur P (2018) Prevention of healthcare-associated infections in low- and middle-income countries: the 'bundle approach.'. Indian J Med Microbiol 36:155–162. https://doi.org/10.4103/ijmm.IJMM_18_152

McLean RJ, Whiteley M, Stickler DJ, Fuqua WC (1997) Evidence of autoinducer activity in naturally occurring biofilms. FEMS Microbiol Lett 154:259–263. https://doi.org/10.1111/j.1574-6968.1997.tb12653.x

Milo S, Thet NT, Liu D et al (2016) An in-situ infection detection sensor coating for urinary catheters. Biosens Bioelectron 81:166–172. https://doi.org/10.1016/j.bios.2016.02.059

Modun BJ, Cockayne A, Finch R, Williams P (1998) The Staphylococcus aureus and Staphylococcus epidermidis transferrin-binding proteins are expressed in vivo during infection. Microbiol Read Engl 144(Pt 4):1005–1012. https://doi.org/10.1099/00221287-144-4-1005

Morris NS, Stickler DJ (1998) Encrustation of indwelling urethral catheters by Proteus mirabilis biofilms growing in human urine. J Hosp Infect 39:227–234. https://doi.org/10.1016/s0195-6701(98)90262-6

Nickel JC, Grant SK, Costerton JW (1985) Catheter-associated bacterium: an experimental study. Urology 26:369–375. https://doi.org/10.1016/0090-4295(85)90185-2

Nickel JC, Downey J, Costerton JW (1992) Movement of pseudomonas aeruginosa along catheter surfaces. A mechanism in pathogenesis of catheter-associated infection. Urology 39:93–98. https://doi.org/10.1016/0090-4295(92)90053-y

Nicolle LE (2005) Catheter-related urinary tract infection. Drugs Aging 22:627–639. https://doi.org/10.2165/00002512-200522080-00001

Noimark S, Dunnill CW, Wilson M, Parkin IP (2009) The role of surfaces in catheter-associated infections. Chem Soc Rev 38:3435–3448. https://doi.org/10.1039/b908260c

Nowatzki PJ, Koepsel RR, Stoodley P et al (2012) Salicylic acid-releasing polyurethane acrylate polymers as anti-biofilm urological catheter coatings. Acta Biomater 8:1869–1880. https://doi.org/10.1016/j.actbio.2012.01.032

O'May GA, Jacobsen SM, Stickler DJ et al (2008) Complicated urinary tract infections due to catheters. Springer, Berlin, Heidelberg, pp 1–43

O'Toole GA (2011) Microtiter dish biofilm formation assay. J Vis Exp JoVE 47:2437. https://doi.org/10.3791/2437

O'Toole GA, Kolter R (1998a) Flagellar and twitching motility are necessary for Pseudomonas aeruginosa biofilm development. Mol Microbiol 30:295–304. https://doi.org/10.1046/j.1365-2958.1998.01062.x

O'Toole GA, Kolter R (1998b) Initiation of biofilm formation in Pseudomonas fluorescens WCS365 proceeds via multiple, convergent signalling pathways: a genetic analysis. Mol Microbiol 28:449–461. https://doi.org/10.1046/j.1365-2958.1998.00797.x

Orsi N (2004) The antimicrobial activity of lactoferrin: current status and perspectives. Biometals Int J Role Met Ions Biol Biochem Med 17:189–196. https://doi.org/10.1023/b:biom.0000027691.86757.e2

Otero A, Palacios F, Akinfiev T, Apalkov A (2010) A low cost device for monitoring the urine output of critical care patients. Sensors 10:10714–10732. https://doi.org/10.3390/s101210714

Page K, Wilson M, Parkin I (2009) Antimicrobial surfaces and their potential in reducing the role of the inanimate environment in the incidence of hospital-acquired infections. J Mater Chem 19: 3819–3831. https://doi.org/10.1039/b818698g

Pawley J (2010) Fundamental limits in confocal microscopy. Scanning, In, pp 20–42

Peel TN (2019) Studying biofilm and clinical issues in orthopedics. Front Microbiol 10:359. https://doi.org/10.3389/fmicb.2019.00359

Pereira S, Nguyen L, Stevermer JJ (2014) PURLs: a simple way to reduce catheter-associated UTIs. J Fam Pract 63:E10–E12

Pezzlo M (1988) Detection of urinary tract infections by rapid methods. Clin Microbiol Rev 1:268–280

Pickard R, Lam T, Maclennan G et al (2012) Types of urethral catheter for reducing symptomatic urinary tract infections in hospitalised adults requiring short-term catheterisation: multicentre randomised controlled trial and economic evaluation of antimicrobial- and antiseptic-impregnated urethral catheters (the CATHETER trial). Health Technol Assess Winch Engl 16:1–197. https://doi.org/10.3310/hta16470

Platt R, Polk BF, Murdock B, Rosner B (1982) Mortality associated with nosocomial urinary-tract infection. N Engl J Med 307:637–642. https://doi.org/10.1056/NEJM198209093071101

Podkovik S, Toor H, Gattupalli M et al (2019) Prevalence of catheter-associated urinary tract infections in neurosurgical intensive care patients – the overdiagnosis of urinary tract infections. Cureus 11:e5494. https://doi.org/10.7759/cureus.5494

Portillo ME, Salvadó M, Trampuz A et al (2013) Sonication versus Vortexing of implants for diagnosis of prosthetic joint infection. J Clin Microbiol 51(2):591–594

Pratt LA, Kolter R (1998) Genetic analysis of Escherichia coli biofilm formation: roles of flagella, motility, chemotaxis and type I pili. Mol Microbiol 30:285–293. https://doi.org/10.1046/j.1365-2958.1998.01061.x

Press MJ, Metlay JP (2013) Catheter-associated urinary tract infection: does changing the definition change quality? Infect Control Hosp Epidemiol Off J Soc Hosp Epidemiol Am 34:313–315. https://doi.org/10.1086/669525

Prüß BM, Besemann C, Denton A, Wolfe AJ (2006) A complex transcription network controls the early stages of biofilm development by Escherichia coli. J Bacteriol 188:3731–3739. https://doi.org/10.1128/JB.01780-05

Radmacher M, Tillmann RW, Fritz M, Gaub HE (1992) From molecules to cells: imaging soft samples with the atomic force microscope. Science 257(5078):1900–1905

Rahmani-Badi A, Sepehr S, Mohammadi P et al (2014) A combination of cis-2-decenoic acid and antibiotics eradicates pre-established catheter-associated biofilms. J Med Microbiol 63:1509–1516. https://doi.org/10.1099/jmm.0.075374-0

Razatos A, Ong Y-L, Sharma MM, Georgiou G (1998) Molecular determinants of bacterial adhesion monitored by atomic force microscopy. Proc Natl Acad Sci 95:11059–11064. https://doi.org/10.1073/pnas.95.19.11059

Reddy ST, Chung KK, McDaniel CJ et al (2011) Micropatterned surfaces for reducing the risk of catheter-associated urinary tract infection: an in vitro study on the effect of sharklet micropatterned surfaces to inhibit bacterial colonization and migration of uropathogenic Escherichia coli. J Endourol 25:1547–1552. https://doi.org/10.1089/end.2010.0611

Reid G, Charbonneau-Smith R, Lam D et al (1992) Bacterial biofilm formation in the urinary bladder of spinal cord injured patients. Paraplegia 30:711–717. https://doi.org/10.1038/sc.1992.138

Riboli DFM, Lyra JC, Silva EP et al (2014) Diagnostic accuracy of semi-quantitative and quantitative culture techniques for the diagnosis of catheter-related infections in newborns and molecular typing of isolated microorganisms. BMC Infect Dis 14:283. https://doi.org/10.1186/1471-2334-14-283

Riley DK, Classen DC, Stevens LE, Burke JP (1995) A large randomized clinical trial of a silver-impregnated urinary catheter: lack of efficacy and staphylococcal superinfection. Am J Med 98:349–356. https://doi.org/10.1016/S0002-9343(99)80313-1

Ronin D, Boyer J, Alban N et al (2021) Current and novel diagnostics for orthopedic implant biofilm infections: a review. APMIS Acta Pathol Microbiol Immunol Scand. https://doi.org/10.1111/apm.13197

Roy RK, Lee K-R (2007) Biomedical applications of diamond-like carbon coatings: a review. J Biomed Mater Res B Appl Biomater 83:72–84. https://doi.org/10.1002/jbm.b.30768

Saint S (2000) Clinical and economic consequences of nosocomial catheter-related bacteriuria. Am J Infect Control 28:68–75. https://doi.org/10.1016/s0196-6553(00)90015-4

Sauer K, Camper AK (2001) Characterization of phenotypic changes in Pseudomonas putida in response to surface-associated growth. J Bacteriol 183:6579–6589. https://doi.org/10.1128/JB.183.22.6579-6589.2001

Sauer K, Camper AK, Ehrlich GD et al (2002) Pseudomonas aeruginosa displays multiple phenotypes during development as a biofilm. J Bacteriol 184:1140–1154. https://doi.org/10.1128/jb.184.4.1140-1154.2002

Schabert FA, Henn C, Engel A (1995) Native Escherichia coli OmpF porin surfaces probed by atomic force microscopy. Science 268:92–94. https://doi.org/10.1126/science.7701347

Schaeffer AJ, Chmiel J (1983) Urethral meatal colonization in the pathogenesis of catheter-associated bacteriuria. J Urol 130:1096–1099. https://doi.org/10.1016/s0022-5347(17)51701-2
Schierholz JM, Beuth J (2001) Implant infections: a haven for opportunistic bacteria. J Hosp Infect 49:87–93. https://doi.org/10.1053/jhin.2001.1052
Schierholz JM, Lucas LJ, Rump A, Pulverer G (1998) Efficacy of silver-coated medical devices. J Hosp Infect 40:257–262. https://doi.org/10.1016/s0195-6701(98)90301-2
Schindelin J, Arganda-Carreras I, Frise E et al (2012) Fiji: an open-source platform for biological-image analysis. Nat Methods 9:676–682. https://doi.org/10.1038/nmeth.2019
Schneider CA, Rasband WS, Eliceiri KW (2012) NIH Image to ImageJ: 25 years of image analysis. Nat Methods 9:671–675. https://doi.org/10.1038/nmeth.2089
Selinummi J, Seppälä J, Yli-Harja O, Puhakka JA (2005) Software for quantification of labeled bacteria from digital microscope images by automated image analysis. BioTechniques 39:859–863. https://doi.org/10.2144/000112018
Semwogerere D, Weeks ER (2005) Confocal microscopy. In: In Encyclopedia of biomaterials and biomedical engineering, biomaterials, biomedical engineering. Informa Healthcare, New York
Senturk S, Ulusoy S, Bosgelmez-Tinaz G, Yagci A (2012) Quorum sensing and virulence of Pseudomonas aeruginosa during urinary tract infections. J Infect Dev Ctries 6:501–507. https://doi.org/10.3855/jidc.2543
Seth JH, Haslam C, Panicker JN (2014) Ensuring patient adherence to clean intermittent self-catheterization. Patient Prefer Adherence 8:191–198. https://doi.org/10.2147/PPA.S49060
Shiga H, Yamane Y, Ito E et al (2000) Mechanical properties of membrane surface of cultured astrocyte revealed by atomic force microscopy. Jpn J Appl Phys 39:3711–3716. https://doi.org/10.1143/JJAP.39.3711
Sievert DM, Ricks P, Edwards JR et al (2013) Antimicrobial-resistant pathogens associated with healthcare-associated infections summary of data reported to the national healthcare safety network at the centers for disease control and prevention, 2009–2010. Infect Control Hosp Epidemiol 34:1–14. https://doi.org/10.1086/668770
Singh PK, Parsek MR, Greenberg EP, Welsh MJ (2002) A component of innate immunity prevents bacterial biofilm development. Nature 417:552–555. https://doi.org/10.1038/417552a
Slobbe L, el Barzouhi A, Boersma E, Rijnders BJA (2009) Comparison of the roll plate method to the sonication method to diagnose catheter colonization and bacteremia in patients with long-term tunnelled catheters: a randomized prospective study. J Clin Microbiol 47:885–888. https://doi.org/10.1128/JCM.00998-08
Soto SM, Smithson A, Horcajada JP et al (2006) Implication of biofilm formation in the persistence of urinary tract infection caused by uropathogenic Escherichia coli. Clin Microbiol Infect Off Publ Eur Soc Clin Microbiol Infect Dis 12:1034–1036. https://doi.org/10.1111/j.1469-0691.2006.01543.x
Staudt C, Horn H, Hempel DC, Neu TR (2004) Volumetric measurements of bacterial cells and extracellular polymeric substance glycoconjugates in biofilms. Biotechnol Bioeng 88:585–592. https://doi.org/10.1002/bit.20241
Stewart PS (2003) New ways to stop biofilm infections. Lancet Lond Engl 361:97. https://doi.org/10.1016/S0140-6736(03)12245-3
Stickler DJ (2014) Clinical complications of urinary catheters caused by crystalline biofilms: something needs to be done. J Intern Med 276:120–129. https://doi.org/10.1111/joim.12220
Subramanian P, Shanmugam N, Sivaraman U et al (2012) Antiobiotic resistance pattern of biofilm-forming uropathogens isolated from catheterised patients in Pondicherry, India. Australas Med J 5:344–348. https://doi.org/10.4066/AMJ.2012.1193
Summers JG (2019) The development and validation of a standard in vitro method to evaluate the efficacy of surface modified urinary catheters. Thesis, Montana State University - Bozeman, College of Engineering
Summers J, Goeres DM (2019) Catheter-associated urinary tract infections: development of a test method for assessing the efficacy of antimicrobial technologies/products. In: Williams DL

(ed) Targeting biofilms in translational research, device development, and industrial sectors. Springer International Publishing, Cham, pp 29–53
Sutrave S, Kikhney J, Schmidt J et al (2019) Effect of daptomycin and vancomycin on Staphylococcus epidermidis biofilms: An in vitro assessment using fluorescence in situ hybridization. PloS One 14:e0221786. https://doi.org/10.1371/journal.pone.0221786
Tambyah PA, Maki DG (2000) Catheter-associated urinary tract infection is rarely symptomatic: a prospective study of 1,497 catheterized patients. Arch Intern Med 160:678–682. https://doi.org/10.1001/archinte.160.5.678
Tambyah PA, Halvorson KT, Maki DG (1999) A prospective study of pathogenesis of catheter-associated urinary tract infections. Mayo Clin Proc 74:131–136. https://doi.org/10.4065/74.2.131
Tarkkanen AM, Allen BL, Williams PH et al (1992) Fimbriation, capsulation, and iron-scavenging systems of Klebsiella strains associated with human urinary tract infection. Infect Immun 60: 1187–1192. https://doi.org/10.1128/iai.60.3.1187-1192.1992
Tran PL, Lowry N, Campbell T et al (2012) An organoselenium compound inhibits staphylococcus aureus biofilms on hemodialysis catheters in vivo. Antimicrob Agents Chemother 56:972–978. https://doi.org/10.1128/AAC.05680-11
Trautner BW (2010) Management of Catheter-Associated Urinary Tract Infection (CAUTI). Curr Opin Infect Dis 23:76–82. https://doi.org/10.1097/QCO.0b013e328334dda8
Trautner BW, Darouiche RO (2004) Catheter-associated infections: pathogenesis affects prevention. Arch Intern Med 164:842–850. https://doi.org/10.1001/archinte.164.8.842
Truong VK, Lapovok R, Estrin YS et al (2010) The influence of nano-scale surface roughness on bacterial adhesion to ultrafine-grained titanium. Biomaterials 31:3674–3683. https://doi.org/10.1016/j.biomaterials.2010.01.071
Vaitkus S, Simoes-Torigoe R, Wong N et al (2021) A comparative study of experimental urinary catheters containing silver and zinc for biofilm inhibition. J Biomater Appl 35:1071–1081. https://doi.org/10.1177/0885328221989553
Vaterrodt A, Thallinger B, Daumann K et al (2016) Antifouling and antibacterial multifunctional polyzwitterion/enzyme coating on silicone catheter material prepared by electrostatic layer-by-layer assembly. Langmuir ACS J Surf Colloids 32:1347–1359. https://doi.org/10.1021/acs.langmuir.5b04303
Vejborg RM, Klemm P (2008) Blocking of bacterial biofilm formation by a fish protein coating. Appl Environ Microbiol 74:3551–3558. https://doi.org/10.1128/AEM.00279-08
Wadhera A, Fung M (2005) Systemic argyria associated with ingestion of colloidal silver. Dermatol Online J 11:12
Wald HL, Epstein AM, Radcliff TA, Kramer AM (2008) Extended use of urinary catheters in older surgical patients: a patient safety problem? Infect Control Hosp Epidemiol 29:116–124. https://doi.org/10.1086/526433
Walker JN, Flores-Mireles AL, Pinkner CL et al (2017) Catheterization alters bladder ecology to potentiate Staphylococcus aureus infection of the urinary tract. Proc Natl Acad Sci U S A 114: E8721–E8730. https://doi.org/10.1073/pnas.1707572114
Wang L, Zhang S, Keatch R et al (2019) In-vitro antibacterial and anti-encrustation performance of silver-polytetrafluoroethylene nanocomposite coated urinary catheters. J Hosp Infect 103:55–63. https://doi.org/10.1016/j.jhin.2019.02.012
Wei Tan C, Chlebicki MP (2016) Urinary tract infections in adults. Singapore Med J 57:485–490. https://doi.org/10.11622/smedj.2016153
White JG, Amos WB, Fordham M (1987) An evaluation of confocal versus conventional imaging of biological structures by fluorescence light microscopy. J Cell Biol 105:41–48. https://doi.org/10.1083/jcb.105.1.41
Wilson ML, Gaido L (2004) Laboratory diagnosis of urinary tract infections in adult patients. Clin Infect Dis 38(8):1150–1158. https://doi.org/10.1086/383029

Wiwanitkit V, Udomsantisuk N, Boonchalermvichian C (2005) Diagnostic value and cost utility analysis for urine Gram stain and urine microscopic examination as screening tests for urinary tract infection. Urol Res 33:220–222. https://doi.org/10.1007/s00240-004-0457-z

Wong SS, Joselevich E, Woolley AT et al (1998) Covalently functionalized nanotubes as nanometre-sized probes in chemistry and biology. Nature 394:52–55. https://doi.org/10.1038/27873

Wright KJ, Seed PC, Hultgren SJ (2005) Uropathogenic Escherichia coli flagella aid in efficient urinary tract colonization. Infect Immun 73:7657–7668. https://doi.org/10.1128/IAI.73.11.7657-7668.2005

Xavier JB, White DC, Almeida JS (2003) Automated biofilm morphology quantification from confocal laser scanning microscopy imaging. Water Sci Technol 47:31–37. https://doi.org/10.2166/wst.2003.0273

Xu Y, Moser C, Al-Soud WA et al (2012) Culture-dependent and -independent investigations of microbial diversity on urinary catheters. J Clin Microbiol 50:3901–3908. https://doi.org/10.1128/JCM.01237-12

Yang X, Beyenal H, Harkin G, Lewandowski Z (2000) Quantifying biofilm structure using image analysis. J Microbiol Methods 39:109–119. https://doi.org/10.1016/s0167-7012(99)00097-4

Yu K, Lo JCY, Yan M et al (2017) Anti-adhesive antimicrobial peptide coating prevents catheter associated infection in a mouse urinary infection model. Biomaterials 116:69–81. https://doi.org/10.1016/j.biomaterials.2016.11.047

Zhang T, Fang HHP (2001) Quantification of extracellular polymeric substances in biofilms by confocal laser scanning microscopy. Biotechnol 23(5):405–409

Zumstein V, Betschart P, Albrich WC et al (2017) Biofilm formation on ureteral stents - Incidence, clinical impact, and prevention. Swiss Med Wkly 147:w14408. https://doi.org/10.4414/smw.2017.14408

Probiotics Action Against Biofilms

Mohammad Yousef Memar, Yalda Rahbar Saadat, Seyyedeh Mina Hejazian, Mohammadreza Ardalan, Elham Ahmadian, and Sepideh Zununi Vahed

Abstract Prevention and treatment of biofilm-related infections are still challenging problems in clinics. However, the recent progress in the understanding of the connection between the immune system and probiotics, which are beneficial live microorganisms, has highlighted their plausible effect on biofilm therapy. The bactericidal effects of probiotics have been reported in different human infections including oral, intestinal, vaginal, wound, and medical device-associated infections. Based on their properties, probiotics administration presents as an appealing option for controlling or preventing the development of biofilms. Further studies are needed for a better understanding of the molecular mechanism, action, and the benefits of probiotics usage. In this chapter, we highlight the impact of probiotics and their products on biofilm formation and eradication.

1 Introduction

Biofilm formation is the ability of microorganisms to resist and persist on a variety of surfaces, including medical devices and living tissues (Uruén et al. 2021). In the process of biofilm formation (Fig. 1), microbial cells become highly resistant to antimicrobial agents and the host immune mediators. It has been reported that most bacterial infections in humans correlate with biofilm and about 50% of nosocomial infections are associated with indwelling medical devices (Paredes et al. 2014). Different studies have revealed that antimicrobial drugs alone are often inefficient in the treatment of biofilm-associated infections (Molina-Manso et al. 2013; Ciofu et al. 2012; Moskowitz et al. 2011).

M. Y. Memar
Infectious and Tropical Diseases Research Center, Tabriz University of Medical Sciences, Tabriz, Iran

Y. R. Saadat · S. M. Hejazian · M. Ardalan · E. Ahmadian · S. Z. Vahed (✉)
Kidney Research Center, Tabriz University of Medical Sciences, Tabriz, Iran
e-mail: zununivahed@tbzmed.ac.ir

K. Richter, K. Nørskov Kragh (eds.), *Antibiofilm Strategies*, Springer Series on Biofilms 11, https://doi.org/10.1007/978-3-031-10992-8_5

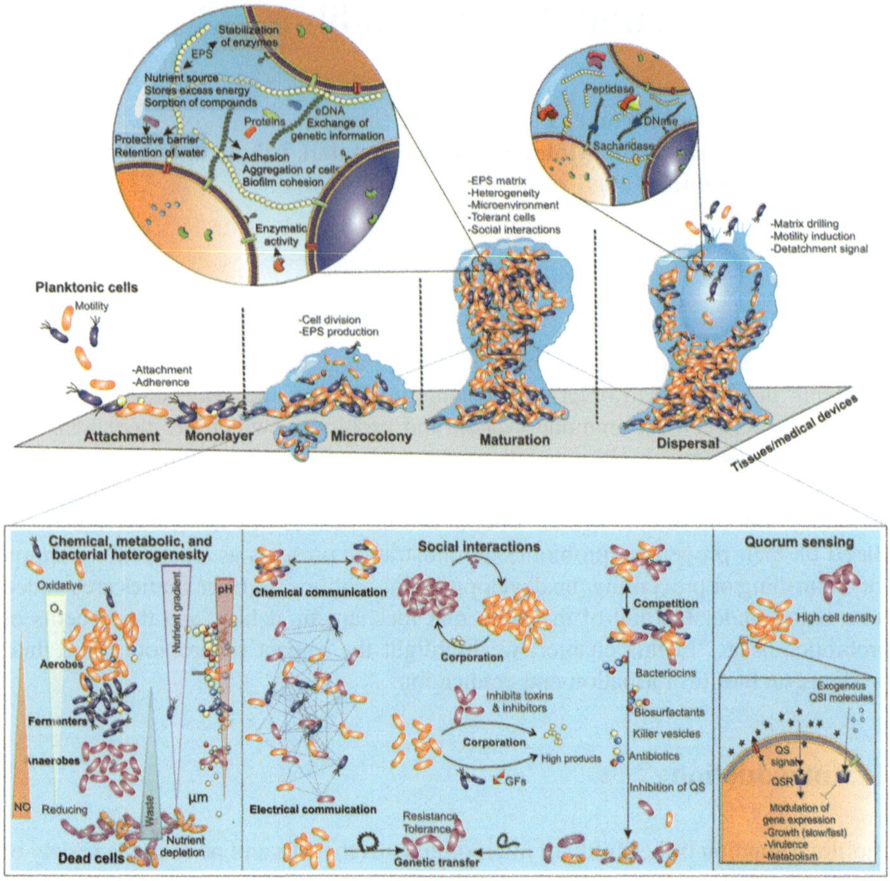

Fig. 1 Distinct stages of biofilm formation. (**a**) (1) Adhesion: microorganisms through cell surface-associated adhesins bind to medical device surfaces or host cells. (2) Initial biofilm formation: microorganisms start to divide and make EPS to enhance their adhesion and to form the matrix for embedding the cells. (3) Maturation of biofilm: the EPS matrix makes a 3D, protective, and multi-functional scaffold, allowing a heterogeneous physical and chemical frame for the synergistic and competitive co-existence of different microorganisms and their social interactions. (4) Dispersal: biofilm cells re-enter the planktonic phase. (**b**) On a small scale, different localized habitats are made by steady gradients of oxygen, pH, nutrition, and waste products. Remodeling of the biofilm community occurs by social connections (competition or cooperation) between different microbial cells. Competition and killing are mediated by growth inhibition strategies (e.g., reduction of nutrients, avoiding QS, production of antibiotics and bacteriocins). Chemical and electrical communications between biofilm cells result in their cooperation. *QS* quorum sensing, *NO* nitric oxide, *GFs* growth factors, *EPS* extracellular polymeric substance. This Figure is adapted from Barzegari et al. (2020) with permission

Probiotics are live microorganisms that have beneficial effects on the host by modulating the functionality and equilibrium of gut microbiota when administered in sufficient amounts (Williams 2010), leading to their worldwide use (Islam 2016). Generally, the health-beneficial effects of probiotics can be achieved by their direct act on the systemic immune system, the gastrointestinal (GI) tract, and the intestinal mucus layer (Rijkers et al. 2010). Probiotics modify the intestinal flora to prevent adhesion and invasion of pathogenic bacteria to the epithelial cells by competing for nutrients and producing antibacterial compounds such as bacteriocins, hydrogen peroxide, and biosurfactants (Gerritsen et al. 2011; Ng et al. 2009). The use of genomics, proteomics, and metabolomics has provided great opportunities in the classification of probiotic strains (Donelli et al. 2013). Lactic acid bacteria, mainly *Bifidobacterium* and *Lactobacillus* species, are the most widely used probiotics (Williams 2010). Probiotics have good efficiency in the prevention and/or treatment of various disorders, such as inflammatory bowel diseases, acute viral gastroenteritis, necrotizing enterocolitis in preterm infants, pediatric post-antibiotic-associated diarrhea, and post-surgical pouchitis (Vuotto et al. 2014; Guandalini et al. 2000). However, owing to the lack of adequate evidence for the pharmacological mechanisms of probiotics, their application in medicine is not yet validated. Several studies have investigated the mechanisms of action of probiotics (Dai et al. 2018). The standards and requirements of the use of these live bio-therapeutic products in humans have also been established (Trush et al. 2020).

Considering the ability of pathogens to produce persistent biofilm-associated infections, the administration of probiotics to inhibit the development of biofilms is an appealing option to be addressed. In this chapter, we highlight the impact of probiotics and probiotic products that can fight against biofilm formation through different mechanisms (Fig. 2). The search strategy includes: searching various databases such as PubMed, ScienceDirect, Google Scholar, and Scopus for English-language articles using the following keywords: probiotics AND biofilm AND ("clinical trials" OR endocarditis OR vagina OR orthopedic OR "dental biofilms" OR plaque OR catheter OR prostheses OR "medical device" OR diarrhea OR infections OR medical OR wound OR medicine OR implant OR "tooth plaque" OR dentistry) up to August 2021.

2 Strategies for Fighting Against Bacterial Biofilm Infections

Antibiofilm strategies can be classified based on the application of medical devices and/or artificial organs in the host (Wu et al. 2015). Long-term and high doses of antimicrobial agents may sometimes eradicate biofilm-related wound infections in chronic diabetes, chronic sinusitis, chronic otitis media, native valve endocarditis, and chronic airway infections (Wu et al. 2015). However, in the presence of a foreign device, its removal is essential for successful treatment. Recently, employing anti-adhesive surfaces with altered physical, chemical, and topographical characteristics seems to be an effective way of preventing surgical site infections and inhibiting

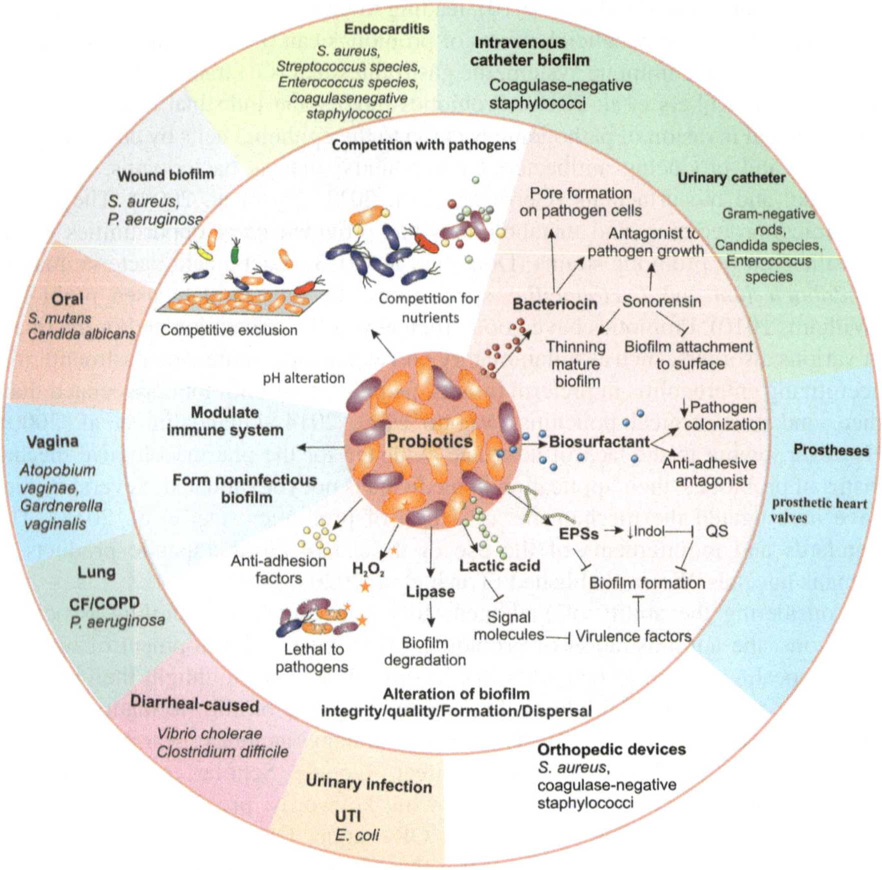

Fig. 2 Anti-biofilm strategies of probiotics and their metabolites. Biofilms can be targeted by probiotics and their derivatives at various stages. Probiotics cells/metabolites compete with pathogens for attachment, adhesion, and nutrition to prevent their adhesion and early phase of biofilm formation. Competitive adhesion of probiotics to medical devices or human tissues inhibits harmful bacterial colonization. Moreover, by targeting the division of pathogenic cells and their EPS production, probiotics can inhibit an early stage of biofilm development. By targeting the social interactions, establishment of hypoxia or low pH microenvironment, probiotics and their derivatives can disrupt the pre-formed pathogenic biofilms. Furthermore, through activation of dispersal mechanisms or remodeling of EPS matrix, probiotics can lead to biofilm dispersion. *UTI* urinary tract infection, *QS* quorum sensing, *EPS* extracellular polymeric substance, *COPD* chronic obstructive pulmonary disease, *CF* cystic fibrosis

biofilm formation on medical devices (Hrynyshyn et al. 2022). Therefore, to control infections, an alternative therapy is urgently required for the prevention, dispersal, and treatment of biofilm infection other than antibiotics (Wu et al. 2015). Several alternative strategies have been developed to combat biofilms including bacteriophage therapy (Rossitto et al. 2018), natural products (Memar et al. 2017), nanoparticle (NPs)-based technology (Memar et al. 2020), matrix-degrading enzyme (Baelo et al. 2015), photodynamic therapy (PDT) (Akhtar et al. 2021), and hyperbaric oxygen therapy (Memar et al. 2019). PDT exploits nontoxic dyes (i.e., photosensitizers such as porphyrins, ruthenium complexes), and visible light to form reactive oxygen species, which in turn, exerts significant antimicrobial activity against a wide variety of microorganisms (Hrynyshyn et al. 2022). Several lines of evidence reported that phytochemicals as natural metabolites of plants, present strong antibiofilm properties (Hrynyshyn et al. 2022). In the case of nanotechnology-based approaches, the broad-spectrum antibacterial properties of silver NPs (AgNPs) have been reported. In line with this, there are products available on the market (e.g., biomedical devices, surgical instruments, contraceptive devices, wound dressings, and bone prostheses) that have different metallic NPs, particularly AgNPs (Hrynyshyn et al. 2022; Singh et al. 2022). The application of probiotics can also provide an interesting approach to fighting against biofilm-related infections.

3 The Mechanisms of Action of Probiotics Against Biofilm-related Infections

There is a growing interest in the application of probiotics and their derivatives because of their potential to either control or treat various disorders such as infections (Santacroce et al. 2019). Probiotics have the potential to inhibit both Gram-positive and Gram-negative bacteria owing to different mechanisms of action (Vieira et al. 2013; Vieira et al. 2016; Sikorska and Smoragiewicz 2013; Huang et al. 2020). However, the precise mechanisms of probiotics that act on the host are not well known. Some of the suggested antimicrobial mechanisms of probiotics are: (1) competitive effects with pathogenic organisms for adhesion and nutrition; (2) production of antimicrobial compounds; (3) improvement of the function of the intestinal barrier; and (4) immunomodulation.

3.1 Competition with Pathogenic Organisms for Adhesion and Nutrition

The intestinal microbiota of humans, including commensal, symbiotic, and pathogenic microorganisms, constitute an intricate microbial ecosystem that plays a significant role in both health and disease (Round and Mazmanian 2009). An accurate balance between the composition and activity of microbiota is an important factor in preserving the health condition of the host. The alterations in the microbiota cause an increased probability of infections or other diseases (Gerritsen et al. 2011). Probiotic bacteria can limit the attachment of pathogenic organisms owing to competitive binding to the host's target cells. In line with this, Sakarya et al. reported that a probiotic yeast (i.e., *Saccharomyces boulardii*) reduces *Helicobacter pylori* adhesion to duodenal epithelial cells through neuraminidase activity (Sakarya and Gunay 2014). This ability is strain specific and is commonly detected in vitro (Lau and Chye 2018; Walsham et al. 2016). In this regard, in a study conducted by Walsham and coworkers, the effects of the *Lactobacillus reuteri* on the adhesion of enteropathogenic *Escherichia coli* (EPEC) to the intestinal epithelium were investigated. *L. reuteri* could diminish EPEC infection by competitive exclusion at the epithelial or mucus level by preventing the cell-to-cell spread and microcolony dispersal that highlight its strain-specific manner (Walsham et al. 2016).

3.2 Secretion of Antimicrobial Substances

It was approved that a broad spectrum of probiotic compounds such as bacteriocins, hydrogen peroxide (H_2O_2), and organic acids have inhibitory effects on a wide range of Gram-negative and Gram-positive bacteria (Gaspar et al. 2018; Ghasemi et al. 2019; Evivie et al. 2019). Organic acids such as lactic acid and acetic acid produced by lactic acid bacteria can change the pH of environmental conditions and prevent the growth of Gram-negative bacteria such as *Salmonella*. Hence, organic acids are considered the key antimicrobial compounds of probiotics that present inhibitory activity against pathogens (Evivie et al. 2019; Mortada et al. 2020). In a study conducted by Evivie et al., it was shown that the organic acids produced by the *L. bulgaricus* KLDS 1.0207 and *S. thermophilus* KLDS 3.1003 strains have the potential to suppress *S. aureus* ATCC 25923 and *E. coli* ATCC 25922 pathogens. Although both strains produced lactic and acetic acid, *L. bulgaricus* KLDS 1.0207 produced higher levels of lactic acid and acetic acid than *S. thermophilus* KLDS 3.1003. Furthermore, both strains could markedly suppress the pathogenicity of *S. aureus* and *E. coli* strains (Evivie et al. 2019). Another study demonstrated the effectiveness of organic acid and probiotic strains (*Enterococcus faecium*, *Bifidobacterium animalis*, *Pediococcus acidilactici*, and *L. reuteri*) in the reduction of *Campylobacter coli* proliferation (Mortada et al. 2020). Acids produced by probiotics enter the target bacterial cytoplasm and decrease intracellular pH or accumulate in the host cell in the ionized form, which are both fatal for bacteria (Russell and Diez-Gonzalez 1997).

Some probiotics can produce bacteriocins, which have a relatively narrow spectrum of antimicrobial activity and mainly affect other closely related bacteria (Hassan et al. 2012). The most common antibacterial mechanisms of bacteriocins (particularly lactic acid bacteria bacteriocins) are formation of pore in the cell membrane and/or inhibition of cell-wall formation, which in turn leads to leakage of low molecular weight compounds (such as ions K^+, H^+, and phosphate) and subsequent dissipation of the proton motive force and cell damage (Hassan et al. 2012).

3.3 Impacting Intestinal Barrier Function

Intestinal barrier integrity is an essential factor in mucosal function. It is reported that the disruption of intestinal barrier integrity is one of the common factors increasing the risk for different gastrointestinal disorders such as infections, inflammatory bowel disease, irritable bowel syndrome, diabetes, and obesity. In vitro, in vivo, and clinical trials have looked into the effects of probiotics on intestinal barrier integrity (Bron et al. 2017). Probiotics can decrease the severity of gastrointestinal infections initiated by some pathogens including *Salmonella enterica* serovar *Typhimurium, Listeria monocytogenes, Yersinia enterocolitica, Citrobacter rodentium*, and *Helicobacter pylori*. It has been suggested that the underlying mechanism is an increase in the concentration of organic acids and a decrease in pH due to dextrose fermentation by probiotic strains. Besides, probiotic therapy could not only boost gut innate defenses but also compete with pathogens in residing adhesion sites (for instance, compete with *H. pylori* for residing in the stomach), and finally, they secrete antibacterial metabolites which exert bactericidal and bacteriostatic activities against a wide range of bacteria (Bujalance et al. 2014; Emara et al. 2016; Asahara et al. 2011). Several possible mechanisms of the modulation by probiotics of intestinal barrier function have been suggested. The butyrate, a short-chain fatty acid (SCFA), production as a secondary metabolite of fermentative processes of the microbiota has been revealed to increase mucus secretion and induce the regulatory T cell (Treg) activities in the intestinal tract (Furusawa et al. 2013). In line with this, Furusawa et al. reported that butyrate stimulates colonic Treg differentiation and enhances colitis induced by adoptive transfer of $CD4^+$ $CD45RB^{hi}$ T cells in $Rag1^{-/-}$ mice. Further analysis proposed that the luminal concentrations of SCFAs correlated positively with the number of Treg cells in the colon. Butyrate treatment of naive T cells under the Treg cell-polarizing conditions could improve histone H3 acetylation in the promoter and conserve non-coding sequence regions of the Foxp3 locus that clarifies the underlying mechanism of butyrate in Treg cell differentiation (Furusawa et al. 2013). Moreover, probiotics can improve intestinal epithelium entirety via controlling genes transcription, which plays an important role in tight junction signaling. It is reported that lactobacilli regulate the expression of genes involved in certain adherence junction molecules such as β-catenin and E-cadherin (Karczewski et al. 2010). Exposure to lactobacilli (i.e., *L. plantarum*) also significantly affects the phosphorylation of adherence junction proteins and increases some protein kinase C levels such as PKCδ, which have a desirable effect on intestinal barrier activity. *L. plantarum* could activate Tool-like receptor 2 (TLR2) signaling in the gut epithelia, which is considered a regulator of epithelial integrity. *L. plantarum* improves epithelial permeability in Caco-2 cells by reducing the dislocation and occluding ZO-1 (Karczewski et al. 2010).

3.4 Immunomodulation

Probiotics can induce an immunomodulatory effect on the GI tract and have the ability to promote the host inflammatory reaction in response to pathogens (Cristofori et al. 2021). After the induction of immune response by commensal microorganisms and probiotics, dendritic cells (DCs) trigger an appropriate reaction, such as the development of Th0 to Treg that inhibits inflammation reaction induced by Th1, Th2, and Th17 cells. For instance, *L. paracasei* CNCM I-4034 stimulates an innate immunity via activation of TLR signaling and various pro-inflammatory cytokines and chemokines. *L. paracasei* CNCM I-4034 could upregulate *TLR9* expression and its upregulation in intestinal epithelia results in an increased state of immune surveillance (Bermudez-Brito et al. 2012). In addition, probiotics can modulate mucosal immunity through the promotion of anti-inflammatory mediators (TGF-β and IL-10) and by reducing the transcription of pro-inflammatory mediators (IFN-γ and TNF) (Corr et al. 2007). Probiotics also alter the immune marker levels, the activity of inflammatory pathways, and the expression of immune-related genes. For example, they can modulate the expression of mitogen-activated protein kinase (MAPK), NF-κB, C-reactive proteins, and pro-inflammatory genes in the intestinal epithelial cells via different mechanisms (Suez et al. 2019; Thomas and Versalovic 2010).

4 Role of Probiotics and their Derivatives in the Eradication of Biofilms

There are a limited number of studies on the antibiofilm activity of probiotics and their effect on planktonic forms of bacteria (Missaoui et al. 2019; Manna et al. 2019; Karadeniz et al. 2021). Various in vitro models have been developed to study the antibiofilm activity of probiotics by attaching the tested organisms to different surfaces (Silva et al. 2020). Two types of studies can be performed to assess the antibiofilm activity of probiotics: a preventive or an interventional study. Most of the time, the findings of in vivo studies are not reproducible in humans. Thus, using 3D cell cultures and human tissues as well as clinical trials is recommended. Further, conducting bioguided studies seems to be practical (Silva et al. 2020). Several animal models have been developed to evaluate the progress of infections in the host. In this regard, BALB/c mice and rabbit models of wound infection of methicillin-resistant *Staphylococcus aureus* have been developed for studying the anti-infective effects of probiotics (Lkhagvadorj et al. 2010). Experimental abutments mimicking dental implants have also been developed for recovering biofilm and studying the nature of biofilms formed around such abutments. The results of the in vivo study indicated that there was considerable plaque growth on the abutments after 2 weeks when hygiene measures were absent (Cortés-Acha et al. 2019). An in vivo study such as the human plasma biofilm model has also been developed to provide a human wound milieu and individual immune competence for studying the effect of probiotics on pathogen biofilms. Employing probiotics, particularly *L. plantarum*, had a great antibacterial effect on various pathogens (*S. aureus*, *S. epidermidis*, *E. faecium*, and *P. aeruginosa*) through the production of glycocalyx, which could eliminate them (Besser et al. 2019a). As previously mentioned, probiotics produce compounds with inhibitory effects on the adhesion of pathogens, including surfactants, bacteriocins, enzymes (amylase and lipase), fatty acids, organic acids, extracellular polysaccharide (EPS), and H_2O_2. Probiotics may also interfere with biofilm integrity, which can result in biofilm elimination.

4.1 The Inhibition of Biofilm Formation by Probiotics

The molecular mechanisms of the inhibitory effect of probiotics on biofilm formation are unknown; however, it has been reported that the expression of genes involved in biofilm formation such as quorum sensing (QS) and bacterial adhesion can play an important role (Deng et al. 2020). QS is a communication system between bacteria that regulates gene expression in response to small compounds known as autoinducers (AIs), which are produced by bacteria. QS regulates important traits in bacterial pathogenesis including the expression of several virulence factors (i.e., proteases, toxins, and adhesins) and related behaviors (i.e., swarming and biofilm formation) (Castillo-Juárez et al. 2015; Deng et al. 2020). As QS plays an important role in the control of biofilm formation by pathogens, its inhibition by QS inhibitors (QSIs) have become anti-infective therapeutic agents (Deng et al. 2020). In line with this, Ouyang et al. reported that quercetin, a natural flavonolon, could decrease the expression levels of QS-related genes (including *lasI*, *lasR*, *rhlI*, and *rhlR*); hence, it was considered a potential inhibitor of QS, biofilm formation, and virulence factors (pyocyanin, protease, and elastase) in *P. aeruginosa* (Ouyang et al. 2016). Bacteriocins, H_2O_2, organic acids, and biosurfactants can act as QS inhibitory agents (Karadeniz et al. 2021; Kiymaci et al. 2018; Yan et al. 2019). It has been reported that the metabolites of *L. casei* and *L. acidophilus* can inhibit *P. aeruginosa* biofilm formation owing to QS inhibition. In vitro assays demonstrated that chloroform extracts of the probiotics could inhibit biofilm biomass, biofilm metabolic activity, the production of elastase, as well as pyocyanin, and reduce QS. Moreover, GC-MS analysis detected diketopiperazines as the major compounds in the extracts (Díaz et al. 2020). *Streptococcus mutans* is a major participant in biofilm formation owing to EPS production on the tooth surface that acts as a binding site, inducing the accumulation of other bacteria on the dental surface and establishment of pathogenic biofilms (Koo et al. 2010). A significant reduction in the expression of genes related to stress responses (*aguD* and *atpD*), signal transduction systems (*comD, vicK, comC*, and *vicR*), and EPS formation (*gtfD, gtfC, sacB* (*ftf*), and *gtfB*) of *S. mutans* has been observed when co-cultured with *Lactobacillus* spp. (Wasfi et al. 2018). EPS-derived from *L. acidophilus* A4 significantly decreases the initial attachment and aggregation of *E. coli* O157: H7 during biofilm formation in vitro owing to a decrease in the expression of genes involved in curli formation (*crl, csgB,* and *csgA*) and chemotaxis (*cheY)* (Kim and Kim 2009).

Probiotics could inhibit in vitro biofilm formation and the adherence of *Porphyromonas gingivalis* to cells by affecting the expression of its virulence-associated factors encoding QS signaling molecules (luxS), proteases (rgpA, kgp, and fsH), and fimbriae (fimA and mfa1) (Ishikawa et al. 2020). *L. acidophilus* LA 5 is reported to be the most effective probiotic against *P. gingivalis* adhesion and biofilm formation via the down-expression of important virulence genes (Ishikawa et al. 2020). Biosurfactants produced by some probiotics such as *Pediococcus acidilactici* and *L. plantarum* can significantly decrease the attachment of bacteria to surfaces and inhibit the biofilm formation of *S. aureus* owing to decreasing expression of genes encoding the signaling molecules (AI-2) in QS and biofilm-related genes (agrA, sortaseA, dltB, icaA, sarA, and cidA) (Yan et al. 2019). It is reported that *L. plantarum* cultures and filtrates can inhibit *P. aeruginosa* QS mediators in vitro. The greatest effect was observed in the exposure of whole culture containing 106 *L. plantarum* CFU/ml (Valdez et al. 2005). *L. plantarum* KU200656 also presents anti-adhesion and antibiofilm activity against *Salmonella typhimurium, E. coli, Listeria monocytogenes,* and *Staphylococcus aureus* owing to its high adhesion ability, and its co-aggregation with these pathogens (Lee et al. 2021b).

Biofilm formation in *Staphylococcus* spp. includes the binding of bacteria to the surfaces, accumulation, maturation, and its release and binding to other sites (Noshak et al. 2019). During the accumulation step, polysaccharide intercellular adhesions (PIAs) are critical for biofilm formation. The genes encoding PIAs are located on the *ica*ADBCR operon, which is regulated by the *icaR* gene (Noshak et al. 2019). As reported, the probiotics (*L. fermentum* TCUESC01 and *L. plantarum* TCUESC02) isolated during the fermentation of fine cocoa, could prevent *S. aureus* growth. Nevertheless, *L. fermentum* TCUESC01 could decrease the biofilm thickness considerably. Further, it could regulate the *icaA* and *icaR* expression levels (increased *icaR* and decreased *icaA* expression). It can be deduced that the decrease in *S. aureus* biofilm upon treatment with *L. fermentum* supernatant might be a result of inhibition of PIA production (Melo et al. 2016). It is reported that *Lactobacillus* strains significantly decrease the adherence ability of *H. pylori* by the synthesis of a biologically active molecule (probably a fixation- and heat-resistant surface-associated molecule) and its release into the medium can decrease the expression of the adhesion-encoding gene *sabA* (De Klerk et al. 2016).

4.2 Probiotics Eradicate the Formed Biofilms

When biofilms are established, the eradication of infection becomes challenging because pathogens in the formed biofilms are protected against host immune mechanisms and antimicrobial drugs (Wu et al. 2015). Probiotics are described to influentially eradicate the biofilm by interfering with the quality/integrity of the biofilm, by completing disrupting the biofilm, and by affecting biofilm microorganisms (Wu et al. 2015). The inhibitory effects of probiotics on microbial biofilms depend on the production of biologically active compounds by probiotics such as EPS, bacteriocins, biosurfactants, organic acids, and H_2O_2 that are responsible for antimicrobial and antibiofilm effects (Barzegari et al. 2020). Bacteriocin and organic acids synthesized by *Lactobacillus* derived from traditional Tunisian fermented food indicate potential antibiofilm activity against *L. monocytogenes* biofilm under cell culture conditions (Ben Slama et al. 2013). In another study, it was reported that *L. plantarum* significantly impaired the biofilm formation of *Klebsiella pneumoniae* (as a multi-resistant opportunistic pathogen). Besides, *L. plantarum* supernatant could modulate *K. pneumoniae* biofilm-related genes expression (reduction of QS-related lsr operon expression and upregulation of type 3 pili structure genes) (Lagrafeuille et al. 2018).

4.3 Probiotic Biofilms Against Pathogenic Biofilms

The biofilm formed by probiotics may be an effective approach against biofilm-forming pathogenic bacteria, as probiotics compete with other microorganisms for sources and surfaces by various mechanisms of action (Gómez et al. 2016). Biofilms can provide more stable probiotics on the mucosal surface of the host, which in turn limits the number of pathogens. The biofilm formation ability of probiotics is a strain-dependent feature that has been described in various studies such as in vitro abiotic surfaces (glass or polystyrene). In this regard, it is reported that the gastric isolate, *L. fermentum* UCO-979C, can produce biofilm (via EPS formation) in vitro on abiotic surfaces as well as on AGS and Caco-2 cells. Furthermore, it could inhibit the adhesion of *H. pylori* in both cell lines (Salas-Jara et al. 2016a).

Biofilms of probiotic *E. coli* Nissle 1917 (EcN) produce DegP, a bifunctional (chaperone and protease) periplasmic protein that inhibits the formation of pathogenic biofilms of enterohemorrhagic *E. coli* (EHEC), *S. aureus*, and *S. epidermidis* by the secreted DegP effect (Fang et al. 2018). Biofilms of EcN on silicone substrates have also been shown to decrease the colonization by pathogenic *Enterococcus faecalis* (Fang et al. 2018). Moreover, it has been elucidated that bacteriocinogenic (*L. sakei* MBSa1, and *L. curvatus* MBSa3) and non-bacteriocinogenic (*L. lactis* 368) probiotic strains could inhibit the biofilm formation of *E. coli* O157: H7, *S. Typhimurium,* as well as *L. monocytogenes*, and decrease the number of aforementioned pathogens (Gómez et al. 2016). *Pediococcus pentosaceus* AC1–2 and *L. curvatus* BSF206 could inhibit the biofilm formation of *S. mutans* by downregulation of the brpA, ftf, gtfB, and gtfA expression involved in biofilm formation (Luan et al. 2022).

Some biosurfactants have been reported to decrease the adhesion of pathogens to glass, silicone rubber, implants, and prostheses (Rodrigues et al. 2004). Nisin-producing probiotic *L. lactis* can significantly suppress and disrupt *K. pneumoniae, Tannerella forsythia,* and *Fusobacterium nucleatum* biofilms and reduce the number of these pathogens in oral biofilms (Radaic et al. 2020). Other beneficial properties have been reported for probiotics in biofilms, including higher levels of resistance to temperature, acidic conditions, and mechanical pressure than their planktonic counterparts (Salas-Jara et al. 2016c).

5 Clinical Applications of Probiotics

Numerous lines of evidence from experimental studies and clinical trials have proposed that probiotics can compete and interfere with biofilm-associated infections in the genitourinary tract, wounds, mouth, skin, and intestine (He et al. 2021; Stanbro et al. 2020; Lorenzo et al. 2020; Buggio et al. 2019; Besser et al. 2019b; Tsai et al. 2021; Del Piano et al. 2012; Vicariotto et al. 2014). Various studies have reported that probiotics may be effective in the treatment and control of urogenital tract infections, particularly in women (Pendharkar et al. 2015; Heczko et al. 2015). Epithelium cell lines of the human urinary bladder were selected under laboratory conditions to determine the adhesion potential of probiotics and their competitive effect on the pathogens' interaction with cells during urogenital tract infections (Chapman et al. 2014). The inhibitory effect of single probiotics (i.e., *L. acidophilus*, *L. fermentum*, *L. plantarum*, and *L. rhamnosus*) and probiotic mixtures against common urinary tract pathogens (i.e., *E. coli* and *E. faecalis*) were investigated by Chapman et al. Their findings indicated that utilizing both single strains and mixtures were considerably effective and result in the reduction of pathogen attachment to the host cells (Chapman et al. 2014). Several clinical trial studies have also demonstrated that prophylaxis by probiotics decreases the frequency of bacterial vaginitis owing to transient colonization of probiotics in the vagina (Ya et al. 2010; Hantoushzadeh et al. 2012). In this regard, Ya and colleagues examined the efficacy of vaginal probiotic capsules for the prevention of recurrent bacterial vaginosis (BV) by employing 120 healthy Chinese women. Short-term administration of one capsule containing *Lactobacillus rhamnosus*, *L acidophilus*, and *Streptococcus thermophilus* and/or one placebo capsule, resulted in the reduction of BV recurrence and *Gardnerella vaginalis* risk after treatment (Ya et al. 2010). In another randomized clinical trial, the efficacy of probiotic yogurt and clindamycin were assessed in the treatment of BV in pregnant women in the third trimester. The results of the study reported that probiotics could exert beneficial effects toward BV during pregnancy (Hantoushzadeh et al. 2012).

A meta-analysis was performed on 50 randomized controlled trials with 3247 participants, mainly children. The results showed that probiotics (*lactobacilli, bifidobacteria*, and other genera) could markedly decrease bleeding-on-probing and gingival index; however, plaque index and caries incidence was not significantly diminished. They could also reduce *S. mutans* to less than 10^4 CFU/ml. Taken all together, although the evidence provided by Gruner and coworkers is inadequate for recommending probiotics in dental caries control, the findings support the beneficial effects of probiotics in gingivitis or periodontitis management (Gruner et al. 2016).

Orally administered probiotics could reduce *Aggregatibacter actinomycetemcomitans* counts at 4 weeks (Sang-Ngoen et al. 2021). Another meta-analysis on 24 studies (20–451 participants aged between 1 and 100 years) indicated that dairy probiotics (in particular, *Bifidobacterium* and *L. rhamnosus*), especially yogurt and milk, could improve salivary and oral parameters. They were effective at elevating salivary pH and decreasing *S. mutans* (Nadelman et al. 2018). The impact of probiotic therapy on oral health was also studied among patients undergoing fixed orthodontic therapy in a systematic review. In this review, 9 clinical trial studies were included, 7 of which concluded that probiotic therapy can decrease the pathogenic counts in saliva and/or in the oral biofilm. Probiotic therapy could diminish halitosis and the scores of gingival index and plaque index citation. However, in one study with daily administration of a probiotic lozenge (including *L. reuteri*) in 85 patients (mean age 15.9 years), no differences in the incidence of white spot lesion formation were detected. Thus, they reported that probiotic therapy does not affect the counts of *S. mutans* (Pietri et al. 2020). A recent meta-analysis on 11 RCTs with 554 patients could find no clear evidence that probiotics can improve oral microbiota and gingival inflammation in patients with plaque-induced gingivitis. Oral administration of probiotics had no significant improvement in the bleeding on probing, the Plaque Index, and the Gingival Index in these patients (Liu et al. 2021). The list of clinical trial studies regarding the effects of probiotics on oral biofilms is presented in Table 1.

Infections related to medical devices are serious challenges owing to their increased usage in the diagnosis, avoidance, and treatment of various diseases for improving the life quality and health care of patients (Khatoon et al. 2018; Azevedo et al. 2017). In recent years, the use of probiotics to inhibit biofilm formation has gained attention in this field through three different antibiofilm strategies. In the first strategy, probiotics and their metabolites (bacteriocins, biosurfactants, EPS, peroxide, carbon dioxide, ethanol, hydrogen sulfide, and organic acids) are added to pre-formed biofilms to disrupt the architecture of pathogens. Walencka et al. reported that *L. acidophilus*-derived surfactants successfully prevent *S. aureus* and *S. epidermidis* biofilms and inhibit the bacterial deposition rate, probably because of interference in the cell-surface hydrophobicity of the pathogenic strains (Walencka et al. 2008). Additionally, the capability of the biosurfactant derived from the *Lactobacillus* sp. CV8LAC, isolated from cabbage, was assessed. The results showed that *Lactobacillus* sp. CV8LAC biosurfactant could considerably inhibit the adhesion of biofilm-forming strains of *Candida albicans* (Fracchia et al. 2010). In another study, it was elucidated that the supernatant of the *L. plantarum* could significantly prevent *P. aeruginosa* pathogenic features, including adhesion, viability, virulence factors, and biofilm formation through the QS mechanism, that made the probiotic strain an acceptable candidate in chronic wound treatment (Ramos et al. 2012). Another strategy is based on the ability of pre-formed probiotic biofilms on the surface of medical devices to hinder the adhesion of pathogens (Salas-Jara et al. 2016b). For instance, in the study by Chen et al., *E. coli* strain Nissle 1917 (EcN) formed stable biofilms on mannoside-presenting silicone substrates and could

Table 1 Clinical trials on the administration of probiotics in oral biofilms

Biofilm former	Participants	Probiotics, dose, duration	Results	References
Orthodontic braces	64	*Streptococcus salivarius* M18, 3.6 × 10^9 CFU/lozenge, 30 days	↓Halitosis level, with minimal effects on dental biofilm microflora, gingival index and plaque index	(Benic et al. 2019)
Fixed orthodontic treatment	45	Probiotic mouthwash[a], 6 months	↓ *Porphyromonas gingivalis*. Mouth wash with probiotic may be applied besides regular brushing to expand periodontal status	(Goyal et al. 2019)
Fixed orthodontic treatment	30	Probiotic mouthwash, 2 × 10^8 CFU/g, 28 days	As a chemical plaque control agent, probiotics can be effective in a similar way to chlorhexidine	(Shah et al. 2019)
Fixed orthodontic treatment	30	*Lactobacillus acidophilus* and *Streptococcus thermophiles*, *Bifidobacteria*, and *Lactobacillus acidophilus*, *L. acidophilus* CFU/g 20 × 10^7, *Bifidobacteria* 5.4 × 10^7/*L. acidophilus* CFU/g 10^6, *S. thermophiles* 35 × 10^4,14 days	Probiotics decrease the *S. mutans* levels in plaque and saliva	(Megha et al. 2019)
Fixed orthodontic treatment	45	Lactic acid bacteria culture, *Lactobacillus acidophilus, Streptococcus thermophiles, Leuconostoc* species, *Lactobacillus* species, I.2 × 100 ml of kefir daily, 42 days, bacteriocin from lactic acid bacteria, 42 days	↓ *Lactobacillus* levels and salivary *S. mutans* and elevates the capacity of salivary buffering	(Alp and Baka 2018)
Fixed orthodontic treatment	50	*Lactobacillus acidophilus*, *Streptococcus salivarius* M18, 3 months	↓ Gingival Index and Plaque Index, and a shift in the oral microbiome composition in favor of beneficial bacteria	(Kaklamanos et al. 2019)

(continued)

Table 1 (continued)

Biofilm former	Participants	Probiotics, dose, duration	Results	References
Tongue coating-associated halitosis	28	*Streptococcus salivarius* K12	No significant differences were seen between the studied groups regarding with TCS, VSC, and OLT	(He et al. 2020)
Halitosis	92	*Weissella cibaria* CMU1 × 10^8 CFU/g, for 8 weeks	Tablets containing *W. cibaria* CMU can be useful for managing bad breath	(Lee et al. 2020)
Gingivitis	92	*Weissella cibaria* CMU1 × 10^8 CFU/g, for 8 weeks	The *W. cibaria* lives and colonizes in the oral cavity and dental plaque and destroys acids, and hence reduces the risk of dental caries developing from acid production in the oral flora	(Kang et al. 2021)
Gingivitis subjects	40	*Lactobacillus reuteri* (ATCC PTA 5289 and DSM-17938), 2 × 10^8 CFU/tablet, 8 weeks	Did not impact clinical parameters. ↓Periodontal pathogens in the subgingival microbiota	(Iniesta et al. 2012)
Periodontal health and oral microbiota	92	*Weissella cibaria* CMU1 × 10^8 CFU/g, for 8 weeks	An improvement was seen in bleeding on probing oral microbiota. *W. cibaria* exerted antimicrobial activity against *Fusobacterium nucleatum*	(Lee et al. 2020)
Chronic periodontitis	41	*Bifidobacterium lactis* HN019 with lozenges as adjuvant to SRP, 10^9 CFUs for 90 days	↓Pro-inflammatory cytokine levels, delayed the recolonization of periodontal pockets	(Invernici et al. 2018)
Chronic periodontitis	30	*Bifidobacterium animalis* subsp. *lactis* HN019, 10^9 CFU, 30 days	*B. lactis* HN019 present the antimicrobial and immunological characteristics and can be applied in patients with chronic periodontitis	(Invernici et al. 2020)
Dental caries	66	*Bifidobacterium lactis Bb12,* 10^6 CFU/ml, 14 days	↓Salivary *Streptococcus mutans* and *Lactobacillus* counts, can modify the oral biofilm	(Zare Javid et al. 2020)

(continued)

Table 1 (continued)

Biofilm former	Participants	Probiotics, dose, duration	Results	References
Type 1 diabetic children with caries and gingivitis	68	*Lactobacillus brevis* CD2, 2×10^9 colonies, 60 days	*L. brevis* CD2 improved gingival health and caries-related risk factors in diabetic children	(Lai et al. 2021)
Oral health	27	*ATCC* 5289 and *Lactobacillus reuteri* DSM 17938, 1×10^8 CFU/table, for 28 days	Gingivitis, bleeding, and plaque indices were insignificantly decreased	(Borrell García et al. 2021)
Dental biofilm and saliva	35	*Lactobacillus rhamnosus* GG, *Lactobacillus casei*, or *Enterococcus faecalis*, for 28 days	Probiotic bacteria could neither integrate nor persist in dental biofilm and saliva but did influence the growth of streptococci in biofilm and saliva	(Arweiler et al. 2020)
Saliva and dental plaque	26	Yogurt containing *B. animalis* subsp. *lactis* DN-173010, 200 g of yogurt per day, 14 days	Beneficial effects in reducing total microbial counts in dental plaque	(Pinto et al. 2014)
Inhibiting *Streptococcus mutans* in saliva	40	15 days	Probiotic could inhibit *S. mutans* growth more than synbiotics	(Duraisamy et al. 2021)
Chronic periodontitis	30	The probiotic lozenges consisted of *L. reuteri* strains ATCC PTA5289 and DSM17938, 1×10^8 CFU of each strain, 12 weeks	*L. reuteri* lozenge administration was a useful adjunct to SRP	(Teughels et al. 2013)
Chronic periodontitis	47	*Lactobacillus rhamnosus SP1*, 2×10^7 CFU/ day, 3 months	Adjunctive use of *L. rhamnosus* had similar microbiological and clinical improvements to the placebo group	(Morales et al. 2018)
Chronic periodontitis	30	SRP+ *lactobacillus reuteri* containing lozenges, 360 days	A helpful supplement in patients with moderately deep pockets	(İnce et al. 2015)
Oral and respiratory tract infections	40	Hyperbiotic PRO-kids ENT[b]	Probiotic supplementation had positive effects against incidence of respiratory and oral tract infections	(Campanella et al. 2018)

(continued)

Table 1 (continued)

Biofilm former	Participants	Probiotics, dose, duration	Results	References
Periodontal patients	51	*L. salivarius* SGL03, 30 days	*L. Salivarius* SGL03 could decrease pocket depth in supragingival plaque	(Nędzi-Góra et al. 2020)
Healthy adult	44	Lactobacilli lozenges contained *L. reuteri* (PTA 5289 and DSM 17938 strains), 10^8 CFU per strain, 12 weeks	Consumption of *L. reuteri* did not affect the total species richness. It induced a shift in composition of oral microbiota, ↓*Prevotella maculosa, Fusobacterium periodicum, N. mucosa, F. nucleatum* subsp. *vincentii, S. anginosus,* and *S. mutans*	(Romani Vestman et al. 2015)
Halitosis	92	*Weissella cibaria* CMU, 1.0 × 10^8 CFU/g, 8 weeks	Oral administration of *W. cibaria* CMU improved oral health-related quality of life and decreased subjective halitosis	(Lee et al. 2021a)
Periodontitis	47	*Lactobacillus rhamnosus* SP1, 2 × 10^7 CFU/day	Adjunct probiotic therapy to SRP failed to exert benefits in patients with periodontitis (stage III)	(Morales et al. 2021)
Peri-implant mucositis	77	*Lactobacillus reuteri*, DSM 17938 and ATCC PTA 5289 strains, 100 million active units of each strain, 30 days	It could prevent and treat peri-implant mucositis. It also improved clinical factors in healthy controls	(Flichy-Fernández et al. 2015)
Mucositis or peri-implantitis	44	*Lactobacillus reuteri* containing 1 × 10^8 living cells of each ATCC PTA 5289 and DSM 17938, 30 days	Could improve the clinical parameters of implants with peri-implantitis or mucositis	(Galofré et al. 2018)
Peri-implant mucositis	49	*Lactobacillus reuteri* strains ATCC PTA 5289and DSM 17938, 2× 10^7 CFU of each strain, 3 months	Probiotic supplements were not different than placebo	(Hallström et al. 2016)

(continued)

Table 1 (continued)

Biofilm former	Participants	Probiotics, dose, duration	Results	References
Experimental gingivitis	18	Lozenges containing *L. reuteri* ATCC PTA5289 and ATCC55730 strains, 10^8 CFU of each strain, 3 weeks twice a day	Probiotic lozenges had no significant impact on the composition of the biofilm, inflammatory reaction, or plaque accumulation during gingivitis	(Hallström et al. 2013)

CFU colony forming units, *OLT* organoleptic test scores, *VSC* volatile sulfur compound levels, *TCS* tongue-coating scores, *SRP* scaling and root planning

[a] *Bifidobacterium bifidum*, *Bifidobacterium longum*, *Lactobacillus rhamnosus*, *Lactobacillus reuteri*, lactilol Fructo-oligosaccharides, and zinc

[b] *Lactobacillus paracasei*, *Lactobacillus sakei*, *Lactobacillus reuteri*, *Streptococcus salivarius* M18, and *Streptococcus salivarius* K12

significantly diminish the colonization of *E. faecalis* by interference in pathogen adhesion (Chen et al. 2017).

The third strategy includes the competition of probiotics and/or their substances with pathogenic cells for attachment (Carvalho et al. 2020). For example, it has been shown that in an in vitro model, *L. rhamnosus* GG could integrate with biofilms formed on saliva-coated hydroxyapatite (HA) disks, and markedly decrease the number of *S. sanguinis* and *C. albicans* (Jiang et al. 2016). After reviewing 15 randomized controlled trials, 3 systematic reviews, and 13 meta-analyses, it can be concluded that *L. rhamnosus* GG could recover the symptoms of gastrointestinal disorders, especially acute gastroenteritis, as well as irritable bowel syndrome, and inhibit antibiotics-associated diarrhea in children (Szajewska and Hojsak 2020).

6 Conclusion and Future Perspectives

The treatment of biofilm-related infections remains a challenging problem in clinics. Recent progress in the understanding of the connection between the immune system and probiotics has posed a plausible effect of probiotics in biofilm infection therapy. Fatty acids, peroxidases, and specific bacteriocins are yielded from probiotics. This bactericidal effect has been reported in different human infections, including oral, intestinal, vaginal, and wound-related infections. The results of different in vitro studies have revealed the adhesion and release of extracellular antibiofilm factors, as well as the growth inhibitory and co-aggregatory effects of probiotics against biofilm formation. Moreover, probiotics have been able to eradicate the mature biofilm, proposing a promising role for them in counteracting biofilm-associated infections.

The current data propose the potential effects of probiotics in counteracting biofilms. However, it should be noted that scant evidence has proved their inhibitory impact against biofilm formation and/or dispersion as commercial drugs. Maintaining the viability of probiotic strains, evaluating their effectiveness, posology, and the diversity of delivery strategies are challenging issues in this context. Taking into account the tight connection between the particular cell type of the host and its related molecular signaling reprogramming with distinct bacterial strains, the strain-specific effect of probiotics should be considered. Each strain of probiotic exhibits unique features. More interestingly, a distinct strain might function differently under different situations (Trush et al. 2020). Thus, the molecular screening of probiotic strains and the competing pathogenic biofilms should be assessed at specific pathophysiological states to yield an ideal strain. Moreover, the clinical administration of probiotics requires preceding safety tests. The capability of probiotic strains to form tight adhesions with epithelial cells and mucus and their co-aggregation with the pathogens are the key criteria in selecting the ideal strain. Additionally, the selected strains should reveal high antimicrobial effects, viability, and stability. Drug resistance and the transfection of resistance-related genes from probiotics to other bacterial communities is an emerging concern in the application of probiotics against biofilms. Also, more studies are required to address the comparative efficacy of monostrain probiotics versus the multistrain ones, although the latter have shown stronger effects in the prohibition of the adhesion and proliferation of pathogenic bacteria (Trush et al. 2020).

A more comprehensive consideration regarding the molecular mechanism of action of probiotics is required to benefit from their usage, especially in the area of the capabilities of the diverse strains, their colonization, and adhesion. Nowadays, growing research focuses on the use of probiotics; however, it is not possible to reach an agreement on the optimal dose/concentration, mode of administration, and required duration for each probiotic strain. An ideal probiotic(s) with the capacity to interfere and compete with biofilm-producing pathogens in the context of specific infection and bacterial targets needs to be identified (Vuotto et al. 2014). It is also important to understand the effectiveness of each strain alone, as well as combining the synergistic effects of probiotics.

Compliance with Ethical Standards The authors declare that there are no conflicts of interest.

References

Akhtar F, Khan AU, Misba L, Akhtar K, Ali A (2021) Antimicrobial and antibiofilm photodynamic therapy against vancomycin resistant Staphylococcus aureus (VRSA) induced infection in vitro and in vivo. Eur J Pharm Biopharm 160:65–76

Alp S, Baka ZM (2018) Effects of probiotics on salivary Streptococcus mutans and Lactobacillus levels in orthodontic patients. Am J Orthod Dentofac Orthop 154:517–523

Arweiler NB, Auschill TM, Heumann C, Hellwig E, Al-Ahmad A (2020) Influence of probiotics on the salivary microflora oral streptococci and their integration into oral biofilm. Antibiotics (Basel) 9(11):803

Asahara T, Shimizu K, Takada T, Kado S, Yuki N, Morotomi M, Tanaka R, Nomoto K (2011) Protective effect of Lactobacillus casei strain Shirota against lethal infection with multi-drug resistant Salmonella enterica serovar Typhimurium DT104 in mice. J Appl Microbiol 110:163–173

Azevedo AS, Almeida C, Melo LF, Azevedo NF (2017) Impact of polymicrobial biofilms in catheter-associated urinary tract infections. Crit Rev Microbiol 43:423–439

Baelo A, Levato R, Julián E, Crespo A, Astola J, Gavaldà J, Engel E, Mateos-Timoneda MA, Torrents E (2015) Disassembling bacterial extracellular matrix with DNase-coated nanoparticles to enhance antibiotic delivery in biofilm infections. J Control Release 209:150–158

Barzegari A, Kheyrolahzadeh K, Hosseiniyan Khatibi SM, Sharifi S, Memar MY, Zununi Vahed S (2020) The battle of probiotics and their derivatives against biofilms. Infect Drug Resist 13:659–672

Ben Slama R, Kouidhi B, Zmantar T, Chaieb K, Bakhrouf A (2013) Anti-listerial and anti-biofilm activities of potential probiotic Lactobacillus strains isolated from Tunisian traditional fermented food. J Food Saf 33:8–16

Benic GZ, Farella M, Morgan XC, Viswam J, Heng NC, Cannon RD, Mei L (2019) Oral probiotics reduce halitosis in patients wearing orthodontic braces: a randomized, triple-blind, placebo-controlled trial. J Breath Res 13:036010

Bermudez-Brito M, Muñoz-Quezada S, Gomez-Llorente C, Matencio E, Bernal MJ, Romero F, Gil A (2012) Human intestinal dendritic cells decrease cytokine release against Salmonella infection in the presence of Lactobacillus paracasei upon TLR activation. PLoS One 7(8):e43197

Besser M, Terberger J, Weber L, Ghebremedhin B, Naumova E, Arnold W, Stuermer E (2019a) Impact of probiotics on pathogen survival in an innovative human plasma biofilm model (hp BIOM). J Transl Med 17:1–9

Besser M, Terberger J, Weber L, Ghebremedhin B, Naumova EA, Arnold WH, Stuermer EK (2019b) Impact of probiotics on pathogen survival in an innovative human plasma biofilm model (hpBIOM). J Transl Med 17:243

Borrell García C, Ribelles Llop M, García Esparza M, Flichy-Fernández AJ, Marqués Martínez L, Izquierdo Fort R (2021) The use of Lactobacillus reuteri DSM 17938 and ATCC PTA 5289 on oral health indexes in a school population: a pilot randomized clinical trial. Int J Immunopathol Pharmacol 35:20587384211031107

Bron PA, Kleerebezem M, Brummer R-J, Cani PD, Mercenier A, Macdonald TT, Garcia-Ródenas CL, Wells JM (2017) Can probiotics modulate human disease by impacting intestinal barrier function? Br J Nutr 117:93–107

Buggio L, Somigliana E, Borghi A, Vercellini P (2019) Probiotics and vaginal microecology: fact or fancy? BMC Womens Health 19:25

Bujalance C, Jiménez-Valera M, Moreno E, Ruiz-López M-D, Lasserrot A, Ruiz-Bravo A (2014) Lack of correlation between in vitro antibiosis and in vivo protection against enteropathogenic bacteria by probiotic lactobacilli. Res Microbiol 165:14–20

Campanella V, Syed J, Santacroce L, Saini R, Ballini A, Inchingolo F (2018) Oral probiotics influence oral and respiratory tract infections in pediatric population: a randomized double-blinded placebo-controlled pilot study. Eur Rev Med Pharmacol Sci 22:8034–8041

Carvalho FM, Teixeira-Santos R, Mergulhão FJM, Gomes LC (2020) The use of probiotics to fight biofilms in medical devices: a systematic review and meta-analysis. Microorganisms 9(1):27

Castillo-Juárez I, Maeda T, Mandujano-Tinoco EA, Tomás M, Pérez-Eretza B, García-Contreras SJ, Wood TK, García-Contreras R (2015) Role of quorum sensing in bacterial infections. World J Clin Cases 3:575

Chapman C, Gibson G, Rowland I (2014) Effects of single-and multi-strain probiotics on biofilm formation and in vitro adhesion to bladder cells by urinary tract pathogens. Anaerobe 27:71–76

Chen Q, Zhu Z, Wang J, Lopez AI, Li S, Kumar A, Yu F, Chen H, Cai C, Zhang L (2017) Probiotic E. coli Nissle 1917 biofilms on silicone substrates for bacterial interference against pathogen colonization. Acta Biomater 50:353–360

Ciofu O, Mandsberg LF, Wang H, Høiby N (2012) Phenotypes selected during chronic lung infection in cystic fibrosis patients: implications for the treatment of Pseudomonas aeruginosa biofilm infections. FEMS Microbiol Immunol 65:215–225

Corr SC, Li Y, Riedel CU, O'toole PW, Hill C, Gahan CG (2007) Bacteriocin production as a mechanism for the antiinfective activity of Lactobacillus salivarius UCC118. Proc Natl Acad Sci 104:7617–7621

Cortés-Acha B, Figueiredo R, Blanc V, Soler-Ollé A, León R, Valmaseda-Castellón E (2019) Development and viability of biofilms grown on experimental abutments mimicking dental implants: an in vivo model. Med Oral Patol Oral Cir Bucal 24:e511

Cristofori F, Dargenio VN, Dargenio C, Miniello VL, Barone M, Francavilla R (2021) Anti-inflammatory and immunomodulatory effects of probiotics in gut inflammation: a door to the body. Front Immunol 12:178

Dai C-E, Li H-L, He X-P, Zheng F-F, Zhu H-L, Liu L-F, Du W (2018) Research advance in metabolism of effective ingredients from traditional Chinese medicines by probiotics. Zhongguo Zhong Yao Za Zhi 43:31–38

De Klerk N, Maudsdotter L, Gebreegziabher H, Saroj SD, Eriksson B, Eriksson OS, Roos S, Lindén S, Sjölinder H, Jonsson A-B (2016) Lactobacilli reduce Helicobacter pylori attachment to host gastric epithelial cells by inhibiting adhesion gene expression. Infect Immun 84:1526–1535

Del Piano M, Anderloni A, Balzarini M, Ballarè M, Carmagnola S, Montino F, Orsello M, Pagliarulo M, Tari R, Soattini L, Sforza F, Mogna L, Mogna G (2012) The innovative potential of Lactobacillus rhamnosus LR06, Lactobacillus pentosus LPS01, Lactobacillus plantarum LP01, and Lactobacillus delbrueckii Subsp. delbrueckii LDD01 to restore the "gastric barrier effect" in patients chronically treated with PPI: a pilot study. J Clin Gastroenterol 46 Suppl:S18–S26

Deng Z, Luo XM, Liu J, Wang H (2020) Quorum sensing, biofilm, and intestinal mucosal barrier: involvement the role of probiotic. Front Cell Infect Microbiol 10:504

Díaz MA, González SN, Alberto MR, Arena ME (2020) Human probiotic bacteria attenuate Pseudomonas aeruginosa biofilm and virulence by quorum-sensing inhibition. Biofouling 36: 597–609

Donelli G, Vuotto C, Mastromarino P (2013) Phenotyping and genotyping are both essential to identify and classify a probiotic microorganism. Microb Ecol Health Dis 24:20105

Duraisamy V, Geethapriya PR, Bharath C, Niveditha RS, John JB (2021) Role of probiotics and synbiotics on inhibiting Streptococcus mutans level in saliva of children: a randomized controlled trial. J Indian Soc Pedod Prev Dent 39:275–278

Emara MH, Elhawari SA, Yousef S, Radwan MI, Abdel-Aziz HR (2016) Emerging role of probiotics in the management of Helicobacter pylori infection: histopathologic perspectives. Helicobacter 21:3–10

Evivie SE, Abdelazez A, Li B, Bian X, Li W, Du J, Huo G, Liu F (2019) In vitro organic acid production and in vivo food pathogen suppression by probiotic S. thermophilus and L. bulgaricus. Front Microbiol 10:782

Fang K, Jin X, Hong SH (2018) Probiotic Escherichia coli inhibits biofilm formation of pathogenic E. coli via extracellular activity of DegP. Sci Rep 8:1–12

Flichy-Fernández AJ, Ata-Ali J, Alegre-Domingo T, Candel-Martí E, Ata-Ali F, Palacio JR, Peñarrocha-Diago M (2015) The effect of orally administered probiotic Lactobacillus reuteri-containing tablets in peri-implant mucositis: a double-blind randomized controlled trial. J Periodontal Res 50:775–785

Fracchia L, Cavallo M, Allegrone G, Martinotti MG (2010) A Lactobacillus-derived biosurfactant inhibits biofilm formation of human pathogenic Candida albicans biofilm producers. Appl Microbiol Biotechnol 2:827–837

Furusawa Y, Obata Y, Fukuda S, Endo TA, Nakato G, Takahashi D, Nakanishi Y, Uetake C, Kato K, Kato T (2013) Commensal microbe-derived butyrate induces the differentiation of colonic regulatory T cells. Nature 504:446–450

Galofré M, Palao D, Vicario M, Nart J, Violant D (2018) Clinical and microbiological evaluation of the effect of Lactobacillus reuteri in the treatment of mucositis and peri-implantitis: a triple-blind randomized clinical trial. J Periodontal Res 53:378–390

Gaspar C, Donders G, Palmeira-De-Oliveira R, Queiroz J, Tomaz C, Martinez-De-Oliveira J, Palmeira-De-Oliveira A (2018) Bacteriocin production of the probiotic Lactobacillus acidophilus KS400. AMB Express 8:1–8

Gerritsen J, Smidt H, Rijkers GT, De Vos WM (2011) Intestinal microbiota in human health and disease: the impact of probiotics. Genes Nutr 6:209–240

Ghasemi A, Moosavi-Nasab M, Setoodeh P, Mesbahi G, Yousefi G (2019) Biosurfactant production by lactic acid bacterium Pediococcus dextrinicus SHU1593 grown on different carbon sources: strain screening followed by product characterization. Sci Rep 9:1–12

Gómez NC, Ramiro JM, Quecan BX, De Melo Franco BD (2016) Use of potential probiotic lactic acid bacteria (LAB) biofilms for the control of Listeria monocytogenes, Salmonella Typhimurium, and Escherichia coli O157: H7 biofilms formation. Front Microbiol 7:863

Goyal N, Shamanna PU, Varughese ST, Abraham R, Antony B, Emmatty R, Paul P (2019) Effects of amine fluoride and probiotic mouthwash on levels of Porphyromonas gingivalis in orthodontic patients: a randomized controlled trial. J Indian Soc Periodontol 23:339–344

Gruner D, Paris S, Schwendicke F (2016) Probiotics for managing caries and periodontitis: systematic review and meta-analysis. J Dent 48:16–25

Guandalini S, Pensabene L, Zikri MA, Dias JA, Casali LG, Hoekstra H, Kolacek S, Massar K, Micetic-Turk D, Papadopoulou A, De Sousa JS, Sandhu B, Szajewska H, Weizman Z (2000) Lactobacillus GG administered in oral rehydration solution to children with acute diarrhea: a multicenter European trial. J Pediatr Gastroenterol Nutr 30:54–60

Hallström H, Lindgren S, Yucel-Lindberg T, Dahlén G, Renvert S, Twetman S (2013) Effect of probiotic lozenges on inflammatory reactions and oral biofilm during experimental gingivitis. Acta Odontol Scand 71:828–833

Hallström H, Lindgren S, Widén C, Renvert S, Twetman S (2016) Probiotic supplements and debridement of peri-implant mucositis: a randomized controlled trial. Acta Odontol Scand 74: 60–66

Hantoushzadeh S, Golshahi F, Javadian P, Khazardoost S, Aram S, Hashemi S, Mirarmandehi B, Borna S (2012) Comparative efficacy of probiotic yoghurt and clindamycin in treatment of bacterial vaginosis in pregnant women: a randomized clinical trial. J Matern Fetal Neonatal Med 25:1021–1024

Hassan M, Kjos M, Nes I, Diep D, Lotfipour F (2012) Natural antimicrobial peptides from bacteria: characteristics and potential applications to fight against antibiotic resistance. J Appl Microbiol 113:723–736

He L, Yang H, Chen Z, Ouyang X (2020) The effect of Streptococcus salivarius K12 on halitosis: a double-blind, randomized, placebo-controlled trial. Probiotics Antimicrob Proteins 12:1321–1329

He Y, Na R, Niu X, Xiao B, Yang H (2021) Lactobacillus rhamnosus and Lactobacillus casei affect various stages of Gardnerella species biofilm formation. Front Cell Infect Microbiol 11:568178

Heczko PB, Tomusiak A, Adamski P, Jakimiuk AJ, Stefański G, Mikołajczyk-Cichońska A, Suda-Szczurek M, Strus M (2015) Supplementation of standard antibiotic therapy with oral probiotics for bacterial vaginosis and aerobic vaginitis: a randomised, double-blind, placebo-controlled trial. BMC Womens Health 15:1–12

Hrynyshyn A, Simoes M, Borges A (2022) Antibiotics (Basel). In: Biofilms in surgical site infections: recent advances and novel prevention and eradication strategies, p 11

Huang F-C, Lu Y-T, Liao Y-H (2020) Beneficial effect of probiotics on Pseudomonas aeruginosa–infected intestinal epithelial cells through inflammatory IL-8 and antimicrobial peptide human beta-defensin-2 modulation. Innate Immun 26:592–600

İnce G, Gürsoy H, İpçi ŞD, Cakar G, Emekli-Alturfan E, Yılmaz S (2015) Clinical and biochemical evaluation of lozenges containing Lactobacillus reuteri as an adjunct to non-surgical periodontal therapy in chronic periodontitis. J Periodontol 86:746–754

Iniesta M, Herrera D, Montero E, Zurbriggen M, Matos AR, Marín MJ, Sánchez-Beltrán MC, Llama-Palacio A, Sanz M (2012) Probiotic effects of orally administered Lactobacillus reuteri-

containing tablets on the subgingival and salivary microbiota in patients with gingivitis. A randomized clinical trial. J Clin Periodontol 39:736–744
Invernici MM, Salvador SL, Silva PHF, Soares MSM, Casarin R, Palioto DB, Souza SLS, Taba M Jr, Novaes AB Jr, Furlaneto FAC, Messora MR (2018) Effects of Bifidobacterium probiotic on the treatment of chronic periodontitis: a randomized clinical trial. J Clin Periodontol 45:1198–1210
Invernici MM, Furlaneto FAC, Salvador SL, Ouwehand AC, Salminen S, Mantziari A, Vinderola G, Ervolino E, Santana SI, Silva PHF, Messora MR (2020) Bifidobacterium animalis subsp lactis HN019 presents antimicrobial potential against periodontopathogens and modulates the immunological response of oral mucosa in periodontitis patients. PLoS One 15:e0238425
Ishikawa KH, Mita D, Kawamoto D, Nicoli JR, Albuquerque-Souza E, Lorenzetti Simionato MR, Mayer MPA (2020) Probiotics alter biofilm formation and the transcription of Porphyromonas gingivalis virulence-associated genes. J Oral Microbiol 12:1805553
Islam SU (2016) Clinical uses of probiotics. Medicine 95(5):e2658
Jiang Q, Stamatova I, Kainulainen V, Korpela R, Meurman JH (2016) Interactions between Lactobacillus rhamnosus GG and oral micro-organisms in an in vitro biofilm model. BMC Microbiol 16:1–11
Kaklamanos EG, Nassar R, Kalfas S, Al Halabi M, Kowash M, Hannawi H, Hussein I, Salami A, Hassan A, Senok AC (2019) A single-centre investigator-blinded randomised parallel group clinical trial to investigate the effect of probiotic strains Streptococcus salivarius M18 and Lactobacillus acidophilus on gingival health of paediatric patients undergoing treatment with fixed orthodontic appliances: study protocol. BMJ Open 9:e030638
Kang MS, Lee DS, Kim M, Lee SA, Nam SH (2021) A randomized, double-blind, placebo-controlled trial to assess the acidogenic potential of dental biofilms through a tablet containing Weissella cibaria CMU. Int J Environ Res Public Health 18(9):4674
Karadeniz DG, Kaskatepe B, Kiymaci ME, Tok KC, Gumustas M, Karaaslan C (2021) Microbial exopolysaccharide production of Streptococcus thermophilus and its antiquorum sensing activity. Arch Microbiol 203(6):1–9
Karczewski J, Troost FJ, Konings I, Dekker J, Kleerebezem M, Brummer R-JM, Wells JM (2010) Regulation of human epithelial tight junction proteins by Lactobacillus plantarum in vivo and protective effects on the epithelial barrier. Am J Physiol Gastrointest Liver Physiol 298:G851–G859
Khatoon Z, Mctiernan CD, Suuronen EJ, Mah TF, Alarcon EI (2018) Bacterial biofilm formation on implantable devices and approaches to its treatment and prevention. Heliyon 4:e01067
Kim Y, Kim SH (2009) Released exopolysaccharide (r-EPS) produced from probiotic bacteria reduce biofilm formation of enterohemorrhagic Escherichia coli O157: H7. Biochem Biophys Res Commun 379:324–329
Kiymaci ME, Altanlar N, Gumustas M, Ozkan SA, Akin A (2018) Quorum sensing signals and related virulence inhibition of Pseudomonas aeruginosa by a potential probiotic strain's organic acid. Microb Pathog 121:190–197
Koo H, Xiao J, Klein M, Jeon J (2010) Exopolysaccharides produced by Streptococcus mutans glucosyltransferases modulate the establishment of microcolonies within multispecies biofilms. J Bacteriol 192(12):3024–3032
Lagrafeuille R, Miquel S, Balestrino D, Vareille-Delarbre M, Chain F, Langella P, Forestier C (2018) Opposing effect of Lactobacillus on in vitro Klebsiella pneumoniae in biofilm and in an in vivo intestinal colonisation model. Benefic Microbes 9:87–100
Lai S, Lingström P, Cagetti MG, Cocco F, Meloni G, Arrica MA, Campus G (2021) Effect of Lactobacillus brevis CD2 containing lozenges and plaque pH and cariogenic bacteria in diabetic children: a randomised clinical trial. Clin Oral Investig 25:115–123
Lau LYJ, Chye FY (2018) Antagonistic effects of Lactobacillus plantarum 0612 on the adhesion of selected foodborne enteropathogens in various colonic environments. Food Control 91:237–247

Lee DS, Lee SA, Kim M, Nam SH, Kang MS (2020) Reduction of halitosis by a tablet containing Weissella cibaria CMU: a randomized, double-blind, placebo-controlled study. J Med Food 23: 649–657

Lee DS, Kim M, Nam SH, Kang MS, Lee SA (2021a) Effects of Oral probiotics on subjective halitosis, oral health, and psychosocial health of college students: a randomized, double-blind. Int J Environ Res Public Health, Placebo-Controlled Study, p 18

Lee JE, Lee NK, Paik HD (2021b) Antimicrobial and anti-biofilm effects of probiotic Lactobacillus plantarum KU200656 isolated from kimchi. Food Sci Biotechnol 30:97–106

Liu J, Liu Z, Huang J, Tao R (2021) Effect of probiotics on gingival inflammation and oral microbiota: a meta-analysis. Oral Dis 28(4):1058–1067

Lkhagvadorj E, Nagata S, Wada M, Bian L, Wang C, Chiba Y, Yamashiro Y, Shimizu T, Asahara T, Nomoto K (2010) Anti-infectious activity of synbiotics in a novel mouse model of methicillin-resistant Staphylococcus aureus infection. Microbiol Immunol 54:265–275

Lorenzo B, Luca S, Antonio M, Alberto DM, Cesare F, Omar C (2020) Effects of probiotics in the management of infected chronic wounds: from cell culture to human studies. Curr Clin Pharmacol 15:193–206

Luan C, Jiang N, Zhou X, Zhang C, Zhao Y, Li Z, Li C (2022) Antibacterial and anti-biofilm activities of probiotic Lactobacillus curvatus BSF206 and Pediococcus pentosaceus AC1-2 against Streptococcus mutans. Microb Pathog 164:105446

Manna S, Ghosh AK, Mandal SM (2019) Curd-peptide based novel hydrogel inhibits biofilm formation, quorum sensing, swimming mortility of multi-antibiotic resistant clinical isolates and accelerates wound healing activity. Front Microbiol 10:951

Megha S, Shalini G, Varsha SA, Abhishek D, Neetu J (2019) Effect of short-term placebo-controlled consumption of probiotic yoghurt and Indian curd on the Streptococcus mutans level in children undergoing fixed interceptive orthodontic therapy. Turk J Orthod 32:16–21

Melo TA, Dos Santos TF, De Almeida ME, Junior LAGF, Andrade EF, Rezende RP, Marques LM, Romano CC (2016) Inhibition of Staphylococcus aureus biofilm by Lactobacillus isolated from fine cocoa. BMC Microbiol 16:1–9

Memar MY, Raei P, Alizadeh N, Aghdam MA, Kafil HS (2017) Carvacrol and thymol: strong antimicrobial agents against resistant isolates. Rev Med Microbiol 28:63–68

Memar MY, Yekani M, Alizadeh N, Baghi HB (2019) Hyperbaric oxygen therapy: antimicrobial mechanisms and clinical application for infections. Biomed Pharmacother 109:440–447

Memar MY, Yekani M, Ghanbari H, Shahi S, Sharifi S, Maleki Dizaj S (2020) Biocompatibility, cytotoxicity and antibacterial effects of meropenem-loaded mesoporous silica nanoparticles against carbapenem-resistant Enterobacteriaceae. Artif Cells Nanomed Biotechnol 48:1354–1361

Missaoui J, Saidane D, Mzoughi R, Minervini F (2019) Fermented seeds ("Zgougou") from Aleppo pine as a novel source of potentially probiotic lactic acid bacteria. Microorganisms 7:709

Molina-Manso D, Del Prado G, Ortiz-Pérez A, Manrubia-Cobo M, Gómez-Barrena E, Cordero-Ampuero J, Esteban J (2013) In vitro susceptibility to antibiotics of staphylococci in biofilms isolated from orthopaedic infections. Int J Antimicrob Agents 41:521–523

Morales A, Gandolfo A, Bravo J, Carvajal P, Silva N, Godoy C, Garcia-Sesnich J, Hoare A, Diaz P, Gamonal J (2018) Microbiological and clinical effects of probiotics and antibiotics on nonsurgical treatment of chronic periodontitis: a randomized placebo- controlled trial with 9-month follow-up. J Appl Oral Sci 26:e20170075

Morales A, Contador R, Bravo J, Carvajal P, Silva N, Strauss FJ, Gamonal J (2021) Clinical effects of probiotic or azithromycin as an adjunct to scaling and root planning in the treatment of stage III periodontitis: a pilot randomized controlled clinical trial. BMC Oral Health 21:12

Mortada M, Cosby D, Shanmugasundaram R, Selvaraj R (2020) In vivo and in vitro assessment of commercial probiotic and organic acid feed additives in broilers challenged with Campylobacter coli. J Appl Poult Res 29:435–446

Moskowitz SM, Emerson JC, Mcnamara S, Shell RD, Orenstein DM, Rosenbluth D, Katz MF, Ahrens R, Hornick D, Joseph PM (2011) Randomized trial of biofilm testing to select antibiotics for cystic fibrosis airway infection. Pediatr Pulmonol 46:184–192

Nadelman P, Magno MB, Masterson D, Da Cruz AG, Maia LC (2018) Are dairy products containing probiotics beneficial for oral health? A systematic review and meta-analysis. Clin Oral Investig 22:2763–2785

Nędzi-Góra M, Wróblewska M, Górska R (2020) The effect of Lactobacillus salivarius SGL03 on clinical and microbiological parameters in periodontal patients. Pol J Microbiol 69:441–451

Ng SC, Hart AL, Kamm MA, Stagg AJ, Knight SC (2009) Mechanisms of action of probiotics: recent advances. Inflamm Bowel Dis 15:300–310

Noshak MA, Rezaee MA, Hasani A, Mirzaii M, Memar MY (2019) Biofilm formation capacity in common SCCmec types of coagulase-negative staphylococci isolated from hospitalized patients and health-care workers in northwest of Iran. Gene Rep 17:100531

Ouyang J, Sun F, Feng W, Sun Y, Qiu X, Xiong L, Liu Y, Chen Y (2016) Quercetin is an effective inhibitor of quorum sensing, biofilm formation and virulence factors in Pseudomonas aeruginosa. J Appl Microbiol 120:966–974

Paredes J, Alonso-Arce M, Schmidt C, Valderas D, Sedano B, Legarda J, Arizti F, Gómez E, Aguinaga A, Del Pozo J (2014) Smart central venous port for early detection of bacterial biofilm related infections. Biomed Microdevices 16:365–374

Pendharkar S, Brandsborg E, Hammarström L, Marcotte H, Larsson P-G (2015) Vaginal colonisation by probiotic lactobacilli and clinical outcome in women conventionally treated for bacterial vaginosis and yeast infection. BMC Infect Dis 15:1–12

Pietri FK, Rossouw PE, Javed F, Michelogiannakis D (2020) Role of probiotics in oral health maintenance among patients undergoing fixed orthodontic therapy: a systematic review of randomized controlled clinical trials. Probiotics Antimicrob Proteins 12:1349–1359

Pinto G, Cenci M, Azevedo M, Epifanio M, Jones M (2014) Effect of yogurt containing Bifidobacterium animalis subsp. lactis DN-173010 probiotic on dental plaque and saliva in orthodontic patients. Caries Res 48:63–68

Radaic A, Ye C, Parks B, Gao L, Kuraji R, Malone E, Kamarajan P, Zhan L, Kapila YL (2020) Modulation of pathogenic oral biofilms towards health with nisin probiotic. J Oral Microbiol 12: 1809302

Ramos AN, Sesto Cabral ME, Noseda D, Bosch A, Yantorno OM, Valdez JC (2012) Antipathogenic properties of Lactobacillus plantarum on Pseudomonas aeruginosa: the potential use of its supernatants in the treatment of infected chronic wounds. Wound Repair Regen 20: 552–562

Rijkers GT, Bengmark S, Enck P, Haller D, Herz U, Kalliomaki M, Kudo S, Lenoir-Wijnkoop I, Mercenier A, Myllyluoma E, Rabot S, Rafter J, Szajewska H, Watzl B, Wells J, Wolvers D, Antoine JM (2010) Guidance for substantiating the evidence for beneficial effects of probiotics: current status and recommendations for future research. J Nutr 140:671s–676s

Rodrigues L, Van Der Mei H, Teixeira JA, Oliveira R (2004) Biosurfactant from Lactococcus lactis 53 inhibits microbial adhesion on silicone rubber. Appl Microbiol Biotechnol 66:306–311

Romani Vestman N, Chen T, Lif Holgerson P, Öhman C, Johansson I (2015) Oral microbiota shift after 12-week supplementation with Lactobacillus reuteri DSM 17938 and PTA 5289; a randomized control trial. PLoS One 10:e0125812

Rossitto M, Fiscarelli EV, Rosati P (2018) Challenges and promises for planning future clinical research into bacteriophage therapy against Pseudomonas aeruginosa in cystic fibrosis. An argumentative review. Front Microbiol 9:775

Round JL, Mazmanian SK (2009) The gut microbiota shapes intestinal immune responses during health and disease. Nat Rev Immunol 9:313–323

Russell JB, Diez-Gonzalez F (1997) The effects of fermentation acids on bacterial growth. Adv Microb Physiol 39:205–234

Sakarya S, Gunay N (2014) Saccharomyces boulardii expresses neuraminidase activity selective for α2, 3-linked sialic acid that decreases Helicobacter pylori adhesion to host cells. APMIS 122: 941–950

Salas-Jara M, Sanhueza E, Retamal-Díaz A, González C, Urrutia H, García A (2016a) Probiotic Lactobacillus fermentum UCO-979C biofilm formation on AGS and Caco-2 cells and Helicobacter pylori inhibition. Biofouling 32:1245–1257

Salas-Jara MJ, Ilabaca A, Vega M, García A (2016b) Microorganisms. In: Biofilm forming lactobacillus: new challenges for the development of probiotics, vol 4

Salas-Jara MJ, Ilabaca A, Vega M, García A (2016c) Biofilm forming Lactobacillus: new challenges for the development of probiotics. Microorganisms 4:35

Sang-Ngoen T, Czumbel LM, Sadaeng W, Mikó A, Németh DI, Mátrai P, Hegyi P, Tóth B, Csupor D, Kiss I, Szabó A, Gerber G, Varga G, Kerémi B (2021) Orally administered probiotics decrease Aggregatibacter actinomycetemcomitans but not other periodontal pathogenic bacteria counts in the oral cavity: a systematic review and meta-analysis. Front Pharmacol 12:682656

Santacroce L, Charitos IA, Bottalico L (2019) A successful history: probiotics and their potential as antimicrobials. Expert Rev Anti-Infect Ther 17:635–645

Shah SS, Nambiar S, Kamath D, Suman E, Unnikrishnan B, Desai A, Mahajan S, Dhawan KK (2019) Comparative evaluation of plaque inhibitory and antimicrobial efficacy of probiotic and chlorhexidine Oral rinses in orthodontic patients: a randomized clinical trial. Int J Dent 2019: 1964158

Sikorska H, Smoragiewicz W (2013) Role of probiotics in the prevention and treatment of methicillin-resistant Staphylococcus aureus infections. Int J Antimicrob Agents 42:475–481

Silva DR, Sardi JDCO, De Souza Pitangui N, Roque SM, Da Silva ACB, Rosalen PL (2020) Probiotics as an alternative antimicrobial therapy: current reality and future directions. J Funct Foods 73:104080

Singh A, Amod A, Pandey P, Bose P, Pingali MS, Shivalkar S, Varadwaj P, Sahoo A, Samanta S (2022) Bacterial biofilm infections, their resistance to antibiotics therapy and current treatment strategies. Biomed Mater 17(2):022003

Stanbro J, Park JM, Bond M, Stockelman MG, Simons MP, Watters C (2020) Topical delivery of lactobacillus culture supernatant increases survival and wound resolution in traumatic Acinetobacter baumannii infections. Probiotics Antimicrob Proteins 12:809–818

Suez J, Zmora N, Segal E, Elinav E (2019) The pros, cons, and many unknowns of probiotics. Nat Med 25:716–729

Szajewska H, Hojsak I (2020) Health benefits of Lactobacillus rhamnosus GG and Bifidobacterium animalis subspecies lactis BB-12 in children. Postgrad Med 132:441–451

Teughels W, Durukan A, Ozcelik O, Pauwels M, Quirynen M, Haytac MC (2013) Clinical and microbiological effects of Lactobacillus reuteri probiotics in the treatment of chronic periodontitis: a randomized placebo-controlled study. J Clin Periodontol 40:1025–1035

Thomas CM, Versalovic J (2010) Probiotics-host communication: modulation of signaling pathways in the intestine. Gut Microbes 1:148–163

Trush EA, Poluektova EA, Beniashvilli AG, Shifrin OS, Poluektov YM, Ivashkin VT (2020) The evolution of human probiotics: challenges and prospects. Probiotics Antimicrob Proteins 12: 1291–1299

Tsai WH, Chou CH, Chiang YJ, Lin CG, Lee CH (2021) Regulatory effects of Lactobacillus plantarum-GMNL6 on human skin health by improving skin microbiome. Int J Med Sci 18: 1114–1120

Uruén C, Chopo-Escuin G, Tommassen J, Mainar-Jaime RC, Arenas J (2021) Biofilms as promoters of bacterial antibiotic resistance and tolerance. Antibiotics 10:3

Valdez J, Peral M, Rachid M, Santana M, Perdigon G (2005) Interference of Lactobacillus plantarum with Pseudomonas aeruginosa in vitro and in infected burns: the potential use of probiotics in wound treatment. Clin Microbiol Infect 11:472–479

Vicariotto F, Mogna L, Del Piano M (2014) Effectiveness of the two microorganisms Lactobacillus fermentum LF15 and Lactobacillus plantarum LP01, formulated in slow-release vaginal tablets,

in women affected by bacterial vaginosis: a pilot study. J Clin Gastroenterol 48(Suppl 1):S106–S112
Vieira AT, Teixeira MM, Martins FDS (2013) The role of probiotics and prebiotics in inducing gut immunity. Front Immunol 4:445
Vieira AT, Rocha VM, Tavares L, Garcia CC, Teixeira MM, Oliveira SC, Cassali GD, Gamba C, Martins FS, Nicoli JR (2016) Control of Klebsiella pneumoniae pulmonary infection and immunomodulation by oral treatment with the commensal probiotic Bifidobacterium longum 51A. Microbes Infect 18:180–189
Vuotto C, Longo F, Donelli G (2014) Probiotics to counteract biofilm-associated infections: promising and conflicting data. Int J Oral Sci 6:189–194
Walencka E, Różalska S, Sadowska B, Różalska B (2008) The influence of Lactobacillus acidophilus-derived surfactants on staphylococcal adhesion and biofilm formation. Folia Microbiol 53:61–66
Walsham AD, Mackenzie DA, Cook V, Wemyss-Holden S, Hews CL, Juge N, Schüller S (2016) Lactobacillus reuteri inhibition of enteropathogenic Escherichia coli adherence to human intestinal epithelium. Front Microbiol 7:244
Wasfi R, Abd El-Rahman OA, Zafer MM, Ashour HM (2018) Probiotic Lactobacillus sp. inhibit growth, biofilm formation and gene expression of caries-inducing Streptococcus mutans. J Cell Mol Med 22:1972–1983
Williams NT (2010) Probiotics. Am J Health Syst Pharm 67:449–458
Wu H, Moser C, Wang H-Z, Høiby N, Song Z-J (2015) Strategies for combating bacterial biofilm infections. Int J Oral Sci 7:1–7
Ya W, Reifer C, Miller LE (2010) Efficacy of vaginal probiotic capsules for recurrent bacterial vaginosis: a double-blind, randomized, placebo-controlled study. Am J Obstet Gynecol 203: 120.e1–120.e6
Yan X, Gu S, Cui X, Shi Y, Wen S, Chen H, Ge J (2019) Antimicrobial, anti-adhesive and anti-biofilm potential of biosurfactants isolated from Pediococcus acidilactici and Lactobacillus plantarum against Staphylococcus aureus CMCC26003. Microb Pathog 127:12–20
Zare Javid A, Amerian E, Basir L, Ekrami A, Haghighizadeh MH, Maghsoumi-Norouzabad L (2020) Effects of the consumption of probiotic yogurt containing Bifidobacterium lactis Bb12 on the levels of Streptococcus mutans and Lactobacilli in saliva of students with initial stages of dental caries: a double-blind randomized controlled trial. Caries Res 54:68–74

Clinical Translation of Biofilm Dispersal Agents

Whitni Redman, Derek Fleming, Karin Sauer, and Kendra Rumbaugh

Abstract The global problem of chronic infection is enormous and increasing. Combined with the challenge that antibiotic resistance already poses, the task of developing new agents that address chronic infection is arduous. As biofilms are a major strategy used by microbes to evade the immune response and antibiotics, developing therapeutics that prevent or destroy biofilms has become a major focus for many researchers. One promising anti-biofilm strategy is to release or disperse cells from the biofilm, which better exposes them to antimicrobial agents. Dispersal can be accomplished with a variety of agents that can initiate either active or passive dispersal, and many of these agents are in preclinical or clinical development. Yet, significant hurdles exist for the clinical translation of biofilm dispersal agents. Regulatory requirements are ambiguous and daunting and can present major road-blocks. There are also real therapeutic risks for dispersing bacteria during active infection, including sepsis and even death. However, proof of concept studies have provided evidence that dispersal can be combined with antibiotics for an effective and safe treatment. As our understanding of dispersal increases, better models are developed, and dispersal agents are refined, it is likely that we will achieve more success in translating the strategy of biofilm dispersal into the clinic.

W. Redman · K. Sauer
Department of Biological Sciences, Binghamton University, Binghamton, NY, USA

Binghamton Biofilm Research Center, Binghamton University, Binghamton, NY, USA

D. Fleming
Division of Clinical Microbiology, Department of Laboratory Medicine and Pathology, Mayo Clinic, Rochester, MN, USA

K. Rumbaugh (✉)
Department of Surgery, Texas Tech University Health Sciences Center, Lubbock, TX, USA

Immunology and Molecular Microbiology, Texas Tech University Health Sciences Center, Lubbock, TX, USA

Burn Center of Research Excellence, Texas Tech University Health Sciences Center, Lubbock, TX, USA
e-mail: Kendra.Rumbaugh@ttuhsc.edu

K. Richter, K. Nørskov Kragh (eds.), *Antibiofilm Strategies*, Springer Series on Biofilms 11, https://doi.org/10.1007/978-3-031-10992-8_6

1 Introduction

Biofilms are communities of microorganisms that are protected by a hydrated matrix composed of exopolysaccharides, extracellular DNA (eDNA), proteins, lipids, and various host components, collectively called the extracellular polymeric substance (EPS) (Flemming et al. 2007). In vitro studies suggest the EPS represents approximately 85% of total biofilm biomass (Flemming et al. 2007). Biofilms can include single or multiple microbial species and develop on biotic and abiotic surfaces in natural, medical, and industrial settings (Lopez et al. 2010). In humans, biofilms account for up to 80% of chronic tissue- and device-related infections, including chronic wound infections, urinary tract infections, lung infections in cystic fibrosis (CF) patients, endocarditis, prosthetic joint infections, and many more (Davies 2003; Omar et al. 2017; Hatt and Rather 2008; Costerton 2001; Elgharably et al. 2016; Gbejuade et al. 2015). As the development of a biofilm is thought to confer tolerance to some antibiotics and protect bacteria against various host defense mechanisms (Flemming and Wingender 2010; Solano et al. 2014), they represent a major challenge for health care professionals. In the current era of increasing antibiotic resistance and dwindling new agents in the pipeline, new strategies are needed to combat these infections. This has led many investigators to propose different strategies for preventing or disrupting biofilms (Fleming and Rumbaugh 2017).

Anti-biofilm agents typically fall into one of three broad categories: agents that prevent bacterial attachment, those that prevent or interrupt biofilm maturation, and those that disrupt or degrade mature biofilms. Prevention of bacterial attachment and/or biofilm maturation are ideal because, if successful, they avoid the biofilm problem altogether. Nevertheless, these prophylactic agents are not realistic for many types of infection. For example, a catheter made from a substrate that resists bacterial attachment might be excellent for preventing catheter-associated biofilm infections, but a different strategy is required to treat a chronic wound that has been infected for months and likely contains mature, polymicrobial biofilm. Halting biofilm maturation allows for generally more effective treatment of less-protected, more metabolically active microbial cells (Ciofu et al. 2017; Wu et al. 2015). However, biofilm maturation is a dynamic and complex process that differs from species to species, and thus clinical application of broadly effective strategies to limit maturation is highly challenging. Therefore, to tackle mature biofilm present at infection sites, strategies that disrupt biofilm and disperse bacterial cells are needed.

In this chapter, we discuss the principles and prospects for biofilm dispersal as a therapeutic strategy. We first give an overview of the different broad processes involved in biofilm dispersal and provide examples of agents that have been shown to induce dispersal. Next, we discuss in greater detail dispersal agents that are in preclinical and clinical development, describing their proposed mechanisms of action and studies that have been performed. Lastly, we provide perspective on the hurdles these agents face for future clinical development.

2 Biofilm Dispersal

Dispersal, the release of microbial cells from a biofilm, can occur by a number of processes (Rumbaugh and Sauer 2020). From the perspective of the microbe, these processes can be either passive or active. For example, the passive release of cells could occur from the shear stress of liquid running over the biofilm, creating a force that causes cells to detach. Conversely, active dispersal is a process initiated by the cells in response to specific environmental cues and can occur via different mechanisms that involve cell signaling pathways (Rumbaugh and Sauer 2020). As biofilms are so difficult to remove, and the microbes within them are so difficult to kill, it is not surprising that therapeutic agents that aim to exploit biofilm dispersal have been developed. In this section, we provide an overview of active and passive biofilm dispersal agents that have promising potential for therapeutic use.

2.1 Active Biofilm Dispersal

Active dispersal requires a signal that leads to a dispersal event (Fig. 1). This signal can be a trigger from the microbes themselves or an environmental cue that induces dispersion (Fleming and Rumbaugh 2017). Mediators of active dispersal include environmental cues, fatty acid signals, peptide signals, nitric oxide (NO), c-di-GMP, Pseudomonas quinolone signal (PQS), and acyl-homoserine lactones (AHLs). It should be noted that many of these agents have complex effects on biofilms and dispersal. While a detailed review of these effects falls beyond this chapter's scope, the multifactorial effects of these agents should be acknowledged.

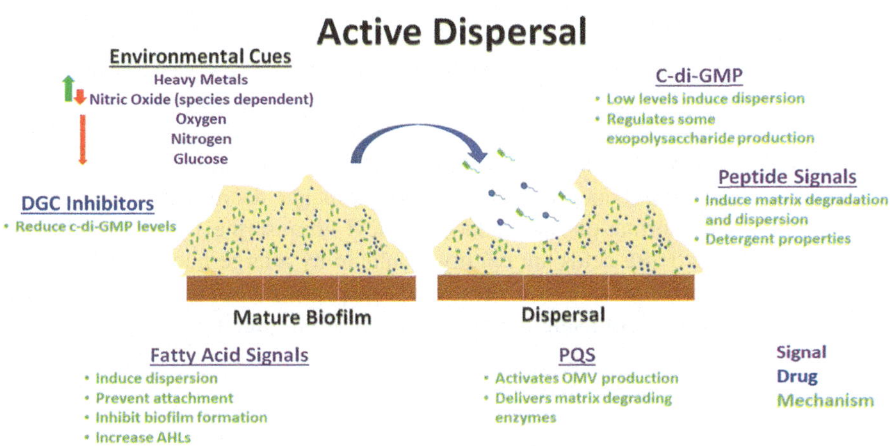

Fig. 1 Examples of active dispersal mediators

2.1.1 Environmental Cues

Environmental cues that induce bacterial dispersal include heavy metals, carbon substrate availability, oxygen depletion, and lack of nutrients. Induced dispersion has yet to be fully understood, but typically a nutrient-depleted environment encourages biofilm formation (Walters 3rd et al. 2003), while the opposite encourages cells to become planktonic (Uppuluri and Lopez-Ribot 2016). For example, a previous study investigated the effect of oxygen depletion on *Pseudomonas putida* biofilms. The depletion of oxygen resulted in an increase in calcium concentrations in the biofilm, inducing catastrophic sloughing events in vitro as compared to biofilms in an oxygen-rich environment (Applegate and Bryers 1991). It is suggested that both the availability, as well as an increase of nutrients in the environment, activate expression of motility-associated genes, e.g., those encoding flagella, and a decrease in the production of adhesion structures such as pili (Sauer et al. 2004).

Some environmental cues induce dispersal as a protective mechanism for the bacteria to escape toxicity. One example of a toxic dispersion signal is an increase in heavy metals. An increase in heavy metals in the environment can result in biofilm dispersal as a protective mechanism through the sensor BdlA, which has a downstream effect on c-di-GMP degradation, therefore inducing dispersal, similar to nutrients and NO dispersion cues (Barraud et al. 2009). Interestingly, other dispersal cues such as hypoxia or low levels of NO both potentiate biofilm dispersal and also affect c-di-GMP levels (Uppuluri and Lopez-Ribot 2016), suggesting c-di-GMP may have an important role in nutrient-induced dispersal.

2.1.2 Fatty Acid Signals

Fatty acids act as bacterial detergents at high concentrations as well as inhibit bacterial colonization at a lower concentration (Kumar et al. 2020; Marques et al. 2015). One prime example of a fatty acid signal is cis-2-decenoic acid (cis-DA) produced by *Pseudomonas aeruginosa*. Cis-DA has been shown to induce dispersion of Gram-positive and Gram-negative bacteria, as well as yeast (Davies and Marques 2009). This fatty acid signaling molecule has also been shown to revert persister cells to their planktonic counterpart, improving antimicrobial susceptibility (Marques et al. 2014). Another group demonstrated that the application of the fatty acids palmitoleic (PoA) and myristoleic (MoA) acids decreased *Acinetobacter baumannii* biofilm formation and reduced *N*-acyl-homoserine lactone production (Nicol et al. 2018). Fatty acids are readily available by a large number of microorganisms and have been shown to effectively disperse biofilms at relatively low levels, making them a promising therapeutic for biofilm infections.

2.1.3 Peptide Signals

Peptide signals, produced mostly by Gram-positive bacteria, have been shown to induce biofilm dispersal. Staphylococci produce two peptide signals, in particular, δ toxin and the autoinducing peptide I (AIP- I). The δ toxin is produced by both *S. aureus and Staphylococcus epidermidis* and is involved in detachment of cells for dispersal (Rudkin et al. 2017). δ toxin falls into the family of phenol-soluble modulins (PSMs) that are small, amphipathic peptides that have detergent-like properties (Otto 2014). *S. aureus* produces AIP-I, which also exhibits dispersal activity. AIP-I is expressed through the *agr* quorum sensing system of *S. aureus* (Canovas et al. 2016). Previous studies have shown the exogenous addition of AIP triggered detachment and increased levels of serine proteases, suggesting a protease-mediated mechanism (Boles and Horswill 2008). In particular, AIP-I upregulates the expression of the proteases aureolysin and Sp1, and both are involved in matrix degradation (Boles and Horswill 2008). Given these findings, peptide signals could be promising dispersal agents as they exhibit the ability to induce biofilm detachment.

2.1.4 Nitric Oxide

NO is a well-known signaling molecule in both prokaryotes and eukaryotes. It has been previously demonstrated that an increase in the level of NO can induce bacterial dispersal (Barraud et al. 2006) by stimulating c-di-GMP phosphodiesterase (PDEs) activity, which inhibits c-di-GMP biosynthesis (Valentini and Filloux 2016). In *P. aeruginosa*, the PDEs DipA and RbdA have been shown to be essential for biofilm dispersion (Roy et al. 2012; Liu et al. 2018a). However, NO may play contradicting roles in biofilm formation and dispersal depending upon the species. For example, NO inhibits biofilm dispersal for *Bacillus subtilis*, but increases motility and reduces biofilm formation for *P. putida* (Schreiber et al. 2011; Liu et al. 2012). Whereas low levels of NO induces *P. aeruginosa* detachment, and high levels enhance biofilm formation (Barraud et al. 2006; Williams and Boon 2019).

2.1.5 Cyclic Dimeric GMP

For over 25 years, c-di-GMP has been recognized as a vital secondary messenger that is associated with biofilm formation, motility, virulence, cellular differentiation, cell cycle regulation and biofilm dispersal (Romling et al. 2013). C-di-GMP is synthesized by diguanylate cyclases (DGC) and has been shown to regulate the biosynthesis of the poly-β-1,6-*N*-acetylglucosamine (PNAG) exopolysaccharide, which is important in the biofilm matrixome of various bacteria including *Yersinia pestis* and *Escherichia coli* (Cho et al. 2020; Yakandawala et al. 2011). While *S. aureus* and *S. epidermidis* also produce PNAG, it is not regulated by c-di-GMP in this genus (Gotz 2002). C-di-GMP also regulates the production of the *P. aeruginosa* exopolysaccharides, Pel and Psl (Colvin et al. 2012). The structures of fimbriae and pili, which assist in the initial attachment of planktonic cells to a surface or each other, are additionally regulated by c-di-GMP (Berne et al. 2018). Lastly, low levels of c-di-GMP have been shown to induce cell lysis, which increases the amount of eDNA available to be utilized for the biofilm matrix, as well as biofilm dispersal (Ha and O'Toole 2015). For these reasons, c-di-GMP is an attractive target for inducing biofilm dispersal and reducing biofilm formation. The earliest described c-di-GMP inhibitor, glycosylated triterpenoid saponin, came from extracts of a garden pea (*Pisum sativum*) and inhibited *Gluconacetobacter xylinus* DGC (Ohana et al. 1998). More recently, the *Vibrio cholerae* small molecules DI-3-N-(4-anilinophenyl)benzamide and DI-10-N-{[(2-phenylethyl) amino] carbonothioyl}benzamide have been described to reduce both *P. aeruginosa* and *V. cholerae* biofilms (Sambanthamoorthy et al. 2012). These compounds act as antagonists to DGC enzymes, and therefore decrease c-di-GMP levels. For these reasons, DGC inhibitors show promising potential as dispersal therapeutic agents.

2.1.6 Pseudomonas Quinolone Signal

PQS is important for cell-cell communication in *Pseudomonas* biofilms. However, the hydrophobicity of the PQS structure can impede dissemination to other bacterial cells. To combat this, PQS has been shown to actively induce outer membrane vesicle (OMV) formation in *P. aeruginosa* (Cooke et al. 2020). OMVs have displayed importance in cell-cell communication, transfer of antibiotic resistance genes, virulence factor delivery, and biofilm development (Schwechheimer and Kuehn 2015). PQS and OMV production is elevated during biofilm dispersal, and this elevation is thought to assist bacterial escape from the biofilm by coordinating a controlled degradation of the matrix (Mashburn and Whiteley 2005). OMVs assist in this process by trafficking matrix-degrading enzymes, such as protease, lipase, and DNase (Cooke et al. 2020). Theoretically, the application of PQS could be used as a therapeutic agent by inducing OMV production, leading to EPS degradation and biofilm dispersal (Cooke et al. 2020).

2.1.7 Acyl-Homoserine Lactones

AHLs are Gram-negative species-specific quorum sensing (QS) signals (Papenfort and Bassler 2016). Depending on the species, AHLs can be involved in the regulation of genes that are required for surface attachment, biofilm development and maintenance, and dispersal (Kjelleberg and Molin 2002; Stanley and Lazazzera 2004). Multiple AHLs have been described to induce dispersion such as 7,8-*cis*-*N*-(tetradecenoyl) homoserine lactone (Puskas et al. 1997), *N*-3-oxo-dodecanoyl homoserine lactone (3-oxo-C12-HSL) (Lade et al. 2014), and *N*-butanoyl-L-homoserine lactone (C4-HSL) (Wood et al. 2018) produced by *Rhodobacter sphaeroides, P. aeruginosa*, and *Serratia marscecens* and *P. aeruginosa*, respectively. In *P. aeruginosa* biofilms, the exogenous addition of C4-HSL induced biofilm dispersion by upregulating the gene *rhlA* that encodes a rhamnolipid production, which is connected to dispersal and negatively regulating the production of exopolysaccharides (Davey et al. 2003; McDougald et al. 2011). When *P. aeruginosa* is deficient in the production of C4-HSL, these biofilms do not undergo central hollowing, a key mechanism in dispersion. Additionally, 3-oxo-C12-HSL was shown to inhibit *S. epidermidis* growth and biofilm formation, allowing Pseudomonas to dominate in co-infections (Singh et al. 2017). Although this approach has been investigated for over 25 years, there has been a lack of translational application of AHLs.

2.2 *Passive Biofilm Dispersal*

Passive dispersal refers to the mechanical break up of biofilm through physical forces such as fluid or sheer force (Fig. 2) (Fleming and Rumbaugh 2017). Pertaining to medical biofilms, passive debridement can be sub-classified into surgical, mechanical, biological, autolytic, and enzymatic debridement (Fig. 2). For the purpose of this chapter, enzymatic debridement will be discussed in further detail. Enzymatic debridement is a popular method of removing infected and/or necrotic tissue, which utilizes enzymatic agents that are safe and effective (Steed 2004). Two currently FDA-approved enzymatic debridement agents contain papain or collagenase, which digest fibrin and collagen or collagen and elastin, respectively (McCallon et al. 2014). While these enzymes target components of the host's extracellular matrix, it is plausible that they disperse microbial biofilm that is present

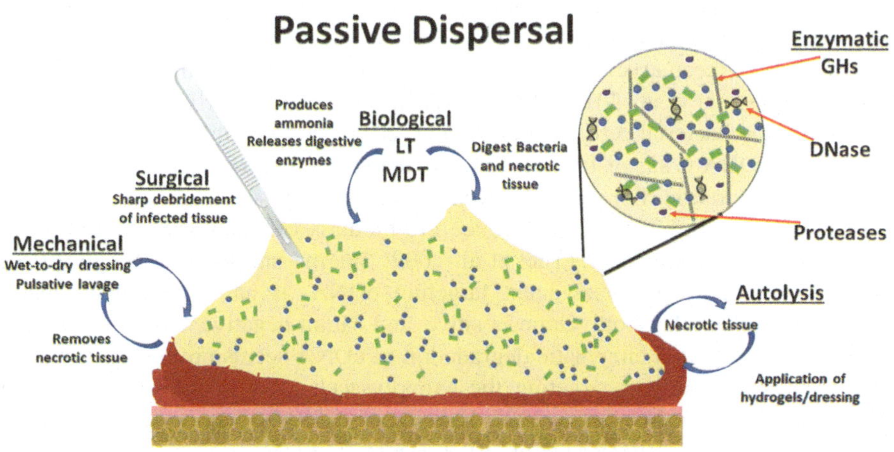

Fig. 2 Examples of passive dispersal mechanisms and agents. *LT* larval therapy, *MDT* maggot debridement therapy

in the tissue. There are other enzymes that specifically target components of the bacterial EPS in order to reduce infection and promote healing, which will be discussed in subsequent sections. As the dispersal of bacteria by the enzymatic breakdown of the bacterial EPS does not rely on the activation of cellular mechanisms within bacterial cells, it is considered to be passive.

2.2.1 Proteases

Proteases that target the extracellular proteins commonly found in the biofilm matrix, and that are utilized to maintain and modify the EPS, have been investigated as dispersal agents (Lasa and Penades 2006; Jiao et al. 2010; Zhang and Bishop 2003). For example, the protease SKS10 produced by *Microbacterium* sp. was shown to degrade *S. aureus* biofilms without inducing cellular cytotoxicity towards host cells (Saggu et al. 2019). In another study, Proteinase K inhibited *S. aureus* biofilm development and enhanced dispersion (Snarr et al. 2017). Ficin, a plant protease, has also been shown to disrupt mature *S. aureus* biofilms (Baidamshina et al. 2017), and a serine protease produced by *Bacillus pumilus* has shown effectiveness against *Serratia marcescens* biofilms (Mitrofanova et al. 2017). Streptococcal cysteine protease, SpeB, exhibited a significant decrease in biofilm formation against Group A Streptococci biofilms (Connolly et al. 2011). These studies are just a representation of proteases that have shown either biofilm degradation, prevention, or both against a variety of bacterial species (Suresh et al. 2019).

2.2.2 DNases

Extracellular DNA is one major component of the biofilm EPS of *P. aeruginosa, Streptococcus intermedius, Streptococcus mutans, Enterococcus faecalis,* and *Staphylococci* (Montanaro et al. 2011). eDNA is vital for the structural and mechanical integrity of the matrix and protects bacteria from antibiotics and the host immune response (Devaraj et al. 2019). The eDNA in the biofilm matrixome comes from lysed host and bacterial cells (Vorkapic et al. 2016), and the utilization of enzymes that cleave eDNA is suggested to lead to an altered biofilm formation that increases antibiotic penetration. For example, DNase one has been shown to impair biofilm formation (Tetz and Tetz 2010), reduce the biomass of established biofilms and improve antibiotic clearance (Tetz et al. 2009; Sharma and Pagedar Singh 2018). Another promising dispersal agent targeting eDNA, is an extracellular DNase (NucB) that degrades both Gram-positive and Gram-negative biofilms (Nijland et al. 2010). Although enzymatic degradation of eDNA seems promising, there are also studies that support targeting the exopolysaccharides in the EPS to induce biofilm dispersal (Redman et al. 2021; Fleming et al. 2017; Fleming and Rumbaugh 2018; Nijland et al. 2010; Ren et al. 2019; Lin et al. 2022).

2.2.3 Glycoside Hydrolases

Glycoside hydrolases (GHs) are a class of enzymes that target the glycosidic linkages between sugars, like those found within the exopolysaccharide structures of the EPS (Fleming and Rumbaugh 2017). The exopolysaccharide component of the matrixome is important for initial adhesion, protection and structure (Limoli et al. 2015). $PelA_h$ and $PslG_h$ are GHs that are encoded in the *P. aeruginosa pel* and *psl* operons, respectively and have previously been shown to be effective in disrupting and preventing the formation of *P. aeruginosa* biofilms (Baker et al. 2015, 2016; Szymanska et al. 2020). $Sph3_h$ is a hydrolase produced by *Aspergillus fumigatus* which disrupts biofilms made by *A. fumigatus* and is also effective cross-kingdom against *P. aeruginosa* biofilms (Snarr et al. 2017; Le Mauff et al. 2019). The GH β-*N*-acetylglucosaminidase (DispersinB), produced by *Actinobacillus actinomycetemcomitans,* disrupts and prevents biofilm formation by *A. actinomycetemcomitans* and *S. epidermidis* (Ramasubbu et al. 2005; Kaplan et al. 2004). In addition to these examples, we have shown that the utilization of cellulase, α-amylase, and xylanase from *Aspergillus niger, B. subtilis,* and *Aspergillus oryzae,* respectively, can degrade *S. aureus, P. aeruginosa* and *E. faecalis* biofilms (Fleming et al. 2017, 2020; Fleming and Rumbaugh 2018). Thus, considering their proven efficacy, and the abundance of different GHs, these enzymes hold substantial therapeutic potential to fight biofilm infections.

3 Biofilm Dispersal Agents in the Pipeline

Given the rise of antibiotic resistance and the recalcitrance of chronic biofilm infections to traditional therapeutic methods, there is a growing need for the clinical application of alternative management strategies, including dispersal agents. However, despite the plethora of preclinical investigations into biofilm dispersal agents, very few have progressed to clinical trials to date, and only one, human recombinant DNase dornase alpha (Pulmozyme), is currently in clinical use for the treatment of CF infections (Kaplan et al. 2012; Lewenza 2013; Martin et al. 2021). That being said, there are several anti-biofilm compounds currently being developed including novel antimicrobial agents such as gallium-based iron analogs (e.g., gallium citrate, also referred to as Panaecin, by Aridis Pharmaceuticals Inc. (Thompson et al. 2015), phage therapy, inhibitors of cell-to-cell signaling, as well as some more unusual ones such as bioenergetic inhibitors, essential oils, small molecule inhibitors, and chelators. An example of a bioenergetic inhibitor is Pravibismane (bismuth-ethanedithiol) (Baker 2020), which received "orphan drug status" for the treatment of CF lung infections from the FDA in 2000 (CysticFibrosisNewsToday 2000). However, only a small subset of these compounds aim to disperse or disassemble biofilms. In this section, we detail examples of compounds that are currently in the pipeline, ranging from the discovery to clinical trial phase. For clinical trials, we discuss those that are both ongoing and recently completed in the last 10 years and provide the results where available. It should be noted that the mechanisms of various proposed anti-biofilm agents are not always completely understood, and thus differentiating between antibacterial and dispersal agents can be challenging. However, we made every effort to discuss the mechanism by which compounds, both in preclinical and clinical phases of investigation, contribute to dispersion or the disassembly of biofilms. As such, any agent found within the ClinicalTrials.gov database with potential dispersal activity as of September 2021 was included (Table 1). Furthermore, repetitive entries into the ClinicalTrials.gov database by the same group, investigating the same agent, and with the same status were not discussed, except for the most recently updated trial.

3.1 Small Molecule Modulators of Dispersion

A compound that is currently being considered as a biofilm dispersal agent is 2-aminoimidazole. The small molecule 2-aminoimidazole is a class of secondary metabolite molecules that is produced by marine sponges belonging to the Calcarea family (Koswatta and Lovely 2011). The brown tube sponge *Agelas conifera* utilizes 2-aminoimidazole to minimize biofilm colonization by marine bacteria such as *Rhodospirillum salexigens (*Keifer et al. 1991*;* Yamada et al. 1997*). Since its first discovery, variants of* 2-aminoimidazole have been reported to re-sensitize bacteria to existing antibiotics (Rogers et al. 2010), as well as inhibit and disperse biofilms made by *P. aeruginosa, S. aureus,* and *Mycobacterium tuberculosis* (Draughn et al. 2017; Richards et al. 2008; Ballard et al. 2008; Jeon et al. 2019). While the mode of action is not fully understood, it appears that 2-aminoimidazole targets the DNA-binding proteins or response regulators of two-component systems involved in virulence, antibiotic resistance responses of bacterial pathogens, and biofilm formation (Draughn et al. 2017). Its effect on biofilms may be two-fold; in *Salmonella*, variants of 2-aminoimidazole strongly reduce the expression of genes (i.e., *csgD*, *csgB*, and *adrA*) involved in the production of the EPS (Robijns et al. 2014). In contrast, studies using *M. tuberculosis* suggest 2-aminoimidazole affects mycobacterial bioenergetics, as a 2-aminoimidazole variant, 2B8, collapsing both components of the proton motive force in a manner similar to other cationic amphiphiles (Jeon et al. 2019). Variants of 2-aminoimidazole, referred to as AGL-110 and AGL-503 by Agile Sciences Inc., are currently being evaluated in a preclinical phase for the treatment of wound infections, CF lung infections, and *P. aeruginosa* infections. Additional products derived from the marine environment have been investigated for antimicrobial/anti-biofilm activity including furanones from the red algae *Delisea pulchra* (Pérez et al. 2016*;* Hentzer et al. 2002). However, due to their relatively high toxicity, none of these potential therapeutic approaches are as of yet in clinical use.

Another compound affecting signaling is the chelator diethylene triamine pentaacetic acid, also referred to as DTPA or pentetic acid. This compound has been

Table 1 Recent promising biofilm dispersal agents for translational use

In clinical use				
Agent	**Product**	**Use and mechanism of action**		**Reference**
DNase (dornase alpha)	Pulmozyme®	Dornase alpha cleaves extracellular DNA, released from both neutrophils and bacteria, weakening the biofilm and dispersing the bacteria. Pulmozyme® is currently used to treat CF.		(Kaplan et al. 2012; Lewenza 2013; Martin et al. 2021)
Ongoing clinical trials				
Agent	**Clinical trial**	**Clinical trial status**	**Mechanism of action**	**ClinicalTrials.gov identifier**
Nitric oxide (topical)	Nitric oxide releasing sinus irrigation (NOSi) to treat recalcitrant chronic rhinosinusitis	Phase 2: Recruiting (as of February 8, 2021)	NO acts as a dispersal signaling molecule by increasing bacterial phosphodiesterase activity, reducing c-di-GMP levels (Barraud et al. 2006).	NCT04163978
CB-280	Study to evaluate the safety of CB-280 in patients with cystic fibrosis	Phase 1: Recruiting (as of March 18,2021	Inhibition of arginase, an enzyme that competes with NO synthase for L-arginine as substrate.	NCT04279769
Essential oils	Effectiveness of Salvadora Persica Miswak in improving plaque control and gingival health	Recruiting (as of January 5, 2021)	Disruption of cell walls and membranes. Possible dispersal at sub-inhibitory concentrations through unknown mechanisms.	NCT04650685
EDTA	Conservative treatment of catheter—Related injections with gentamicin/EDTA (CATH-GE)	Phase 1: Recruiting (as of March 10, 2021)	Chelation of minerals needed for biofilm maintenance.	NCT04789928
Next science	Clinical assessment of next science wound gels in healing below the knee amputation surgical wound compared to SOC	Not yet recruiting (as of August 26, 2019)	Destabilization of the exopolysaccharide matrix through the chelation of calcium (Miller et al. 2014).	NCT04053946

(continued)

Table 1 (continued)

TRL1068	Study to evaluate safety and activity of TRL1068 in prosthetic joint infections	Phase 1; recruiting (as of October 5, 2021)	Targets a highly conserved epitope on the DNABII family of bacterial DNA-binding proteins, resulting disruption of the EPS matrix.	NCT04763759
Trials completed in the last 10 years				
Agent	**Clinical trial**	**Outcome**	**Mechanism of action**	**ClinicalTrials.gov identifier**
Nitric oxide (inhaled)	RATNO, reducing antibiotic tolerance using nitric oxide in CF	Phase 2: Significant reduction in *P. aeruginosa* biofilm in CF patients treated with NO + antibiotics vs. antibiotics alone (Cathie et al. 2014).	NO acts as a dispersal signaling molecule by increasing bacterial phosphodiesterase activity, reducing c-di-GMP levels (Barraud et al. 2006).	NCT02295566
Nitric oxide (inhaled)	Efficacy and safety of inhaled nitric oxide (NO) in cystic fibrosis (CF) patients	Phase 2: Terminated for the safety of cystic fibrosis clinical trial subjects due to Covid-19		NCT02498535
Nitric oxide (inhaled)	Safety and tolerability of inhaled nitric oxide in patients with cystic fibrosis	Phase 1: Safe and tolerable in terms of O_2 saturation and forced expiratory volume in 1 second.		NCT00570349
OligoG CF-5/20	Safety and efficacy of inhaled OligoG CF-5/20 for the treatment cystic fibrosis	Phase 1: No results posted	Destruction of polymer networks by disrupting DNA-Ca2 + -DNA bridges and the biofilm matrix.	NCT00970346
Essential oils	Immediate effect and substantively of the essential oils without alcohol on the Oral biofilm in situ	Phase 4: Improved efficacy over 0.2% chlorhexidine against dental plaques over 7 hours (Quintas et al. 2015b).	Disruption of cell walls and membranes. Possible dispersal at sub-inhibitory concentrations through unknown mechanisms.	NCT02946814
Essential oils	Antiplaque effect of essential oils with and without alcohol on an in situ model of Oral biofilm growth	Phase 4: EOs showed similar Efficacy to 0.2% chlorhexidine to reducing bacterial viability in dental plaque, but reduced efficacy in reducing biofilm biomass (Quintas et al. 2015a).		NCT02946801

Essential oils	Essential oils with and without alcohol: Substantively and antiplaque effect	Phase 4: EOs were effective at reducing bacterial viability and biofilm biomass after 4 days on dental plaques (Quintas et al. 2017).		NCT03146390
Ferumoxytol	Evaluation of anti-biofilm effects of Ferumoxytol (Feraheme) using an in situ biofilm demineralization model	Phase 1: Unknown	EPS matrix degradation through glucan structure cleavage (Liu et al. 2018b).	NCT03678012
Tea polyphenols	Effectiveness of Black tea mouth-wash on plaque formation (Blacktea)	Phase 3: Unknown	Disruption of amyloid curli fibers and cellulose synthesis (Shinde et al. 2021).	NCT02545244
N-acetylcysteine	A novel catheter lock solution for treatment of Tunneled Hemodialysis catheter-associated Bacteremia	Phase 1: Treatment was successfully for 83% of cases within 90 days of follow-up (Aslam et al. 2008).	Possible direct effect on the matrix (chelation of calcium and magnesium or interaction with crucial components in the matrix (Blasi et al. 2016).	NCT00614679
AR-105 (Aerucin®)	Adjunctive therapeutic treatment with human monoclonal antibody AR-105 (Aerucin®) in P. Aeruginosa pneumonia	Phase 2: 20% improvement in clinical cure rate in patients treated with AR-105 vs. placebo in a sub-group of patients.	Targets alginate, a *P. aeruginosa*-produced cell surface polysaccharide involved in surface adhesion, biofilm formation, and protection against the human immune system.	NCT03027609

reported to suppress the transcription of the elastase-encoding *lasB* gene and levels of the PQS, a molecule that mediates *P. aeruginosa* QS. This compound is being considered by Hoth Therapeutics Inc. in combination with antibiotic gentamicin as a conjunctive therapy (referred to as BioLexa) for the treatment of atopic dermatitis (ClinicalTrials.gov Identifier: NCT04544943) and the treatment of wounds and diabetic foot ulcers (preclinical phase). Three other chelators of interest are ethylenediaminetetraacetic acid (EDTA), "Next Science," and N-acetylcysteine. EDTA has been shown to induce biofilm dispersal of *P. aeruginosa* biofilms, likely by sequestering key minerals (magnesium, calcium, and iron) necessary for biofilm maintenance (Banin et al. 2006). A clinical study is currently underway (recruitment phase) investigating the efficacy of EDTA and gentamicin against central venous catheter biofilm infections (ClinicalTrials.gov Identifier: NCT04789928). "Next Science," is a proprietary agent with antimicrobial and biofilm dispersal properties. Next Science Ltd's approach, referred to as XBIO Technology, is to deconstruct the structure of biofilms based on the removal of the metal ions that hold the EPS together, thus separating the polymers into separate units. Metal ion removal is combined with surfactants and osmotic stress to induce further imbalance, collapsing biofilms and rendering biofilm bacteria susceptible to eradication.(Bounds et al. 2021; Miller et al. 2014). A clinical trial is underway testing the efficacy of Next Science against surgical site infections (ClinicalTrials.gov Identifier: NCT04053946). N-acetylcysteine is a mucolytic agent that is used clinically to treat patients with chronic bronchopulmonary disorders. N-acetylcysteine has been studied extensively for the treatment of biofilm infections, especially respiratory-associated biofilm infections, and is thought to induce biofilm dispersal by chelating calcium and magnesium, and by interacting directly with the EPS matrix (Blasi et al. 2016). A clinical trial was performed investigating the potential of N-acetylcysteine as a catheter lock solution where indwelling catheters were filled with an intraluminal solution of antimicrobial and anticoagulant solution (Shah et al. 2002) (ClinicalTrials.gov Identifier: NCT00614679). In phase one of the trial, treatment was successful for 83% of the patients within 90 days of follow-up (Aslam et al. 2008), with no further results posted to date.

Rather than jamming the bacterial communication system, cis-DA, produced by *P. aeruginosa*, has been reported as a cross-kingdom dispersion signaling molecule, capable of inducing *P. aeruginosa* and other Gram-negative and Gram-positive bacteria and fungi to undergo a physiologically-mediated dispersion response, resulting in the dis-aggregation of surface-associated microbial populations and biofilms (Davies and Marques 2009; Davies 2007). While this small molecule is currently not being developed for the treatment of biofilm-related infections, Compounds Ltd. is considering it for various applications relating to water treatment, considering that the incorporation of cis-DA into bromine-based water treatment greatly improved the effectiveness of biofilm control compared to the brominated biocide acting alone (Rodensky et al. 2020). A similar but somewhat smaller effect was observed in chlorine-based water treatment (Rodensky et al. 2020).

3.2 Modulators of c-di-GMP

Previous studies demonstrated that a decrease in c-di-GMP level coincided with dispersion of established *P. aeruginosa* biofilms grown in vitro and on infected silicone implants in the peritoneal cavity of mice (Christensen et al. 2013; Andersen et al. 2021b). Several research groups have since used high throughput methods to detect compounds that reduce the levels of c-di-GMP in a number of different bacterial species (Andersen et al. 2021a; Sambanthamoorthy et al. 2012) or used rationally designed peptides to sequester c-di-GMP (Hee et al. 2020). While none of these compounds appear to be in clinical use as of yet, azathioprine may emerge as a front runner in the modulation of c-di-GMP levels and thus, dispersion. Azathioprine is an immunosuppressive drug that is already in clinical use. The drug has been demonstrated in biofilms formed by *E. coli* to inhibit biosynthesis of c-di-GMP by interfering with intracellular nucleotide pool availability and to suppress the expression of extracellular matrix proteins (Antoniani et al. 2013).

Another compound that affects c-di-GMP levels to induce biofilm dispersion and is currently undergoing clinical trials is NO. NO has long been recognized as a vital signaling compound in eukaryote biology including humans. This highly water soluble and diffusible molecule has been identified as a signaling molecule that induces biofilm dispersion, with dispersion coinciding with increased phosphodiesterase activity and in turn, reduced levels of c-di-GMP (Barraud et al. 2006; Li et al. 2013). While NO is toxic at high levels, NO donors posed no toxicity towards bacterial cells at the low dosage required to induce biofilm dispersion (Cai and Webb 2020). There are currently three approaches to using NO therapeutically in bacterial infection, (i) delivering NO donor prodrugs, (ii) direct inhalation of NO, and (ii) increasing endogenous NO formation.

NO donor prodrugs are designed to release NO at specific sites, e.g., upon cleavage of the prodrug by enzymes such as glycosidases and β-lactamases (Yang et al. 2021). NO donor prodrugs include sulfathiazole [93], and cephalosporin-3′-diazeniumdiolates, with NO being released following cleavage of its β-lactam moiety by bacterial β-lactamases (Soren et al. 2020; Barraud et al. 2012), and spermine NONOate (N-[4-[1-(3-aminopropyl)-2-hydroxy-2-nitrosohydrazino]butyl]-1,3-propanediamine, S150) that spontaneously dissociates in a pH-dependent manner (Cai and Webb 2020; Marvasi et al. 2014). These NO donor prodrugs were shown to induce significant dispersal of biofilms formed by *P. aeruginosa, E. coli and Salmonella enterica* in vitro (Cai and Webb 2020; Marvasi et al. 2014), improve dispersal when co-administered with tobramycin, and nearly completely eradicate biofilms when combined with colistin (Soren et al. 2020). Gaseous NO requires inhalation via a nasal cannula. Results from a small pilot study in CF patients with *P. aeruginosa* infection, using direct inhalation of gaseous NO, indicated lower numbers of bacterial aggregates in the sputum when compared with controls when used as adjunctive therapy to antibiotics (Howlin et al. 2017). Additionally, two clinical trials using inhaled NO have been performed to evaluate its safety and tolerability when administered at 160 ppb (clinicaltrials.gov,

NCT02498535) and at 20 and 40 ppm (NCT00570349) in CF patients, and one has been conducted to investigate the ability of NO to reduce antibiotic tolerance in CF patients (NCT02295566). In the latter, it was reported that a significant reduction in *P. aeruginosa* biofilm occurred in CF patients treated with inhaled NO + IV antibiotics (ceftazidime and tobramycin) compared to antibiotics alone (Cathie et al. 2014). Another clinical trial currently recruiting aims to test the efficacy of topical NO-releasing sinus irrigation in treating biofilm-associated chronic sinusitis (NCT04163978).

An alternative approach for NO therapy is to increase endogenous NO production by the inhibition of arginase, an enzyme that competes with NO synthase for L-arginine as substrate. A phase 1B randomized, double-blind, placebo-controlled trial investigating a compound called CB-280, a potent arginase inhibitor, is currently being conducted (NCT04279769). Taken together, these ongoing studies illustrate both the promise and versatility of using NO to tackle many different types of biofilm infections.

3.3 Compounds Targeting the Protective Biofilm Matrix

The majority of dispersal agents that are currently being pursued target the biofilm matrix. Strategies include the use of matrix-degrading enzymes, but primarily focus on inhibiting the function of enzymes and components present in the biofilm matrix to disassemble biofilms. To date, the only such agent in clinical use is the DNase Pulmozyme, which works by cleaving the extracellular DNA produced by both neutrophils and bacteria. This leads to a disruption of the mucus and biofilm EPS that builds up in chronic CF lung infections and increases the activity of antibiotics against the resident microbes (Martin et al. 2021). In addition to its uses in CF, DNase has been and is currently being investigated for numerous other anti-biofilm applications, including the treatment of clogged tympanostomy tubes (Chan et al. 2018), the treatment of ventilator-associated lung infections (131), and others. The use of other matrix-degrading enzymes (including alginate lyase) as a biofilm disruptors has so far not been commercialized, despite advances in enzyme engineering (Jang et al. 2016); however one compound has been somewhat successful. The hydrolase *DispersinB, produced* by *A. actinomycetemcomitans* degrades the polysaccharide, poly(1,6)-N-acetyl-D-glucosamine (PNAG), by hydrolyzing β(1,6) glycosidic linkages. This enzyme is effective at disrupting the biofilms made by multiple Gram-negative and -positive bacterial species, including many leading pathogens (*S. aureus, S. epidermidis, Enterococci, Acinetobacter baumannii, Klebsiella pneumoniae,* and *P. aeruginosa*) (Waryah et al. 2017; Izano et al. 2007; Kaplan et al. 2004; Gawande et al. 2014b; Itoh et al. 2005). DispersinB has also shown efficacy in some in vivo models (Gawande et al. 2014a; Kaplan et al. 2018; Darouiche et al. 2009), and commercialization of the agent is being pursued by Kane Biotech, Inc., for the treatment of animals. In 2019, the company announced its focus on health strategies with the development of a wound care hydrogel containing the enzyme DispersinB (KaneBiotech 2019).

An example of a nonenzymatic agent that potentially targets the EPS matrix is Ferumoxytol, an FDA-approved, carbohydrate-coated, superparamagnetic iron oxide nanoparticle that is used to treat iron deficiency anemia (Schwenk 2010). However, it has also been shown to disrupt oral biofilms by binding to EPS components and generating free radicals from hydrogen peroxide, causing both bacterial cell membrane disruption and EPS matrix degradation (Liu et al. 2018b). A clinical trial was performed investigating the efficacy of ferumoxytol against cariogenic dental plaques (ClinicalTrials.gov Identifier: NCT03678012), with no results posted to date.

Previous findings demonstrated that bacteriophages, which have been engineered to express a biofilm-degrading enzyme during infection, can simultaneously attack the bacterial cells in the biofilm and the biofilm matrix. Use of these bacteriophages resulted in a significant removal of established biofilms relative to nonenzymatic bacteriophage treatment (Lu and Collins 2007). EnBiotix is taking advantage of this two-prong approach with their engineered bacteriophage platform. Their lead compound, EBX-003 (drug ID# 311748), enables the delivery of a biofilm-degrading enzyme payload to biofilms via optimized natural bacteriophages and is being developed for hospital-acquired pneumonia, ventilator-associated pneumonia (clinical phase II), and the treatment of prosthetic joint infections.

3.4 Inhibiting the Function of Enzymes and Components Present in the Biofilm Matrix

Approaches to inhibit the function of biofilm matrix components entails disrupting the interactions between matrix components and blocking/removing matrix components. A compound capable of disrupting the interactions between matrix components is OligoG CF-5/20, a low-molecular-weight alginate oligosaccharide ($Mn = 3200\,g\,mol^{-1}$) derived from the marine algae *Laminaria hyperborea* with a G content >85%. OligoG CF-5/20 has been shown to diffuse into biofilms and disrupt polymer networks by disrupting DNA-Ca^{2+}-DNA bridges and the biofilm matrix, resulting in the disassembly of Pseudomonal biofilms (Oakley et al. 2021; Powell et al. 2018). Overall, OligoG treatment significantly reduced the structural quantities of EPS polysaccharides and eDNA, while enhancing antibiotic efficacy against established biofilms. Additionally, the oligomer has been shown to alter sputum viscoelasticity. AlgiPharma AS is currently testing inhaled OligoG CF-5/20 alginate oligomer therapy in phase IIb/III clinical trials for the treatment of CF patients (NCT00970346, completed phase I).

Recent findings suggest that rather than degrading matrix components via enzymes such as DNAse, proteinases, or hydrolases, to disassemble biofilms, the same result can be achieved by using antibodies that target matrix components. For example, Novotny et al. targeted the DNABII family of bacterial DNA-binding proteins, which serve as critical structural components within the eDNA scaffold of biofilms formed by all bacterial species tested to date (Novotny et al. 2020). Using chimeric peptide immunogens or humanized monoclonal antibodies, they demonstrated that this approach could disassemble the biofilms formed by the respiratory tract pathogen non-typeable *Haemophilus influenzae*. Moreover, anti-DNABII-treated biofilm cells were more vulnerable to the effects of antibiotics and the immune system (Goodman and Bakaletz 2021; Novotny et al. 2020). The humanized monoclonal antibody CMTX-101 by Clarametyx Biosciences Inc. targets a region of DNABII binding proteins to induce the disassembly of biofilms. The antibody is currently in preclinical development for hospital-acquired pneumonia. Trellis Bioscience Inc. is likewise focusing on an antibody targeting DNABII, specifically a highly conserved epitope on the DNABII family of bacterial DNA-binding proteins that includes histone-like (HU) and integration host factor (IHF) proteins of clinically relevant Gram-positive and Gram-negative bacteria. The respective high-affinity and broad-spectrum native human antibody, TRL1068 (drug ID# 231794), was originally discovered using single B-lymphocyte screening technology (Estellés et al. 2016). The DNABII epitope bound by TRL1068 has no homologs in the human proteome. Clinical trials are currently underway to evaluate TRL1068 for the treatment of prosthetic joint infections (NCT04763759). In contrast to targeting a broadly present matrix component, the fully human IgG1 monoclonal antibody, AR-105 (Aerucin), by Aridis Pharmaceuticals targets *P. aeruginosa* alginate, a cell surface polysaccharide involved in surface adhesion, biofilm formation, and protection against the human immune system. Clinical phase II testing of AR-105 for the treatment of ventilator-associated pneumonia has been completed (NCT03027609), and trial results are available (AridisPharmaceuticals 2019). While AR-105 did not perform significantly better than placebo in 158 patients, a post hoc analysis found a 20% improvement in clinical cure rate in a subgroup of patients.

3.5 Targeting the Outer Membrane

Lectins are bacterial outer membrane proteins that allow bacteria to adhere to host tissues (Nilsson 2003). The adhesion of *P. aeruginosa* to the lung epithelial cell surface is mediated by two specific lectins, LecA and LecB (Chemani et al. 2009) which are also involved in *P. aeruginosa* biofilm formation (Pang et al. 2019; Tielker et al. 2005). In turn, galactose derivatives capable of binding to the LecA and LecB ligand nitrophenylfucoside prevented biofilm formation and partially dispersed wild-type biofilms (Diggle et al. 2006; Johansson et al. 2008). Not surprisingly, the *P. aeruginosa* lectins have become targets for anti-biofilm treatment strategies (Grishin et al. 2015). The University of Konstanz is currently pursuing two small molecules, small molecule one (drug ID# 275317, cinnamide derivative) and small molecule two (drug ID# 275294, trimethylphenylsulfonamide), which are capable of inhibiting lectin B for the treatment of *P. aeruginosa* infections.

Similarly, ProclaRx LLC has made use of antibodies to inhibit bacterial outer membrane proteins with the goal of disrupting the physical barrier which protects biofilm bacteria. In addition, monoclonal antibodies for the treatment of secretory otitis media (PCL 1402, PCL 1440, drug ID# 329648) and sinusitis (PCL 1404, PCL 1550, drug ID# 331939) are currently in preclinical phases.

3.6 Essential Oils

Various essential oils (EOs) have also been the subject of past and present clinical trials. Many preclinical studies have shown EOs to be promising candidates for anti-biofilm applications. It is believed that the majority of EOs act by disrupting bacterial cell membranes and cell walls (El-Tarabily et al. 2021), and some studies have hinted that they may also display activity as dispersal agents (Papa et al. 2020; Jain and Parihar 2018; Gao et al. 2020). However, the exact mechanisms, especially in respect to individual EOs, requires further study. Another type of natural product shown to exhibit biofilm dispersal activity is certain tea polyphenols. These compounds synergistically target amyloid curli fibers and cellulose synthesis to disrupt biofilms (Shinde et al. 2021). For example, a clinical trial was undertaken to investigate the efficacy of black tea mouthwash on dental plaque formation (ClinicalTrials.gov Identifier: NCT02545244), with no results posted to date. Another group has found through multiple trials (ClinicalTrials.gov Identifiers: NCT02946814, NCT02946801, NCT03146390) that essential oils reduce bacterial viability and biofilm biomass of, and improve the activity of chlorhexidine against, dental plaques (Quintas et al. 2015a, b, 2017).

To conclude, although preclinical research into biofilm dispersal agents is expansive, therapies that have advanced to clinical trials are sparse. While many promising agents are in development, more incentives and resources need to be allocated towards human testing.

4 Hurdles for Development

As the prevalence of biofilm-associated infections increases worldwide, the development of novel therapeutic agents is crucial. However, there is currently a lack of preclinical and clinical guidelines. For example, the US Food & Drug Administration (FDA) lacks clear guidelines for anti-biofilm and biofilm dispersal agents. This is in part due to the lack of requirements set to determine an effective "anti-biofilm" or "biofilm-reducing" agent. One of the main questions that needs to be defined for preclinical screening of these therapeutics is, "what is a clinically relevant reduction of biofilm?" Does this include a reduction of overall metabolically active cells, a reduction in biomass, or both? Establishing what "effective" means clinically is one large hurdle in itself. Additionally, dispersal agents do not have antimicrobial activity themselves, but are instead utilized in conjunction with an antimicrobial to kill the cells released from the biofilm. Thus their efficacy cannot be based on traditional measures for anti-infective agents.

4.1 Regulatory Obstacles

As dispersal agents fall into multiple categories already outlined on the FDA website, this can lead to confusion on what is necessary to move a promising agent from preclinical screening to human clinical trials. These categories include wound topical agents and antibacterial therapies for patients with unmet medical need for the treatment of serious bacterial diseases. Additionally, there are further federal requirements that need to be met depending upon how the dispersal agent is applied. For example, the FDA recommendations outlined for "wound topical" agents require proving that subcutaneous injections of the agent do not induce fibroplasia and stroma formation (Group 2001). Topical application of the agent should also be conducted to measure contraction and re-epithelization of the skin. Additionally, histological sections of infections treated with "wound topical" agents should be analyzed for lesions, infiltration of inflammatory cells, and acanthosis, all of which could indicate toxicity to the host (Mitsuishi et al. 2015). Lastly, a defined toxic dose should be determined through the application of exaggerated concentrations to a wound in vivo. The FDA recommendations also suggest conducting studies to ensure the application of the agent does not induce carcinogenic or reproductive and developmental toxicity as well. Lastly, cutaneous irritation and sensitization should be examined for the topical application of treatments. As many novel agents continue to be designed, it is imperative that investigators performing preclinical studies while keeping FDA guidelines in mind to ensure efficacy and safety prior to moving to clinical trials.

Dispersal agents may also fall into the category of "Antibacterial Therapies for Patients With Unmet Medical Need for the Treatment of Serious Bacterial Diseases" as these agents meet the requirement of having "some other characteristic that has a potential to lead to enhanced effectiveness" (Research 2017). This particular guideline set for preclinical screening lacks safety protocols. Instead, the recommendations are highly focused on in vitro and in vivo efficacy against targeted bacterial pathogens as well as describing the mechanism of action and potential mechanisms of resistance to the drug. This program is classified as a streamlined development program, expediting the transition into clinical trials.

Not only has there been a rise in the discovery of anti-biofilm and biofilm dispersal agents, but devices with antimicrobial and biofilm prevention have also been advanced, further revealing the lack of federal guidelines. These newly developed devices tend to fall into a gray zone within the guidelines as they often qualify as both a device and a drug. Any coating, dressing, gel, or medical device is currently under the jurisdiction of device Protecting and Promoting Public Health (PMOA) under the FDA, while any agent that "prevents," "reduces," or "treats" is considered a drug and therefore falls under the drug PMOA jurisdiction. This leads to the daunting gray area of a medical device used in combination with a drug. For example, consider a newly developed catheter (medical device) coated with zinc (medical device) that prevents biofilm formation. The zinc coating, originally labeled as a device, now is acting as a drug, therefore making the catheter+zinc coating a combination product, which lacks a clear jurisdiction in the FDA. As more potential therapeutics are developed, it is imperative that guidelines be set quickly to move life-saving discoveries to bedside practice. A specific drug development pathway for anti-biofilm agents, under the antimicrobial therapies umbrella, would be desirable. This pathway would include suggested models and benchmarks for the effective demonstration of preventing or treating biofilm in different types of infection (e.g., wound, implants, lung, etc. . . .) either by single agents or combination projects (e.g., dispersal agent/antibiotic combination).

4.2 Therapeutic Risks and Rewards

In addition to better guidelines and goalposts for efficacy, better animal models to test biofilm dispersion are needed. With few exceptions, biofilm dispersal has not been definitively demonstrated in vivo. More commonly, dispersal agents are shown to disperse biofilms in an in vitro assay and then subsequent experiments performed in an in vivo model reveal that the same agent decreases bacterial load. Therefore, dispersal in vivo is assumed, rather than demonstrated. However, decreases in bacterial load could occur for a variety of reasons. For example, a dispersal agent could inhibit bacterial growth in comparison to a control application, or perhaps the dispersal agent stimulates the immune response above that of the control. It is difficult to say definitively that the bacterial load was reduced because the bacteria were dispersed from the biofilm and subsequently killed. Thus more models of dispersal are needed for different types of biofilm infection in order to better understand the dynamics of dispersal, e.g., how quickly does it happen, how much of the population is dispersed and how does dispersal affect antibiotic susceptibility.

In vivo models for biofilm dispersal are important both to test preclinical efficacy of experimental agents and also to gain a better understanding of how the host responds to the release of bacteria from a biofilm. For example, the therapeutic risks of biofilm dispersal during infection have been proposed and demonstrated experimentally (Watters et al. 2016; Fleming and Rumbaugh 2017)' (Wille and Coenye 2020). Fleming et al. tested the efficacy of glycoside hydrolases to disperse *P. aeruginosa* from a mouse wound infection (Fleming and Rumbaugh 2018). With in vivo imaging, the investigators were able to visualize bioluminescent *P. aeruginosa* appearing outside of the infection site shortly after treatment with the dispersal agent, but not after application of the vehicle control. This mass dispersion of a high number of *P. aeruginosa* quickly caused bloodstream infection and death. However, the investigators also showed that antibiotics given in conjunction with the dispersal agent prevented bacteremia and greater therapeutic efficacy than the antibiotic alone. Thus more models of dispersion are needed for different types of biofilm infection in order to better understand how both the host and the bacteria respond to dispersal. There are many gaps in knowledge to be filled, including how many cells can safely be dispersed without causing sepsis? Does the risk of systemic infection differ when bacteria are dispersed from different locations (e.g., wound vs. lung). Are there changes in the bacterial phenotype of dispersed cells that make them more of a threat to the host (e.g., more virulent or less susceptible to antibiotics). Clearly there are inherent risks to dispersing bacteria from infection sites. However, these risks are likely dependent on bacterial load, the species and motility of bacteria being dispersed, the location of the infection and the overall health of the host. While it has been demonstrated that risks can be augmented by the administration of dispersal agents adjunctive to antibiotics, there are still a lot of unknowns pertaining to dispersing bacteria in vivo. However, there is a promise of great reward for gaining a better understanding of dispersal. With dwindling options for new anti-infective agents, dispersal may provide us with an effective way to extend the life and increase the efficacy of already existing antibiotics.

5 Conclusion and Future Perspectives

The growing need to generate antimicrobial agents with novel mechanisms of action has been a driving force for the development of anti-biofilm agents. The most effective method for fighting biofilm-associated infections is to prevent the biofilm from ever establishing. However, the unfortunate reality is that many patients require treatment for infections that already have mature biofilm present. Proof of principle studies has shown that dispersing microbes from biofilm infections is a promising method of making them more susceptible to conventional antibiotic treatments. These methods of dispersion have included environmental signals, manipulating intracellular signals, inhibitors, and matrix-degrading enzymes. While more studies and better in vivo models are needed to gain a better understanding of the risks and rewards of biofilm dispersal, there are many agents already in the pipeline. Agents with proven or hypothetical dispersal activity are in preclinical development, and some are even in clinical trials. Thus, the burden is now on researchers to expand our understanding of the consequences, whether they be positive or negative, that dispersal has on the host. This will require better imaging and tracking of dispersed cells in vivo, an understanding of phenotypic changes that dispersal induces, and a thorough assessment of the host response to dispersal. As with all novel therapeutics, there is risk associated with biofilm dispersal. However, the promise of being able to better treat the 80% of chronic infections that involve biofilm is worth this risk.

Acknowledgments This work was supported by grants from the National Institutes of Health (R21 AI137462-01A1 to KPR and R01AI150761-01A1 to KS), the Ted Nash Long Life Foundation, The

Jasper L. and Jack Denton Wilson Foundation and the Department of Defense (MIDRP W0318_19_NM_PP) to KPR.

Compliance with Ethical Standards The authors have no conflicts of interest to disclose.

References

Andersen JB, Hultqvist LD, Jansen CU, Jakobsen TH, Nilsson M, Rybtke M, Uhd J, Fritz BG, Seifert R, Berthelsen J (2021a) Identification of small molecules that interfere with c-di-GMP signaling and induce dispersal of *Pseudomonas aeruginosa* biofilms. NPJ Biofilms Microbiomes 7:1–13

Andersen JB, Kragh KN, Hultqvist LD, Rybtke M, Nilsson M, Jakobsen TH, Givskov M, Tolker-Nielsen T (2021b) Induction of native c-di-GMP phosphodiesterases leads to dispersal of *Pseudomonas aeruginosa* biofilms. Antimicrob Agents Chemother 65:e02431–e02420

Antoniani D, Rossi E, Rinaldo S, Bocci P, Lolicato M, Paiardini A, Raffaelli N, Cutruzzolà F, Landini P (2013) The immunosuppressive drug azathioprine inhibits biosynthesis of the bacterial signal molecule cyclic-di-GMP by interfering with intracellular nucleotide pool availability. Appl Microbiol Biotechnol 97:7325–7336

Applegate DH, Bryers JD (1991) Effects of carbon and oxygen limitations and calcium concentrations on biofilm removal processes. Biotechnol Bioeng 37:17–25

Aridis Pharmaceuticals (2019). https://www.biospace.com/article/releases/aridis-pharmaceuticals-reports-phase-2-clinical-trial-results-of-ar-105-for-the-treatment-of-ventilator-associated-pneumonia-caused-by-pseudomonas-aeruginosa/

Aslam S, Trautner BW, Ramanathan V, Darouiche RO (2008) Pilot trial of N-acetylcysteine and tigecycline as a catheter-lock solution for treatment of hemodialysis catheter-associated bacteremia. Infect Control Hosp Epidemiol 29:894–897

Baidamshina DR, Trizna EY, Holyavka MG, Bogachev MI, Artyukhov VG, Akhatova FS, Rozhina EV, Fakhrullin RF, Kayumov AR (2017) Targeting microbial biofilms using Ficin, a nonspecific plant protease. Sci Rep 7:46068

Baker B (2020) 1289. Pravibismane is a potent, broad spectrum anti-infective small molecule that rapidly disrupts bacterial bioenergetics and halts bacterial growth. Open Forum Infect Dis 7: S659–SS60

Baker P, Whitfield GB, Hill PJ, Little DJ, Pestrak MJ, Robinson H, Wozniak DJ, Howell PL (2015) Characterization of the Pseudomonas aeruginosa glycoside hydrolase PslG reveals that its levels are critical for Psl polysaccharide biosynthesis and biofilm formation. J Biol Chem 290:28374–28387

Baker P, Hill PJ, Snarr BD, Alnabelseya N, Pestrak MJ, Lee MJ, Jennings LK, Tam J, Melnyk RA, Parsek MR, Sheppard DC, Wozniak DJ, Howell PL (2016) Exopolysaccharide biosynthetic glycoside hydrolases can be utilized to disrupt and prevent Pseudomonas aeruginosa biofilms. Sci Adv 2:e1501632

Ballard TE, Richards JJ, Wolfe AL, Melander C (2008) Synthesis and antibiofilm activity of a second-generation reverse-amide oroidin library: a structure–activity relationship study. Chem Eur J 14:10745–10761

Banin E, Brady KM, Peter Greenberg E (2006) Chelator-induced dispersal and killing of Pseudomonas aeruginosa cells in a biofilm. Appl Environ Microbiol 72:2064–2069

Barraud N, Hassett DJ, Hwang SH, Rice SA, Kjelleberg S, Webb JS (2006) Involvement of nitric oxide in biofilm dispersal of Pseudomonas aeruginosa. J Bacteriol 188:7344–7353

Barraud N, Schleheck D, Klebensberger J, Webb JS, Hassett DJ, Rice SA, Kjelleberg S (2009) Nitric oxide signaling in Pseudomonas aeruginosa biofilms mediates phosphodiesterase activity, decreased cyclic di-GMP levels, and enhanced dispersal. J Bacteriol 191:7333–7342

Barraud N, Kardak BG, Yepuri NR, Howlin RP, Webb JS, Faust SN, Kjelleberg S, Rice SA, Kelso MJ (2012) Cephalosporin-3′-diazeniumdiolates: targeted NO-donor prodrugs for dispersing bacterial biofilms. Angew Chem Int Ed 51:9057–9060

Berne C, Ellison CK, Ducret A, Brun YV (2018) Bacterial adhesion at the single-cell level. Nat Rev Microbiol 16:616–627

Blasi F, Page C, Rossolini GM, Pallecchi L, Matera MG, Rogliani P, Cazzola M (2016) The effect of N-acetylcysteine on biofilms: implications for the treatment of respiratory tract infections. Respir Med 117:190–197

Boles BR, Horswill AR (2008) Agr-mediated dispersal of Staphylococcus aureus biofilms. PLoS Pathog 4:e1000052

Bounds K, Colmer-Hamood JA, Myntti M, Jeter RM, Hamood AN (2021) The influence of a biofilm-dispersing wound gel on the wound healing process. Int Wound J 19(3):553–572

Cai Y-m, Webb JS (2020) Optimization of nitric oxide donors for investigating biofilm dispersal response in *Pseudomonas aeruginosa* clinical isolates. Appl Microbiol Biotechnol 104:8859–8869

Canovas J, Baldry M, Bojer MS, Andersen PS, Grzeskowiak PK, Stegger M, Damborg P, Olsen CA, Ingmer H (2016) Cross-talk between Staphylococcus aureus and other Staphylococcal species via the agr quorum sensing system. Front Microbiol 7:1733

Cathie K, Howlin R, Carroll M, Clarke S, Connett G, Cornelius V, Daniels T, Duignan C, Hall-Stoodley L, Jefferies J, Kelso M, Kjelleberg S, Legg J, Pink S, Rogers G, Salib R, Stoodley P, Sukhtankar P, Webb J, Faust S (2014) G385 RATNO–reducing antibiotic tolerance using nitric oxide in cystic fibrosis: report of a proof of concept clinical trial. Arch Dis Child 99:A159

Chan KH, Allen GC, Kelley PE, Streubel SO, Friedman NR, Yoon P, Gao D, Ruiz AG, Jung TTK (2018) Dornase alfa ototoxic effects in animals and efficacy in the treatment of clogged tympanostomy tubes in children: a preclinical study and a randomized clinical trial. JAMA Otolaryngol Head Neck Surg 144:776–780

Chemani C, Imberty A, de Bentzmann S, Pierre M, Wimmerová M, Guery BP, Faure K (2009) Role of LecA and LecB lectins in *Pseudomonas aeruginosa*-induced lung injury and effect of carbohydrate ligands. Infect Immun 77:2065–2075

Cho KH, Tryon RG, Kim JH (2020) Screening for Diguanylate cyclase (DGC) inhibitors mitigating bacterial biofilm formation. Front Chem 8:264

Christensen LD, van Gennip M, Rybtke MT, Hong W, Chiang W-C, Alhede M, Høiby N, Nielsen TE, Givskov M, Tolker-Nielsen T (2013) Clearance of *Pseudomonas aeruginosa* foreign-body biofilm infections through reduction of the cyclic di-GMP level in the bacteria. Infect Immun 81:2705–2713

Ciofu O, Rojo-Molinero E, Macià MD, Antonio O (2017) Antibiotic treatment of biofilm infections. APMIS 125:304–319

Colvin KM, Irie Y, Tart CS, Urbano R, Whitney JC, Ryder C, Howell PL, Wozniak DJ, Parsek MR (2012) The Pel and Psl polysaccharides provide Pseudomonas aeruginosa structural redundancy within the biofilm matrix. Environ Microbiol 14:1913–1928

Connolly KL, Roberts AL, Holder RC, Reid SD (2011) Dispersal of Group A streptococcal biofilms by the cysteine protease SpeB leads to increased disease severity in a murine model. PLoS One 6:e18984

Cooke AC, Florez C, Dunshee EB, Lieber AD, Terry ML, Light CJ, Schertzer JW (2020) Pseudomonas quinolone signal-induced outer membrane vesicles enhance biofilm dispersion in Pseudomonas aeruginosa. mSphere 5(6):1

Costerton JW (2001) Cystic fibrosis pathogenesis and the role of biofilms in persistent infection. Trends Microbiol 9:50–52

CysticFibrosisNewsToday (2000). https://cysticfibrosisnewstoday.com/2020/06/01/pravibismane-granted-fda-orphan-drug-designation-for-cf-lung-infections/

Darouiche RO, Mansouri MD, Gawande PV, Madhyastha S (2009) Antimicrobial and antibiofilm efficacy of triclosan and DispersinB combination. J Antimicrob Chemother 64:88–93

Davey ME, Caiazza NC, O'Toole GA (2003) Rhamnolipid surfactant production affects biofilm architecture in Pseudomonas aeruginosa PAO1. J Bacteriol 185:1027–1036

Davies D (2003) Understanding biofilm resistance to antibacterial agents. Nat Rev Drug Discov 2: 114–122

Davies D (2007) Induction of a physiological dispersion response in bacterial cells in a biofilm, WO2008143889A1. In Google Patents

Davies DG, Marques CN (2009) A fatty acid messenger is responsible for inducing dispersion in microbial biofilms. J Bacteriol 191:1393–1403

Devaraj A, Buzzo JR, Mashburn-Warren L, Gloag ES, Novotny LA, Stoodley P, Bakaletz LO, Goodman SD (2019) The extracellular DNA lattice of bacterial biofilms is structurally related to Holliday junction recombination intermediates. Proc Natl Acad Sci U S A 116:25068–25077

Diggle SP, Stacey RE, Dodd C, Cámara M, Williams P, Winzer K (2006) The galactophilic lectin, LecA, contributes to biofilm development in *Pseudomonas aeruginosa*. Environ Microbiol 8: 1095–1104

Draughn G, Logan CL, Allen PA, Routh MR, Stone KR, Kirker LB, Schuchman RM, Linder KE, Baynes RE, James G, Melander C, Pollard A, Cavanagh J (2017) Evaluation of a 2-aminoimidazole variant as adjuvant treatment for dermal bacterial infections. Drug Des Devel Ther 11:153–162

Elgharably H, Hussain ST, Shrestha NK, Blackstone EH, Pettersson GB (2016) Current hypotheses in cardiac surgery: biofilm in infective endocarditis. Semin Thorac Cardiovasc Surg 28:56–59

El-Tarabily KA, El-Saadony MT, Alagawany M, Arif M, Batiha GE, Khafaga AF, Elwan HAM, Elnesr SS, Abd ME, El-Hack. (2021) Using essential oils to overcome bacterial biofilm formation and their antimicrobial resistance. Saudi J Biol Sci 28:5145–5156

Estellés A, Woischnig A-K, Liu K, Stephenson R, Lomongsod E, Nguyen D, Zhang J, Heidecker M, Yang Y, Simon RJ (2016) A high-affinity native human antibody disrupts biofilm from *Staphylococcus aureus* bacteria and potentiates antibiotic efficacy in a mouse implant infection model. Antimicrob Agents Chemother 60:2292–2301

Fleming D, Rumbaugh KP (2017) Approaches to dispersing medical biofilms. Microorganisms 5: 15

Fleming D, Rumbaugh K (2018) The consequences of biofilm dispersal on the host. Sci Rep 8: 10738

Fleming D, Chahin L, Rumbaugh K (2017) Glycoside hydrolases degrade Polymicrobial bacterial biofilms in wounds. Antimicrob Agents Chemother 61(2):e01998–e01916

Fleming D, Redman WK, Welch GS, Mdluli NV, Rouchon CN, Frank KL, Rumbaugh KP (2020) Utilizing glycoside hydrolases to improve the quantification and visualization of biofilm bacteria. Biofilms 2:100037

Flemming HC, Wingender J (2010) The biofilm matrix. Nat Rev Microbiol 8:623–633

Flemming HC, Neu TR, Wozniak DJ (2007) The EPS matrix: the "house of biofilm cells". J Bacteriol 189:7945–7947

Gao Z, Zhong W, Chen K, Tang P, Guo J (2020) Chemical composition and anti-biofilm activity of essential oil from Citrus medica L. var. sarcodactylis Swingle against Listeria monocytogenes. Ind Crop Prod 144:112036

Gawande PV, Clinton AP, LoVetri K, Yakandawala N, Rumbaugh KP, Madhyastha S (2014a) Antibiofilm efficacy of DispersinB((R)) wound spray used in combination with a silver wound dressing. Microbiol Insights 7:9–13

Gawande PV, Leung KP, Madhyastha S (2014b) Antibiofilm and antimicrobial efficacy of DispersinB(R)-KSL-W peptide-based wound gel against chronic wound infection associated bacteria. Curr Microbiol 68:635–641

Gbejuade HO, Lovering AM, Webb JC (2015) The role of microbial biofilms in prosthetic joint infections. Acta Orthop 86:147–158

Goodman SD, Bakaletz LO (2021) Antibody compositions for disrupting biofilms. In Google Patents

Gotz F (2002) Staphylococcus and biofilms. Mol Microbiol 43:1367–1378

Grishin AV, Krivozubov MS, Karyagina AS, Gintsburg AL (2015) *Pseudomonas aeruginosa* lectins as targets for novel antibacterials. Acta Nat 7:29–41

Group, F. D. A. Wound Healing Clinical Focus (2001) Guidance for industry: chronic cutaneous ulcer and burn wounds-developing products for treatment. Wound Repair Regen 9:258–268

Ha DG, O'Toole GA (2015) C-di-GMP and its effects on biofilm formation and dispersion: a pseudomonas aeruginosa review. Microbiol Spectr 3:MB-0003-2014

Hatt JK, Rather PN (2008) Role of bacterial biofilms in urinary tract infections. Curr Top Microbiol Immunol 322:163–192

Hee C-S, Habazettl J, Schmutz C, Schirmer T, Jenal U, Grzesiek S (2020) Intercepting second-messenger signaling by rationally designed peptides sequestering c-di-GMP. Proc Natl Acad Sci 117:17211–17220

Hentzer M, Riedel K, Rasmussen TB, Heydorn A, Andersen JB, Parsek MR, Rice SA, Eberl L, Molin S, Hoiby N, Kjelleberg S, Givskov M (2002) Inhibition of quorum sensing in *Pseudomonas aeruginosa* biofilm bacteria by a halogenated furanone compound. Microbiology 148: 87–102

Howlin RP, Cathie K, Hall-Stoodley L, Cornelius V, Duignan C, Allan RN, Fernandez BO, Barraud N, Bruce KD, Jefferies J, Kelso M, Kjelleberg S, Rice SA, Rogers GB, Pink S, Smith C, Sukhtankar PS, Salib R, Legg J, Carroll M, Daniels T, Feelisch M, Stoodley P, Clarke SC, Connett G, Faust SN, Webb JS (2017) Low-dose nitric oxide as targeted anti-biofilm adjunctive therapy to treat chronic *Pseudomonas aeruginosa* infection in cystic fibrosis. Mol Ther 25:2104–2116

Itoh Y, Wang X, Hinnebusch BJ, Preston JF 3rd, Romeo T (2005) Depolymerization of beta-1,6-N-acetyl-D-glucosamine disrupts the integrity of diverse bacterial biofilms. J Bacteriol 187:382–387

Izano EA, Wang H, Ragunath C, Ramasubbu N, Kaplan JB (2007) Detachment and killing of Aggregatibacter actinomycetemcomitans biofilms by dispersin B and SDS. J Dent Res 86:618–622

Jain A, Parihar DK (2018) Antibacterial, biofilm dispersal and antibiofilm potential of alkaloids and flavonoids of Curcuma. Biocatal Agric Biotechnol 16:677–682

Jang CH, Piao YL, Huang X, Yoon EJ, Park SH, Lee K, Zhan C-G, Cho H (2016) Modeling and re-engineering of Azotobacter vinelandii alginate lyase to enhance its catalytic efficiency for accelerating biofilm degradation. PLoS One 11:e0156197

Jeon AB, Ackart DF, Li W, Jackson M, Melander RJ, Melander C, Abramovitch RB, Chicco AJ, Basaraba RJ, Obregón-Henao A (2019) 2-aminoimidazoles collapse mycobacterial proton motive force and block the electron transport chain. Sci Rep 9:1513

Jiao Y, Cody GD, Harding AK, Wilmes P, Schrenk M, Wheeler KE, Banfield JF, Thelen MP (2010) Characterization of extracellular polymeric substances from acidophilic microbial biofilms. Appl Environ Microbiol 76:2916–2922

Johansson EMV, Crusz SA, Kolomiets E, Buts L, Kadam RU, Cacciarini M, Bartels K-M, Diggle SP, Cámara M, Williams P (2008) Inhibition and dispersion of *Pseudomonas aeruginosa* biofilms by glycopeptide dendrimers targeting the fucose-specific lectin LecB. Chem Biol 15: 1249–1257

KaneBiotech (2019). https://ir.kanebiotech.com/press-releases/detail/210/kane-biotech-positions-dispersinb-for-wound-care

Kaplan JB, Ragunath C, Velliyagounder K, Fine DH, Ramasubbu N (2004) Enzymatic detachment of Staphylococcus epidermidis biofilms. Antimicrob Agents Chemother 48:2633–2636

Kaplan JB, LoVetri K, Cardona ST, Madhyastha S, Sadovskaya I, Jabbouri S, Izano EA (2012) Recombinant human DNase I decreases biofilm and increases antimicrobial susceptibility in staphylococci. J Antibiot 65:73–77

Kaplan JB, Mlynek KD, Hettiarachchi H, Alamneh YA, Biggemann L, Zurawski DV, Black CC, Bane CE, Kim RK, Granick MS (2018) Extracellular polymeric substance (EPS)-degrading enzymes reduce staphylococcal surface attachment and biocide resistance on pig skin in vivo. PLoS One 13:e0205526

Keifer PA, Schwartz RE, Koker MES, Hughes Jr RG, Rittschof D, Rinehart KL (1991) Bioactive bromopyrrole metabolites from the Caribbean sponge *Agelas conifera*. J Org Chem 56:2965–2975

Kjelleberg S, Molin S (2002) Is there a role for quorum sensing signals in bacterial biofilms? Curr Opin Microbiol 5:254–258

Koswatta PB, Lovely CJ (2011) Structure and synthesis of 2-aminoimidazole alkaloids from Leucetta and Clathrina sponges. Nat Prod Rep 28:511–528

Kumar P, Lee JH, Beyenal H, Lee J (2020) Fatty acids as Antibiofilm and Antivirulence agents. Trends Microbiol 28:753–768

Lade H, Paul D, Kweon JH (2014) N-acyl homoserine lactone-mediated quorum sensing with special reference to use of quorum quenching bacteria in membrane biofouling control. Biomed Res Int 2014:162584

Lasa I, Penades JR (2006) Bap: a family of surface proteins involved in biofilm formation. Res Microbiol 157:99–107

Le Mauff F, Bamford NC, Alnabelseya N, Zhang Y, Baker P, Robinson H, Codee JDC, Howell PL, Sheppard DC (2019) Molecular mechanism of Aspergillus fumigatus biofilm disruption by fungal and bacterial glycoside hydrolases. J Biol Chem 294:10760–10772

Lewenza S (2013) Extracellular DNA-induced antimicrobial peptide resistance mechanisms in Pseudomonas aeruginosa. Front Microbiol 4:21

Li Y, Heine S, Entian M, Sauer K, Frankenberg-Dinkel N (2013) NO-induced biofilm dispersion in *Pseudomonas aeruginosa* is mediated by a MHYT-domain coupled phosphodiesterase. J Bacteriol 195:3531–3542

Limoli DH, Jones CJ, Wozniak DJ (2015) Bacterial extracellular polysaccharides in biofilm formation and function. Microbiol Spectr 3:3

Lin Y, Zhou X, Li Y (2022) Strategies for Streptococcus mutans biofilm dispersal through extracellular polymeric substances disruption. Mol Oral Microbiol 37:1–8

Liu P, Huang Q, Chen W (2012) Heterologous expression of bacterial nitric oxide synthase gene: a potential biological method to control biofilm development in the environment. Can J Microbiol 58:336–344

Liu C, Liew CW, Wong YH, Tan ST, Poh WH, Manimekalai MSS, Rajan S, Xin L, Liang ZX, Gruber G, Rice SA, Lescar J (2018a) Insights into biofilm dispersal regulation from the crystal structure of the PAS-GGDEF-EAL region of RbdA from Pseudomonas aeruginosa. J Bacteriol 200:e00515–e00517

Liu Y, Naha PC, Hwang G, Kim D, Huang Y, Simon-Soro A, Jung H-I, Ren Z, Li Y, Gubara S, Alawi F, Zero D, Hara AT, Cormode DP, Koo H (2018b) Topical ferumoxytol nanoparticles disrupt biofilms and prevent tooth decay in vivo via intrinsic catalytic activity. Nat Commun 9: 2920

Lopez D, Vlamakis H, Kolter R (2010) Biofilms. Cold Spring Harb Perspect Biol 2:a000398

Lu TK, Collins JJ (2007) Dispersing biofilms with engineered enzymatic bacteriophage. Proc Natl Acad Sci U S A 104:11197–11202

Marques CN, Morozov A, Planzos P, Zelaya HM (2014) The fatty acid signaling molecule cis-2-decenoic acid increases metabolic activity and reverts persister cells to an antimicrobial-susceptible state. Appl Environ Microbiol 80:6976–6991

Marques CN, Davies DG, Sauer K (2015) Control of biofilms with the fatty acid signaling molecule cis-2-decenoic acid. Pharmaceuticals (Basel) 8:816–835

Martin I, Waters V, Grasemann H (2021) Approaches to targeting bacterial biofilms in cystic fibrosis airways. Int J Mol Sci 22(4):2155

Marvasi M, Chen C, Carrazana M, Durie IA, Teplitski M (2014) Systematic analysis of the ability of nitric oxide donors to dislodge biofilms formed by *Salmonella enterica* and *Escherichia coli* O157: H7. AMB Express 4:1–11

Mashburn LM, Whiteley M (2005) Membrane vesicles traffic signals and facilitate group activities in a prokaryote. Nature 437:422–425

McCallon SK, Weir D, Lantis JC 2nd (2014) Optimizing wound bed preparation with collagenase enzymatic debridement. J Am Coll Clin Wound Spec 6:14–23
McDougald D, Rice SA, Barraud N, Steinberg PD, Kjelleberg S (2011) Should we stay or should we go: mechanisms and ecological consequences for biofilm dispersal. Nat Rev Microbiol 10: 39–50
Miller KG, Tran PL, Haley CL, Kruzek C, Colmer-Hamood JA, Myntti M, Hamood AN (2014) Next science wound gel technology, a novel agent that inhibits biofilm development by gram-positive and gram-negative wound pathogens. Antimicrob Agents Chemother 58:3060–3072
Mitrofanova O, Mardanova A, Evtugyn V, Bogomolnaya L, Sharipova M (2017) Effects of bacillus serine proteases on the bacterial biofilms. Biomed Res Int 2017:8525912
Mitsuishi M, Oshikata T, Kumabe S, Kobayashi A, Katoku K, Kanno T, Hamamura M, Tsuchitani M (2015) Histological dermal changes caused by preparation and application procedures in percutaneous dose toxicity studies in dogs, rabbits and rats. J Toxicol Pathol 28:1–9
Montanaro L, Poggi A, Visai L, Ravaioli S, Campoccia D, Speziale P, Arciola CR (2011) Extracellular DNA in biofilms. Int J Artif Organs 34:824–831
Nicol M, Alexandre S, Luizet JB, Skogman M, Jouenne T, Salcedo SP, De E (2018) Unsaturated fatty acids affect quorum sensing communication system and inhibit motility and biofilm formation of Acinetobacter baumannii. Int J Mol Sci 19:214
Nijland R, Hall MJ, Burgess JG (2010) Dispersal of biofilms by secreted, matrix degrading, bacterial DNase. PLoS One 5:e15668
Nilsson CL (2003) Lectins: proteins that interpret the sugar code. Anal Chem 75:348–A-53 A
Novotny LA, Goodman SD, Bakaletz LO (2020) Targeting a bacterial DNABII protein with a chimeric peptide immunogen or humanised monoclonal antibody to prevent or treat recalcitrant biofilm-mediated infections. EBioMedicine 59:102867
Oakley JL, Weiser R, Powell LC, Forton J, Mahenthiralingam E, Rye PD, Hill KE, Thomas DW, Pritchard MF (2021) Phenotypic and genotypic adaptations in *Pseudomonas aeruginosa* biofilms following long-term exposure to an alginate oligomer therapy. mSphere 6:e01216–e01220
Ohana P, Delmer DP, Carlson RW, Glushka J, Azadi P, Bacic T, Benziman M (1998) Identification of a novel triterpenoid saponin from Pisum sativum as a specific inhibitor of the diguanylate cyclase of Acetobacter xylinum. Plant Cell Physiol 39:144–152
Omar A, Wright JB, Schultz G, Burrell R, Nadworny P (2017) Microbial biofilms and chronic wounds. Microorganisms 5(1):9
Otto M (2014) Staphylococcus aureus toxins. Curr Opin Microbiol 17:32–37
Pang Z, Raudonis R, Glick BR, Lin T-J, Cheng Z (2019) Antibiotic resistance in *Pseudomonas aeruginosa:* mechanisms and alternative therapeutic strategies. Biotechnol Adv 37:177–192
Papa R, Garzoli S, Vrenna G, Sabatino M, Sapienza F, Relucenti M, Donfrancesco O, Fiscarelli EV, Artini M, Selan L, Ragno R (2020) Essential oils biofilm modulation activity, chemical and machine learning analysis. Application on Staphylococcus aureus isolates from cystic fibrosis patients. Int J Mol Sci 21:9258
Papenfort K, Bassler BL (2016) Quorum sensing signal-response systems in gram-negative bacteria. Nat Rev Microbiol 14:576–588
Pérez MJ, Falqué E, Domínguez H (2016) Antimicrobial action of compounds from marine seaweed. Mar Drugs 14:52
Powell LC, Pritchard MF, Ferguson EL, Powell KA, Patel SU, Rye PD, Sakellakou S-M, Buurma NJ, Brilliant CD, Copping JM, Menzies GE, Lewis PD, Hill KE, Thomas DW (2018) Targeted disruption of the extracellular polymeric network of *Pseudomonas aeruginosa* biofilms by alginate oligosaccharides. NPJ Biofilms Microbiomes 4:13–13
Puskas A, Greenberg EP, Kaplan S, Schaefer AL (1997) A quorum-sensing system in the free-living photosynthetic bacterium Rhodobacter sphaeroides. J Bacteriol 179:7530–7537
Quintas V, Prada-López I, Donos N, Suárez-Quintanilla D, Tomás I (2015a) Antiplaque effect of essential oils and 0.2% chlorhexidine on an in situ model of oral biofilm growth: a randomised clinical trial. PLoS One 10:e0117177

Quintas V, Prada-López I, Prados-Frutos JC, Tomás I (2015b) In situ antimicrobial activity on oral biofilm: essential oils vs. 0.2% chlorhexidine. Clin Oral Investig 19:97–107

Quintas V, Prada-López I, Carreira MJ, Suárez-Quintanilla D, Balsa-Castro C, Tomás I (2017) In situ antibacterial activity of essential oils with and without alcohol on oral biofilm: a randomized clinical trial. Front Microbiol 8:2162

Ramasubbu N, Thomas LM, Ragunath C, Kaplan JB (2005) Structural analysis of dispersin B, a biofilm-releasing glycoside hydrolase from the periodontopathogen Actinobacillus actinomycetemcomitans. J Mol Biol 349:475–486

Redman WK, Welch GS, Rumbaugh KP (2021) Assessing biofilm dispersal in murine wounds. J Vis Exp 174:62136

Ren Z, Kim D, Paula AJ, Hwang G, Liu Y, Li J, Daniell H, Koo H (2019) Dual-targeting approach degrades biofilm matrix and enhances bacterial killing. J Dent Res 98:322–330

Research, Center for Drug Evaluation (2017) Antibacterial therapies for patients with an unmet medical need for the treatment of serious bacterial diseases. Center for Drug Evaluation and Research

Richards JJ, Eric Ballard T, Melander C (2008) Inhibition and dispersion of *Pseudomonas aeruginosa* biofilms with reverse amide 2-aminoimidazole oroidin analogues. Org Biomol Chem 6:1356–1363

Robijns SCA, Roberfroid S, Van Puyvelde S, De Pauw B, Santamaría EU, De Weerdt A, De Coster D, Kim Hermans SCJ, Keersmaecker D, Vanderleyden J (2014) A GFP promoter fusion library for the study of *Salmonella* biofilm formation and the mode of action of biofilm inhibitors. Biofouling 30:605–625

Rodensky M, Zolkov C, Davies DG (2020) Method and composition for water treatment, WO2020240559A1. Google Patents

Rogers SA, Huigens III RW, Cavanagh J, Melander C (2010) Synergistic effects between conventional antibiotics and 2-aminoimidazole-derived antibiofilm agents. Antimicrob Agents Chemother 54:2112–2118

Romling U, Galperin MY, Gomelsky M (2013) Cyclic di-GMP: the first 25 years of a universal bacterial second messenger. Microbiol Mol Biol Rev 77:1–52

Roy AB, Petrova OE, Sauer K (2012) The phosphodiesterase DipA (PA5017) is essential for Pseudomonas aeruginosa biofilm dispersion. J Bacteriol 194:2904–2915

Rudkin JK, McLoughlin RM, Preston A, Massey RC (2017) Bacterial toxins: offensive, defensive, or something else altogether? PLoS Pathog 13:e1006452

Rumbaugh KP, Sauer K (2020) Biofilm dispersion. Nat Rev Microbiol 18:571–586

Saggu SK, Jha G, Mishra PC (2019) Enzymatic degradation of biofilm by metalloprotease from microbacterium sp. SKS10. Front Bioeng Biotechnol 7:192

Sambanthamoorthy K, Sloup RE, Parashar V, Smith JM, Kim EE, Semmelhack MF, Neiditch MB, Waters CM (2012) Identification of small molecules that antagonize diguanylate cyclase enzymes to inhibit biofilm formation. Antimicrob Agents Chemother 56:5202–5211

Sauer K, Cullen MC, Rickard AH, Zeef LA, Davies DG, Gilbert P (2004) Characterization of nutrient-induced dispersion in Pseudomonas aeruginosa PAO1 biofilm. J Bacteriol 186:7312–7326

Schreiber F, Beutler M, Enning D, Lamprecht-Grandio M, Zafra O, Gonzalez-Pastor JE, de Beer D (2011) The role of nitric-oxide-synthase-derived nitric oxide in multicellular traits of Bacillus subtilis 3610: biofilm formation, swarming, and dispersal. BMC Microbiol 11:111

Schwechheimer C, Kuehn MJ (2015) Outer-membrane vesicles from gram-negative bacteria: biogenesis and functions. Nat Rev Microbiol 13:605–619

Schwenk MH (2010) Ferumoxytol: a new intravenous iron preparation for the treatment of iron deficiency anemia in patients with chronic kidney disease. Pharmacotherapy 30:70–79

Shah CB, Mittelman MW, Costerton JW, Parenteau S, Pelak M, Arsenault R, Mermel LA (2002) Antimicrobial activity of a novel catheter lock solution. Antimicrob Agents Chemother 46: 1674–1679

Sharma K, Pagedar Singh A (2018) Antibiofilm effect of DNase against single and mixed species biofilm. Foods 7(3):42

Shinde S, Lee LH, Chu T (2021) Inhibition of biofilm formation by the synergistic action of EGCG-S and antibiotics. Antibiotics (Basel) 10(2):102

Singh PK, Yadav VK, Kalia M, Dohare S, Sharma D, Agarwal V (2017) Pseudomonas aeruginosa auto inducer3-oxo-C12-HSL exerts bacteriostatic effect and inhibits Staphylococcus epidermidis biofilm. Microb Pathog 110:612–619

Snarr BD, Baker P, Bamford NC, Sato Y, Liu H, Lehoux M, Gravelat FN, Ostapska H, Baistrocchi SR, Cerone RP, Filler EE, Parsek MR, Filler SG, Howell PL, Sheppard DC (2017) Microbial glycoside hydrolases as antibiofilm agents with cross-kingdom activity. Proc Natl Acad Sci U S A 114:7124–7129

Solano C, Echeverz M, Lasa I (2014) Biofilm dispersion and quorum sensing. Curr Opin Microbiol 18:96–104

Soren O, Rineh A, Silva DG, Cai Y, Howlin RP, Allan RN, Feelisch M, Davies JC, Connett GJ, Faust SN (2020) Cephalosporin nitric oxide-donor prodrug DEA-C3D disperses biofilms formed by clinical cystic fibrosis isolates of *Pseudomonas aeruginosa*. J Antimicrob Chemother 75:117–125

Stanley NR, Lazazzera BA (2004) Environmental signals and regulatory pathways that influence biofilm formation. Mol Microbiol 52:917–924

Steed DL (2004) Debridement. Am J Surg 187:71S–74S

Suresh MK, Biswas R, Biswas L (2019) An update on recent developments in the prevention and treatment of Staphylococcus aureus biofilms. Int J Med Microbiol 309:1–12

Szymanska M, Karakulska J, Sobolewski P, Kowalska U, Grygorcewicz B, Bottcher D, Bornscheuer UT, Drozd R (2020) Glycoside hydrolase (PelAh) immobilization prevents Pseudomonas aeruginosa biofilm formation on cellulose-based wound dressing. Carbohydr Polym 246:116625

Tetz VV, Tetz GV (2010) Effect of extracellular DNA destruction by DNase I on characteristics of forming biofilms. DNA Cell Biol 29:399–405

Tetz GV, Artemenko NK, Tetz VV (2009) Effect of DNase and antibiotics on biofilm characteristics. Antimicrob Agents Chemother 53:1204–1209

Thompson MG, Truong-Le V, Alamneh YA, Black CC, Anderl J, Honnold CL, Pavlicek RL, Abu-Taleb R, Wise MC, Hall ER (2015) Evaluation of gallium citrate formulations against a multidrug-resistant strain of *Klebsiella pneumoniae* in a murine wound model of infection. Antimicrob Agents Chemother 59:6484–6493

Tielker D, Hacker S, Loris R, Strathmann M, Wingender J, Wilhelm S, Rosenau F, Jaeger K-E (2005) Pseudomonas aeruginosa lectin LecB is located in the outer membrane and is involved in biofilm formation. Microbiology 151:1313–1323

Uppuluri P, Lopez-Ribot JL (2016) Go forth and colonize: dispersal from clinically important microbial biofilms. PLoS Pathog 12:e1005397

Valentini M, Filloux A (2016) Biofilms and cyclic di-GMP (c-di-GMP) signaling: lessons from Pseudomonas aeruginosa and other bacteria. J Biol Chem 291:12547–12555

Vorkapic D, Pressler K, Schild S (2016) Multifaceted roles of extracellular DNA in bacterial physiology. Curr Genet 62:71–79

Walters MC 3rd, Roe F, Bugnicourt A, Franklin MJ, Stewart PS (2003) Contributions of antibiotic penetration, oxygen limitation, and low metabolic activity to tolerance of Pseudomonas aeruginosa biofilms to ciprofloxacin and tobramycin. Antimicrob Agents Chemother 47:317–323

Waryah CB, Wells K, Ulluwishewa D, Chen-Tan N, Gogoi-Tiwari J, Ravensdale J, Costantino P, Gokcen A, Vilcinskas A, Wiesner J, Mukkur T (2017) In vitro antimicrobial efficacy of tobramycin against Staphylococcus aureus biofilms in combination with or without DNase I and/or Dispersin B: a preliminary investigation. Microb Drug Resist 23:384–390

Watters C, Fleming D, Bishop D, Rumbaugh KP (2016) Host responses to biofilm. Prog Mol Biol Transl Sci 142:193–239

Wille J, Coenye T (2020) Biofilm dispersion: the key to biofilm eradication or opening Pandora's box? Biofilms 2:100027
Williams DE, Boon EM (2019) Towards understanding the molecular basis of nitric oxide-regulated group behaviors in pathogenic bacteria. J Innate Immun 11:205–215
Wood TL, Gong T, Zhu L, Miller J, Miller DS, Yin B, Wood TK (2018) Rhamnolipids from Pseudomonas aeruginosa disperse the biofilms of sulfate-reducing bacteria. NPJ Biofilms Microbiomes 4:22
Wu H, Moser C, Wang H-Z, Høiby N, Song Z-J (2015) Strategies for combating bacterial biofilm infections. Int J Oral Sci 7:1–7
Yakandawala N, Gawande PV, LoVetri K, Cardona ST, Romeo T, Nitz M, Madhyastha S (2011) Characterization of the poly-beta-1,6-N-acetylglucosamine polysaccharide component of Burkholderia biofilms. Appl Environ Microbiol 77:8303–8309
Yamada A, Kitamura H, Yamaguchi K, Fukuzawa S, Kamijima C, Yazawa K, Kuramoto M, Wang G-Y-S, Fujitani Y, Uemura D (1997) Development of chemical substances regulating biofilm formation. Bull Chem Soc Jpn 70:3061–3069
Yang Y, Huang Z, Li L-L (2021) Advanced nitric oxide donors: chemical structure of NO drugs NO nanomedicines and biomedical applications. Nanoscale 13:444–459
Zhang X, Bishop PL (2003) Biodegradabilitys of biofilm extracellular polymeric substances. Chemosphere 50:63–69

Chemical Regulation of *Pseudomonas aeruginosa* Biofilm Formation as a Potential Treatment for Persistent Infections

Rasmus Kristensen, Tim Tolker-Nielsen, and Tim Holm Jakobsen

Abstract At the end of the last century, bacterial biofilms were recognised as being a common cause of persistent infections. This finding encouraged intense research in biofilms, revealing that the biofilm life form is an essential factor for bacteria to develop persistent infections, which is because of an increased resistance to the immune system and conventional antibiotics. However, the increased scientific focus on studying the underlying mechanisms of biofilm formation has not yet led to the development of efficient therapeutic applications to eradicate biofilm infections, and treatment of these infections remains a critical struggle for physicians. The bacterial cell-to-cell quorum-sensing signalling system and the intracellular cyclic-di-GMP signalling system have gained increasing attention as potential targets for the treatment of persistent biofilm infections. Here, we describe these two systems in relation to virulence and regulation of biofilm formation in *Pseudomonas aeruginosa*. We also describe the potential use of small molecules to chemically interfere with these systems for treatment purposes, and we review the scientific development and novel findings of compounds that have been identified as potential modulators of the two regulatory processes.

1 Introduction

Over the last decades, it has been generally accepted that bacteria exist not only as solitary single cells but also in the form of biofilms (Tolker-Nielsen 2015). In biofilms, the bacteria are located in densely packed microcolonies concealed in a self-produced matrix of exopolymers. In this life mode, the bacteria attain the highest levels of tolerance to the currently existing assortment of antibiotics and the immune system (Ciofu and Tolker-Nielsen 2019; Hall and Mah 2017). Due to their ability to tolerate high doses of antibiotics, biofilms are a common cause of persistent

R. Kristensen · T. Tolker-Nielsen · T. H. Jakobsen (✉)
Costerton Biofilm Center, Department of Immunology and Microbiology, Faculty of Health and Medical Sciences, University of Copenhagen, Copenhagen, Denmark
e-mail: tholm@sund.ku.dk

K. Richter, K. Nørskov Kragh (eds.), *Antibiofilm Strategies*, Springer Series on Biofilms 11, https://doi.org/10.1007/978-3-031-10992-8_7

infections, hospitalisations, patient suffering and reduced quality of life (Costerton et al. 1999). For example, it is estimated that around 60–70% of hospital-acquired infections are caused by bacterial biofilms (Bryers 2008). Since our current treatment regimens in many cases fail to cure biofilm-based infections, new anti-biofilm drugs and novel treatment strategies are urgently needed.

Exploring the concept of alternative therapies towards antibiotic-resistant infections has gained increasing momentum over the last couple of decades, as the development of new antibiotics has stagnated, and cases of antibiotic-resistant nosocomial infections are on an all-time high (Jernigan et al. 2020). A possible solution to the biofilm infection problem is to use novel drugs to target non-essential components of cell signalling that control biofilm formation and virulence. Combination treatments with such novel drugs and the current classes of antibiotics are expected to result in improved infection clearance.

The primary focus in this chapter is to explore the current strategies targeting (1) the cell-to-cell communication system called quorum sensing (QS), and (2) cyclic-di-GMP (c-di-GMP) mediated secondary messenger signalling, as a means of combating acute and chronic infections by disrupting these non-essential signalling pathways. In this chapter, we will give an introduction to the molecular biology of the QS and c-di-GMP signalling systems and provide examples of the commonly used targets and modulators that have been developed to perturb these systems.

Although QS and c-di-GMP signalling can be found across the bacterial phylum (Manefield and Turner 2002; Ryjenkov et al. 2005), this review will mainly focus on the Gram-negative opportunistic pathogen *Pseudomonas aeruginosa*. This bacterium is involved in a variety of biofilm-based infections, such as cystic fibrosis (CF) pneumonia, chronic wound infections, catheter-associated urinary tract infections, and ventilator-associated pneumonia. Hence, *P. aeruginosa* is considered a model organism for studying biofilm formation and chronic infections (Tolker-Nielsen 2014). Therefore, most of the current knowledge regarding the molecular mechanisms of microbial biofilm formation has originated from work with this bacterium (Tolker-Nielsen 2014).

2 Quorum Sensing

Host colonisation is a difficult task for bacterial pathogens. For it to be successful, it often relies on counteracting the human immune response by synchronising the production of virulence factors. This level of community-wide coordination is accomplished by the cell-to-cell communication QS systems, in which the production of virulence factors in the bacteria is regulated in a cell-density-dependent manner.

QS was first found to be a vital part of controlling the bioluminescent properties of the Gram-negative marine bacterium *Vibrio fischeri* (Nealson and Hastings 1979), through the LuxI-mediated synthesis of the signal molecule *N*-acyl homoserine lactone (AHL) and its interaction with the native receptor protein LuxR. It was observed that *V. fischeri* lacked bioluminescent properties under low cell density condition. However, when the community had reached a certain density, genes related to bioluminescence were found to be upregulated, thereby generating light. Further research into the topic revealed the role that community size plays in the regulation of QS-related genes. In general, an individual cell synthesising AHL cannot reach the signal molecule threshold by itself and thereby cannot self-activate the QS-related genes. Only when the bacterial population reaches a certain size does the surrounding concentration of signal molecules become high enough to form a complex with the QS receptor leading to the binding of DNA and transcription initiation (Kaplan and Greenberg 1985). An additional effect of this is the creation of a positive feedback loop, leading to the additional synthesis of signal molecules. Although there exist homologues to the *luxI/luxR* system among Gram-negative bacteria, the AHL molecules are species-specific, resulting from a variation in the attached side groups of the core AHL molecule, thus avoiding communication jamming in multi-species microbial communities (Gray and Garey 2001).

The main QS system of *P. aeruginosa* consists of two *luxI-luxR* homologues placed in a hierarchical formation, with *lasI-lasR* being the main regulator, followed by the second system *rhlI-rhlR*. The Rhl system is positively regulated by the Las system (Latifi et al. 1995; Pearson et al. 1994). Both systems are composed of a signal synthetase LasI/RhlI and receptor proteins LasR/RhlR, with each of the signal synthetases producing unique signal molecules. LasR forms a complex with *N*-3-oxo-dodecanoyl-L-homoserine lactone (3-oxo-C12-HSL) leading to activation of *rhlI* and the synthesis of *N*-butyryl-L-homoserine lactone (C4-HSL) (Pearson et al. 1994).

Sputum samples from patients with late-stage CF often contain *P. aeruginosa lasR* mutants (Hoffman et al. 2009). However, this does not mean that the entire QS network of *P. aeruginosa* is dismantled, as shown by Lee et al. (2013) and (Pesci et al. (1999) and their studies on an additional QS network that is interwoven with the *las* and *rhl* QS circuit. They further observed induction of *rhl*-related genes in a *lasR* mutant strain and found that additional supplementation of the native signal molecule does not increase the expression of these genes. This occurs as a result of the activity of an additional system, employing its unique signal molecule, 2-heptyl-3-hydroxy-4-quinolone (PQS), which is regarded as a mediator between the *las* and *rhl* system.

Depending on the growth conditions, it is estimated that up to 6% of *P. aeruginosa*'s genome is regulated by the QS signal system (Schuster et al. 2003; Hentzer et al. 2003). Among these genes is a plethora of virulence factors that aid in countering the immune response and, more importantly, in a biofilm context, components providing crucial structure integrity to the biofilm.

Extracellular DNA (eDNA) is one of the core building blocks of biofilms that provide durability and tolerance towards positively charged antibiotics, such as aminoglycosides and anti-microbial peptides (Whitchurch et al. 2002; Chiang et al. 2013; Allesen-Holm et al. 2006; Lewenza 2013). In addition, secondary metabolites, such as phenazines, enhance the integrity of biofilms by intercalating with eDNA (Das et al. 2015), and pyocyanin can contribute to the release of DNA through the formation of reactive oxygen species which impair cell membranes (Das and Manefield 2013). Although the generation of eDNA by *P. aeruginosa* depends on the PQS system (Tahrioui et al. 2019; Allesen-Holm et al. 2006; Soh et al. 2021), it has been proposed that *P. aeruginosa* biofilms can be stabilised through the release of additional eDNA during cell lysis of immune cells (Walker et al. 2005). Recently, the twin-arginine translocation (tat) pathway was reported to be involved in eDNA release as well as production of phenazines, rhamnolipid and micro-vesicles through PQS signalling. It was shown that *tat* mutants were unable to translocate the Rieske subunit of the cytochrome bc1 complex leading to defective PQS signalling (Soh et al. 2021).

Two important virulence factors for host colonisation, pyocyanin and rhamnolipid regulated by the rhl/pqs system can mediate lysis of eukaryotic cells. Pyocyanin (5-methyl-1[5H]-phenazinone) is a secondary metabolite that belongs to a class of molecules known as phenazines, that is often easily recognisable through their vibrant colours (Lau et al. 2004). Genes related to phenazine synthesis are upregulated by PQS, leading to the production of a precursor molecule that will eventually become one of three phenazines. Pyocyanin affects eukaryotic cells by crossing the cell membrane and creating intracellular reactive oxygen species, such as hydrogen peroxide which damages the mitochondria (Virreira Winter and Zychlinsky 2018; Managò et al. 2015) and causes cell lysis (Das and Manefield 2012). Rhamnolipid is a bio-surfactant that is produced under the regulation of the rhl system in *P. aeruginosa*. It has multiple roles in the pathogenesis of *P. aeruginosa*, as it exhibits haemolytic properties (Johnson and Boese-Marrazzo 1980), aids in bacterial motility and plays various roles in biofilm formation (Pamp and Tolker-Nielsen 2007; Davey et al. 2003). The reported haemolytic effect that rhamnolipid has on eukaryotic cells, such as polymorphonuclear leukocytes (PMNs) (Jensen et al. 2007), has led to the coinage of the term "rhamnolipid shielding" which is used to describe rhamnolipid-mediated protection of biofilms against the activity of PMNs (Alhede et al. 2009).

3 c-di-GMP Signalling

Moshe Benziman and colleagues were the first to describe that the synthesis of cellulose by *Gluconacetobacter xylinus* is controlled by the intracellular secondary messenger bis(3′→5′)-cyclic dimeric guanylic acid (c-di-GMP) (Ross et al. 1987). However, it was not until the beginning of this century that it became clear that most Gram-negative bacteria and several Gram-positive bacteria employ c-di-GMP signalling to regulate whether they form a biofilm or assume a planktonic single-cell lifestyle (Jenal et al. 2017; Ryjenkov et al. 2005).

The planktonic bacterial lifestyle is characterised by a low intracellular concentration of c-di-GMP, whereas an elevated level of c-di-GMP leads to the production of adhesins and exopolysaccharides and thereby biofilms. The concentration of c-di-

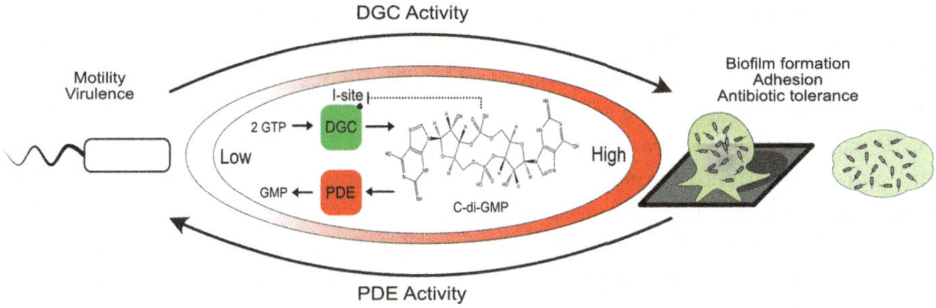

Fig. 1 The intracellular concentration of c-di-GMP drives the biofilm cycle. High c-di-GMP levels are associated with biofilm bacteria, whereas low c-di-GMP levels are associated with planktonic bacteria. In response to environmental cues, the synthesis (DGC activity) or degradation (PDE activity) of c-di-GMP occurs, causing transitions between the planktonic and biofilm lifestyles. An elevated cellular level of c-di-GMP drives planktonic bacteria to form surface-attached biofilms or aggregates, whereas a reduction in the c-di-GMP level causes dispersal of biofilm bacteria into the planktonic mode of life. *GMP* guanosine monophosphate, *DGC* diguanylate cyclase, *PDE* phosphodiesterase

GMP remains in a constant flux as the cells react to various cues produced by the surrounding environment and adjust the metabolism of c-di-GMP to cope with such changes (Barraud et al. 2009; Schleheck et al. 2009). Unlike the community-wide cooperation observed in the QS system, the nature of c-di-GMP regulation is "self-centred" as individual cells respond to the local changes in the environment. In response to external cues such as carbon starvation or a lack of oxygen, the bacteria in a biofilm can initiate a dispersal process that is characterised by a decrease in their c-di-GMP content and a down-regulation of the synthesis of adhesins and exopolysaccharides, as well as enzymatic cleavage of biofilm matrix components (Gjermansen et al. 2005, 2006, 2010; Rybtke et al. 2015). Biofilm dispersal can be a localised event within the biofilm in which a sub-population of the inhabiting bacteria disperses and displays enhanced motility to escape unsuitable living conditions, resulting in the release of these cells from the biofilm and their return to a planktonic state (Yu et al. 2015).

The metabolism of the c-di-GMP molecule relies on the opposing activities of two families of enzymes, namely, diguanylate cyclases (DGCs) and c-di-GMP phosphodiesterases (PDEs). The synthesis of c-di-GMP occurs through the binding of two GTP molecules in an antiparallel manner to the GGDEF domain of DGCs, which results in the dimerization of the GTP molecules into a c-di-GMP molecule via condensation (Chan et al. 2004; Chen et al. 2012). In some cases, feedback inhibition of DGCs occurs through the binding of c-di-GMP to an inhibition site (I-site) located upstream of the GGDEF domain (Chan et al. 2004) (Fig. 1). C-di-GMP can be hydrolysed by PDEs containing either an HD-GYP or EAL domain into different molecules depending on the motif of the domain. In some cases, an enzyme can contain both a GGDEF and an EAL domain, hence acting as a hybrid protein that can utilise both catalytic processes. However, only one domain can be active at a time (Hay et al. 2009). In addition to their catalytic domains, DGCs and PDEs have regulatory domains through which they are believed to regulate the bacterial lifestyle (planktonic versus biofilm) in response to environmental cues (Jenal et al. 2017). DNA sequence analysis suggests that most bacterial species possess multiple DGCs and PDEs, and the available evidence suggest that a specific output of distinct c-di-GMP pathways is achieved through local signalling in which specific DGC, PDE, and c-di-GMP effector/target components are arranged in large complexes or c-di-GMP signalling modules (Hengge 2021).

While the intracellular concentration of c-di-GMP is determined by the activity of DGCs and PDEs receiving the appropriate cues, the role of the c-di-GMP effectors is to translate the information into physical adaptation (Ryan et al. 2012). The c-di-GMP effectors are very diverse in structure, ranging from non-coding RNA called riboswitches, to DNA binding transcriptional regulators and biosynthetic enzymes, with the common trend of having no sequence or structure similarity of their c-di-GMP binding domains (Tamayo 2019; Skotnicka et al. 2020; Chou and Galperin 2016). Valentini and Filloux (2016) provided a comprehensive overview of the c-di-GMP effectors discovered in *P. aeruginosa* and their role in the biofilm cycle. Among the most important effectors for biofilm formation are those related to the synthesis of biofilm matrix components. For example, in *P. aeruginosa* cells with low c-di-GMP content, the synthesis of the biofilm matrix components Pel polysaccharide and CdrA adhesin is repressed through binding of FleQ to the promoters of the respective genes (Hickman and Harwood 2008; Baraquet et al. 2012). In case a shift occurs towards an elevated bacterial c-di-GMP content, FleQ binds c-di-GMP and shifts its activity from being a repressor of transcription to being an activator of transcription (Baraquet et al. 2012). The synthesis of Pel and CdrA then promotes biofilm formation.

Genomic analyses have shown that the conserved GGDEF and EAL domains are the most abundant motives among bacteria. Moreover, c-di-GMP has been found to be the central biofilm regulator in virtually all Gram-negative bacteria investigated to date (Ryjenkov et al. 2005; Jenal et al. 2017). However, c-di-GMP is not produced by humans or other mammals, suggesting that c-di-GMP signalling is a promising anti-biofilm drug target.

4 Chemical Modulation of Quorum Sensing and c-di-GMP Signalling

Generally, the development of QS inhibitors (QSIs) and c-di-GMP modulators has in most cases been initiated by screening for compounds that affect a target in the respective signalling pathways. In most cases, this has been performed using high-throughput screening, in which a chemical compound library consisting of thousands of different chemical structures is tested against purified target proteins, or in vitro with reporter strains. Identified compounds capable of modulating these systems originate from natural sources, such as food and plant extracts, or are synthetic small molecules from either random compound libraries or libraries specifically synthesised on the basis of the native signal molecules, or from previous studies focusing on the structure-activity relationship (SAR). One often-used method for screening compounds for biological activity is to construct live reporter bacteria with a promoter of interest fused to a reporter gene encoding, for example green fluorescence protein (GFP) or β-galactosidase. Several *P. aeruginosa* QS reporter strains have been published, such as *lasB-lacZ* (Pearson et al. 1994), *lasB-gfp* (Hentzer et al. 2002), *rhlA-gfp* (Yang et al. 2009) and *pqsA-gfp* (Yang et al. 2007). A transcriptional c-di-GMP reporter which consists of the DGC *cdrA* promoter fused to *gfp* has also been constructed (Rybtke et al. 2012). In one of the versions, the reporter strain is constructed in a *wspF* mutant background which exhibits a high level of c-di-GMP for better detection and it has mutation in *pelA* and *pslBCD* to avoid clumping during growth (Rybtke et al. 2012).

Because QSIs and c-di-GMP modulating compounds are not designed to effectively kill bacteria an effective treatment of biofilm infections would presumably be a combination treatment with antibiotics. The use of conventional antibiotics would be effective in eradicating the dispersed bacteria after treatment with a c-di-GMP modulating compound or a less tolerant biofilm after treatment with a QS inhibitor.

5 Quorum Sensing Inhibition

With regard to targeting the QS system to diminish the virulence of biofilm-forming pathogens, there are three main targets for which QSIs are developed. While the approaches differ in their mechanism of action, the main goal is the same, namely to prevent the QS system from reaching the signal molecule threshold.

Targeting the synthesis of AHL can be achieved by interfering with the formation of AHL by binding the precursor *S*-adenosyl-L-methionine (SAM), thereby preventing the LuxI synthases from catalysing SAM and an acyl carrier protein (ACP) into AHL. In the same vein, an alternative strategy is to target the lasI activity pocket and block the overall process (Lidor et al. 2015).

Silencing the signal by degrading the AHL molecule through enzymatic reactions has been achieved using AHLases. These enzymes originate from bacteria and fungi and are believed to prevent competing bacteria from gaining a fitness advantage by jamming the signal network (Kusada et al. 2019; Dong et al. 2000). AHLases consist of different classes, such as AHL-lactonase, AHL-acylase, paraoxonase that break the AHL molecule and oxidoreductases that reduce the carbonyl to a hydroxyl group (Chen et al. 2013). The first lactonase to be described was the AiiA lactonase isolated from the soil bacterium *Bacillus* sp. 240B1 (Dong et al. 2000). And later the AiiM lactonase was discovered and reported to be an effective inhibitor of QS in *P. aeruginosa* and to reduce pneumonia infection in a mouse model (Migiyama et al. 2013). Several paraoxonases (PON) have been discovered in different mammals. PON have strong degradation activity of especially long acylated chains of

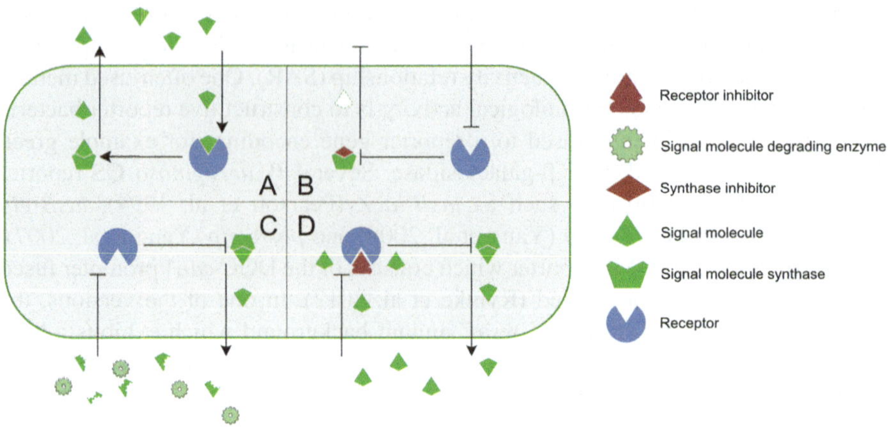

Fig. 2 Different approaches to target the QS system (A) A normally functioning QS system. The signal molecule is synthesised by the synthase, and the concentration increases at a constant rate, leading to a threshold being reached by binding signal molecules to their native receptors, resulting in the upregulation of QS-related genes and the synthase enzyme. (B) The synthase inhibitor binds to the catalytic area of the signal synthase, thereby preventing the synthesis of signal molecules. (C) Enzymes degrades the signal molecules, leading to no reaction with the signal receptor. (D) A non-productive signal is produced through the binding of inhibitor molecules to the signal receptor, blocking the native signal molecules from activating the QS threshold

AHLs, such as the 3-oxo-C12-HSL in *P. aeruginosa* (Ozer et al. 2005; Teiber et al. 2008).

Blocking the native signal molecule from binding its native receptor protein, thus resulting in a non-productive signal, is the strategy that has yielded the largest number of QSI candidates. Among these candidates are compounds mimicking the structure of the native AHL. However, the affinity with which the QSI binds to the receptor protein does not correlate with the structural resemblance of the native signal molecules, as compounds whose structure is not reminiscent of the native AHL still demonstrate a potent inhibitory effect (Fig. 2).

The first generation of QSIs that spurred the ongoing search originated from the discovery of halogenated furanones produced by the marine macro-alga *Delisea pulchra* (Hentzer et al. 2003). Among these furanones was the modified C-30, a compound that is still currently being used as a positive control of QS inhibitory effect. However, C-30 cannot be used medically as it is known to be carcinogenic (Janecki et al. 2002). Since then, an overwhelming number of new QSI candidates with diverse structures have been described each year. The ongoing research into new QSI candidates with a therapeutic potential has yielded a plethora of compounds from different sources, such as compounds derived from aquatic organisms or from plants as reviewed by Bouyahya et al. (2017) and Zhao et al. (2019).

A large number of QSI candidates consists of flavonoids, a secondary metabolite found in all types of plants. However, the mechanism of action of many of these compounds against the QS system is still unclear. Recently, Paczkowski et al. (2017)

pinpointed the inhibitory mechanism of molecules derived from a screening of 60,000s compounds of diverse origin, as well as two previous QSI flavonoids, baicalein and quercetin. They concluded that the flavonoids bind to the LasR ligand-binding domain in a non-competitive manner, which prevents the LasR protein from binding DNA, leading to a lack of expression of downstream genes. The results of the SAR study performed by the Paczkowski group showed that the binding of the QSI flavonoids to the transcription factor depends on the flavone A-ring backbone containing two hydroxyl moieties.

Ajoene, a sulphur-containing compound isolated from the crude extract of garlic is just one of many cases in which a QSI compound has been isolated from a food source (Jakobsen et al. 2012a; Kim et al. 2015). In vivo treatments with ajoene against *P. aeruginosa* have been found to cause a decrease in the production of rhamnolipids at the same level as a *ΔrhlA* mutant (Jakobsen et al. 2012b). The same tendency was observed in a PMN killing assay, as no significant cell lysis was detected in the presence of the QSI (Jakobsen et al. 2012b). Moreover, combined treatment with the antibiotic tobramycin and ajoene generated a higher level of sensitivity towards tobramycin, leading to a much more effective clearing of infection in vivo. A follow-up study was performed using ajoene analogues generated from a SAR study. These analogues were similar to the sulphur-rich ajoene and were highly effective as QS inhibitors (Fong et al. 2017). Transcriptomic analysis of *P. aeruginosa* treated with the original ajoene compound did not narrow down the potential target, as the expression of none of the QS master controls was significantly changed. Additional studies aiming at identifying the target led to the discovery of ajoene's QS inhibitory effect being through modulation of the synthesis of the small RNAs *rsmY* and *rsmZ*, which in turn regulates the synthesis of AHL (Jakobsen et al. 2017). Similarly, the QSI *N*-decanoyl-L-homoserine benzyl ester affects the QS circuit by upregulating the gene responsible for the LuxR-type receptor QscR (QS control repressor) (Weng et al. 2014; Yang et al. 2012). This research into alternative targets located upstream of the QS system has opened the door for unconventional approaches of searching for QSIs that do not rely on interfering with the main part of the QS network.

Different studies using single or double QS mutants of the signal synthetases (*lasI, rhlI*) or regulators (*lasR, rhlR or pqsR*) in different in vivo models have revealed a reduction in virulence and faster clearing of infection compared to infection with a wild-type *P. aeruginosa* (Rumbaugh et al. 1999; Smith et al. 2002; Pearson et al. 2000; Wu et al. 2001; Christensen et al. 2007; Xiao et al. 2006). These results indicate that an infection with a *P. aeruginosa* strain with an impaired or a non-functional QS system restores the innate immune system, hence leading to clearance of the infecting bacteria. Using a murine foreign body infection model, Christensen et al. confirmed the hypothesis of the synergistic effect of antibiotics and QSIs for treating infections (Christensen et al. 2012). The QSIs were C-30, iberin (a compound isolated from horseradish), and ajoene. While all three compounds increased the sensitivity of the bacteria towards tobramycin, the authors found that the synergistic effect diminished at a later stage of the infection.

This suggests that QSI treatment can be applied as a preventive measure rather than using it for the treatment of late-stage chronic infections.

The preferred target of QSIs research was previously the las system, however, it has been given a lower priority mainly due to the appearance of *lasI/lasR* mutants in clinical isolates from patients with CF. To steer clear of this conundrum, research have shifted focus towards QSI therapies targeting the PQS and the rhl system. As the biosynthesis of the PQS signal molecule requires a cascade of enzymes, the range in which inhibitors can interact with a potential target is more diverse than that of traditional QSIs, making it a much more attractive point of focus. Among the more recent PQS inhibitors are a series of thiazole-containing quinazolinones that inhibit the activity of the PQS receptor PqsR at a potency value of $IC_{50} = 300$ nm by generating hydrophobic interactions between the compounds' alkyl chain and the ligand-binding domain of the receptor (Grossman et al. 2020).

As mentioned, mutations in *lasI/lasR* are frequent, especially in patients with late-stage CF. Therefore, many researchers are already aware of the possibly limited duration in which QSIs can be applied as a therapy (Hamood et al. 1996; Cabrol et al. 2003; Bjarnsholt et al. 2010; Smith et al. 2006). However, the rhl and PQS parts of the QS system have not been reported to mutate to the same degree, allowing the production of a multitude of QS-regulated virulence factors, including rhamnolipid. Suggestions for why las mutants increase during infections are metabolic advantages and increasing fitness potential (D'Argenio et al. 2007; Smith et al. 2006). An alternative explanation is that mutants evolve in a mixed population acting as social cheats (Diggle et al. 2007; Sandoz et al. 2007). According to this concept, mutants take advantage of QS proficient bacteria producing expensive products for the benefit of the entire population. The notion of a functional rhl and PQS system that has been shown to control some of the most important QS-regulated features of biofilm infections, together with research revealing the presence of mixed populations, points to a prolonged use of QS inhibitors for infection control.

For an additional perspective of the chemistry of these modulators, please refer to (Welsh and Blackwell 2016; Qvortrup et al. 2019) as well as to QSIs targeting other bacterial species (Majik et al. 2020). For a more in-depth analysis of the inhibitors targeting different parts of the PQS biosynthesis cascade and the receptor of PQS, please refer to (Schütz and Empting 2018).

6 Interference with c-di-GMP Signalling

Research on c-di-GMP modulators has resulted in a handful of compounds that can inhibit biofilm formation and disperse preformed biofilms (Qvortrup et al. 2019). The core strategies explored so far have mainly focused on modulating DGC activity through the binding of small molecules to the I-site or the catalytic site, thereby inhibiting the synthesis of c-di-GMP signal molecules (Qvortrup et al. 2019).

The idea of enhancing antibiotic treatment through the co-treatment of infections with compounds that disperse biofilms originated from the discovery of how nitric oxide (NO) can disperse biofilms of *P. aeruginosa* (Barraud et al. 2006). The compound sodium nitroprusside (SNP) releases NO, leading to the dispersal of *P. aeruginosa* biofilms (Barraud et al. 2006), and is often used as a positive control when testing the anti-biofilm capabilities of other compounds against *P. aeruginosa*. The available evidence suggests that NO stimulates the activity of the PDE NbdA thereby leading to the hydrolysis of c-di-GMP (Li et al. 2013).

The first proof of concept regarding induction of biofilm dispersal by PDE expression in vivo as a viable strategy to counter biofilm infections came in 2013 (Christensen et al. 2013). It was found that biofilms of a *P. aeruginosa* strain containing an inducible PDE from *Escherichia coli* cleared in a murine biofilm infection model when the PDE activity was induced.

The c-di-GMP lowering compound ebselen was discovered through radiation-based high-throughput screening and has been demonstrated to have a unique mechanism of covalently modifying cysteine residues located near the I-site of the DGCs WspR and PelD (Lieberman et al. 2014). Although it has been shown to reduce biofilm formation of *P. aeruginosa* PA14, the dispersal of preformed biofilms has not been demonstrated (Lieberman et al. 2014). Ebselen has since reached phase III of clinical trials for treatment against brain blood clots and acute hearing loss, hence sparking renewed interest in the possibility of using ebselen analogues to combat biofilm-related infections. Kim et al. recently demonstrated a reduction in the synthesis of alginate in response to treatment with ebselen (Kim et al. 2021). They determined that the mechanism of action is through an interaction between the compound and the PilZ domain of the effector Alg44. They also found that the inhibitory effect is due to the modification of cysteine residues in the PilZ domain of Alg44, hence rendering c-di-GMP unable to bind the domain and initiate synthesis of alginate by Alg44 and the other proteins in the alginate synthase.

Sambanthamoorthy et al. screened 66,000 compounds of natural origin and identified the compound DI-3, which was found to reduce the c-di-GMP content in *Vibrio cholerae* and *P. aeruginosa* (Sambanthamoorthy et al. 2012). Although DI-3 exhibits an activity against *V. cholerae* and *P. aeruginosa* biofilm formation, similar to ebselen no dispersal of pre-formed biofilms was accomplished.

As mentioned previously, Rybtke et al. constructed genetically modified *P. aeruginosa* bacteria that report their internal level of c-di-GMP by expressing a c-di-GMP-regulated fluorescent reporter gene fusion *cdrA-gfp* (Rybtke et al. 2012). Subsequently, Andersen et al. used these reporter bacteria to perform high-throughput screening of a library consisting of 50,000 chemical compounds and identified a number of chemically distinct molecules that decrease the c-di-GMP level in *P. aeruginosa* (Andersen et al. 2021a). A highly active lead compound called H6-335-P1 was developed in a SAR study. The compound was reported to prevent *P. aeruginosa* biofilm formation and dismantle established *P. aeruginosa* biofilms in vitro and in a murine biofilm infection model (Andersen et al. 2021b; Jansen et al. 2021). In addition, it was demonstrated that the bacteria that are dispersed from biofilms can be efficiently killed when conventional antibiotics are co-administered with H6-335-P1 (Andersen et al. 2021a). Evidence was presented that H6-335-P1 activates the PDE BifA, which is only produced by pseudomonads

(Andersen et al. 2021a). Accordingly, the anti-biofilm activity of H6-335-P1 is specific for pseudomonads, and it does not affect biofilms of other bacteria (Andersen et al. 2021a).

7 Conclusions and Perspectives

Since the discovery of penicillin in 1928 by Alexander Fleming, pathogens have proven to be the masters of survival and adaption to specified treatments, a battle that does not seem to have an end in sight. In addition, to the general increase in antibiotic resistance, is the complexity and difficulty of treating biofilm infections as compared to acute infections which often are treatable with a single antibiotic. The heterogeneity, diverse growth patterns and the matrix components of biofilms complicate the treatment with the current assortment of antibiotics. There is a general increase in the number of persistent infections as a result of an increasingly ageing population as well as a rise in lifestyle-related diseases. Therefore, designing alternative strategies for treating biofilm infections has become more important than ever. The development of QSIs used in combined therapy with current antibiotics has been well researched in in vitro as well as in different animal models (Liu et al. 2022; Schütz et al. 2021; Rezzoagli et al. 2020; Brackman et al. 2011). It has nearly been two decades since the discovery of C-30 and its QSI capabilities, and research has since then expanded our knowledge of the astonishing variety in the structural skeleton of these molecules and their mechanisms of action. Despite the amount of research into QS inhibitors over the years as a potential treatment, it has not succeeded in the development of actual medical products. Natural sources have been shown to be a favourable venue for identifying active inhibitors of QS. However, when studying extracts from natural sources, the active molecule is seldom reported which complicates the assessment of the potential active molecules for further investigations. In general, it is difficult to evaluate the identified QS inhibitors as future drug candidates. Hopefully, more in-depth studies will be published in the future so as to promote this potential treatment method to clinical trials. The research in potential modulation of c-di-GMP with small compounds is less studied and as previously described, studies of the first well-investigated and promising candidates have recently been published.

Currently, there are no active clinical trials on QSIs or modulators of c-di-GMP signalling on ClinicalTrials.gov or the EU Clinical Trials Register. Therefore, it is important to investigate whether drugs that have been approved for medical use and were not originally intended for the treatment of biofilm infections, can be regarded as possible future candidates. Among these drugs are the analgesic paracetamol (Seleem et al. 2021), the anti-cancer drug doxorubicin (Groizeleau et al. 2016), and the immunosuppressant azathioprine (Antoniani et al. 2013).

Given the history of bacterial infection treatment, there will always be a risk of the emergence of mutants, which may hamper the effectiveness of new treatment options (Maeda et al. 2012). Many studies have revealed resistance towards the

QSI C-30 in *P. aeruginosa* when it is grown under selective pressure generated by a minimal medium supplemented with adenosine as the sole carbon source. Mutations were observed in genes related to the regulation of efflux pumps, *mexR*, and *nalC* that are repressors of the *mexAB-oprM* multi-drug resistance operon, which have previously been linked to resistance towards antibiotics. Mutations in efflux pump genes such as *mexEF-oprD* have been linked to a decrease in virulence factor production, as efflux diminishes the intracellular concentration of AHL, thus rendering the signal molecule threshold unreachable (Bové et al. 2021). At the time of publishing this chapter, no published studies have shown resistance mechanisms towards modulators of the c-di-GMP content. However, mutants will undoubtedly emerge, but selection might be less severe compared to traditional antibiotics that target metabolic functions, thereby killing or hampering the growth and reproduction of the bacteria.

It is important to continue and also upscale the search and investigations for potential novel compounds useable as treatment of biofilm infections. The importance of this research is emphasised by the fact that The World Health Organization (WHO), United Nations (UN) and Center for Disease Control and Prevention (CDC) predict that the number of antibiotic-resistant bacterial infections will rise to unmanageable levels in all parts of the world if adequate and appropriate countermeasures are not taken.

Compliance with Ethical Standards **Funding:** This study was funded by Independent Research Fund Denmark (grant no. 9040-00023B).

Disclosure of Interests: All authors declare they have no conflict of interest.

References

Alhede M, Bjarnsholt T, Jensen PO, Phipps RK, Moser C, Christophersen L, Christensen LD, van Gennip M, Parsek M, Hoiby N, Rasmussen TB, Givskov M (2009) Pseudomonas aeruginosa recognizes and responds aggressively to the presence of polymorphonuclear leukocytes. Microbiology (Reading) 155(Pt 11):3500–3508. https://doi.org/10.1099/mic.0.031443-0

Allesen-Holm M, Barken KB, Yang L, Klausen M, Webb JS, Kjelleberg S, Molin S, Givskov M, Tolker-Nielsen T (2006) A characterization of DNA release in Pseudomonas aeruginosa cultures and biofilms. Mol Microbiol 59(4):1114–1128. https://doi.org/10.1111/j.1365-2958.2005.05008.x

Andersen JB, Hultqvist LD, Jansen CU, Jakobsen TH, Nilsson M, Rybtke M, Uhd J, Fritz BG, Seifert R, Berthelsen J, Nielsen TE, Qvortrup K, Givskov M, Tolker-Nielsen T (2021a) Identification of small molecules that interfere with c-di-GMP signaling and induce dispersal of Pseudomonas aeruginosa biofilms. NPJ Biofilms Microbiomes 7(1):59. https://doi.org/10.1038/s41522-021-00225-4

Andersen JB, Kragh KN, Hultqvist LD, Rybtke M, Nilsson M, Jakobsen TH, Givskov M, Tolker-Nielsen T (2021b) Induction of native c-di-GMP phosphodiesterases leads to dispersal of Pseudomonas aeruginosa biofilms. Antimicrob Agents Chemother. https://doi.org/10.1128/AAC.02431-20

Antoniani D, Rossi E, Rinaldo S, Bocci P, Lolicato M, Paiardini A, Raffaelli N, Cutruzzola F, Landini P (2013) The immunosuppressive drug azathioprine inhibits biosynthesis of the bacterial signal molecule cyclic-di-GMP by interfering with intracellular nucleotide pool availability. Appl Microbiol Biotechnol 97(16):7325–7336. https://doi.org/10.1007/s00253-013-4875-0

Baraquet C, Murakami K, Parsek MR, Harwood CS (2012) The FleQ protein from Pseudomonas aeruginosa functions as both a repressor and an activator to control gene expression from the pel operon promoter in response to c-di-GMP. Nucleic Acids Res 40(15):7207–7218. https://doi.org/10.1093/nar/gks384

Barraud N, Hassett DJ, Hwang S-H, Rice SA, Kjelleberg S, Webb JS (2006) Involvement of nitric oxide in biofilm dispersal of Pseudomonas aeruginosa. J Bacteriol 188(21):7344–7353. https://doi.org/10.1128/JB.00779-06

Barraud N, Schleheck D, Klebensberger J, Webb JS, Hassett DJ, Rice SA, Kjelleberg S (2009) Nitric oxide signaling in Pseudomonas aeruginosa biofilms mediates phosphodiesterase activity, decreased cyclic di-GMP levels, and enhanced dispersal. J Bacteriol 191(23):7333–7342. https://doi.org/10.1128/JB.00975-09

Bjarnsholt T, Jensen PO, Jakobsen TH, Phipps R, Nielsen AK, Rybtke MT, Tolker-Nielsen T, Givskov M, Hoiby N, Ciofu O, Scandinavian Cystic Fibrosis Study Consortium (2010) Quorum sensing and virulence of Pseudomonas aeruginosa during lung infection of cystic fibrosis patients. PLoS One 5(4):e10115. https://doi.org/10.1371/journal.pone.0010115

Bouyahya A, Dakka N, Et-Touys A, Abrini J, Bakri Y (2017) Medicinal plant products targeting quorum sensing for combating bacterial infections. Asian Pac J Trop Med 10(8):729–743. https://doi.org/10.1016/j.apjtm.2017.07.021

Bové M, Bao X, Sass A, Crabbé A, Coenye T (2021) The quorum-sensing inhibitor furanone C-30 rapidly loses its tobramycin-potentiating activity against Pseudomonas aeruginosa biofilms during experimental evolution. Antimicrob Agents Chemother 65(7):e0041321. https://doi.org/10.1128/aac.00413-21

Brackman G, Cos P, Maes L, Nelis HJ, Coenye T (2011) Quorum sensing inhibitors increase the susceptibility of bacterial biofilms to antibiotics in vitro and in vivo. Antimicrob Agents Chemother 55(6):2655–2661. https://doi.org/10.1128/aac.00045-11

Bryers JD (2008) Medical biofilms. Biotechnol Bioeng 100(1):1–18. https://doi.org/10.1002/bit.21838

Cabrol S, Olliver A, Pier GB, Andremont A, Ruimy R (2003) Transcription of quorum-sensing system genes in clinical and environmental isolates of Pseudomonas aeruginosa. J Bacteriol 185(24):7222–7230. https://doi.org/10.1128/JB.185.24.7222-7230.2003

Chan C, Paul R, Samoray D, Amiot NC, Giese B, Jenal U, Schirmer T (2004) Structural basis of activity and allosteric control of diguanylate cyclase. Proc Natl Acad Sci U S A 101(49): 17084–17089. https://doi.org/10.1073/pnas.0406134101

Chen MW, Kotaka M, Vonrhein C, Bricogne G, Rao F, Chuah MLC, Svergun D, Schneider G, Liang Z-X, Lescar J (2012) Structural insights into the regulatory mechanism of the response regulator RocR from Pseudomonas aeruginosa in cyclic Di-GMP signaling. J Bacteriol 194(18): 4837–4846. https://doi.org/10.1128/JB.00560-12

Chen F, Gao Y, Chen X, Yu Z, Li X (2013) Quorum quenching enzymes and their application in degrading signal molecules to block quorum sensing-dependent infection. Int J Mol Sci 14(9): 17477–17500. https://doi.org/10.3390/ijms140917477

Chiang WC, Nilsson M, Jensen PO, Hoiby N, Nielsen TE, Givskov M, Tolker-Nielsen T (2013) Extracellular DNA shields against aminoglycosides in Pseudomonas aeruginosa biofilms. Antimicrob Agents Chemother 57(5):2352–2361. https://doi.org/10.1128/AAC.00001-13

Chou S-H, Galperin MY (2016) Diversity of cyclic Di-GMP-binding proteins and mechanisms. J Bacteriol 198(1):32–46. https://doi.org/10.1128/JB.00333-15

Christensen LD, Moser C, Jensen PO, Rasmussen TB, Christophersen L, Kjelleberg S, Kumar N, Hoiby N, Givskov M, Bjarnsholt T (2007) Impact of Pseudomonas aeruginosa quorum sensing on biofilm persistence in an in vivo intraperitoneal foreign-body infection model. Microbiology (Reading) 153(Pt 7):2312–2320. https://doi.org/10.1099/mic.0.2007/006122-0

Christensen LD, van Gennip M, Jakobsen TH, Alhede M, Hougen HP, Hoiby N, Bjarnsholt T, Givskov M (2012) Synergistic antibacterial efficacy of early combination treatment with tobramycin and quorum-sensing inhibitors against Pseudomonas aeruginosa in an intraperitoneal foreign-body infection mouse model. J Antimicrob Chemother 67(5): 1198–1206. https://doi.org/10.1093/jac/dks002

Christensen LD, van Gennip M, Rybtke MT, Wu H, Chiang W-C, Alhede M, Høiby N, Nielsen TE, Givskov M, Tolker-Nielsen T (2013) Clearance of Pseudomonas aeruginosa foreign-body biofilm infections through reduction of the cyclic Di-GMP level in the bacteria. Infect Immun 81(8):2705–2713. https://doi.org/10.1128/IAI.00332-13

Ciofu O, Tolker-Nielsen T (2019) Tolerance and resistance of pseudomonas aeruginosa biofilms to antimicrobial agents-how P. aeruginosa can escape antibiotics. Front Microbiol 10:913. https://doi.org/10.3389/fmicb.2019.00913

Costerton JW, Stewart PS, Greenberg EP (1999) Bacterial biofilms: a common cause of persistent infections. Science 284(5418):1318–1322

D'Argenio DA, Wu M, Hoffman LR, Kulasekara HD, Deziel E, Smith EE, Nguyen H, Ernst RK, Larson Freeman TJ, Spencer DH, Brittnacher M, Hayden HS, Selgrade S, Klausen M, Goodlett DR, Burns JL, Ramsey BW, Miller SI (2007) Growth phenotypes of Pseudomonas aeruginosa lasR mutants adapted to the airways of cystic fibrosis patients. Mol Microbiol 64(2):512–533. https://doi.org/10.1111/j.1365-2958.2007.05678.x

Das T, Manefield M (2012) Pyocyanin promotes extracellular DNA release in Pseudomonas aeruginosa. PLoS One 7(10):e46718–e46718. https://doi.org/10.1371/journal.pone.0046718

Das T, Manefield M (2013) Phenazine production enhances extracellular DNA release via hydrogen peroxide generation in Pseudomonas aeruginosa. Commun Integr Biol 6(3):e23570. https://doi.org/10.4161/cib.23570

Das T, Kutty SK, Tavallaie R, Ibugo AI, Panchompoo J, Sehar S, Aldous L, Yeung AW, Thomas SR, Kumar N, Gooding JJ, Manefield M (2015) Phenazine virulence factor binding to extracellular DNA is important for Pseudomonas aeruginosa biofilm formation. Sci Rep 5:8398. https://doi.org/10.1038/srep08398

Davey ME, Caiazza NC, O'Toole GA (2003) Rhamnolipid surfactant production affects biofilm architecture in Pseudomonas aeruginosa PAO1. J Bacteriol 185(3):1027–1036

Diggle SP, Griffin AS, Campbell GS, West SA (2007) Cooperation and conflict in quorum-sensing bacterial populations. Nature 450(7168):411–414. https://doi.org/10.1038/nature06279

Dong YH, Xu JL, Li XZ, Zhang LH (2000) AiiA, an enzyme that inactivates the acylhomoserine lactone quorum-sensing signal and attenuates the virulence of Erwinia carotovora. Proc Natl Acad Sci U S A 97(7):3526–3531. https://doi.org/10.1073/pnas.060023897

Fong J, Yuan M, Jakobsen TH, Mortensen KT, Delos Santos MM, Chua SL, Yang L, Tan CH, Nielsen TE, Givskov M (2017) Disulfide bond-containing ajoene analogues as novel quorum sensing inhibitors of Pseudomonas aeruginosa. J Med Chem 60(1):215–227. https://doi.org/10.1021/acs.jmedchem.6b01025

Gjermansen M, Ragas P, Sternberg C, Molin S, Tolker-Nielsen T (2005) Characterization of starvation-induced dispersion in Pseudomonas putida biofilms. Environ Microbiol 7(6): 894–906. https://doi.org/10.1111/j.1462-2920.2005.00775.x

Gjermansen M, Ragas P, Tolker-Nielsen T (2006) Proteins with GGDEF and EAL domains regulate Pseudomonas putida biofilm formation and dispersal. FEMS Microbiol Lett 265(2): 215–224. https://doi.org/10.1111/j.1574-6968.2006.00493.x

Gjermansen M, Nilsson M, Yang L, Tolker-Nielsen T (2010) Characterization of starvation-induced dispersion in Pseudomonas putida biofilms: genetic elements and molecular mechanisms. Mol Microbiol 75(4):815–826. https://doi.org/10.1111/j.1365-2958.2009.06793.x. MMI6793 [pii]

Gray KM, Garey JR (2001) The evolution of bacterial LuxI and LuxR quorum sensing regulators. Microbiology (Reading) 147(Pt 8):2379–2387. https://doi.org/10.1099/00221287-147-8-2379

Groizeleau J, Rybtke M, Andersen JB, Berthelsen J, Liu Y, Yang L, Nielsen TE, Kaever V, Givskov M, Tolker-Nielsen T (2016) The anti-cancerous drug doxorubicin decreases the c-di-GMP content in Pseudomonas aeruginosa but promotes biofilm formation. Microbiology (Reading) 162(10):1797–1807. https://doi.org/10.1099/mic.0.000354

Grossman S, Soukarieh F, Richardson W, Liu R, Mashabi A, Emsley J, Williams P, Cámara M, Stocks MJ (2020) Novel quinazolinone inhibitors of the Pseudomonas aeruginosa quorum sensing transcriptional regulator PqsR. Eur J Med Chem 208:112778. https://doi.org/10.1016/j.ejmech.2020.112778

Hall CW, Mah TF (2017) Molecular mechanisms of biofilm-based antibiotic resistance and tolerance in pathogenic bacteria. FEMS Microbiol Rev 41(3):276–301. https://doi.org/10.1093/femsre/fux010

Hamood AN, Griswold J, Colmer J (1996) Characterization of elastase-deficient clinical isolates of Pseudomonas aeruginosa. Infect Immun 64(8):3154–3160. https://doi.org/10.1128/iai.64.8.3154-3160.1996

Hay ID, Remminghorst U, Rehm BH (2009) MucR, a novel membrane-associated regulator of alginate biosynthesis in Pseudomonas aeruginosa. Appl Environ Microbiol 75(4):1110–1120. https://doi.org/10.1128/AEM.02416-08. AEM.02416-08 [pii]

Hengge R (2021) High-specificity local and global c-di-GMP signaling. Trends Microbiol. https://doi.org/10.1016/j.tim.2021.02.003

Hentzer M, Riedel K, Rasmussen TB, Heydorn A, Andersen JB, Parsek MR, Rice SA, Eberl L, Molin S, Hoiby N, Kjelleberg S, Givskov M (2002) Inhibition of quorum sensing in Pseudomonas aeruginosa biofilm bacteria by a halogenated furanone compound. Microbiology (Reading) 148(Pt 1):87–102. https://doi.org/10.1099/00221287-148-1-87

Hentzer M, Wu H, Andersen JB, Riedel K, Rasmussen TB, Bagge N, Kumar N, Schembri MA, Song Z, Kristoffersen P, Manefield M, Costerton JW, Molin S, Eberl L, Steinberg P, Kjelleberg S, Høiby N, Givskov M (2003) Attenuation of Pseudomonas aeruginosa virulence by quorum sensing inhibitors. EMBO J 22(15):3803–3815. https://doi.org/10.1093/emboj/cdg366

Hickman JW, Harwood CS (2008) Identification of FleQ from Pseudomonas aeruginosa as a c-di-GMP-responsive transcription factor. Mol Microbiol 69(2):376–389. https://doi.org/10.1111/j.1365-2958.2008.06281.x

Hoffman LR, Kulasekara HD, Emerson J, Houston LS, Burns JL, Ramsey BW, Miller SI (2009) Pseudomonas aeruginosa lasR mutants are associated with cystic fibrosis lung disease progression. J Cyst Fibros 8(1):66–70. https://doi.org/10.1016/j.jcf.2008.09.006

Jakobsen TH, Bragason SK, Phipps RK, Christensen LD, van Gennip M, Alhede M, Skindersoe M, Larsen TO, Høiby N, Bjarnsholt T, Givskov M (2012a) Food as a source for quorum sensing inhibitors: iberin from horseradish revealed as a quorum sensing inhibitor of Pseudomonas aeruginosa. Appl Environ Microbiol 78(7):2410–2421. https://doi.org/10.1128/AEM.05992-11

Jakobsen TH, van Gennip M, Phipps RK, Shanmugham MS, Christensen LD, Alhede M, Skindersoe ME, Rasmussen TB, Friedrich K, Uthe F, Jensen PØ, Moser C, Nielsen KF, Eberl L, Larsen TO, Tanner D, Høiby N, Bjarnsholt T, Givskov M (2012b) Ajoene, a sulfur-rich molecule from garlic, inhibits genes controlled by quorum sensing. Antimicrob Agents Chemother 56(5):2314–2325. https://doi.org/10.1128/AAC.05919-11

Jakobsen TH, Warming AN, Vejborg RM, Moscoso JA, Stegger M, Lorenzen F, Rybtke M, Andersen JB, Petersen R, Andersen PS, Nielsen TE, Tolker-Nielsen T, Filloux A, Ingmer H, Givskov M (2017) A broad range quorum sensing inhibitor working through sRNA inhibition. Sci Rep 7(1):9857–9857. https://doi.org/10.1038/s41598-017-09886-8

Janecki T, Błaszczyk E, Studzian K, Rózalski M, Krajewska U, Janecka A (2002) New stereocontrolled synthesis and biological evaluation of 5-(1′-hydroxyalkyl)-3-methylidenetetrahydro-2-furanones as potential cytotoxic agents. J Med Chem 45(5):1142–1145. https://doi.org/10.1021/jm011019v

Jansen CU, Uhd J, Andersen JB, Hultqvist LD, Jakobsen TH, Nilsson M, Nielsen TE, Givskov M, Tolker-Nielsen T, Qvortrup KM (2021) SAR study of 4-arylazo-3,5-diamino-1H-pyrazoles: identification of small molecules that induce dispersal of Pseudomonas aeruginosa biofilms. RSC Med Chem. https://doi.org/10.1039/D1MD00275A

Jenal U, Reinders A, Lori C (2017) Cyclic di-GMP: second messenger extraordinaire. Nat Rev Microbiol 15(5):271–284. https://doi.org/10.1038/nrmicro.2016.190

Jensen P, Bjarnsholt T, Phipps R, Rasmussen TB, Calum H, Christoffersen L, Moser C, Williams P, Pressler T, Givskov M, Høiby N (2007) Rapid necrotic killing of polymorphonuclear leukocytes is caused by quorum-sensing-controlled production of rhamnolipid by Pseudomonas aeruginosa. Microbiology (Reading) 153(Pt 5):1329–1338. https://doi.org/10.1099/mic.0.2006/003863-0

Jernigan JA, Hatfield KM, Wolford H, Nelson RE, Olubajo B, Reddy SC, McCarthy N, Paul P, McDonald LC, Kallen A, Fiore A, Craig M, Baggs J (2020) Multidrug-resistant bacterial infections in U.S. hospitalized patients, 2012-2017. N Engl J Med 382(14):1309–1319. https://doi.org/10.1056/NEJMoa1914433

Johnson MK, Boese-Marrazzo D (1980) Production and properties of heat-stable extracellular hemolysin from Pseudomonas aeruginosa. Infect Immun 29(3):1028–1033. https://doi.org/10.1128/iai.29.3.1028-1033.1980

Kaplan HB, Greenberg EP (1985) Diffusion of autoinducer is involved in regulation of the Vibrio fischeri luminescence system. J Bacteriol 163(3):1210–1214. https://doi.org/10.1128/jb.163.3.1210-1214.1985

Kim H-S, Lee S-H, Byun Y, Park H-D (2015) 6-Gingerol reduces Pseudomonas aeruginosa biofilm formation and virulence via quorum sensing inhibition. Sci Rep 5:8656–8656. https://doi.org/10.1038/srep08656

Kim SK, Ngo HX, Dennis EK, Thamban Chandrika N, DeShong P, Garneau-Tsodikova S, Lee VT (2021) Inhibition of Pseudomonas aeruginosa alginate synthesis by ebselen oxide and its analogues. ACS Infect Dis 7(6):1713–1726. https://doi.org/10.1021/acsinfecdis.1c00045

Kusada H, Zhang Y, Tamaki H, Kimura N, Kamagata Y (2019) Novel N-acyl homoserine lactone-degrading bacteria isolated from penicillin-contaminated environments and their quorum-quenching activities. Front Microbiol 10:455–455. https://doi.org/10.3389/fmicb.2019.00455

Latifi A, Winson MK, Foglino M, Bycroft BW, Stewart GS, Lazdunski A, Williams P (1995) Multiple homologues of LuxR and LuxI control expression of virulence determinants and secondary metabolites through quorum sensing in Pseudomonas aeruginosa PAO1. Mol Microbiol 17(2):333–343. https://doi.org/10.1111/j.1365-2958.1995.mmi_17020333.x

Lau GW, Hassett DJ, Ran H, Kong F (2004) The role of pyocyanin in Pseudomonas aeruginosa infection. Trends Mol Med 10(12):599–606. https://doi.org/10.1016/j.molmed.2004.10.002

Lee J, Wu J, Deng Y, Wang J, Wang C, Wang J, Chang C, Dong Y, Williams P, Zhang LH (2013) A cell-cell communication signal integrates quorum sensing and stress response. Nat Chem Biol 9(5):339–343. https://doi.org/10.1038/nchembio.1225

Lewenza S (2013) Extracellular DNA-induced antimicrobial peptide resistance mechanisms in Pseudomonas aeruginosa. Front Microbiol 4:21–21. https://doi.org/10.3389/fmicb.2013.00021

Li Y, Heine S, Entian M, Sauer K, Frankenberg-Dinkel N (2013) NO-induced biofilm dispersion in Pseudomonas aeruginosa is mediated by an MHYT domain-coupled phosphodiesterase. J Bacteriol 195(16):3531–3542. https://doi.org/10.1128/JB.01156-12

Lidor O, Al-Quntar A, Pesci EC, Steinberg D (2015) Mechanistic analysis of a synthetic inhibitor of the Pseudomonas aeruginosa LasI quorum-sensing signal synthase. Sci Rep 5:16569. https://doi.org/10.1038/srep16569

Lieberman OJ, Orr MW, Wang Y, Lee VT (2014) High-throughput screening using the differential radial capillary action of ligand assay identifies ebselen as an inhibitor of diguanylate cyclases. ACS Chem Biol 9(1):183–192. https://doi.org/10.1021/cb400485k

Liu J, Hou JS, Chang YQ, Peng LJ, Zhang XY, Miao ZY, Sun PH, Lin J, Chen WM (2022) New Pqs quorum sensing system inhibitor as an antibacterial synergist against multidrug-resistant Pseudomonas aeruginosa. J Med Chem 65(1):688–709. https://doi.org/10.1021/acs.jmedchem.1c01781

Maeda T, García-Contreras R, Pu M, Sheng L, Garcia LR, Tomás M, Wood TK (2012) Quorum quenching quandary: resistance to antivirulence compounds. ISME J 6(3):493–501. https://doi.org/10.1038/ismej.2011.122

Majik MS, Gawas UB, Mandrekar VK (2020) Next generation quorum sensing inhibitors: Accounts on structure activity relationship studies and biological activities. Bioorg Med Chem 28(21):115728. https://doi.org/10.1016/j.bmc.2020.115728

Managò A, Becker KA, Carpinteiro A, Wilker B, Soddemann M, Seitz AP, Edwards MJ, Grassmé H, Szabò I, Gulbins E (2015) Pseudomonas aeruginosa pyocyanin induces neutrophil death via mitochondrial reactive oxygen species and mitochondrial acid sphingomyelinase. Antioxid Redox Signal 22(13):1097–1110. https://doi.org/10.1089/ars.2014.5979

Manefield M, Turner SL (2002) Quorum sensing in context: out of molecular biology and into microbial ecology. Microbiology (Reading) 148(Pt 12):3762–3764. https://doi.org/10.1099/00221287-148-12-3762

Migiyama Y, Kaneko Y, Yanagihara K, Morohoshi T, Morinaga Y, Nakamura S, Miyazaki T, Hasegawa H, Izumikawa K, Kakeya H, Kohrogi H, Kohno S (2013) Efficacy of AiiM, an N-acylhomoserine lactonase, against Pseudomonas aeruginosa in a mouse model of acute pneumonia. Antimicrob Agents Chemother 57(8):3653–3658. https://doi.org/10.1128/aac.00456-13

Nealson KH, Hastings JW (1979) Bacterial bioluminescence: its control and ecological significance. Microbiol Rev 43(4):496–518. https://doi.org/10.1128/mr.43.4.496-518.1979

Ozer EA, Pezzulo A, Shih DM, Chun C, Furlong C, Lusis AJ, Greenberg EP, Zabner J (2005) Human and murine paraoxonase 1 are host modulators of Pseudomonas aeruginosa quorum-sensing. FEMS Microbiol Lett 253(1):29–37. https://doi.org/10.1016/j.femsle.2005.09.023

Paczkowski JE, Mukherjee S, McCready AR, Cong JP, Aquino CJ, Kim H, Henke BR, Smith CD, Bassler BL (2017) Flavonoids suppress pseudomonas aeruginosa virulence through allosteric inhibition of quorum-sensing receptors. J Biol Chem 292(10):4064–4076. https://doi.org/10.1074/jbc.M116.770552

Pamp SJ, Tolker-Nielsen T (2007) Multiple roles of biosurfactants in structural biofilm development by Pseudomonas aeruginosa. J Bacteriol 189(6):2531–2539. https://doi.org/10.1128/JB.01515-06

Pearson JP, Gray KM, Passador L, Tucker KD, Eberhard A, Iglewski BH, Greenberg EP (1994) Structure of the autoinducer required for expression of Pseudomonas aeruginosa virulence genes. Proc Natl Acad Sci U S A 91(1):197–201. https://doi.org/10.1073/pnas.91.1.197

Pearson JP, Feldman M, Iglewski BH, Prince A (2000) Pseudomonas aeruginosa cell-to-cell signaling is required for virulence in a model of acute pulmonary infection. Infect Immun 68(7):4331–4334. https://doi.org/10.1128/IAI.68.7.4331-4334.2000

Pesci EC, Milbank JB, Pearson JP, McKnight S, Kende AS, Greenberg EP, Iglewski BH (1999) Quinolone signaling in the cell-to-cell communication system of Pseudomonas aeruginosa. Proc Natl Acad Sci U S A 96(20):11229–11234

Qvortrup K, Hultqvist LD, Nilsson M, Jakobsen TH, Jansen CU, Uhd J, Andersen JB, Nielsen TE, Givskov M, Tolker-Nielsen T (2019) Small molecule anti-biofilm agents developed on the basis of mechanistic understanding of biofilm formation. Front Chem 7:742. https://doi.org/10.3389/fchem.2019.00742

Rezzoagli C, Archetti M, Mignot I, Baumgartner M, Kümmerli R (2020) Combining antibiotics with antivirulence compounds can have synergistic effects and reverse selection for antibiotic resistance in Pseudomonas aeruginosa. PLoS Biol 18(8):e3000805. https://doi.org/10.1371/journal.pbio.3000805

Ross P, Weinhouse H, Aloni Y, Michaeli D, Weinberger-Ohana P, Mayer R, Braun S, de Vroom E, van der Marel GA, van Boom JH, Benziman M (1987) Regulation of cellulose synthesis in Acetobacter xylinum by cyclic diguanylic acid. Nature 325(6101):279–281. https://doi.org/10.1038/325279a0

Rumbaugh KP, Griswold JA, Iglewski BH, Hamood AN (1999) Contribution of quorum sensing to the virulence of Pseudomonas aeruginosa in burn wound infections. Infect Immun 67(11):5854–5862. https://doi.org/10.1128/IAI.67.11.5854-5862.1999

Ryan RP, Tolker-Nielsen T, Dow JM (2012) When the PilZ don't work: effectors for cyclic di-GMP action in bacteria. Trends Microbiol 20(5):235–242. https://doi.org/10.1016/j.tim.2012.02.008

Rybtke MT, Borlee BR, Murakami K, Irie Y, Hentzer M, Nielsen TE, Givskov M, Parsek MR, Tolker-Nielsen T (2012) Fluorescence-based reporter for gauging cyclic di-GMP levels in Pseudomonas aeruginosa. Appl Environ Microbiol 78(15):5060–5069. https://doi.org/10.1128/AEM.00414-12

Rybtke M, Berthelsen J, Yang L, Hoiby N, Givskov M, Tolker-Nielsen T (2015) The LapG protein plays a role in Pseudomonas aeruginosa biofilm formation by controlling the presence of the CdrA adhesin on the cell surface. Microbiologyopen 4(6):917–930. https://doi.org/10.1002/mbo3.301

Ryjenkov DA, Tarutina M, Moskvin OV, Gomelsky M (2005) Cyclic diguanylate is a ubiquitous signaling molecule in bacteria: insights into biochemistry of the GGDEF protein domain. J Bacteriol 187(5):1792–1798. https://doi.org/10.1128/JB.187.5.1792-1798.2005. 187/5/1792 [pii]

Sambanthamoorthy K, Sloup RE, Parashar V, Smith JM, Kim EE, Semmelhack MF, Neiditch MB, Waters CM (2012) Identification of small molecules that antagonize diguanylate cyclase enzymes to inhibit biofilm formation. Antimicrob Agents Chemother 56(10):5202–5211. https://doi.org/10.1128/aac.01396-12

Sandoz KM, Mitzimberg SM, Schuster M (2007) Social cheating in Pseudomonas aeruginosa quorum sensing. Proc Natl Acad Sci U S A 104(40):15876–15881. https://doi.org/10.1073/pnas.0705653104

Schleheck D, Barraud N, Klebensberger J, Webb JS, McDougald D, Rice SA, Kjelleberg S (2009) Pseudomonas aeruginosa PAO1 preferentially grows as aggregates in liquid batch cultures and disperses upon starvation. PLoS One 4(5):e5513. https://doi.org/10.1371/journal.pone.0005513

Schuster M, Lostroh CP, Ogi T, Greenberg EP (2003) Identification, timing, and signal specificity of *Pseudomonas aeruginosa* quorum-controlled genes: a transcriptome analysis. J Bacteriol 185(7):2066–2079. https://doi.org/10.1128/JB.185.7.2066-2079.2003

Schütz C, Empting M (2018) Targeting the Pseudomonas quinolone signal quorum sensing system for the discovery of novel anti-infective pathoblockers. Beilstein J Org Chem 14:2627–2645. https://doi.org/10.3762/bjoc.14.241

Schütz C, Ho DK, Hamed MM, Abdelsamie AS, Röhrig T, Herr C, Kany AM, Rox K, Schmelz S, Siebenbürger L, Wirth M, Börger C, Yahiaoui S, Bals R, Scrima A, Blankenfeldt W, Horstmann JC, Christmann R, Murgia X, Koch M, Berwanger A, Loretz B, Hirsch AKH, Hartmann RW, Lehr CM, Empting M (2021) A new PqsR inverse agonist potentiates tobramycin efficacy to eradicate Pseudomonas aeruginosa biofilms. Adv Sci (Weinh) 8(12):e2004369. https://doi.org/10.1002/advs.202004369

Seleem NM, Atallah H, Abd El Latif HK, Shaldam MA, El-Ganiny AM (2021) Could the analgesic drugs, paracetamol and indomethacin, function as quorum sensing inhibitors? Microb Pathog 158:105097. https://doi.org/10.1016/j.micpath.2021.105097

Skotnicka D, Steinchen W, Szadkowski D, Cadby IT, Lovering AL, Bange G, Søgaard-Andersen L (2020) CdbA is a DNA-binding protein and c-di-GMP receptor important for nucleoid organization and segregation in Myxococcus xanthus. Nat Commun 11(1):1791. https://doi.org/10.1038/s41467-020-15628-8

Smith RS, Harris SG, Phipps R, Iglewski B (2002) The Pseudomonas aeruginosa quorum-sensing molecule N-(3-oxododecanoyl)homoserine lactone contributes to virulence and induces inflammation in vivo. J Bacteriol 184(4):1132–1139. https://doi.org/10.1128/jb.184.4.1132-1139.2002

Smith EE, Buckley DG, Wu Z, Saenphimmachak C, Hoffman LR, D'Argenio DA, Miller SI, Ramsey BW, Speert DP, Moskowitz SM, Burns JL, Kaul R, Olson MV (2006) Genetic adaptation by Pseudomonas aeruginosa to the airways of cystic fibrosis patients. Proc Natl Acad Sci U S A 103(22):8487–8492. https://doi.org/10.1073/pnas.0602138103

Soh EY, Smith F, Gimenez MR, Yang L, Vejborg RM, Fletcher M, Halliday N, Bleves S, Heeb S, Camara M, Givskov M, Hardie KR, Tolker-Nielsen T, Ize B, Williams P (2021) Disruption of the Pseudomonas aeruginosa Tat system perturbs PQS-dependent quorum sensing and biofilm maturation through lack of the Rieske cytochrome bc1 sub-unit. PLoS Pathog 17(8):e1009425. https://doi.org/10.1371/journal.ppat.1009425

Tahrioui A, Duchesne R, Bouffartigues E, Rodrigues S, Maillot O, Tortuel D, Hardouin J, Taupin L, Groleau MC, Dufour A, Déziel E, Brenner-Weiss G, Feuilloley M, Orange N, Lesouhaitier O, Cornelis P, Chevalier S (2019) Extracellular DNA release, quorum sensing, and PrrF1/F2 small RNAs are key players in Pseudomonas aeruginosa tobramycin-enhanced biofilm formation. NPJ Biofilms Microbiomes 5(1):15. https://doi.org/10.1038/s41522-019-0088-3

Tamayo R (2019) Cyclic diguanylate riboswitches control bacterial pathogenesis mechanisms. PLoS Pathog 15(2):e1007529. https://doi.org/10.1371/journal.ppat.1007529

Teiber JF, Horke S, Haines DC, Chowdhary PK, Xiao J, Kramer GL, Haley RW, Draganov DI (2008) Dominant role of paraoxonases in inactivation of the Pseudomonas aeruginosa quorum-sensing signal N-(3-oxododecanoyl)-L-homoserine lactone. Infect Immun 76(6):2512–2519. https://doi.org/10.1128/iai.01606-07

Tolker-Nielsen T (2014) Pseudomonas aeruginosa biofilm infections: from molecular biofilm biology to new treatment possibilities. APMIS Suppl 138:1–51. https://doi.org/10.1111/apm.12335

Tolker-Nielsen T (2015) Biofilm development. Microbiol Spectr 3(2):MB-0001-2014. https://doi.org/10.1128/microbiolspec.MB-0001-2014

Valentini M, Filloux A (2016) Biofilms and cyclic di-GMP (c-di-GMP) signaling: lessons from Pseudomonas aeruginosa and other bacteria. J Biol Chem 291(24):12547–12555. https://doi.org/10.1074/jbc.R115.711507

Virreira Winter S, Zychlinsky A (2018) The bacterial pigment pyocyanin inhibits the NLRP3 inflammasome through intracellular reactive oxygen and nitrogen species. J Biol Chem 293(13):4893–4900. https://doi.org/10.1074/jbc.RA117.001105

Walker TS, Tomlin KL, Worthen GS, Poch KR, Lieber JG, Saavedra MT, Fessler MB, Malcolm KC, Vasil ML, Nick JA (2005) Enhanced Pseudomonas aeruginosa biofilm development mediated by human neutrophils. Infect Immun 73(6):3693–3701. https://doi.org/10.1128/iai.73.6.3693-3701.2005

Welsh MA, Blackwell HE (2016) Chemical probes of quorum sensing: from compound development to biological discovery. FEMS Microbiol Rev 40(5):774–794. https://doi.org/10.1093/femsre/fuw009

Weng L-X, Yang Y-X, Zhang Y-Q, Wang L-H (2014) A new synthetic ligand that activates QscR and blocks antibiotic-tolerant biofilm formation in Pseudomonas aeruginosa. Appl Microbiol Biotechnol 98(6):2565–2572. https://doi.org/10.1007/s00253-013-5420-x

Whitchurch CB, Tolker-Nielsen T, Ragas PC, Mattick JS (2002) Extracellular DNA required for bacterial biofilm formation. Science 295(5559):1487. https://doi.org/10.1126/science.295.5559.1487

Wu H, Song Z, Givskov M, Doring G, Worlitzsch D, Mathee K, Rygaard J, Hoiby N (2001) Pseudomonas aeruginosa mutations in lasI and rhlI quorum sensing systems result in milder chronic lung infection. Microbiology (Reading) 147(Pt 5):1105–1113. https://doi.org/10.1099/00221287-147-5-1105

Xiao G, Deziel E, He J, Lepine F, Lesic B, Castonguay MH, Milot S, Tampakaki AP, Stachel SE, Rahme LG (2006) MvfR, a key Pseudomonas aeruginosa pathogenicity LTTR-class regulatory protein, has dual ligands. Mol Microbiol 62(6):1689–1699. https://doi.org/10.1111/j.1365-2958.2006.05462.x

Yang L, Barken KB, Skindersoe ME, Christensen AB, Givskov M, Tolker-Nielsen T (2007) Effects of iron on DNA release and biofilm development by Pseudomonas aeruginosa. Microbiology (Reading) 153(Pt 5):1318–1328. https://doi.org/10.1099/mic.0.2006/004911-0

Yang L, Rybtke MT, Jakobsen TH, Hentzer M, Bjarnsholt T, Givskov M, Tolker-Nielsen T (2009) Computer-aided identification of recognized drugs as Pseudomonas aeruginosa quorum-sensing inhibitors. Antimicrob Agents Chemother 53(6):2432–2443. https://doi.org/10.1128/AAC.01283-08

Yang YX, Xu ZH, Zhang YQ, Tian J, Weng LX, Wang LH (2012) A new quorum-sensing inhibitor attenuates virulence and decreases antibiotic resistance in Pseudomonas aeruginosa. J Microbiol 50(6):987–993. https://doi.org/10.1007/s12275-012-2149-7

Yu S, Su T, Wu H, Liu S, Wang D, Zhao T, Jin Z, Du W, Zhu M-J, Chua SL, Yang L, Zhu D, Gu L, Ma LZ (2015) PslG, a self-produced glycosyl hydrolase, triggers biofilm disassembly by disrupting exopolysaccharide matrix. Cell Res 25(12):1352–1367. https://doi.org/10.1038/cr.2015.129

Zhao J, Li X, Hou X, Quan C, Chen M (2019) Widespread existence of quorum sensing inhibitors in marine bacteria: potential drugs to combat pathogens with novel strategies. Mar Drugs 17(5):275. https://doi.org/10.3390/md17050275

Bacteriophages for the Treatment of Biofilm-Associated Infections

David R. Cameron, Luca G. Valente, Melissa Pitton, Josef Prazak, and Yok-Ai Que

Abstract Bacteriophages (phages) are viruses that specifically infect and lyse bacteria. Contrary to the theme of the current book, the therapeutic application of phages is not a "new" anti-biofilm strategy at all. The process, known as phage therapy, has been investigated for just as long as antibiotic therapy although its use is less widespread. As we approach what some describe as the post-antibiotic era, phage therapy has re-emerged as an intriguing antibiotic alternative or complementary treatment strategy. The relationship between phages and their bacterial prey, however, can be volatile with each party equipped with strategies to thwart the other. This is particularly true in biofilms, where phages may penetrate the matrix and lyse bacteria within, and bacteria may conversely evolve to overcome phage infection. Given this relationship, various strategies have been investigated to maximize phage treatment efficacy including the formulation of multiphage treatment cocktails, the exploitation of phage-antibiotic synergies, and the use of targeted phage genome engineering. Currently, phage therapy is being used increasingly as a salvage

D. R. Cameron (✉)
Department of Intensive Care Medicine, Inselspital, Bern University Hospital, University of Bern, Bern, Switzerland

Department for Biomedical Research, University of Bern, Bern, Switzerland

Debiopharm, Lausanne, Switzerland
e-mail: david.cameron@debiopharm.com

L. G. Valente
Department of Intensive Care Medicine, Inselspital, Bern University Hospital, University of Bern, Bern, Switzerland

Institute for Infectious Diseases, University of Bern, Bern, Switzerland

M. Pitton · J. Prazak
Department of Intensive Care Medicine, Inselspital, Bern University Hospital, University of Bern, Bern, Switzerland

Y.-A. Que
Department of Intensive Care Medicine, Inselspital, Bern University Hospital, University of Bern, Bern, Switzerland

Department for Biomedical Research, University of Bern, Bern, Switzerland

K. Richter, K. Nørskov Kragh (eds.), *Antibiofilm Strategies*, Springer Series on Biofilms 11, https://doi.org/10.1007/978-3-031-10992-8_8

therapy in humans following unsuccessful antibiotic treatments. The next big frontier for phage therapy against biofilm infections, however, is to prove efficacy in randomized controlled clinical trials.

1 Introduction

Bacteriophages, or simply phages, are viruses that can lyse bacterial cells. It is this property that has seen them considered for the treatment of bacterial infections for over 100 years (d'Herelle 1917). "Phage therapy" continues to be used in countries such as Russia, Poland, and Georgia (Międzybrodzki et al. 2018). However, the efficacy of phage therapy in early clinical studies was questionable (Eaton and Bayne-Jones 1934), and this coincided with phages being shelved in most countries following the advancement of antibiotics, most notably penicillin. Phage therapy is currently being reconsidered as a potential answer to the ongoing antibiotic resistance crisis. A particularly appealing property of phage therapy is its potential to disrupt biofilms that are refractory to antibacterial chemotherapy. In the current chapter, we will discuss the advantages of phage therapy for the treatment of biofilm-associated infections, as well as some of the limitations. We will discuss innovative approaches that are being investigated to maximize the usefulness of phages for therapy. Finally, we will provide an update as to the clinical outcomes of phage therapy when applied in human biofilm-associated infections.

2 Biofilm Basics

In nature, bacterial cells are often encased within biofilms. The biofilm matrix is a largely self-produced conglomeration of polysaccharides, proteins, nucleic acids, and lipids collectively referred to as extracellular polymeric substances (EPS, Fig. 1a) (Flemming and Wingender 2010). The assembly and disassembly of biofilm is regulated by integrated molecular networks and often involves cell-to-cell communication via the exchange of quorum sensing molecules (Parsek and Greenberg 2005). Notably, the exact composition and regulation of biofilm varies within and across bacterial species.

A great many bacterial infections afflicting human patients are associated with biofilms; they form on native tissues such as the teeth (plaque), skin (i.e., burn wounds), heart, bone, and lungs, as well as on indwelling foreign material including catheters (Fig. 1b), endotracheal tubes and prosthetic joints or heart valves. Biofilms are particularly problematic because they are difficult to eradicate with conventional therapies. The matrix produces a physical barrier that retards the diffusion of various immune factors and some antibiotics (Lewis 2001). Oxygen and nutrient gradients exist across the matrix and this contributes to physiological heterogeneity for cells within the biofilm; in regions where oxygen and nutrients may be limited, bacteria

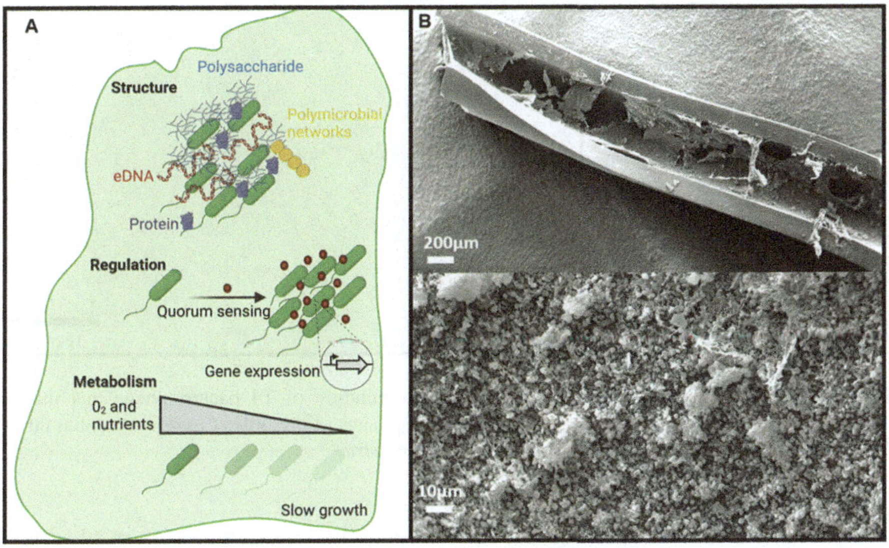

Fig. 1 Bacterial biofilm. (**a**) Schematic illustration of a bacterial biofilm. Structural elements of the matrix are typically self-made polymers known as extracellular polymeric substances (EPS). Cells at high density in biofilms often communicate using quorum-sensing systems. The metabolic activity of cells throughout the biofilm is not uniform and often depends on oxygen and nutrient availability. eDNA, extracellular deoxyribonucleic acid. (**b**) Transmission electron micrographs of *Staphylococcus epidermidis* biofilm formed on the lumen of an indwelling venous catheter

grow slowly and a subpopulation of these cells are not efficiently killed by antibiotics (often referred to as "persister cells") (Stewart and Franklin 2008).

Collectively, an effective anti-biofilm agent should be able to penetrate deep within the biofilm and kill cells in a range of metabolic states. If the agent can disrupt the structural elements of the biofilm (i.e., degrade EPS), or interfere with quorum sensing and biofilm regulation, it would likely be advantageous. The natural properties of bacteriophages, as will be discussed, fulfill at least some of these requirements, suggesting that they may be effective for the treatment of biofilm-associated infections.

3 Bacteriophage Basics

Bacteriophages are natural viral predators of bacteria. They are the most abundant replicative biological entity on the planet and they play an enormous role in shaping the biosphere (Dion et al. 2020; Mushegian 2020; Hendrix et al. 1999). In the environment, while estimates vary widely, roughly 20% of prokaryotes in the ocean are the subject of viral lysis each day (Suttle 2007). In humans, phages are particularly abundant in the gut and may influence gut health by modulating the

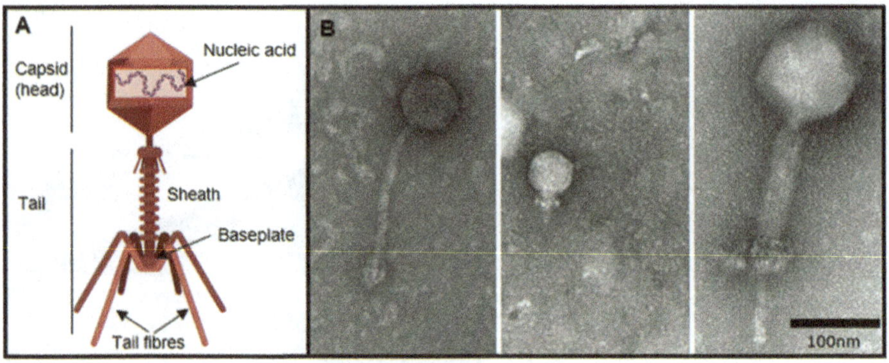

Fig. 2 Bacteriophage structure. (**a**) Schematic representation of T4 bacteriophage that infects *E. coli*. (**b**) Scanning electron micrographs of bacteriophages of the order *Caudovirales* that infect *S. epidermidis*. Left to right: *siphovirus*, *podovirus*, *myovirus*

microbiome, and/or interacting with the host-immune system (Barr et al. 2013; Norman et al. 2015). Aside from abundance, phages are impressively diverse in terms of genomics, physical structure and function (Dion et al. 2020). Phages can be either tailed, non-tailed or filamentous, and they can possess dsDNA, ssDNA, dsRNA, or ssRNA genomes. The model phage T4 that infects *Escherichia coli* is presented schematically in Fig. 2a and representative images of staphylococcal phages (Valente et al. 2021b) from the order *Caudovirales* are presented in Fig. 2b. The tropism of a given phage is largely dictated by the bacterial surface receptors for which it interacts (Rakhuba et al. 2010). Some phages have relatively wide host-ranges, capable of infecting isolates from a few closely related bacterial species. Others have narrow host-ranges, only capable of infecting a limited number of isolates from within a single species. Thus, in comparison to classical antibiotics, phages have narrow-spectrum therapeutic potential.

The two prototypical phage lifecycles are the "lytic" cycle, and the "lysogenic" cycle. As the name suggests, the endpoint of the lytic cycle is lysis of the bacterial cell. In contrast, the lysogenic cycle involves the integration of phage DNA into bacterial genetic elements producing a bacterial "lysogen." The phage is then replicated along with the bacterial DNA and vertically transmitted to daughter cells. A "virulent phage" is obligatorily lytic, for example, the T4-like phages that infect *E. coli*. "Temperate" phages are those that can lysogenize bacteria and following an induction event exercise a lytic cycle, an example of which is lambda phage. An additional phage state worth noting is the "chronic" lifestyle, whereby phage progeny can be persistently released without causing bacterial cell lysis. Filamentous phages, such as M13, utilize a chronic phage lifestyle.

Temperate phages contribute to the spread of genetic material between bacterial strains and species (known as horizontal gene transfer [HGT]) and can include undesirable genes such as those coding for virulence factors and/or antibiotic resistance (Haaber et al. 2016; Waldor and Mekalanos 1996). Additionally, upon treatment, lysogens will likely form which prevent future infection by the same or similar phages. It is for these reasons that temperate phages are not typically considered for phage therapy. Similarly, in the context of human infection, phages capable of chronic release, especially the filamentous phages of *Pseudomonas aeruginosa*, contribute to disease by promoting biofilm formation and interacting with the immune system (Sweere et al. 2019; Secor et al. 2015). Thus, obligatorily lytic/virulent phages are those most commonly used for therapeutic purposes.

4 Phage-Biofilm Interactions

The efficacy of phages for the removal of biofilms (without considering pharmacology) will be mostly dictated by the diffusion or active penetration of phages through the matrix, degradation or disruption of the matrix, and killing of cells within the matrix. The biofilm matrix, however, is a major barrier for phage infection. Biofilms have evolved to restrict phage penetration and can harbor cells that avoid phage lysis (Fig. 3).

Movement of Phages Through the Biofilm A growing number of studies have shown that phages can diffuse through biofilms (Gonzalez et al. 2018; Briandet et al. 2008; Vilas Boas et al. 2016). However, numerous factors will dictate the efficacy with which this occurs. For example, the composition of a biofilm matrix will

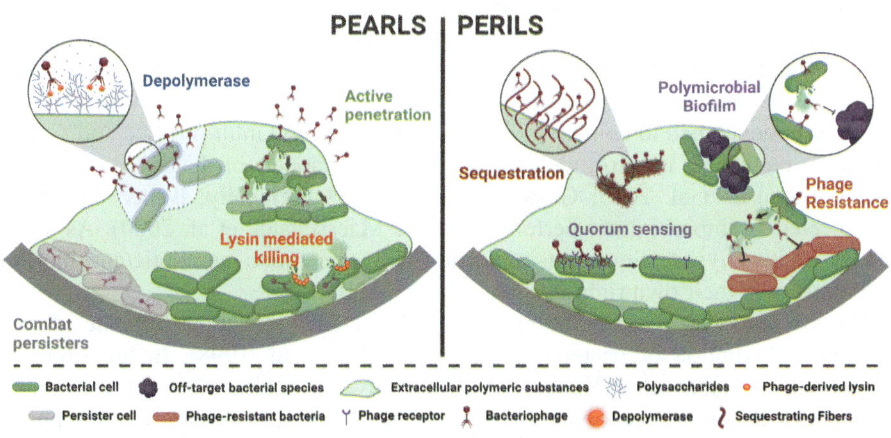

Fig. 3 The pearls and perils of using bacteriophages to combat biofilms. Desirable phage properties are presented on the left and include active penetration through the biofilm matrix, enzymes that can disrupt components of the matrix and a potential to interact with persister cells. Limitations are presented on the right. Phage treatment efficacy can be restricted by phage resistance and/or the presence of bacterial species for which the phage cannot infect. The matrix can contain bacterially derived products that limit infection, and/or the bacteria may sequester phages. Finally, bacteria can communicate using quorum sensing to downregulate the presentation of phage receptors, limiting phage infectivity

influence diffusivity. At least for *E. coli*, the production of amyloid fibers both limits phage mobility, and shields individual cells from infection by physically sequestering phages (Vidakovic et al. 2018). Importantly, the amyloid fiber network is only established following extended biofilm incubation (>60 h) highlighting that the "age" of a biofilm will also dictate phage treatment efficacy (Vidakovic et al. 2018). This presents a possible issue when combatting well-established biofilms in chronic infection settings. Additionally, the thickness, and density with which the biofilm is packed can restrict phage diffusion twofold; if densely packed, large phages in particular may not be able to penetrate through available pores, and if cells are sparsely available, an epidemic of phage infection is unlikely to be sustainable (Simmons et al. 2018; Hu et al. 2010; Vidakovic et al. 2018).

Aside from passive diffusion, phages should be able to move through biofilms by actively lysing cells facilitating access to bacteria in adjacent and perhaps deeper layers of the biofilm. This mode of movement, however, is disrupted by the presence of bacteria that are not susceptible to the specific action of a given phage. Here, phage-resistant bacteria, and/or off-target bacteria can provide a physical barrier that shields susceptible populations, facilitating long-term persistence of the bacteria within the biofilm (Simmons et al. 2020).

Disruption of the Matrix Phages can code for a range of enzymes capable of disrupting biofilms. A key example is peptidoglycan (or murein) hydrolases termed "lysins." Upon release from the cytoplasm, soluble phage lysins break bonds within peptidoglycan facilitating the release of phage progeny from infected bacterial cells. Conversely, lysins can be virion-associated and breakdown peptidoglycan "from without," in order to introduce phage DNA into uninfected cells (Moak and Molineux 2004). Lysins are dually effective as they lyse bacterial cells and this in turn destabilizes and disperses the biofilm matrix (Meng et al. 2011). Additionally, some have shown activity against persister cells (Briers et al. 2014). Purified lysins have been developed to combat biofilms formed by both Gram-positive and Gram-negative pathogens in vitro, and in various animal models of disease (Lood et al. 2015; Entenza et al. 2005; Gilmer et al. 2013). Additionally, lysins have been engineered to improve their performance (De Maesschalck et al. 2020). As such, phage-derived lysins alone are interesting next-generation antibiotic/anti-biofilm alternatives in their own right (Czaplewski et al. 2016).

Depolymerases are an additional class of enzymes that may facilitate phage-mediated disruption of bacterial biofilms (Hughes et al. 1998). In this context, depolymerases that can degrade capsular polysaccharides (Guo et al. 2017), exopolysaccharides or lipopolysaccharides are of interest as they expose phage receptors facilitating phage infection, and they can reduce resistance to innate immunity (Majkowska-Skrobek et al. 2018). A drawback for depolymerases as therapeutic agents is that they are often highly specific, targeting only a limited set of substrates (Hughes et al. 1998). This is not ideal within complex biofilms consisting of a range of polymeric substances. Notwithstanding, when added exogenously to established biofilms of staphylococci, depolymerase Dpo7 reduced biomass in vitro (Gutierrez et al. 2015). The exact contribution of phage-derived depolymerases for biofilm biomass removal in vivo is currently unclear.

While phages and phage-derived enzymes have shown potential for biofilm disruption, it is important to note that they can also have the reverse effect, whereby bacteria under phage predation may be triggered to produce more biofilm (Hansen et al. 2019). Some examples include the production of larger aggregates (as shown for *E. coli* (Lacqua et al. 2006) and *P. aeruginosa* (Henriksen et al. 2019)), the evolution of the mucoid phenotype of *Pseudomonas fluorescens* that results in the overproduction of alginate (Scanlan and Buckling 2012), and the induction of the SOS response in *S. aureus*, which was associated with the production of thicker biofilms (Fernandez et al. 2017).

Killing Cells in the Biofilm Many studies have shown that bacteria in biofilms can be lysed by phages (Szafranski et al. 2017; Waters et al. 2017). The two biological entities, however, are engaged in a "co-evolutionary arms race," whereby bacteria evolve to resist phage infection, and phages subsequently evolve to overcome phage resistance (Hampton et al. 2020). In the arms race analogy, the biofilm represents the battlefield. In the context of therapy, phage resistance is similar to antibiotic resistance as it negatively affects treatment efficacy. Three main mechanisms of bacterial resistance to phage have been identified in vitro and include (1) spontaneous mutation or phase variation of surface receptors preventing adsorption of phages to their bacterial host, (2) specific cleavage of incoming phage DNA by bacterial restriction modification systems, or (3) bacterial adaptive immunity mediated by CRISPR-Cas systems (Clustered Regularly Interspaced Short Palindromic Repeats, coupled to CRISPR-associated genes) (Oechslin 2018).

The metabolism of bacterial cells within a biofilm is not uniform. Certainly, antibiotics are less effective against cells that are exercising limited metabolism, as they typically target active processes (i.e., cell wall biosynthesis, transcription, translation) (Cameron et al. 2018; Shan et al. 2017; Conlon et al. 2016). Importantly, there is some evidence to suggest that phages can interact with slow-growing persister cells of bacteria. Analysis of gene expression for the model pairing of *E. coli* and temperate phage lambda revealed that while lytic phage pathways are suppressed in dormant cells, they are activated upon resuscitation (in other words, resumption of bacterial growth), leading to cell lysis (Pearl et al. 2008). The relevance of this finding in the context of phage therapy, however, is unclear because temperate phages are not typically considered for therapeutic purposes. Where virulent phages are concerned, the data are conflicting. At least for *S. aureus*, Tkhilaishvili et al. reported that staphylococcal phage Sb-1 could kill persister cells (Tkhilaishvili et al. 2018). In contrast, Berryhill and colleagues reported that PYO^{Sa}, which comes from the same source as Sb-1 and is thus likely to be highly similar, did not kill *S. aureus* cells in the stationary phase, which is a growth phase known to harbor a high proportion of persister cells (Conlon et al. 2016; Berryhill et al. 2021). In fact, phage titers decreased when in combination with stationary cells suggesting dormant cells may actually act as a phage sink. A possible advantage of using phage-derived endolysins for therapy against biofilms is that their lytic capacity does not seem to depend on an active bacterial metabolism suggesting they may be effective against slow-growing cells within the biofilm (Briers et al. 2014; Gutierrez et al. 2014). More research is warranted to understand the interaction between phages and phage-derived enzymes, and bacterial persister cells.

In addition to killing cells in various metabolic states, phages must overcome coordinated physiological processes of cellular communities within the biofilm that are mediated via the exchange of quorum-sensing molecules. The paradigmatic classes of quorum-sensing molecules are the acyl-homoserine lactones (AHL) of Gram-negative bacteria, and the autoinducing peptides (AIP) of Gram-positive bacteria (Federle and Bassler 2003). In the context of the former, *N*-acyl-L-homoserine lactone has been found to be a mediator of anti-phage defence (Hoyland-Kroghsbo et al. 2013). Exogenous addition of AHL to *E. coli* reduces the presentation of phage receptors on the surface of the bacteria, which in turn improves survival of *E. coli* under phage attack (Hoyland-Kroghsbo et al. 2013). Similar quorum-mediated phage-receptor downregulation has been shown for additional Gram-negatives including *Vibrio cholerae* (Hoque et al. 2016). In contrast, a role for AIP in phage-bacteria interactions for Gram-positive bacteria is yet to be defined.

5 Overcoming the Limitations of Using Phages for Therapy

The desired outcome when developing an anti-biofilm strategy is the eradication of the bacteria that is producing the biofilm. A caveat for phage therapy is that phages and bacteria naturally coexist, and host bacteria is required for phage replication. It is not in the best interest of the phage population to eradicate all of the susceptible bacteria. Thus, pushing native phages to eliminate their host presents a challenge. Innovative strategies to address this limitation are being investigated including the application of multi-phage cocktails, the use of engineered phages with designer properties, and the exploitation of phage-antibiotic synergies (PAS).

Phage Cocktails The therapeutic potential of an individual phage is limited by its host-range. To increase the usefulness of a phage treatment product, it is commonplace to combine multiple distinct phages into "phage cocktails." This has the dual purpose or broadening the spectrum of activity and mitigating the threat of phage resistance. For the former, multi-phage cocktails have revealed improved therapeutic efficacy against mono- and multi-species biofilms in experimental in vitro models (Alves et al. 2014; Lehman and Donlan 2015). For the later, it is theoretically unlikely that a bacterial cell will simultaneously evolve to overcome the action of multiple distinct phages. In support of this theory, resistance of *S. aureus* to four individual phages emerged within 24 h in vitro, yet no resistance was detected when the four phages were used in combination (Prazak et al. 2019).

Cocktail design can be seemingly straightforward for some bacterial pathogens. *S. aureus* phages, for example, can have comparatively broad intraspecies host-ranges, and combining a small number of phages (i.e., four) can produce a therapeutic that can lyse >90% of isolates tested (Prazak et al. 2019). In contrast, combining ≥10 phages does not ensure widespread infectivity against some troublesome gram-negative pathogens including *E. coli* and *P. aeruginosa* (Jault et al. 2018; Sarker et al. 2012). Further, combining too many different phages in a cocktail is undesirable as the production and maintenance of the therapeutic product becomes complex, and pharmacokinetics/pharmacodynamic outcomes become difficult to predict (Gordillo Altamirano and Barr 2020). The most effective phage cocktails are likely to be those that are designed rationally, with each component providing a complementary characteristic (i.e., choosing phages that target different receptors).

Phage Engineering Primitive phage therapy involves the use of naturally occurring phages that are isolated from the environment, purified, and then given to patients. To expand the functionality of native phages, they can be evolved by serial passage with different bacterial hosts in a process referred to as "phage training." However, the next generation of phage therapy will most likely be realized with the aid of targeted phage engineering.

Phages can be engineered to alter the function of existing genes within the phage genome. A prototypical example involves manipulating phage encoded receptor binding proteins (RBPs), particularly tail fibers, which facilitates host-range tailoring and has the added benefit of mitigating the emergence of phage resistance (Yehl et al. 2019).

Phages can be engineered to harbor genes that provide desirable properties. Genes coding for enzymes that target structural elements of the biofilm, that quench quorum-sensing and disrupt biofilm formation, and those that aide in killing of cells within the biofilm have all been investigated (Lu and Collins 2007; Pei and Lamas-Samanamud 2014; Lemon et al. 2019; Lu and Collins 2009). Phages have also been exploited to deliver the RNA-guided nuclease Cas9 that can target both virulence and antibiotic resistance genes, and to deliver gold nanoparticles that can kill bacteria via photothermal ablation (Bikard et al. 2014; Citorik et al. 2014; Peng et al. 2020). Conversely, phage genes with undesirable properties in the context of therapy, such as lysogeny modules, can be removed to generate pseudo virulent phages from temperate phage backbones (Dedrick et al. 2019).

As our capacity to engineer phage genomes advances, any number of distinct phages may serve as chasses to carry diverse cargo with anti-biofilm properties; the possibilities are almost limitless.

Phage-Antibiotic Combinations Given the proven long-term efficacy of antibiotics for the treatment of bacterial infections, it is unlikely that phage therapy alone will replace antibiotics, particularly in critical care situations. Indeed, in recently published case reports/series, the vast majority performed phage therapy as an adjunct to standard-of-care treatments (Valente et al. 2021a; Dedrick et al. 2019, 2021). Thus, understanding exactly how phages and antibiotics behave in combination is paramount.

Phage-antibiotic synergy (PAS) is defined as the enhanced lytic efficacy of bacteriophages when combined with sub-lethal concentrations of antibiotics

(Comeau et al. 2007). In the specific context of biofilms, PAS has been described for both Gram-positive and Gram-negative pathogens (Ryan et al. 2012; Chaudhry et al. 2017; Dickey and Perrot 2019). In some instances, sequential treatments may be more effective than simultaneous treatments. However, the utility of PAS is not without caveats. Some phage-antibiotic combinations are unlikely to be effective; for instance, combining phages with linezolid, a bacteriostatic protein synthesis inhibitor, has the dual effect of inhibiting both bacterial and phage replication, which likely impairs phage efficacy (Prazak et al. 2021). Additionally, some specific phage-antibiotic combinations can reveal synergies or antagonisms depending purely upon the concentration of each agent applied (Dickey and Perrot 2019; Gu Liu et al. 2020).

In addition to the synergistic killing of biofilms, combining phages with antibiotics has the potential to force evolutionary trade-offs, whereby the induction of phage resistance may consequently impair bacterial pathogenesis or sensitize bacteria to antibiotics. To facilitate infection, phages first interact with specific receptors at the bacterial cell surface. As described above, phage resistance can arise quite quickly if a mutation emerges that is significant enough to destabilize the phage-receptor interaction. A common receptor for gram-negative bacteria is lipopolysaccharide (LPS), which is a key virulence determinant. Thus, an LPS mutation may evolve during therapy that hinders phage infection; however, it may come at the cost of pathogenicity. Similarly, efflux pumps that extrude antibiotics can serve as phage receptor proteins, and pump mutations resulting in phage resistance can improve the efficacy of traditional antibiotics (Chan et al. 2016). However, antagonistic mutations, whereby antibiotic sensitivity decreases, can also emerge (Burmeister et al. 2020; Moulton-Brown and Friman 2018).

6 Phages in Action

Phage therapy has been used to treat patients afflicted by biofilm-associated infections for over 100 years. Most of the clinical experience comes from countries of the former Soviet Union, with institutes reporting promising results (Międzybrodzki et al. 2018). The clinical application of phage therapy in most countries, however, is limited to expanded access or compassionate use as it has yet to be authorized by western medical frameworks. This is at least partly because the phage therapeutic process has not proven to be efficacious when scrutinized in randomized controlled clinical trials (RCT).

Compassionate Use Multiple clinical teams around the world are currently investigating the use of phages as a salvage therapy following unsuccessful antibiotic application in complex infection settings (Fig. 4, left side). Phage therapy in humans has followed one of two contrasting frameworks; a one-size-fits-all approach, where the same therapeutic is used for multiple patients, or a personalized approach where a treatment product is tailored to a single patient. Phages have been delivered to

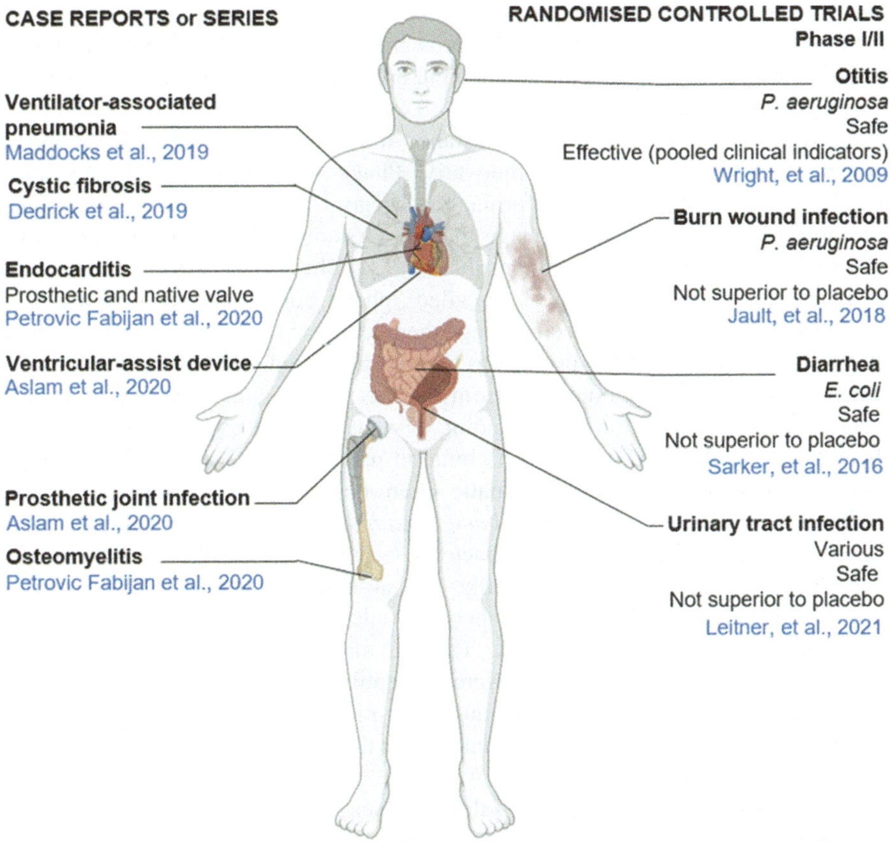

Fig. 4 Infection settings targeted using phage therapy in recent clinical studies. A single reference is provided in blue for each setting. The reference list is not exhaustive

patients in various ways including intravenously (Petrovic Fabijan et al. 2020), topically (Rose et al. 2014) and via nebulization (Maddocks et al. 2019); however, it remains unclear what effects each modality has on phage treatment efficacy.

S. aureus is an attractive target for the one-size-fits-all approach as staphylococcal phages are available with broad intraspecies host-ranges. In Australia, the Westmead Bacteriophage Therapy Team applied a multiphage cocktail (AB-SA01) systematically for the treatment of 13 patients with staphylococcal biofilm-associated infections such as endocarditis and osteomyelitis. Intravenous therapy was safe, with no adverse reactions. Although this single-arm study was not designed to assess efficacy, phage therapy was associated with positive patient outcomes in eight of the 13 cases (62%) (Petrovic Fabijan et al. 2020). Further, only one case was complicated by phage resistance, which was detected in vitro. The major benefits of a one-size-fits-all approach are that it better fits the typical regulatory framework for a therapeutic, production is simpler (i.e., producing one cocktail is more efficient than producing many), and the product can be applied to patients relatively quickly. The drawback for the one-size-fits-all approach is that it is unlikely to be effective in situations where the infective bacteria is inherently resistant to the phages within the pre-established product.

Indeed, the problem of phage resistance mandates personalization for many human infections. The Center for Innovative Phage Applications and Therapeutics (IPATH) at the University of California, San Diego, USA, has explored and refined the application of personalized phage therapy (Aslam et al. 2020). Their pipeline involves seven steps: (1) patient consultation, (2) collection of infective organism, (3) phage susceptibility testing (a so-called "phage hunt"), (4) manufacture of a rationally designed, patient-specific phage therapeutic, (5) application for regulatory approval, (6) shipping of the product to treating physician, and finally (7) administration. Summarizing the outcomes from the first ten uses revealed important and incremental advances in the understanding of what might be expected from personalized phage therapy. First, the most common pathogens for which phage therapy was requested included the problematic Gram-negatives *P. aeruginosa*, *E. coli*, *Klebsiella pneumoniae* and *Acinetobacter baumannii* (~35% of identified cases, collectively), nontuberculosis mycobacteria (~18% collectively), and *S. aureus* (~15%). Secondly, the process typically took an extended period of time (median of 170.5 days, range 28–386). As such, chronic, biofilm and device associated infections were commonly targeted. Thirdly, matching of the bacteria causing infection to a phage(s) for which it were susceptible was not altogether proficient. Of the 99 phage hunts initiated, an appropriate phage-bacteria pairing was established in only 47 cases (47%). Finally, once the pipeline reached the application stage in human subjects, the phages were safe, and therapy was associated with positive patient outcomes in seven of the ten cases (70%).

In previous sections, we described the interaction between bacteria and phages without mention of the third crucial player; the human host. The potential for antibody-mediated phage neutralization has been recognized since the 1930s (Eaton and Bayne-Jones 1934). More recently, sustained intravenous application of phages produced a specific and potent antibody response that was associated with therapeutic failure for a patient with *M. abscessus* infection (Dedrick et al. 2021). Thus, in addition to overcoming the inherent limitations of using phages for therapy (i.e., non-phage susceptibility for the infective strain), the immunocompetency of the patient needs to be considered, particularly when planning long-course, systemic phage therapy.

Randomized, Double-Blinded, Controlled Clinical Trials One of the next key hurdles in the pathway of widespread phage implementation is to prove its efficacy in RCTs that have been appropriately designed to satisfy medical regulatory bodies.

Aside from a single placebo-controlled trial assessing phage therapy for the treatment of chronic otitis due to *P. aeruginosa,* which reported efficacy based on "pooled patient- and physician-reported clinical indicators" (Wright et al. 2009), each of the three most recent phage therapy RCTs failed to show superiority when phages were compared to control treatment groups (Jault et al. 2018; Sarker et al.

2016; Leitner et al. 2021) (Fig. 4, right side). The most relevant trial as it pertains to the treatment of biofilm-associated infection was "Phagoburn," which was a phase I/II multicenter clinical trial aimed at evaluating phage therapy for the treatment of burn wounds infected with *P. aeruginosa* (NCT02116010) (Jault et al. 2018). It assessed a one-size-fits-all approach using a cocktail composed of twelve different anti-*Pseudomonas* phages. The results of the trial were disappointing. Patients receiving phage therapy took longer to reach the primary endpoint (reduced bacterial burden) compared to the standard-of-care (144 h vs. 47 h, $p = 0.018$), and fewer patients reached the endpoint after 7 days of treatment (50% vs. 85%, $p = 0.0917$). Correlates for the poor outcome were poor stability of the therapeutic product (patients received ~10,000-times fewer phages than intended), and bacterial resistance to the phage cocktail at the beginning of therapy.

7 Outlook

Positive results from compassionate case reports and small case series have fuelled a growing enthusiasm for the consideration of phage therapy for broader clinical usage. However, phage therapy is unlikely to be used widely until it is proved efficacious in the framework of evidence-based medicine. While results from recent RCTs were disappointing, each provided important insights into the hurdles that need to be overcome in future trials, be that biological (i.e., limitations of phages), or administrative/regulatory. Regarding phage biology, advancing our understanding of phage pharmacokinetics and pharmacodynamics when applied to biofilms established in vivo, as well as the interaction between bacteria, phages, and antibiotics, will likely help guide us toward improved efficacy. Regarding regulation, innovative clinical trials need to be designed to integrate the somewhat unique requirements of personalized phage therapy, in order to evaluate its therapeutic potential.

Acknowledgments Writing of this piece was supported by Swiss Heart Foundation grant #FF20114 awarded to DRC, and Swiss National Foundation Grants # 320030_176216 and # CR31I3_166124 awarded to YAQ.

Illustrations were generated using https://biorender.com.

References

Alves DR, Gaudion A, Bean JE, Perez Esteban P, Arnot TC, Harper DR, Kot W, Hansen LH, Enright MC, Jenkins AT (2014) Combined use of bacteriophage K and a novel bacteriophage to reduce *Staphylococcus aureus* biofilm formation. Appl Environ Microbiol 80(21):6694–6703. https://doi.org/10.1128/AEM.01789-14

Aslam S, Lampley E, Wooten D, Karris M, Benson C, Strathdee S, Schooley RT (2020) Lessons learned from the first 10 consecutive cases of intravenous bacteriophage therapy to treat multidrug-resistant bacterial infections at a single center in the United States. Open Forum Infect Dis 7(9):ofaa389. https://doi.org/10.1093/ofid/ofaa389

Barr JJ, Auro R, Furlan M, Whiteson KL, Erb ML, Pogliano J, Stotland A, Wolkowicz R, Cutting AS, Doran KS, Salamon P, Youle M, Rohwer F (2013) Bacteriophage adhering to mucus provide a non-host-derived immunity. Proc Natl Acad Sci U S A 110(26):10771–10776. https://doi.org/10.1073/pnas.1305923110

Berryhill BA, Huseby DL, McCall IC, Hughes D, Levin BR (2021) Evaluating the potential efficacy and limitations of a phage for joint antibiotic and phage therapy of *Staphylococcus aureus* infections. Proc Natl Acad Sci U S A 118(10). https://doi.org/10.1073/pnas.2008007118

Bikard D, Euler CW, Jiang W, Nussenzweig PM, Goldberg GW, Duportet X, Fischetti VA, Marraffini LA (2014) Exploiting CRISPR-Cas nucleases to produce sequence-specific antimicrobials. Nat Biotechnol 32(11):1146–1150. https://doi.org/10.1038/nbt.3043

Briandet R, Lacroix-Gueu P, Renault M, Lecart S, Meylheuc T, Bidnenko E, Steenkeste K, Bellon-Fontaine MN, Fontaine-Aupart MP (2008) Fluorescence correlation spectroscopy to study diffusion and reaction of bacteriophages inside biofilms. Appl Environ Microbiol 74(7): 2135–2143. https://doi.org/10.1128/AEM.02304-07

Briers Y, Walmagh M, Grymonprez B, Biebl M, Pirnay JP, Defraine V, Michiels J, Cenens W, Aertsen A, Miller S, Lavigne R (2014) Art-175 is a highly efficient antibacterial against multidrug-resistant strains and persisters of *Pseudomonas aeruginosa*. Antimicrob Agents Chemother 58(7):3774–3784. https://doi.org/10.1128/AAC.02668-14

Burmeister AR, Fortier A, Roush C, Lessing AJ, Bender RG, Barahman R, Grant R, Chan BK, Turner PE (2020) Pleiotropy complicates a trade-off between phage resistance and antibiotic resistance. Proc Natl Acad Sci U S A 117(21):11207–11216. https://doi.org/10.1073/pnas.1919888117

Cameron DR, Shan Y, Zalis EA, Isabella V, Lewis K (2018) A genetic determinant of persister cell formation in bacterial pathogens. J Bacteriol 200(17). https://doi.org/10.1128/JB.00303-18

Chan BK, Sistrom M, Wertz JE, Kortright KE, Narayan D, Turner PE (2016) Phage selection restores antibiotic sensitivity in MDR *Pseudomonas aeruginosa*. Sci Rep 6:26717. https://doi.org/10.1038/srep26717

Chaudhry WN, Concepción-Acevedo J, Park T, Andleeb S, Bull JJ, Levin BR (2017) Synergy and order effects of antibiotics and phages in killing *Pseudomonas aeruginosa* biofilms. PLoS One 12(1):e0168615. https://doi.org/10.1371/journal.pone.0168615

Citorik RJ, Mimee M, Lu TK (2014) Sequence-specific antimicrobials using efficiently delivered RNA-guided nucleases. Nat Biotechnol 32(11):1141–1145. https://doi.org/10.1038/nbt.3011

Comeau AM, Tetart F, Trojet SN, Prere MF, Krisch HM (2007) Phage-antibiotic synergy (PAS): beta-lactam and quinolone antibiotics stimulate virulent phage growth. PLoS One 2(8):e799. https://doi.org/10.1371/journal.pone.0000799

Conlon BP, Rowe SE, Gandt AB, Nuxoll AS, Donegan NP, Zalis EA, Clair G, Adkins JN, Cheung AL, Lewis K (2016) Persister formation in *Staphylococcus aureus* is associated with ATP depletion. Nat Microbiol 1:16051. https://doi.org/10.1038/nmicrobiol.2016.51

Czaplewski L, Bax R, Clokie M, Dawson M, Fairhead H, Fischetti VA, Foster S, Gilmore BF, Hancock RE, Harper D, Henderson IR, Hilpert K, Jones BV, Kadioglu A, Knowles D, Olafsdottir S, Payne D, Projan S, Shaunak S, Silverman J, Thomas CM, Trust TJ, Warn P, Rex JH (2016) Alternatives to antibiotics-a pipeline portfolio review. Lancet Infect Dis 16(2): 239–251. https://doi.org/10.1016/S1473-3099(15)00466-1

Dedrick RM, Guerrero-Bustamante CA, Garlena RA, Russell DA, Ford K, Harris K, Gilmour KC, Soothill J, Jacobs-Sera D, Schooley RT, Hatfull GF, Spencer H (2019) Engineered bacteriophages for treatment of a patient with a disseminated drug-resistant *Mycobacterium abscessus*. Nat Med 25(5):730–733. https://doi.org/10.1038/s41591-019-0437-z

Dedrick RM, Freeman KG, Nguyen JA, Bahadirli-Talbott A, Smith BE, Wu AE, Ong AS, Lin CT, Ruppel LC, Parrish NM, Hatfull GF, Cohen KA (2021) Potent antibody-mediated neutralization limits bacteriophage treatment of a pulmonary *Mycobacterium abscessus* infection. Nat Med. https://doi.org/10.1038/s41591-021-01403-9

De Maesschalck V, Gutierrez D, Paeshuyse J, Lavigne R, Briers Y (2020) Advanced engineering of third-generation lysins and formulation strategies for clinical applications. Crit Rev Microbiol 46(5):548–564. https://doi.org/10.1080/1040841X.2020.1809346

d'Herelle F (1917) On an invisible microbe antagonistic to dysentery bacilli. Note by M. F. d'Herelle, presented by M. Roux. Comptes Rendus Academie des Sciences 1917; 165:373–5. Bacteriophage 1(1):3–5. https://doi.org/10.4161/bact.1.1.14941

Dickey J, Perrot V (2019) Adjunct phage treatment enhances the effectiveness of low antibiotic concentration against *Staphylococcus aureus* biofilms in vitro. PLoS One 14(1):e0209390. https://doi.org/10.1371/journal.pone.0209390

Dion MB, Oechslin F, Moineau S (2020) Phage diversity, genomics and phylogeny. Nat Rev Microbiol 18(3):125–138. https://doi.org/10.1038/s41579-019-0311-5

Eaton M, Bayne-Jones S (1934) Bacteriophage therapy: review of the principles and results of the use of bacteriophage in the treatment of infections. JAMA 103(23):1769–1776. https://doi.org/10.1001/jama.1934.72750490003007

Entenza JM, Loeffler JM, Grandgirard D, Fischetti VA, Moreillon P (2005) Therapeutic effects of bacteriophage Cpl-1 lysin against Streptococcus pneumoniae endocarditis in rats. Antimicrob Agents Chemother 49(11):4789–4792. https://doi.org/10.1128/AAC.49.11.4789-4792.2005

Federle MJ, Bassler BL (2003) Interspecies communication in bacteria. J Clin Invest 112(9): 1291–1299. https://doi.org/10.1172/JCI20195

Fernandez L, Gonzalez S, Campelo AB, Martinez B, Rodriguez A, Garcia P (2017) Low-level predation by lytic phage phiIPLA-RODI promotes biofilm formation and triggers the stringent response in Staphylococcus aureus. Sci Rep 7:40965. https://doi.org/10.1038/srep40965

Flemming HC, Wingender J (2010) The biofilm matrix. Nat Rev Microbiol 8(9):623–633. https://doi.org/10.1038/nrmicro2415

Gilmer DB, Schmitz JE, Euler CW, Fischetti VA (2013) Novel bacteriophage lysin with broad lytic activity protects against mixed infection by *Streptococcus pyogenes* and methicillin-resistant *Staphylococcus aureus*. Antimicrob Agents Chemother 57(6):2743–2750. https://doi.org/10.1128/AAC.02526-12

Gonzalez S, Fernandez L, Gutierrez D, Campelo AB, Rodriguez A, Garcia P (2018) Analysis of different parameters affecting diffusion, propagation and survival of staphylophages in bacterial biofilms. Front Microbiol 9:2348. https://doi.org/10.3389/fmicb.2018.02348

Gordillo Altamirano FL, Barr JJ (2020) Unlocking the next generation of phage therapy: the key is in the receptors. Curr Opin Biotechnol 68:115–123. https://doi.org/10.1016/j.copbio.2020.10.002

Gu Liu C, Green SI, Min L, Clark JR, Salazar KC, Terwilliger AL, Kaplan HB, Trautner BW, Ramig RF, Maresso AW (2020) Phage-antibiotic synergy is driven by a unique combination of antibacterial mechanism of action and stoichiometry. mBio 11(4). https://doi.org/10.1128/mBio.01462-20

Guo Z, Huang J, Yan G, Lei L, Wang S, Yu L, Zhou L, Gao A, Feng X, Han W, Gu J, Yang J (2017) Identification and characterization of Dpo42, a novel depolymerase derived from the Escherichia coli phage vB_EcoM_ECOO78. Front Microbiol 8:1460. https://doi.org/10.3389/fmicb.2017.01460

Gutierrez D, Ruas-Madiedo P, Martinez B, Rodriguez A, Garcia P (2014) Effective removal of staphylococcal biofilms by the endolysin LysH5. PLoS One 9(9):e107307. https://doi.org/10.1371/journal.pone.0107307

Gutierrez D, Briers Y, Rodriguez-Rubio L, Martinez B, Rodriguez A, Lavigne R, Garcia P (2015) Role of the pre-neck appendage protein (Dpo7) from phage vB_SepiS-phiIPLA7 as an anti-biofilm agent in staphylococcal species. Front Microbiol 6:1315. https://doi.org/10.3389/fmicb.2015.01315

Haaber J, Leisner JJ, Cohn MT, Catalan-Moreno A, Nielsen JB, Westh H, Penades JR, Ingmer H (2016) Bacterial viruses enable their host to acquire antibiotic resistance genes from neighbouring cells. Nat Commun 7:13333. https://doi.org/10.1038/ncomms13333

Hampton HG, Watson BNJ, Fineran PC (2020) The arms race between bacteria and their phage foes. Nature 577(7790):327–336. https://doi.org/10.1038/s41586-019-1894-8

Hansen MF, Svenningsen SL, Roder HL, Middelboe M, Burmolle M (2019) Big impact of the tiny: bacteriophage-bacteria interactions in biofilms. Trends Microbiol 27(9):739–752. https://doi.org/10.1016/j.tim.2019.04.006

Hendrix RW, Smith MC, Burns RN, Ford ME, Hatfull GF (1999) Evolutionary relationships among diverse bacteriophages and prophages: all the world's a phage. Proc Natl Acad Sci U S A 96(5): 2192–2197. https://doi.org/10.1073/pnas.96.5.2192

Henriksen K, Rorbo N, Rybtke ML, Martinet MG, Tolker-Nielsen T, Hoiby N, Middelboe M, Ciofu O (2019) *P. aeruginosa* flow-cell biofilms are enhanced by repeated phage treatments but can be eradicated by phage-ciprofloxacin combination. Pathog Dis 77(2). https://doi.org/10.1093/femspd/ftz011

Hoque MM, Naser IB, Bari SM, Zhu J, Mekalanos JJ, Faruque SM (2016) Quorum regulated resistance of *Vibrio cholerae* against environmental bacteriophages. Sci Rep 6:37956. https://doi.org/10.1038/srep37956

Hoyland-Kroghsbo NM, Maerkedahl RB, Svenningsen SL (2013) A quorum-sensing-induced bacteriophage defense mechanism. mBio 4(1):e00362–e00312. https://doi.org/10.1128/mBio.00362-12

Hu J, Miyanaga K, Tanji Y (2010) Diffusion properties of bacteriophages through agarose gel membrane. Biotechnol Prog 26(5):1213–1221. https://doi.org/10.1002/btpr.447

Hughes KA, Sutherland IW, Jones MV (1998) Biofilm susceptibility to bacteriophage attack: the role of phage-borne polysaccharide depolymerase. Microbiology (Reading) 144(Pt 11): 3039–3047. https://doi.org/10.1099/00221287-144-11-3039

Jault P, Leclerc T, Jennes S, Pirnay JP, Que YA, Resch G, Rousseau AF, Ravat F, Carsin H, Le Floch R, Schaal JV, Soler C, Fevre C, Arnaud I, Bretaudeau L, Gabard J (2018) Efficacy and tolerability of a cocktail of bacteriophages to treat burn wounds infected by *Pseudomonas aeruginosa* (PhagoBurn): a randomised, controlled, double-blind phase 1/2 trial. Lancet Infect Dis 19(1):35–45. https://doi.org/10.1016/s1473-3099(18)30482-1

Lacqua A, Wanner O, Colangelo T, Martinotti MG, Landini P (2006) Emergence of biofilm-forming subpopulations upon exposure of Escherichia coli to environmental bacteriophages. Appl Environ Microbiol 72(1):956–959. https://doi.org/10.1128/AEM.72.1.956-959.2006

Lehman SM, Donlan RM (2015) Bacteriophage-mediated control of a two-species biofilm formed by microorganisms causing catheter-associated urinary tract infections in an in vitro urinary catheter model. Antimicrob Agents Chemother 59(2):1127–1137. https://doi.org/10.1128/AAC.03786-14

Leitner L, Ujmajuridze A, Chanishvili N, Goderdzishvili M, Chkonia I, Rigvava S, Chkhotua A, Changashvili G, McCallin S, Schneider MP, Liechti MD, Mehnert U, Bachmann LM, Sybesma W, Kessler TM (2021) Intravesical bacteriophages for treating urinary tract infections in patients undergoing transurethral resection of the prostate: a randomised, placebo-controlled, double-blind clinical trial. Lancet Infect Dis 21(3):427–436. https://doi.org/10.1016/S1473-3099(20)30330-3

Lemon DJ, Kay MK, Titus JK, Ford AA, Chen W, Hamlin NJ, Hwang YY (2019) Construction of a genetically modified T7Select phage system to express the antimicrobial peptide 1018. J Microbiol 57(6):532–538. https://doi.org/10.1007/s12275-019-8686-6

Lewis K (2001) Riddle of biofilm resistance. Antimicrob Agents Chemother 45(4):999–1007. https://doi.org/10.1128/AAC.45.4.999-1007.2001

Lood R, Winer BY, Pelzek AJ, Diez-Martinez R, Thandar M, Euler CW, Schuch R, Fischetti VA (2015) Novel phage lysin capable of killing the multidrug-resistant gram-negative bacterium Acinetobacter baumannii in a mouse bacteremia model. Antimicrob Agents Chemother 59(4): 1983–1991. https://doi.org/10.1128/AAC.04641-14

Lu TK, Collins JJ (2007) Dispersing biofilms with engineered enzymatic bacteriophage. Proc Natl Acad Sci U S A 104(27):11197–11202. https://doi.org/10.1073/pnas.0704624104

Lu TK, Collins JJ (2009) Engineered bacteriophage targeting gene networks as adjuvants for antibiotic therapy. Proc Natl Acad Sci U S A 106(12):4629–4634. https://doi.org/10.1073/pnas.0800442106

Maddocks S, Fabijan AP, Ho J, Lin RCY, Ben Zakour NL, Dugan C, Kliman I, Branston S, Morales S, Iredell JR (2019) Bacteriophage therapy of ventilator-associated pneumonia and empyema caused by *Pseudomonas aeruginosa*. Am J Respir Crit Care Med 200(9):1179–1181. https://doi.org/10.1164/rccm.201904-0839LE

Majkowska-Skrobek G, Latka A, Berisio R, Squeglia F, Maciejewska B, Briers Y, Drulis-Kawa Z (2018) Phage-borne depolymerases decrease Klebsiella pneumoniae resistance to innate defense mechanisms. Front Microbiol 9:2517. https://doi.org/10.3389/fmicb.2018.02517

Meng X, Shi Y, Ji W, Meng X, Zhang J, Wang H, Lu C, Sun J, Yan Y (2011) Application of a bacteriophage lysin to disrupt biofilms formed by the animal pathogen *Streptococcus suis*. Appl Environ Microbiol 77(23):8272–8279. https://doi.org/10.1128/AEM.05151-11

Międzybrodzki R, Hoyle N, Zhvaniya F, Łusiak-Szelachowska M, Weber-Dąbrowska B, Łobocka M, Borysowski J, Alavidze Z, Kutter E, Górski A, Gogokhia L (2018) Current updates from the long-standing phage research centers in Georgia, Poland, and Russia. In: Harper DR, Abedon ST, Burrowes BH, McConville ML (eds) Bacteriophages: biology, technology, therapy. Springer, Cham, pp 1–31. https://doi.org/10.1007/978-3-319-40598-8_31-1

Moak M, Molineux IJ (2004) Peptidoglycan hydrolytic activities associated with bacteriophage virions. Mol Microbiol 51(4):1169–1183. https://doi.org/10.1046/j.1365-2958.2003.03894.x

Moulton-Brown CE, Friman VP (2018) Rapid evolution of generalized resistance mechanisms can constrain the efficacy of phage-antibiotic treatments. Evol Appl 11(9):1630–1641. https://doi.org/10.1111/eva.12653

Mushegian AR (2020) Are there 10(31) virus particles on earth, or more, or fewer? J Bacteriol 202(9). https://doi.org/10.1128/JB.00052-20

Norman JM, Handley SA, Baldridge MT, Droit L, Liu CY, Keller BC, Kambal A, Monaco CL, Zhao G, Fleshner P, Stappenbeck TS, McGovern DP, Keshavarzian A, Mutlu EA, Sauk J, Gevers D, Xavier RJ, Wang D, Parkes M, Virgin HW (2015) Disease-specific alterations in the enteric virome in inflammatory bowel disease. Cell 160(3):447–460. https://doi.org/10.1016/j.cell.2015.01.002

Oechslin F (2018) Resistance development to bacteriophages occurring during bacteriophage therapy. Viruses 10(7):doi:10.3390/v10070351

Parsek MR, Greenberg EP (2005) Sociomicrobiology: the connections between quorum sensing and biofilms. Trends Microbiol 13(1):27–33. https://doi.org/10.1016/j.tim.2004.11.007

Pearl S, Gabay C, Kishony R, Oppenheim A, Balaban NQ (2008) Nongenetic individuality in the host-phage interaction. PLoS Biol 6(5):e120. https://doi.org/10.1371/journal.pbio.0060120

Pei R, Lamas-Samanamud GR (2014) Inhibition of biofilm formation by T7 bacteriophages producing quorum-quenching enzymes. Appl Environ Microbiol 80(17):5340–5348. https://doi.org/10.1128/AEM.01434-14

Peng H, Borg RE, Dow LP, Pruitt BL, Chen IA (2020) Controlled phage therapy by photothermal ablation of specific bacterial species using gold nanorods targeted by chimeric phages. Proc Natl Acad Sci U S A 117(4):1951–1961. https://doi.org/10.1073/pnas.1913234117

Petrovic Fabijan A, Lin RCY, Ho J, Maddocks S, Ben Zakour NL, Iredell JR, Westmead Bacteriophage Therapy T (2020) Safety of bacteriophage therapy in severe *Staphylococcus aureus* infection. Nat Microbiol 5(3):465–472. https://doi.org/10.1038/s41564-019-0634-z

Prazak J, Iten M, Cameron DR, Save J, Grandgirard D, Resch G, Goepfert C, Leib SL, Takala J, Jakob SM, Que YA, Haenggi M (2019) Bacteriophages Improve Outcomes in Experimental *Staphylococcus aureus* Ventilator-associated Pneumonia. Am J Respir Crit Care Med 200(9):1126–1133. https://doi.org/10.1164/rccm.201812-2372OC

Prazak J, Valente L, Iten M, Federer L, Grandgirard D, Soto S, Resch G, Leib SL, Jakob SM, Haenggi M, Cameron DR, Que YA (2021) Benefits of aerosolized phages for the treatment of pneumonia due to methicillin-resistant *Staphylococcus aureus* (MRSA): an experimental study in rats. J Infect Dis. https://doi.org/10.1093/infdis/jiab112

Rakhuba DV, Kolomiets EI, Dey ES, Novik GI (2010) Bacteriophage receptors, mechanisms of phage adsorption and penetration into host cell. Pol J Microbiol 59(3):145–155

Rose T, Verbeken G, Vos DD, Merabishvili M, Vaneechoutte M, Lavigne R, Jennes S, Zizi M, Pirnay JP (2014) Experimental phage therapy of burn wound infection: difficult first steps. Int J Burns Trauma 4(2):66–73

Ryan EM, Alkawareek MY, Donnelly RF, Gilmore BF (2012) Synergistic phage-antibiotic combinations for the control of Escherichia coli biofilms in vitro. FEMS Immunol Med Microbiol 65(2):395–398. https://doi.org/10.1111/j.1574-695X.2012.00977.x

Sarker SA, McCallin S, Barretto C, Berger B, Pittet AC, Sultana S, Krause L, Huq S, Bibiloni R, Bruttin A, Reuteler G, Brussow H (2012) Oral T4-like phage cocktail application to healthy adult volunteers from Bangladesh. Virology 434(2):222–232. https://doi.org/10.1016/j.virol.2012.09.002

Sarker SA, Sultana S, Reuteler G, Moine D, Descombes P, Charton F, Bourdin G, McCallin S, Ngom-Bru C, Neville T, Akter M, Huq S, Qadri F, Talukdar K, Kassam M, Delley M, Loiseau C, Deng Y, El Aidy S, Berger B, Brussow H (2016) Oral phage therapy of acute bacterial diarrhea with two coliphage preparations: a randomized trial in children from Bangladesh. EBioMedicine 4:124–137. https://doi.org/10.1016/j.ebiom.2015.12.023

Scanlan PD, Buckling A (2012) Co-evolution with lytic phage selects for the mucoid phenotype of *Pseudomonas fluorescens* SBW25. ISME J 6(6):1148–1158. https://doi.org/10.1038/ismej.2011.174

Secor PR, Sweere JM, Michaels LA, Malkovskiy AV, Lazzareschi D, Katznelson E, Rajadas J, Birnbaum ME, Arrigoni A, Braun KR, Evanko SP, Stevens DA, Kaminsky W, Singh PK, Parks WC, Bollyky PL (2015) Filamentous bacteriophage promote biofilm assembly and function. Cell Host Microbe 18(5):549–559. https://doi.org/10.1016/j.chom.2015.10.013

Shan Y, Brown Gandt A, Rowe SE, Deisinger JP, Conlon BP, Lewis K (2017) ATP-dependent persister formation in *Escherichia coli*. mBio 8(1). https://doi.org/10.1128/mBio.02267-16

Simmons M, Drescher K, Nadell CD, Bucci V (2018) Phage mobility is a core determinant of phage-bacteria coexistence in biofilms. ISME J 12(2):531–543. https://doi.org/10.1038/ismej.2017.190

Simmons EL, Bond MC, Koskella B, Drescher K, Bucci V, Nadell CD (2020) Biofilm structure promotes coexistence of phage-resistant and phage-susceptible bacteria. mSystems 5(3). https://doi.org/10.1128/mSystems.00877-19

Stewart PS, Franklin MJ (2008) Physiological heterogeneity in biofilms. Nat Rev Microbiol 6(3):199–210. https://doi.org/10.1038/nrmicro1838

Suttle CA (2007) Marine viruses—major players in the global ecosystem. Nat Rev Microbiol 5(10):801–812. https://doi.org/10.1038/nrmicro1750

Sweere JM, Van Belleghem JD, Ishak H, Bach MS, Popescu M, Sunkari V, Kaber G, Manasherob R, Suh GA, Cao X, de Vries CR, Lam DN, Marshall PL, Birukova M, Katznelson E, Lazzareschi DV, Balaji S, Keswani SG, Hawn TR, Secor PR, Bollyky PL (2019) Bacteriophage trigger antiviral immunity and prevent clearance of bacterial infection. Science 363(6434). https://doi.org/10.1126/science.aat9691

Szafranski SP, Winkel A, Stiesch M (2017) The use of bacteriophages to biocontrol oral biofilms. J Biotechnol 250:29–44. https://doi.org/10.1016/j.jbiotec.2017.01.002

Tkhilaishvili T, Lombardi L, Klatt AB, Trampuz A, Di Luca M (2018) Bacteriophage Sb-1 enhances antibiotic activity against biofilm, degrades exopolysaccharide matrix and targets persisters of *Staphylococcus aureus*. Int J Antimicrob Agents 52(6):842–853. https://doi.org/10.1016/j.ijantimicag.2018.09.006

Valente L, Prazak J, Que YA, Cameron DR (2021a) Progress and pitfalls of bacteriophage therapy in critical care: a concise definitive review. Crit Care Explor 3(3):e0351. https://doi.org/10.1097/CCE.0000000000000351

Valente LG, Pitton M, Fürholz M, Oberhaensli S, Bruggmann R, Leib SL, Jakob SM, Resch G, Que Y-A, Cameron DR (2021b) Isolation and characterization of bacteriophages from the human skin microbiome that infect *Staphylococcus epidermidis*. FEMS Microbes 2. https://doi.org/10.1093/femsmc/xtab003

Vidakovic L, Singh PK, Hartmann R, Nadell CD, Drescher K (2018) Dynamic biofilm architecture confers individual and collective mechanisms of viral protection. Nat Microbiol 3(1):26–31. https://doi.org/10.1038/s41564-017-0050-1

Vilas Boas D, Almeida C, Sillankorva S, Nicolau A, Azeredo J, Azevedo NF (2016) Discrimination of bacteriophage infected cells using locked nucleic acid fluorescent in situ hybridization (LNA-FISH). Biofouling 32(2):179–190. https://doi.org/10.1080/08927014.2015.1131821

Waldor MK, Mekalanos JJ (1996) Lysogenic conversion by a filamentous phage encoding cholera toxin. Science 272(5270):1910–1914. https://doi.org/10.1126/science.272.5270.1910

Waters EM, Neill DR, Kaman B, Sahota JS, Clokie MRJ, Winstanley C, Kadioglu A (2017) Phage therapy is highly effective against chronic lung infections with Pseudomonas aeruginosa. Thorax 72(7):666–667. https://doi.org/10.1136/thoraxjnl-2016-209265

Wright A, Hawkins CH, Anggard EE, Harper DR (2009) A controlled clinical trial of a therapeutic bacteriophage preparation in chronic otitis due to antibiotic-resistant *Pseudomonas aeruginosa*; a preliminary report of efficacy. Clin Otolaryngol 34(4):349–357. https://doi.org/10.1111/j.1749-4486.2009.01973.x

Yehl K, Lemire S, Yang AC, Ando H, Mimee M, Torres MT, de la Fuente-Nunez C, Lu TK (2019) Engineering phage host-range and suppressing bacterial resistance through phage tail fiber mutagenesis. Cell 179(2):459–469 e459. https://doi.org/10.1016/j.cell.2019.09.015

Hyperbaric Oxygen Treatment May Advance the Outcome of Antibiotic Treatment of Biofilm Infections

Peter Østrup Jensen, Signe Agnete Møller, Simone Stjernekilde, Pernille Olsen, Claus Moser, Franziska Angelika Schwartz, Christian Johann Lerche, Niels Høiby, Ole Hyldegaard, Daniel Faurholt-Jepsen, and Mette Kolpen

Abstract The failing eradication of bacterial biofilm by antibiotic treatment is a major clinical challenge for patients with chronic infections. The effect of re-sensitizing infectious biofilms to antibiotics by supplemental O_2 has recently been explained by the pathophysiological conditions of the infectious microenvironment. In this chapter, we argue that multifactorial restrictions of O_2, caused by O_2 depletion within the infectious microenvironment, determine the tolerance of bacterial phenotypes ranging from the endogenous formation of persistent subpopulations within biofilms to the induction of tolerant biofilm by the host response. Consequently, we describe how reoxygenation may re-sensitize microorganisms growing in biofilms to antibiotics. To further emphasize the importance of O_2 distribution we demonstrate the significance of O_2 availability for the outcome of susceptibility testing.

P. Ø. Jensen · N. Høiby
Department of Clinical Microbiology, Rigshospitalet, Copenhagen, Denmark

Costerton Biofilm Center, Department of Immunology and Microbiology, University of Copenhagen, Copenhagen, Denmark

S. A. Møller · S. Stjernekilde · P. Olsen · C. Moser · F. A. Schwartz · C. J. Lerche · M. Kolpen (✉)
Department of Clinical Microbiology, Rigshospitalet, Copenhagen, Denmark
e-mail: Mette.Kolpen@regionh.dk

O. Hyldegaard
Department of Anesthesia, Hyperbaric Unit, Rigshospitalet, University of Copenhagen, Copenhagen, Denmark

D. Faurholt-Jepsen
Department of Infectious Diseases, Rigshospitalet, Copenhagen, Denmark

K. Richter, K. Nørskov Kragh (eds.), *Antibiofilm Strategies*, Springer Series on Biofilms 11, https://doi.org/10.1007/978-3-031-10992-8_9

1 Introduction

The most extensively examined biofilm infection is the chronic *Pseudomonas aeruginosa* pneumonia in patients with cystic fibrosis (CF). Even though CF patients who have acquired lung infection with *P. aeruginosa* biofilm receive early antibiotic treatments and are subsequently intensively treated with antibiotics, the infection persists (Høiby et al. 2017). This persistence is not completely due to the development of conventional antibiotic resistance related to genetic changes nor by the lack of susceptibility to antibiotics judged by standard testing, nor by deficient penetration of the antibiotics into the lungs or the biofilms (Høiby et al. 2011; Cao et al. 2015, 2016). Lately, however, evidence has been obtained that suggests the lack of O_2 may contribute to the poor outcome of antibiotic treatment in CF patients with chronic *P. aeruginosa* pneumonia as well as other groups of patients with biofilm-related infections., The beneficial potential of supplying oxygen together with antibiotic treatment of infectious biofilm is therefore reviewed.

2 Internal Microenvironment

The chemical microenvironment inside the biofilm is determined by multiple factors involving bacterial metabolic activity, solute exchange with the surroundings, diffusion properties, and restricted bacterial mobility in the exopolymeric matrix of biofilms. The outcome of these interacting mechanisms is the creation of chemical gradients from the surface to the center of biofilms which in turn generates physiological gradients with distinct localization of adapted subpopulations. Interestingly, the distribution of molecular oxygen (O_2) may stratify the subpopulations in biofilms resulting in metabolically active peripheral subpopulations with consumption of O_2 and inner subpopulations that are metabolically inactive due to depletion of O_2 (Jensen et al. 2017). In in vitro biofilm grown at atmospheric O_2, the O_2 consumption by the metabolically active peripheral subpopulation may limit the penetration depth of O_2 to 50 μm, beneath which metabolically inactive subpopulations are located (Walters et al. 2003). The susceptibility of bacteria to most antibiotics is strongly determined by the metabolic activity of the bacteria. In this respect, high metabolic activity may result in increased growth which is related to enhanced susceptibility to beta-lactams, quinolones, and aminoglycosides in biofilm as well as in planktonic cultures (Cozens et al. 1986; Tuomanen et al. 1986; Evans et al. 1991), but high metabolic activity may even increase the susceptibility to antibiotics in non-growing cells (Lopatkin et al. 2019). The metabolic activity may be driven by aerobic respiration, which renders several bacterial species susceptible to antibiotics (Lobritz et al. 2015). In biofilm, the contribution of aerobic respiration to the susceptibility to antibiotics has been evidenced by removing O_2 (Borriello et al. 2004) and by applying potassium cyanide (KCN) for blocking the exploitation of O_2 as an electron acceptor by the terminal oxidases (Fig. 1).

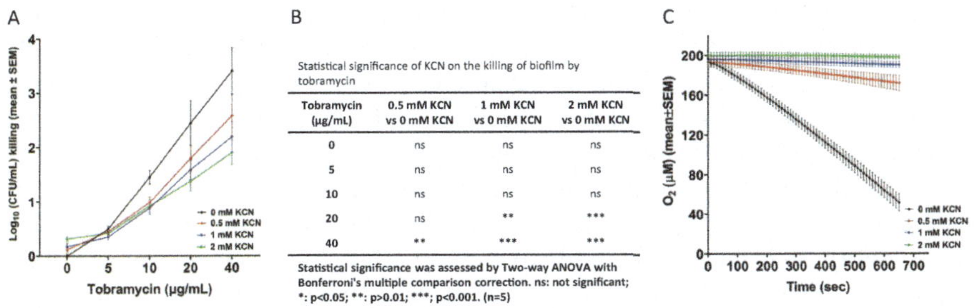

Statistical significance of KCN on the killing of biofilm by tobramycin

Tobramycin (μg/mL)	0.5 mM KCN vs 0 mM KCN	1 mM KCN vs 0 mM KCN	2 mM KCN vs 0 mM KCN
0	ns	ns	ns
5	ns	ns	ns
10	ns	ns	ns
20	ns	**	***
40	**	***	***

Statistical significance was assessed by Two-way ANOVA with Bonferroni's multiple comparison correction. ns: not significant; *: p<0.05; **: p>0.01; ***; p<0.001. (n=5)

Fig. 1 Effect of aerobic respiration on the susceptibility of *P. aeruginosa* biofilms to tobramycin. (**a**) and (**b**) novel demonstration of cyanide-induced reduced susceptibility to tobramycin of *P. aeruginosa* biofilm. *P. aeruginosa* (PAO1) was embedded in agarose and grown in microtiter-plates at atmospheric conditions for 3 days as previously described (Møller et al. 2019) before treatment. Briefly, overnight cultures of PAO1 were adjusted to an optical density at 600 nm of 0.4 and then diluted ×100 in Lurai-Bertani (LB) medium supplemented with 2% 2-hydroxyethylagarose (Sigma-Aldrich, Brøndby, Denmark) to achieve a cell loading of approximately 10^6 CFU/mL. 50 μL agar suspensions were loaded into each well in a flat-bottom microtiter plate (Nunclon™ Delta Surface; Thermo Fischer Scientific, Roskilde, Denmark) and 50 μL of LB medium was added as supernatant on top of the solid gels. The microtiter plates were sealed with Parafilm (Bemis, Neenah, WI, USA) and lid, and incubated at 37°C for 3 days to establish mature agarose-embedded bacterial aggregates. The supernatant was replaced daily to ensure adequate growth. Inhibition of aerobic respiration was initiated on biofilms after 3 days by replacing the supernatant with 50 μL of LB with potassium cyanide (KCN) (final concentration: 0, 0.5, 1, and 2 mM KCN). Fifteen minutes after aerobic inhibition the supernatant was replaced with an LB with tobramycin (final concentration: 0, 5, 10, 20, and 40 mM μg/mL). The microtiter plates were incubated for 90 min at 37°C. At termination of the experiments, the supernatant was removed, and the agarose-embedded aggregates were placed in 2.95 mL phosphate-buffered saline (PBS) (Substrate Department, Panum Institute, Copenhagen, Denmark) before homogenizing for 20 s (SilentCrusher M; Heidolph, Schwabach, Germany). Quantitative bacteriology was performed by standard tenfold dilutions in 0.9% NaCl and plating on Blue plates followed by CFU determination after incubation overnight at 37°C. (**a**) Impact of inhibiting aerobic respiration with potassium cyanide (KCN) on the killing of *P. aeruginosa* biofilms by tobramycin. (**b**) Statistical evaluation of the impact of KCN on the killing of *P. aeruginosa* biofilms. Statistical significance was evaluated by a two-way ANOVA test followed by Bonferroni's multiple comparison correction. (**c**) Inhibition of aerobic respiration by KCN was demonstrated by micro-respirometry according to a modification of our previous study (Kolpen et al. 2010). Briefly, the impact of KCN on O_2 consumption by *P. aeruginosa* (PAO1) was determined using overnight cultures diluted ten times in LB. Measuring of O_2 concentration and time was performed in respiration vials (OXVIAL4, Pyro-Science) with a magnet. During O_2 measurement the samples were stirred at 450 rpm. The O_2 concentration was recorded with separate adaptor rings (ADVIAL4, Pyro-Science) fixated to optical fibers (SPFIB-BARE, Pyro-Science) which connects the respiration vials to the optical O_2 meter (FireStingO2, FSO2-4, Pyro-Science). This was connected to the Pure Oxygen Logger software (v.3.206 with FireStingO2 (Firmware 3.07), Pyro-Science) which measured the O_2 concentration in continuous mode for each second. Factory calibrations were used, and the settings salinity 0.9 g/L and 1013 mbar were used. Micro respirometry experiments for PAO1 were performed in triplicates in an incubation room at 37°C. Each overnight culture was pretreated with 0–2 mM of KCN for 15 min before thorough oxygenation by manually shaking and initiation of O_2 measurements

The ability of O_2 to increase metabolic activity resulting in antibiotic-susceptible physiology (Walters et al. 2003) may link the internal distribution of O_2 to the stratified susceptibility in biofilm. In the absence of O_2, respiration using alternative electron acceptors, such as nitrate, as well as oxidative phosphorylation may also increase susceptibility to aminoglycosides (Borriello et al. 2006). The potential of O_2 and alternative electron acceptors to enhance the killing efficacy of major types of antibiotics is not a unique property of the biofilm mode of growth since the susceptibility of planktonic cultures of the same bacterial species display similar responses to the availability of electron acceptors (Brochmann et al. 2014; Dwyer et al. 2014). At present, however, it is not known if biofilms are stratified in zones with high or low susceptibility when O_2 is absent from the surrounding microenvironment and alternative electron acceptors are used.

The impact of bacterial metabolism on the susceptibility to antibiotics may result from the ability of low metabolic activity to decrease the uptake of antibiotics and downregulate the expression of antibiotic targets (Van Acker and Coenye 2017). These mechanisms are likely contributing to the reduced bactericidal efficacy of antibiotics when metabolism is reduced due to lack of O_2. However, the presence of O_2 may induce additional killing arising from formation of cytotoxic levels of reactive oxygen species (ROS) induced by the antibiotics (Brochmann et al. 2014; Dwyer et al. 2014). The induction of cytotoxic levels of ROS by antibiotics is associated with increased production of NADH by the tricarboxylic acid cycle (TCA) (Kohanski et al. 2007). Consequently, part of the impact of O_2 on the susceptibility to antibiotics may result from aerobic bacterial respiration stimulating the TCA cycle, which enhances the effect of antibiotic treatment (Crabbe et al. 2019). In accordance, O_2-depletion may further increase bacterial tolerance to antibiotics by inducing a shift from the TCA cycle to anaerobic glycolysis and by preventing ROS formation. Moreover, several enzymes of the TCA cycle are activated by O_2 (Mitchell and Dawes 1982).

In biofilm, the significance of the accumulation of cytotoxic ROS is correlated to the downregulation of the TCA cycle in surviving cells from *Burkholderia cenocepacia* biofilm treated with tobramycin (Van Acker et al. 2013). The importance of cytotoxic ROS is further evidenced by the connection between deficient antioxidative defense and increased susceptibility of *P. aeruginosa* biofilm to ciprofloxacin (Jensen et al. 2014) and of *Staphylococcus aureus* biofilm to daptomycin (El Haj et al. 2021). Thus, it is evident that the internal concentration of O_2 in biofilms is a key determinant of their susceptibility to antibiotics and that the metabolic activity of biofilms consumes and thereby affects the distribution of O_2 in biofilms.

3 External Microenvironment

The direct correlation of the O_2 concentration at the surface of the biofilm to the distribution of O_2 inside the biofilm (Stewart 2003) suggests that the O_2 concentration in the external microenvironment may directly influence its susceptibility to antibiotics. During infections, the pathophysiological conditions—such as inflammation—consume and thereby reduce the O_2 concentration in the external microenvironment (Jensen et al. 2017) causing a lack of O_2 with the potential to provide an explanation for the unsuccessful outcome of antibiotic therapy. Lack of O_2 in the external microenvironment may result from the concerted O_2 consumption of host cells and the infectious biofilm (Wu et al. 2018) and it is important to distinguish O_2 consumption by the host response from O_2 consumption by the infectious biofilm. It is, however, also important to determine if O_2 consumption by the host response prevents the aerobic bacterial metabolism that may be connected to susceptibility to antibiotics. The possibility to disentangle the O_2 consumption by the host response from that of the infectious biofilms has been demonstrated in freshly expectorated sputum samples from CF patients with chronic *P. aeruginosa* lung infection (Kolpen et al. 2010, 2014). In CF patients with chronic *P. aeruginosa* lung infection, lack of O_2 has been shown directly in endobronchial secretions in the lungs (Worlitzsch et al. 2002). By examining sputum samples from CF patients with chronic *P. aeruginosa* lung infection with micro-respirometry it was shown that the O_2 consumption is dominated by activated human polymorphonuclear leukocytes (PMNs) while the consumption from bacterial aerobic respiration is diminutive (Kolpen et al. 2010, 2014).

Major contributions of the host response to chronic *P. aeruginosa* biofilm infections are provided by PMNs and macrophages through pattern recognition receptors (PRRs). The PRRs of PMNs and macrophages recognize and bind pathogen-associated molecular patterns (PAMPs) expressed on *P. aeruginosa* and biofilm-associated molecular patterns (BAMPs) present in the biofilm matrix. The binding of BAMPs and PAMPs to PRRs activates the PMNs and macrophages (Moser et al. 2021). The activation of PMNs and macrophages includes stimulation of the respiratory burst in PMNs resulting in strong consumption of O_2 for the formation of ROS (Kolpen et al. 2010) and nitric oxide (NO) (Kolpen et al. 2014).

The innate immune response further comprises secretion of proteases by PMNs and release of pro-inflammatory cytokines by macrophages. During the maturation of the adaptive immune response, the T-cells and the B-cells reside distantly from the biofilms. Upon activation, T-cells secrete pro-inflammatory cytokines that attract and activate PMNs and stimulate production of IgG from plasma cells leading to immune-complex mediated stimulation of the PMNs and activation of the classical complement pathway (Moser et al. 2017). Accordingly, the persistence of chronic biofilm infections enables the adaptive immune response to mature and to accelerate the local inflammation causing collateral tissue damage but without eliminating the infectious biofilm (Jensen et al. 2010). The depletion of O_2 by activated PMNs restricts the amount of O_2 available for bacterial aerobic respiration, which impacts bacterial growth as evidenced by the inverse correlation of the growth of *P. aeruginosa* to the amount of PMNs surrounding the biofilms in endobronchial secretions from CF patients with chronic lung infection (Kragh et al. 2014). Slow bacterial growth resulting from O_2 depletion may also apply to pathogens other than *P. aeruginosa* (Jensen et al. 2017). Various technologies have indicated slow growth of *Stenotrophomonas maltophilia* (Kolpen et al. 2015), *Achromobacter xylosoxidans* (DePas et al. 2016) and *Staphylococcus aureus* (Kopf et al. 2016) in endobronchial secretions from CF patients with chronic lung infection. Due to the relation of slow bacterial growth with low susceptibility to antibiotics, O_2 consumption the PMNs may determine the recalcitrance of chronic biofilm infections to the intense antibiotic treatment given to CF patients. In contrast to the slowly growing bacteria in the biofilms of difficult-to-treat chronic lung infections, biofilms with fast-growing bacteria have recently been reported to dominate acute lung infections, which are much more susceptible to antibiotics (Kolpen et al. 2022). This further strengthens the importance of the growth rate for the susceptibility of infectious biofilm to antibiotics.

In chronic wounds, the prevalence of bacterial biofilms may exceed 80% (Malone et al. 2017). The influence of bacterial biofilms on the deficient healing of chronic wounds has long been appreciated (Bjarnsholt et al. 2008; James et al. 2008) and confirmed experimentally (Seth et al. 2012; Trøstrup et al. 2013; Watters et al. 2013). The microenvironment of chronic wounds also contains zones with low O_2 as suggested by luminescence lifetime imaging (Meier et al. 2014), transcriptomics and steep O_2 gradients (James et al. 2016) as well as by the high abundance of anaerobes (Dowd et al. 2008) and the presence of anaerobic metabolites (Debats et al. 2006). The mechanisms of O_2 depletion in infected wounds are far from clarified, but the accumulation of PMNs is increased in wounds with biofilm infection (Fazli et al. 2011; Trøstrup et al. 2013) and the concerted action of biofilm and the summoned PMNs may account for the steep O_2 gradients in chronic wounds similar to the lungs of biofilm-infected CF patients (Wu et al. 2018). The possibility of ongoing depletion of O_2 by PMNs is proposed by the connection between the extent of the respiratory burst and the bacterial load in infected wounds (Belotsky et al. 1990). Lack of O_2 may significantly delay wound healing (Hunt et al. 1969; Gottrup et al. 2017; Frykberg et al. 2020), but the impact of hypoxia on the efficacy of antibiotic treatment in chronic wounds remains to be proven.

The influence of pathophysiological structural changes, such as edema, necrosis, and poor blood supply, may also cause hypoxia (Gottrup et al. 2017) with the potential to induce tolerance to antibiotics. Nonetheless, the enhanced effect of antibiotic treatments of numerous types of biofilm infections by additional O_2 suggests that lack of O_2 in the microenvironment of biofilms may provide protection against antibiotics. The types of biofilm infections that have been sensitized to antibiotics by supplemental O_2 include endocarditis (Ozkan et al. 2016; Lerche et al. 2017), pneumonia (Haft 1991), osteomyelitis (Yu et al. 2011), brain abscesses (Kutlay et al. 2008; Bartek et al. 2016) and device-related infections (Bartek et al. 2018). The recent discovery of formation of thick biofilms in necrotizing soft-tissue infections (NSTI) (Siemens et al. 2016) has been proposed to provide an explanation for the beneficial outcome of adjuvant HBOT of NSTI patients due to the alleviation of hypoxia by HBOT resulting in restored susceptibility of biofilm to antibiotics (Hedetoft et al. 2021a).

4 Reversing Lack of O_2 by HBOT Improves Treatment Outcome

To overcome hypoxia-induced antibiotic tolerance in biofilms, alleviation of O_2 limitation may activate bacterial aerobic respiration and thus increase the susceptibility of pathogens to several antibiotics targeting metabolically active bacteria. In accordance with this, addition of extra O_2 by hyperbaric O_2 treatment (HBOT) significantly enhances the bactericidal activity of several antibiotics on in vitro biofilm (Kolpen et al. 2016; Lerche et al. 2017, Møller et al. 2019) (Fig. 2) and it has been shown to enhance the antibiotic efficacy during experimental in vivo biofilm infections (Haft 1991; Ozkan et al. 2016, Lerche et al. 2017). The bactericidal activity of colistin on *P. aeruginosa* biofilm, however, is decreased by HBOT (Fig. 2). The protective effect of HBOT against colistin treatment of biofilms may result from O_2-dependent increased formation of ATP, which fuels active efflux of colistin and removal of drug targets by LPS-modifications (Pamp et al. 2008). In addition, the bactericidal activity of colistin on *P. aeruginosa* does not involve the formation of ROS, which requires the presence of O_2 (Brochmann et al. 2014).

Hyperoxia by HBOT is a treatment modality in which patients breathe 100% O_2 while exposed to increased atmospheric pressure (ATA) up to 2.0–2.8 ATA to enhance the amount of O_2 dissolved in the body tissues. Gases, like O_2, are compressible and related to volume, temperature, and pressure. The amount of dissolved O_2 is proportional to the O_2 partial pressure at a specific temperature, according to Henry's law (Trayhurn 2019). Therefore, standard therapy of HBOT exploits Henry's law by increasing the pressure and reducing the volume of gas-filled spaces according to Boyle's law (Thom 2011). During HBOT, the arterial O_2 tension typically exceeds 2000 mmHg, and levels of 200–400 mmHg occur in

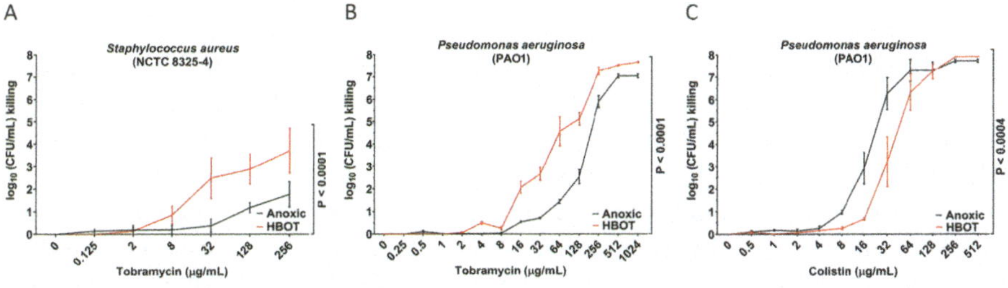

Fig. 2 Effects of HBOT on the killing of 3-day old anoxic *P. aeruginosa* and *S. aureus* biofilms by common antibiotics. Anoxic agarose-embedded bacterial biofilms were cultured for 3 days (Lerche et al. 2017; Møller et al. 2019). Antibiotics were added and HBOT was applied to one part of the biofilms at 2.8 ATA for 90 min. The other part of the samples was left in the anaerobic bench for 90 min. (**a**) HBOT increases killing by tobramycin in *S. aureus* (NCTC 8325-4) biofilms ($p < 0.0001$, $n = 3$) (Lerche et al. 2017). (**b**) HBOT increases killing by tobramycin in PAO1 biofilms ($p < 0.0001$, $n = 3$) (Møller et al. 2019). (**c**) Novel test where HBOT decreases killing by colistin in PAO1 biofilms ($p = 0.0004$, $n = 3$) Statistical significance was evaluated by two-way ANOVA test followed by Bonferroni's multiple comparison correction

tissues (Thom 1989; Choudhury et al. 2018) as compared to the 100 mmHg in arteries of healthy people at sea level (Neuman and Thom2008). Traditionally, the rationale for the use of HBOT for especially necrotizing skin and soft-tissue infections is based on retro- and prospective clinical and preclinical data showing a bacteriostatic effect on anaerobic bacterial growth and reduction in the production of bacterial toxins (Moon 2019). However, this view is simplified, and recent preclinical data suggest that HBOT in combination with certain types of antibiotics contributes to the achievement of infection control, as bacteria are subject to metabolic adaptations to the biofilm environment in which O_2 is involved (Sønderholm et al. 2017; Jensen et al. 2019). Kolpen et al. demonstrated that reoxygenation by HBOT in an agarose *P. aeruginosa* biofilm model with slow-growing bacterial subpopulations in O_2-free zones leads to increased susceptibility to antibiotics (Møller et al. 2019). Reoxygenation was established with HBOT, where biofilm was exposed to an atmosphere of 100% O_2 at 2.8 ATA for 90 min (Fig. 3a).

In combination with tobramycin treatment, HBOT enhanced the killing of clinical *P. aeruginosa* isolates from CF patients grown as biofilm more than six orders of magnitude (Møller et al. 2019) and the killing of PAO1 biofilm more than two orders of magnitude in combination with ciprofloxacin (Kolpen et al. 2016). HBOT also reduced the amount of tobramycin needed to achieve the clinically relevant biofilm bactericidal concentration (BBC) by more than 50% (Møller et al. 2019). The possibility that HBOT enhances the susceptibility to antibiotics due to induction of aerobic respiration is evident from serial O_2-microprofiles demonstrating a fast consumption of the added O_2 immediately after 90 min of HBOT of *P. aeruginosa* biofilm (Møller et al. 2019) (Fig. 3a). Interestingly, the ability to reoxygenate anoxic zones of biofilm infections by HBOT is demonstrated by the O_2-microprofiles in freshly expectorated sputum, with ongoing O_2 consumption by PMNs, from CF patients with chronic *P. aeruginosa* lung infection (Fig. 3b). In addition to increasing the susceptibility of biofilm to antibiotics, adjuvant HBOT may improve the host response by reducing the production of pro-inflammatory cytokines (Lerche et al. 2017; Hedetoft et al. 2021b) and by shedding of endothelial adhesion molecules potentially resulting in decreased extravasation of neutrophils thereby providing protection from collateral damage caused by the excessive inflammatory response (Hedetoft et al. 2021c). The inflammation may also be ameliorated by HBOT by driving macrophages from the pro-inflammatory subtype (M1) to the resolving and anti-inflammatory type (M2) (Geng et al. 2015; Han et al. 2017). Besides controlling inflammation, HBOT may by itself enhance the bacterial clearing by the host response to levels resembling the outcome of antibiotic treatment (Mader et al. 1978). This augmented bacterial clearing by the host response may in part result from the potentiation of the neutrophil bactericidal capacity by extra O_2 (Mader et al. 1980; Schwartz et al. 2021). The O_2-dependent part of neutrophil bactericidal activity is mediated by the formation of ROS that act by oxidation and accordingly HBOT impacts oxidative stress markers in patients with necrotizing soft-tissue infection (Hedetoft et al. 2021d).

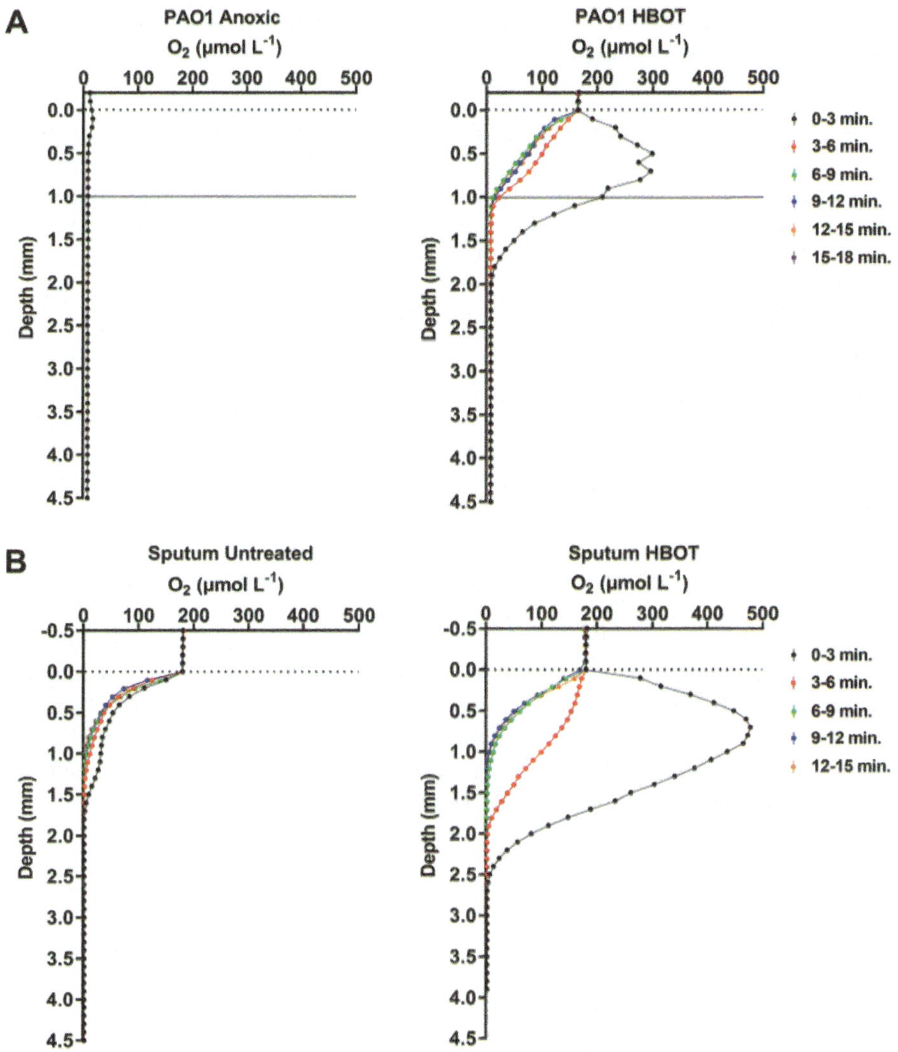

Fig. 3 Oxygenation of the infectious microenvironment by HBOT. Vertical O_2-profiles demonstrating the effect of HBOT on the penetration of O_2 in 3-day-old anoxic *P. aeruginosa* biofilm and in freshly expectorated sputum from a CF patient with chronic *P. aeruginosa* lung infection. HBOT was applied at 2.8 ATA for 90 min. The surface of the samples is at 0 µm. (**a**) *P. aeruginosa* (PAO1) biofilms were embedded in agarose and overlayered with 1 mm of medium to prevent drying out. The surface of the agarose embedded PAO1 biofilm is at 1 µm. Left: Anoxic *P. aeruginosa* biofilm before HBOT. The profiling was performed at atmospheric conditions immediately after leaving the anaerobic chamber. Right: O_2-profiling in *P. aeruginosa* biofilm immediately after HBOT showed that oxygenation of biofilm by HBOT may exceed 200 µM O_2. These high amounts of O_2 were reduced within few minutes, indicating bacterial aerobic respiration (Møller et al. 2019). (**b**) Novel O_2-profiling in untreated sputum from a cystic fibrosis patient with chronic *P. aeruginosa* lung infection. A freshly expectorated sputum sample from a CF patient were divided in two. In one part O_2 profiling was performed immediately at normoxic conditions. The other part was treated with HBOT for 90 min followed by O_2 profiling. The O_2 profiles were recorded measured several times according to Møller et al. 2019. Briefly, O_2 profiling was accomplished with a retractable needle-

5 Mimicking the Infectious Microenvironment Impacts Susceptibility Testing

As argued, lack of O_2 may contribute to the low antibiotic susceptibility of sensitive bacteria in infectious biofilm and supplementation with O_2 may restore susceptibility. It was recently warranted to consider the infectious microenvironment for obtaining relevant information on antibiotic susceptibility testing (Bjarnsholt et al. 2022). The opposing effects of O_2 depletion on the response of *P. aeruginosa* biofilm to ciprofloxacin and colistin (Fig. 4) is in strong support of the relevance of the infectious microenvironment when performing susceptibility testing on clinical isolates. In addition, the reverse outcome of O_2 depletion on the susceptibility to ciprofloxacin and colistin is in line with the effect of replenishing the deficiency of O_2 by HBOT (Fig. 2) and further emphasizes the relevance of the microenvironment. In particular, low O_2 availability is a common condition in the CF paranasal sinuses (Aanaes et al. 2011) and endobronchial mucus (Worlitzsch et al. 2002), in abscesses (Hays and Mandell 1974) and in infectious biofilm (Høiby et al. 2010). The presence of O_2 during culture of clinical samples may also profoundly affect the growth of the isolated bacteria (Kvich et al. 2016), which may have acquired deficient ability to grow in the presence of O_2 during growth in biofilms (Kvich et al. 2019). Clinical relevant biofilm susceptibility testing, however, is challenged by the difficulties of isolating and growing the relevant clinical isolates under conditions mimicking the location of the biofilm infection adhering to artificial surfaces (the Calgary biofilm susceptibility testing) or embedded in mucus or in the tissues where the biofilms are very small (Høiby et al. 2019). This problem may be solved by oxygenating the infectious microenvironment with HBOT in order to induce susceptibility to bacterial physiology.

Fig. 3 (continued) type O_2 microsensor (OXR60-UHS, PyroScience, GmbH) mounted on a heavy stand (HS1, PyroScience) in a motorized PC-controlled profiling setup (Micromanipulator Mu1, PyroScience). A metal rack heather (Dri-Block®DB-3D, Techne, Staffordshire, UK) was placed on top of the heavy stand and set for 37°C to keep the microtiter plate at 37°C. The microsensor (tip diameter 50–70 μm) was connected to a FireSting O_2 fiber-optic oxygen meter (PyroScience) allowing logging of O_2 profiles using the dedicated software Profix (Version 4.51, PyroScience). O_2 profiling was done by movement of sensor in vertical steps of 100 μm into the sputum sample. Profiling was done for 2 s at each step in a top-down direction. The dashed line indicates the surface of the sputum sample. Left: Serial O_2 profiling before HBOT. The O_2 penetration approximates 1.5 mm. Right: O_2 profiling in sputum immediately after HBOT demonstrated oxygenation to O_2 concentrations far above atmospheric O_2 and an increased penetration depth exceeding 2.5 mm. The O_2 levels were quickly depleted to levels resembling the sample before HBOT, indicating ongoing O_2 consumption within the sample.

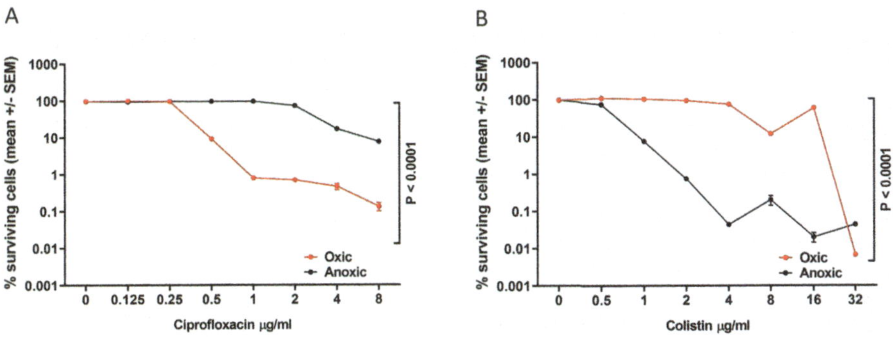

Fig. 4 Impacts of the presence of oxygen on susceptibility testing. Novel demonstration of the opposing effect of the availability of O_2 on the outcome of treatment with ciprofloxacin or colistin of *P. aeruginosa* (PAO1) biofilm. Overnight cultures of PAO1 in LB were adjusted to an optical density at 600 nm of 0.4 and then diluted ×100 in ABTG minimal media supplemented with trace metals and 0.5% (v/w) glucose (ABTG) and 1 mM sodium nitrate (NO_3^-) to achieve a cell loading of ~10^6 CFU/mL. Two hundred microliter culture were loaded into each well in a flat bottom microtiter plate (Nunclon™ Delta Surface; Thermo Fischer Scientific, Roskilde, Denmark). The microtiter plates were sealed with Parafilm (Bemis, Neenah, WI, USA) and lid, and incubated at 37°C at normoxic conditions or at anoxic conditions in an anoxomat (Concept 400 Work Station, Ruskinn Nordic, Bridgend, UK) for 3 days to establish mature biofilms. The density of biofilms after 3 days was 8×10^6 CFU/mL and 3×10^6 CFU/mL in normoxic and anoxic conditions, respectively. Susceptibility testing of the biofilms was initiated after 3 days by replacing the supernatant with a 200 μL colistin solution in ABTG medium supplemented with 1 mM NO_3^- in twofold dilutions from 0 to 32 μg/mL or a 200 μL ciprofloxacin solution in ABTG medium supplemented with 1 mM NO_3^- in twofold dilutions from 0 to 8 μg/mL. The microtiter plates were incubated for 3 h at 37°C at normoxic conditions or at anoxic conditions. The supernatant was removed and 250 μL 0.9% saline was added to each well and sonicated for 5 min. Quantitative bacteriology was performed by standard tenfold dilutions in 0.9% NaCl and plating on Blue plates followed by CFU determination after incubation overnight at 37°C. Effect of oxic (red line) and anoxic (black line) treatment-conditions on the survival of PAO1 biofilms grown in microtiter plates for 3 days before treatment with (**a**) ciprofloxacin ($n = 3$) and (**b**) colistin ($n = 3$) after treatment for 3 h. Statistical significance was evaluated by a two-way ANOVA test followed by Bonferroni's multiple comparison correction

6 Oxygen Therapy as an Adjunctive to Antibiotic Treatment of Bacterial Biofilms

The clinical effects of HBOT, on uncomplicated as well as severe, complicated infections, are described in pro- and retrospective case–control studies (Thom 2011; Memar et al. 2019), whereas randomized, controlled trials are still lacking (Huang et al. 2020). To elucidate the relevance of O_2 therapy as an adjunctive to antibiotic therapy of persistent bacterial biofilm infections we included 30 studies for analysis based on a systematic Medline search (Fig. 5). After evaluation, 26 articles were found to fulfill the inclusion criteria of studies that examine the antibiotic susceptibility of biofilm-forming pathogens, when O_2 is supplied. These consisted of 13 experimental in vitro studies (Table 1), 4 experimental animal studies

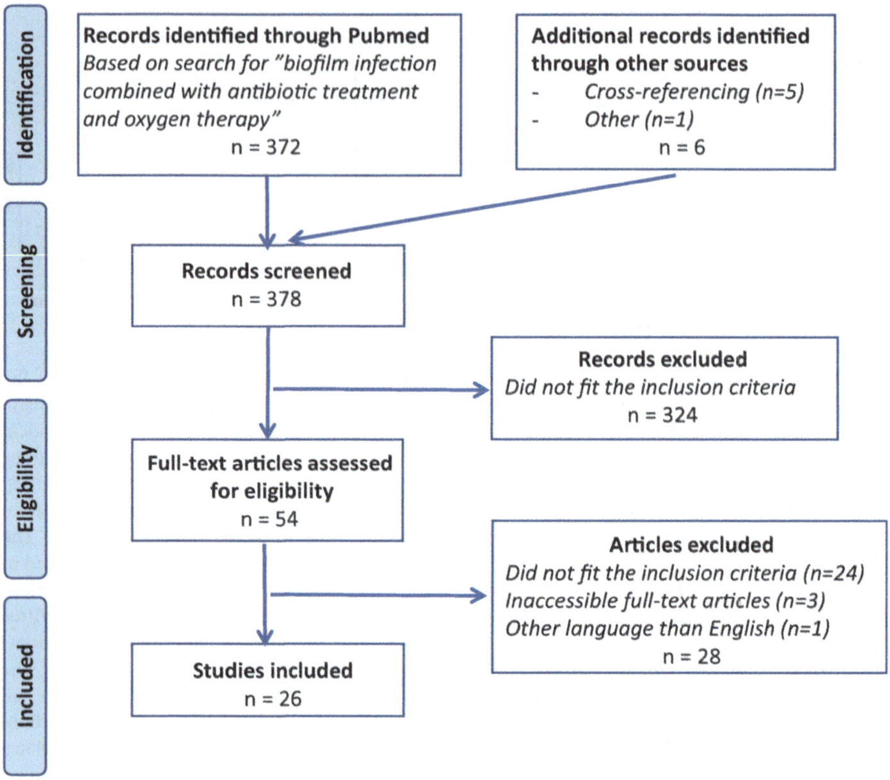

Fig. 5 Flow diagram showing search strategy. A strategical search was performed on Medline, based on the entries "biofilm infection" combined with "antibiotic" and "supplementary oxygen." Inclusion of the screened studies was restricted to studies that examine the antibiotic susceptibility of biofilm-forming pathogens, when O_2 is supplied. Both studies providing O_2 under hyperbaric pressure and under normobaric pressure were included. There was no restriction according to year of publication or study design; however, most of the studies included were performed in vitro. All non-English written studies were excluded. Studies concerning gangrene were excluded due to lack of biofilm like properties

(Table 2), and 9 human studies (Table 3). The search was completed 8th of March 2021. Thirteen in vitro studies examined the effect of additional O_2 and its influence on antibiotic susceptibility of biofilm formed by bacterial pathogens that are capable of biofilm formation (Table 1). In 12 out of the 13 studies, supplemental O_2 was found to reinforce the antibiotic effect (Jass et al. 1995, Jass and Lappin-Scott 1996, Stewart et al. 1999, Wattanakaroon and Stewart 2000, Borriello et al. 2004, Reighard and Schoenfisch 2015, Gupta et al. 2016; Kolpen et al. 2016, 2020; Lerche et al. 2017; Møller et al. 2019). Supplemental O_2 was applied by HBOT in seven studies, by normobaric pressure in seven studies, and by employing an electric current to generate O_2 from water electrolysis in three studies. In all studies, supplemental O_2 increased the susceptibility to antibiotics, but with the exception that *P. aeruginosa*

Table 1 In vitro biofilm studies on effect of extra O_2 on the bactericidal efficacy of antibiotics

Reference	Pathogen	Antibiotics	Oxygen	Summary of outcome
Kolpen et al. (2016)	*Pseudomonas aeruginosa*	Ciprofloxacin	HBOT (2.8 ATM, 90 min) vs. normoxia	HBOT enhanced antibiotic efficacy
Møller et al. (2019)	*Pseudomonas aeruginosa*	Tobramycin	HBOT (2.8 ATM, 90 min) vs. anoxia	HBOT enhanced antibiotic efficacy
	Pseudomonas aeruginosa	Ciprofloxacin	HBOT (2.8 ATM, 90 min) vs. anoxia	HBOT enhanced antibiotic efficacy
Lerche et al. (2017)	*Staphylococcus aureus*	Tobramycin	HBOT (2.8 ATM, 90 min) vs. anoxia	HBOT enhanced antibiotic efficacy
Stewart et al. (1999)	*Pseudomonas aeruginosa*	Tobramycin	Normobaric (21% O_2 vs. 100% O_2)	100% O_2 enhanced antibiotic efficacy
Borriello et al. (2004)	*Pseudomonas aeruginosa*	More than5[a]	Normoxia vs. anoxia	Normoxia enhanced antibiotic efficacy
Reighard and Schoenfisch (2015)	*Pseudomonas aeruginosa*	Tobramycin	Normoxia vs. anoxia	Normoxia enhanced antibiotic efficacy
Kolpen et al. (2016)	*Pseudomonas aeruginosa*	Colistin	Normoxia vs. anoxia	Normoxia enhanced antibiotic efficacy
Kolpen et al. (2020)	*Mycobacterium abscessus*	Amikacin	Normoxia vs. anoxia	Normoxia enhanced antibiotic efficacy
Gupta et al. (2016)	*Staphylococcus aureus, Klebsiella pneumonia, Pseudomonas aeruginosa*	More than 10[b]	Normobaric (normoxia vs. hyperoxia) (95–99% O_2)	Hyperoxia enhanced antibiotic efficacy
Jass et al. (1995)	*Pseudomonas aeruginosa*	Tobramycin	Electrical current 9 mA/cm^2 12 h	Electrical current enhanced antibiotic efficacy

(continued)

Table 1 (continued)

Reference	Pathogen	Antibiotics	Oxygen	Summary of outcome
Jass and Lappin-Scott (1996)	*Pseudomonas aeruginosa*	Ciprofloxacin Polymyxin B Piperacillin	Electrical current 9 mA/ cm^2 12 h	Electrical current enhanced efficacy of ciprofloxacin and Poly-myxin B
Wattanakaroon and Stewart (2000)	*Streptococcus gordonii*	Tobramycin	Electrical current 9 mA/ cm^2 12 h	Electrical current enhanced antibiotic efficacy

HBOT: hyperbaric oxygen therapy, ATA: atmospheres absolute
Normoxia: 21% O_2 at 1 atmosphere
Anoxia: 0% O_2 at 1 atmosphere
[a]Carbenicillin, ceftazidime, chloramphenicol, ciprofloxacin, tetracycline, tobramycin
[b]Meropenem, doripenem, ciprofloxacin, rifampicin, ampicillin, piperacillin/tazobactam, vancomycin, tetracycline, tigecycline, gentamycin, amikacin, tobramycin, kanamycin, azithromycin

biofilms were less susceptible to colistin in aerobic conditions compared to anaerobic conditions (Kolpen et al. 2016).

Four experimental in vivo studies examined the effect of additional O_2 and its influence on antibiotic susceptibility of representative clinical infections. In three studies supplemental O_2 was found to reduce the bacterial burden and to improve the survival rate in pulmonary infection and infective endocarditis (Luongo et al. 1999; Kikuchi et al. 2009; Lerche et al. 2017). However, one study reported no enhancement of the antibiotic therapy by HBOT. Probably, because these two antibiotics (daptomycin and rifampicin) are independent of oxygen in their mechanism of action (Miller et al. 2016) and because of the effective treatment by this combination (Jørgensen et al. 2017). Jørgensen et al. also measured bone resorption and formation makers, which were significantly increased in the HBOT group compared to the non-HBOT group. However, no apparent adverse events of HBOT were observed (Jørgensen et al. 2017). The duration, number of treatments, timing, and breathing pressure of 100% O_2 differed between the animal studies, which may explain the diversities. In particular, timing of HBOT is crucial since early application of HBOT after induction of sepsis resulted in improved survival and decreased levels of pro-inflammatory cytokines while delayed HBOT after induction of sepsis did not change the levels of pro-inflammatory cytokines and survival (Baernthsen et al. 2017; Halbach et al. 2019). Multiple HBOT, however, did not significantly improve survival further than one early HBOT (Halbach et al. 2019).

Five retrospective studies (Morrey et al. 1979, Chen et al. 2003, 2004, Narozny et al. 2006; Bartek et al. 2018) and four case studies (Shupak et al. 1989; Chong et al. 2007; Fernandes et al. 2013; Vestita et al. 2017) examined the effect of HBOT on antibiotic susceptibility (Table 3). A total of 90 patients with bacterial infections

Table 2 Biofilm studies in animals on the effect of extra O_2 on the bactericidal efficacy of antibiotics

Reference	Infection Pathogen	Antibiotics	Oxygen	Summary of outcome
Luongo et al. (1999)	**Subcutaneous infection Pulmonary infection** *Pseudomonas aeruginosa*	Amikacin	HBOT 100% O_2, 3 ATA, 55 min daily, 8 days	HBOT increased therapeutic effect of amikacin and improved survival rate of pulmonary infected rats to 100%
Kikuchi et al. (2009)	**Pulmonary infection** *Pseudomonas aeruginosa*	Azithromycin	NORMOBAR Hyperoxia, 90% O_2, 60 h	Azithromycin and hyperoxia significantly reduced bacterial burden and dissemination
Lerche et al. (2017)	**Infective endocarditis** *Staphylococcus aureus*	Tobramycin	HBOT 100% O_2, 2.8 ATA, 90 min, twice/day, 4 days	HBOT treatment significantly reduced bacterial load and the clinical condition of the rat was improved
Jørgensen et al. (2017)	**Implant-associated osteomyelitis** *Staphylococcus aureus*	Rifampicin in combination with Daptomycin	HBOT 100% O_2, 3 ATA, 60 min daily, 14 days	Cure rate and bacterial load in bone and on implants were unaffected by HBOT

HBOT: hyperbaric oxygen therapy, ATA: atmospheres absolute
NORMOBAR: 21% O_2 at 1 atmosphere

treated with HBOT and antibiotics were identified. Seventy-seven patients out of 90 (85%) had a successful outcome of the treatment with no reported relapses. In 4 out of 90 (4%) patients, failure of treatment was reported (Chen et al. 2003, 2004). In 9 out of 90 patients, infection recurred after treatment (10%) (Morrey et al. 1979; Bartek et al. 2018). The spectrum of antibiotics across the studies is broad, and no consistent trend was found since several patients were given a combination of two or more antibiotics. Only two patients reported side effects from HBOT (Bartek et al. 2018). Even though the side effects of HBOT are generally few and relatively benign when proper attentions are paid to patient evaluation, it is still strongly recommended to carefully select the patients eligible for HBOT. In particular, application of HBOT to patients with lung tissue damage should be taken into careful consideration, as serious complications such as pulmonary barotrauma could arise during the treatment (Heyboer et al. 2017). However, HBOT is used regularly for many other chronic wound care patients among some of which have pulmonary disease, without any major adverse events and pulmonary barotrauma is a rare event although anecdotal reports exist (Camporesi 2014; Hadanny et al. 2016; Jokinen-Gordon et al. 2017; Hadanny et al. 2019). The occurrence of pulmonary side effects, however, is likely to be reported from two ongoing randomized controlled trials on the effect of HBOT on COVID-19 infected patients requiring supplemental oxygen therapy, including HBOT (Boet et al. 2021; Kjellberg et al. 2021).

Table 3 Biofilm-related infection studies on effect of HBOT on the bactericidal efficacy of antibiotics

Reference	Infection Pathogen	Antibiotics	Oxygen	Type of study	Summary of outcome
Morrey et al. (1979)	**Chronic osteomyelitis** *Staphyloccus*. sp ($n = 11$) *Proteus mirabilis* ($n = 1$) *Serratia marascens* ($n = 1$) Mixed infection ($n = 24$) Anaerobic ($n = 3$)	Types of antibiotics not reported	2.4 ATA, 90 min, 9–208 times	Preliminary report $n = 40$	In 34 patients the infection was eradicated. In six patients the infection relapsed
Shupak et al. (1989)	**Malign necrotizing external otitis** *Pseudomonas aeruginosa*	Azlocillin	2.5ATA, 90 min, 27 times	Case study $n = 1$	Eradication of *Pseudomonas* infection
Chen et al. (2003)	**Chronic refractory osteomyelitis tibia** *Staphylococcus* sp. ($n = 7$) *Enterococcus* ($n = 1$) *Serratia marascens* ($n = 1$) Mixed infection ($n = 5$)	Vancomycin, oxacillin, ampicillin, ciproxin, cefazolin, piperacillin, gentamycin	2.5 ATA, 120 min, 30–60 times	Retrospective $n = 14$	11/14 patients recovered from infection
Chen et al. (2004)	**Chronic refractory osteomyelitis femur**	Vancomycin, Pipril, gentamycin, cefamezine, fortum, ampicillin, amikacin	2.5ATA, 120 min, 5 days/week, 18–75 treatments	Retrospective $n = 13$	Complete eradication of infection in 12/13 patients

	Staphylococcus sp. ($n = 7$) *Pseudomonas aeruginosa* ($n = 1$) Mixed infection ($n = 5$)				
Narozny et al. (2006)	**Malign external otitis** *Pseudomonas aeruginosa* ($n = 3$) *Staphylococcus* sp. ($n = 1$) Mixed infection ($n = 2$)	Flouroqinolones, III generation cephalosporins, Semisynthetic and polienic penicillin	2.5 ATA, 70 min daily, 14–45 treatments	Retrospective $n = 7$	7/7 patients recovered from infection
Chong et al. (2007)	**Keratitis** *Pseudomonas aeruginosa*	Cephalothin, ciprofloxacin, tobramycin, moxifloxacin, Ticarcillin	2.0 ATA, 90 min, 6 days	Case study $n = 1$	Improved efficiency of antibiotic treatment and the woman's visual acuity recovered
Fernandes et al. (2013)	**Perichondritis of the pinna** *Escherichia coli*	Imipenem	3.0 ATA, 120 min daily, 10 times	Case study $n = 1$	Eradication of infection with no relapse. Otoscopy normal
Vestita et al. (2017)	**Leg ulcer** *Pseudomonas aeruginosa*	Meropenem, tobramycin, daptomycin	2.5 ATA, 60 min, 16 times	Case study $n = 1$	Eradication of infection
Bartek et al. (2018)	**Hardware related infection** *Staphylococcus aureus* ($n = 8$) *Propionibacterium acnes* ($n = 2$) *Enterobacter* ($n = 1$)	Clindamycin, dicloxacillin, flucloxacillin	2.0–2.8 ATM, 1–2 times daily, 28–73 treatments	Retrospective $n = 12$	All patients (12) achieved infection resolution. Three patients had reoccurring infection

ATA: atmospheres absolute

References

Aanaes K, Rickelt LF, Johansen HK, von Buchwald C, Pressler T, Høiby N, Jensen PØ (2011) Decreased mucosal oxygen tension in the maxillary sinuses in patients with cystic fibrosis. J Cyst Fibros 10:114–120

Baernthsen NF, Hansen MB, Wahl AM, Simonsen U, Hyldegaard O (2017) Treatment with 24 h-delayed normo- and hyperbaric oxygenation in severe sepsis induced by cecal ligation and puncture in rats. J Inflamm (Lond) 14:27

Bartek J Jr, Jakola AS, Skyrman S, Forander P, Alpkvist P, Schechtmann G, Glimaker M, Larsson A, Lind F, Mathiesen T (2016) Hyperbaric oxygen therapy in spontaneous brain abscess patients: a population-based comparative cohort study. Acta Neurochir 158:1259–1267

Bartek J Jr, Skyrman S, Nekludov M, Mathiesen T, Lind F, Schechtmann G (2018) Hyperbaric oxygen therapy as adjuvant treatment for hardware-related infections in neuromodulation. Stereotact Funct Neurosurg 96:100–107

Belotsky SM, Guzu EV, Karlov VA, Dikovskaya ES, Filjukova OB, Snastina TI (1990) Wound tissue respiratory burst and local microbial inflammation. Inflammation 14:663–668

Bjarnsholt T, Kirketerp-Møller K, Jensen PØ, Madsen KG, Phipps R, Krogfelt K, Høiby N, Givskov M (2008) Why chronic wounds will not heal: a novel hypothesis. Wound Repair Regen 16:2–10

Bjarnsholt T, Whiteley M, Rumbaugh KP, Stewart PS, Jensen PØ, Frimodt-Moller N (2022) The importance of understanding the infectious microenvironment. Lancet Infect Dis 22(3):e88–e92

Boet S, Katznelson R, Castellucci LA, Fergusson D, Gonevski M, Clarke H, Etherington C, Burns JK, Gagné S, McIsaac DI, Djaiani G, Pollock NW, Proulx L, Sun LY, Thavorn K, Wherrett C, Thompson C, Moffett S, Blatteau JE, Louge P, Pignel R, Roth V, MacDonald JR, Taljaard M, Booth RA (2021) Protocol for a multicentre randomized controlled trial of normobaric versus hyperbaric oxygen therapy for hypoxemic COVID-19 patients. medRxiv. https://doi.org/10.1101/2020.07.15.20154609

Borriello G, Werner E, Roe F, Kim AM, Ehrlich GD, Stewart PS (2004) Oxygen limitation contributes to antibiotic tolerance of *Pseudomonas aeruginosa* in biofilms. Antimicrob Agents Chemother 48:2659–2664

Borriello G, Richards L, Ehrlich GD, Stewart PS (2006) Arginine or nitrate enhances antibiotic susceptibility of *Pseudomonas aeruginosa* in biofilms. Antimicrob Agents Chemother 50:382–384

Brochmann RP, Toft A, Ciofu O, Briales A, Kolpen M, Hempel C, Bjarnsholt T, Høiby N, Jensen PØ (2014) Bactericidal effect of colistin on planktonic *Pseudomonas aeruginosa* is independent of hydroxyl radical formation. Int J Antimicrob Agents 43:140–147

Camporesi EM (2014) Side effects of hyperbaric oxygen therapy. Undersea Hyperb Med 41:253–257

Cao B, Christophersen L, Thomsen K, Sonderholm M, Bjarnsholt T, Jensen PØ, Høiby N, Moser C (2015) Antibiotic penetration and bacterial killing in a *Pseudomonas aeruginosa* biofilm model. J Antimicrob Chemother 70:2057–2063

Cao B, Christophersen L, Kolpen M, Jensen PØ, Sneppen K, Høiby N, Moser C, Sams T (2016) Diffusion retardation by binding of tobramycin in an alginate biofilm model. PLoS One 11:e0153616

Chen CE, Shih ST, Fu TH, Wang JW, Wang CJ (2003) Hyperbaric oxygen therapy in the treatment of chronic refractory osteomyelitis: a preliminary report. Chang Gung Med J 26:114–121

Chen CE, Ko JY, Fu TH, Wang CJ (2004) Results of chronic osteomyelitis of the femur treated with hyperbaric oxygen: a preliminary report. Chang Gung Med J 27:91–97

Chong R, Ayer CJ, Francis IC, Coroneo MT, Wolfers DL (2007) Adjunctive hyperbaric oxygen in *Pseudomonas keratitis*. Br J Ophthalmol 91:560–561

Choudhury A, Young G, Reyad B, Shah N, Rahman R (2018) Can we improve the prescribing and delivery of oxygen on a respiratory ward in accordance with new British Thoracic Society oxygen guidelines? BMJ Open Qual 7:e000371

Cozens RM, Tuomanen E, Tosch W, Zak O, Suter J, Tomasz A (1986) Evaluation of the bactericidal activity of beta-lactam antibiotics on slowly growing bacteria cultured in the chemostat. Antimicrob Agents Chemother 29:797–802

Crabbe A, Jensen PØ, Bjarnsholt T, Coenye T (2019) Antimicrobial tolerance and metabolic adaptations in microbial biofilms. Trends Microbiol 27(10):850–863

Debats IB, Booi D, Deutz NE, Buurman WA, Boeckx WD, van der Hulst RR (2006) Infected chronic wounds show different local and systemic arginine conversion compared with acute wounds. J Surg Res 134:205–214

DePas WH, Starwalt-Lee R, Van Sambeek L, Ravindra Kumar S, Gradinaru V, Newman DK (2016) Exposing the three-dimensional biogeography and metabolic states of pathogens in cystic fibrosis sputum via hydrogel embedding, clearing, and rRNA labeling. mBio 7

Dowd SE, Sun Y, Secor PR, Rhoads DD, Wolcott BM, James GA, Wolcott RD (2008) Survey of bacterial diversity in chronic wounds using pyrosequencing, DGGE, and full ribosome shotgun sequencing. BMC Microbiol 8:43

Dwyer DJ, Belenky PA, Yang JH et al (2014) Antibiotics induce redox-related physiological alterations as part of their lethality. Proc Natl Acad Sci U S A 111:E2100–E2109

El Haj C, Lichtenberg M, Nielsen KL, Bjarnsholt T, Jensen PØ (2021) Catalase protects biofilm of *Staphylococcus aureus* against daptomycin activity. Antibiotics (Basel) 10

Evans DJ, Allison DG, Brown MR, Gilbert P (1991) Susceptibility of *Pseudomonas aeruginosa* and Escherichia coli biofilms towards ciprofloxacin: effect of specific growth rate. J Antimicrob Chemother 27:177–184

Fazli M, Bjarnsholt T, Kirketerp-Møller K, Jørgensen A, Andersen CB, Givskov M, Tolker-Nielsen T (2011) Quantitative analysis of the cellular inflammatory response against biofilm bacteria in chronic wounds. Wound Repair Regener 19:387–391

Fernandes FL, Lavor M, de Carvalho GM, Guimaraes AC (2013) Hyperbaric oxygen in the treatment of perichondritis of the pinna. Diving Hyperb Med 43:166–167

Frykberg RG, Franks PJ, Edmonds M et al (2020) A multinational, multicenter, randomized, double-blinded, placebo-controlled trial to evaluate the efficacy of cyclical topical wound oxygen (TWO2) therapy in the treatment of chronic diabetic foot ulcers: the TWO2 study. Diabetes Care 43:616–624

Geng M, Zhou L, Liu X, Li P (2015) Hyperbaric oxygen treatment reduced the lung injury of type II decompression sickness. Int J Clin Exp Pathol 8:1797–1803

Gottrup F, Dissemond J, Baines C, Frykberg R, Jensen PO, Kot J, Kroger K, Longobardi P (2017) Use of oxygen therapies in wound healing. J Wound Care 26:S1–S43

Gupta S, Laskar N, Kadouri DE (2016) Evaluating the effect of oxygen concentrations on antibiotic sensitivity, growth, and biofilm formation of human pathogens. Microbiol Insights 9:37–46

Hadanny A, Meir O, Bechor Y, Fishlev G, Bergan J, Efrati S (2016) The safety of hyperbaric oxygen treatment—retrospective analysis in 2,334 patients. Undersea Hyperb Med 43:113–122

Hadanny A, Zubari T, Tamir-Adler L, Bechor Y, Fishlev G, Lang E, Polak N, Bergan J, Friedman M, Efrati S (2019) Hyperbaric oxygen therapy effects on pulmonary functions: a prospective cohort study. BMC Pulm Med 19:148

Haft CT Jr (1991) The potentiation of the aminoglycoside tobramycin by hyperbaric oxygen therapy in an animal model of chronic *Pseudomonas aeruginosa* pneumonia (abstract). Undersea Biomed Res 18

Halbach JL, Prieto JM, Wang AW et al (2019) Early hyperbaric oxygen therapy improves survival in a model of severe sepsis. Am J Physiol Regul Integr Comp Physiol 317:R160–R168

Han CH, Zhang PX, Xu WG, Li RP, Xu JJ, Liu WW (2017) Polarization of macrophages in the blood after decompression in mice. Med Gas Res 7:236–240

Hays RC, Mandell GL (1974) PO2, pH, and redox potential of experimental abscesses. Proc Soc Exp Biol Med Soc Exp Biol Med 147:29–30

Hedetoft M, Bennett MH, Hyldegaard O (2021a) Adjunctive hyperbaric oxygen treatment for necrotising soft-tissue infections: A systematic review and meta-analysis. Diving Hyperb Med 51:34–43

Hedetoft M, Garred P, Madsen MB, Hyldegaard O (2021b) Hyperbaric oxygen treatment is associated with a decrease in cytokine levels in patients with necrotizing soft-tissue infection. Physiol Rep 9:e14757

Hedetoft M, Moser C, Jensen PØ, Vinkel J, Hyldegaard O (2021c) Soluble ICAM-1 is modulated by hyperbaric oxygen treatment and correlates with disease severity and mortality in patients with necrotizing soft-tissue infection. J Appl Physiol 130:729–736

Hedetoft M, Jensen PØ, Moser C, Vinkel J, Hyldegaard O (2021d) Hyperbaric oxygen treatment impacts oxidative stress markers in patients with necrotizing soft-tissue infection. J Invest Med 69:1330–1338

Heyboer M 3rd, Sharma D, Santiago W, McCulloch N (2017) Hyperbaric oxygen therapy: side effects defined and quantified. Adv Wound Care (New Rochelle) 6:210–224

Høiby N, Ciofu O, Bjarnsholt T (2010) *Pseudomonas aeruginosa* biofilms in cystic fibrosis. Future Microbiol 5:1663–1674

Høiby N, Ciofu O, Johansen HK, Song ZJ, Moser C, Jensen PØ, Molin S, Givskov M, Tolker-Nielsen T, Bjarnsholt T (2011) The clinical impact of bacterial biofilms. Int J Oral Sci 3:55–65

Høiby N, Bjarnsholt T, Moser C, Jensen PØ, Kolpen M, Qvist T, Aanaes K, Pressler T, Skov M, Ciofu O (2017) Diagnosis of biofilm infections in cystic fibrosis patients. APMIS: Acta Pathologica, Microbiologica, et Immunologica Scandinavica 125:339–343

Høiby N, Henneberg KA, Wang H, Stavnsbjerg C, Bjarnsholt T, Ciofu O, Johansen UR, Sams T (2019) Formation of *Pseudomonas aeruginosa* inhibition zone during tobramycin disk diffusion is due to transition from planktonic to biofilm mode of growth. Int J Antimicrob Agents 53:564–573

Huang ET, Savaser DJ, Heyboer Iii M (2020) Arterial insufficiencies: hyperbaric oxygen therapy for selected problem wounds. Undersea Hyperb Med 47:491–530

Hunt TK, Zederfeldt B, Goldstick TK (1969) Oxygen and healing. Am J Surg 118:521–525

James GA, Swogger E, Wolcott R, Pulcini E, Secor P, Sestrich J, Costerton JW, Stewart PS (2008) Biofilms in chronic wounds. Wound Repair Regener 16:37–44

James GA, Zhao AG, Usui M et al (2016) Microsensor and transcriptomic signatures of oxygen depletion in biofilms associated with chronic wounds. Wound Repair Regener 24(2):373–383

Jass J, Lappin-Scott HM (1996) The efficacy of antibiotics enhanced by electrical currents against *Pseudomonas aeruginosa* biofilms. J Antimicrob Chemother 38:987–1000

Jass J, Costerton JW, Lappin-Scott HM (1995) The effect of electrical currents and tobramycin on *Pseudomonas aeruginosa* biofilms. J Ind Microbiol 15:234–242

Jensen PØ, Givskov M, Bjarnsholt T, Moser C (2010) The immune system vs. *Pseudomonas aeruginosa* biofilms. FEMS Immunol Med Microbiol 59:292–305

Jensen PO, Briales A, Brochmann RP, Wang H, Kragh KN, Kolpen M, Hempel C, Bjarnsholt T, Høiby N, Ciofu O (2014) Formation of hydroxyl radicals contributes to the bactericidal activity of ciprofloxacin against *Pseudomonas aeruginosa* biofilms. Pathogens Dis 70:440–443

Jensen PØ, Kolpen M, Kragh KN, Kühl M (2017) Micro-environmental characteristics and physiology of biofilms in chronic infections of CF patients are strongly affected by the host immune response. APMIS 125(4):276–288

Jensen PØ, Møller SA, Lerche CJ, Moser C, Bjarnsholt T, Ciofu O, Faurholt-Jepsen D, Høiby N, Kolpen M (2019) Improving antibiotic treatment of bacterial biofilm by hyperbaric oxygen therapy: Not just hot air. Biofilm 1:100008

Jokinen-Gordon H, Barry RC, Watson B, Covington DS (2017) A retrospective analysis of adverse events in hyperbaric oxygen therapy (2012-2015): lessons learned from 1.5 million treatments. Adv Skin Wound Care 30:125–129

Jørgensen NP, Hansen K, Andreasen CM, Pedersen M, Fuursted K, Meyer RL, Petersen E (2017) Hyperbaric oxygen therapy is ineffective as an adjuvant to daptomycin with rifampicin treatment in a murine model of *Staphylococcus aureus* in implant-associated osteomyelitis. Microorganisms 5

Kikuchi Y, Tateda K, Fuse ET, Matsumoto T, Gotoh N, Fukushima J, Takizawa H, Nagase T, Standiford TJ, Yamaguchi K (2009) Hyperoxia exaggerates bacterial dissemination and lethality in *Pseudomonas aeruginosa* pneumonia. Pulm Pharmacol Ther 22:333–339

Kjellberg A, Douglas J, Pawlik MT et al (2021) Randomised, controlled, open label, multicentre clinical trial to explore safety and efficacy of hyperbaric oxygen for preventing ICU admission, morbidity and mortality in adult patients with COVID-19. BMJ Open 11:e046738

Kohanski MA, Dwyer DJ, Hayete B, Lawrence CA, Collins JJ (2007) A common mechanism of cellular death induced by bactericidal antibiotics. Cell 130:797–810

Kolpen M, Hansen CR, Bjarnsholt T et al (2010) Polymorphonuclear leucocytes consume oxygen in sputum from chronic *Pseudomonas aeruginosa* pneumonia in cystic fibrosis. Thorax 65:57–62

Kolpen M, Bjarnsholt T, Moser C, Hansen CR, Rickelt LF, Kühl M, Hempel C, Pressler T, Høiby N, Jensen PØ (2014) Nitric oxide production by polymorphonuclear leucocytes in infected cystic fibrosis sputum consumes oxygen. Clin Exp Immunol 177:310–319

Kolpen M, Kragh KN, Bjarnsholt T, Line L, Hansen CR, Dalbøge CS, Hansen N, Kühl M, Høiby N, Jensen PØ (2015) Denitrification by cystic fibrosis pathogens—*Stenotrophomonas maltophilia* is dormant in sputum. Int J Med Microbiol IJMM 305:1–10

Kolpen M, Appeldorff CF, Brandt S et al (2016) Increased bactericidal activity of colistin on *Pseudomonas aeruginosa* biofilms in anaerobic conditions. Pathogens Dis 74(1):ftv086

Kolpen M, Jensen PØ, Qvist T, Kragh KN, Ravnholt C, Fritz BG, Johansen UR, Bjarnsholt T, Høiby N (2020) Biofilms of *Mycobacterium abscessus* complex can be sensitized to antibiotics by disaggregation and oxygenation. Antimicrob Agents Chemother 64

Kolpen M, Kragh KN, Enciso JB et al (2022) Bacterial biofilms predominate in both acute and chronic human lung infections. Thorax:217576

Kopf SH, Sessions AL, Cowley ES, Reyes C, Van Sambeek L, Hu Y, Orphan VJ, Kato R, Newman DK (2016) Trace incorporation of heavy water reveals slow and heterogeneous pathogen growth rates in cystic fibrosis sputum. Proc Natl Acad Sci U S A 113:E110–E116

Kragh KN, Alhede M, Jensen PØ et al (2014) Polymorphonuclear leukocytes restrict growth of *Pseudomonas aeruginosa* in the lungs of cystic fibrosis patients. Infect Immun 82:4477–4486

Kutlay M, Colak A, Yildiz S, Demircan N, Akin ON (2008) Stereotactic aspiration and antibiotic treatment combined with hyperbaric oxygen therapy in the management of bacterial brain abscesses. Neurosurgery 62(Suppl 2):540–546

Kvich L, Jensen PØ, Justesen US, Bjarnsholt T (2016) Incidence of *Propionibacterium acnes* in initially culture-negative thioglycollate broths-a prospective cohort study at a Danish University Hospital. Clin Microbiol Infect 22:941–945

Kvich L, Fritz B, Crone S, Kragh KN, Kolpen M, Sønderholm M, Andersson M, Koch A, Jensen PØ, Bjarnsholt T (2019) Oxygen restriction generates difficult-to-culture *P. aeruginosa*. Front Microbiol 10:1992

Lerche CJ, Christophersen LJ, Kolpen M et al (2017) Hyperbaric oxygen therapy augments tobramycin efficacy in experimental *Staphylococcus aureus* endocarditis. Int J Antimicrob Agents 50(3):406–412

Lobritz MA, Belenky P, Porter CB, Gutierrez A, Yang JH, Schwarz EG, Dwyer DJ, Khalil AS, Collins JJ (2015) Antibiotic efficacy is linked to bacterial cellular respiration. Proc Natl Acad Sci U S A 112:8173–8180

Lopatkin AJ, Stokes JM, Zheng EJ, Yang JH, Takahashi MK, You L, Collins JJ (2019) Bacterial metabolic state more accurately predicts antibiotic lethality than growth rate. Nat Microbiol 4: 2109–2117

Luongo C, Imperatore F, Matera MG, Mangoni G, Marmo M, Baroni A, Catalanotti P, Rossi F, Filippelli A (1999) Effect of hyperbaric oxygen therapy in experimental subcutaneous and pulmonary infections due to *Pseudomonas aeruginosa*. Undersea Hyperb Med 26:21–25

Mader JT, Guckian JC, Glass DL, Reinarz JA (1978) Therapy with hyperbaric oxygen for experimental osteomyelitis due to *Staphylococcus aureus* in rabbits. J Infect Dis 138:312–318

Mader JT, Brown GL, Guckian JC, Wells CH, Reinarz JA (1980) A mechanism for the amelioration by hyperbaric oxygen of experimental staphylococcal osteomyelitis in rabbits. J Infect Dis 142: 915–922

Malone M, Bjarnsholt T, McBain AJ, James GA, Stoodley P, Leaper D, Tachi M, Schultz G, Swanson T, Wolcott RD (2017) The prevalence of biofilms in chronic wounds: a systematic review and meta-analysis of published data. J Wound Care 26:20–25

Memar MY, Yekani M, Alizadeh N, Baghi HB (2019) Hyperbaric oxygen therapy: Antimicrobial mechanisms and clinical application for infections. Biomed Pharmacother 109:440–447

Miller WR, Bayer AS, Arias CA (2016) Mechanism of action and resistance to daptomycin in Staphylococcus aureus and Enterococci. Cold Spring Harb Perspect Med 6

Mitchell CG, Dawes EA (1982) The role of oxygen in the regulation of glucose metabolism, transport and the tricarboxylic acid cycle in *Pseudomonas aeruginosa*. J Gen Microbiol 128:49–59

Møller SA, Jensen PØ, Høiby N, Ciofu O, Kragh KN, Bjarnsholt T, Kolpen M (2019) Hyperbaric oxygen treatment increases killing of aggregating *Pseudomonas aeruginosa* isolates from cystic fibrosis patients. J Cyst Fibros 18(5):657–664

Moon R (2019) Hyperbaric oxygen therapy indications. Best Publishing, Palm Beach

Morrey BF, Dunn JM, Heimbach RD, Davis J (1979) Hyperbaric oxygen and chronic osteomyelitis. Clin Orthop Relat Res 121–127

Moser C, Pedersen HT, Lerche CJ, Kolpen M, Line L, Thomsen K, Høiby N, Jensen PØ (2017) Biofilms and host response—helpful or harmful. APMIS 125:320–338

Moser C, Jensen PØ, Thomsen K, Kolpen M, Rybtke M, Lauland AS, Trøstrup H, Tolker-Nielsen T (2021) Immune responses to *Pseudomonas aeruginosa* biofilm infections. Front Immunol 12: 625597

Narozny W, Kuczkowski J, Stankiewicz C, Kot J, Mikaszewski B, Przewozny T (2006) Value of hyperbaric oxygen in bacterial and fungal malignant external otitis treatment. Eur Arch Otorhinolaryngol 263:680–684

Neuman TS, Thom SR (2008) Physiology and medicine of hyperbaric oxygen therapy. Saunders Elsevier, Philadelphia, PA

Ozkan MT, Vural A, Cicek OF, Yener AU, Ozcan S, Toman H, Unver A, Sacar M (2016) Is hyperbaric oxygen or ozone effective in experimental endocarditis? J Surg Res 202:66–70

Pamp SJ, Gjermansen M, Johansen HK, Tolker-Nielsen T (2008) Tolerance to the antimicrobial peptide colistin in *Pseudomonas aeruginosa* biofilms is linked to metabolically active cells, and depends on the pmr and mexAB-oprM genes. Mol Microbiol 68:223–240

Reighard KP, Schoenfisch MH (2015) Antibacterial action of nitric oxide-releasing chitosan oligosaccharides against *Pseudomonas aeruginosa* under aerobic and anaerobic conditions. Antimicrob Agents Chemother 59:6506–6513

Schreml S, Meier RJ, Kirschbaum M et al (2014) Luminescent dual sensors reveal extracellular pH-gradients and hypoxia on chronic wounds that disrupt epidermal repair. Theranostics 4:721–735

Schwartz FA, Lerche CJ, Christophersen L, Jensen PØ, Laulund AS, Woetmann A, Høiby N, Moser C (2021) Distinct contribution of hyperbaric oxygen therapy to human neutrophil function and antibiotic efficacy against *Staphylococcus aureus*. APMIS 129:566–573

Seth AK, Geringer MR, Galiano RD, Leung KP, Mustoe TA, Hong SJ (2012) Quantitative comparison and analysis of species-specific wound biofilm virulence using an in vivo, rabbit-ear model. J Am Coll Surg 215:388–399

Shupak A, Greenberg E, Hardoff R, Gordon C, Melamed Y, Meyer WS (1989) Hyperbaric oxygenation for necrotizing (malignant) otitis externa. Arch Otolaryngol Head Neck Surg 115:1470–1475

Siemens N, Chakrakodi B, Shambat SM et al (2016) Biofilm in group A streptococcal necrotizing soft tissue infections. JCI Insight 1:e87882

Sønderholm M, Bjarnsholt T, Alhede M, Kolpen M, Jensen PØ, Kühl M, Kragh KN (2017) The consequences of being in an infectious biofilm: microenvironmental conditions governing antibiotic tolerance. Int J Mol Sci 18

Stewart PS (2003) Diffusion in biofilms. J Bacteriol 185:1485–1491

Stewart PS, Wattanakaroon W, Goodrum L, Fortun SM, McLeod BR (1999) Electrolytic generation of oxygen partially explains electrical enhancement of tobramycin efficacy against *Pseudomonas aeruginosa* biofilm. Antimicrob Agents Chemother 43:292–296

Thom SR (1989) Hyperbaric oxygen therapy. J Intensive Care Med 4:58–74

Thom SR (2011) Hyperbaric oxygen: its mechanisms and efficacy. Plast Reconstr Surg 127(Suppl 1):131S–141S

Trayhurn P (2019) Oxygen—a critical, but overlooked, nutrient. Front Nutr 6:10

Trøstrup H, Thomsen K, Christophersen LJ, Hougen HP, Bjarnsholt T, Jensen PØ, Kirkby N, Calum H, Høiby N, Moser C (2013) *Pseudomonas aeruginosa* biofilm aggravates skin inflammatory response in BALB/c mice in a novel chronic wound model. Wound Repair Regener 21: 292–299

Tuomanen E, Cozens R, Tosch W, Zak O, Tomasz A (1986) The rate of killing of *Escherichia coli* by beta-lactam antibiotics is strictly proportional to the rate of bacterial growth. J Gen Microbiol 132:1297–1304

Van Acker H, Coenye T (2017) The role of reactive oxygen species in antibiotic-mediated killing of bacteria. Trends Microbiol 25(6):456–466

Van Acker H, Sass A, Bazzini S et al (2013) Biofilm-grown *Burkholderia cepacia* complex cells survive antibiotic treatment by avoiding production of reactive oxygen species. PLoS One 8: e58943

Vestita M, Giudice G, Bonamonte D (2017) Hyperbaric oxygen therapy in the management of severe leg ulcers from prolidase deficiency. BMJ Case Rep. https://doi.org/10.1136/bcr-2016-217329

Walters MC 3rd, Roe F, Bugnicourt A, Franklin MJ, Stewart PS (2003) Contributions of antibiotic penetration, oxygen limitation, and low metabolic activity to tolerance of *Pseudomonas aeruginosa* biofilms to ciprofloxacin and tobramycin. Antimicrob Agents Chemother 47:317–323

Wattanakaroon W, Stewart PS (2000) Electrical enhancement of *Streptococcus gordonii* biofilm killing by gentamicin. Arch Oral Biol 45:167–171

Watters C, DeLeon K, Trivedi U, Griswold JA, Lyte M, Hampel KJ, Wargo MJ, Rumbaugh KP (2013) *Pseudomonas aeruginosa* biofilms perturb wound resolution and antibiotic tolerance in diabetic mice. Med Microbiol Immunol 202:131–141

Worlitzsch D, Tarran R, Ulrich M et al (2002) Effects of reduced mucus oxygen concentration in airway *Pseudomonas infections* of cystic fibrosis patients. J Clin Invest 109:317–325

Wu Y, Klapper I, Stewart PS (2018) Hypoxia arising from concerted oxygen consumption by neutrophils and microorganisms in biofilms. Pathogens Dis 76

Yu WK, Chen YW, Shie HG, Lien TC, Kao HK, Wang JH (2011) Hyperbaric oxygen therapy as an adjunctive treatment for sternal infection and osteomyelitis after sternotomy and cardiothoracic surgery. J Cardiothorac Surg 6:141

Cold Plasma Therapy as a Physical Antibiofilm Approach

Adrian I. Abdo, Thomas Schmitt-John, and Katharina Richter

Abstract The emergence and spread of antimicrobial resistance have led to a dire need for innovative, alternative intervention strategies for infectious diseases, such as cold plasma technology. Plasma, known as the "fourth state of matter," is a reactive mix of excited and fragmented gas molecules, ions, free electrons and radicals. Cold plasma is generated by a strong electric field ionising a gas and can be applied safely onto human tissue. Early studies at the turn of the century showed cold plasma has antimicrobial activity towards several types of bacteria and fungi, ushering in plasma medicine as a potential strategy to prevent and treat infections without using antimicrobial drugs. As a result, medical plasma devices have been developed for first applications in wound care and dentistry. However, these plasma devices have only passed through small, randomised controlled trials in diabetic foot ulcers and chronic leg ulcers, while dental applications of plasma devices are limited to in vitro and ex vivo experiments. These early studies showed promising results, thereby opening new avenues of plasma technology for medical therapies. This book chapter summarises the current state of knowledge concerning how plasma medicine could revolutionise biofilm infection control.

A. I. Abdo
Richter Lab, Department of Surgery, Basil Hetzel Institute for Translational Health Research & The Queen Elizabeth Hospital, Adelaide, SA, Australia

Adelaide Medical School, University of Adelaide, Adelaide, SA, Australia

T. Schmitt-John
Plasmatreat GmbH, Steinhagen, Germany

K. Richter (✉)
Richter Lab, Department of Surgery, Basil Hetzel Institute for Translational Health Research & The Queen Elizabeth Hospital, Adelaide, SA, Australia

Adelaide Medical School, University of Adelaide, Adelaide, SA, Australia

Institute for Photonics and Advanced Sensing, University of Adelaide, Adelaide, SA, Australia
e-mail: katharina.richter@adelaide.edu.au

K. Richter, K. Nørskov Kragh (eds.), *Antibiofilm Strategies*, Springer Series on Biofilms 11, https://doi.org/10.1007/978-3-031-10992-8_10

1 Introduction

Biofilms are complicit to causing 60–80% of chronic infections by abetting persistent and recurrent infections (de la Fuente-Nunez et al. 2013). These biofilms are microbial fortresses that provide physical and chemical protection against the external environment and the host immune system, while they endow resident microbes with tolerance against antibiotic therapy by reducing contact with antibiotics. Effectively, this exposes microbes to sublethal concentrations of antibiotics in biofilms, which promotes antimicrobial resistance. Additionally, the heterogenous microbial population that forms in biofilms lead to the emergence of persister cells that have suppressed metabolic activity and can withstand harsh antibiotic therapies. To make a bad situation worse, antibiotic resistance genes can be transferred between microorganisms, both within species and between species if the biofilm is polymicrobial. These traits of biofilms frequently lead to therapeutic failure of best medical care, therefore, driving research into alternatives to antibiotic therapies to treat infections. Besides surface cleaning (removal of dust, particles and organic compounds) and activation of technical surfaces, plasma has a proven antibacterial effect and thus, has been used for disinfection and sterilisation purposes and was tested for medical use, particularly in dermatology and dentistry (Bernhardt et al. 2019; Borges et al. 2021).

Plasma is referred to as the "fourth state of matter" after solid, liquid and gas, regardless of being the most common state of matter in the universe in the form of stellar material, but manifests naturally on earth as lightning, auroras and fire. Plasma is ionised gas with an approximately neutral ("quasi-neutral") overall charge, but consists of a reactive mix of excited and fragmented gas ions, free electrons and radicals and emitting electromagnetic radiation in the form of heat and UV light. Gas can be converted to plasma either by extreme high temperatures ("hot" plasma) or by the influence of intense electric fields ("cold" or "non-thermal" plasma). In the lab, plasma is produced by a microwave, radiofrequency, DC or AC electric discharge that ionises an operator gas, usually argon (Ar), helium (He), oxygen (O_2), nitrogen (N_2), air, or a mixture of a noble gas with O_2 or air by direct electron impact ionisation (Bruggeman et al. 2017). The various technical solutions for electrical discharge and the geometry of atmospheric pressure plasma sources were extensively reviewed (Tendero et al. 2006).

The field of plasma medicine has emerged as a result of holistic scientific amalgamation of physics, chemistry and biology, involving collaboration between plasma physicists and engineers, health scientists, microbiologists, chemists and clinicians. During its research infancy at the turn of the century, cold plasma was observed to be antimicrobial to several types of bacteria and fungi on surfaces and in aqueous solutions, while sparing mammalian cells (Stoffels et al. 2003). In the last decade, research in plasma medicine has surged, particularly in the fields of dentistry and dermatology to treat microbial infections. Cold plasma also has great potential to replace other dangerous and damaging forms of sterilisation of medical surfaces and devices (e.g. ultraviolet/gamma radiation, nuclear radiation, incineration, ethylene oxide) which are harmful to humans and the environment. It was best said by von Woedtke et al., "In the 20th century, laser was a physical technology that was integrated very successfully into medicine[. . .] Plasma has the chance to repeat this successful integration of a physical technology into medicine to be in the fore of scientific and technological development in medicine of the 21st century" (von Woedtke et al. 2020). Not until recently has the first generation of plasma medical devices passed from clinical trial to government-approved medical therapy. However, approval is currently restricted to skin treatment (Bernhardt et al. 2019).

This chapter is an introduction to the research and applications of cold plasma in biofilm-related infections and treatment with selected cornerstone and landmark papers, but it is not exhaustive. As such, the chapter will refer to further reading in other comprehensive reviews when required.

2 Cold Plasma

In plasma medicine, cold plasma can also refer to "hot" plasma that is applied in pulsed bursts to avoid heating of tissue. As a helpful rule, as long as the therapeutic level of plasma exposure does not heat the tissue of the applied area >40°C at any time during the procedure, it can be classified as cold plasma therapy (von Woedtke et al. 2020). The main types of cold plasma discharges are corona discharge, microwave discharge, gliding arc, dielectric barrier discharge (DBD) and plasma jet. Detailed summaries of the physical principles and properties of these discharge plasma have been reviewed in detail (Bruggeman et al. 2017). Of these, DBD and plasma jets are the most applied in plasma medicine.

DBD plasma is formed by the interference of a dielectric material, for example, glass, plastic or rubber, between two alternating current electrodes (Bruggeman et al. 2017). By virtue of this design, human tissue can act as the dielectric material when applying a DBD for treatment. A plasma jet is not technically a type of discharge. The plasma is formed by one of the other aforementioned modes, including DBD, which is then blown outside the plasma-forming electrode by an ancillary gas into a jet of active particles (Scholtz et al. 2015). In plasma medicine, this makes the application of plasma in a jet much more precise and portable than a DBD, while DBDs might have an advantage when larger areas are to be treated.

3 Biofilm Destruction by Plasma

Plasma and plasma-activated liquids (PAL) are highly effective at sterilising surfaces inoculated with planktonic bacteria, but microbes in a biofilm are naturally more tolerant to sterilisation by plasma and PAL (Flynn et al. 2015). The biofilm imposes tolerance against plasma, which depends on the maturity and thickness of the biofilm (Ermolaeva et al. 2011) and on the microbe itself. Microbial tolerance to an antimicrobial strategy is not unique to protecting against plasma therapy of course, conventional antimicrobial therapies also have to overcome the chemical and physical barriers of the biofilm. Therefore, as biofilms form a protective layer against the external environment, higher concentrations of conventional antimicrobial therapies are required to diffuse further into the biofilm at lethal concentrations. As protection against plasma and PAL is dependent on biofilm thickness, analogously, higher sterilisation time, or multiple repeated sterilisations are required to overcome the biofilm (Mai-Prochnow et al. 2014). This needs to be considered when taking actions like sterilising a wound to prevent infection versus treating an established biofilm-related infection to minimise the potential risks against the benefits of sterilisation time.

Plasma has been reported to effectively eradicate a wide selection of bacterial biofilms, including *Chromobacterium violaceum*, *Bacillus cereus*, *Escherichia coli*, *Candida albicans* and the "ESKAPE" pathogens *Enterococci* and *Staphylococci* spp., *Klebsiella pneumoniae*, *Acinetobacter baumannii*, *Pseudomonas aeruginosa*, and *Enterobacter* spp. (Abdo et al. 2021; Flynn et al. 2015; Vandervoort et al. 2008; Joaquin et al. 2009; Koban et al. 2010; Alkawareek et al. 2012; Fricke et al. 2012; Matthes et al. 2013; Daeschlein et al. 2014; Modic et al. 2017). A comparison of several in vitro studies investigating the biofilm eradication efficacy of "ESKAPE" pathogens and *E. coli* by counting the colony-forming units (CFU) of the remaining viable microbes is summarised in Table 1. In general, Gram-negative bacterial biofilms were more susceptible to plasma and PAL than Gram-positive bacteria (Alkawareek et al. 2012; Zhao et al. 2020a). This can be observed in studies that compared the CFU reduction of Gram-positive (*Staphylococci* spp., *Bacillus subtilis*, *Listeria monocytogenes*) and Gram-negative (*P. aeruginosa*, *E. coli*, *Enterobacter* spp., *K. pneumoniae* and *A. baumannii*) bacteria biofilms, which are tabulated in Table 1 (Flynn et al. 2015; Matthes et al. 2013; Modic et al. 2017; Mai-Prochnow et al. 2016). For instance, 24-h single-species biofilms of Gram-negative *P. aeruginosa* had biofilm eradication levels of 3.3–4.1 log CFU, respectively, compared to 1.0–2.2 CFU reductions of Gram-positive *Staphylococcus epidermidis* after treatment with cold plasma (Matthes et al. 2013; Mai-Prochnow et al. 2016). Addition of O_2 to noble gas plasmas generally enhanced biofilm killing. He with 0.5% O_2 plasma jets reduced biofilm CFU counts to non-detectable levels in almost all "ESKAPE" pathogens, while *A. baumannii* was reduced by 4–5 log CFU in 24-h biofilm models (Flynn et al. 2015; Alkawareek et al. 2012; Flynn et al. 2019). These CFU reductions of "ESKAPE" pathogens with highly pure noble gas plasmas achieved more modest 0.5–3.5 log CFU reductions (Matthes et al. 2013; Mai-Prochnow et al. 2016; Xu et al. 2015), with the exception of *P. aeruginosa* biofilm eradicated with the Ar kINPen 09 (4.1 log CFU reduction (Matthes et al. 2013)). Air surface barrier discharge (SBD) plasma were also potent antimicrobial agents, achieving 2.7–4.8 log CFU reductions of 24-h biofilms of "ESKAPE" pathogen (Modic et al. 2017; Salgado et al. 2021). However, air DBD plasma was significantly more powerful than SBD, achieving 4.1 log and > 5 log CFU reductions of 48-h *S. aureus* and *E. coli* biofilms, respectively (Ziuzina et al. 2015). Many in vitro and ex vivo studies investigating cold plasma therapy on *C. albicans*,

Table 1 Summary of in vitro studies of cold plasma treatment against "ESKAPE" biofilms that were measured by $\log_{10}$ CFU reduction. All biofilms are single-species, unless stated otherwise

Ref.	Plasma treatment	Exposure time (min)	Biofilm model	Species	(Log_{10} CFU) Untreated control	Reduction after plasma treatment
Alkawareek et al. (2012)	He + 0.5% O_2 plasma jet	2	48-h biofilm on CBD	*S. aureus*	5.2	ND
		8		*E. coli*	7.2	ND
		10		*P. aeruginosa*	6.1	ND
Matthes et al. (2013)	Ar kINPen 09	2.5	24-h biofilms on PS culture plates	*P. aeruginosa*	NA	–4.1
	Ar + 1% O_2 kINPen 09					–4.6
	Ar kINPen 09	5		*S. epidermidis*		–3.1
	Ar + 1% O_2 kINPen 09					–2.2
Flynn et al. (2015)	He + 0.5% O_2 plasma jet	6	24-h biofilm on PS culture plates	*S. aureus*	6.5	ND
		2		*K. pneumoniae*	5.7	ND
				A. baumannii	5.7	–4.0
		4		*P. aeruginosa*	6.5	ND
				E. cloacae	5.9	ND
Xu et al. (2015)	He plasma jet	10	12-h biofilm on borosilicate slices	*S. aureus*	6.6	–3.0
Ziuzina et al. (2015)	Air DBD	5	48-h biofilm in PS culture plates	*E. coli*	5.4	ND
				S. aureus	6.3	–4.1
Mai-Prochnow et al. (2016)	Ar kINPen Med	10	24-h biofilm on SS coupons	*Pseudomonas aeruginosa*	4.5	–3.5
				Enterobacter cloacae	5.2	–3.3
				Bacillus subtilis	3.3	–0.5
				Staphylococcus epidermidis	5.0	–1.0
			24-h dual species biofilm on SS coupons	*P. aeruginosa*	5.6	–2.6
				S. epidermidis	3.5	–1.4

(continued)

Table 1 (continued)

Ref.	Plasma treatment	Exposure time (min)	Biofilm model	Species	(Log_{10} CFU) Untreated control	Reduction after plasma treatment
Modic et al. (2017)	Air SBD	4	24-h biofilm in a CDC Biofilm Reactor	*P. aeruginosa*	6.8	–4.2
				S. aureus	6.3	–3.7
			24-h multispecies biofilm in a CDC Biofilm Reactor	*P. aeruginosa*	6.4	–2.8
				S. aureus	5.5	–2.7
				K. pneumoniae	6.3	–4.0
				E. faecalis	5.5	–3.1
Flynn et al. (2019)	He + 0.5% O_2 plasma jet	9	24-h biofilm on CBD	*A. baumannii*	6.0	–5.0
			48-h biofilm on CBD			–3.9
			72-h biofilm on CBD			–4.2
Salgado et al. (2021)	Air SBD	3	24-h biofilm on PP CDC coupons	*E. coli*	NA	–4.8

Abbreviations: *PP* polypropylene, *PS* polystyrene, *SS* stainless steel, *CBD* calgary biofilm device, *SBD* surface barrier discharge, *DCD* dielectric barrier discharge, *ND* not detectable, *NA* not available

Enterococcus faecalis, *Streptococci* spp., *Lactobacilli* spp. and other dental pathogens have also been comprehensively summarised in a recent systematic review (Jungbauer et al. 2021), so we have not included them in Table 1, but we summarise many of these studies in Sect. 6.

Some studies have also observed bacteria being inactivated, and suggested these bacteria switch to a viable but non-culturable (VBNC) state, instead of actually being destroyed by the plasma (Vandervoort et al. 2008; Joaquin et al. 2009). The induction of a VBNC phenotype using cold plasma would make it an ineffective treatment. However, cold plasma inducing a VBNC phenotype was empirically based on morphological data by both studies, which is controversial for identifying a VBNC phenotype. Scientific experimentation with validated assays of VBNC phenotype are still required. Although cold plasma can be applied widely as an antimicrobial strategy, the effectiveness and mechanisms of action against different kinds of bacteria depend on the type of plasma (discharge, gas, flow rate, power), working distance, treatment time, the relevant environmental conditions (humidity, temperature, surface application), whether microbes are in planktonic or biofilm form, and the maturity of the biofilm (Scholtz et al. 2015; Puligundla and Mok 2017). For this reason, comparing results between studies using different plasma exposures, biofilm models and environmental conditions is challenging. Consequently, standardised experimental testing needs to be established, while environmental conditions are replicated as close as possible to real-world applications.

3.1 Biofilm Damage by Reactive Oxygen and Nitrogen Species (RONS)

It has become clear in recent years that RONS are pivotal to the destruction of essential components of the biofilm. However, the destruction of these biofilm components is not necessarily evidence of microbicidal activity, since the modification of the EPS, proteins and eDNA of the biofilm can afford protection to the residing microbes by being a sink for RONS. Therefore, researchers and clinicians should be vigilant and ensure enough plasma RONS is produced to overwhelm the biofilm and damage the microbial cells.

High purity noble working gases produce higher levels of reactive nitrogen species (RNS), but by spiking the noble working gas with other molecular gases (O_2, air, H_2O) there is a large shift in the production of reactive oxygen species (ROS), with a lower proportion of RNS being produced (Lietz and Kushner 2018). This enables the tailoring of particular reactive species in the plasma to cater the effective killing of different microbial species. Cold plasma generates a milieu of RONS (Fig. 1), including atomic oxygen (O), superoxide ($O_2^{\bullet -}$), hydroxyl radicals ($^{\bullet}OH$), singlet oxygen ($^1O_2$), ozone (O_3), hydrogen peroxide (H_2O_2), nitric oxide ($NO^{\bullet}$), nitrite/nitrate (NO_2^-/NO_3^-), dinitrogen pentoxide (N_2O_5) and peroxynitrite ($ONOO^-$) (Salgado et al. 2021; Kondeti et al. 2018; Lu et al. 2016; Nicol et al. 2020).

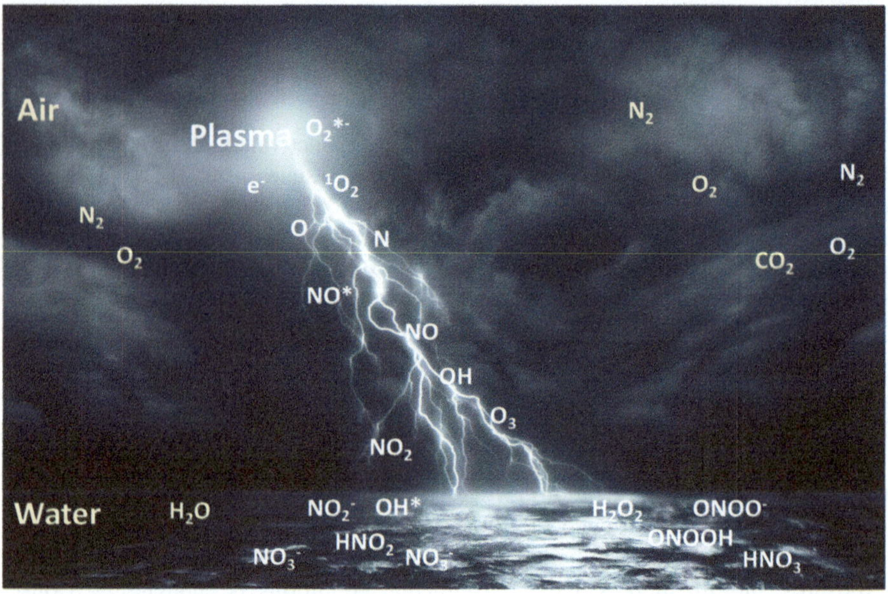

Fig. 1 Lightning strike through the air into the ocean depicting plasma discharge and plasma activation of the air and water to form PAW. Lightning bolt (plasma discharge) in the air generates RONS from atmospheric gases. These gaseous RONS diffuse into the water, while the discharge at the air–liquid interphase also forms RONS to generate plasma-activated water

These species react with the complex biological components of the biofilm, such as the water, ions, polysaccharides, eDNA, lipids and proteins, the composition of which can be different between bacterial species. So unlike antibiotics that have a specific mode of action against microbes, that microbes can then adapt to and resist (Davies and Davies 2010), cold plasma produces a plethora of reactive species and particles that inflict damage in a variety of ways. This difference makes cold plasma advantageous to conventional antimicrobials since microbes cannot induce any single response to resist or reverse damage by cold plasma.

Precise mechanisms for how plasma-derived RONS damage biofilm are not well characterised, since the vast majority of studies investigated plasma-derived RONS on planktonic forms. However, the presence of antioxidants and scavengers of free radicals like ascorbic acid and α-tocopherol during plasma exposure completely prevented antimicrobial killing, lipid peroxidation and DNA oxidation in *E. coli* (Joshi et al. 2011), providing indirect evidence for the importance of RONS. Alginate, a major polysaccharide of *P. aeruginosa* biofilms, performs essential functions including cell-cell adhesion, biofilm structure stability and desiccation resistance (Limoli et al. 2015), also reacts with free radicals and ROS.

The ability of plasma to etch the biofilm is another important process of physically removing the source of infection that should also be categorised as antibiofilm activity. However, PAL do not cause an etching effect. Since it is reasonable to assume all biofilms contain persister cells, physically removing biofilms from the host is sometimes easier than killing the cells in the biofilm. Etching of *S. aureus* biofilms was caused by He plasma, which was also associated with time-dependent accumulation of plasma-derived intracellular ROS (Xu et al. 2015). Meanwhile, 7-day *C. albicans* biofilms required the presence of O_2 in the plasma gas to effectively etch biofilms (Fricke et al. 2012), lending credence to oxygen-centred radicals and oxidants being needed to cause etching in some biofilm species. Etching using a plasma jet is limited in large biofilm infection areas, since the etching area is maximum at the plume's axis, then rapidly diminishes further than 1 mm radial distance from the axis (Fricke et al. 2012).

3.2 Ultraviolet Radiation

The contribution of UV light on the microbicidal or antibiofilm activity of plasma is still contentious. While vacuum plasma does emit significant UV light at the microbicidal wavelengths (UV-C; 200–280 nm), vacuum plasma is not practical in sterilisation or medical device applications and most atmospheric cold plasma does not emit germicidal UV-C radiation (Laroussi and Leipold 2004; Birmingham 2004; Kieft et al. 2004; Taghizadeh et al. 2015). Exposing planktonic *S. aureus* and *E. coli* samples to DBD plasma in sterilisation bags to block UV contact, or *Bacillus globigii* and *E. coli* behind quartz to block UV did not alter the bacterial killing effect of plasma (Kelly-Wintenberg et al. 1998; Herrmann et al. 1999; Dobrynin et al. 2009). Also, DBD plasma humidity (which contributes to the formation of RONS) and not UV-C was found to strongly correlate to *B. subtilis* spore inactivation (Kogelheide et al. 2020). Neither does UV radiation alone significantly contribute to biofilm etching, shown with *C. albicans* (Fricke et al. 2012). Some studies refute this and show UV can prominently contribute to microbial killing. UV from Ar plasma sources killed *B. subtilis* and *Aspergilius niger* spores, and planktonic MRSA and *E. coli*, but still retained killing power when UV was excluded (Trompeter et al. 2002; Lee et al. 2005; Boudam et al. 2006; Armand et al. 2019).

The conflicting data regarding UV, the overwhelming data implicating RONS in killing microbes and that cold atmospheric plasma does not emit microbicidal UV-C indicate that UV has a minor role in plasma disinfection. This was supported recently in an *E. coli* biofilm study. UV and RONS produced by plasma synergistically contributed to a biofilm killing of 3 log CFU reduction, though RONS was the major contributor compared to UV (*cf.* 2.5 log and 0.5 log CFU reduction, respectively) when the contribution of plasma-derived RONS and UV were experimentally isolated (Salgado et al. 2021). However, UV may contribute to biofilm destruction through means other than the conventional killing of bacteria through direct DNA, membrane or intracellular protein damage. Unfortunately, most studies omit investigating this effect and instead focus on direct bacterial damage by UV-C. Regardless, perhaps one way plasma-derived UV could amplify biofilm killing is through photolysis of H_2O_2, NO_2^-, HNO_2 and HNO_3 also produced by plasma to generate the more potent and destructive $^{\bullet}OH$ radical (Kondeti et al. 2018; Park et al. 2019).

Plasma-derived UV may also photolytically generate atomic oxygen closer to the plasma source and O_3 further from the plasma source to accelerate bacterial killing and etch biofilms (Schneider et al. 2011). This evidence supports that UV may contribute to biological damage by generating more potent RONS through photolysis, but this is yet to be reported in biofilms.

3.3 Quorum Sensing Disruption

Biofilms are like a close-knit, gated community of gregarious microbes that communicate and coordinate behaviour. This communication is mediated by quorum sensing (QS), where extracellular signalling molecules called autoinducers transcriptionally regulate several important cellular processes, including biofilm formation, in Gram-positive and Gram-negative bacteria (Whiteley et al. 2017). Disrupting QS by destruction of autoinducers is an attractive prospect for suppressing biofilm formation, and is termed "quorum quenching." The interactions of cold plasma with bacterial signalling pathways and virulence factors are not well understood, but a couple of studies showed cold plasma has some abiotic quorum quenching activity. In a *P. aeruginosa* biofilm model, cold plasma treatment inactivated the *las-rhl* QS signalling molecule *N*-acyl homoserine lactone, significantly inhibiting the release of QS-regulated virulence factors pyocyanin and elastase Las B (Ziuzina et al. 2015; Flynn et al. 2016). However, a single cold plasma treatment did not reduce biofilm forming capacity (Ziuzina et al. 2015). Interestingly, the *las-rhl* QS system in *P. aeruginosa* has long been known to regulate the expression of catalase and superoxide dismutase (Hassett et al. 1999), which may reduce the risk of plasma tolerance discussed later in the chapter. Overall, the QS disruption effects of cold plasma therapy indicate that the effect is more likely an anti-virulence rather than an antibiofilm activity. Notably, this has only been demonstrated in *P. aeruginosa* biofilm models and the response of other microbial species is still required.

4 Plasma-Activated Liquids

Aqueous solutions, typically water or saline, that have been exposed to plasma are often referred to as "plasma-activated" liquids (PAL). It is unfortunately easy to make comparison with "snake oils" or "cure-all elixirs" due to the ambiguity of the term, but fortunately, plasma-activation of several different liquids and plasma sources has been extensively characterised in the chemical composition and physical properties that make them antimicrobial. PAL can be a direct application when the plasma is sterilising a solution contaminated with microbes, but this is an application to water treatment and not plasma medicine. Therefore, this section will describe PAL when being applied to surfaces or tissue to kill microbes. As the plasma source is not present during the application of PAL, only the aqueous long-lived chemical species of the liquid are present and UV radiation and electric fields emitted from plasma do not play a role.

4.1 Physicochemical Properties of Plasma-Activated Liquids

When aqueous liquids are subjected to plasma, a plethora of short- and long-lived reactive oxygen and nitrogen species (RONS) are produced. PAL have complex chemistry, depending on several factors: plasma source (microwave, DBD, arc discharge), position of the plasma (in liquid or air above the liquid), liquid composition (water, saline, media), plasma operating gas, plasma power and voltage, pulse duration and more (Schnabel et al. 2020; Zhou et al. 2020). Just how complex is it? Well, the latest computational models include 53 species involved in 624 reactions in the plasma region and 33 species involved in 109 reactions in the liquid region (Liu et al. 2015). In Fig. 1 is summarised the key RONS that are experimentally measured. Short-lived $^{\bullet}OH$, $^{1}O_2$, $O_2^{\bullet-}$ and $NO^{\bullet}$ are formed first and are highly reactive (chemical lifetime [*t*] of 10^{-9} to 1 s), which then form secondary RONS like (1) H_2O_2 (t = several minutes) from $^{\bullet}OH$ by the spontaneous dismutation of $O_2^{\bullet-}$, (2) $ONOO^-$ ($t < 1$ s) by the reaction of $O_2^{\bullet-}$ and $NO^{\bullet}$, or NO_2 and $^{\bullet}OH$, and (3) O_2NOOH by the reaction of $ONOO^-$ and H_2O_2 (t = minutes) (Aboubakr et al. 2016; Ikawa et al. 2016; Zhang et al. 2018; Zhou et al. 2018, 2019; Tarabova et al. 2019). Eventually, long-lived species are produced, O_3 (t = several minutes), NO_2^- (t = several days), NO_3^- (t = several years), and nitric acid (HNO_3; t = several years) (Zhou et al. 2019; Liu et al. 2021).

The consequence of HNO_3, $ONOO^-$ and H_2O_2 formation in PAL is a large pH decrease with concurrent increase in oxidoreductive potential (ORP) and electrical conductivity (Thirumdas et al. 2018; Wang and Salvi 2021). Studies that experimentally simulated plasma-activated water (PAW) by using HCl or HNO_3 had comparable bactericidal effect (Wang and Salvi 2021), showing that pH and ORP are integral properties dictating the antimicrobial effect (Zhang et al. 2013; Tian et al. 2015). PAW can reach highly acidic pH levels below 2.0 depending on the plasma setup (Shen et al. 2016). In addition, pH, ORP and electrical conductivity do not seem to change during storage while many of the RONS decomposed (Wang and Salvi 2021; Shen et al. 2016). The ORP of the solution is proportional to the oxidising capacity, with PAW having a substantially higher ORP reported between up to 550 mV (Shen et al. 2016; Ma et al. 2015) compared to water (approximately 250 mV). The significant increases in conductivity do not have any direct effect on the antimicrobial activity of PAL but can be another indicator of the formation of RONS. This indicates that the microbicidal activity of PAW is generally determined by the pH and ORP of the solution. Otherwise, some nuanced considerations for generating PAL may be considered. For example, anaerobic microorganisms may be more susceptible than aerobic microorganisms to PAL with higher ROS content, in which activation with a plasma gas enriched with pure O_2, H_2O or air would be preferred.

4.2 Antibiofilm Activity of Plasma-Activated Liquids

PAW has exhibited potent antimicrobial activity in a broad spectrum of planktonic microorganisms, including *Staphylococcus* spp., *E. coli*, *E. faecalis*, *P. aeruginosa* and *C. albicans* (Abdo et al. 2021; Kondeti et al. 2018; Tian et al. 2015; Kamgang-Youbi et al. 2009; Naitali et al. 2010; Laurita et al. 2015; Zhang et al. 2016; Li et al. 2019; Royintarat et al. 2019; Zhao et al. 2020b). The antibacterial efficacy of PAW on planktonic *S. aureus* is well documented. PAW increased intracellular DNA leakage from planktonic *S. aureus* cells over treatment time (Tian et al. 2015; Zhang et al. 2016). The effect of storage temperatures (25, 4, −20 and −80 °C) on the bactericidal efficacy of PAW against *S. aureus* were also investigated over a 30-day period and found that 30 days storage at −80 °C maintained a considerable 3.5 log CFU reduction, but 4 °C storage after 30 days was only capable of 1 log CFU reduction of *S. aureus* following exposure to the PAW (Shen et al. 2016). When *E. faecalis* biofilm were exposed to PAW, there was prolonged downregulation of quorum sensing virulence related genes *cylR1*, *cylA*, *gelE* and *sprE* (Li et al. 2019), indicating the biofilm-forming process was disrupted and the infection may be more easily detected and eradicated by the host immune system. These findings support the use of PAW as an alternative disinfectant, but most of these studies were performed on planktonic bacteria and direct effects of PAW on the biofilm forms are scarcer. However, one recent study showed that PAW had a significant antibiofilm effect, as well as a deactivation effect on *S. aureus* biofilm post-treatment with PAW (Xu et al. 2020). In this study, PAW was tested in two ways; (1) the water was plasma-activated for between 5 and 60 min, with *S. aureus* biofilms being treated for 30 min, and (2) water was activated for 30 min, with *S. aureus* biofilms being treated for 5–60 min. Interestingly, the effectiveness of both treatment methods were equivalent, showing a >99.99% killing of *S. aureus* in biofilms and >99% reduction in the cultivability of *S. aureus* to produce new biofilms (Xu et al. 2020). Studies like this warrant being replicated in other relevant pathogenic biofilms.

PAW has significant bactericidal effect on oral pathogens. There was a reduction of *Staphylococcus mutans*, *Actinomyces viscosus* and *Porphyromonas gingivalis* to non-detectable levels in less than 1 min of PAW treatment associated with significant cell damage demonstrated by rapid protein and DNA leakage from the cells (Li et al. 2017). This shows promise that PAW can be an effective antimicrobial mouthwash, but data is still limited and the in vitro design of these studies were not applied to an oral experimental model. To our knowledge, there is still lack of in vitro and in vivo studies investigating the safety of PAW on cells of the tongue or gingiva, or the dentin and enamel of teeth to assure PAW is safe on these tissues for acute treatment and for repeated usages for preventative care. The ideal PAW mouth wash would be prophylactic against oral infection while keeping healthy oral microbiota in balance and be safe if swallowed.

5 Plasma and Plasma-Activated Liquid Cleaning and Coating of Medical Devices and Surfaces

Sterilisation is important for clinical and dental surfaces and devices. Disposable instruments are increasingly being used, but the frequent refurbishment and sterilisation of a wide variety of devices cannot be avoided and thus, cleaning and sterilisation devices are required. Besides heat sterilisers such as autoclaves, a number of alternative sterilisation procedures have been established. However, a fast, reliable, and material-friendly sterilisation procedure could improve safety and facilitate the workflow in dentistry and thus, plasma sterilisation is worth considering. The market approval of several plasma sterilisation devices, mostly based on low-pressure plasma or atmospheric pressure plasma technology, drives the use of plasma for sterilisation of medical devices, instruments and materials. However, these devices predominantly use H_2O_2 as a sterilising agent in addition to plasma and are mainly dedicated to the sterilisation of temperature-sensitive reusable medical products (Scholtz et al. 2015; Kramer et al. 2019).

The reactive plasma species interfere and react with surfaces and thereby cause chemical surface modifications. Plasma is already widely used in medical device manufacturing for surface activation or coating via plasma-enhanced vapour deposition (PEVD) of polymer surfaces for improved bonding, sealing or printing (Tendero et al. 2006). The introduction of functional groups, such as hydroxyl- or amino groups leads to a shift in surface energy to more hydrophilic properties and increased surface roughness to enhance the bonding of antimicrobial compounds. For instance, cold plasma can facilitate binding of antibiofilm compounds like acrylic acid and chitosan to otherwise non-adhesive polymeric hydrocarbon surface like polylactic acid (PLA) (Muro-Fraguas et al. 2020; Stoleru et al. 2016), an eco-friendly, non-toxic, biodegradable polymer useful for sutures and implants. Similarly, hernia repair surgeries see a high rate of mesh infection leading to hernia repair failure (Whitehead-Clarke and Windsor 2021), so antimicrobials such as levofloxacin, polydopamine and cyclodextrins have been bonded to polypropylene, a common polymer for hernia repair meshes (Pankaj et al. 2017; Sanbhal et al. 2018a; Sanbhal et al. 2018b; Sanbhal et al. 2019; Saitaer et al. 2019). Evidence that this reduces incidence of surgical site infection is still pending. Thus, atmospheric pressure plasma is widely used for industrial surface treatment with the aim of cleaning and improvement of processes like bonding and printing.

Cold plasma has been investigated to combat healthcare-associated infections through disinfection of surfaces touched frequently by healthcare workers and patients, commonly contaminated with methicillin-resistant *S. aureus* (MRSA), vancomycin-resistant *E. faecalis* (VREF), *E. coli*, *K. pneumoniae*, *A. baumannii* and *P. aeruginosa* (Otter et al. 2011). Surfaces found in the clinical environment (linoleum flooring, polyurethane mattress fabric, polypropylene plastic, powder-coated steel, stainless steel) were inoculated with MRSA, *E. coli*, VREF, *A. baumannii* and *C. difficile* spores before being cold plasma-sterilised for up to 90 s. A cold plasma jet applied for 90 s produced the most significant reductions of up to 5 log CFU in MRSA and VREF, and up to 3 log CFU of *E. coli*, *A. baumannii* and *C. difficile* on all surfaces (Cahill et al. 2014; Claro et al. 2015; Lis et al. 2018). In dental unit waterlines which can be heavily contaminated with bacteria and fungi, plasma-decontamination of the waterlines resulted in a significant 98% reduction in microbial contamination compared to untreated control after just 1 week, which was sustained until the end of the trial of 16 weeks (Noopan et al. 2019). *E. faecalis* biofilms were also significantly reduced by PAW in addition to disturbing new biofilm formation (Li et al. 2019). These evidence indicate that flushing waterlines with PAW may be a practical aseptic solution for dental waterlines. Conventional waterline cleaners such as peracetic acid or chlorine compounds corrode waterlines (Puttaiah et al. 1998; Zhang et al. 2019) and are potentially harmful to the patient. Therefore, it should be determined whether PAW is inert to the waterlines and safe for human consumption, as it may be a viable replacement for conventional corrosive compounds. These levels of reduction would significantly reduce the risk of surface transmission of pathogens; however, the practicality and safety of using plasma-based decontamination strategies in a dynamic healthcare setting needs to be conducted for efficacy.

The advantages of using PAL over conventional detergents or disinfectants are multifaceted and extend beyond medicine. As accessibility to plasma devices become widely available, PAL can be economically produced to-demand and on-site. PAL is also environmentally friendly compared to industrially-produced disinfectants and detergents, which need large factories and produce significant wastes that harm the environment. PAL can easily be produced, stored and applied as liquid, spray or fog (Wang and Salvi 2021) and has been assessed as an alternative for disinfection of medical devices. For instance, 5-min circulation of PAW through dental instrument waterlines reduced 5-day old *E. faecalis* biofilms to undetectable levels, without accelerating waterline corrosion (Pan et al. 2017). In another setting, after being submerged in solutions of 10^7 CFU mL^{-1} *E. coli*, *K. pneumoniae*, *A. baumannii* or *P. aeruginosa* planktonic suspensions for 15 min, duodenal endoscopes were completely decontaminated after 20–30 min immersion in PAW without modifying the surface or composition of the endoscope (Balan et al. 2018). Sterilising medical devices with PAW has been demonstrated in vitro, but whether this translates to reduced medical device-associated infection is yet to be determined.

6 Plasma Medicine in Dental Infections

Complex polymicrobial biofilms called "dental plaque" form on teeth and are not necessarily harmful to the host. They can provide an optimum environment for the microorganism in exchange for aiding digestion of complex nutrients and nitrates (Marsh et al. 2015). However, dysbiosis of the plaque can progress to cause caries and periodontal disease (gum disease). This involves inflammation of the gums that can lead to tooth detachment. It should be acknowledged that dental/periodontal infections are usually due to dysbiosis of the existing biofilms in the mouth rather than via an external pathogen, as the resident biofilms intrinsically protect the mouth by resisting external colonisation of pathogenic microbes (Marsh 2003). Increasing oral hygiene procedures like brushing, flossing and tooth picking in combination with the use of fluoridated toothpaste, combined with lifestyle changes like altering diet, starting (or quitting) smoking, which can change the chemical or compositional properties of the saliva and gums can promote the growth of a particular microorganism in the biofilm most suited to that new environment.

The use of cold plasma has been studied in several applications of dentistry including tooth-whitening, disinfection of teeth and cavities, plaque removal and strengthening composite restorations in vitro. In 2021, the first systematic review, which included 55 studies over the last two decades regarding the efficacy of cold plasma on dental pathogens in vitro has been published (Jungbauer et al. 2021). The review includes a great summary of the included studies spanning investigation of several types of operator gases, plasma modes/devices, treatment times (<1 to >10 min) and dental caries and periodontal pathogens including *Streptococci* and *Lactobacilli* spp., *E. faecalis*, *P. gingivalis*, *Aggregatibacter actinomycetemcomitans* and *C. albicans*. Unfortunately, only a few animal studies and no clinical trials in humans have been performed, which warrants further investigation.

6.1 Dental Caries

Dental caries (tooth decay) are a progressive degradation of teeth caused by an acidic environment in the mouth (Silk 2014). The acidic environment can be caused by the catabolism of sugars by oral aciduric and acidogenic microbes, primarily *L. acidophilus* and *S. mutans*, leading to dyshomeostasis of the oral microbiome (Caufield et al. 2015; Lamont et al. 2018). This explains why the incidence of dental caries is strongly correlated with the consumption of free sugars in our diet, even when fluoride in drinking water is used to strengthen teeth (Sheiham and James 2014). *L. acidophilus* and *S. mutans* were not detectable on tooth enamel analogues such as hydroxyapatite disks following treatment with Ar plasma brushes for less than 1 min, which was more effective in the absence of O_2 (Yang et al. 2011; Blumhagen et al. 2014). Plasma jet and DBD devices (including self-constructed, kINPen and PlasmaDerm) also effectively removed *Streptococcus* spp. biofilms on several surfaces by as much as 5.8 log CFU (Koban et al. 2011; Huang et al. 2013; Preissner et al. 2016). However, the plasma exposure was between 5 and 10 min, whereas *Lactobacillus rhamnosus* only had a 1 log CFU reduction after 1 min treatment with the kINPen Med or PlasmaDerm (Hertel et al. 2018). After 2-min exposure to He DBD plasma on tooth slices, *S. mutans* biofilms could not be detected (Molnar et al. 2013). Interestingly, the volume DBD Ar plasma more effectively reduced *S. mutans* biofilm and saliva biofilm CFU counts on Ti disks by 5.4 log and 5.7 log CFU, respectively, compared to 0.1% chlorhexidine rinse (3.4 log and 1.5 log reduction, respectively) (Koban et al. 2011). Although Ar plasma jet alone was effective at killing *S. mutans* on a tooth slice, it could also be further enhanced by combining therapy with gold nanoparticles, likely by intensifying the damage at the cell membrane (Park et al. 2016). Thereby, combined therapy could be a sound strategy to achieve strong antibiofilm efficacy while reducing treatment time.

In a self-controlled Sprague-Dawley rat model, an Ar and 1% O_2 plasma brush was applied once to the left molar of Sprague-Dawley rats for 2 min, while the right molar was untreated (Hong et al. 2019). Six months later, there was significantly less decayed surfaces, and the caries rate was 40% lower in the lower molars and 20% lower in the upper molars of the cold plasma-treated side (Hong et al. 2019). This is potentially clinically significant, due to this being attributed to only a single application of plasma treatment in 6 months, but no trials in human caries have been conducted to translate this finding.

6.2 *Root Canal Infections (*E. faecalis*)*

Compared to other in-house plasma jet designs using He or He/O_2, the kINPen using Ar may not be as effective on *E. faecalis* biofilms, with some studies finding a non-significant 0.5 log CFU reduction (Hufner et al. 2017; Ballout et al. 2018), or 1.5–3.2 log CFU reduction (Hufner et al. 2017; Du et al. 2012; Schaudinn et al. 2013; Herbst et al. 2015). Ar plasma with 2% O_2 instead of He plasma caused complete eradication of *E. faecalis* biofilm in a root canal in vitro model (Pan et al. 2013; Li et al. 2015). Some studies found these plasma jets were more effective at killing *E. faecalis* biofilm than 2% chlorhexidine and calcium hydroxide, but not more effective than 1–6% hypochlorite (bleach) (Hufner et al. 2017; Ballout et al. 2018; Schaudinn et al. 2013; Pan et al. 2013). In comparison, DBD plasma with He or He/O_2 were generally more effective than plasma jet designs. He gas plasma reduced *E. faecalis* biofilm by 2.1 log CFU, which was significantly augmented to 5.1 log CFU reduction with He/0.5% O_2 plasma (Armand et al. 2019). Sectional analysis of root canal found that the coronal and apical root canal showed a 3.2 and 3.0 log CFU reduction, while the middle root canal had no detectable *E. faecalis* following 5 min plasma treatment (Ureyen Kaya et al. 2014). Additionally, DBD plasma was better than photodynamic or O_3 therapy and was comparable to hypochlorite in disinfecting root canals (Armand et al. 2019; Ureyen Kaya et al. 2014).

6.3 *Periodontal Disease Bacteria*

Plasma jets and DBD devices have been applied to periodontal pathogens *P. gingivalis* and *A. actinomycetemcomitans* in vitro. Agar plates inoculated with *A. actinomycetemcomitans* were completely sterilised with an O_2 plasma jet in just 1 min, while N_2 and 1:1 N_2:O_2 plasma jets also significantly eradicated the bacteria, with Ar plasma jet having no effect (Abonti et al. 2016). In addition to that, a 2-min application of air DBD plasma was shown to reduce *A. actinomycetemcomitans* biofilms by 2 log CFU reduction and pre-application of plasma before inoculating with the bacteria significantly prevented adhesion and growth of new biofilm (Liguori et al. 2017). Plasma jets were also effective at eradicating *P. gingivalis*. Though the kINPen Ar plasma jet could remove 1.3 log CFU of *P. gingivalis* on Ti disks after 3 min treatment, this was not as effective as 0.2% chlorhexidine (3 log CFU reduction) (Carreiro et al. 2019). Otherwise, plasma jets utilising either air or He were much more successful at eradicating *P. gingivalis* to undetectable levels at 5–6 min application on agar and Ti disk (Yang et al. 2018; Lee et al. 2019).

6.4 C. albicans

It is estimated that approximately 50% of the healthy human population carry oral *C. albicans* commensally, but dyshomeostasis of the oral microbiome can lead to oral candidiasis, more commonly known as oral "thrush" (Hebecker et al. 2014). The kINPen plasma jet has been tested on *C. albicans* on a number of surfaces in vitro. With Ar operator gas, there was a 2 log CFU reduction of biofilms in culture wells following 5-min plasma exposure (Handorf et al. 2018) and approximately 1.5 log reduction in *C. albicans* and *C. albicans/S. aureus* dual-biofilm on acrylic resin surfaces after 1-min plasma exposure, while causing no harm to infected oral epithelial cells (Delben et al. 2016). The kINPen did not effectively remove biofilms on Ti surfaces with Ar alone (0.5 log CFU reduction) but was somewhat improved with Ar/1% O_2 plasma (1 log CFU reduction) (Koban et al. 2010). Ar/1% O_2 kINPen plasma jet applied to 7-day *C. albicans* biofilms on root canals in vitro resulted in a 2.1 and 2.7 log CFU reduction from 6- and 12-min plasma exposures, respectively (Kerlikowski et al. 2020). Further, combined plasma and antiseptic (0.1% octenidine or 2% chlorhexidine) therapy was not more beneficial than plasma therapy alone (Kerlikowski et al. 2020). DBD Ar plasma sources effectively removed *C. albicans* biofilm by between 2.3 and 5.9 log CFU with between 1- and 10-min plasma therapy, respectively on Ti discs (Koban et al. 2010) and 2.8–4.1 log CFU between 5- and 10-min plasma therapy, respectively on polymethylmethacrylate (Matthes et al. 2015). There was no additional benefit to an Ar DBD plasma device with the addition of 1% O_2 (Koban et al. 2010; Matthes et al. 2015). In an in vivo murine model of oral candidiasis, He plasma jet therapy for 5 min on 2 consecutive days did not cause a significant reduction in CFU, but there was significantly reduced tongue tissue invasion (Borges et al. 2018). In general, DBD and plasma jet sources have been effective against *C. albicans* biofilm; however, no commercial DBD plasma devices have been tested against *C. albicans*.

7 Plasma Medicine in Dermatological Infections

Cold plasma therapy exhibited powerful antimicrobial effect on many prominent skin and wound pathogens, including *E. coli*, MRSA, *S. epidermidis*, *Streptococcus* spp. and *P. aeruginosa* (Daeschlein et al. 2012, 2014). Although almost all studies focus on patient applications, plasma could be a useful hand hygiene tool for surgeons as well, to mitigate potential surgical site infections upstream of the patient (Daeschlein et al. 2012), with the added advantage of reducing the use of alcohol-based disinfectants that can irritate broken or sensitive skin.

Plasma in dermatology should not be confused with plasma cosmetics. Although they are applied to the skin in both cases, the difference is in the intended purpose where plasma medicine is applied to harmed or destroyed skin tissue to aid wound healing and skin disinfection, and plasma cosmetics are applied "for cleansing, beautifying, promoting attractiveness, or altering the appearance" according to the U.S. Food and Drug Administration (FD & C Act, sec. 201(i)) (FDA n.d.). However, these two fields are not always mutually exclusive. Hypothetically, getting dermatological plasma treatment may have secondary benefits of beautifying, and plasma cosmetics may create an antimicrobial environment on the skin. These topics and implications are discussed in greater detail elsewhere (von Woedtke et al. 2020).

7.1 *Plasma Therapy in Wound Healing*

Biofilms in wounds lead to persistent and recurrent infections that prevent healing. Typically, chronic wounds are caused by opportunistic pathogens as a secondary morbidity to other chronic diseases or conditions, e.g., diabetes mellitus leading to complications of diabetic foot ulcers, which increases hospital stay, likelihood of amputation and death (Bakker et al. 2016). Plasma treatment supports wound healing by both disinfecting wounds and stimulating healthy human cells by modulating local inflammation, which is described later in this section. The exception to this is the plasma and PAL being utilised to induce apoptosis in cancer cells by causing DNA and mitochondrial damage through intracellular generation of RONS (Azzariti et al. 2019; Nguyen et al. 2016; Jo et al. 2020; Girard et al. 2016). Plasma-based cancer therapy is outside the scope of this chapter but has been extensively reviewed elsewhere (Yan et al. 2017; Semmler et al. 2020).

The DBD device PlasmaDerm® VU-2010 (CINOGY Technologies GmbH), the kINPen® MED atmospheric pressure plasma jet (INP Greifswald/neoplas tools GmbH) and the MicroPlaSter (Adtec Ltd.) are commercial medical plasma products that have undergone the first prospective randomised controlled trials (RCT) to treat infections in chronic wounds in the last decade (Table 2). As the first and most researched and developed plasma-based medical device, a review of the Physics and chemistry of the kINPen is available from Reuter et al. (Reuter et al. 2018). Pleasingly, plasma application was highly tolerable, with no observed adverse

Table 2 Clinical trials of cold plasma devices to treat chronic infected wounds

Plasma device	Ref.	Trial description	Participants and treatment groups	Results and conclusion
kINPen® Med plasma jet	Stratmann et al. (2020)	P, SB, RPCT in diabetic foot ulcers	$n = 45$. Wound care with plasma ($n = 33$) or placebo ($n = 32$) applied for 30 s cm^{-2} wound once daily for 5 days, followed by 3 more treatments every 2nd day	NSD in bacterial load, reduction of clinical infection or time to infection between groups. Plasma therapies lead to significantly more wound closure compared with placebo. 26% greater wound closure in plasma compared to placebo ($p = 0.03$)
kINPen® Med plasma jet	Ulrich et al. (2015)	P, RCT in chronic leg ulcers	$n = 16$. Wound care alone ($n = 6$), with plasma ($n = 6$) or both ($n = 4$). 2-week plasma therapy for 1 min cm^{-2}, 3 times per week	Bacterial load was significantly reduced immediately after plasma therapy compared to immediately before therapy, but recolonised quickly. NSD in bacterial load after 2-week plasma therapy
Bioplasma Jet	Chuangsuwanich et al. (2016)	P, RCT in pressure ulcers	$n = 42$. Wound care with plasma ($n = 23$) or without plasma ($n = 19$). Plasma applied once weekly for 8 weeks	Bacterial load reduced after 1 week in plasma treatment. The wound size and wound base were significantly improved after 2 plasma treatments
PlasmaDerm® VU-2010 DBD	Brehmer et al. (2015)	Two-armed, open, P, RCT in chronic venous leg ulcers	$n = 14$. Wound care with plasma ($n = 7$) or without plasma ($n = 7$). Plasma applied for 45 s cm^{-2} ulcer, 3 times per week for 8 weeks + 4-week follow-up	88.2 ± 51.1 mm^2 increase in bacteria-free area of lesions ($p = 0.0313$). NSD in ulcer lesion size between groups at the end of trial
MicroPlaSter plasma torch	Isbary et al. (2010)	P, RCT in chronic infected skin wounds	$n = 36$. Standard wound care with or without plasma in different wounds (self-controls). Daily 5 min cold plasma treatment. Treatment	67% reduction in bacteria load in plasma-treated wounds compared to 35% in control wounds ($p < 0.0001$)

(continued)

Table 2 (continued)

Plasma device	Ref.	Trial description	Participants and treatment groups	Results and conclusion
			period not constant (range: 1–54 treatments)	
MicroPlaSter plasma torch	Isbary et al. (2012)		$n = 24$. Standard wound care with once daily 2-min plasma therapy in different wounds (self-controls). Treatment period not constant (range: 1–71 treatments)	74% reduction in bacteria load in plasma-treated wounds compared to 50% in control wounds ($p <$ 0.008)
Housemade He plasma jet	Mirpour et al. (2020)	P, DB, RCT in diabetic foot ulcers	$n = 44$. Standard care with ($n = 22$) or without ($n = 22$) cold plasma. Plasma applied for 5 min, 3 times per week for 3 weeks	After 3 weeks, wound closure was accelerated significantly with plasma therapy (61%) vs. standard therapy alone (31%; $p < 0.05$). Bacterial load was significantly reduced immediately after plasma therapy compared to immediately before therapy but would recolonise by the next week

Abbreviations: *P* prospective, *SB* single-blinded, *DB* double-blinded, *RCT* randomised controlled trial, *RPCT* randomised placebo-controlled trial, *NSD* no significant difference

events related to the use of plasma devices. The kINPen MED and PlasmaDerm VU-2010 significantly reduced bacterial load in chronic leg ulcers greater than standard wound care, which was concluded based on a combined 30 participants in the trials (Brehmer et al. 2015; Ulrich et al. 2015). Similarly, two MicroPlaSter clinical trials with a combined 60 participants, ranging from 2 to 5-min plasma treatments significantly reduced the bacterial load of chronic infected skin wounds far more effectively than standard wound care in other wounds of the same patient (self-controls) (Isbary et al. 2010; Isbary et al. 2012). A Bioplasma jet device applied to 23 participants with pressure ulcers significantly reduced bacterial load after one use (Chuangsuwanich et al. 2016). However, two recent trials in diabetic foot ulcers, one using the kINPen MED and the other an inhouse He plasma jet design, with combined 89 participants did not get the desired reduction in bacterial load at the end of the trials (Stratmann et al. 2020; Mirpour et al. 2020). Interestingly, bacterial load was significantly lower immediately after plasma therapy compared to immediately before, but due to the advanced disease states of participants with diabetic foot ulcers, bacteria were able to quickly recolonise (Ulrich et al. 2015; Stratmann et al. 2020; Mirpour et al. 2020). This is not surprising, as uncontrolled diabetes mellitus causes impaired healing and immune responses to infection (reviewed by Daryabor et al. 2020).

Overall, these studies indicate that cold plasma therapy could be a well-tolerated conjunctive therapy to standard wound care. Currently, the available research is small-scale, single-centred, short-term studies, but they are robustly designed prospective RCT, with studies including placebo groups and participants with multiple wounds being self-controls. Therefore, these results are encouraging and justify conducting larger, multicentred prospective RCT. Further, there are no clinical studies on long-term prospective consequences of plasma on the skin. This is understandable given the youth of the technology in clinical trials and the advanced age of participants that usually qualify for recruitment in these studies, but this requires investigation. There are difficulties in assessing the true antimicrobial efficacy of plasma therapy in RCT, since participants in plasma and control groups are commonly on antibiotic therapy. Larger studies would also allow logistic regression analysis with adequate statistical power to determine if cold plasma is an independently effective antimicrobial tool in the clinic.

Regardless of the potential antimicrobial activity of cold plasma therapy, it goes further than antibiofilm and antisepsis to aid in wound healing (Fig. 2). Traditionally in surgery, hot plasma has been used to thermally cauterise wounds. However, cold plasma has also demonstrated the capability to augment wound healing by promoting platelet activation, aggregation and fibrin polymerisation to accelerate blood coagulation (Kalghatgi et al. 2007; Fridman et al. 2008), scarring the wound without the need to cauterise. Fibroblast differentiation into myofibroblasts is also induced by cold plasma, promoting matrix synthesis as a framework for wound closure while keratinocyte and fibroblast migration and proliferation are also induced to close the wound (Arndt et al. 2018a). While the mechanisms are not yet clear, it is apparent that vascular perfusion is improved at the site of plasma exposure on the skin, demonstrated by the increased partial pressure O_2 (pO_2) and consequent blood flow, which lasted for over 1 h after the 5-min plasma therapy ceased (Collet et al. 2014; Heuer et al. 2015; Kisch et al. 2016). This was due to localised increase in NO (Heuer et al. 2015; Shekhter et al. 2005), decreasing skin pH to inhibit microbial activity and being a potent vasodilator and modulator of inflammation (Carpenter and Schoenfisch 2012) to accelerated wound closure (Shekhter et al. 2005). This mechanism is supported by a study in BALB/c mice demonstrating with direct plasma jet application onto burn wounds, that cold plasma promoted neovascularization, which increased endothelial nitric oxide synthase (eNOS) expression (Duchesne et al. 2019). In concert with this, inflammatory cytokines and pro-angiogenic factors are released by keratinocytes and fibroblasts to promote inflammation and message vascular endothelial cells to grow new blood vessels in the wounded area to increase immune cell infiltration and nutrient transport for tissue remodelling (Arndt et al. 2015, 2018b). Therefore, application of cold plasma is not

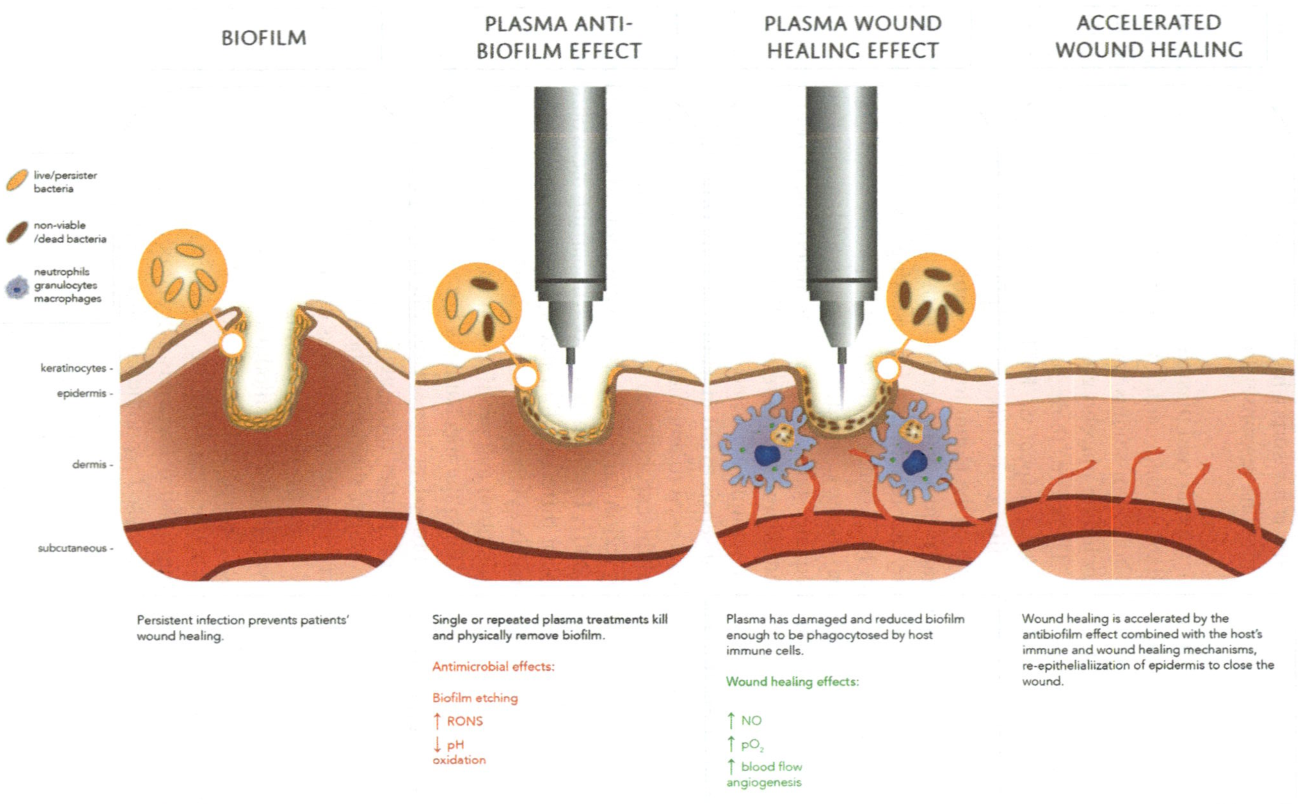

Fig. 2 Concepts for antibiofilm activity and wound healing promoted by plasma. Bacteria populate the wound and form a biofilm to adhere to tissue and shelter from the host immune system. The biofilm infection causes unresolved inflammation that prevents the wound from healing. Direct cold plasma treatment onto the wound etches the biofilm with RONS and the consequent pH reduction of the skin. These RONS also penetrate the biofilm to directly kill residing microbial

cells to halt further biofilm development. Meanwhile, plasma actively promotes localised angiogenic response and vascularises the wound site through the production of NO. Biofilm etching allows immune cells to eradicate the source of infection by phagocytosis, while the increased blood flow increases oxygenation and nutrition transport into the wound site to rebuild and remodel the tissue. Finally, inflammation is resolved and re-epithelialisation of the skin closes the wound. Illustration by Animate Your Science restricted to its antimicrobial therapy and in fact actively participates in healing the wound itself.

When commercial plasma devices were assessed for healing chronic wound healing capacity in the clinic, their effectiveness is less clear (Table 2). Isbary and colleagues also assessed the wound healing capability of the MicroPlaSter in 70 participants but wound healing was not significantly different between plasma and control wounds in the same patient (Isbary et al. 2013). In contrast, skin graft donor sites of 34 participants stimulated by the MicroPlaSter β for 2 min, 5 days a week, saw significantly improved wound healing after the second treatment day onwards compared to the placebo group (Heinlin et al. 2013). Evidently, these two studies are very different clinical settings, Isbary and colleagues performed cold plasma therapy on infected wounds (Isbary et al. 2013), while Heinlin and colleagues on disinfected skin graft sites (Heinlin et al. 2013). Hence, wound healing would still be hampered by the chronic infection, regardless of wound healing effects by plasma therapy.

8 Challenges to Biofilm Destruction by Cold Plasma

Heterogeneity in biofilm EPS composition, thickness, O_2 content, nutrient gradients, microbial species, persister cells and location of the biofilm (tissue surface, medical device) pose a challenge to test cold plasma therapy. As biofilms mature, the effectiveness of cold plasma treatment will likely diminish, so cold plasma as a preventative treatment or up to the early stages of biofilm development will be most effective. Cold plasma properties and therapeutic treatment periods need to be optimised to adequately cover differences in these biofilm characteristics, which requires a thorough fundamental understanding of cold plasma physicochemical properties and biological responses. Alternatively, adjuvant treatments of cold plasma with antibiotics or antiseptics could magnify the antibiofilm effect of both. This has been demonstrated to work in 5-day-old *S. aureus* and *P. aeruginosa* biofilm treated with cold plasma combined with either daptomycin or ciprofloxacin, respectively, or with chlorhexidine in vitro (Brun et al. 2018). Similarly, cold plasma combined with mild H_2O_2 adjuvant treatment was more effective than either adjuvant alone in *L. monocytogenes* and *Salmonella typhimurium* (Govaert et al. 2019). These evidence give hope that cold plasma can also extend the life of our current antimicrobial therapies.

8.1 Plasma Tolerance

A recent review has proposed putative potential challenges to cold plasma, such as preconditioning and attenuation of RONS to eDNA and EPS and plasma-induced persistence and tolerance (Gilmore et al. 2018). Compared to antibiotic resistance, the likelihood of generalised plasma resistance is low due to the diverse chemistry of plasma which can be altered in several parameters. Regardless, repeated suboptimal plasma doses have led to the emergence of some *P. aeruginosa* cells in vitro that have a higher tolerance to cold plasma and plasma-derived H_2O_2 (Mai-Prochnow et al. 2015). Similarly, *E. coli* was adapted to tolerate PAW after strains were preconditioned to H_2O_2 exposure (Rodriguez et al. 2020).

A few studies have determined whether or not plasma tolerance could arise if biofilms were allowed time to recover after repeated exposure to plasma. This has been performed by applying plasma to an initial bacteria colony or biofilm, then a surviving colony is re-cultured and exposed to plasma again several times, with a CFU count being performed after each plasma treatment to determine plasma tolerance by a diminution of the reduction factor of plasma treatment over repeated applications. The first study to test this applied four cycles of a small micro-discharge plasma to *E. coli* and *Enterococcus mundtii* colonies as Gram-positive and Gram-negative representatives, respectively. After four plasma exposure regiments, no plasma tolerance was detected (Zimmermann et al. 2012). Later, this was tested in a similar study in *S. aureus* biofilms that were treated with the kINPen09 plasma jet every hour over 6 h, which did not show evidence of plasma tolerance after each cold plasma regiment (Matthes et al. 2014). However, the short 1-h rest time between cold plasma applications may not be enough time to observe potential adaptation to the repeated cold plasma exposure. Fortunately, this was more recently expanded to daily 2-min plasma treatments for 7 days, which consistently led to a 2.5 log and 3 log CFU reduction of *S. aureus* and *P. aeruginosa* biofilm, respectively, showing no sign of plasma tolerance (Brun et al. 2018). Although this evidence does show positive signs for the longevity of plasma treatment as an antibiofilm therapy, these studies are nonetheless obtained through very short-term experiments on bacterial populations, and the potential adaptation of plasma tolerance over longer timespans is still warranted.

Bacteria are armed with a host of antioxidant mechanisms, including the glutathione reductase system, thioredoxin reductase system, catalase and superoxide dismutase (Cheng et al. 2020). Microbial cell responses to cold plasma exposure are an important issue that is still relatively unexplored, but some responses have been documented. Cold plasma exposure to *E. coli* activated the SOS regulon and an array of genes involved in DNA and oxidative damage repair, protein binding and folding, iron metabolism, transport proteins and transcription regulators (Sharma et al. 2009). Additionally, housekeeping genes were downregulated to maximise energy and nutrient availability for cellular repair (Sharma et al. 2009). These mechanisms are induced to counteract the increased oxidative stress and associated damage to the biofilm and cellular metabolic systems imposed by plasma-derived RONS, so cold plasma therapy needs to be applied with enough intensity and frequency to exhaust and overwhelm these protective mechanisms. Evidently, we cannot let a low likelihood lead to complacency, as recurrence of infection with plasma tolerance is possible.

9 Conclusions and Future Perspectives

There are significant conceptual advantages to the application of cold plasma compared to conventional antimicrobial drugs. For instance, cold plasma (and the RONS and other reactive constituents) can be applied directly to the local site of interest for a determinant amount of time, without affecting other tissue. This is not the case for conventional antimicrobials drugs, which (1) have to be absorbed, (2) usually disperse systemically through the body, (3) get metabolised, and (4) have the compound and metabolic byproducts remain in the body after its usefulness. These aspects of conventional antimicrobials can lead to toxicity or contraindicate the use of these drugs in patients with renal or hepatic impairment.

In dentistry, animal and clinical studies are still required to translate the large body of in vitro evidence for the effectiveness of cold plasma against oral pathogens and its side-effects on the oral microbiome. More is known in dermatology, and although small, RCT of cold plasma could significantly improve wound healing, its effectiveness at reducing bacterial load in wounds is mostly acute. However, these clinical studies are small and only in older patients with chronic wounds, some of which are secondary to underlying illnesses that impair immune function, such as diabetes mellitus. In contrast, clinical trials for cold plasma therapy surgical wounds are absent but could be useful as a prophylactic strategy. Another intriguing prospect is the application of cold plasma therapy for medical emergencies and wound care in space as an "'energy-based' and 'dry' technology that can replace perishable drugs" (Laroussi 2020).

Although plasma medicine is a promising alternative to antibiotics, we should still learn from our experiences regarding antibiotics by harbouring the same concerns for plasma antimicrobial therapy. Similar to microbial adaptations leading to antimicrobial resistance and tolerance, we must wonder, can inadequate plasma or PAL exposure coax microbes to produce more adaptive biofilms or induce some cells into a persister state? Or select for tolerance or resistance to RONS generated by the plasma? So far, this has not been reported in the scientific community, which is encouraging, but further studies to investigate the expansive range of cold plasma parameters and the phenotypic changes in plasma-treated bacteria are required to answer this. Ultimately, we do not want to end up looking for an alternative to plasma antimicrobial medicine in the future, like we are today for antibiotics.

Acknowledgements Work in the laboratory and for this review was supported by the National Health and Medical Research Council (GNT1163634, GNT2004036), the University of Adelaide and The Hospital Research Foundation, Australia. The authors kindly acknowledge Animate Your Science (www.animateyour.science) for professional graphics.

References

Abdo A, Schmitt-John T, Richter K (2021) Plasma-activated water as a new weapon against multidrug-resistant bacteria. EMJ Microbiol Infect Dis 2(1):38–39. ECCMID 2021 Abstract Review No. AR9

Abonti TR, Kaku M, Kojima S, Sumi H, Kojima S, Yamamoto T, Yashima Y, Miyahara H, Okino A, Kawata T, Tanne K, Tanimoto K (2016) Irradiation effects of low temperature multi gas plasma jet on oral bacteria. Dent Mater J 35(5):822–828. https://doi.org/10.4012/dmj.2016-062

Aboubakr HA, Gangal U, Youssef MM, Goyal SM, Bruggeman PJ (2016) Inactivation of virus in solution by cold atmospheric pressure plasma: identification of chemical inactivation pathways. J Phys D Appl Phys 49(20). https://doi.org/10.1088/0022-3727/49/20/204001

Alkawareek MY, Algwari QT, Gorman SP, Graham WG, O'Connell D, Gilmore BF (2012) Application of atmospheric pressure nonthermal plasma for the *in vitro* eradication of bacterial biofilms. FEMS Immunol Med Microbiol 65(2):381–384. https://doi.org/10.1111/j.1574-695X.2012.00942.x

Armand A, Khani M, Asnaashari M, AliAhmadi A, Shokri B (2019) Comparison study of root canal disinfection by cold plasma jet and photodynamic therapy. Photodiagnosis Photodyn Ther 26:327–333. https://doi.org/10.1016/j.pdpdt.2019.04.023

Arndt S, Landthaler M, Zimmermann JL, Unger P, Wacker E, Shimizu T, Li YF, Morfill GE, Bosserhoff AK, Karrer S (2015) Effects of cold atmospheric plasma (CAP) on ß-defensins, inflammatory cytokines, and apoptosis-related molecules in keratinocytes *in vitro* and *in vivo*. PLoS One 10(3):e0120041. https://doi.org/10.1371/journal.pone.0120041

Arndt S, Schmidt A, Karrer S, von Woedtke T (2018a) Comparing two different plasma devices kINPen and Adtec SteriPlas regarding their molecular and cellular effects on wound healing. Clin Plasma Med 9:24–33. https://doi.org/10.1016/j.cpme.2018.01.002

Arndt S, Unger P, Berneburg M, Bosserhoff AK, Karrer S (2018b) Cold atmospheric plasma (CAP) activates angiogenesis-related molecules in skin keratinocytes, fibroblasts and endothelial cells and improves wound angiogenesis in an autocrine and paracrine mode. J Dermatol Sci 89(2): 181–190. https://doi.org/10.1016/j.jdermsci.2017.11.008

Azzariti A, Iacobazzi RM, Di Fonte R, Porcelli L, Gristina R, Favia P, Fracassi F, Trizio I, Silvestris N, Guida G, Tommasi S, Sardella E (2019) Plasma-activated medium triggers cell death and the presentation of immune activating danger signals in melanoma and pancreatic cancer cells. Sci Rep 9(1):4099. https://doi.org/10.1038/s41598-019-40637-z

Bakker K, Apelqvist J, Lipsky BA, Van Netten JJ, International Working Group on the Diabetic Foot (2016) The 2015 IWGDF guidance documents on prevention and management of foot problems in diabetes: development of an evidence-based global consensus. Diabetes Metab Res Rev 32(Suppl 1):2–6. https://doi.org/10.1002/dmrr.2694

Balan GG, Rosca I, Ursu EL, Doroftei F, Bostanaru AC, Hnatiuc E, Nastasa V, Sandru V, Stefanescu G, Trifan A, Mares M (2018) Plasma-activated water: a new and effective alternative for duodenoscope reprocessing. Infect Drug Resist 11:727–733. https://doi.org/10.2147/IDR.S159243

Ballout H, Hertel M, Doehring J, Kostka E, Hartwig S, Paris S, Preissner S (2018) Effects of plasma jet, dielectric barrier discharge, photodynamic therapy and sodium hypochlorite on infected curved root canals. J Biophotonics 11(3). https://doi.org/10.1002/jbio.201700186

Bernhardt T, Semmler ML, Schafer M, Bekeschus S, Emmert S, Boeckmann L (2019) Plasma medicine: Applications of cold atmospheric pressure plasma in dermatology. Oxid Med Cell Longev 2019:3873928. https://doi.org/10.1155/2019/3873928

Birmingham JG (2004) Mechanisms of bacterial spore deactivation using ambient pressure non-thermal discharges. IEEE Trans Plasma Sci 32(4):1526–1531. https://doi.org/10.1109/tps.2004.832609

Blumhagen A, Singh P, Mustapha A, Chen M, Wang Y, Yu QS (2014) Plasma deactivation of oral bacteria seeded on hydroxyapatite disks as tooth enamel analogue. Am J Dent 27(2):84–90

Borges AC, Lima GMG, Nishime TMC, Gontijo AVL, Kostov KG, Koga-Ito CY (2018) Amplitude-modulated cold atmospheric pressure plasma jet for treatment of oral candidiasis: *In vivo* study. PLoS One 13(6):e0199832. https://doi.org/10.1371/journal.pone.0199832

Borges AC, Kostov KG, Pessoa RS, de Abreu GMA, Lima GMG, Figueira LW, Koga-Ito CY (2021) Applications of cold atmospheric pressure plasma in dentistry. Appl Sci 11(5). https://doi.org/10.3390/app11051975

Boudam MK, Moisan M, Saoudi B, Popovici C, Gherardi N, Massines F (2006) Bacterial spore inactivation by atmospheric-pressure plasmas in the presence or absence of UV photons as obtained with the same gas mixture. J Phys D Appl Phys 39(16):3494–3507. https://doi.org/10.1088/0022-3727/39/16/s07

Brehmer F, Haenssle HA, Daeschlein G, Ahmed R, Pfeiffer S, Gorlitz A, Simon D, Schon MP, Wandke D, Emmert S (2015) Alleviation of chronic venous leg ulcers with a hand-held dielectric barrier discharge plasma generator (PlasmaDerm((R)) VU-2010): results of a monocentric, two-armed, open, prospective, randomized and controlled trial (NCT01415622). J Eur Acad Dermatol Venerol 29(1):148–155. https://doi.org/10.1111/jdv.12490

Bruggeman PJ, Iza F, Brandenburg R (2017) Foundations of atmospheric pressure non-equilibrium plasmas. Plasma Sources Sci Technol 26(12). https://doi.org/10.1088/1361-6595/aa97af

Brun P, Bernabe G, Marchiori C, Scarpa M, Zuin M, Cavazzana R, Zaniol B, Martines E (2018) Antibacterial efficacy and mechanisms of action of low power atmospheric pressure cold plasma: membrane permeability, biofilm penetration and antimicrobial sensitization. J Appl Microbiol 125(2):398–408. https://doi.org/10.1111/jam.13780

Cahill OJ, Claro T, O'Connor N, Cafolla AA, Stevens NT, Daniels S, Humphreys H (2014) Cold air plasma to decontaminate inanimate surfaces of the hospital environment. Appl Environ Microbiol 80(6):2004–2010. https://doi.org/10.1128/AEM.03480-13

Carpenter AW, Schoenfisch MH (2012) Nitric oxide release: part II. Ther Appl Chem Soc Rev 41(10):3742–3752. https://doi.org/10.1039/c2cs15273h

Carreiro AFP, Delben JA, Guedes S, Silveira EJD, Janal MN, Vergani CE, Pushalkar S, Duarte S (2019) Low-temperature plasma on peri-implant-related biofilm and gingival tissue. J Periodontol 90(5):507–515. https://doi.org/10.1002/JPER.18-0366

Caufield PW, Schon CN, Saraithong P, Li Y, Argimon S (2015) Oral Lactobacilli and dental caries: A model for niche adaptation in humans. J Dent Res 94(9 Suppl):110S–118S. https://doi.org/10.1177/0022034515576052

Cheng J-H, Lv X, Pan Y, Sun D-W (2020) Foodborne bacterial stress responses to exogenous reactive oxygen species (ROS) induced by cold plasma treatments. Trends Food Sci Technol 103:239–247. https://doi.org/10.1016/j.tifs.2020.07.022

Chuangsuwanich A, Assadamongkol T, Boonyawan D (2016) The healing effect of low-temperature atmospheric-pressure plasma in pressure ulcer: A randomized controlled trial. Int J Low Extrem Wounds 15(4):313–319. https://doi.org/10.1177/1534734616665046

Claro T, Cahill OJ, O'Connor N, Daniels S, Humphreys H (2015) Cold-air atmospheric pressure plasma against *Clostridium difficile* spores: a potential alternative for the decontamination of hospital inanimate surfaces. Infect Control Hosp Epidemiol 36(6):742–744. https://doi.org/10.1017/ice.2015.39

Collet G, Robert E, Lenoir A, Vandamme M, Darny T, Dozias S, Kieda C, Pouvesle JM (2014) Plasma jet-induced tissue oxygenation: potentialities for new therapeutic strategies. Plasma Sources Sci Technol 23(1). https://doi.org/10.1088/0963-0252/23/1/012005

Daeschlein G, Scholz S, Ahmed R, von Woedtke T, Haase H, Niggemeier M, Kindel E, Brandenburg R, Weltmann KD, Juenger M (2012) Skin decontamination by low-temperature atmospheric pressure plasma jet and dielectric barrier discharge plasma. J Hosp Infect 81(3): 177–183. https://doi.org/10.1016/j.jhin.2012.02.012

Daeschlein G, Napp M, von Podewils S, Lutze S, Emmert S, Lange A, Klare I, Haase H, Gümbel D, von Woedtke T, Jünger M (2014) *In vitro* susceptibility of multidrug resistant skin and wound pathogens against low temperature atmospheric pressure plasma jet (APPJ) and dielectric barrier discharge plasma (DBD). Plasma Processes Polym 11(2):175–183. https://doi.org/10.1002/ppap.201300070

Daryabor G, Atashzar MR, Kabelitz D, Meri S, Kalantar K (2020) The effects of type 2 diabetes mellitus on organ metabolism and the immune system. Front Immunol 11:1582. https://doi.org/10.3389/fimmu.2020.01582

Davies J, Davies D (2010) Origins and evolution of antibiotic resistance. Microbiol Mol Biol Rev 74(3):417–433. https://doi.org/10.1128/MMBR.00016-10

de la Fuente-Nunez C, Reffuveille F, Fernandez L, Hancock RE (2013) Bacterial biofilm development as a multicellular adaptation: antibiotic resistance and new therapeutic strategies. Curr Opin Microbiol 16(5):580–589. https://doi.org/10.1016/j.mib.2013.06.013

Delben JA, Zago CE, Tyhovych N, Duarte S, Vergani CE (2016) Effect of atmospheric-pressure cold plasma on pathogenic oral biofilms and *in vitro* reconstituted oral epithelium. PLoS One 11(5):e0155427. https://doi.org/10.1371/journal.pone.0155427

Dobrynin D, Fridman G, Friedman G, Fridman A (2009) Physical and biological mechanisms of direct plasma interaction with living tissue. New J Phys 11(11). https://doi.org/10.1088/1367-2630/11/11/115020

Du T, Ma J, Yang P, Xiong Z, Lu X, Cao Y (2012) Evaluation of antibacterial effects by atmospheric pressure nonequilibrium plasmas against *Enterococcus faecalis* biofilms *in vitro*. J Endod 38(4):545–549. https://doi.org/10.1016/j.joen.2011.10.021

Duchesne C, Banzet S, Lataillade JJ, Rousseau A, Frescaline N (2019) Cold atmospheric plasma modulates endothelial nitric oxide synthase signalling and enhances burn wound neovascularisation. J Pathol 249(3):368–380. https://doi.org/10.1002/path.5323

Ermolaeva SA, Varfolomeev AF, Chernukha MY, Yurov DS, Vasiliev MM, Kaminskaya AA, Moisenovich MM, Romanova JM, Murashev AN, Selezneva II, Shimizu T, Sysolyatina EV, Shaginyan IA, Petrov OF, Mayevsky EI, Fortov VE, Morfill GE, Naroditsky BS, Gintsburg AL (2011) Bactericidal effects of non-thermal argon plasma *in vitro*, in biofilms and in the animal model of infected wounds. J Med Microbiol 60(Pt 1):75–83. https://doi.org/10.1099/jmm.0.020263-0

FDA (n.d.) FDA authority over cosmetics: How cosmetics are not FDA-approved, but are FDA-regulated. https://www.fda.gov/cosmetics/cosmetics-laws-regulations/fda-authorityover-cosmetics-how-cosmetics-are-not-fda-approved-are-fda-regulated. Accessed 15 June 2021

Flynn PB, Higginbotham S, Alshraiedeh NH, Gorman SP, Graham WG, Gilmore BF (2015) Bactericidal efficacy of atmospheric pressure non-thermal plasma (APNTP) against the ESKAPE pathogens. Int J Antimicrob Agents 46(1):101–107. https://doi.org/10.1016/j.ijantimicag.2015.02.026

Flynn PB, Busetti A, Wielogorska E, Chevallier OP, Elliott CT, Laverty G, Gorman SP, Graham WG, Gilmore BF (2016) Non-thermal plasma exposure rapidly attenuates bacterial AHL-dependent quorum sensing and virulence. Sci Rep 6:26320. https://doi.org/10.1038/srep26320

Flynn PB, Graham WG, Gilmore BF (2019) *Acinetobacter baumannii* biofilm biomass mediates tolerance to cold plasma. Lett Appl Microbiol 68(4):344–349. https://doi.org/10.1111/lam.13122

Fricke K, Koban I, Tresp H, Jablonowski L, Schroder K, Kramer A, Weltmann KD, von Woedtke T, Kocher T (2012) Atmospheric pressure plasma: a high-performance tool for the efficient removal of biofilms. PLoS One 7(8):e42539. https://doi.org/10.1371/journal.pone.0042539

Fridman G, Friedman G, Gutsol A, Shekhter AB, Vasilets VN, Fridman A (2008) Applied plasma medicine. Plasma Processes Polym 5(6):503–533. https://doi.org/10.1002/ppap.200700154

Gilmore BF, Flynn PB, O'Brien S, Hickok N, Freeman T, Bourke P (2018) Cold plasmas for biofilm control: Opportunities and challenges. Trends Biotechnol 36(6):627–638. https://doi.org/10.1016/j.tibtech.2018.03.007

Girard PM, Arbabian A, Fleury M, Bauville G, Puech V, Dutreix M, Sousa JS (2016) Synergistic effect of H_2O_2 and NO_2 in cell death induced by cold atmospheric He plasma. Sci Rep 6:29098. https://doi.org/10.1038/srep29098

Govaert M, Smet C, Verheyen D, Walsh JL, Van Impe JFM (2019) Combined effect of cold atmospheric plasma and hydrogen peroxide treatment on mature *Listeria monocytogenes* and *Salmonella typhimurium* biofilms. Front Microbiol 10:2674. https://doi.org/10.3389/fmicb.2019.02674

Handorf O, Weihe T, Bekeschus S, Graf AC, Schnabel U, Riedel K, Ehlbeck J (2018) Nonthermal plasma jet treatment negatively affects the viability and structure of *Candida albicans* SC5314 biofilms. Appl Environ Microbiol 84(21). https://doi.org/10.1128/aem.01163-18

Hassett DJ, Ma JF, Elkins JG, McDermott TR, Ochsner UA, West SEH, Huang CT, Fredericks J, Burnett S, Stewart PS, McFeters G, Passador L, Iglewski BH (1999) Quorum sensing in Pseudomonas aeruginosa controls expression of catalase and superoxide dismutase genes and mediates biofilm susceptibility to hydrogen peroxide. Mol Microbiol 34(5):1082–1093. https://doi.org/10.1046/j.1365-2958.1999.01672.x

Hebecker B, Naglik JR, Hube B, Jacobsen ID (2014) Pathogenicity mechanisms and host response during oral *Candida albicans* infections. Expert Rev Anti Infect Ther 12(7):867–879. https://doi.org/10.1586/14787210.2014.916210

Heinlin J, Zimmermann JL, Zeman F, Bunk W, Isbary G, Landthaler M, Maisch T, Monetti R, Morfill G, Shimizu T, Steinbauer J, Stolz W, Karrer S (2013) Randomized placebo-controlled human pilot study of cold atmospheric argon plasma on skin graft donor sites. Wound Repair Regen 21(6):800–807. https://doi.org/10.1111/wrr.12078

Herbst SR, Hertel M, Ballout H, Pierdzioch P, Weltmann KD, Wirtz HC, Abu-Sirhan S, Kostka E, Paris S, Preissner S (2015) Bactericidal efficacy of cold plasma at different depths of infected root canals *in vitro*. Open Dent J 9:486–491. https://doi.org/10.2174/1874210601509010486

Herrmann HW, Henins I, Park J, Selwyn GS (1999) Decontamination of chemical and biological warfare (CBW) agents using an atmospheric pressure plasma jet (APPJ). Phys Plasmas 6(5): 2284–2289. https://doi.org/10.1063/1.873480

Hertel M, Schwill-Engelhardt J, Gerling T, Weltmann K-D, Imiolczyk SM, Hartwig S, Preissner S (2018) Antibacterial efficacy of plasma jet, dielectric barrier discharge, chlorhexidine, and silver diamine fluoride varnishes in caries lesions. Plasma Med 8(1):73–82. https://doi.org/10.1615/PlasmaMed.2018024767

Heuer K, Hoffmanns MA, Demir E, Baldus S, Volkmar CM, Rohle M, Fuchs PC, Awakowicz P, Suschek CV, Oplander C (2015) The topical use of non-thermal dielectric barrier discharge (DBD): nitric oxide related effects on human skin. Nitric Oxide 44:52–60. https://doi.org/10.1016/j.niox.2014.11.015

Hong Q, Dong X, Chen M, Sun H, Hong L, Wang Y, Li H, Yu Q (2019) An *in vitro* and *in vivo* study of plasma treatment effects on oral biofilms. J Oral Microbiol 11(1):1603524. https://doi.org/10.1080/20002297.2019.1603524

Huang WK, Weng CC, Liao JD, Wang YC, Chuang SF (2013) Capillary-tube-based micro-plasma system for disinfecting dental biofilm. Int J Radiat Biol 89(5):364–370. https://doi.org/10.3109/09553002.2013.756594

Hufner A, Steffen H, Holtfreter B, Schluter R, Duske K, Matthes R, von Woedtke T, Weltmann KD, Kocher T, Jablonowski L (2017) Effects of non-thermal atmospheric pressure plasma and sodium hypochlorite solution on *Enterococcus faecalis* biofilm: An investigation in extracted teeth. Plasma Processes Polym 14(3):1600064. https://doi.org/10.1002/ppap.201600064

Ikawa S, Tani A, Nakashima Y, Kitano K (2016) Physicochemical properties of bactericidal plasma-treated water. J Phys D Appl Phys 49(42):425401. https://doi.org/10.1088/0022-3727/49/42/425401

Isbary G, Morfill G, Schmidt HU, Georgi M, Ramrath K, Heinlin J, Karrer S, Landthaler M, Shimizu T, Steffes B, Bunk W, Monetti R, Zimmermann JL, Pompl R, Stolz W (2010) A first prospective randomized controlled trial to decrease bacterial load using cold atmospheric argon plasma on chronic wounds in patients. Br J Dermatol 163(1):78–82. https://doi.org/10.1111/j.1365-2133.2010.09744.x

Isbary G, Heinlin J, Shimizu T, Zimmermann JL, Morfill G, Schmidt HU, Monetti R, Steffes B, Bunk W, Li Y, Klaempfl T, Karrer S, Landthaler M, Stolz W (2012) Successful and safe use of 2 min cold atmospheric argon plasma in chronic wounds: results of a randomized controlled trial. Br J Dermatol 167(2):404–410. https://doi.org/10.1111/j.1365-2133.2012.10923.x

Isbary G, Stolz W, Shimizu T, Monetti R, Bunk W, Schmidt HU, Morfill GE, Klämpfl TG, Steffes B, Thomas HM, Heinlin J, Karrer S, Landthaler M, Zimmermann JL (2013) Cold atmospheric argon plasma treatment may accelerate wound healing in chronic wounds: Results of an open retrospective randomized controlled study *in vivo*. Clin Plasma Med 1(2):25–30. https://doi.org/10.1016/j.cpme.2013.06.001

Jo A, Joh HM, Chung TH, Chung JW (2020) Anticancer effects of plasma-activated medium produced by a microwave-excited atmospheric pressure argon plasma jet. Oxid Med Cell Longev 2020:4205640. https://doi.org/10.1155/2020/4205640

Joaquin JC, Kwan C, Abramzon N, Vandervoort K, Brelles-Marino G (2009) Is gas-discharge plasma a new solution to the old problem of biofilm inactivation? Microbiology 155(3): 724–732. https://doi.org/10.1099/mic.0.021501-0

Joshi SG, Cooper M, Yost A, Paff M, Ercan UK, Fridman G, Friedman G, Fridman A, Brooks AD (2011) Nonthermal dielectric-barrier discharge plasma-induced inactivation involves oxidative DNA damage and membrane lipid peroxidation in Escherichia coli. Antimicrob Agents Chemother 55(3):1053–1062. https://doi.org/10.1128/AAC.01002-10

Jungbauer G, Moser D, Muller S, Pfister W, Sculean A, Eick S (2021) The antimicrobial effect of cold atmospheric plasma against dental pathogens—a systematic review of i*n vitro* studies. Antibiotics (Basel) 10(2). https://doi.org/10.3390/antibiotics10020211

Kalghatgi SU, Fridman G, Cooper M, Nagaraj G, Peddinghaus M, Balasubramanian M, Vasilets VN, Gutsol AF, Fridman A, Friedman G (2007) Mechanism of blood coagulation by nonthermal atmospheric pressure dielectric barrier discharge plasma. IEEE Trans Plasma Sci 35(5): 1559–1566. https://doi.org/10.1109/tps.2007.905953

Kamgang-Youbi G, Herry JM, Meylheuc T, Brisset JL, Bellon-Fontaine MN, Doubla A, Naitali M (2009) Microbial inactivation using plasma-activated water obtained by gliding electric discharges. Lett Appl Microbiol 48(1):13–18. https://doi.org/10.1111/j.1472-765X.2008.02476.x

Kelly-Wintenberg K, Montie TC, Brickman C, Roth JR, Carr AK, Sorge K, Wadsworth LC, Tsai PPY (1998) Room temperature sterilization of surfaces and fabrics with a one atmosphere uniform glow discharge plasma. J Ind Microbiol Biotechnol 20(1):69–74. https://doi.org/10.1038/sj.jim.2900482

Kerlikowski A, Matthes R, Pink C, Steffen H, Schluter R, Holtfreter B, Weltmann KD, von Woedtke T, Kocher T, Jablonowski L (2020) Effects of cold atmospheric pressure plasma and disinfecting agents on *Candida albicans* in root canals of extracted human teeth. J Biophotonics 13(12):e202000221. https://doi.org/10.1002/jbio.202000221

Kieft IE, Laan EPvd, Stoffels E (2004) Electrical and optical characterization of the plasma needle. New J Phys 6:149–149. https://doi.org/10.1088/1367-2630/6/1/149

Kisch T, Helmke A, Schleusser S, Song J, Liodaki E, Stang FH, Mailaender P, Kraemer R (2016) Improvement of cutaneous microcirculation by cold atmospheric plasma (CAP): Results of a controlled, prospective cohort study. Microvasc Res 104:55–62. https://doi.org/10.1016/j.mvr.2015.12.002

Koban I, Matthes R, Hübner N-O, Welk A, Meisel P, Holtfreter B, Sietmann R, Kindel E, Weltmann K-D, Kramer A, Kocher T (2010) Treatment of Candida albicans biofilms with low-temperature plasma induced by dielectric barrier discharge and atmospheric pressure plasma jet. New J Phys 12(7). https://doi.org/10.1088/1367-2630/12/7/073039

Koban I, Holtfreter B, Hubner NO, Matthes R, Sietmann R, Kindel E, Weltmann KD, Welk A, Kramer A, Kocher T (2011) Antimicrobial efficacy of non-thermal plasma in comparison to chlorhexidine against dental biofilms on titanium discs *in vitro* - proof of principle experiment. J Clin Periodontol 38(10):956–965. https://doi.org/10.1111/j.1600-051X.2011.01740.x

Kogelheide F, Voigt F, Hillebrand B, Moeller R, Fuchs F, Gibson AR, Awakowicz P, Stapelmann K, Fiebrandt M (2020) The role of humidity and UV-C emission in the inactivation of *B. subtilis* spores during atmospheric-pressure dielectric barrier discharge treatment. J Phys D Appl Phys 53(29). https://doi.org/10.1088/1361-6463/ab77cc

Kondeti VSSK, Phan CQ, Wende K, Jablonowski H, Gangal U, Granick JL, Hunter RC, Bruggeman PJ (2018) Long-lived and short-lived reactive species produced by a cold atmospheric pressure plasma jet for the inactivation of *Pseudomonas aeruginosa* and *Staphylococcus aureus*. Free Radic Biol Med 124:275–287. https://doi.org/10.1016/j.freeradbiomed.2018.05.083

Kramer B, Hasse D, Guist S, Schmitt-John T, Muranyi P (2019) Inactivation of bacterial endospores on surfaces by plasma processed air. J Appl Microbiol 128(4):920–933. https://doi.org/10.1111/jam.14528

Lamont RJ, Koo H, Hajishengallis G (2018) The oral microbiota: dynamic communities and host interactions. Nat Rev Microbiol 16(12):745–759. https://doi.org/10.1038/s41579-018-0089-x

Laroussi M (2020) Cold plasma in medicine and healthcare: The new frontier in low temperature plasma applications. Front Phys 8. https://doi.org/10.3389/fphy.2020.00074

Laroussi M, Leipold F (2004) Evaluation of the roles of reactive species, heat, and UV radiation in the inactivation of bacterial cells by air plasmas at atmospheric pressure. Int J Mass Spectrom 233(1-3):81–86. https://doi.org/10.1016/j.ijms.2003.11.016

Laurita R, Barbieri D, Gherardi M, Colombo V, Lukes P (2015) Chemical analysis of reactive species and antimicrobial activity of water treated by nanosecond pulsed DBD air plasma. Clin Plasma Med 3(2):53–61. https://doi.org/10.1016/j.cpme.2015.10.001

Lee K-Y, Joo Park B, Hee Lee D, Lee I-S, Hyun SO, Chung K-H, Park J-C (2005) Sterilization of Escherichia coli and MRSA using microwave-induced argon plasma at atmospheric pressure. Surf Coat Technol 193(1–3):35–38. https://doi.org/10.1016/j.surfcoat.2004.07.034

Lee JY, Kim KH, Park SY, Yoon SY, Kim GH, Lee YM, Rhyu IC, Seol YJ (2019) The bactericidal effect of an atmospheric-pressure plasma jet on Porphyromonas gingivalis biofilms on sandblasted and acid-etched titanium discs. J Periodontal Implant Sci 49(5):319–329. https://doi.org/10.5051/jpis.2019.49.5.319

Li Y, Sun K, Ye G, Liang Y, Pan H, Wang G, Zhao Y, Pan J, Zhang J, Fang J (2015) Evaluation of cold plasma treatment and safety in disinfecting 3-week root canal *Enterococcus faecalis* biofilm *in vitro*. J Endod 41(8):1325–1330. https://doi.org/10.1016/j.joen.2014.10.020

Li Y, Pan J, Ye G, Zhang Q, Wang J, Zhang J, Fang J (2017) *In vitro* studies of the antimicrobial effect of non-thermal plasma-activated water as a novel mouthwash. Eur J Oral Sci 125(6): 463–470. https://doi.org/10.1111/eos.12374

Li Y, Pan J, Wu D, Tian Y, Zhang J, Fang J (2019) Regulation of *Enterococcus faecalis* biofilm formation and quorum sensing related virulence factors with ultra-low dose reactive species produced by plasma activated water. Plasma Chem Plasma Process 39(1):35–49. https://doi.org/10.1007/s11090-018-9930-2

Lietz AM, Kushner MJ (2018) Molecular admixtures and impurities in atmospheric pressure plasma jets. J Appl Phys 124(15). https://doi.org/10.1063/1.5049430

Liguori A, Cochis A, Stancampiano A, Laurita R, Azzimonti B, Sorrentino R, Varoni EM, Petri M, Colombo V, Gherardi M, Rimondini L (2017) Cold atmospheric plasma treatment affects early bacterial adhesion and decontamination of soft reline palatal obturators. Clin Plasma Med 7–8: 36–45. https://doi.org/10.1016/j.cpme.2017.08.001

Limoli DH, Jones CJ, Wozniak DJ (2015) Bacterial extracellular polysaccharides in biofilm formation and function. Microbiol Spectr 3(3). https://doi.org/10.1128/microbiolspec.MB-0011-2014

Lis KA, Kehrenberg C, Boulaaba A, von Kockritz-Blickwede M, Binder S, Li Y, Zimmermann JL, Pfeifer Y, Ahlfeld B (2018) Inactivation of multidrug-resistant pathogens and *Yersinia enterocolitica* with cold atmospheric-pressure plasma on stainless-steel surfaces. Int J Antimicrob Agents 52(6):811–818. https://doi.org/10.1016/j.ijantimicag.2018.08.023

Liu ZC, Liu DX, Chen C, Li D, Yang AJ, Rong MZ, Chen HL, Kong MG (2015) Physicochemical processes in the indirect interaction between surface air plasma and deionized water. J Phys D Appl Phys 48(49). https://doi.org/10.1088/0022-3727/48/49/495201

Liu K, Ren W, Ran C, Zhou R, Tang W, Zhou R, Yang Z, Ostrikov K (2021) Long-lived species in plasma-activated water generated by an AC multi-needle-to-water discharge: effects of gas flow on chemical reactions. J Phys D Appl Phys 54(6). https://doi.org/10.1088/1361-6463/abc211

Lu X, Naidis GV, Laroussi M, Reuter S, Graves DB, Ostrikov K (2016) Reactive species in non-equilibrium atmospheric-pressure plasmas: Generation, transport, and biological effects. Phys Rep 630:1–84. https://doi.org/10.1016/j.physrep.2016.03.003

Ma R, Wang G, Tian Y, Wang K, Zhang J, Fang J (2015) Non-thermal plasma-activated water inactivation of food-borne pathogen on fresh produce. J Hazard Mater 300:643–651. https://doi.org/10.1016/j.jhazmat.2015.07.061

Mai-Prochnow A, Murphy AB, McLean KM, Kong MG, Ostrikov KK (2014) Atmospheric pressure plasmas: infection control and bacterial responses. Int J Antimicrob Agents 43(6): 508–517. https://doi.org/10.1016/j.ijantimicag.2014.01.025

Mai-Prochnow A, Bradbury M, Ostrikov K, Murphy AB (2015) *Pseudomonas aeruginosa* biofilm response and resistance to cold atmospheric pressure plasma Is linked to the redox-active molecule phenazine. PLoS One 10(6):e0130373. https://doi.org/10.1371/journal.pone.0130373

Mai-Prochnow A, Clauson M, Hong J, Murphy AB (2016) Gram positive and Gram negative bacteria differ in their sensitivity to cold plasma. Sci Rep 6:38610. https://doi.org/10.1038/srep38610

Marsh PD (2003) Are dental diseases examples of ecological catastrophes? Microbiology (Reading) 149(Pt 2):279–294. https://doi.org/10.1099/mic.0.26082-0

Marsh PD, Head DA, Devine DA (2015) Dental plaque as a biofilm and a microbial community—Implications for treatment. J Oral Biosci 57(4):185–191. https://doi.org/10.1016/j.job.2015.08.002

Matthes R, Koban I, Bender C, Masur K, Kindel E, Weltmann K-D, Kocher T, Kramer A, Hübner N-O (2013) Antimicrobial efficacy of an atmospheric pressure plasma jet against biofilms of *Pseudomonas aeruginosa* and *Staphylococcus epidermidis*. Plasma Processes Polym 10(2): 161–166. https://doi.org/10.1002/ppap.201100133

Matthes R, Assadian O, Kramer A (2014) Repeated applications of cold atmospheric pressure plasma does not induce resistance in *Staphylococcus aureus* embedded in biofilms. GMD Hyg. Infect Control 9(3):Doc17. https://doi.org/10.3205/dgkh000237

Matthes R, Jablonowski L, Koban I, Quade A, Hubner NO, Schlueter R, Weltmann KD, von Woedtke T, Kramer A, Kocher T (2015) *In vitro* treatment of *Candida albicans* biofilms on denture base material with volume dielectric barrier discharge plasma (VDBD) compared with common chemical antiseptics. Clin Oral Investig 19:2319–2326. https://doi.org/10.1007/s00784-015-1463-y

Mirpour S, Fathollah S, Mansouri P, Larijani B, Ghoranneviss M, Mohajeri Tehrani M, Amini MR (2020) Cold atmospheric plasma as an effective method to treat diabetic foot ulcers: A randomized clinical trial. Sci Rep 10(1):10440. https://doi.org/10.1038/s41598-020-67232-x

Modic M, McLeod NP, Sutton JM, Walsh JL (2017) Cold atmospheric pressure plasma elimination of clinically important single- and mixed-species biofilms. Int J Antimicrob Agents 49(3): 375–378. https://doi.org/10.1016/j.ijantimicag.2016.11.022

Molnar I, Papp J, Simon A, Anghel SD (2013) Deactivation of *Streptococcus mutans* biofilms on a tooth surface using He dielectric barrier discharge at atmospheric pressure. Plasma Sci Technol 15(6):535–541. https://doi.org/10.1088/1009-0630/15/6/09

Muro-Fraguas I, Sainz-Garcia A, Lopez M, Rojo-Bezares B, Mugica-Vidal R, Sainz-Garcia E, Toledano P, Saenz Y, Gonzalez-Marcos A, Alba-Elias F (2020) Antibiofilm coatings through atmospheric pressure plasma for 3D printed surgical instruments. Surf Coat Technol 399: 126163. https://doi.org/10.1016/j.surfcoat.2020.126163

Naitali M, Kamgang-Youbi G, Herry JM, Bellon-Fontaine MN, Brisset JL (2010) Combined effects of long-living chemical species during microbial inactivation using atmospheric plasma-treated water. Appl Environ Microbiol 76(22):7662–7664. https://doi.org/10.1128/AEM.01615-10

Nguyen NH, Park HJ, Yang SS, Choi KS, Lee JS (2016) Anti-cancer efficacy of nonthermal plasma dissolved in a liquid, liquid plasma in heterogeneous cancer cells. Sci Rep 6:29020. https://doi.org/10.1038/srep29020

Nicol MJ, Brubaker TR, Honish BJ 2nd, Simmons AN, Kazemi A, Geissel MA, Whalen CT, Siedlecki CA, Bilen SG, Knecht SD, Kirimanjeswara GS (2020) Antibacterial effects of low-temperature plasma generated by atmospheric-pressure plasma jet are mediated by reactive oxygen species. Sci Rep 10(1):3066. https://doi.org/10.1038/s41598-020-59652-6

Noopan S, Unchui P, Techotinnakorn S, Ampornaramveth RS (2019) Plasma sterilization effectively reduces bacterial contamination in dental unit waterlines. Int J Dentist 2019:5720204. https://doi.org/10.1155/2019/5720204

Otter JA, Yezli S, French GL (2011) The role played by contaminated surfaces in the transmission of nosocomial pathogens. Infect Control Hosp Epidemiol 32(7):687–699. https://doi.org/10.1086/660363

Pan J, Sun K, Liang Y, Sun P, Yang X, Wang J, Zhang J, Zhu W, Fang J, Becker KH (2013) Cold plasma therapy of a tooth root canal infected with *Enterococcus faecalis* biofilms *in vitro*. J Endod 39(1):105–110. https://doi.org/10.1016/j.joen.2012.08.017

Pan J, Li YL, Liu CM, Tian Y, Yu S, Wang KL, Zhang J, Fang J (2017) Investigation of cold atmospheric plasma-activated water for the dental unit waterline system contamination and safety evaluation *in vitro*. Plasma Chem Plasma Process 37(4):1091–1103. https://doi.org/10.1007/s11090-017-9811-0

Pankaj SK, Bueno-Ferrer C, Misra NN, O'Neill L, Bourke P, Cullen PJ (2017) Effects of cold plasma on surface, thermal and antimicrobial release properties of chitosan film. J Renew Mater 5(1):14–20. https://doi.org/10.7569/jrm.2016.634105

Park JH, Kim M, Shiratani M, Cho AE, Choi EH, Attri P (2016) Variation in structure of proteins by adjusting reactive oxygen and nitrogen species generated from dielectric barrier discharge jet. Sci Rep 6:35883. https://doi.org/10.1038/srep35883

Park S, Park JY, Choe W (2019) Origin of hydroxyl radicals in a weakly ionized plasma-facing liquid. Chem Eng J 378. https://doi.org/10.1016/j.cej.2019.122163

Preissner S, Wirtz HC, Tietz AK, Abu-Sirhan S, Herbst SR, Hartwig S, Pierdzioch P, Schmidt-Westhausen AM, Dommisch H, Hertel M (2016) Bactericidal efficacy of tissue tolerable plasma on microrough titanium dental implants: An *in-vitro*-study. J Biophotonics 9(6):637–644. https://doi.org/10.1002/jbio.201500189

Puligundla P, Mok C (2017) Potential applications of nonthermal plasmas against biofilm-associated micro-organisms in vitro. J Appl Microbiol 122(5):1134–1148. https://doi.org/10.1111/jam.13404

Puttaiah R, Karpay RI, Fabre C, Sherman LR, Nemeth JF, Mills SE, Plamondon TJ (1998) Dental unit water line treatment with sodium hypochlorite and acetic acid. Microchem J 59(2):333–340. https://doi.org/10.1006/mchj.1998.1589

Reuter S, von Woedtke T, Weltmann KD (2018) The kINPen-a review on physics and chemistry of the atmospheric pressure plasma jet and its applications. J Phys D Appl Phys 51(233001). https://doi.org/10.1088/1361-6463/aab3ad

Rodriguez C, Wandell RJ, Zhang Z, Neurohr JM, Tang Y, Rhodes R, Kinsey ST, Locke BR (2020) *Escherichia coli* survival in plasma-treated water and in a gas–liquid plasma reactor. Plasma Processes Polym 17(12). https://doi.org/10.1002/ppap.202000099

Royintarat T, Seesuriyachan P, Boonyawan D, Choi EH, Wattanutchariya W (2019) Mechanism and optimization of non-thermal plasma-activated water for bacterial inactivation by underwater plasma jet and delivery of reactive species underwater by cylindrical DBD plasma. Curr Appl Phys 19(9):1006–1014. https://doi.org/10.1016/j.cap.2019.05.020

Saitaer X, Sanbhal N, Qiao Y, Li Y, Gao J, Brochu G, Guidoin R, Khatri A, Wang L (2019) Polydopamine-inspired surface modification of polypropylene hernia mesh devices via cold oxygen plasma: antibacterial and drug release properties. Coatings 9(3). https://doi.org/10.3390/coatings9030164

Salgado BAB, Fabbri S, Dickenson A, Hasan MI, Walsh JL (2021) Surface barrier discharges for *Escherichia coli* biofilm inactivation: Modes of action and the importance of UV radiation. PLoS One 16(3):e0247589. https://doi.org/10.1371/journal.pone.0247589

Sanbhal N, Mao Y, Sun G, Xu RF, Zhang Q, Wang L (2018a) Surface modification of polypropylene mesh devices with cyclodextrin via cold plasma for hernia repair: Characterization and antibacterial properties. Appl Surf Sci 439:749–759. https://doi.org/10.1016/j.apsusc.2017.12.192

Sanbhal N, Mao Y, Sun G, Li Y, Peerzada M, Wang L (2018b) Preparation and characterization of antibacterial polypropylene meshes with covalently incorporated beta-cyclodextrins and captured antimicrobial agent for hernia repair. Polymers (Basel) 10(1). https://doi.org/10.3390/polym10010058

Sanbhal N, Li Y, Khatri A, Peerzada M, Wang L (2019) Chitosan cross-linked bio-based antimicrobial polypropylene meshes for hernia repair loaded with levofloxacin HCl via cold oxygen plasma. Coatings 9(3). https://doi.org/10.3390/coatings9030168

Schaudinn C, Jaramillo D, Freire MO, Sedghizadeh PP, Nguyen A, Webster P, Costerton JW, Jiang C (2013) Evaluation of a nonthermal plasma needle to eliminate ex vivo biofilms in root canals of extracted human teeth. Int Endod J 46(10):930–937. https://doi.org/10.1111/iej.12083

Schnabel U, Handorf O, Stachowiak J, Boehm D, Weit C, Weihe T, Schäfer J, Below H, Bourke P, Ehlbeck J (2020) Plasma-functionalized water: from bench to prototype for fresh-cut lettuce. Food Eng Rev 13(1):115–135. https://doi.org/10.1007/s12393-020-09238-9

Schneider S, Lackmann JW, Narberhaus F, Bandow JE, Denis B, Benedikt J (2011) Separation of VUV/UV photons and reactive particles in the effluent of a He/O2atmospheric pressure plasma jet. J Phys D Appl Phys 44(37). https://doi.org/10.1088/0022-3727/44/37/379501

Scholtz V, Pazlarova J, Souskova H, Khun J, Julak J (2015) Nonthermal plasma-A tool for decontamination and disinfection. Biotechnol Adv 33(6):1108–1119. https://doi.org/10.1016/j.biotechadv.2015.01.002

Semmler ML, Bekeschus S, Schafer M, Bernhardt T, Fischer T, Witzke K, Seebauer C, Rebl H, Grambow E, Vollmar B, Nebe JB, Metelmann HR, Woedtke TV, Emmert S, Boeckmann L (2020) Molecular mechanisms of the efficacy of cold atmospheric pressure plasma (CAP) in cancer treatment. Cancers (Basel) 12(2). https://doi.org/10.3390/cancers12020269

Sharma A, Collins G, Pruden A (2009) Differential gene expression in *Escherichia coli* following exposure to nonthermal atmospheric pressure plasma. J Appl Microbiol 107(5):1440–1449. https://doi.org/10.1111/j.1365-2672.2009.04323.x

Sheiham A, James WPT (2014) A reappraisal of the quantitative relationship between sugar intake and dental caries: the need for new criteria for developing goals for sugar intake. BMC Public Health 14. https://doi.org/10.1186/1471-2458-14-863

Shekhter AB, Serezhenkov VA, Rudenko TG, Pekshev AV, Vanin AF (2005) Beneficial effect of gaseous nitric oxide on the healing of skin wounds. Nitric Oxide 12(4):210–219. https://doi.org/10.1016/j.niox.2005.03.004

Shen J, Tian Y, Li Y, Ma R, Zhang Q, Zhang J, Fang J (2016) Bactericidal effects against *S. aureus* and physicochemical properties of plasma activated water stored at different temperatures. Sci Rep 6:28505. https://doi.org/10.1038/srep28505

Silk H (2014) Diseases of the mouth. Prim Care 41(1):75–90. https://doi.org/10.1016/j.pop.2013.10.011

Stoffels E, Kieft IE, Sladek REJ (2003) Superficial treatment of mammalian cells using plasma needle. J Phys D Appl Phys 36(23):2908–2913. https://doi.org/10.1088/0022-3727/36/23/007

Stoleru E, Dumitriu RP, Munteanu BS, Zaharescu T, Tănase EE, Mitelut A, Ailiesei G-L, Vasile C (2016) Novel procedure to enhance PLA surface properties by chitosan irreversible immobilization. Appl Surf Sci 367:407–417. https://doi.org/10.1016/j.apsusc.2016.01.200

Stratmann B, Costea TC, Nolte C, Hiller J, Schmidt J, Reindel J, Masur K, Motz W, Timm J, Kerner W, Tschoepe D (2020) Effect of cold atmospheric plasma therapy vs standard therapy

placebo on wound healing in patients with diabetic foot ulcers: A randomized clinical trial. JAMA Netw Open 3(7):e2010411. https://doi.org/10.1001/jamanetworkopen.2020.10411
Taghizadeh L, Brackman G, Nikiforov A, van der Mullen J, Leys C, Coenye T (2015) Inactivation of biofilms using a low power atmospheric pressure argon plasma jet; the role of entrained nitrogen. Plasma Processes Polym 12(1):75–81. https://doi.org/10.1002/ppap.201400074
Tarabova B, Lukes P, Hammer MU, Jablonowski H, von Woedtke T, Reuter S, Machala Z (2019) Fluorescence measurements of peroxynitrite/peroxynitrous acid in cold air plasma treated aqueous solutions. Phys Chem Chem Phys 21(17):8883–8896. https://doi.org/10.1039/c9cp00871c
Tendero C, Tixier C, Tristant P, Desmaison J, Leprince P (2006) Atmospheric pressure plasmas: A review. Spectrochim Acta Part B Atomic Spectrosc 61(1):2–30. https://doi.org/10.1016/j.sab.2005.10.003
Thirumdas R, Kothakota A, Annapure U, Siliveru K, Blundell R, Gatt R, Valdramidis VP (2018) Plasma activated water (PAW): Chemistry, physico-chemical properties, applications in food and agriculture. Trends Food Sci Technol 77:21–31. https://doi.org/10.1016/j.tifs.2018.05.007
Tian Y, Ma R, Zhang Q, Feng H, Liang Y, Zhang J, Fang J (2015) Assessment of the physicochemical properties and biological effects of water activated by non-thermal plasma above and beneath the water surface. Plasma Processes Polym 12(5):439–449. https://doi.org/10.1002/ppap.201400082
Trompeter FJ, Neff WJ, Franken O, Heise M, Neiger M, Shuhai L, Pietsch GJ, Saveljew AB (2002) Reduction of *Bacillus subtilis* and *Aspergillus niger* spores using nonthermal atmospheric gas discharges. IEEE Trans Plasma Sci 30(4):1416–1423. https://doi.org/10.1109/tps.2002.804182
Ulrich C, Kluschke F, Patzelt A, Vandersee S, Czaika VA, Richter H, Bob A, von Hutten J, Painsi C, Hüge R, Kramer A, Assadian O, Lademann J, Lange-Asschenfeldt B (2015) Clinical use of cold atmospheric pressure argon plasma in chronic leg ulcers: A pilot study. J Wound Care 24(5):196, 198–200, 202–193. https://doi.org/10.12968/jowc.2015.24.5.196
Ureyen Kaya B, Kececi AD, Guldas HE, Cetin ES, Ozturk T, Oksuz L, Bozduman F (2014) Efficacy of endodontic applications of ozone and low-temperature atmospheric pressure plasma on root canals infected with Enterococcus faecalis. Lett Appl Microbiol 58(1):8–15. https://doi.org/10.1111/lam.12148
Vandervoort KG, Abramzon N, Brelles-Marino G (2008) Plasma interactions with bacterial biofilms as visualized through atomic force microscopy. IEEE Trans Plasma Sci 36(4):1296–1297. https://doi.org/10.1109/tps.2008.917953
von Woedtke T, Emmert S, Metelmann H-R, Rupf S, Weltmann K-D (2020) Perspectives on cold atmospheric plasma (CAP) applications in medicine. Phys Plasmas 27(7):070601. https://doi.org/10.1063/5.0008093
Wang Q, Salvi D (2021) Evaluation of plasma-activated water (PAW) as a novel disinfectant: Effectiveness on *Escherichia coli* and *Listeria innocua*, physicochemical properties, and storage stability. LWT Food Sci Technol 149. https://doi.org/10.1016/j.lwt.2021.111847
Whitehead-Clarke T, Windsor A (2021) Surgical site infection: The scourge of abdominal wall reconstruction. Surg Infect (Larchmt) 22(4):357–362. https://doi.org/10.1089/sur.2020.325
Whiteley M, Diggle SP, Greenberg EP (2017) Progress in and promise of bacterial quorum sensing research. Nature 551(7680):313–320. https://doi.org/10.1038/nature24624
Xu Z, Shen J, Zhang Z, Ma J, Ma R, Zhao Y, Sun Q, Qian S, Zhang H, Ding L, Cheng C, Chu PK, Xia W (2015) Inactivation effects of non-thermal atmospheric-pressure helium plasma jet on *Staphylococcus aureus* biofilms. Plasma Processes Polym 12(8):827–835. https://doi.org/10.1002/ppap.201500006
Xu Z, Zhou X, Yang W, Zhang Y, Ye Z, Hu S, Ye C, Li Y, Lan Y, Shen J, Ye X, Yang F, Cheng C (2020) *In vitro* antimicrobial effects and mechanism of air plasma-activated water on *Staphylococcus aureus* biofilm. Plasma Processes Polym 17(8). https://doi.org/10.1002/ppap.201900270
Yan DY, Sherman JH, Keidar M (2017) Cold atmospheric plasma, a novel promising anti-cancer treatment modality. Oncotarget 8(9):15977–15995. https://doi.org/10.18632/oncotarget.13304

Yang B, Chen JR, Yu QS, Li H, Lin MS, Mustapha A, Hong LA, Wang Y (2011) Oral bacterial deactivation using a low-temperature atmospheric argon plasma brush. J Dent 39(1):48–56. https://doi.org/10.1016/j.jdent.2010.10.002

Yang Y, Guo J, Zhou X, Liu Z, Wang C, Wang K, Zhang J, Wang Z (2018) A novel cold atmospheric pressure air plasma jet for peri-implantitis treatment: An *in vitro* study. Dent Mater J 37(1):157–166. https://doi.org/10.4012/dmj.2017-030

Zhang Q, Liang Y, Feng H, Ma R, Tian Y, Zhang J, Fang J (2013) A study of oxidative stress induced by non-thermal plasma-activated water for bacterial damage. Appl Phys Lett 102(20): 203701. https://doi.org/10.1063/1.4807133

Zhang Q, Ma R, Tian Y, Su B, Wang K, Yu S, Zhang J, Fang J (2016) Sterilization efficiency of a novel electrochemical disinfectant against *Staphylococcus aureus*. Environ Sci Technol 50(6): 3184–3192. https://doi.org/10.1021/acs.est.5b05108

Zhang X, Zhou R, Bazaka K, Liu Y, Zhou R, Chen G, Chen Z, Liu Q, Yang S, Ostrikov KK (2018) Quantification of plasma produced OH radical density for water sterilization. Plasma Processes Polym 15(6):1700241. https://doi.org/10.1002/ppap.201700241

Zhang H, Tian Y, Kang M, Chen C, Song Y, Li H (2019) Effects of chlorination/chlorine dioxide disinfection on biofilm bacterial community and corrosion process in a reclaimed water distribution system. Chemosphere 215:62–73. https://doi.org/10.1016/j.chemosphere.2018.09.181

Zhao YM, Patange A, Sun DW, Tiwari B (2020a) Plasma-activated water: Physicochemical properties, microbial inactivation mechanisms, factors influencing antimicrobial effectiveness, and applications in the food industry. Compr Rev Food Sci Food Saf 19(6):3951–3979. https://doi.org/10.1111/1541-4337.12644

Zhao YM, Ojha S, Burgess CM, Sun DW, Tiwari BK (2020b) Inactivation efficacy of plasma-activated water: influence of plasma treatment time, exposure time and bacterial species. Int J Food Sci Technol 56(2):721–732. https://doi.org/10.1111/ijfs.14708

Zhou R, Zhou R, Prasad K, Fang Z, Speight R, Bazaka K, Ostrikov K (2018) Cold atmospheric plasma activated water as a prospective disinfectant: the crucial role of peroxynitrite. Green Chem 20(23):5276–5284. https://doi.org/10.1039/c8gc02800a

Zhou R, Zhou R, Wang P, Luan B, Zhang X, Fang Z, Xian Y, Lu X, Ostrikov KK, Bazaka K (2019) Microplasma bubbles: Reactive vehicles for biofilm dispersal. ACS Appl Mater Interfaces 11(23):20660–20669. https://doi.org/10.1021/acsami.9b03961

Zhou R, Zhou R, Wang P, Xian Y, Mai-Prochnow A, Lu X, Cullen PJ, Ostrikov K, Bazaka K (2020) Plasma-activated water: generation, origin of reactive species and biological applications. J Phys D Appl Phys 53(30):303001. https://doi.org/10.1088/1361-6463/ab81cf

Zimmermann JL, Shimizu T, Schmidt HU, Li YF, Morfill GE, Isbary G (2012) Test for bacterial resistance build-up against plasma treatment. New J Phys 14(7). https://doi.org/10.1088/1367-2630/14/7/073037

Ziuzina D, Boehm D, Patil S, Cullen PJ, Bourke P (2015) Cold plasma inactivation of bacterial biofilms and reduction of quorum sensing regulated virulence factors. PLoS One 10(9): e0138209. https://doi.org/10.1371/journal.pone.0138209

Antibodies Against Biofilms: Mechanisms and Applications

Avia Watson, Harsh Maan, Ilana Kolodkin-Gal, and Natalia T. Freund

Abstract Biofilms shield microbes from the surrounding environment, enabling them to spread and colonialize. Therefore, they represent a significant economic and medical burden. Antibodies are extensively used in the clinic to treat various diseases, such as cancer and autoimmune disorders, and were suggested as a potential solution to cope with biofilms. The major advantages of antibody therapy are in providing immediate protection which remains in the blood for weeks to months, high specificity with low toxicity while not harming the gut microbiota, making antibodies an appealing therapeutic regiment. Several promising anti-biofilm antibodies are in pre-clinical and clinical evaluations, heralding the beginning of a new era in biofilm therapeutics.

1 Introduction to Antibodies and B Cells

Antibodies (also known as immunoglobulins) are naturally produced proteins that exhibit exquisite specificity and diversity, and therefore hold limitless possibilities as reagents, diagnostic tools, therapeutics, and immunomodulators. Antibodies can also be used as guides for vaccine design and as correlates of protection. After reviewing the main traits of antibodies in the following chapter, we will discuss antibodies against biofilm components that show anti-biofilm activity in vitro, ex vivo, or in vivo and the application of antibodies as potential treatments against biofilms.

A. Watson · N. T. Freund (✉)
Department of Clinical Microbiology and Immunology, Sackler Faculty of Medicine, Tel Aviv University, Tel Aviv, Israel
e-mail: nfreund@tauex.tau.ac.il

H. Maan
Department of Molecular Genetics, Weizmann Institute of Science, Rehovot, Israel

I. Kolodkin-Gal
Department of Molecular Genetics, Weizmann Institute of Science, Rehovot, Israel

Department of Plant Pathology and Microbiology, The Robert H. Smith Faculty of Agriculture, Food & Environment, The Hebrew University of Jerusalem, Jerusalem, Israel

K. Richter, K. Nørskov Kragh (eds.), *Antibiofilm Strategies*, Springer Series on Biofilms 11, https://doi.org/10.1007/978-3-031-10992-8_11

1.1 Antibody Structure and Functional Classification

Several reviews describe the structure of an antibody (Janeway et al. 2001b; Chiu et al. 2019). Antibodies are large Y-shaped proteins with a molecular weight of approximately 150 kDa. There are several ways to classify antibodies. Structurally, a single molecule is comprised of four peptide chains; two identical heavy chains, and two identical light chains connected to each other through a series of disulfide bonds. Functionally, an antibody can be divided into three regions; two identical antigen-binding domains (called Fab domains, together F(ab)2), and a fragment crystallizable (Fc) domain that is responsible for antibody effector functions. The Fc domain is connected to the Fab parts by a flexible "hinge" region (Fig. 1a, b). Another genetic and mechanistic classification of an antibody is into "variable" and

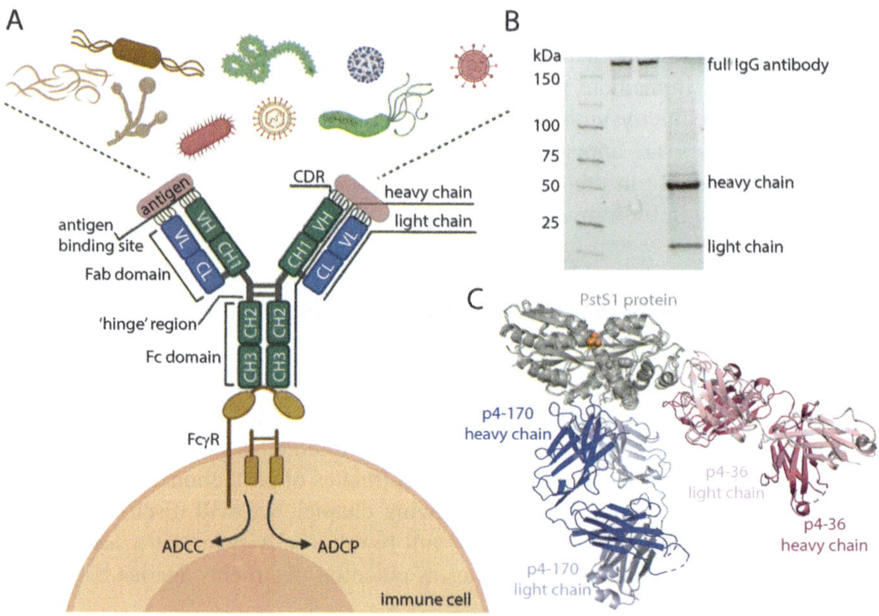

Fig. 1 IgG antibody classification. (**a**) Schematic illustration of an antibody molecule, IgG isotype. Two identical heavy chains (teal) and two identical light chains (blue) form two antigen-binding domains and are linked by a "hinge" region. Within the Fab antigen-binding site, the three complementary-determining regions (CDRs). The Fc domain is responsible for activating other immune cells, that in turn can activate antibody-dependent cell cytotoxicity (ADCC) or cell phagocytosis (ADCP) through an FcγR (indicated in mustard). Illustration was created with https://BioRender.com. (**b**) Protein gel demonstrating the size of a fully human IgG antibody. Lanes from left to right: 1—Marker, 2,3—Two different IgG molecules, 4—Unpaired heavy and light chains of an IgG molecule. (**c**) Atomic structure obtained by X-ray crystallography of two Fab fragments of two different antibodies (p4-36 mAb (PDB: 7DM1) in magenta shades, and p4-170 mAb (PDB: 7DM2) in blue shades) binding two non-overlapping sites on Mycobacterium tuberculosis antigen PstS1 (depicted in gray, with phosphate atom is in orange), (PDB: 1PC3). The two structures are super-imposed on one another. Illustration was created using PyMOL software

"constant" domains. The variable fragments are located in the N-terminal part of each chain and define the antibody's specificity. The variable fragments are termed VH and VL, for the variable part of the heavy and light chains, respectively. These domains are created via a unique somatic DNA editing and recombination process that takes place in the bone marrow during the early stages of B cell development. In this process, termed "V(D)J recombination," a functional VDJ cassette of the heavy chain is joined with a functional VJ cassette of the light chain to form the complete antibody binding site (Janeway et al. 2001a). Within the binding site, there is a six-loop finger-like structure contributed by the three complementary-determining regions (CDRs) of the heavy and light chains, representing the most variable parts of the VH and VL. The constant domain is located at the C-terminus of the light and heavy chains (CL and CH, respectively). The C-terminus of the heavy chain also determines whether the antibody is in its membrane-bound or soluble form. The constant parts of the heavy chain also determine the antibody isotype, which corresponds to the plethora of functions the antibody fulfills. In humans, antibodies can assume one of the following isotypes IgM, IgD, IgA, IgE, or IgG, with IgG being the most prevalent in human serum. The light chains do not define the antibody isotype and are encoded by two different genes, becoming either Kappa (κ) or Lambda (λ) light chain types. The mechanism by which antibodies neutralize or eliminate pathogens is related to both the Fab and Fc functional regions and are described in the next paragraphs, as well as in these reviews (Forthal 2014; van Erp et al. 2019; Li and Javid 2018; Lu et al. 2018).

1.2 Antibodies Are Naturally Produced by B Cells

B lymphocytes (B cells) development is the basis for antibodies (Akkaya et al. 2020; Gitlin et al. 2015). B cells originate from progenitor cells derived from hematopoietic stem cells in the bone marrow and fetal liver. In the bone marrow, B cells rearrange their VH and VL gene segments to display a surface antibody, often termed as B cell receptor (BCR). BCRs first rearrange into IgM and IgD isotypes. B cells are negatively selected for reactivity to self-antigens within the bone marrow, a crucial step for preventing autoimmunity (Wardemann et al. 2003). Mature naïve B cells exit the bone marrow and circulate in the body until they encounter an antigen that fits their BCR binding site profile. The naive B cell's encounter with an antigen results in a sequence of events. First, the antigen:BCR complex is internalized and the antigen is degraded by the proteosome into peptides which are presented on B cell MHC Class II complexes to T helper cells. The B cell then becomes activated and is directed to the T:B border, where it proliferates and differentiates. New immune structures are formed within the secondary lymphoid organs, named germinal centers (GCs), where B cells are improved by mutations and selected by T cells for the B cells carrying the highest affinity antibodies. The GC-graduating B cells become short-lived or long-lived plasma cells secreting antigen-specific monoclonal antibodies (mAbs), and memory B cells that are the basis for immunological memory and vaccines (Gitlin et al. 2015; Victora and Nussenzweig 2012). Naturally produced antibodies can be isolated using a "baiting" strategy where target-specific B cells are single-cell sorted, and their V(D)J sequences (both VH and VL) are amplified by two rounds of polymerase chain reactions (PCR) and subsequently cloned, to generate human mAbs (Tiller et al. 2008; Freund et al. 2015; Liao et al. 2013).

1.3 Monoclonal Antibodies Can Be Produced Synthetically upon Request

Besides being naturally produced, mAbs can also be synthetically produced using various techniques (Lu et al. 2020). The first successful attempt was by fusing spleen cells from immunized mice and a myeloma cell line resulting in hybridomas that secrete antigen-specific mAbs (Kohler and Milstein 1975). However, clinical use of murine mAbs revealed several limitations such as their incompatibility with human complement and cellular systems, as well as an anti-therapy response. To decrease toxicity, the murine antibody content was reduced using genetic engineering techniques. Combination of the murine variable and the human constant domains or grafting the murine CDRs within the binding site to a human antibody backbone results in chimeric (Boulianne et al. 1984) or humanized mAbs (Jones et al. 1986) respectively. Alternatively, human synthetic mAbs can be generated by incorporating human antibody gene fragments, such as single-chain fragment variable (scFv) or Fab, on the surface of filamentous phages creating phage-display peptide libraries allowing the selection of strong binders (McCafferty et al. 1990). Another technology is target-immunization of transgenic mice which involves genetic manipulations, replacing the endogenous murine Ig genes with their corresponding human genes, allowing the transgenic mice to produce human antibodies upon immunization (Lonberg et al. 1994). Antibodies can be further engineered by mutagenesis design to improve their affinity, half-life, stability, and Fc effector function (Presta 2008).

2 Mechanisms of Antibody-Mediated Functions

Antibodies have many functions by which they eliminate pathogens, kill infected or cancerous cells and either promote, or downregulate inflammation. While the antibody's activity is dependent on both its Fab and Fc parts, some functions rely more on one or the other.

2.1 Fab-Mediated Pathogen Blocking

Physical blocking is the most straightforward way by which antibodies neutralize pathogens. This is largely observed in viral infections, where antibodies bind to critical sites of the viral envelope, thus preventing the viral proteins from binding to their corresponding cellular receptors, subsequently blocking the infection (Mor et al. 2021; Robbiani et al. 2020; Scheid et al. 2011). Antibodies can also bind to viral sites that block conformational rearrangements needed for membrane fusion, subsequent to the primary attachment step (Huang et al. 2012). Whether by blocking the attachment step or the subsequent rearrangements and fusion, it is mostly the F (ab')2 binding that prevents viral entry to the host cells. These mechanisms have been defined structurally. For example, the neutralizing antibody VRC01 binds to the HIV-1 gp120 fragment, thereby preventing the binding to its target CD4 receptor on human cells (Zhou et al. 2010). Similarly, antibodies that bind the receptor-binding domain (RBD) located on the SARS Coronavirus 2 (SARS-CoV-2) prevent the entry of viruses through angiotensin-converting enzyme 2 (ACE2) to the host cells (Barnes et al. 2020). Leading examples of antiviral neutralizing antibodies are the clinically approved mAbs, anti-Ebola virus, Ansuvimab-zykl (mAb114) that targets the EBOV glycoprotein (Corti et al. 2016), and anti-SARS Coronavirus mAbs, Casirivimab and Imdevimab by Regeneron Pharmaceuticals that target the SARS-CoV-2 Spike protein RBD (Razonable et al. 2021; Baum et al. 2020), Bamlanivimab plus Etesevimab by Lilly (Dougan et al. 2021), and Sotrovimab by GSK (Cathcart et al. 2021). More examples that are still in the clinical stages of development are anti-Zika (Wang et al. 2017a), anti-HBV (Gao et al. 2017; Wang et al. 2020), and anti-HIV-1 mAbs (Caskey et al. 2016, 2019).

Toxins released by pathogenic bacteria, such as *Clostridium difficile* and *Bacillus anthracis*, are classical targets for antibodies, which prevent toxin-mediated killing of the host cells (Navalkele and Chopra 2018; Greig 2016). Two antibacterial toxin mAbs have been clinically approved, the human obiltoxaximab, treating inhalational anthrax due to *Bacillus anthracis* (Greig 2016) and bezlotoxumab, a humanized mAb, targeting *Clostridium difficile* toxin B, and preventing its attachment to the host cells (Navalkele and Chopra 2018). Antibodies can also inhibit parasites, such as *Plasmodium falciparum,* that cause malaria (Kisalu et al. 2018).

2.2 Fc-Mediated Pathogen Blocking

Direct blocking of the pathogen is not always possible. In many cases, pathogens utilize several cellular receptors for host cell entry, and thus blocking one pathogenic protein might not be enough to prevent infection. Additionally, antibodies have other roles, such as mediating the destruction of cancerous and infected cells by marking those cells to be destroyed by other components of the immune system. The antibody's Fc region links antibodies, which are considered humoral adaptive components, with other cellular and innate arms of the immune system. When the antibody's F(ab)2 domains bind their target, they create an "immune complex". The Fc region is stabilized and can now be recognized by a collection of Fc receptors (FcRs) on the host cell surface of various immune cells (Ravetch and Clynes 1998). This binding of immune complexes to cells through the FcR can lead to a cascade that will trigger cellular activation or inhibition. One of the leading examples for such antibody-mediated cell activation is **antibody-dependent cell cytotoxicity (ADCC)**, which is mediated through interactions of the Fc domain of an antibody-antigen complex with FcγRs on the cell surface of innate immune cells such as natural killer (NK) cells, neutrophils, monocytes, and macrophages. These cells can kill virally infected or cancerous cells by a massive release of cytotoxins to the adjacent cell. However, these immune cells themselves lack specificity and cannot tell harmful from normal cells. Another mechanism that results in opsonization of the immune complex is relevant when large pathogens, such as bacteria are captured by antibodies. **Antibody-dependent cellular phagocytosis (ADCP)** is an important mechanism by which antibody-bacteria complexes are internalized by phagocytic cells, directed to the lysosome, followed by degradation and antigen processing and presentation to T cells. ADCP was found to be much faster and 4000-fold more efficient than regular opsonization that is not antibody-mediated. Antibodies are small compared to cells, but in some cases, with the help of their Fc, they can kill large cells that are either infected or cancer transformed. This is done through **complement-dependent cytotoxicity (CDC)**, where antibodies activate a complement cascade resulting in the assembly of proteins around the Fc region and the formation of pores in the cell membrane by the membrane attack complex (MAC). FcγRs can also be inhibitory, resulting in a negative feedback loop that aims to balance inflammation and immunity (Vogelpoel et al. 2015).

3 Antibacterial Antibodies in the Clinic and Biotech

Currently, there are only a handful of clinically approved mAbs for the therapy of infectious diseases. However, many promising antibodies are in various stages of clinical research and might become clinically available in the near future (Lu et al. 2020). Unlike viruses that encode a limited number of antigens that are produced by the host cells, bacteria make their own genetic material and proteins, which usually requires encoding hundreds to thousands of different proteins, which are then post-translationally modified in a unique fashion. Therefore, the complex nature of bacteria presents a far bigger challenge for mAbs therapy. For many years, antibiotics served as the optimal "magic bullet" for treating bacteria, presenting little need for alternative approaches. With the presence of antibiotic-resistant bacteria and the need for new antibacterial approaches, several antibacterial antibodies are in development. A case in point is the two fully human mAbs isolated from an active tuberculosis-infected patient, both targeting two non-overlapping sites on the phosphate-binding *Mycobacterium tuberculosis* protein, PstS1 (Fig. 1c). The mAbs inhibit *Mycobacterium tuberculosis* both in vitro and in vivo in an Fc-dependent manner (Watson et al. 2021). MAB1, a murine mAb targeting BamA, the main component of the *Escherichia coli (E-coli)* β-barrel assembly machine, promotes *E-coli* killing (Storek et al. 2018). Another example is the anti-*Salmonella Typhimurium* (*S. Typhimurium*) mAb-449, which is protective in mice by promoting phagocytosis (Aribam et al. 2016). Other examples are mAbs against *Yersinia pestis*, the causative agent of the bubonic plague (Xiao et al. 2010), anti-*Vibrio cholerae* (Charles et al. 2020) and anti-tetanus toxin (Wang et al. 2021).

4 Antibodies as Weapons Against Biofilm

In addition to the genetic and structural complexity that microbial organisms present, many of them add another layer of complexity by forming composite differentiated multicellular communities known as **biofilms**. The biofilm structure provides many benefits to its community members. In vitro, the 3D structure of the biofilm was suggested to relieve metabolic stress. For example, channels formed below the ridges and wrinkles within the colony may facilitate the diffusion of fluids, nutrients, and oxygen (Wilking et al. 2013; Dietrich et al. 2013; Bloom-Ackermann et al. 2016; Kolodkin-Gal et al. 2013). Furthermore, cells located in different areas of the colony are exposed to different levels of oxygen, nutrients, and quorum-sensing molecules, allowing biofilm cells to display heterogeneous capacities and to endure stressful environments (Stewart and Franklin 2008; Monds and O'Toole 2009; Serra and Hengge 2014; Hassanov et al. 2018; Asally et al. 2012; Liu et al. 2015). In vivo, preliminary evidence suggests the potential for similar metabolic differentiation, frequently associated with polymicrobial biofilms. Growth in a murine thigh abscess model showed that *Streptococcus gordonii* produces L-lactate, which serves as a carbon source for *Aggregatibacter actinomycetemcomitans* and thus increases its pathogenicity (Ramsey et al. 2011). Furthermore, biofilms of *Bacillus subtilis* formed over their host *Arabidopsis thaliana* are differentiated to motile cells, matrix producers and non-matrix producers, a division of labor which is affected by plant exudations and cues (Beauregard et al. 2013; Allard-Massicotte et al. 2016).

Bacterial biofilms are clinically important, being inherently associated with many persistent and chronic bacterial infections (Costerton et al. 1999). Biofilm-forming bacteria present a major medical problem by contaminating biological, and non-biological surfaces. When formed on biological surfaces, biofilms can be the causative agents of various chronic and recurrent infections such as otitis media (OM), pneumonia, keratitis as well as burn and wound-related infections (Costerton et al. 1999). When, biofilms are formed on non-biological surfaces, for example medical devices, such as heart valves, catheters, and implants, they can lead to serious consequences requiring hospitalization, prolonged treatment, and sometimes the replacement of the medical device (Romling et al. 2014; Wi and Patel 2018). Moreover, biofilms can be the cause of antibiotics tolerance and resistance (Bryers 2008; Hill et al. 2005; Lewis 2007), with bacteria in biofilm being up to 1000 times more resistant than planktonic (free-living) bacteria (Bryers 2008; Ceri et al. 1999), and with an increased spread of antibiotic resistance genes throughout the biofilms (Flores-Vargas et al. 2021). The mechanisms supporting this phenotypic resistance are still poorly understood (Costerton et al. 1999; Oppenheimer-Shaanan et al. 2013). Due to the growing problem of biofilm-forming bacteria and the need for new strategies to disrupt biofilms, antibodies were suggested as a potential solution. Antibodies can disrupt biofilms by either preventing biofilm formation or by inducing an active degradation of already established biofilms. Similar to antiviral and antibacterial antibodies, the main challenge for designing anti-biofilm antibodies is finding a biofilm structural component that is conserved among multiple species and can be readily targeted. In the following paragraphs, we will review several antibodies that exhibit anti-biofilm functions and discuss their functionality in vitro and in vivo, and their potential benefits for proceeding to clinical trials. One should note, that in some cases antibodies can also exhibit detrimental effects. Studies reported that by promoting bacterial attachment and induction of key steps in biofilm formation antibodies can sometimes serve as a double edge sword (Sheu et al. 2007; Lam et al. 2014). Therefore, careful selection of antibodies and their targets is crucial.

One key step in biofilm formation is extracellular matrix (ECM) generation. Like multicellular eukaryotes, bacterial cells in a biofilm are surrounded by self-produced polymers that constitute the ECM, which binds them to each other and to the surface. The ECM is comprised of extracellular polymeric substance (EPS), which vary in composition among different microbes. EPS can be functionally divided into **cell surface associated components** such as proteins and polysaccharides decorating the cell wall, bacterial pili and adhesins, and to **secreted extracellular components,** such as nucleic acids binding proteins, exopolysaccharides, and proteins contributing to the matrix architecture (Fig. 2) (Flemming et al. 2016; Karygianni et al. 2020).

5 Anti-biofilm Antibodies Targeting Bacterial Cell Surface Associated Components

Bacterial adherence is the first step in biofilm formation, enabling microbial transformation from a free state to a sessile form. Bacteria express numerous inducible biofilm proteins on their membrane and cell wall, which function as adhesins and promote attachment to biotic and abiotic surfaces. In the following paragraphs, we will review the different adhesins, other proteins and polysaccharides that can be targeted by antibodies which are summarized in Table 1.

Type IV pili (T4P) are long and uniquely thin appendages and are composed of dynamic filaments that rapidly polymerize and depolymerize from a pool of pilin subunits such as the pilus structural protein PilA (Craig et al. 2019). Frequently, these adhesins provide the reversible attachment that is an important component of successful biofilm formation (Craig et al. 2019) and are therefore appealing targets for antibodies. Rabbit polyclonal anti-PilA antibodies inhibit biofilm formation by

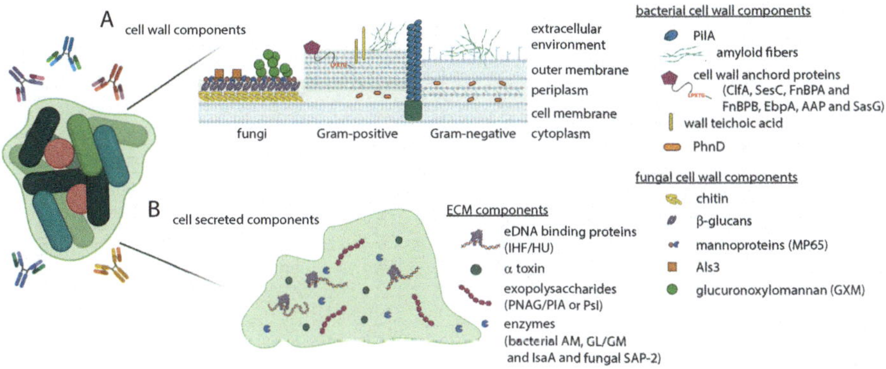

Fig. 2 Bacterial and fungal targets for anti-biofilm antibodies. A schematic illustration of the molecular composition of Biofilms divided into (**a**) cell wall and (**b**) cell secreted components. (**a**) A schematic cell wall structure presentation of fungi, Gram-positive and Gram-negative bacteria and the main components mentioned in the chapter embedded in their cell wall and being targets for anti-biofilm antibodies. Gram-positive and Gram-negative bacteria comprise with pilin subunit **PilA**, the building block of the filamentous Type IV pili, **amyloid** structures that extend from the cell wall into the biofilm matrix, and **PhnD** periplasmic protein. Proteins that are typical to Gram-positive bacteria are: wall teichoic acid (**WTA**) and the cell wall anchored (CWA) proteins such as the fibrinogen binding proteins, clumping factor A (**ClfA**), surface protein C (**SesC**), fibronectin-binding protein A and B (**FnBPA** or **FnBPB**), Endocarditis biofilm-associated pilus A subunit (**EbpA**), accumulation associated protein (**AAP**) or the surface protein G (**SasG**). Antibodies can also be directed against fungal cellular components proteins such as **Als3**, **MP65**, and the polysaccharide glucuronoxylomannan (**GXM**). (**b**) A schematic presentation of the biofilm ECM and the bacterial and fungal secreted components within it. Those components are the target for anti-biofilm antibodies. ECM components that mentioned in this chapter in a bacterial context are: eDNA binding proteins (**IHF/HU**), exopolysaccharides such as **PNAG/PIA** and **PsI**, extracellular enzymes like amidase (**AM**), glucosaminidase (**GL** or **GM**) and **IsaA** or toxins such as **α-toxin** hemolysin. ECM components mentioned in fungal context are aspartic proteases (**SAP-2**) and the secreted and cellular **β-glucans**. Illustration was created with https://BioRender.com

non-typeable *Haemophilus influenza* (NTHI), and disrupt already established biofilm in vitro, showing reduced biomass accompanied by transiently released bacteria. In the chinchilla NTHI OM infection model, immunization with PilA results in an increased recovery, accompanied with reduced levels of free bacteria and mucosal biofilms. A synergistic effect was obtained following dual immunization with PilA and the DNA binding protein, integration host factor (IHF) (see below) (Novotny et al. 2015). An additional adhesin that can be targeted by antibodies is the outer membrane protein P5 (OMP P5) that present in pathogenic bacteria such as *Helicobacter pylori* (Oleastro and Menard 2013). Rabbit polyclonal antibodies targeting both PilA and OMP P5 improve the inhibition of early biofilm formed by either NTHI alone, or by both NTHI and *Moraxella catarrhalis* (Mcat) (Mokrzan et al. 2016, 2018), indicating that antibodies are able to inhibit early stages of biofilm formation by preferentially targeting more than one biofilm component.

Table 1 Antibodies against bacterial biofilms

Target group	Specific target	Antibodies	Origin	Biofilm species	References
Antibodies targeting surface elements					
Adhesins	PilA	polyclonal	rabbit	NTHI	Mokrzan et al. (2016); Novotny et al. (2015)
				NTHI + Mcat	Mokrzan et al. (2018)
	OMP P5	polyclonal	rabbit	NTHI	Mokrzan et al. (2016)
				NTHI + Mcat	Mokrzan et al. (2018)
	ClfA	12-9 mAb	murine	*S. aureus* or MRSA	Hall et al. (2003)
		T1-2 mAb	humanized	MRSA	Domanski et al. (2005)
	SecS	Polyclonal	Rabbit	*S. epidermidis*	Shahrooei et al. (2009); Shahrooei et al. (2012)
	FnBPA	Polyclonal	Rabbit	MRSA	O'Neill et al. (2008)
	AAP	Polyclonal	Rabbit	*S. epidermidis*	Rohde et al. (2005); Sun et al. (2005)
		12C6 mAb	Murine		Sun et al. (2005)
		12A1 mAb			
		3C1 mAb			
		18B6 mAb			Hu et al. (2011)
	SasG	Polyclonal anti-SasG	Rabbit	*S. aureus*	Belyi et al. (2018)
	Cna	9G7 mAb	Murine	*S. aureus*	Visai et al. (2000)
Other cell associated proteins	Amyloid-b	3H3 mAb	Human	*S. Typhimurium*	Levites et al. (2015); Tursi et al. (2020)
	PhnD	Polyclonal	Rabbit	*S. epidermidis*	Lam et al. (2014)
Polysaccharides	WTA	Antibody-antibiotic conjugate	Human	MRSA	Lehar et al. (2015)
		4497-IgG		*S. aureus*	Lisanne de Vor (2021)

Antibodies targeting extracellular secreted elements					
DNABII proteins	$IHF_{E\text{-}coli}$	Polyclonal or IgG	Rabbit	NTHI or *E-coli*	Brockson et al. (2014); Goodman et al. (2011)
	α and β-tip IHF	Polyclonal	Rabbit	NTHI	Novotny et al. (2019)
	β-tip IHF	Fab	Murine/ humanized	NTHI, *S. aureus*, *P. aeruginosa*, Mcat or *Burkholderia cenocepacia*	Novotny et al. (2020)
	α and β-tip IHF	$IhfA5_{NTHI}$ mAb	Murine		Novotny et al. (2016)
		mIhfB4NTHI mAb			
	IHF/HU	TRL1068 mAb	human	MRSA, *Acinetobacter baumannii*, *S. aureus* or *P. aeruginosa*	Estelles et al. (2016); Xiong et al. (2017)
Exopolysaccharides	PNAG	Polyclonal	Rabbit	*S. epidermidis*	Franca et al. (2013)
		F598 mAb	Human	*S. aureus*	Cerca et al. (2007); Kelly-Quintos et al. (2006)
		SAR279356 mAb			Sanofi (2012)
	Psl	Cam-003 mAb	Human	*P. aeruginosa*	DiGiandomenico et al. (2012)
		PsI0096 mAb			Morris et al. (2021)
Enzymes	AM	Polyclonal	Rabbit/ murine	*S. aureus or S. epidermidis*	McCarthy et al. (2016); Nair et al. (2015)
	GM	1C11 mAb	Murine	MRSA	Varrone et al. (2014)
	IsaA	UK-66P	Murine	*S. aureus*	Lorenz et al. (2011)
		1D9 mAb	Human		van den Berg et al. (2015)
	SdgA and SdgB	rF1 mAb	Human	*S. aureus*	Lisanne de Vor (2021)
Toxins	β-toxin	2A3 mAb	Chimeric	*S. aureus*	Tkaczyk et al. (2012)
		LC10	Human		Hua et al. (2014)
		MEDI4893	Human	MRSA	Anderson et al. (2018)
Antibodies targeting surface and extracellular elements					
Exopolysaccharides + T3S	PsI + PcrV	BiS4aPa (MEDI3902) bispecific	Human	*P. aeruginosa*	DiGiandomenico et al. (2014)

Another large group of adhesins commonly found in Gram-positive Staphylococci are the **cell wall anchored (CWA) proteins**. Microbial surface component recognizing adhesive matrix molecules (MSCRAMM) are the most common CWA protein group that are responsible for adhering to various mammalian ECM proteins such as fibrinogen and collagen, thus promoting adhesion and colonization to host tissues as well as medical devices (Foster et al. 2014; Foster 2019). Some of the leading fibrinogen binding MSCRAMM proteins that are targeted by antibodies are the homologous proteins clumping factor A (ClfA) and surface protein C (SesC), as well as fibronectin-binding proteins A and B (FnBPA or FnBPB) and Endocarditis biofilm-associated pilus A subunit (EbpA). In the next paragraphs, we will describe some of the existing mAbs against CWA proteins.

Murine anti-ClfA 12-9 mAb, binds a conserved epitope on multiple *Staphylococcus aureus* (*S. aureus*) strains and inhibits ClfA binding to fibrinogen. Moreover, 12-9 mAb inhibits adhesion and promotes detachment of biofilm cells in vitro, and prophylactically increases survival in infected mice (Hall et al. 2003). The humanized 12-9 version, T1-2 mAb, induces phagocytosis, and prophylactically reduces the bacterial load in the rabbit catheterized endocarditis methicillin-resistant *S. aureus* (MRSA) model. Co-application with vancomycin further reduces colony forming unit (CFU) bacteria (Domanski et al. 2005). Despite its promising results as an anti-biofilm agent, when T1-2 mAb (Tefibazumab) was clinically evaluated for treating *S. aureus* bacteraemia in a phase II clinical trial, it did not show significant activity (Weems et al. 2006). Another fibrinogen binding adhesin, SesC, is highly expressed in biofilm-associated cells and conserved among clinical isolates of *Staphylococcus epidermidis* (*S. epidermidis*) (Shahrooei et al. 2012). Rabbit polyclonal serum against SesC was able to inhibit both bacterial attachment and biofilm formation in a rat catheter model (Shahrooei et al. 2009), and in a jugular vein-catheterization, it reduced catheter *S. epidermidis* CFU and dissemination by opsonophagocytosis killing (Shahrooei et al. 2012). Upon SesC rat immunization, catheter adhering *S. epidermidis* were significantly reduced (Shahrooei et al. 2012). *S. aureus* displays two CWA fibronectin-binding adhesins FnBPA and FnBPB, that promote adhesion to fibrinogen and biofilm formation in acidic pH or with glucose. Polyclonal anti-FnBPA inhibits biofilm formation in four different MRSA strains (O'Neill et al. 2008). The interactions of EbpA and fibrinogen play an important role in catheter-associated urinary tract infections (CAUTIs) by *Enterococcus faecalis* (*E. faecalis*) which accumulate on fibrinogen in a catheter mice model. Immunizing mice against EbpA or its N-terminal domain significantly reduces bacterial load in both catheter and bladder *E. faecalis* infections (Flores-Mireles et al. 2014).

Another group of CWA proteins are adhesins that contain structurally identical G5 domains, characterized by five conserved glycine residues separated by an E region sequence (Foster et al. 2014). Examples for such G5-E adhesins are the *S. epidermidis* accumulation associated protein (AAP) or the *S. aureus* surface protein G (SasG). Besides adhesion, AAP and SasG promote bacterial aggregation which further support biofilm development (Foster et al. 2014). Rabbit anti-AAP sera and three murine anti-AAP mAbs, 12C6, 12A1 and 3C1, inhibit *S. epidermidis* biofilm formation with up to an 87% inhibition when using the three murine mAbs as a mix (Rohde et al. 2005; Sun et al. 2005). Another murine anti-AAP mAb, 18B6, was found to reduce *S. epidermidis* biofilm density, and disrupt biofilm arrangement and formation (Rohde et al. 2005; Hu et al. 2011). However, Lam et al, demonstrated that rabbit anti-AAP or anti-SesC (target previously described) polyclonal antibodies allowed the initial bacterial attachment and aggregation, leading to increased biofilms, in contrast to anti-SasG antibodies which inhibited *S. aureus* biofilm formation in vitro and in support with the initial findings above (Belyi et al. 2018; Lam et al. 2014).

5.1 Antibodies Directed Against Other Cell-Associated Proteins

Amyloids are traditionally associated with several degenerative human diseases such as Alzheimer and Parkinson that result from misfolded amyloid structures (Duyckaerts 2013). However, bacteria can also form amyloid structures, with a generic structural similarity. Bacterial amyloid proteins possess a conserved cross β-sheet structure and can serve as a strong scaffold for the ECM. Amyloids are highly resistant to chemical, proteolytic, and enzymatic degradation, and therefore provide a strong scaffold for the community, acting almost like a shield under stress (Nicastro and Tukel 2019; Hufnagel et al. 2013). Functional amyloids are the major components of the proteinous ECM. These proteins form fibrils with amyloid-like structures and their assemblage is important for biofilm integrity and functionality. Known biofilm-associated amyloids include the *E-coli and S. Typhimurium* Curli CsgA/B proteins (Chapman et al. 2002), *S. aureus* phenol soluble modulins (Schwartz et al. 2012), and *Pseudomonas aeruginosa* (*P. aeruginosa*) FapB/C (Lembre et al. 2014).

Since most of these proteins are both common and conserved, and their assembly is important for biofilm integrity, several antibodies were assessed for their ability to target fiber-forming proteins that comprise the biofilm. One example is the A11 antibody that recognizes fiber-forming intermediate protein aggregates as diverse as amyloid-β (Aβ), synuclein, islet amyloid polypeptide, insulin, prions, lysozyme, and curli (Kayed et al. 2003; O'Nuallain and Wetzel 2002). Another example is the human anti-Aβ 3H3 mAb, isolated from memory B cells of a healthy donor (Levites et al. 2015), which successfully inhibits *S. Typhimurium* curli fibrilization and biofilm formation in vitro. 3H3 mAb also reduces the density and alters the integrity of already established biofilms, by enhanced uptake of bacteria and curli fibers by macrophages. A combination of mAb 3H3 with ampicillin was effective in destroying *S. Typhimurium* pre-established biofilms in a catheter mouse model (Tursi et al. 2020).

Another group of targets for antibodies are the many transporters that are expressed on the bacterial cell walls and membranes. The phosphonate ABC transporter substrate binding protein, PhnD, a periplasmic *E-coli* protein (Metcalf and

Wanner 1993), conserved across the coagulase-negative *Staphylococci* and *S. aureus*, can also be targeted by antibodies. Polyclonal rabbit anti-PhnD antibodies inhibit *S. epidermidis* biofilm formation, preventing initial adherence and subsequent aggregation through neutrophil opsonophagocytosis (Lam et al. 2014).

5.2 Antibodies Directed Against Cell-Associated Polysaccharides

The cell wall consists of various components: peptidoglycan, anionic polymers like teichoic acids (Coley et al. 1978; Kawai et al. 2011) and teichuronic acids (Soldo et al. 1999). Peptidoglycans, with their mesh-like structure, are built from disaccharide precursors with covalently linked stem peptides (*N*-acetylmuramic acid-(β-1,4)-*N*-acetylglucosamine with an L-Ala-γ-D-Glu-*meso*-A_2pm-D-Ala-D-Ala pentapeptide linked to the first saccharide). Wall teichoic acid (WTA) is a glycopolymer made of polyglycerol or ribitol phosphate units linked to the peptidoglycan layer. WTA forms a physical barrier protecting the cells from antimicrobials. Furthermore, teichoic acids were recently shown to be essential for biofilm formation (Bucher et al. 2015; Sumrall et al. 2020), which makes them appealing targets for antibodies, targeting both the bacteria and bacterial biofilm (Weidenmaier and Peschel 2008). In a study published by Lehar et al., 40 mAbs were isolated from B cells of *S. aureus* convalescent patients. The anti-β-WTA mAb, demonstrated antibacterial activity and MRSA killing within macrophages, endothelial and epithelial cells when conjugated to a rifamycin derivative that was cleaved upon phagocytosis (Lehar et al. 2015). In regard to *S. aureus* biofilms, an anti-β-WTA 4497-IgG was observed to be co-localized in *S. aureus* infected implant sites in mice (Lisanne de Vor et al. 2021), suggesting that this mAb could exhibit an anti-biofilm activity. However, further evaluation is necessary.

6 Anti-biofilm Antibodies Targeting Secreted Bacterial Extracellular Components

The most extensively researched anti-biofilm antibodies, also summarized in Table 1, target secreted ECM elements rather than cell wall components. Secreted polymers include extracellular DNA (eDNA), eDNA binding proteins, and exopolysaccharides that maintain biofilm stability and structural architecture, enzymes, and toxins.

6.1 Antibodies Directed Against DNA Binding Proteins

eDNA plays an important role in biofilm formation in various bacteria type, such as *P. aeruginosa*, *Staphylococci* and *Streptococcus*. eDNA maintains the biofilm structural organization, being a nutrient source for resident microbes and stimulating antimicrobial resistance (Karygianni et al. 2020). As such, DNases are sometimes used to disrupt biofilms (Shields et al. 2013). The DNABII family proteins, IHF and histone-like protein (HU) stabilize eDNA through binding, bending, and compacting the double-stranded DNA and therefore are crucial for biofilm formation. Several approaches have attempted to generate antibodies targeting IHF/HU proteins, thereby disrupting eDNA biofilms. Both rabbit polyclonal and murine monoclonal anti-IHF antibodies disrupt the biofilm layer by reducing its height, thickness, and biomass, an effect that was further enhanced with the addition of amoxicillin. This combination induced bacterial killing, as the bacteria are more accessible to drugs when no longer shielded by the biofilm (Brockson et al. 2014; Goodman et al. 2011). Biochemical mapping revealed that anti-IHF inhibiting antibodies are directed against a distinct structure within the IHF tip (Brockson et al. 2014). Anti-IHF tip mAbs effectively disrupt biofilm in a chinchilla NTHI OM model, as well as in *P. aeruginosa*-formed biofilm in a mouse lung infection model. In the latter model, a more dramatic protection was achieved when the mAbs were combined with tobramycin antibiotic therapy (Novotny et al. 2016; Novotny et al. 2019). Moreover, a humanized version of an anti-tip Fab was able to induce anti-inflammatory responses exhibited by the induction of IL-10 and IL-13 cytokines (Novotny et al. 2020).

Several human anti-IHF/HU mAbs were generated from human B cells (Ryser et al. 2019; Harriman et al. 2009; Estelles et al. 2016). The TRL1068 mAb targets a conserved IHF/HU epitope, common to a large group of Gram-negative and Gram-positive bacterial species. Like murine mAbs, combining TRL1068 with the antibiotics daptomycin, vancomycin or imipenem is extremely efficient in preventing biofilm formation and treating pre-established biofilm (Estelles et al. 2016; Xiong et al. 2017). Currently, TRL1068 mAb (Identifier: NCT04763759) is being clinically tested for safety and efficacy in eliminating biofilms in prosthetic joint infections, making its resident bacteria more susceptible to antibiotic regiments. IHF has also been tested as an immunogen, resulting in reduced biomass of *H. influenzae* biofilm in the middle ear and delayed OM signs in a chinchilla model (Novotny et al. 2020; Goodman et al. 2011). The general development of therapeutic anti-biofilm antibodies targeting IHF/HU proteins is described in Fig. 3.

6.2 Antibodies Directed Against Exopolysaccharides

Exopolysaccharides generated by bacteria have been recognized for decades to significantly impact bacterial virulence and promote capsule formation (Taylor and

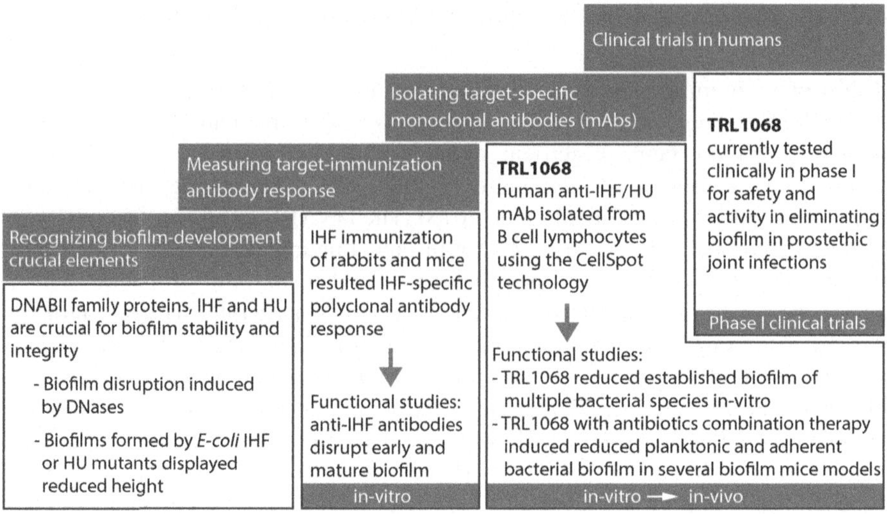

Fig. 3 Targeting DNABII proteins, step-by-step model for the general experimental processes influencing the development of anti-biofilm mAb therapy. **First step,** evaluating the role of DNABII proteins, IHF and HU in biofilms. **Second step**, measuring the polyclonal antibody response following IHF immunization and its effect on biofilms in vitro. **Third step**, generating anti-IHF/HU human mAb (TRL1068) and testing its function on biofilms in vitro. Next, moving on to another level when testing TRL1068 mAb in vivo in biofilm murine models. Moreover, TRL1068 mAb was tested in combination with antibiotic therapy in vivo for additive effect. **Fourth step**, Clinical evaluation of TRL1068 mAb in humans

Roberts 2005). Various genetic analyses have provided strong evidence that biofilm exopolysaccharides play a fundamental structural role in different bacterial species. Bacterial strains mutated in exopolysaccharide production display severe defects in biofilm formation and in achieving complex biofilm architecture. Common bacterial exopolysaccharides include cellulose (Ross et al. 1991; Ude et al. 2006; Serra et al. 2013; Jonas et al. 2007; Solano et al. 2002) and the *Staphylococcal* polysaccharide intercellular adhesin (PIA) (Rohde et al. 2010). PIA-related polymers are produced by *S. epidermidis* and *S. aureus* (Mack et al. 1996; Cramton et al. 1999), and several Gram-negative bacteria (Wang et al. 2004; Darby et al. 2002; Bobrov et al. 2008; Kaplan et al. 2004; Izano et al. 2007; Parise et al. 2007). Due to their significance for biofilm formation and structural integrity, exopolysaccharides were targeted by various anti-biofilm mAbs. One example is the polysaccharide-*N*-acetylated glucosamine (PNAG), also known as PIA, the main exopolysaccharide located on the staphylococcal biofilm surface. PNAG/PIA mediates adhesion, bacterial aggregation and biofilm formation and thus plays an important role in biofilm architecture (Cerca et al. 2007; Sadovskaya et al. 2005). PNAG/PIA saccharides are produced and regulated by the (*ica*) ADCB operon associated with biofilm formation in bacteria and fungi, providing a universal potential target for antibody-based therapy and vaccination (Arciola et al. 2001; Cywes-Bentley et al. 2013). Several attempts were made in this direction. Rabbit polyclonal anti-PNAG antibodies were able to penetrate the biofilm biomass and mediate bacterial killing (Cerca et al. 2006). A case in point is the human anti-PNAG mAb, F598, isolated from an *S. aureus* convalescent patient. F598 mAb showed complement-dependent efficacy in vitro (Franca et al. 2013), and an increasing survival from an *S. aureus* lethal dose in a mice infection model (Cerca et al. 2007; Kelly-Quintos et al. 2006). Structural analysis revealed that F598 binds multiple PNAG GlcNAc residues in an anchored binding mechanism (Soliman et al. 2018). When tested clinically to prevent *Neisseria gonorrhoeae* urethral infection, F598 mAb failed to show the expected therapeutic effect (Identifier: NCT03222401). Testing F598 mAb (SAR279356) clinically in mechanically ventilated patients for preventing infections was terminated due to the difficulty in recruiting patients (Identifier: NCT01389700) (Sanofi 2012).

Pel and Psl are two exopolysaccharides produced by *P. aeruginosa* (Colvin et al. 2012). Psl is composed of mannose, rhamnose, and glucose, and is synthesized by the gene products of the *pslA-O* operon (Friedman and Kolter 2004b; Matsukawa and Greenberg 2004; Jackson et al. 2004; Byrd et al. 2009). Pel is a glucose-rich polymer, and its synthesis is mediated by the *pelA-G* operon (Friedman and Kolter 2004a; Colvin et al. 2011). In addition to its structural role, Psl is specifically important for surface attachment (Colvin et al. 2012; Ma et al. 2006) and both exopolysaccharides were shown to be essential for virulence and host colonization in mice (Mann and Wozniak 2012). Thus, Pel and Psl represent an additional interesting target for anti-biofilm antibodies. Screening *P. aeruginosa* with human single-chain variable region (scFv) phage display libraries, revealed three *P. aeruginosa* PsI epitopes targeted by antibodies: the scFv with the greatest activity, Cam-003 mAb which inhibits *P. aeruginosa* attachment to human pulmonary epithelial cells in vitro, and when converted to IgG, prophylactically protects through FcγR and opsonophagocytic killing in acute lethal pneumonia, corneal, thermal injury, and ocular keratitis mice infection models of *P. aeruginosa* infection (DiGiandomenico et al. 2012). Combination therapy of mAbs targeting three non-overlapping epitopes, together with human neutrophils and meropenem antibiotics have reduced the in vitro biomass of *P. aeruginosa* biofilm (Ray et al. 2017). Cam-003 or its affinity optimized version, PsI0096 mAb, showed a significantly increased binding to biofilms formed by *P. aeruginosa* cystic fibrosis isolates (Morris et al. 2021). However, although the promising prophylactic results, in vivo applications as a therapeutic had no protective effect (Kugadas et al. 2019). BiS4αPa (MEDI3902) is a promising bispecific mAb constructed with PsI0096 anti-Psl mAb and V2L2-MD anti-PcrV mAb (DiGiandomenico et al. 2014). Thus, targeting Psl and PcrV proteins simultaneously. PcrV is an important component of *P. aeruginosa* type III secretion system (T3S), located in the bacterial cytoplasm but also extracellularly and involved in cytotoxicity of the host cells. BiS4αPa mAb enhanced protection in *P. aeruginosa* bacteremia, pneumonia, or thermal injury mouse models (DiGiandomenico et al. 2014). Despite encouraging Phase I clinical data (Ali et al. 2019) single dose of MEDI3902 in Phase II was not sufficient to reduce *P. aeruginosa* pneumonia in mechanically ventilated patients compared to placebo (Chastre et al. 2020).

6.3 Antibodies Directed Against Extracellular Enzymes

Another important biofilm component is extracellular enzymes essential for bacterial growth and virulence. For example, the cell wall autolysin, Atl, is the most predominant peptidoglycan hydrolase in Staphylococci. It is cleaved into two functional extracellular enzymes, an *N*-acetylmuramoyl-L-alanine amidase (AM) and endo-β-*N*-acetylglucosaminidase (GL or GM) and is involved in bacterial adherence and biofilm formation (McCarthy et al. 2016). Polyclonal anti-AM antibodies reduced biofilm formation on polystyrene plate and shown protection in vivo by activating neutrophils and complement-mediated bacterial killing (McCarthy et al. 2016; Nair et al. 2015). Murine anti-GM 1C11 mAb prophylactically reduced MRSA infections in an implant-associated osteomyelitis murine model mediated by MRSA internalization by macrophages (Varrone et al. 2014). An additional enzyme for biofilm-mediated infection is a conserved extracellular and cell wall glycosylase, immunodominant staphylococcal antigen A (IsaA) (Sakata et al. 2005). IsaA promotes growth among *S. aureus* strains and induces elevated blood anti-IsaA antibody responses in infected individuals (Lorenz et al. 2000). A murine anti-IsaA, UK-66P mAb, induces bacterial killing by neutrophils and has protected against infection in a catheter-related sepsis mice model (Lorenz et al. 2011). In *S. aureus* bacteremia mice model another anti-IsaA mAb, 1D9, increased survival (van den Berg et al. 2015).

6.4 Antibodies Directed to Toxins

As previously mentioned, toxins are successful targets for immunotherapy for many bacterial infections including tetanus, botulism, and anthrax. A case in point is the α-toxin hemolysin (HLA), a pore-forming toxin required for cell-cell interactions during biofilm formation (Caiazza and O'Toole 2003; Anderson et al. 2012). α-toxin can be targeted by neutralizing antibodies. The human anti-α-toxin 2A3 mAb inhibits cell lysis, reduces *S. aureus* skin infected lesions and the infection severity of *S. aureus* clinical isolates (Tkaczyk et al. 2012). Prophylactic administration of an affinity optimized version of 2A3, named LC10 mAb, has increased survival in an *S. aureus* pneumonia mice model, showing mild alveolar inflammation with decreased proinflammatory cytokines, an effect that was further increased when mAb therapy was combined with vancomycin or linezolid antibiotics (Hua et al. 2014). When used as a prophylaxis, the next-generation LC10 mAb, MEDI4893 (suvratoxumab, with a longer half-life), prevents biofilm formation in an MRSA porcine vaginal mucosa neutralization assay (Anderson et al. 2018). The crystal structure of MEDI4893 and α-toxin revealed that MEDI4893 binds the rim domain of α-toxin preventing interactions with the host cells (Oganesyan et al. 2014). Despite encouraging MEDI4893 Phase I clinical trial data (Yu et al. 2017) (Identifier: NCT01769417), Phase II evaluation showed similar *S. aureus* occurrence in the treated and placebo groups (Francois et al. 2021) (identifier: NCT02296320). Interestingly, targeting α-toxin and ClfA simultaneously results in a better outcome that was also correlated with reduced ex vivo CFU in the joints or bone tissue from the implant surgical leg (Wang et al. 2017b). These results indicate that combination therapy mAbs, targeting different antigens simultaneously, might be advantageous compared to targeting only one component at a time.

Another indication that targeting more than one antigen will be advantageous, comes from the fact that so far, all efforts to develop vaccine against *S. aureus* based on single antigen were less effective clinically, despite being promising targets. Two prominent examples are Merck's V710 vaccine that contain iron surface determinant B (IsdB) which promotes *S. aureus* adherence and Nabi's StaphVAX that contain capsular polysaccharides type 5 and 8 (CP5/CP8) conjugated with the carrier *P. aeruginosa* exoprotein A (Bagnoli et al. 2012; Clegg et al. 2021).

7 Antibodies Against Fungi-Associated Biofilm

Fungi are eukaryotes naturally present in the environment and in the flora of humans and animals. Pathogenic fungi mainly affect immunosuppressed individuals, such as HIV-1 infected, or individuals following surgical interventions. The most studied infection-causing fungi are *Candida*, *Aspergillus*, and *Cryptococcus* which result in approximately 1.7 million deaths annually worldwide (Kainz et al. 2020). Similar to bacteria, fungi form biofilms on medical devices, with complex ECM structures comprised of cell-associated and secreted components that vary in different fungi types. Various factors can influence the chemistry of fungal biofilm matrices such as the biofilm composition and the conditions under which biofilms are formed. The most noticeable differences in the matrix composition appear at the protein level (Mitchell et al. 2016). For instance, in *Candida albicans* (*C. albicans*) about 55% of the dry weight of the extracellular matrix was proteinaceous, far more than its carbohydrate content (25%) (Zarnowski et al. 2014). Matrix carbohydrates were primarily composed of arabinose, mannose, glucose, and xylose (Pierce et al. 2017; Zarnowski et al. 2014). Lipids and eDNA were also present in the fungal biofilm matrix.

Extensive use of antifungal drugs has resulted in the emergence of drug-resistant strains, making treatment ineffective (Perlin et al. 2015), similar to bacterial infections. While this chapter focuses on antibodies against bacterial biofilms, several attempts to generate antibodies targeting fungal biofilms were described as well (Casadevall and Pirofski 2007). In the following paragraphs, we will review anti-biofilm antibodies against the two most common pathogenic fungi *C. albicans* and *Cryptococcus neoformans* (*C. neoformans*), summarized in Table 2.

C. albicans and its biofilm extracts were used in immunizing rabbits and mice to generate anti-biofilm antibodies. Rabbit polyclonal antibodies reduce *C. albicans* viability, metabolic activity, and early biofilm formation (Carrano et al. 2019), whereas murine 7D7 mAb, targeting an unknown 47.2 kDa protein, inhibit

Table 2 Antibodies against fungal biofilms

Target group	Specific target	Antibodies	Origin	Biofilm species	References
Antibodies targeting surface elements					
C. albicans structures	*C. albicans* blastopores	Polyclonal	Rabbit	*C. albicans*	Carrano et al. (2019)
Proteins	Unknown 47.2 kDa protein	7D7 mAb	Murine	*C. albicans*	Mishra et al. (2015)
	Als3	3A-5 mAb 113 mAb	Murine	*C. albicans*	Coleman et al. (2009)
	MP65	ADR3-6 DAb	Human	*C. albicans*	De Bernardis et al. (2007)
	CR3-RP	Polyclonal or IgGs	Rabbit	*C. albicans*	Bujdakova et al. (2010); Dekkerova et al. (2019); Bujdakova et al. (2008)
	Stress mannoprotein	C7 mAb	Murine	*C. albicans*	Moragues et al. (2003)
Polysaccharides	GXM	18B7 mAb 12A1 mAb	Murine	*C. neoformans*	Martinez and Casadevall (2005); Martinez et al. (2006b)
Antibodies targeting extracellular secreted elements					
Enzyme	SAP-2	ADR4-6	Human	*C. albicans*	De Bernardis et al. (2007)
Antibodies targeting surface and extracellular elements					
Polysaccharides	β-glucan	2G8 mAb	Murine	*C. albicans*	Torosantucci et al. (2009)
Polysaccharide and enzyme	MP65 + SAP-2	bispecific	Human	*C. albicans*	De Bernardis et al. (2007)

C. albicans adherence and mature biofilms (Mishra et al. 2015). Antibodies against specific known targets such as the murine antibodies, 3-A5 and 113 mAbs, directed at *C. albicans* cell surface protein adhesin, Als3, inhibit *C. albicans* aggregation in vitro and in vivo as well as adherence to human endothelial and epithelial cells (Coleman et al. 2009). Phase I clinical trial using NDV-3, an Als3 candidate vaccine (NCT01926028) for the prevention of recurrent vulvovaginal candidiasis, showed elevated antibody responses only in patients with no recurrent infection (Uppuluri et al. 2018).

Cell surface mannoproteins are also targeted by mAbs. Mannoprotein MP65 is the major *C. albicans* cell wall component, highly represented in the biofilm matrix and plays an important role in biofilm development and maintenance (Sandini et al. 2011). Variable domain antibody (DAb) ADR3-6, an anti-MP65, has shown in vitro and in vivo protection, *C. albicans* adherence inhibition with rapid fungal clearance in a *C. albicans* vaginal rat infection model (De Bernardis et al. 2007). MAbs targeting the *C. albicans* receptor 3-related protein (CR3-RP) were also described. These mAbs act by mimicking the mammalian neutrophil receptor MAC-1 (CD11b/CD18) and thus inhibiting opsonization and phagocytosis. Several studies have demonstrated that anti-CR3-RP IgGs inhibit *C. albicans* adherence and biofilm formation to epithelial cells and polyethylene coverslips (Bujdakova et al. 2008, 2010; Dekkerova et al. 2019), and had some activity in vivo (Chupacova et al. 2018).

Other antifungal antibodies are directed against surface polysaccharides. The polysaccharide capsule of *C. neoformans* is composed of glucuronoxylomannan (GXM) that upon *cap59* deletion, the acapsular *C. neoformans* fails to form biofilm, which highlights GXM's contribution to virulence. Two murine anti-GXM mAbs, 18B7 and 12A1, inhibit in vitro *C. neoformans* biofilm formation (Martinez and Casadevall 2005). 18B7 mAb together with amphotericin reduce free *C. neoformans* metabolic activity. However, this combination had an opposite effect on pre-established biofilms (Martinez et al. 2006b). Radioimmunotherapy (RIT) is a highly targeted delivery system approach that was evaluated and approved for cancer therapy (Green and Press 2017) and can be exploited to deliver cytotoxic radiation directly to the infection site. Radiolabelled 18B7 mAb greatly damaged free *C. neoformans* rather than the biofilm, with at most, 85% and 50% reduction in metabolic activity, respectively (Martinez et al. 2006a). 18B7 mAb is the first mAb therapy against fungal disease evaluated in humans. A phase I clinical trial for the safety and pharmacokinetics in HIV-infected patients pre-treated effectively for cryptococcal meningitis shows a transient decline in serum cryptococcal antigens (Larsen et al. 2005).

Similar to bacterial biofilms, fungal biofilms are also comprised of secreted components that support biofilm structure and function. For example, fungal growth is accompanied by β-glucan secretion, an ECM exopolysaccharide also found in the cell wall and its associated proteins. β-glucan acts as a shield, preventing antifungal drugs to reach their cellular targets through a drug sequestering mechanism and are involved in cell wall assembly and growth (Mitchell et al. 2016). The anti-β-glucan, 2G8 mAb mediates protection during *C. albicans* infections, increases mice survival in lethal dose infections, accelerates fungal clearance in a self-healing rat vaginitis model, and inhibits growth and adherence to a carcinoma cell line (Torosantucci et al. 2009). In addition to exopolysaccharides, fungal enzymes such as the secreted aspartic proteases (SAP-2) promote fungal cell adhesion (Naglik et al. 2003). The anti-(SAP)-2 variable domain antibody (DAb), ADR4-6, has shown in vitro and in vivo protection and *C. albicans* adherence inhibition with *C. albicans* rapid clearance in a vaginal rat infection model. As observed for bacterial biofilm, targeting several antigens might result in better inhibition. Bispecific DAbs targeting both mannoprotein, MP65 and (SAP)-2, were more potent than mono-specific DAbs, demonstrating further accelerated clearance of *C. albicans* (De Bernardis et al. 2007). Other secreted extracellular components, such as nucleic acids and lipids, are also present in fungal biofilms. However, to our knowledge mAbs against these secreted components were not tested in fungal biofilm assays.

8 Conclusions

Although currently, there is no available licensed antibody-based therapy or vaccine for biofilm-associated infections, there are several potentially interesting anti-biofilm antibodies with in vitro and in vivo efficacy. Antibodies have gained much clinical focus during the last years, due to their unique traits such as safety, specificity, and stability. However, antibody therapy has also several disadvantages that were not discussed here, such as the high production costs and very specific preservation conditions. Further research on the functions of antibodies in the battle against pathogenic microbes is needed, and it is still unclear whether antibodies by themselves can neutralize bacteria and fungi as effectively as antiviral antibodies that already exist in the clinic and act against viruses. To our knowledge so far, human clinical trials mostly used antibodies against single biofilm component, which failed to show significant therapeutic effect. Therefore, antibodies exhibiting similar mechanisms of action for biofilm treatment in humans might show limited effect as well. Furthermore, most of the antibody therapy presented in this chapter was greatly enhanced by the addition of an antibiotic drug. Therefore, anti-biofilm antibodies can act as "smart missiles" that can penetrate through the biofilm shield and deliver antibacterial drugs, which in turn can kill the biofilm-forming bacteria. Another approach that shows enhanced anti-biofilm activity is targeting multiple components at once. This "cocktail" approach is frequently used in medicine and can address issues like drug escape, resulting in better outcomes when facing the highly complex structure of the biofilm matrix. While there is still a long way to go for marketed anti-biofilm antibodies, using these approaches might improve the scope of clinical outcomes of anti-biofilm antibodies in humans and bring us a step forward toward an effective therapy.

Acknowledgments A Watson is supported by a PhD Scholarship from the Tel Aviv University Center for Combatting Pandemics. N T Freund is supported by ISF grants # 1422/18 and #3711/20. Illustartions were created with https://BioRender.com (agreement number: NC23SO1RCH (Figure 1) and WH23SO3TMJ (Figure 2)) and PyMOL software (public available version).

References

Akkaya M, Kwak K, Pierce SK (2020) B cell memory: building two walls of protection against pathogens. Nat Rev Immunol 20(4):229–238. https://doi.org/10.1038/s41577-019-0244-2

Ali SO, Yu XQ, Robbie GJ, Wu Y, Shoemaker K, Yu L, DiGiandomenico A, Keller AE, Anude C, Hernandez-Illas M, Bellamy T, Falloon J, Dubovsky F, Jafri HS (2019) Phase 1 study of MEDI3902, an investigational anti-Pseudomonas aeruginosa PcrV and Psl bispecific human monoclonal antibody, in healthy adults. Clin Microbiol Infect 25(5):629 e621–629 e626. https://doi.org/10.1016/j.cmi.2018.08.004

Allard-Massicotte R, Tessier L, Lecuyer F, Lakshmanan V, Lucier JF, Garneau D, Caudwell L, Vlamakis H, Bais HP, Beauregard PB (2016) Bacillus subtilis early colonization of arabidopsis

thaliana roots involves multiple chemotaxis receptors. mBio 7(6). https://doi.org/10.1128/mBio.01664-16
Anderson MJ, Lin YC, Gillman AN, Parks PJ, Schlievert PM, Peterson ML (2012) Alpha-toxin promotes Staphylococcus aureus mucosal biofilm formation. Front Cell Infect Microbiol 2:64. https://doi.org/10.3389/fcimb.2012.00064
Anderson MJ, Schaaf E, Breshears LM, Wallis HW, Johnson JR, Tkaczyk C, Sellman BR, Sun J, Peterson ML (2018) Alpha-toxin contributes to biofilm formation among Staphylococcus aureus wound isolates. Toxins (Basel) 10(4). https://doi.org/10.3390/toxins10040157
Arciola CR, Baldassarri L, Montanaro L (2001) Presence of icaA and icaD genes and slime production in a collection of staphylococcal strains from catheter-associated infections. J Clin Microbiol 39(6):2151–2156. https://doi.org/10.1128/JCM.39.6.2151-2156.2001
Aribam SD, Harada T, Elsheimer-Matulova M, Iwata T, Kanehira K, Hikono H, Matsui H, Ogawa Y, Shimoji Y, Eguchi M (2016) Specific monoclonal antibody overcomes the Salmonella enterica serovar Typhimurium's adaptive mechanisms of intramacrophage survival and replication. PLoS One 11(3):e0151352. https://doi.org/10.1371/journal.pone.0151352
Asally M, Kittisopikul M, Rue P, Du Y, Hu Z, Cagatay T, Robinson AB, Lu H, Garcia-Ojalvo J, Suel GM (2012) Localized cell death focuses mechanical forces during 3D patterning in a biofilm. Proc Natl Acad Sci U S A 109(46):18891–18896. https://doi.org/10.1073/pnas.1212429109
Bagnoli F, Bertholet S, Grandi G (2012) Inferring reasons for the failure of Staphylococcus aureus vaccines in clinical trials. Front Cell Infect Microbiol 2:16. https://doi.org/10.3389/fcimb.2012.00016
Barnes CO, West AP Jr, Huey-Tubman KE, Hoffmann MAG, Sharaf NG, Hoffman PR, Koranda N, Gristick HB, Gaebler C, Muecksch F, Lorenzi JCC, Finkin S, Hagglof T, Hurley A, Millard KG, Weisblum Y, Schmidt F, Hatziioannou T, Bieniasz PD, Caskey M, Robbiani DF, Nussenzweig MC, Bjorkman PJ (2020) Structures of human antibodies bound to SARS-CoV-2 spike reveal common epitopes and recurrent features of antibodies. Cell. https://doi.org/10.1016/j.cell.2020.06.025
Baum A, Fulton BO, Wloga E, Copin R, Pascal KE, Russo V, Giordano S, Lanza K, Negron N, Ni M, Wei Y, Atwal GS, Murphy AJ, Stahl N, Yancopoulos GD, Kyratsous CA (2020) Antibody cocktail to SARS-CoV-2 spike protein prevents rapid mutational escape seen with individual antibodies. Science 369(6506):1014–1018. https://doi.org/10.1126/science.abd0831
Beauregard PB, Chai Y, Vlamakis H, Losick R, Kolter R (2013) Bacillus subtilis biofilm induction by plant polysaccharides. Proc Natl Acad Sci U S A 110(17):E1621–E1630. https://doi.org/10.1073/pnas.1218984110
Belyi Y, Rybolovlev I, Polyakov N, Chernikova A, Tabakova I, Gintsburg A (2018) Staphylococcus aureus surface protein G is an immunodominant protein and a possible target in an anti-biofilm drug development. Open Microbiol J 12:94–106. https://doi.org/10.2174/1874285801812010094
Bloom-Ackermann Z, Steinberg N, Rosenberg G, Oppenheimer-Shaanan Y, Pollack D, Ely S, Storzi N, Levy A, Kolodkin-Gal I (2016) Toxin-antitoxin systems eliminate defective cells and preserve symmetry in Bacillus subtilis biofilms. Environ Microbiol 18(12):5032–5047. https://doi.org/10.1111/1462-2920.13471
Bobrov AG, Kirillina O, Forman S, Mack D, Perry RD (2008) Insights into Yersinia pestis biofilm development: topology and co-interaction of Hms inner membrane proteins involved in exopolysaccharide production. Environ Microbiol 10(6):1419–1432. https://doi.org/10.1111/j.1462-2920.2007.01554.x
Boulianne GL, Hozumi N, Shulman MJ (1984) Production of functional chimaeric mouse/human antibody. Nature 312(5995):643–646. https://doi.org/10.1038/312643a0
Brockson ME, Novotny LA, Mokrzan EM, Malhotra S, Jurcisek JA, Akbar R, Devaraj A, Goodman SD, Bakaletz LO (2014) Evaluation of the kinetics and mechanism of action of anti-integration host factor-mediated disruption of bacterial biofilms. Mol Microbiol 93(6):1246–1258. https://doi.org/10.1111/mmi.12735

Bryers JD (2008) Medical biofilms. Biotechnol Bioeng 100(1):1–18. https://doi.org/10.1002/bit.21838

Bucher T, Oppenheimer-Shaanan Y, Savidor A, Bloom-Ackermann Z, Kolodkin-Gal I (2015) Disturbance of the bacterial cell wall specifically interferes with biofilm formation. Environ Microbiol Rep 7(6):990–1004. https://doi.org/10.1111/1758-2229.12346

Bujdakova H, Paulovicova E, Borecka-Melkusova S, Gasperik J, Kucharikova S, Kolecka A, Lell C, Jensen DB, Wurzner R, Chorvat D, Pichova I (2008) Antibody response to the 45 kDa Candida albicans antigen in an animal model and potential role of the antigen in adherence. J Med Microbiol 57(Pt 12):1466–1472. https://doi.org/10.1099/jmm.0.2008/001479-0

Bujdakova H, Paulovicova E, Paulovicova L, Simova Z (2010) Participation of the Candida albicans surface antigen in adhesion, the first phase of biofilm development. FEMS Immunol Med Microbiol 59(3):485–492. https://doi.org/10.1111/j.1574-695X.2010.00713.x

Byrd MS, Sadovskaya I, Vinogradov E, Lu H, Sprinkle AB, Richardson SH, Ma L, Ralston B, Parsek MR, Anderson EM, Lam JS, Wozniak DJ (2009) Genetic and biochemical analyses of the Pseudomonas aeruginosa Psl exopolysaccharide reveal overlapping roles for polysaccharide synthesis enzymes in Psl and LPS production. Mol Microbiol 73(4):622–638. https://doi.org/10.1111/j.1365-2958.2009.06795.x

Caiazza NC, O'Toole GA (2003) Alpha-toxin is required for biofilm formation by Staphylococcus aureus. J Bacteriol 185(10):3214–3217. https://doi.org/10.1128/JB.185.10.3214-3217.2003

Carrano G, Paulone S, Lainz L, Sevilla MJ, Blasi E, Moragues MD (2019) Anti-Candida albicans germ tube antibodies reduce in vitro growth and biofilm formation of C. albicans. Rev Iberoam Micol 36(1):9–16. https://doi.org/10.1016/j.riam.2018.07.005

Casadevall A, Pirofski LA (2007) Antibody-mediated protection through cross-reactivity introduces a fungal heresy into immunological dogma. Infect Immun 75(11):5074–5078. https://doi.org/10.1128/IAI.01001-07

Caskey M, Klein F, Nussenzweig MC (2016) Broadly neutralizing antibodies for HIV-1 prevention or immunotherapy. N Engl J Med 375(21):2019–2021. https://doi.org/10.1056/NEJMp1613362

Caskey M, Klein F, Nussenzweig MC (2019) Broadly neutralizing anti-HIV-1 monoclonal antibodies in the clinic. Nat Med 25(4):547–553. https://doi.org/10.1038/s41591-019-0412-8

Cathcart AL, Havenar-Daughton C, Lempp FA, Ma D, Schmid M, Agostini ML, Guarino B, Di Iulio J, Rosen L, Tucker H, Dillen J, Subramanian S, Sloan B, Bianchi S, Wojcechowskyj J, Zhou J, Kaiser H, Chase A, Montiel-Ruiz M, Czudnochowski N, Cameroni E, Ledoux S, Colas C, Soriaga L, Telenti A, Hwang S, Snell G, Virgin HW, Corti D, Hebner CM (2021) The dual function monoclonal antibodies VIR-7831 and VIR-7832 demonstrate potent in vitro and in vivo activity against SARS-CoV-2. bioRxiv. https://doi.org/10.1101/2021.03.09.434607

Cerca N, Jefferson KK, Oliveira R, Pier GB, Azeredo J (2006) Comparative antibody-mediated phagocytosis of Staphylococcus epidermidis cells grown in a biofilm or in the planktonic state. Infect Immun 74(8):4849–4855. https://doi.org/10.1128/IAI.00230-06

Cerca N, Jefferson KK, Maira-Litran T, Pier DB, Kelly-Quintos C, Goldmann DA, Azeredo J, Pier GB (2007) Molecular basis for preferential protective efficacy of antibodies directed to the poorly acetylated form of staphylococcal poly-N-acetyl-beta-(1-6)-glucosamine. Infect Immun 75(7):3406–3413. https://doi.org/10.1128/IAI.00078-07

Ceri H, Olson ME, Stremick C, Read RR, Morck D, Buret A (1999) The Calgary Biofilm Device: new technology for rapid determination of antibiotic susceptibilities of bacterial biofilms. J Clin Microbiol 37(6):1771–1776. https://doi.org/10.1128/JCM.37.6.1771-1776.1999

Chapman MR, Robinson LS, Pinkner JS, Roth R, Heuser J, Hammar M, Normark S, Hultgren SJ (2002) Role of Escherichia coli curli operons in directing amyloid fiber formation. Science 295(5556):851–855. https://doi.org/10.1126/science.1067484

Charles RC, Kelly M, Tam JM, Akter A, Hossain M, Islam K, Biswas R, Kamruzzaman M, Chowdhury F, Khan AI, Leung DT, Weil A, LaRocque RC, Bhuiyan TR, Rahman A, Mayo-Smith LM, Becker RL, Vyas JM, Faherty CS, Nickerson KP, Giffen S, Ritter AS, Waldor MK, Xu P, Kovac P, Calderwood SB, Kauffman RC, Wrammert J, Qadri F, Harris JB, Ryan ET

(2020) Humans surviving cholera develop antibodies against Vibrio cholerae O-specific polysaccharide that inhibit pathogen motility. mBio 11(6). https://doi.org/10.1128/mBio.02847-20

Chastre J, François B, Bourgeois M, Komnos A, Ferrer R, Rahav G, De Schryver N, Lepape A, Koksal I, Luyt C-E, Garcia MS, Torres A, Holland TL, Holland TL, Ali O, Shoemaker K, Ren P, Ruzin A, Jiang Y, Colbert S, Vandamme D, Bellamy T, Reisner C, Dubovsky F, Jafri HS (2020) 635. Efficacy, pharmacokinetics (PK), and safety profile of MEDI3902, an anti-Pseudomonas aeruginosa bispecific human monoclonal antibody in mechanically ventilated intensive care unit patients; results of the phase 2 EVADE study conducted by the Public-Private COMBACTE-MAGNET Consortium in the Innovative Medicines Initiative (IMI) Program. Open Forum Infectious Diseases 7(Suppl_1):S377–S378. https://doi.org/10.1093/ofid/ofaa439.829

Chiu ML, Goulet DR, Teplyakov A, Gilliland GL (2019) Antibody structure and function: the basis for engineering therapeutics. Antibodies (Basel) 8(4). https://doi.org/10.3390/antib8040055

Chupacova J, Borghi E, Morace G, Los A, Bujdakova H (2018) Anti-biofilm activity of antibody directed against surface antigen complement receptor 3-related protein-comparison of Candida albicans and Candida dubliniensis. Pathog Dis 76(1). https://doi.org/10.1093/femspd/ftx127

Clegg J, Soldaini E, McLoughlin RM, Rittenhouse S, Bagnoli F, Phogat S (2021) Staphylococcus aureus vaccine research and development: the past, present and future, including novel therapeutic strategiesincluding novel therapeutic strategies. Front Immunol 12:705360. https://doi.org/10.3389/fimmu.2021.705360

Coleman DA, Oh SH, Zhao X, Zhao H, Hutchins JT, Vernachio JH, Patti JM, Hoyer LL (2009) Monoclonal antibodies specific for Candida albicans Als3 that immunolabel fungal cells in vitro and in vivo and block adhesion to host surfaces. J Microbiol Methods 78(1):71–78. https://doi.org/10.1016/j.mimet.2009.05.002

Coley J, Tarelli E, Archibald AR, Baddiley J (1978) The linkage between teichoic acid and peptidoglycan in bacterial cell walls. FEBS Lett 88(1):1–9

Colvin KM, Gordon VD, Murakami K, Borlee BR, Wozniak DJ, Wong GC, Parsek MR (2011) The pel polysaccharide can serve a structural and protective role in the biofilm matrix of Pseudomonas aeruginosa. PLoS Pathog 7(1):e1001264. https://doi.org/10.1371/journal.ppat.1001264

Colvin KM, Irie Y, Tart CS, Urbano R, Whitney JC, Ryder C, Howell PL, Wozniak DJ, Parsek MR (2012) The Pel and Psl polysaccharides provide Pseudomonas aeruginosa structural redundancy within the biofilm matrix. Environ Microbiol 14(8):1913–1928. https://doi.org/10.1111/j.1462-2920.2011.02657.x

Corti D, Misasi J, Mulangu S, Stanley DA, Kanekiyo M, Wollen S, Ploquin A, Doria-Rose NA, Staupe RP, Bailey M, Shi W, Choe M, Marcus H, Thompson EA, Cagigi A, Silacci C, Fernandez-Rodriguez B, Perez L, Sallusto F, Vanzetta F, Agatic G, Cameroni E, Kisalu N, Gordon I, Ledgerwood JE, Mascola JR, Graham BS, Muyembe-Tamfun JJ, Trefry JC, Lanzavecchia A, Sullivan NJ (2016) Protective monotherapy against lethal Ebola virus infection by a potently neutralizing antibody. Science 351(6279):1339–1342. https://doi.org/10.1126/science.aad5224

Costerton JW, Stewart PS, Greenberg EP (1999) Bacterial biofilms: a common cause of persistent infections. Science 284(5418):1318–1322. https://doi.org/10.1126/science.284.5418.1318

Craig L, Forest KT, Maier B (2019) Type IV pili: dynamics, biophysics and functional consequences. Nat Rev Microbiol 17(7):429–440. https://doi.org/10.1038/s41579-019-0195-4

Cramton SE, Gerke C, Schnell NF, Nichols WW, Gotz F (1999) The intercellular adhesion (ica) locus is present in Staphylococcus aureus and is required for biofilm formation. Infect Immun 67(10):5427–5433. https://doi.org/10.1128/IAI.67.10.5427-5433.1999

Cywes-Bentley C, Skurnik D, Zaidi T, Roux D, Deoliveira RB, Garrett WS, Lu X, O'Malley J, Kinzel K, Zaidi T, Rey A, Perrin C, Fichorova RN, Kayatani AK, Maira-Litran T, Gening ML, Tsvetkov YE, Nifantiev NE, Bakaletz LO, Pelton SI, Golenbock DT, Pier GB (2013) Antibody to a conserved antigenic target is protective against diverse prokaryotic and eukaryotic pathogens. Proc Natl Acad Sci U S A 110(24):E2209–E2218. https://doi.org/10.1073/pnas.1303573110

Darby C, Hsu JW, Ghori N, Falkow S (2002) Caenorhabditis elegans: plague bacteria biofilm blocks food intake. Nature 417(6886):243–244. https://doi.org/10.1038/417243a

De Bernardis F, Liu H, O'Mahony R, La Valle R, Bartollino S, Sandini S, Grant S, Brewis N, Tomlinson I, Basset RC, Holton J, Roitt IM, Cassone A (2007) Human domain antibodies against virulence traits of Candida albicans inhibit fungus adherence to vaginal epithelium and protect against experimental vaginal candidiasis. J Infect Dis 195(1):149–157. https://doi.org/10.1086/509891

Dekkerova J, Lopez-Ribot JL, Bujdakova H (2019) Activity of anti-CR3-RP polyclonal antibody against biofilms formed by Candida auris, a multidrug-resistant emerging fungal pathogen. Eur J Clin Microbiol Infect Dis 38(1):101–108. https://doi.org/10.1007/s10096-018-3400-x

Dietrich LE, Okegbe C, Price-Whelan A, Sakhtah H, Hunter RC, Newman DK (2013) Bacterial community morphogenesis is intimately linked to the intracellular redox state. J Bacteriol 195(7):1371–1380. https://doi.org/10.1128/JB.02273-12

DiGiandomenico A, Warrener P, Hamilton M, Guillard S, Ravn P, Minter R, Camara MM, Venkatraman V, Macgill RS, Lin J, Wang Q, Keller AE, Bonnell JC, Tomich M, Jermutus L, McCarthy MP, Melnick DA, Suzich JA, Stover CK (2012) Identification of broadly protective human antibodies to Pseudomonas aeruginosa exopolysaccharide Psl by phenotypic screening. J Exp Med 209(7):1273–1287. https://doi.org/10.1084/jem.20120033

DiGiandomenico A, Keller AE, Gao C, Rainey GJ, Warrener P, Camara MM, Bonnell J, Fleming R, Bezabeh B, Dimasi N, Sellman BR, Hilliard J, Guenther CM, Datta V, Zhao W, Gao C, Yu XQ, Suzich JA, Stover CK (2014) A multifunctional bispecific antibody protects against Pseudomonas aeruginosa. Sci Transl Med 6(262):262ra155. https://doi.org/10.1126/scitranslmed.3009655

Domanski PJ, Patel PR, Bayer AS, Zhang L, Hall AE, Syribeys PJ, Gorovits EL, Bryant D, Vernachio JH, Hutchins JT, Patti JM (2005) Characterization of a humanized monoclonal antibody recognizing clumping factor A expressed by Staphylococcus aureus. Infect Immun 73(8):5229–5232. https://doi.org/10.1128/IAI.73.8.5229-5232.2005

Dougan M, Nirula A, Azizad M, Mocherla B, Gottlieb RL, Chen P, Hebert C, Perry R, Boscia J, Heller B, Morris J, Crystal C, Igbinadolor A, Huhn G, Cardona J, Shawa I, Kumar P, Adams AC, Van Naarden J, Custer KL, Durante M, Oakley G, Schade AE, Holzer TR, Ebert PJ, Higgs RE, Kallewaard NL, Sabo J, Patel DR, Dabora MC, Klekotka P, Shen L, Skovronsky DM, BLAZE-1 Investigators (2021) Bamlanivimab plus etesevimab in mild or moderate Covid-19. N Engl J Med. https://doi.org/10.1056/NEJMoa2102685

Duyckaerts C (2013) Neurodegenerative lesions: seeding and spreading. Rev Neurol (Paris) 169(10):825–833. https://doi.org/10.1016/j.neurol.2013.07.018

Estelles A, Woischnig AK, Liu K, Stephenson R, Lomongsod E, Nguyen D, Zhang J, Heidecker M, Yang Y, Simon RJ, Tenorio E, Ellsworth S, Leighton A, Ryser S, Gremmelmaier NK, Kauvar LM (2016) A high-affinity native human antibody disrupts biofilm from Staphylococcus aureus bacteria and potentiates antibiotic efficacy in a mouse implant infection model. Antimicrob Agents Chemother 60(4):2292–2301. https://doi.org/10.1128/AAC.02588-15

Flemming HC, Wingender J, Szewzyk U, Steinberg P, Rice SA, Kjelleberg S (2016) Biofilms: an emergent form of bacterial life. Nat Rev Microbiol 14(9):563–575. https://doi.org/10.1038/nrmicro.2016.94

Flores-Mireles AL, Pinkner JS, Caparon MG, Hultgren SJ (2014) EbpA vaccine antibodies block binding of Enterococcus faecalis to fibrinogen to prevent catheter-associated bladder infection in mice. Sci Transl Med 6(254):254ra127. https://doi.org/10.1126/scitranslmed.3009384

Flores-Vargas G, Bergsveinson J, Lawrence JR, Korber DR (2021) Environmental biofilms as reservoirs for antimicrobial resistance. Front Microbiol 12:766242. https://doi.org/10.3389/fmicb.2021.766242

Forthal DN (2014) Functions of antibodies. Microbiol Spectr 2(4):AID-0019-2014. https://doi.org/10.1128/microbiolspec.AID-0019-2014

Foster TJ (2019) The MSCRAMM family of cell-wall-anchored surface proteins of gram-positive cocci. Trends Microbiol 27(11):927–941. https://doi.org/10.1016/j.tim.2019.06.007

Foster TJ, Geoghegan JA, Ganesh VK, Hook M (2014) Adhesion, invasion and evasion: the many functions of the surface proteins of Staphylococcus aureus. Nat Rev Microbiol 12(1):49–62. https://doi.org/10.1038/nrmicro3161

Franca A, Vilanova M, Cerca N, Pier GB (2013) Monoclonal antibody raised against PNAG has variable effects on static S. epidermidis biofilm accumulation in vitro. Int J Biol Sci 9(5): 518–520. https://doi.org/10.7150/ijbs.6102

Francois B, Jafri HS, Chastre J, Sanchez-Garcia M, Eggimann P, Dequin PF, Huberlant V, Vina Soria L, Boulain T, Bretonniere C, Pugin J, Trenado J, Hernandez Padilla AC, Ali O, Shoemaker K, Ren P, Coenjaerts FE, Ruzin A, Barraud O, Timbermont L, Lammens C, Pierre V, Wu Y, Vignaud J, Colbert S, Bellamy T, Esser MT, Dubovsky F, Bonten MJ, Goossens H, Laterre PF, COMBACTE Consortium and the SAATELLITE Study Group (2021) Efficacy and safety of suvratoxumab for prevention of Staphylococcus aureus ventilator-associated pneumonia (SAATELLITE): a multicentre, randomised, double-blind, placebo-controlled, parallel-group, phase 2 pilot trial. Lancet Infect Dis. https://doi.org/10.1016/S1473-3099(20)30995-6

Freund NT, Horwitz JA, Nogueira L, Sievers SA, Scharf L, Scheid JF, Gazumyan A, Liu C, Velinzon K, Goldenthal A, Sanders RW, Moore JP, Bjorkman PJ, Seaman MS, Walker BD, Klein F, Nussenzweig MC (2015) A new glycan-dependent CD4-binding site neutralizing antibody exerts pressure on HIV-1 in vivo. PLoS Pathog 11(10):e1005238. https://doi.org/10.1371/journal.ppat.1005238

Friedman L, Kolter R (2004a) Genes involved in matrix formation in Pseudomonas aeruginosa PA14 biofilms. Mol Microbiol 51(3):675–690. https://doi.org/10.1046/j.1365-2958.2003.03877.x

Friedman L, Kolter R (2004b) Two genetic loci produce distinct carbohydrate-rich structural components of the Pseudomonas aeruginosa biofilm matrix. J Bacteriol 186(14):4457–4465. https://doi.org/10.1128/JB.186.14.4457-4465.2004

Gao Y, Zhang TY, Yuan Q, Xia NS (2017) Antibody-mediated immunotherapy against chronic hepatitis B virus infection. Hum Vaccin Immunother 13(8):1768–1773. https://doi.org/10.1080/21645515.2017.1319021

Gitlin AD, Mayer CT, Oliveira TY, Shulman Z, Jones MJ, Koren A, Nussenzweig MC (2015) Humoral immunity. T cell help controls the speed of the cell cycle in germinal center B cells. Science 349(6248):643–646. https://doi.org/10.1126/science.aac4919

Goodman SD, Obergfell KP, Jurcisek JA, Novotny LA, Downey JS, Ayala EA, Tjokro N, Li B, Justice SS, Bakaletz LO (2011) Biofilms can be dispersed by focusing the immune system on a common family of bacterial nucleoid-associated proteins. Mucosal Immunol 4(6):625–637. https://doi.org/10.1038/mi.2011.27

Green DJ, Press OW (2017) Whither radioimmunotherapy: to be or not to be? Cancer Res 77(9): 2191–2196. https://doi.org/10.1158/0008-5472.CAN-16-2523

Greig SL (2016) Obiltoxaximab: first global approval. Drugs 76(7):823–830. https://doi.org/10.1007/s40265-016-0577-0

Hall AE, Domanski PJ, Patel PR, Vernachio JH, Syribeys PJ, Gorovits EL, Johnson MA, Ross JM, Hutchins JT, Patti JM (2003) Characterization of a protective monoclonal antibody recognizing Staphylococcus aureus MSCRAMM protein clumping factor A. Infect Immun 71(12): 6864–6870. https://doi.org/10.1128/IAI.71.12.6864-6870.2003

Harriman WD, Collarini EJ, Sperinde GV, Strandh M, Fatholahi MM, Dutta A, Lee Y, Mettler SE, Keyt BA, Ellsworth SL, Kauvar LM (2009) Antibody discovery via multiplexed single cell characterization. J Immunol Methods 341(1-2):135–145. https://doi.org/10.1016/j.jim.2008.11.009

Hassanov T, Karunker I, Steinberg N, Erez A, Kolodkin-Gal I (2018) Novel antibiofilm chemotherapies target nitrogen from glutamate and glutamine. Sci Rep 8(1):7097. https://doi.org/10.1038/s41598-018-25401-z

Hill D, Rose B, Pajkos A, Robinson M, Bye P, Bell S, Elkins M, Thompson B, Macleod C, Aaron SD, Harbour C (2005) Antibiotic susceptabilities of Pseudomonas aeruginosa isolates derived from patients with cystic fibrosis under aerobic, anaerobic, and biofilm conditions. J Clin Microbiol 43(10):5085–5090. https://doi.org/10.1128/JCM.43.10.5085-5090.2005

Hu J, Xu T, Zhu T, Lou Q, Wang X, Wu Y, Huang R, Liu J, Liu H, Yu F, Ding B, Huang Y, Tong W, Qu D (2011) Monoclonal antibodies against accumulation-associated protein affect EPS biosynthesis and enhance bacterial accumulation of Staphylococcus epidermidis. PLoS One 6(6):e20918. https://doi.org/10.1371/journal.pone.0020918

Hua L, Hilliard JJ, Shi Y, Tkaczyk C, Cheng LI, Yu X, Datta V, Ren S, Feng H, Zinsou R, Keller A, O'Day T, Du Q, Cheng L, Damschroder M, Robbie G, Suzich J, Stover CK, Sellman BR (2014) Assessment of an anti-alpha-toxin monoclonal antibody for prevention and treatment of Staphylococcus aureus-induced pneumonia. Antimicrob Agents Chemother 58(2):1108–1117. https://doi.org/10.1128/AAC.02190-13

Huang J, Ofek G, Laub L, Louder MK, Doria-Rose NA, Longo NS, Imamichi H, Bailer RT, Chakrabarti B, Sharma SK, Alam SM, Wang T, Yang Y, Zhang B, Migueles SA, Wyatt R, Haynes BF, Kwong PD, Mascola JR, Connors M (2012) Broad and potent neutralization of HIV-1 by a gp41-specific human antibody. Nature 491(7424):406–412. https://doi.org/10.1038/nature11544

Hufnagel DA, Tukel C, Chapman MR (2013) Disease to dirt: the biology of microbial amyloids. PLoS Pathog 9(11). https://doi.org/10.1371/journal.ppat.1003740

Izano EA, Sadovskaya I, Vinogradov E, Mulks MH, Velliyagounder K, Ragunath C, Kher WB, Ramasubbu N, Jabbouri S, Perry MB, Kaplan JB (2007) Poly-N-acetylglucosamine mediates biofilm formation and antibiotic resistance in Actinobacillus pleuropneumoniae. Microb Pathog 43(1):1–9. https://doi.org/10.1016/j.micpath.2007.02.004

Jackson KD, Starkey M, Kremer S, Parsek MR, Wozniak DJ (2004) Identification of psl, a locus encoding a potential exopolysaccharide that is essential for Pseudomonas aeruginosa PAO1 biofilm formation. J Bacteriol 186(14):4466–4475. https://doi.org/10.1128/JB.186.14.4466-4475.2004

Janeway CA, Travers P, Walport M, Shlomchik MJ (2001a) The generation of diversity in immunoglobulins. In: Immunobiology: the immune system in health and disease, 5th edn. Garland, New York. https://www.ncbi.nlm.nih.gov/books/NBK27140/

Janeway CA, Travers P, Walport M, Shlomchik MJ (2001b) The structure of a typical antibody molecule. In: Immunobiology: the immune system in health and disease, 5th edn. Garland, New York

Jonas K, Tomenius H, Kader A, Normark S, Romling U, Belova LM, Melefors O (2007) Roles of curli, cellulose and BapA in Salmonella biofilm morphology studied by atomic force microscopy. BMC Microbiol 7:70. https://doi.org/10.1186/1471-2180-7-70

Jones PT, Dear PH, Foote J, Neuberger MS, Winter G (1986) Replacing the complementarity-determining regions in a human antibody with those from a mouse. Nature 321(6069):522–525. https://doi.org/10.1038/321522a0

Kainz K, Bauer MA, Madeo F, Carmona-Gutierrez D (2020) Fungal infections in humans: the silent crisis. Microb Cell 7(6):143–145. https://doi.org/10.15698/mic2020.06.718

Kaplan JB, Velliyagounder K, Ragunath C, Rohde H, Mack D, Knobloch JK, Ramasubbu N (2004) Genes involved in the synthesis and degradation of matrix polysaccharide in Actinobacillus actinomycetemcomitans and Actinobacillus pleuropneumoniae biofilms. J Bacteriol 186(24): 8213–8220. https://doi.org/10.1128/JB.186.24.8213-8220.2004

Karygianni L, Ren Z, Koo H, Thurnheer T (2020) Biofilm matrixome: extracellular components in structured microbial communities. Trends Microbiol 28(8):668–681. https://doi.org/10.1016/j.tim.2020.03.016

Kawai Y, Marles-Wright J, Cleverley RM, Emmins R, Ishikawa S, Kuwano M, Heinz N, Bui NK, Hoyland CN, Ogasawara N, Lewis RJ, Vollmer W, Daniel RA, Errington J (2011) A widespread family of bacterial cell wall assembly proteins. EMBO J 30(24):4931–4941. https://doi.org/10.1038/emboj.2011.358

Kayed R, Head E, Thompson JL, McIntire TM, Milton SC, Cotman CW, Glabe CG (2003) Common structure of soluble amyloid oligomers implies common mechanism of pathogenesis. Science 300(5618):486–489. https://doi.org/10.1126/science.1079469

Kelly-Quintos C, Cavacini LA, Posner MR, Goldmann D, Pier GB (2006) Characterization of the opsonic and protective activity against Staphylococcus aureus of fully human monoclonal antibodies specific for the bacterial surface polysaccharide poly-N-acetylglucosamine. Infect Immun 74(5):2742–2750. https://doi.org/10.1128/IAI.74.5.2742-2750.2006

Kisalu NK, Idris AH, Weidle C, Flores-Garcia Y, Flynn BJ, Sack BK, Murphy S, Schon A, Freire E, Francica JR, Miller AB, Gregory J, March S, Liao HX, Haynes BF, Wiehe K, Trama AM, Saunders KO, Gladden MA, Monroe A, Bonsignori M, Kanekiyo M, Wheatley AK, McDermott AB, Farney SK, Chuang GY, Zhang B, Kc N, Chakravarty S, Kwong PD, Sinnis P, Bhatia SN, Kappe SHI, Sim BKL, Hoffman SL, Zavala F, Pancera M, Seder RA (2018) A human monoclonal antibody prevents malaria infection by targeting a new site of vulnerability on the parasite. Nat Med 24(4):408–416. https://doi.org/10.1038/nm.4512

Kohler G, Milstein C (1975) Continuous cultures of fused cells secreting antibody of predefined specificity. Nature 256(5517):495–497. https://doi.org/10.1038/256495a0

Kolodkin-Gal I, Elsholz AK, Muth C, Girguis PR, Kolter R, Losick R (2013) Respiration control of multicellularity in Bacillus subtilis by a complex of the cytochrome chain with a membrane-embedded histidine kinase. Genes Dev 27(8):887–899. https://doi.org/10.1101/gad.215244.113

Kugadas A, Geddes-McAlister J, Guy E, DiGiandomenico A, Sykes DB, Mansour MK, Mirchev R, Gadjeva M (2019) Frontline Science: Employing enzymatic treatment options for management of ocular biofilm-based infections. J Leukoc Biol 105(6):1099–1110. https://doi.org/10.1002/JLB.4HI0918-364RR

Lam H, Kesselly A, Stegalkina S, Kleanthous H, Yethon JA (2014) Antibodies to PhnD inhibit staphylococcal biofilms. Infect Immun 82(9):3764–3774. https://doi.org/10.1128/IAI.02168-14

Larsen RA, Pappas PG, Perfect J, Aberg JA, Casadevall A, Cloud GA, James R, Filler S, Dismukes WE (2005) Phase I evaluation of the safety and pharmacokinetics of murine-derived anticryptococcal antibody 18B7 in subjects with treated cryptococcal meningitis. Antimicrob Agents Chemother 49(3):952–958. https://doi.org/10.1128/AAC.49.3.952-958.2005

Lehar SM, Pillow T, Xu M, Staben L, Kajihara KK, Vandlen R, DePalatis L, Raab H, Hazenbos WL, Morisaki JH, Kim J, Park S, Darwish M, Lee BC, Hernandez H, Loyet KM, Lupardus P, Fong R, Yan D, Chalouni C, Luis E, Khalfin Y, Plise E, Cheong J, Lyssikatos JP, Strandh M, Koefoed K, Andersen PS, Flygare JA, Wah Tan M, Brown EJ, Mariathasan S (2015) Novel antibody-antibiotic conjugate eliminates intracellular S. aureus. Nature 527(7578):323–328. https://doi.org/10.1038/nature16057

Lembre P, Di Martino P, Vendrely C (2014) Amyloid peptides derived from CsgA and FapC modify the viscoelastic properties of biofilm model matrices. Biofouling 30(4):415–426. https://doi.org/10.1080/08927014.2014.880112

Levites Y, O'Nuallain B, Puligedda RD, Ondrejcak T, Adekar SP, Chen C, Cruz PE, Rosario AM, Macy S, Mably AJ, Walsh DM, Vidal R, Solomon A, Brown D, Rowan MJ, Golde TE, Dessain SK (2015) A human monoclonal IgG that binds abeta assemblies and diverse amyloids exhibits anti-amyloid activities in vitro and in vivo. J Neurosci 35(16):6265–6276. https://doi.org/10.1523/JNEUROSCI.5109-14.2015

Lewis K (2007) Persister cells, dormancy and infectious disease. Nat Rev Microbiol 5(1):48–56. https://doi.org/10.1038/nrmicro1557

Li H, Javid B (2018) Antibodies and tuberculosis: finally coming of age? Nat Rev Immunol 18(9):591–596. https://doi.org/10.1038/s41577-018-0028-0

Liao HX, Lynch R, Zhou T, Gao F, Alam SM, Boyd SD, Fire AZ, Roskin KM, Schramm CA, Zhang Z, Zhu J, Shapiro L, Program NCS, Mullikin JC, Gnanakaran S, Hraber P, Wiehe K, Kelsoe G, Yang G, Xia SM, Montefiori DC, Parks R, Lloyd KE, Scearce RM, Soderberg KA, Cohen M, Kamanga G, Louder MK, Tran LM, Chen Y, Cai F, Chen S, Moquin S, Du X, Joyce MG, Srivatsan S, Zhang B, Zheng A, Shaw GM, Hahn BH, Kepler TB, Korber BT, Kwong PD,

Mascola JR, Haynes BF (2013) Co-evolution of a broadly neutralizing HIV-1 antibody and founder virus. Nature 496(7446):469–476. https://doi.org/10.1038/nature12053

Lisanne de Vor BVD, van Kessel KPM, Kavanaugh JS, de Haas CJC, Aerts PC, Viveen MC, Boel ECH, Fluit AC, Kwiecinski JM, Krijger GC, Ramakers RM, Beekman FJ, Dadachova E, Lam MGEH, Vogely HC, van der Wal BCH, van Strijp JAG, Horswill AR, Weinans H, Rooijakkers SHM (2021) Human monoclonal antibodies against Staphylococcus aureus surface antigens recognize in vitro biofilm and in vivo implant infections. bioRxiv. https://doi.org/10.1101/2021.02.09.429966

Liu J, Prindle A, Humphries J, Gabalda-Sagarra M, Asally M, Lee DY, Ly S, Garcia-Ojalvo J, Suel GM (2015) Metabolic co-dependence gives rise to collective oscillations within biofilms. Nature 523(7562):550–554. https://doi.org/10.1038/nature14660

Lonberg N, Taylor LD, Harding FA, Trounstine M, Higgins KM, Schramm SR, Kuo CC, Mashayekh R, Wymore K, McCabe JG et al (1994) Antigen-specific human antibodies from mice comprising four distinct genetic modifications. Nature 368(6474):856–859. https://doi.org/10.1038/368856a0

Lorenz U, Ohlsen K, Karch H, Hecker M, Thiede A, Hacker J (2000) Human antibody response during sepsis against targets expressed by methicillin resistant Staphylococcus aureus. FEMS Immunol Med Microbiol 29(2):145–153. https://doi.org/10.1111/j.1574-695X.2000.tb01517.x

Lorenz U, Lorenz B, Schmitter T, Streker K, Erck C, Wehland J, Nickel J, Zimmermann B, Ohlsen K (2011) Functional antibodies targeting IsaA of Staphylococcus aureus augment host immune response and open new perspectives for antibacterial therapy. Antimicrob Agents Chemother 55(1):165–173. https://doi.org/10.1128/AAC.01144-10

Lu LL, Suscovich TJ, Fortune SM, Alter G (2018) Beyond binding: antibody effector functions in infectious diseases. Nat Rev Immunol 18(1):46–61. https://doi.org/10.1038/nri.2017.106

Lu RM, Hwang YC, Liu IJ, Lee CC, Tsai HZ, Li HJ, Wu HC (2020) Development of therapeutic antibodies for the treatment of diseases. J Biomed Sci 27(1):1. https://doi.org/10.1186/s12929-019-0592-z

Ma L, Jackson KD, Landry RM, Parsek MR, Wozniak DJ (2006) Analysis of Pseudomonas aeruginosa conditional psl variants reveals roles for the psl polysaccharide in adhesion and maintaining biofilm structure postattachment. J Bacteriol 188(23):8213–8221. https://doi.org/10.1128/JB.01202-06

Mack D, Fischer W, Krokotsch A, Leopold K, Hartmann R, Egge H, Laufs R (1996) The intercellular adhesin involved in biofilm accumulation of Staphylococcus epidermidis is a linear beta-1,6-linked glucosaminoglycan: purification and structural analysis. J Bacteriol 178(1):175–183. https://doi.org/10.1128/jb.178.1.175-183.1996

Mann EE, Wozniak DJ (2012) Pseudomonas biofilm matrix composition and niche biology. FEMS Microbiol Rev 36(4):893–916. https://doi.org/10.1111/j.1574-6976.2011.00322.x

Martinez LR, Casadevall A (2005) Specific antibody can prevent fungal biofilm formation and this effect correlates with protective efficacy. Infect Immun 73(10):6350–6362. https://doi.org/10.1128/IAI.73.10.6350-6362.2005

Martinez LR, Bryan RA, Apostolidis C, Morgenstern A, Casadevall A, Dadachova E (2006a) Antibody-guided alpha radiation effectively damages fungal biofilms. Antimicrob Agents Chemother 50(6):2132–2136. https://doi.org/10.1128/AAC.00120-06

Martinez LR, Christaki E, Casadevall A (2006b) Specific antibody to Cryptococcus neoformans glucurunoxylomannan antagonizes antifungal drug action against cryptococcal biofilms in vitro. J Infect Dis 194(2):261–266. https://doi.org/10.1086/504722

Matsukawa M, Greenberg EP (2004) Putative exopolysaccharide synthesis genes influence Pseudomonas aeruginosa biofilm development. J Bacteriol 186(14):4449–4456. https://doi.org/10.1128/JB.186.14.4449-4456.2004

McCafferty J, Griffiths AD, Winter G, Chiswell DJ (1990) Phage antibodies: filamentous phage displaying antibody variable domains. Nature 348(6301):552–554. https://doi.org/10.1038/348552a0

McCarthy H, Waters EM, Bose JL, Foster S, Bayles KW, O'Neill E, Fey PD, O'Gara JP (2016) The major autolysin is redundant for Staphylococcus aureus USA300 LAC JE2 virulence in a murine device-related infection model. FEMS Microbiol Lett 363(9). https://doi.org/10.1093/femsle/fnw087

Metcalf WW, Wanner BL (1993) Mutational analysis of an Escherichia coli fourteen-gene operon for phosphonate degradation using TnphoA' elements. J Bacteriol 175(11):3430–3442. https://doi.org/10.1128/jb.175.11.3430-3442.1993

Mishra NN, Ali S, Shukla PK (2015) A monoclonal antibody against 47.2 kDa cell surface antigen prevents adherence and affects biofilm formation of Candida albicans. World J Microbiol Biotechnol 31(1):11–21. https://doi.org/10.1007/s11274-014-1760-7

Mitchell KF, Zarnowski R, Andes DR (2016) Fungal super glue: the biofilm matrix and its composition, assembly, and functions. PLoS Pathog 12(9):e1005828. https://doi.org/10.1371/journal.ppat.1005828

Mokrzan EM, Ward MO, Bakaletz LO (2016) Type IV pilus expression is upregulated in nontypeable haemophilus influenzae biofilms formed at the temperature of the human nasopharynx. J Bacteriol 198(19):2619–2630. https://doi.org/10.1128/JB.01022-15

Mokrzan EM, Novotny LA, Brockman KL, Bakaletz LO (2018) Antibodies against the majority subunit (PilA) of the type IV pilus of nontypeable Haemophilus influenzae disperse Moraxella catarrhalis from a dual-species biofilm. mBio 9(6). https://doi.org/10.1128/mBio.02423-18

Monds RD, O'Toole GA (2009) The developmental model of microbial biofilms: ten years of a paradigm up for review. Trends Microbiol 17(2):73–87. https://doi.org/10.1016/j.tim.2008.11.001

Mor M, Werbner M, Alter J, Safra M, Chomsky E, Lee JC, Hada-Neeman S, Polonsky K, Nowell CJ, Clark AE, Roitburd-Berman A, Ben-Shalom N, Navon M, Rafael D, Sharim H, Kiner E, Griffis ER, Gershoni JM, Kobiler O, Leibel SL, Zimhony O, Carlin AF, Yaari G, Dessau M, Gal-Tanamy M, Hagin D, Croker BA, Freund NT (2021) Multi-clonal SARS-CoV-2 neutralization by antibodies isolated from severe COVID-19 convalescent donors. PLoS Pathog 17(2): e1009165. https://doi.org/10.1371/journal.ppat.1009165

Moragues MD, Omaetxebarria MJ, Elguezabal N, Sevilla MJ, Conti S, Polonelli L, Ponton J (2003) A monoclonal antibody directed against a Candida albicans cell wall mannoprotein exerts three anti-C. albicans activities. Infect Immun 71(9):5273–5279. https://doi.org/10.1128/IAI.71.9.5273-5279.2003

Morris AJ, Jackson L, Cw Yau Y, Reichhardt C, Beaudoin T, Uwumarenogie S, Guttman KM, Lynne Howell P, Parsek MR, Hoffman LR, Nguyen D, DiGiandomenico A, Guttman DS, Wozniak DJ, Waters VJ (2021) The role of Psl in the failure to eradicate Pseudomonas aeruginosa biofilms in children with cystic fibrosis. NPJ Biofilms Microbiomes 7(1):63. https://doi.org/10.1038/s41522-021-00234-3

Naglik JR, Challacombe SJ, Hube B (2003) Candida albicans secreted aspartyl proteinases in virulence and pathogenesis. Microbiol Mol Biol Rev 67(3):400–428., Table of contents. https://doi.org/10.1128/MMBR.67.3.400-428.2003

Nair N, Vinod V, Suresh MK, Vijayrajratnam S, Biswas L, Peethambaran R, Vasudevan AK, Biswas R (2015) Amidase, a cell wall hydrolase, elicits protective immunity against Staphylococcus aureus and S. epidermidis. Int J Biol Macromol 77:314–321. https://doi.org/10.1016/j.ijbiomac.2015.03.047

Navalkele BD, Chopra T (2018) Bezlotoxumab: an emerging monoclonal antibody therapy for prevention of recurrent Clostridium difficile infection. Biologics 12:11–21. https://doi.org/10.2147/BTT.S127099

Nicastro L, Tukel C (2019) Bacterial amyloids: the link between bacterial infections and autoimmunity. Trends Microbiol 27(11):954–963. https://doi.org/10.1016/j.tim.2019.07.002

Novotny LA, Jurcisek JA, Ward MO Jr, Jordan ZB, Goodman SD, Bakaletz LO (2015) Antibodies against the majority subunit of type IV Pili disperse nontypeable Haemophilus influenzae biofilms in a LuxS-dependent manner and confer therapeutic resolution of experimental otitis media. Mol Microbiol 96(2):276–292. https://doi.org/10.1111/mmi.12934

Novotny LA, Jurcisek JA, Goodman SD, Bakaletz LO (2016) Monoclonal antibodies against DNA-binding tips of DNABII proteins disrupt biofilms in vitro and induce bacterial clearance in vivo. EBioMedicine 10:33–44. https://doi.org/10.1016/j.ebiom.2016.06.022

Novotny LA, Goodman SD, Bakaletz LO (2019) Redirecting the immune response towards immunoprotective domains of a DNABII protein resolves experimental otitis media. NPJ Vaccines 4:43. https://doi.org/10.1038/s41541-019-0137-1

Novotny LA, Goodman SD, Bakaletz LO (2020) Targeting a bacterial DNABII protein with a chimeric peptide immunogen or humanised monoclonal antibody to prevent or treat recalcitrant biofilm-mediated infections. EBioMedicine 59:102867. https://doi.org/10.1016/j.ebiom.2020.102867

Oganesyan V, Peng L, Damschroder MM, Cheng L, Sadowska A, Tkaczyk C, Sellman BR, Wu H, Dall'Acqua WF (2014) Mechanisms of neutralization of a human anti-alpha-toxin antibody. J Biol Chem 289(43):29874–29880. https://doi.org/10.1074/jbc.M114.601328

Oleastro M, Menard A (2013) The role of Helicobacter pylori outer membrane proteins in adherence and pathogenesis. Biology (Basel) 2(3):1110–1134. https://doi.org/10.3390/biology2031110

O'Neill E, Pozzi C, Houston P, Humphreys H, Robinson DA, Loughman A, Foster TJ, O'Gara JP (2008) A novel Staphylococcus aureus biofilm phenotype mediated by the fibronectin-binding proteins, FnBPA and FnBPB. J Bacteriol 190(11):3835–3850. https://doi.org/10.1128/JB.00167-08

O'Nuallain B, Wetzel R (2002) Conformational Abs recognizing a generic amyloid fibril epitope. Proc Natl Acad Sci U S A 99(3):1485–1490. https://doi.org/10.1073/pnas.022662599

Oppenheimer-Shaanan Y, Steinberg N, Kolodkin-Gal I (2013) Small molecules are natural triggers for the disassembly of biofilms. Trends Microbiol. https://doi.org/10.1016/j.tim.2013.08.005

Parise G, Mishra M, Itoh Y, Romeo T, Deora R (2007) Role of a putative polysaccharide locus in Bordetella biofilm development. J Bacteriol 189(3):750–760. https://doi.org/10.1128/JB.00953-06

Perlin DS, Shor E, Zhao Y (2015) Update on antifungal drug resistance. Curr Clin Microbiol Rep 2(2):84–95. https://doi.org/10.1007/s40588-015-0015-1

Pierce CG, Vila T, Romo JA, Montelongo-Jauregui D, Wall G, Ramasubramanian A, Lopez-Ribot JL (2017) The Candida albicans biofilm matrix: composition, structure and function. J Fungi (Basel) 3(1):doi:10.3390/jof3010014

Presta LG (2008) Molecular engineering and design of therapeutic antibodies. Curr Opin Immunol 20(4):460–470. https://doi.org/10.1016/j.coi.2008.06.012

Ramsey MM, Rumbaugh KP, Whiteley M (2011) Metabolite cross-feeding enhances virulence in a model polymicrobial infection. PLoS Pathog 7(3):e1002012. https://doi.org/10.1371/journal.ppat.1002012

Ravetch JV, Clynes RA (1998) Divergent roles for Fc receptors and complement in vivo. Annu Rev Immunol 16:421–432. https://doi.org/10.1146/annurev.immunol.16.1.421

Ray VA, Hill PJ, Stover CK, Roy S, Sen CK, Yu L, Wozniak DJ, DiGiandomenico A (2017) Anti-Psl targeting of Pseudomonas aeruginosa biofilms for neutrophil-mediated disruption. Sci Rep 7(1):16065. https://doi.org/10.1038/s41598-017-16215-6

Razonable RR, Pawlowski C, O'Horo JC, Arndt LL, Arndt R, Bierle DM, Borgen MD, Hanson SN, Hedin MC, Lenehan P, Puranik A, Seville MT, Speicher LL, Tulledge-Scheitel SM, Venkatakrishnan AJ, Wilker CG, Badley AD, Ganesh R (2021) Casirivimab-Imdevimab treatment is associated with reduced rates of hospitalization among high-risk patients with mild to moderate coronavirus disease-19. EClinicalMedicine. https://doi.org/10.1016/j.eclinm.2021.101102

Robbiani DF, Gaebler C, Muecksch F, Lorenzi JCC, Wang Z, Cho A, Agudelo M, Barnes CO, Gazumyan A, Finkin S, Hagglof T, Oliveira TY, Viant C, Hurley A, Hoffmann HH, Millard KG, Kost RG, Cipolla M, Gordon K, Bianchini F, Chen ST, Ramos V, Patel R, Dizon J, Shimeliovich I, Mendoza P, Hartweger H, Nogueira L, Pack M, Horowitz J, Schmidt F, Weisblum Y, Michailidis E, Ashbrook AW, Waltari E, Pak JE, Huey-Tubman KE,

Koranda N, Hoffman PR, West AP Jr, Rice CM, Hatziioannou T, Bjorkman PJ, Bieniasz PD, Caskey M, Nussenzweig MC (2020) Convergent antibody responses to SARS-CoV-2 in convalescent individuals. Nature. https://doi.org/10.1038/s41586-020-2456-9

Rohde H, Burdelski C, Bartscht K, Hussain M, Buck F, Horstkotte MA, Knobloch JK, Heilmann C, Herrmann M, Mack D (2005) Induction of Staphylococcus epidermidis biofilm formation via proteolytic processing of the accumulation-associated protein by staphylococcal and host proteases. Mol Microbiol 55(6):1883–1895. https://doi.org/10.1111/j.1365-2958.2005.04515.x

Rohde H, Frankenberger S, Zahringer U, Mack D (2010) Structure, function and contribution of polysaccharide intercellular adhesin (PIA) to Staphylococcus epidermidis biofilm formation and pathogenesis of biomaterial-associated infections. Eur J Cell Biol 89(1):103–111. https://doi.org/10.1016/j.ejcb.2009.10.005

Romling U, Kjelleberg S, Normark S, Nyman L, Uhlin BE, Akerlund B (2014) Microbial biofilm formation: a need to act. J Intern Med 276(2):98–110. https://doi.org/10.1111/joim.12242

Ross P, Mayer R, Benziman M (1991) Cellulose biosynthesis and function in bacteria. Microbiol Rev 55(1):35–58. https://doi.org/10.1128/mr.55.1.35-58.1991

Ryser S, Tenorio E, Estelles A, Kauvar LM (2019) Human antibody repertoire frequently includes antibodies to a bacterial biofilm associated protein. PLoS One 14(7):e0219256. https://doi.org/10.1371/journal.pone.0219256

Sadovskaya I, Vinogradov E, Flahaut S, Kogan G, Jabbouri S (2005) Extracellular carbohydrate-containing polymers of a model biofilm-producing strain, Staphylococcus epidermidis RP62A. Infect Immun 73(5):3007–3017. https://doi.org/10.1128/IAI.73.5.3007-3017.2005

Sakata N, Terakubo S, Mukai T (2005) Subcellular location of the soluble lytic transglycosylase homologue in Staphylococcus aureus. Curr Microbiol 50(1):47–51. https://doi.org/10.1007/s00284-004-4381-9

Sandini S, Stringaro A, Arancia S, Colone M, Mondello F, Murtas S, Girolamo A, Mastrangelo N, De Bernardis F (2011) The MP65 gene is required for cell wall integrity, adherence to epithelial cells and biofilm formation in Candida albicans. BMC Microbiol 11:106. https://doi.org/10.1186/1471-2180-11-106

Sanofi (2012) A randomized, double-blind, placebo-controlled trial to assess the pharmacokinetics, pharmacodynamics, and safety of a single dose of SAR279356 in patients hospitalized in intensive care unit and on mechanical ventilation. https://www.sanofi.com/en/science-and-innovation/clinical-trials-and-results/our-disclosure-commitments/pharma/-/media/Project/One-Sanofi-Web/Websites/Global/Sanofi-COM/Home/common/docs/clinical-study-results/PKD11791_summary.pdf.

Scheid JF, Mouquet H, Ueberheide B, Diskin R, Klein F, Oliveira TY, Pietzsch J, Fenyo D, Abadir A, Velinzon K, Hurley A, Myung S, Boulad F, Poignard P, Burton DR, Pereyra F, Ho DD, Walker BD, Seaman MS, Bjorkman PJ, Chait BT, Nussenzweig MC (2011) Sequence and structural convergence of broad and potent HIV antibodies that mimic CD4 binding. Science 333(6049):1633–1637. https://doi.org/10.1126/science.1207227

Schwartz K, Syed AK, Stephenson RE, Rickard AH, Boles BR (2012) Functional amyloids composed of phenol soluble modulins stabilize Staphylococcus aureus biofilms. PLoS Pathog 8(6):e1002744. https://doi.org/10.1371/journal.ppat.1002744

Serra DO, Hengge R (2014) Stress responses go three dimensional—the spatial order of physiological differentiation in bacterial macrocolony biofilms. Environ Microbiol 16(6):1455–1471. https://doi.org/10.1111/1462-2920.12483

Serra DO, Richter AM, Hengge R (2013) Cellulose as an architectural element in spatially structured Escherichia coli biofilms. J Bacteriol 195(24):5540–5554. https://doi.org/10.1128/JB.00946-13

Shahrooei M, Hira V, Stijlemans B, Merckx R, Hermans PW, Van Eldere J (2009) Inhibition of Staphylococcus epidermidis biofilm formation by rabbit polyclonal antibodies against the SesC protein. Infect Immun 77(9):3670–3678. https://doi.org/10.1128/IAI.01464-08

Shahrooei M, Hira V, Khodaparast L, Khodaparast L, Stijlemans B, Kucharikova S, Burghout P, Hermans PW, Van Eldere J (2012) Vaccination with SesC decreases Staphylococcus epidermidis biofilm formation. Infect Immun 80(10):3660–3668. https://doi.org/10.1128/IAI.00104-12

Sheu SM, Sheu BS, Yang HB, Lei HY, Wu JJ (2007) Anti-Lewis X antibody promotes Helicobacter pylori adhesion to gastric epithelial cells. Infect Immun 75(6):2661–2667. https://doi.org/10.1128/IAI.01689-06

Shields RC, Mokhtar N, Ford M, Hall MJ, Burgess JG, ElBadawey MR, Jakubovics NS (2013) Efficacy of a marine bacterial nuclease against biofilm forming microorganisms isolated from chronic rhinosinusitis. PLoS One 8(2):e55339. https://doi.org/10.1371/journal.pone.0055339

Solano C, Garcia B, Valle J, Berasain C, Ghigo JM, Gamazo C, Lasa I (2002) Genetic analysis of Salmonella enteritidis biofilm formation: critical role of cellulose. Mol Microbiol 43(3): 793–808. https://doi.org/10.1046/j.1365-2958.2002.02802.x

Soldo B, Lazarevic V, Pagni M, Karamata D (1999) Teichuronic acid operon of Bacillus subtilis 168. Mol Microbiol 31(3):795–805

Soliman C, Walduck AK, Yuriev E, Richards JS, Cywes-Bentley C, Pier GB, Ramsland PA (2018) Structural basis for antibody targeting of the broadly expressed microbial polysaccharide poly-N-acetylglucosamine. J Biol Chem 293(14):5079–5089. https://doi.org/10.1074/jbc.RA117.001170

Stewart PS, Franklin MJ (2008) Physiological heterogeneity in biofilms. Nat Rev Microbiol 6(3): 199–210. https://doi.org/10.1038/nrmicro1838

Storek KM, Auerbach MR, Shi H, Garcia NK, Sun D, Nickerson NN, Vij R, Lin Z, Chiang N, Schneider K, Wecksler AT, Skippington E, Nakamura G, Seshasayee D, Koerber JT, Payandeh J, Smith PA, Rutherford ST (2018) Monoclonal antibody targeting the beta-barrel assembly machine of Escherichia coli is bactericidal. Proc Natl Acad Sci U S A 115(14): 3692–3697. https://doi.org/10.1073/pnas.1800043115

Sumrall ET, Keller AP, Shen Y, Loessner MJ (2020) Structure and function of Listeria teichoic acids and their implications. Mol Microbiol 113(3):627–637. https://doi.org/10.1111/mmi.14472

Sun D, Accavitti MA, Bryers JD (2005) Inhibition of biofilm formation by monoclonal antibodies against Staphylococcus epidermidis RP62A accumulation-associated protein. Clin Diagn Lab Immunol 12(1):93–100. https://doi.org/10.1128/CDLI.12.1.93-100.2005

Taylor CM, Roberts IS (2005) Capsular polysaccharides and their role in virulence. Contrib Microbiol 12:55–66. https://doi.org/10.1159/000081689

Tiller T, Meffre E, Yurasov S, Tsuiji M, Nussenzweig MC, Wardemann H (2008) Efficient generation of monoclonal antibodies from single human B cells by single cell RT-PCR and expression vector cloning. J Immunol Methods 329(1–2):112–124. https://doi.org/10.1016/j.jim.2007.09.017

Tkaczyk C, Hua L, Varkey R, Shi Y, Dettinger L, Woods R, Barnes A, MacGill RS, Wilson S, Chowdhury P, Stover CK, Sellman BR (2012) Identification of anti-alpha toxin monoclonal antibodies that reduce the severity of Staphylococcus aureus dermonecrosis and exhibit a correlation between affinity and potency. Clin Vaccine Immunol 19(3):377–385. https://doi.org/10.1128/CVI.05589-11

Torosantucci A, Chiani P, Bromuro C, De Bernardis F, Palma AS, Liu Y, Mignogna G, Maras B, Colone M, Stringaro A, Zamboni S, Feizi T, Cassone A (2009) Protection by anti-beta-glucan antibodies is associated with restricted beta-1,3 glucan binding specificity and inhibition of fungal growth and adherence. PLoS One 4(4):e5392. https://doi.org/10.1371/journal.pone.0005392

Tursi SA, Puligedda RD, Szabo P, Nicastro LK, Miller AL, Qiu C, Gallucci S, Relkin NR, Buttaro BA, Dessain SK, Tukel C (2020) Salmonella Typhimurium biofilm disruption by a human antibody that binds a pan-amyloid epitope on curli. Nat Commun 11(1):1007. https://doi.org/10.1038/s41467-020-14685-3

Ude S, Arnold DL, Moon CD, Timms-Wilson T, Spiers AJ (2006) Biofilm formation and cellulose expression among diverse environmental Pseudomonas isolates. Environ Microbiol 8(11): 1997–2011. https://doi.org/10.1111/j.1462-2920.2006.01080.x

Uppuluri P, Singh S, Alqarihi A, Schmidt CS, Hennessey JP Jr, Yeaman MR, Filler SG, Edwards JE, Ibrahim AS (2018) Human anti-Als3p antibodies are surrogate markers of NDV-3A vaccine efficacy against recurrent vulvovaginal candidiasis. Front Immunol 9:1349. https://doi.org/10.3389/fimmu.2018.01349

van den Berg S, Bonarius HP, van Kessel KP, Elsinga GS, Kooi N, Westra H, Bosma T, van der Kooi-Pol MM, Koedijk DG, Groen H, van Dijl JM, Buist G, Bakker-Woudenberg IA (2015) A human monoclonal antibody targeting the conserved staphylococcal antigen IsaA protects mice against Staphylococcus aureus bacteremia. Int J Med Microbiol 305(1):55–64. https://doi.org/10.1016/j.ijmm.2014.11.002

van Erp EA, Luytjes W, Ferwerda G, van Kasteren PB (2019) Fc-mediated antibody effector functions during respiratory syncytial virus infection and disease. Front Immunol 10:548. https://doi.org/10.3389/fimmu.2019.00548

Varrone JJ, de Mesy Bentley KL, Bello-Irizarry SN, Nishitani K, Mack S, Hunter JG, Kates SL, Daiss JL, Schwarz EM (2014) Passive immunization with anti-glucosaminidase monoclonal antibodies protects mice from implant-associated osteomyelitis by mediating opsonophagocytosis of Staphylococcus aureus megaclusters. J Orthop Res 32(10):1389–1396. https://doi.org/10.1002/jor.22672

Victora GD, Nussenzweig MC (2012) Germinal centers. Annu Rev Immunol 30:429–457. https://doi.org/10.1146/annurev-immunol-020711-075032

Visai L, Xu Y, Casolini F, Rindi S, Hook M, Speziale P (2000) Monoclonal antibodies to CNA, a collagen-binding microbial surface component recognizing adhesive matrix molecules, detach Staphylococcus aureus from a collagen substrate. J Biol Chem 275(51):39837–39845. https://doi.org/10.1074/jbc.M005297200

Vogelpoel LT, Baeten DL, de Jong EC, den Dunnen J (2015) Control of cytokine production by human fc gamma receptors: implications for pathogen defense and autoimmunity. Front Immunol 6:79. https://doi.org/10.3389/fimmu.2015.00079

Wang X, Preston JF 3rd, Romeo T (2004) The pgaABCD locus of Escherichia coli promotes the synthesis of a polysaccharide adhesin required for biofilm formation. J Bacteriol 186(9): 2724–2734. https://doi.org/10.1128/JB.186.9.2724-2734.2004

Wang Q, Yan J, Gao GF (2017a) Monoclonal antibodies against Zika virus: therapeutics and their implications for vaccine design. J Virol 91(20). https://doi.org/10.1128/JVI.01049-17

Wang Y, Cheng LI, Helfer DR, Ashbaugh AG, Miller RJ, Tzomides AJ, Thompson JM, Ortines RV, Tsai AS, Liu H, Dillen CA, Archer NK, Cohen TS, Tkaczyk C, Stover CK, Sellman BR, Miller LS (2017b) Mouse model of hematogenous implant-related Staphylococcus aureus biofilm infection reveals therapeutic targets. Proc Natl Acad Sci U S A 114(26):E5094–E5102. https://doi.org/10.1073/pnas.1703427114

Wang Q, Michailidis E, Yu Y, Wang Z, Hurley AM, Oren DA, Mayer CT, Gazumyan A, Liu Z, Zhou Y, Schoofs T, Yao KH, Nieke JP, Wu J, Jiang Q, Zou C, Kabbani M, Quirk C, Oliveira T, Chhosphel K, Zhang Q, Schneider WM, Jahan C, Ying T, Horowitz J, Caskey M, Jankovic M, Robbiani DF, Wen Y, de Jong YP, Rice CM, Nussenzweig MC (2020) A combination of human broadly neutralizing antibodies against hepatitis B virus HBsAg with distinct epitopes suppresses escape mutations. Cell Host Microbe 28 (2):335-349:e336. https://doi.org/10.1016/j.chom.2020.05.010

Wang Y, Wu C, Yu J, Lin S, Liu T, Zan L, Li N, Hong P, Wang X, Jia Z, Li J, Wang Y, Zhang M, Yuan X, Li C, Xu W, Zheng W, Wang X, Liao HX (2021) Structural basis of tetanus toxin neutralization by native human monoclonal antibodies. Cell Rep 35(5):109070. https://doi.org/10.1016/j.celrep.2021.109070

Wardemann H, Yurasov S, Schaefer A, Young JW, Meffre E, Nussenzweig MC (2003) Predominant autoantibody production by early human B cell precursors. Science 301(5638): 1374–1377. https://doi.org/10.1126/science.1086907

Watson A, Li H, Ma B, Weiss R, Bendayan D, Abramovitz L, Ben-Shalom N, Mor M, Pinko E, Bar Oz M, Wang Z, Du F, Lu Y, Rybniker J, Dahan R, Huang H, Barkan D, Xiang Y, Javid B, Freund NT (2021) Human antibodies targeting a Mycobacterium transporter protein mediate protection against tuberculosis. Nat Commun 12(1):602. https://doi.org/10.1038/s41467-021-20930-0

Weems JJ Jr, Steinberg JP, Filler S, Baddley JW, Corey GR, Sampathkumar P, Winston L, John JF, Kubin CJ, Talwani R, Moore T, Patti JM, Hetherington S, Texter M, Wenzel E, Kelley VA, Fowler VG Jr (2006) Phase II, randomized, double-blind, multicenter study comparing the safety and pharmacokinetics of tefibazumab to placebo for treatment of Staphylococcus aureus bacteremia. Antimicrob Agents Chemother 50(8):2751–2755. https://doi.org/10.1128/AAC.00096-06

Weidenmaier C, Peschel A (2008) Teichoic acids and related cell-wall glycopolymers in Gram-positive physiology and host interactions. Nat Rev Microbiol 6(4):276–287. https://doi.org/10.1038/nrmicro1861

Wi YM, Patel R (2018) Understanding biofilms and novel approaches to the diagnosis, prevention, and treatment of medical device-associated infections. Infect Dis Clin North Am 32(4):915–929. https://doi.org/10.1016/j.idc.2018.06.009

Wilking JN, Zaburdaev V, De Volder M, Losick R, Brenner MP, Weitz DA (2013) Liquid transport facilitated by channels in Bacillus subtilis biofilms. Proc Natl Acad Sci U S A 110(3):848–852. https://doi.org/10.1073/pnas.1216376110

Xiao X, Zhu Z, Dankmeyer JL, Wormald MM, Fast RL, Worsham PL, Cote CK, Amemiya K, Dimitrov DS (2010) Human anti-plague monoclonal antibodies protect mice from Yersinia pestis in a bubonic plague model. PLoS One 5(10):e13047. https://doi.org/10.1371/journal.pone.0013047

Xiong YQ, Estelles A, Li L, Abdelhady W, Gonzales R, Bayer AS, Tenorio E, Leighton A, Ryser S, Kauvar LM (2017) A human biofilm-disrupting monoclonal antibody potentiates antibiotic efficacy in rodent models of both Staphylococcus aureus and Acinetobacter baumannii infections. Antimicrob Agents Chemother 61(10). https://doi.org/10.1128/AAC.00904-17

Yu XQ, Robbie GJ, Wu Y, Esser MT, Jensen K, Schwartz HI, Bellamy T, Hernandez-Illas M, Jafri HS (2017) Safety, tolerability, and pharmacokinetics of MEDI4893, an investigational, extended-half-life, anti-Staphylococcus aureus alpha-toxin human monoclonal antibody, in healthy adults. Antimicrob Agents Chemother 61(1). https://doi.org/10.1128/AAC.01020-16

Zarnowski R, Westler WM, Lacmbouh GA, Marita JM, Bothe JR, Bernhardt J, Lounes-Hadj Sahraoui A, Fontaine J, Sanchez H, Hatfield RD, Ntambi JM, Nett JE, Mitchell AP, Andes DR (2014) Novel entries in a fungal biofilm matrix encyclopedia. mBio 5(4):e01333-01314. https://doi.org/10.1128/mBio.01333-14

Zhou T, Georgiev I, Wu X, Yang ZY, Dai K, Finzi A, Kwon YD, Scheid JF, Shi W, Xu L, Yang Y, Zhu J, Nussenzweig MC, Sodroski J, Shapiro L, Nabel GJ, Mascola JR, Kwong PD (2010) Structural basis for broad and potent neutralization of HIV-1 by antibody VRC01. Science 329(5993):811–817. https://doi.org/10.1126/science.1192819

Host Defense Peptides: Multifront Attack on Biofilms

Melanie Dostert, Lucas Pedraz, and Robert E. W. Hancock

Abstract Short, cationic, amphipathic peptides are found as natural antimicrobial agents among all life forms. These natural peptides and their synthetic analogs, collectively named host defense peptides, have been extensively studied for their use as antimicrobial drugs. In the treatment of biofilm infections, host defense peptides can orchestrate a multifront attack, taking advantage of their antimicrobial, immunomodulatory, and antibiofilm activities. In this chapter, we will discuss the different activities that host defense peptides present and their potential in the treatment of biofilms infections. We focus primarily on the underlying molecular mechanisms of action and target pathways of these peptides, since understanding their action at a molecular level is becoming increasingly important to facilitate the design and development of improved peptides. We present biofilm-specific mechanisms and direct antimicrobial and immunomodulatory mechanisms that are relevant to the treatment of biofilm infections.

1 Introduction

Antibiotic therapy frequently fails against biofilms, which account for more than 65% of clinical infections (Hardie 2020). Clinical treatment of recalcitrant biofilm infections commonly involves invasive procedures (surgical removal of infected tissues or medical devices) and/or combination therapies with different antibiotic classes (Hilchie et al. 2013; Dostert et al. 2021). While many novel strategies to combat biofilm infections are being explored (Lynch and Abbanat 2010; Wu et al. 2015; Roy et al. 2018), we still lack reliable, clinically effective antibiofilm

Melanie Dostert and Lucas Pedraz contributed equally with all other contributors.

M. Dostert · L. Pedraz · R. E. W. Hancock (✉)
Centre for Microbial Diseases and Immunity Research, Department of Microbiology and Immunology, University of British Columbia, Vancouver, BC, Canada
e-mail: bob@hancocklab.com

K. Richter, K. Nørskov Kragh (eds.), *Antibiofilm Strategies*, Springer Series on Biofilms 11, https://doi.org/10.1007/978-3-031-10992-8_12

therapies, and none have been approved to date. A promising approach may be found in ancient, natural antimicrobial compounds: the host defense peptides (HDPs).

In general, HDPs are short, cationic, amphipathic peptides produced by virtually all life forms as a defense against pathogens (Zasloff 2002; Hancock and Sahl 2006; Hancock et al. 2021). While antimicrobial peptides of microbial origin were already known, animal HDPs were first discovered in insects (cecropins A and B in silk moths) (Steiner et al. 1981) and later expanded to vertebrates in the late 1980s (magainins in *Xenopus* frogs) (Zasloff 1987). HDPs are today considered to be an integral part of the innate immune response of all organisms (Hancock and Scott 2000; Zasloff 2002; Haney et al. 2017a). In mammals, HDPs are produced by epithelial cells in exposed body tissues (Gallo and Hooper 2012) and by innate immune cells, which store them in granules or secrete them (Faurschou and Borregaard 2003). Physiologically, HDPs present immunomodulatory activities and moderate direct antimicrobial effects.

As therapeutic agents, HDPs present broad-spectrum activity and are less prone to induce resistance than conventional antibiotics (Zasloff 2002; Hancock and Sahl 2006; Di Luca et al. 2014). Of particular interest to combat biofilm infections is that some HDPs are known to display antibiofilm activity, thanks to their capacity to target biofilm-specific pathways (Pletzer and Hancock 2016; de la Fuente-Núñez et al. 2016; Dostert et al. 2021). This activity, first discovered for human cathelicidin LL-37 (Overhage et al. 2008), can appear combined with antimicrobial and immunomodulatory activities within the same molecule. To be used as drugs, rather than purifying them from natural sources, HDPs are obtained by chemical synthesis, often solid-phase peptide synthesis (Behrendt et al. 2016; Haney et al. 2017a), or obtained from the recombinant production of their inactive precursors (Bommarius et al. 2010; Haney et al. 2017a). These techniques allow for the development of modified synthetic HDPs with improved antimicrobial potency. Today, more than 5000 natural and synthetic HDPs are listed on the Data Repository of Antimicrobial Peptides (DRAMP) (Kang et al. 2019).

Multiple reviews by us and others discuss HDPs and their potential as therapeutic agents (Haney et al. 2017a; Magana et al. 2020; Mookherjee et al. 2020). In this chapter, we focus on the use of HDPs in the treatment of biofilm infections, detailing the different activities involved. We place special emphasis on mechanisms and target pathways that are relevant to combat biofilms.

2 Main Body

2.1 General Features and Terminology of Host Defense Peptides

As HDPs were initially appreciated for their broad-spectrum activity against planktonic bacteria, they are often referred to as antimicrobial peptides (AMPs). However, nowadays, it is recognized that these peptides display a broad range of biological activities, including antifungal, antiviral, antiparasitic, anticancer, immunomodulatory, and antibiofilm activities (Fig. 1, Dostert et al. 2019; Di Somma et al. 2020; Mookherjee et al. 2020). The last two functions are considered pivotal for the anti-infective capacity of these peptides, and in particular, immunomodulatory activity might be their most critical natural function. In this chapter, we use the broader term HDPs, which reflects their multifunctionality and acknowledges their origin (Haney et al. 2019b). Other terms describe subsets of HDPs: for instance, antibiofilm peptides (ABPs) are HDPs with antibiofilm activity at a concentration below their planktonic MIC (Overhage et al. 2008; Haney et al. 2019b)). Likewise, innate defense regulators (IDRs) are synthetic HDPs that can control infections by modulating host immune responses without strongly affecting the microbial burden (Steinstraesser et al. 2012; Haney et al. 2019b).

HDPs are often called "highly conserved in evolution" (Bommarius et al. 2010; Di Luca et al. 2014; de la Fuente-Núñez et al. 2016; Dostert et al. 2019). However, this claim refers to their general role and structural properties; their primary sequences vary widely, even among mammals (Patil et al. 2004; Hancock and Sahl 2006). Natural HDPs are gene-encoded and produced as inactive pro-peptides. After translation, they are activated through one or more proteolytic events (Hancock and Sahl 2006; Haney et al. 2017a). The final, active peptides range in size between 12 and 50 amino acids and present an abundance of hydrophobic residues (often more than 50%) and a positive net charge (usually +2 to +9) due to the presence of several cationic residues (Zasloff 2002; Hancock and Sahl 2006). Consequently, HDPs present amphipathic secondary structures, which are often only adopted in the presence of biological membranes (Nguyen et al. 2011). The specific secondary structure that HDPs form has traditionally been used to classify them into four categories: α-helical, β-stranded, (single) looped, and extended (Jenssen et al. 2006; Haney et al. 2017a). A small number of anionic antimicrobial peptides that work in conjunction with ions such as zinc are known (Harris et al. 2009), but those are considered beyond the focus of this chapter, as are larger proteins with antimicrobial activity (Gallo and Hooper 2012).

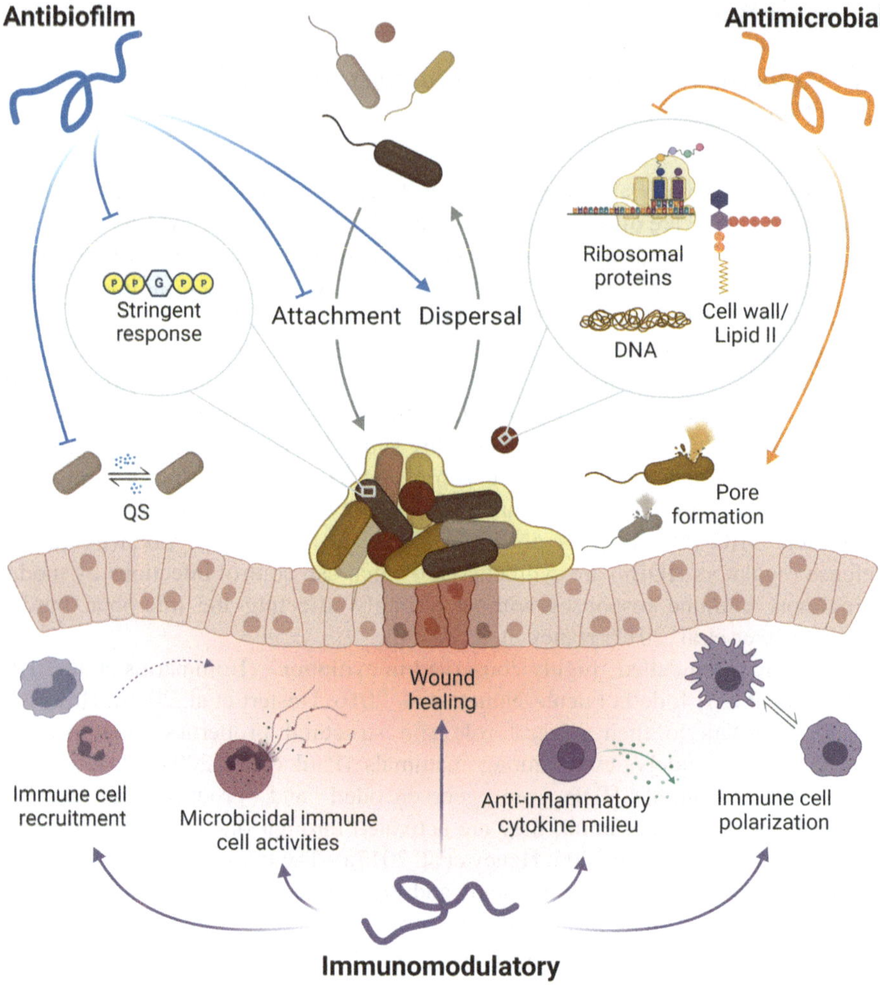

Fig. 1 Overview of HDP activities and associated mechanisms of action. Non-exhaustive summary of antibiofilm (blue), antimicrobial (orange), and immunomodulatory (purple) mechanisms of action of HDPs. Regular arrowheads represent positive actions (induction, upregulation, activation) and flathead arrows represent negative actions (repression, downregulation, inactivation). Created with BioRender.com

2.2 *HDPs Orchestrate Multifront Attacks on Biofilms*

In the context of treating biofilm infections, the multifunctionality of HDPs offers unique opportunities, which make them exceptional drug candidates. First, their ability to directly target biofilm-specific pathways is critical for successful antibiofilm therapy (Dostert et al. 2019). Second, considering that biofilm and planktonic growth are not mutually exclusive and that the disruption of biofilms might, at least partially, promote planktonic growth, the direct antimicrobial activity of HDPs, possibly in synergy with antibiotics, is crucial to preventing chronic biofilm infections from turning into systemic ones, causing issues like sepsis. Finally, immunomodulatory HDPs can control the harmful chronic inflammation that often accompanies biofilm infections and also redirect the host immune response to successfully clear microbial infections and overcome biofilm immune evasion mechanisms (Dostert et al. 2021).

Recently, less studied activities relevant to biofilm infections, such as eradication of mixed-species biofilms and persister cells, have been reported for HDPs (Chung et al. 2017; Casciaro et al. 2020; Scheper et al. 2021; Pati et al. 2021). Despite the therapeutic challenge that interactions between different microbial species pose, the synthetic HDP DRGN-1 reduced the bacterial burden of both *Pseudomonas aeruginosa* and *Staphylococcus aureus*, while promoting wound healing in a murine biofilm model (Chung et al. 2017; Batoni et al. 2021). In an even more complex porcine ex vivo skin biofilm model of *P. aeruginosa*, *S. aureus,* and *C. albicans*, hydrogels containing chitosan and the natural HDP epsilon-poly-L-lysine strongly reduced biofilm thickness and bacterial counts of especially *P. aeruginosa* (Pati et al. 2021). While antibiotics fail in killing persisters, the amphibian HDP temporin G reduced biofilm-derived *S. aureus* persisters by up to 80% (Casciaro et al. 2020). Moreover, the synthetic LL-37 derivative SAAP-148 and the frog HDP pexiganan completely eradicated persister cells derived from methicillin-resistant *S. aureus* (MRSA) biofilms formed on various substrates (Scheper et al. 2021).

Despite their exceptional anti-infective capacity, identifying HDPs that (1) present activity profiles relevant for the treatment of biofilm infections, (2) retain their activity in vivo, and (3) simultaneously display low cytotoxicity against human cells and in relevant animal models is not an easy task. In the quest for the most potent anti-infective HDP sequence, various approaches have been applied over the years. Initially, HDPs were mainly isolated from natural sources and subsequently optimized using single point mutations (Haney and Hancock 2013). Since then, further studies have vindicated the value of systematic amino acid replacements. Studies performing systematic substitutions in the synthetic bactenecin derivative 1018 demonstrated that the sequence requirements for different HDP activities might be distinct but partially overlapping and that multiple activities can be optimized at once (Fig. 1, Haney et al. 2015, 2017b, 2019a). With the rise of bioinformatics, genome mining approaches identifying novel natural HDPs in less studied species, as well as computer-aided de novo design and sequence optimization, have started to shape the field (Haney et al. 2018; Cardoso et al. 2019; Oshiro et al. 2019; Santana et al. 2021).

Beyond the design of more potent HDP sequences, other strategies aim to overcome the limitations of HDPs that impede their clinical application. The lack of translation from potent in vitro activity to effective biofilm eradication in vivo has led to models mimicking relevant in vivo conditions (Dostert et al. 2019). For example, exposure to proteases, extreme pH, and high concentrations of salts and host proteins can strongly affect peptide activity (Dostert et al. 2019). To improve peptide stability under these conditions, HDPs have been cyclized, synthesized using non-natural amino acids, hybridized with natural or synthetic protein domains, and modified with various side chains (Haney et al. 2019a; Etayash et al. 2020; Maystrenko et al. 2020; Koeninger et al. 2021; Lakshmaiah Narayana et al. 2021). Furthermore, the development of more complex screening models can improve the identification of HDPs with promising activity profiles, as recently demonstrated using a human skin organoid model (de Breij et al. 2018; Wu et al. 2021). Other strategies, like peptide fragmentation, formulation, and synergy between antibiotics and peptides, which address cytotoxicity against human cells, high production costs, as well as peptide stability and activity in vivo, are active research areas in the field of HDPs (Pletzer and Hancock 2018; Dostert et al. 2019).

In the following sections, we will detail the mechanisms of action that contribute to the extraordinary therapeutic potential of HDPs against biofilm infections. Unraveling the mechanisms that drive the anti-infective capacity of HDPs can guide researchers to determine favorable HDP activity profiles (Fig. 1). Furthermore, it offers the opportunity to design more effective HDPs by improving their interactions with their molecular targets.

2.3 Diversity of HDP-Mediated Antibiofilm Activities

HDPs can act during different phases of the biofilm life cycle: they can inhibit biofilm formation, eradicate established biofilms, and induce biofilm dispersal (Hancock et al. 2021). One of the first reports of HDP-mediated antibiofilm activity demonstrated that human LL-37 and bovine indolicidin inhibit biofilms at concentrations below those required for their respective antimicrobial activities, suggesting that the molecular mechanisms underlying these two activities are distinct (Overhage et al. 2008; Hancock et al. 2021). Since biofilm inhibitory activity can completely prevent biofilm growth, the development of HDP-coated surfaces for medical devices and implants is an active area of research (Dostert et al. 2019). Even though biofilm inhibition is critical to prevent the initiation of biofilm communities, it cannot completely explain how treatment of established biofilm infections can remove most bacteria from surfaces, since dead bacteria should remain attached to such surfaces (de la Fuente-Núñez Pimentel-Filho et al. 2014). Biofilm dispersal and eradication are arguably more important for the clearance of pre-formed biofilms. Induction of dispersal offers the opportunity to revert highly antibiotic resistant biofilm bacteria to a more susceptible state. However, increased release of planktonic bacteria from biofilms harbors the risk of turning chronic infections into systemic infections; therefore, HDPs triggering dispersal should be either antimicrobial or combined with antimicrobial agents. Probably the most straightforward HDP-mediated antibiofilm activity is biofilm eradication. Even though identifying HDPs able to completely eradicate biofilms is challenging, substantial progress has recently been made (Wu et al. 2021).

HDP-mediated antibiofilm mechanisms have mainly been studied in the context of biofilm inhibition and it remains to be determined if mechanisms underlying eradication and induction of dispersal of pre-formed biofilms are the result of interactions with the same or different molecular targets. Nevertheless, the investigation of biofilm inhibitory mechanisms clearly highlights a prominent role for bacterial regulators as molecular targets. Therefore, we will next provide a brief overview of regulatory pathways driving biofilm growth before discussing biofilm-specific HDP mechanisms.

2.4 Regulatory Pathways Driving Biofilm Growth

Bacteria in biofilms exhibit distinct gene expression profiles compared to their planktonic counterparts (Dostert et al. 2019). A global regulatory gene network is believed to orchestrate this reversible growth adaptation (Dostert et al. 2019). To date, evidence suggests that nucleotide signaling pathways, quorum sensing (QS), and two-component systems (TCS) play important roles in establishing and maintaining biofilms (Passos da Silva et al. 2017; Dostert et al. 2019). The stringent response and cyclic diguanylate monophosphate (c-di-GMP) mediated regulation are universal, nucleotide-mediated signaling pathways involved in the biofilm growth state, thus representing promising targets for broad-spectrum biofilm-specific therapies (Purcell and Tamayo 2016; Dostert et al. 2021). While c-di-GMP promotes biofilm formation at high concentrations and motility and virulence at low concentrations, the stringent response has been linked to biofilm resistance and the formation of persister cells (Dostert et al. 2021). Furthermore, bacterial communication via several QS systems can contribute to biofilm growth by inducing the production of matrix components, such as rhamnolipids and extracellular DNA (Passos da Silva et al. 2017). However, the importance of QS systems in the formation and development of bacterial biofilms seems to be partially environment dependent (Passos da Silva et al. 2017). Another important group of regulators associated with biofilm growth are TCSs, which generally include two interacting proteins: a sensor kinase and a response regulator. Various TCSs have been shown to play important roles in every stage of the biofilm life cycle (Francis et al. 2017). Interestingly, certain TCS genes have also been linked to bacterial HDP resistance (Francis et al. 2017). Prokaryotic genomes encode a few hundred different TCSs with variation between individual species, and some collaborate in complex multikinase networks, which simultaneously integrate several signals and might be interchangeable (Francis et al. 2017; Gumerov et al. 2020). Therefore, TCSs represent challenging molecular targets for the design of antibiofilm therapies.

Even though many research groups have identified key players in biofilm regulation, we still lack a holistic understanding of the global regulatory network. Especially surprising is the involvement of only a limited number of known one-component systems, the most prominent group of bacterial regulators, that combine the sensor and output domains in the same protein (Ulrich et al. 2005).

2.5 *Biofilm-Specific Mechanisms of Action and Target Pathways of HDPs*

2.5.1 Effects on the Stringent Response

When antibiofilm activity was first discovered in HDPs (Overhage et al. 2008), the mechanism of action was largely unknown. Changes in the transcriptome provoked by subinhibitory peptide treatment were studied in detail (Overhage et al. 2008; de la Fuente-Núñez et al. 2012) but the actions performed by the peptide to cause those changes remained unexplained. The first time the antibiofilm mechanism of an HDP was linked to a specific molecular target was to the stringent response (de la Fuente-Núñez Pimentel-Filho et al. 2014).

The stringent response is a highly conserved stress response mechanism used by virtually all bacterial species (and plant chloroplasts) to react to different environmental inputs, such as amino acid, fatty acid, iron, or phosphate starvation (Potrykus and Cashel 2008; Römling and Balsalobre 2012). Upon detection of those sources of stress, the stringent response dramatically alters the transcriptome, diverting resources from growth and division to stress adaptations and survival mechanisms (Potrykus and Cashel 2008; Traxler et al. 2011; Dalebroux and Swanson 2012). At a molecular level, the stringent response is mediated by the nucleotide alarmones guanosine tetraphosphate and guanosine pentaphosphate, collectively referred to as (p)ppGpp (Dalebroux and Swanson 2012). In *Escherichia coli*, where the stringent response was first studied in detail, the synthesis of (p)ppGpp is mediated by two homologous enzymes: the synthase RelA and the synthase/hydrolase SpoT. RelA interacts with ribosomal proteins and can activate the synthesis of alarmones when translation is stopped due to unloaded tRNAs causing a "stalled ribosome" (Battesti and Bouveret 2009; Dalebroux and Swanson 2012). SpoT responds to other stimuli, such as fatty acid starvation, and can balance the production or degradation of alarmones (Battesti and Bouveret 2009; Dalebroux and Swanson 2012). Other bacterial species use different RelA/SpoT homologs (RSHs) to produce (p)ppGpp, in some cases encoding a single enzyme (e.g., *S. aureus*) (Atkinson et al. 2011). Small, non-RSH (p)ppGpp synthases/hydrolases have also been described in species such as *Vibrio cholerae* (Atkinson et al. 2011), but the alarmones remain the universal link between all mechanisms. Above a certain threshold, (p)ppGpp triggers the stringent response mainly through direct interaction with the RNA polymerase. The interaction with the alarmone has two effects: directly influencing the transcription of specific promoters and modulating the apoenzyme affinity for different sigma factors (thus displacing the exponential factor RpoD in favor of others, such as RpoS) (Potrykus and Cashel 2008; Römling and Balsalobre 2012). (p)ppGpp has also been demonstrated to act via other mechanisms, including cross-talk with second messengers, such as c-di-GMP (Dalebroux and Swanson 2012; Römling and Balsalobre 2012). The final effects of the changes in the transcriptome include the reduction of growth-related functions, such as protein and cell wall synthesis, and the activation of pathways related to stress resistance, biofilm formation,

Table 1 Biofilm-specific mechanisms of HDPs. Peptides presenting the mechanisms of action listed are classified into two categories: proven, if the mechanism has been associated with at least one molecular interaction involving the HDP, and proposed (indicated with an asterisk), if no such interaction has been described

Molecular target	HDP	Species	References
1. Stringent response			
(p)ppGpp binding/ degradation	1018;DJK-5/ DJK-6;1018M	*P. aeruginosa* *S. aureus*	de la Fuente-Núñez et al. (2014, 2015); Jiale et al. (2021)
Downregulation of *spoT* transcription	DJK-5*	*P. aeruginosa*	Pletzer et al. (2017)
2. Extracellular matrix and surface attachment			
Prevention of curli polymerization	LL-37	*E. coli*	Kai-Larsen et al. (2010)
Matrix anti-assembly	Capsicumicine	*S. epidermidis*	Von Borowski et al. (2020)
Alteration of surface hydrophobicity	Bovicin HC5; Nisin	*S. aureus*	Pimentel-Filho et al. (2014)
Peptidoglycan binding, surface interference	P1	*S. mutants*	Ansari et al. (2017)
3. Motility systems			
Stimulation of twitching motility	LL-37*;1037*;1018*	*P. aeruginosa*	Overhage et al. (2008); de la Fuente-Núñez et al. (2012); Wilkinson et al. (2021)
Inhibition of swarming motility	1037*;1018*	*P. aeruginosa*	de la Fuente-Núñez et al. (2012); Wilkinson et al. (2021)
4. Other pathways			
Deregulation of quorum sensing	LL-37*;1018*; Lactoferrin*/LF-derivatives*	*P. aeruginosa* *S. aureus*	Overhage et al. (2008); Cao et al. (2020); Ramamourthy and Vogel (2021)
Inhibition of polysaccharide intercellular adhesin (PIA) biosynthesis	β-defensin 3*	*S. aureus* *S. epidermidis*	Zhu et al. (2013)

antimicrobial resistance, antibiotic synthesis, and sporulation (Magnusson et al. 2005; Römling and Balsalobre 2012). The link between the stringent response and production of biofilm is complex and only partially understood, but it is considered a universal mechanism, as it has been described in multiple species of Gram-negative and Gram-positive bacteria that a reduction of (p)ppGpp levels leads to a clear decrease in biofilm formation (Balzer and McLean 2002; Åberg et al. 2006; Römling and Balsalobre 2012; de la Fuente-Núñez Pimentel-Filho et al. 2014).

The first HDP to be linked to the stringent response was peptide 1018 (de la Fuente-Núñez Pimentel-Filho et al. 2014) (Table 1; also called IDR-1018), which was demonstrated to reduce the intracellular levels of (p)ppGpp, inhibiting biofilm formation and stimulating biofilm dispersal in *P. aeruginosa*. Through co-precipitation and NMR experiments, it was demonstrated that HDP 1018 directly binds (p)ppGpp and triggers its degradation (de la Fuente-Núñez Pimentel-Filho et al. 2014). The ubiquity of (p)ppGpp among bacteria suggests a potentially universal antibiofilm mechanism of action. This mechanism was later extended to the highly efficient D-enantiomeric peptide DJK-5 (de la Fuente-Nuñez et al. 2015), which also demonstrated strong antibiofilm activity in vivo against *P. aeruginosa* and *S. aureus* (Mansour et al. 2016; Pletzer et al. 2017). More recently, a modified version of HDP 1018 with improved affinity for (p)ppGpp was designed, named 1018 M, and applied to treat MRSA biofilms (Jiale et al. 2021). Further insights suggest that these peptides may also use other mechanisms beyond direct (p)ppGpp degradation. These mechanisms include further impairing the stringent response by downregulating the expression of *spoT*, which was shown in a mouse skin infection model (Pletzer et al. 2017) (Table 1), as well as other mechanisms described below. However, it remains unclear if these additional effects are evidence of an additional mechanism of action or consequences of the altered stringent response (Wilkinson et al. 2021).

2.5.2 Other Antibiofilm Mechanisms

Other mechanisms have been described, and more remain only partially explained (Batoni et al. 2016; Raheem and Straus 2019; Di Somma et al. 2020). The proposed targets include QS, motility, biosynthesis of matrix polymers, and surface adhesion/detachment (Table 1).

QS systems are used by bacteria to trigger population density-dependent responses, including biofilm formation. The disruption of this form of communication, dubbed quorum quenching, is thus often proposed as a potential mechanism for antibiofilm therapies (Jakobsen et al. 2013; Paluch et al. 2020). The molecular basis of QS is universal: bacteria synthesize and release signal molecules called autoinducers (AIs), which can penetrate all cells in the population so that the total environmental concentration of AI is representative of the number of cells synthesizing it. Bacteria also encode AI receptors, responsible for eliciting a particular response above a certain AI concentration threshold (Gazit et al. 1996; Jakobsen et al. 2013). However, the specific molecules used as AIs, as well as the receptor systems, are generally species-specific, which restricts the usefulness of QS as a target for antibiofilm drugs (Jakobsen et al. 2013; Paluch et al. 2020). The antibiofilm mechanism of several HDPs has been correlated with disrupted QS systems. This is the case for human cathelicidin LL-37, which was associated with the downregulation of the two major QS systems of *P. aeruginosa* (Overhage et al. 2008). Similar effects have been suggested for HDP 1018 on *Streptococcus mutants* (Cao et al. 2020), as well as more recently for dimeric lactoferrin-derived peptides on *P. aeruginosa* (Ramamourthy and Vogel 2021). Nevertheless, no direct link between these HDPs and the QS systems has been discovered. It remains to be proven whether the alteration of QS elements is part of a yet undiscovered antibiofilm mechanism, part of the stringent response-dependent action, or a consequence of biofilm elimination rather than its cause.

Motility systems have also been proposed to be the target of numerous antibiofilm peptides. *P. aeruginosa* treated with LL-37 displays upregulated twitching motility, which reduces surface attachment (Overhage et al. 2008). HDPs 1018 and 1037 were later demonstrated to not only promote twitching in *P. aeruginosa* but also inhibit swarming, a form of surface motility (de la Fuente-Núñez et al. 2012; Wilkinson et al. 2021). Other pathways proposed as potential targets of antibiofilm peptides include the biosynthesis of polysaccharide intercellular adhesin in *Staphylococcus sp.*, which is repressed when treated with human β-defensin 3 (Zhu et al. 2013). While the exact targets of these peptides are still unknown, all represent strong subinhibitory antibiofilm effects, suggesting that additional antibiofilm mechanisms remain to be discovered.

Finally, some antibiofilm peptides have been proposed to inhibit biofilm formation or cause biofilm dispersal by affecting surface attachment and bacterial interactions with extracellular matrix components. Pimentel-Filho et al. explain the antibiofilm activity of the bacteriocins nisin and bovicin HC5 toward *S. aureus* through a purely biophysical mechanism, in which the peptides would reduce bacterial surface attachment by decreasing the hydrophobicity of the surface (Pimentel-Filho Pimentel-Filho et al. 2014). Synthetic peptide P1, derived from a tick antifreeze glycoprotein, has demonstrated strong antibiofilm effects against exclusively Gram-positive bacteria, both in vitro and in vivo (Heisig et al. 2014). Given its capacity to bind the peptidoglycan pentapeptide (Abraham et al. 2017), it has been proposed that peptide P1 inhibits biofilm formation and causes biofilm dispersion by sticking to the bacterial surface, interfering with the secretion of extracellular polymers and their interactions (Ansari et al. 2017). A recent study proposed a related mechanism called matrix anti-assembly for the synthetic peptide capsicumicine (Von Borowski et al. 2020), which is proposed to interact with elements of the extracellular matrix in biofilms of *S. aureus* and *S. epidermidis*, disrupting the assembly process without affecting the viability of the cells. Finally, an antibiofilm mechanism has also been described for human cathelicidin LL-37, which can prevent the polymerization of curli amyloid fibers, involved in surface adhesion and biofilm formation, by binding the major curli subunit, CsgA (Kai-Larsen et al. 2010). These mechanisms based on extracellular interactions offer another promising avenue for antibiofilm peptides. However, as they rely on the biophysical properties of the peptides that are highly susceptible to the environment, it remains to be demonstrated whether their mode of action can be replicated in vivo (Haney et al. 2019b).

2.6 Direct Antimicrobial Activity of HDPs

Despite the importance of specific antibiofilm mechanisms in combatting biofilm infections with HDPs, their direct antimicrobial activity remains relevant for planktonic bacteria released as a consequence of biofilm disruption. Moreover, some studies detailing HDP modes of action showcase how their mechanism against planktonic bacteria can equally apply to their antibiofilm activities (Okuda et al. 2013; Luca et al. 2013; Libardo et al. 2017; Kim et al. 2018). Here, we explore the mechanisms behind the direct antimicrobial activity of HDPs, focusing specifically on bacteria. Other antimicrobial activities (antiviral, antifungal, antiparasitic) are considered out of the scope of this chapter but have been discussed in detail elsewhere (Jenssen et al. 2006; Mookherjee et al. 2020).

All HDPs must initially interact with the surface of cells (Hancock and Sahl 2006; Fjell et al. 2011). Most HDPs are unstructured in solution and adopt their characteristic conformations in the presence of biological membranes or membrane mimetics (Nguyen et al. 2011). This transformation was initially regarded as switching between inactive and active states as part of a single membrane-targeting mechanism. However, this change in structure is now considered a manifestation of the conformational plasticity of HDPs, serving as a first step both for membrane interaction mechanisms and for translocating across membranes and reaching the cytoplasm (Henninot et al. 2018; Sun et al. 2018). The initial interaction between the HDP and the bacterial surface is electrostatic, with the positive residues of the peptide binding either lipopolysaccharide (LPS) in Gram-negative bacteria or teichoic and lipoteichoic acids (LTA) in Gram-positive bacteria (Scott et al. 1999; Hancock 2001; Haney et al. 2019b). In the former, it is known that the accumulation of peptides on the outer membrane gradually increases its permeability in a process termed self-promoted uptake (Hancock 2001). There is little consensus about the mechanisms HDPs employ to penetrate the membrane further or to translocate across it and reach the cytoplasm. However, it is known that the hydrophobicity of the peptide, its charge, and the position of the charged residues play a role (Haney et al. 2019b).

When HDPs reach the cytoplasmic membrane, one potential antimicrobial mechanism is non-specific disruption of the membrane integrity, explained via pore or non-pore forming models (Haney et al. 2017a; Kumar et al. 2018). Classical pore models include the barrel-stave model, in which HDPs insert perpendicularly to the membrane, aligning their hydrophobic sections forming barrel-like channels stabilized by peptide–peptide interactions (Kumar et al. 2018). While commonly discussed, this structure has only been demonstrated for the fungal peptide alamethicin (Laver 1994; Haney et al. 2017a). The toroidal pore model proposes the formation of transient pores, where HDPs gradually curve the lipid bilayer until pores are generated, lined with peptides interacting with the polar regions of the phospholipids (Wimley 2010; Kumar et al. 2018). The toroidal model has been proposed for numerous HDPs, such as magainin Q or melittin (Yang et al. 2001; Lee et al. 2016; Kumar et al. 2018), but there is only modest evidence to support its significance under physiological conditions (Sengupta et al. 2008; Haney et al. 2017a, 2019b). The aggregate model explains the disruption of the membrane in a less formal manner through the formation of transient lipid-peptide aggregates that permit membrane depolarization (Wu et al. 1999). Non-pore models, such as the carpet and detergent models, explain the disruption of the membrane as the effect of a critical concentration of peptide gradually accumulated on the membrane surface (Shai 2002; Haney et al. 2017a; Kumar et al. 2018). These last models have been suggested for several HDPs, including cecropins (Gazit et al. 1996) and human cathelicidin LL-37 (Shai 2002).

Membrane-targeting models dominated the first decades of research on HDPs, as it was initially thought that the single mechanism of action of all HDPs was directly killing bacteria by interacting with and destabilizing the cytoplasmic membrane (Fig. 1, Shai 1999; Zasloff 2002; Hancock et al. 2021). However, it has now become apparent that these early HDP studies were mostly conducted under conditions not representative of those found in vivo. Peptides were assayed in dilute media without physiological concentrations of monovalent cations, divalent cations, and polyanions (heparin, mucins) (Hancock and Sahl 2006) and at peptide concentrations much higher than their MIC, or with very high peptide-lipid ratios on model membranes (Wu et al. 1999; Wimley 2010; Haney et al. 2019b). Today, there is an increasing focus on studying the bactericidal or bacteriostatic effects of HDPs in vivo or under physiologically relevant conditions (Hancock and Sahl 2006). Membrane-targeting mechanisms demonstrated in vivo have been suggested as the basis for the activity of several HDPs against biofilms (Fig. 1). This is the case for the amphibian HDP esculentin (Luca et al. 2013) or peptide Hp1404, obtained from scorpion venom, and its synthetic analogs (Kim et al. 2018).

Besides non-specific membrane disruption, other antimicrobial mechanisms started to be considered after demonstrating that certain HDPs can kill bacterial cells by interacting with intracellular bacterial molecules (Park et al. 1998; Wu et al. 1999). Some HDPs, such as buforin-2 (Park et al. 1998) and indolicidin (Hsu et al. 2005), are considered to target DNA or RNA. These potential mechanisms have been inferred using standard nucleic acid-interaction techniques such as the Electrophoretic Mobility Shift Assay (EMSA). DNA binding, followed by Cu^{2+}-dependent DNA cleavage, has been proposed as the primary mechanism underlying the antimicrobial activity of fish HDP piscidin 3. However, not only intracellular DNA but also the extracellular DNA in the matrix is affected, leading to the eradication of *P. aeruginosa* biofilms (Libardo et al. 2017). Other HDPs perform their action by binding to cytoplasmic protein targets; this is the case of Bac71–35, which can bind to ribosomal proteins (Mardirossian et al. 2014). These mechanisms can be studied using protein–protein interaction techniques, like co-precipitation (Raheem and Straus 2019). HDPs can also act on specific membrane targets: lantibiotics, such as nisin, bind the peptidoglycan precursor lipid II and thus block the biosynthesis of the cell wall (Martin and Breukink 2007). While HDPs were traditionally studied in the context of their capacity to disrupt membrane function, the importance of intracellular targets for HDPs is progressively more recognized (Haney et al. 2019b; Raheem and Straus 2019).

2.7 *Immunomodulatory Mechanisms of HDPs Relevant for the Treatment of Biofilm Infections*

Beyond their direct antimicrobial and antibiofilm activities, HDPs have extensively been studied for their ability to shape immune responses (Mookherjee et al. 2020). A diverse repertoire of immunomodulatory activities enables HDPs to promote clearance of infections (Mookherjee et al. 2020). Since immune evasion and progressive chronic inflammation, due to excessive and prolonged neutrophil infiltration, are considered key characteristics of biofilm infections, HDP-mediated modulation of host responses can offer the opportunity to improve therapeutic success and patient outcome (Dostert et al. 2021; Moser et al. 2021). Host–biofilm interactions have been primarily studied in the context of chronic lung infections and chronic wounds, which typically involve *P. aeruginosa* and *S. aureus* (Moser et al. 2021). However, most studies, including those characterizing HDPs, mainly focus on the host response to planktonic bacteria. Here, we will highlight immunomodulatory activities relevant to our current understanding of host responses to biofilm infections.

First, HDPs have been shown to modulate the secretion of cytokines from various epithelial and immune cell types (Mookherjee et al. 2020; Alford et al. 2020). While the production of pro-inflammatory mediators by these cells in response to LPS is well established, other microbial components such as LTA, mycobacterial cord factor, zymosan, and flagellin can cause similar effects (Filewod et al. 2009; Wu et al. 2017; Yang et al. 2019; Mookherjee et al. 2020). In biofilms, bacterial molecules are proposed to be less accessible to immune recognition due to shielding by the biofilm matrix, LPS modification, and loss of flagella (González et al. 2018; Dostert et al. 2021; Moser et al. 2021). However, microbial components might still escape biofilms, as demonstrated by the fact that LPS, extracellular DNA, and exopolysaccharides have recently been isolated from *P. aeruginosa* biofilm supernatants (Ciszek-Lenda et al. 2019). Importantly, these biofilm-derived microbial molecules induced the production of pro-inflammatory cytokines interleukin 6 (IL-6), IL-12B, and tumor necrosis factor alpha (TNFα) in macrophages and neutrophils (Ciszek-Lenda et al. 2019). Therefore, HDP-mediated suppression of pro-inflammatory cytokine secretion and enhancement of anti-inflammatory cytokine secretion might reduce damage to host tissues by suppressing local inflammatory responses that cause pathology and favor bacterial clearance in the context of biofilm-associated chronic inflammation.

Early studies suggested that the sequestration of microbial factors (e.g., binding of HDPs to LPS), interferes with signaling, e.g., by preventing the interaction with their cognate receptors, leading to the suppression of pro-inflammatory cytokine secretion (Mookherjee et al. 2020; Alford et al. 2020; Kumar and Shin 2020). However, in this situation we would expect complete inhibition of downstream effects mediated by the binding of microbial signatures (also termed pathogen-associated molecular patterns) to their receptors. Instead, transcriptional profiling and immunological experiments have revealed that the inhibition is selective, e.g., reducing certain pro-inflammatory cytokines, such as TNF-α, while enhancing the production of pro-inflammatory chemokines, like monocyte chemoattractant protein-1 (Mookherjee et al. 2006, 2009; Hilchie et al. 2013). Also, the addition of HDPs 30 minutes after LPS, when receptors have already been engaged and downstream signaling pathways and transcription factors have been activated, still allows for selective HDP modulation of these effects (Mookherjee et al. 2006). Indeed, comparisons between different HDPs suggest that the sequestration of microbial factors alone cannot explain their ability to modulate cytokine production (Scheenstra et al. 2019; Wang et al. 2021). The frog cathelicidin PN-CATH1 suppressed LPS-induced production of pro-inflammatory cytokines in mouse peritoneal macrophages even though it cannot bind to LPS (Wang et al. 2021). Even more conclusively, synthetic HDP 1002 suppressed pro-inflammatory responses in a sterile mouse ear inflammation model in the absence of any immunogenic, microbial molecule (Wu et al. 2017). The same peptide profoundly reduced pro-inflammatory cytokines and selectively modulated gene expression in a mouse model of P. aeruginosa infection (Wuerth et al. 2019). Accordingly, other studies have demonstrated that HDPs interact with several intracellular proteins, indicating they enter eukaryotic cells to modulate immune responses (Mookherjee et al. 2020).

HDP-mediated changes of the cytokine milieu at the site of infection can affect recruitment, proliferation, activation, and polarization of leukocytes (Mookherjee et al. 2020). Recruitment of immune cells is usually driven by chemokines, a subgroup of cytokines acting as chemoattractants (Mookherjee et al. 2020). Thus, the treatment of infections in a mouse model led to the induction of multiple chemokines and recruitment of monocytes and neutrophils to the site of infections (Nijnik et al. 2010). Similarly, synthetic LL-37 derived C10-KR8d, in a catheter-associated *S. aureus* biofilm model, induced the production of macrophage inflammatory protein-1α and IL-17A, consistent with the recruitment of neutrophils and macrophages (Lakshmaiah Narayana et al. 2021). Intriguingly, C10-KR8d cleared the biofilm infection highlighting the complexity of host–biofilm interactions. Alternatively, the engineered HDP WLUB2 suppressed the transcription of two chemokines in a murine pneumonia model, which coincided with a reduction of leukocyte infiltrates and bacterial burden in the lung (Chen et al. 2018a).

Beyond inducing chemokine secretion, HDPs can themselves act as weak chemoattractants, promoting migration of various immune, epithelial, and fibroblast cells by interacting with cell surface receptors (van Harten et al. 2018). HDP-mediated migration and proliferation of fibroblasts and keratinocytes plays an important role in wound healing, which is especially relevant in the context of chronic wounds (Chung et al. 2017; Wu et al. 2018). For instance, frog cathelicidin-NV promoted migration and proliferation of keratinocytes, as well as proliferation, differentiation, and collagen production in fibroblasts in vitro, which correlated with potent wound healing activity in vivo (Wu et al. 2018). Interestingly, a study comparing natural cathelicidins demonstrated that some HDPs might enhance immune cell recruitment by combining indirect and direct mechanisms (Xie et al. 2020). Bovine BMAP-34, murine CRAMP, and human LL-37 promoted chemotaxis at 1 μM after 45 minutes, while increased chemokine secretion was observed at higher concentrations after 6 hours (Xie et al. 2020).

Moreover, these three HDPs increased bactericidal neutrophil functions, such as the production of intracellular reactive oxygen species and the formation of extracellular traps, in vitro (Xie et al. 2020). Intriguingly, LL-37 in combination with bovine BMAP-27, an HDP with potent antimicrobial and antibiofilm activity but no ability to modulate neutrophil functions, reduced tissue damage, and augmented bacterial clearance in a murine pulmonary *P. aeruginosa* infection model when compared to single peptide treatments (Xie et al. 2020). It remains to be determined whether this combined HDP therapy would prove similarly effective against biofilm infections. Pro-inflammatory neutrophil functions have been proposed to be less beneficial in the context of biofilm infections since they can cause tissue damage to the host or provide material, such as host nucleic acid, that might be incorporated into the matrix and thus enhance biofilm growth (Dostert et al. 2021). However, LL-37 might be able to antagonize inflammation by inhibiting pyroptosis, a form of cell death stimulated during microbial infections, which leads to the release of intracellular molecules, such as DNA and pro-inflammatory mediators (Hu et al. 2016; Qiu et al. 2019).

Besides modulation of bactericidal immune cell functions, HDPs can also affect immune cell polarization (Pena et al. 2013; Hilchie et al. 2013; Mookherjee et al. 2020). Depending on the cytokine environment, macrophages and T helper (T_h) cells can polarize into different types of effector cells, which are categorized based on their pro-inflammatory (T_h1 and M1) and wound healing or anti-inflammatory (T_h2 and M2) functions (González et al. 2018). During biofilm infections, the balance between T_h1 and T_h2, as well as M1 and M2 responses, has been suggested to be skewed (González et al. 2018; Moser et al. 2021). This imbalance prevents effective clearance of the infections and instead causes damage to the host. Interestingly, T cell immune responses to biofilm infections seem to be pathogen-dependent, highlighting the complexity of host–biofilm interactions (González et al. 2018). While chronic infections with *P. aeruginosa* in cystic fibrosis patients correlated with predominantly anti-inflammatory T_h2 immune responses, pro-inflammatory T_h1 effector cells have been shown to drive immune responses to *S. aureus* biofilms (González et al. 2018; Moser et al. 2021). Experiments with mouse strains displaying biased T cell responses have demonstrated that T_h1 responder C3H/HeN mice and T_h2 responder BALB/c mice more often spontaneously clear biofilm infections caused by *P. aeruginosa* in the lung and *S. aureus* both in a subcutaneous footpad and a prosthetic implant infection model (Moser et al. 1997; Nippe et al. 2011; Prabhakara et al. 2011). These observations could suggest that HDP-mediated modulation of T cell responses might prove beneficial in treating biofilm infections. For example, human HDP LL-37 indirectly promoted T cell polarization to T_h1 effector cells via differentiation of human monocyte derived dendritic cells in vitro (Davidson et al. 2004). However, only a limited number of studies have explored the capacity of HDPs to modulate T cell polarization.

HDP-mediated polarization of macrophages has attracted more attention and might also be a promising approach since *S. aureus* biofilm infections have been associated with predominant anti-inflammatory M2 macrophage responses (Lyu et al. 2017; Chen et al. 2018b; González et al. 2018). Interestingly, local administration of pro-inflammatory M1 macrophages alleviated bacterial burden in a murine catheter-associated model (Hanke et al. 2013). Also, direct polarization of human blood-derived monocytes in the presence of synthetic bactenecin derivative 1018 led to macrophages in an intermediary M1-M2 state, which displayed increased transcription of genes associated with wound healing (Pena et al. 2013).

HDPs display a variety of immunomodulatory activities that offer opportunities to restore immune homeostasis to successfully clear biofilm infections. Modulation of cytokine secretion is probably the best characterized immunomodulatory activity of HDPs. Since biofilms are typically accompanied by excessive and prolonged inflammation, the suppression of pro-inflammatory cytokine production and induction of anti-inflammatory cytokine production are a strong research focus. However, it is worth noting that potent anti-infective HDPs have also been shown to inversely modulate individual cytokines or immune cell functions (Yang et al. 2019; Lakshmaiah Narayana et al. 2021). Additionally, anti-inflammatory macrophage and T cell responses fail to clear biofilm infections in some cases (González et al. 2018), and thus this might be considered an ancillary activity that could assist HDPs in resolving biofilm infections and associated pathologies. Evidence suggests that host–biofilm interactions are complex, pathogen-dependent, and potentially context-specific. Overall, we need to improve our understanding of host–biofilm interactions to define desirable host-directed activities to treat or assist in the resolution of biofilm infections and guide HDP research.

3 Conclusion and Future Perspectives

Since antibiotic therapy often fails against biofilm infections, even in the absence of acquired multidrug resistance, there is an urgent need for new therapeutic options addressing biofilm-specific challenges. HDPs are promising drug candidates developed for their broad-spectrum antimicrobial activity and low potential to develop resistance. However, especially given the rather disappointing clinical results for HDPs to date (Fjell et al. 2011; Hilchie et al. 2013), there is likely even greater potential for the independently determined activities directed at biofilms and the host immune system. The interplay of all three activities offers the opportunity to simultaneously attack biofilm infections through multiple, independent mechanisms.

Translating those promising characteristics to clinical applications depends on our capacity to go beyond the physiological roles of natural HDPs, both through the design of optimized peptide sequences and the development of drug combinations that exploit HDP-HDP and HDP-antibiotic synergies. The future of these therapies hinges on our understanding of how HDPs work, and especially how they combat biofilms. In this chapter, we have highlighted the mechanisms of action of HDPs that are specifically relevant to the treatment of biofilm infections. Today this is a very active area of research, and new mechanisms are continuously being discovered and expanded upon. Understanding these mechanisms will assist the process of HDP optimization and allow us to develop effective drug combination therapies.

Other obstacles will need to be addressed for HDPs to become common in clinical practice. It has become clear that peptides screened and developed under standard lab conditions often fail to retain their activity in vivo. Currently, we are gradually moving toward performing peptide screening, optimization, and mechanistic studies under conditions that mimic in vivo environments. It is worth mentioning that other HDP caveats, such as stability and precipitation, may be less problematic when treating biofilm infections, given the prevalence of HDPs as topical treatments of skin and wound biofilms, or as surface coatings on medical devices.

4 Compliance with Ethical Standards

All research summarized here was done under specific ethical certification for use of animals and biologicals. Our own research in the area covered here is partially supported by a grant from the Canadian Institutes for Health Research (FDN-154287) and Nanomedicines Innovation Network (NMIN) grant (GR015743). LP received a postdoctoral fellowship from the Ramón Areces Foundation (Life and Matter Sciences). REWH holds a Canada Research Chair in Health and Genomics and a University of British Columbia Killam Professorship. The peptide research of REWH has been filed for patent protection by the University of British Columbia and licensed to ABT Innovations Inc., in which REWH holds an ownership position.

References

Åberg A, Shingler V, Balsalobre C (2006) (p)ppGpp regulates type 1 fimbriation of *Escherichia coli* by modulating the expression of the site-specific recombinase FimB. Mol Microbiol 60: 1520–1533. https://doi.org/10.1111/j.1365-2958.2006.05191.x

Abraham NM, Liu L, Jutras BL et al (2017) Pathogen-mediated manipulation of arthropod microbiota to promote infection. Proc Natl Acad Sci U S A 114:E781–E790. https://doi.org/10.1073/pnas.1613422114

Alford MA, Baquir B, Santana FL et al (2020) Cathelicidin host defense peptides and inflammatory signaling: striking a balance. Front Microbiol 11:1902. https://doi.org/10.3389/fmicb.2020.01902

Ansari JM, Abraham NM, Massaro J et al (2017) Anti-biofilm activity of a self-aggregating peptide against *Streptococcus mutans*. Front Microbiol 8:488. https://doi.org/10.3389/fmicb.2017.00488

Atkinson GC, Tenson T, Hauryliuk V (2011) The RelA/SpoT homolog (RSH) superfamily: distribution and functional evolution of ppGpp synthetases and hydrolases across the tree of life. PLoS One 6:e23479. https://doi.org/10.1371/journal.pone.0023479

Balzer GJ, McLean RJC (2002) The stringent response genes relA and spoT are important for *Escherichia coli* biofilms under slow growth conditions. Can J Microbiol 48:675–680. https://doi.org/10.1139/w02-060

Batoni G, Maisetta G, Esin S (2016) Antimicrobial peptides and their interaction with biofilms of medically relevant bacteria. Biochim Biophys Acta 1858:1044–1060. https://doi.org/10.1016/j.bbamem.2015.10.013

Batoni G, Maisetta G, Esin S (2021) Therapeutic potential of antimicrobial peptides in polymicrobial biofilm-associated infections. Int J Mol Sci 22:1–24. https://doi.org/10.3390/ijms22020482

Battesti A, Bouveret E (2009) Bacteria possessing two RelA/SpoT-like proteins have evolved a specific stringent response involving the acyl carrier protein-SpoT interaction. J Bacteriol 191: 616–624. https://doi.org/10.1128/JB.01195-08

Behrendt R, White P, Offer J (2016) Advances in Fmoc solid-phase peptide synthesis. J Pept Sci 22: 4–27. https://doi.org/10.1002/psc.2836

Bommarius B, Jenssen H, Elliott M et al (2010) Cost-effective expression and purification of antimicrobial and host defense peptides in *Escherichia coli*. Peptides 31:1957–1965. https://doi.org/10.1016/j.peptides.2010.08.008

Cao Y, Yin H, Wang W et al (2020) Killing *Streptococcus mutans* in mature biofilm with a combination of antimicrobial and antibiofilm peptides. Amino Acids 52:1–14. https://doi.org/10.1007/s00726-019-02804-4

Cardoso MH, Orozco RQ, Rezende SB et al (2019) Computer-aided design of antimicrobial peptides: are we generating effective drug candidates? Front Microbiol 10:3097. https://doi.org/10.3389/fmicb.2019.03097

Casciaro B, Loffredo MR, Cappiello F et al (2020) The antimicrobial peptide temporin G: anti-biofilm, anti-persister activities, and potentiator effect of tobramycin efficacy against *Staphylococcus aureus*. Int J Mol Sci 21:1–13. https://doi.org/10.3390/ijms21249410

Chen C, Deslouches B, Montelaro RC, Di YP (2018a) Enhanced efficacy of the engineered antimicrobial peptide WLBU2 via direct airway delivery in a murine model of *Pseudomonas aeruginosa* pneumonia. Clin Microbiol Infect 24:547.e1–547.e8. https://doi.org/10.1016/j.cmi.2017.08.029

Chen S, Lu Z, Wang F, Wang Y (2018b) Cathelicidin-WA polarizes *E. coli* K88-induced M1 macrophage to M2-like macrophage in RAW264.7 cells. Int Immunopharmacol 54:52–59. https://doi.org/10.1016/j.intimp.2017.10.013

Chung EMC, Dean SN, Propst CN et al (2017) Komodo dragon-inspired synthetic peptide DRGN-1 promotes wound-healing of a mixed-biofilm infected wound. npj Biofilms Microbiomes 3:1–13. https://doi.org/10.1038/s41522-017-0017-2

Ciszek-Lenda M, Strus M, Walczewska M et al (2019) *Pseudomonas aeruginosa* biofilm is a potent inducer of phagocyte hyperinflammation. Inflamm Res 68:397–413. https://doi.org/10.1007/s00011-019-01227-x

Dalebroux ZD, Swanson MS (2012) ppGpp: magic beyond RNA polymerase. Nat Rev Microbiol 10:203–212. https://doi.org/10.1038/nrmicro2720

Davidson DJ, Currie AJ, Reid GSD et al (2004) The cationic antimicrobial peptide LL-37 modulates dendritic cell differentiation and dendritic cell-induced T cell polarization. J Immunol 172:1146–1156. https://doi.org/10.4049/jimmunol.172.2.1146

de Breij A, Riool M, Cordfunke RA et al (2018) The antimicrobial peptide SAAP-148 combats drug-resistant bacteria and biofilms. Sci Transl Med 10:eaan4044. https://doi.org/10.1126/scitranslmed.aan4044

de la Fuente-Núñez C, Korolik V, Bains M et al (2012) Inhibition of bacterial biofilm formation and swarming motility by a small synthetic cationic peptide. Antimicrob Agents Chemother 56: 2696–2704. https://doi.org/10.1128/AAC.00064-12

de la Fuente-Núñez C, Reffuveille F, Haney EF et al (2014) Broad-spectrum anti-biofilm peptide that targets a cellular stress response. PLoS Pathog 10:e1004152. https://doi.org/10.1371/journal.ppat.1004152

de la Fuente-Nuñez C, Reffuveille F, Mansour SC et al (2015) D-enantiomeric peptides that eradicate wild-type and multi-drug resistant biofilms and protect against lethal Pseudomonas aeruginosa infections. Chem Biol 22:196–205. https://doi.org/10.1016/j.chembiol.2015.01.002

de la Fuente-Núñez C, Cardoso MH, de Souza Cândido E et al (2016) Synthetic antibiofilm peptides. Biochim Biophys Acta 1858:1061–1069. https://doi.org/10.1016/j.bbamem.2015.12.015
Di Luca M, Maccari G, Nifosì R (2014) Treatment of microbial biofilms in the post-antibiotic era: prophylactic and therapeutic use of antimicrobial peptides and their design by bioinformatics tools. Pathog Dis 70:257–270. https://doi.org/10.1111/2049-632X.12151
Di Somma A, Moretta A, Canè C et al (2020) Antimicrobial and antibiofilm peptides. Biomolecules 10:1–15. https://doi.org/10.3390/biom10040652
Dostert M, Belanger CR, Hancock REW (2019) Design and assessment of anti-biofilm peptides: steps toward clinical application. J Innate Immun 11:193–204. https://doi.org/10.1159/000491497
Dostert M, Trimble MJ, Hancock REW (2021) Antibiofilm peptides: overcoming biofilm-related treatment failure. RSC Adv 11:2718–2728. https://doi.org/10.1039/D0RA09739J
Etayash H, Pletzer D, Kumar P et al (2020) Cyclic derivative of host-defense peptide IDR-1018 improves proteolytic stability, suppresses inflammation, and enhances *in vivo* activity. J Med Chem 63:9228–9236. https://doi.org/10.1021/acs.jmedchem.0c00303
Faurschou M, Borregaard N (2003) Neutrophil granules and secretory vesicles in inflammation. Microbes Infect 5:1317–1327. https://doi.org/10.1016/j.micinf.2003.09.008
Filewod NCJ, Pistolic J, Hancock REW (2009) Low concentrations of LL-37 alter IL-8 production by keratinocytes and bronchial epithelial cells in response to proinflammatory stimuli. FEMS Immunol Med Microbiol 56:233–240. https://doi.org/10.1111/j.1574-695X.2009.00571.x
Fjell CD, Hiss JA, Hancock REW, Schneider G (2011) Designing antimicrobial peptides: form follows function. Nat Rev Drug Discov 11:37–51. https://doi.org/10.1038/nrd3591
Francis VI, Stevenson EC, Porter SL (2017) Two-component systems required for virulence in *Pseudomonas aeruginosa*. FEMS Microbiol Lett 364:104. https://doi.org/10.1093/femsle/fnx104
Gallo RL, Hooper LV (2012) Epithelial antimicrobial defence of the skin and intestine. Nat Rev Immunol 12:503–516. https://doi.org/10.1038/nri3228
Gazit E, Miller IR, Biggin PC et al (1996) Structure and orientation of the mammalian antibacterial peptide cecropin P1 within phospholipid membranes. J Mol Biol 258:860–870. https://doi.org/10.1006/jmbi.1996.0293
González JF, Hahn MM, Gunn JS (2018) Chronic biofilm-based infections: skewing of the immune response. Pathog Dis 76:23. https://doi.org/10.1093/femspd/fty023
Gumerov VM, Ortega DR, Adebali O et al (2020) MiST 3.0: an updated microbial signal transduction database with an emphasis on chemosensory systems. Nucleic Acids Res 48: D459–D464. https://doi.org/10.1093/nar/gkz988
Hancock REW (2001) Cationic peptides: effectors in innate immunity and novel antimicrobials. Lancet Infect Dis 1:156–164. https://doi.org/10.1016/S1473-3099(01)00092-5
Hancock REW, Sahl H-G (2006) Antimicrobial and host-defense peptides as new anti-infective therapeutic strategies. Nat Biotechnol 24:1551–1557. https://doi.org/10.1038/nbt1267
Hancock REW, Scott MG (2000) The role of antimicrobial peptides in animal defenses. Proc Natl Acad Sci U S A 97:8856–8861. https://doi.org/10.1073/pnas.97.16.8856
Hancock REW, Alford MA, Haney EF (2021) Antibiofilm activity of host defence peptides: complexity provides opportunities. Nat Rev Microbiol 19(12):786–797. https://doi.org/10.1038/s41579-021-00585-w
Haney EF, Hancock REW (2013) Peptide design for antimicrobial and immunomodulatory applications. Biopolym Peptide Sci 100:572–583. https://doi.org/10.1002/bip.22250
Haney EF, Mansour SC, Hilchie AL et al (2015) High throughput screening methods for assessing antibiofilm and immunomodulatory activities of synthetic peptides. Peptides 71:276–285. https://doi.org/10.1016/j.peptides.2015.03.015
Haney EF, Mansour SC, Hancock REW (2017a) Antimicrobial peptides: an introduction. Methods Mol Biol 1548:3–22

Haney EF, Wu B (Catherine), Lee K et al (2017b) Aggregation and its influence on the immunomodulatory activity of synthetic innate defense regulator peptides. Cell Chem Biol 24:969–980. e4. https://doi.org/10.1016/j.chembiol.2017.07.010

Haney EF, Brito-Sánchez Y, Trimble MJ et al (2018) Computer-aided discovery of peptides that specifically attack bacterial biofilms. Sci Rep 8:1871. https://doi.org/10.1038/s41598-018-19669-4

Haney EF, Barbosa SC, Baquir B, Hancock REW (2019a) Influence of non-natural cationic amino acids on the biological activity profile of innate defense regulator peptides. J Med Chem 62: 10294–10304. https://doi.org/10.1021/acs.jmedchem.9b01344

Haney EF, Straus SK, Hancock REW (2019b) Reassessing the host defense peptide landscape. Front Chem 7:43. https://doi.org/10.3389/fchem.2019.00043

Hanke ML, Heim CE, Angle A et al (2013) Targeting macrophage activation for the prevention and treatment of *Staphylococcus aureus* biofilm infections. J Immunol 190:2159–2168. https://doi.org/10.4049/jimmunol.1202348

Hardie KR (2020) Antimicrobial resistance: the good, the bad, and the ugly. Emerg Top life Sci 4: 129–136. https://doi.org/10.1042/ETLS20190194

Harris F, Dennison SR, Phoenix DA (2009) Anionic antimicrobial peptides from eukaryotic organisms. Curr Protein Pept Sci 10:585–606. https://doi.org/10.2174/138920309789630589

Heisig M, Abraham NM, Liu L et al (2014) Antivirulence properties of an antifreeze protein. Cell Rep 9:417–424. https://doi.org/10.1016/j.celrep.2014.09.034

Henninot A, Collins JC, Nuss JM (2018) The current state of peptide drug discovery: back to the future? J Med Chem 61:1382–1414. https://doi.org/10.1021/acs.jmedchem.7b00318

Hilchie AL, Wuerth K, Hancock REW (2013) Immune modulation by multifaceted cationic host defense (antimicrobial) peptides. Nat Chem Biol 9:761–768. https://doi.org/10.1038/nchembio.1393

Hsu CH, Chen C, Jou ML et al (2005) Structural and DNA-binding studies on the bovine antimicrobial peptide, indolicidin: evidence for multiple conformations involved in binding to membranes and DNA. Nucleic Acids Res 33:4053–4064. https://doi.org/10.1093/nar/gki725

Hu Z, Murakami T, Suzuki K et al (2016) Antimicrobial cathelicidin peptide LL-37 inhibits the pyroptosis of macrophages and improves the survival of polybacterial septic mice. Int Immunol 28:245–253. https://doi.org/10.1093/intimm/dxv113

Jakobsen TH, Bjarnsholt T, Jensen PØ et al (2013) Targeting quorum sensing in *Pseudomonas aeruginosa* biofilms: current and emerging inhibitors. Future Microbiol 8:901–921. https://doi.org/10.2217/FMB.13.57

Jenssen H, Hamill P, Hancock REW (2006) Peptide antimicrobial agents. Clin Microbiol Rev 19: 491–511. https://doi.org/10.1128/CMR.00056-05

Jiale Z, Jian J, Xinyi T et al (2021) Design of a novel antimicrobial peptide 1018M targeted ppGpp to inhibit MRSA biofilm formation. AMB Express 11:49. https://doi.org/10.1186/s13568-021-01208-6

Kai-Larsen Y, Lüthje P, Chromek M et al (2010) Uropathogenic *Escherichia coli* modulates immune responses and its curli fimbriae interact with the antimicrobial peptide LL-37. PLoS Pathog 6:1–16. https://doi.org/10.1371/journal.ppat.1001010

Kang X, Dong F, Shi C et al (2019) DRAMP 2.0, an updated data repository of antimicrobial peptides. Sci Data 6:1–10. https://doi.org/10.1038/s41597-019-0154-y

Kim MK, Kang HK, Ko SJ et al (2018) Mechanisms driving the antibacterial and antibiofilm properties of Hp1404 and its analogue peptides against multidrug-resistant *Pseudomonas aeruginosa*. Sci Rep 8:1763. https://doi.org/10.1038/s41598-018-19434-7

Koeninger L, Osbelt L, Berscheid A et al (2021) Curbing gastrointestinal infections by defensin fragment modifications without harming commensal microbiota. Commun Biol 4:47. https://doi.org/10.1038/s42003-020-01582-0

Kumar SD, Shin SY (2020) Antimicrobial and anti-inflammatory activities of short dodecapeptides derived from duck cathelicidin: plausible mechanism of bactericidal action and endotoxin neutralization. Eur J Med Chem 204:112580. https://doi.org/10.1016/j.ejmech.2020.112580

Kumar P, Kizhakkedathu JN, Straus SK (2018) Antimicrobial peptides: diversity, mechanism of action and strategies to improve the activity and biocompatibility *in vivo*. Biomol Ther 8:4. https://doi.org/10.3390/biom8010004

Lakshmaiah Narayana J, Golla R, Mishra B et al (2021) Short and robust anti-infective lipopeptides engineered based on the minimal antimicrobial peptide KR12 of human LL-37. ACS Infect Dis 7:1795–1808. https://doi.org/10.1021/acsinfecdis.1c00101

Laver DR (1994) The barrel-stave model as applied to alamethicin and its analogs reevaluated. Biophys J 66:355–359. https://doi.org/10.1016/s0006-3495(94)80784-2

Lee T-H, Hall KN, Aguilar M-I (2016) Antimicrobial peptide structure and mechanism of action: a focus on the role of membrane structure. Curr Top Med Chem 16:25–39. https://doi.org/10.2174/1568026615666150703121700

Libardo MDJ, Bahar AA, Ma B et al (2017) Nuclease activity gives an edge to host-defense peptide piscidin 3 over piscidin 1, rendering it more effective against persisters and biofilms. FEBS J 284:3662–3683. https://doi.org/10.1111/febs.14263

Luca V, Stringaro A, Colone M et al (2013) Esculentin(1-21), an amphibian skin membrane-active peptide with potent activity on both planktonic and biofilm cells of the bacterial pathogen *Pseudomonas aeruginosa*. Cell Mol Life Sci 70:2773–2786. https://doi.org/10.1007/s00018-013-1291-7

Lynch AS, Abbanat D (2010) New antibiotic agents and approaches to treat biofilm-associated infections. Expert Opin Ther Pat 20:1373–1387. https://doi.org/10.1517/13543776.2010.505923

Lyu J, Bian T, Chen B et al (2017) β-defensin 3 modulates macrophage activation and orientation during acute inflammatory response to *Porphyromonas gingivalis* lipopolysaccharide. Cytokine 92:48–54. https://doi.org/10.1016/j.cyto.2016.12.015

Magana M, Pushpanathan M, Santos AL et al (2020) The value of antimicrobial peptides in the age of resistance. Lancet Infect Dis 20:e216–e230. https://doi.org/10.1016/S1473-3099(20)30327-3

Magnusson LU, Farewell A, Nyström T (2005) ppGpp: A global regulator in *Escherichia coli*. Trends Microbiol 13:236–242. https://doi.org/10.1016/j.tim.2005.03.008

Mansour SC, Pletzer D, de la Fuente-Núñez C et al (2016) Bacterial abscess formation is controlled by the stringent stress response and can be targeted therapeutically. EBioMedicine 12:219–226. https://doi.org/10.1016/j.ebiom.2016.09.015

Mardirossian M, Grzela R, Giglione C et al (2014) The host antimicrobial peptide Bac71-35 binds to bacterial ribosomal proteins and inhibits protein synthesis. Chem Biol 21:1639–1647. https://doi.org/10.1016/j.chembiol.2014.10.009

Martin NI, Breukink E (2007) Expanding role of lipid II as a target for lantibiotics. Future Microbiol 2:513–525. https://doi.org/10.2217/17460913.2.5.513

Maystrenko A, Feng Y, Akhtar N, Li J (2020) The addition of a synthetic LPS-targeting domain improves serum stability while maintaining antimicrobial, antibiofilm, and cell stimulating properties of an antimicrobial peptide. Biomolecules 10:1–22. https://doi.org/10.3390/biom10071014

Mookherjee N, Brown KL, Bowdish DME et al (2006) Modulation of the TLR-mediated inflammatory response by the endogenous human host defense peptide LL-37. J Immunol 176:2455–2464. https://doi.org/10.4049/jimmunol.176.4.2455

Mookherjee N, Hamill P, Gardy J et al (2009) Systems biology evaluation of immune responses induced by human host defence peptide LL-37 in mononuclear cells. Mol BioSyst 5:483–496. https://doi.org/10.1039/b813787k

Mookherjee N, Anderson MA, Haagsman HP, Davidson DJ (2020) Antimicrobial host defence peptides: functions and clinical potential. Nat Rev Drug Discov 19:311–332. https://doi.org/10.1038/s41573-019-0058-8

Moser C, Johansen HK, Song Z et al (1997) Chronic *Pseudomonas aeruginosa* lung infection is more severe in Th2 responding BALB/c mice compared to Th1 responding C3H/HeN mice. APMIS 105:838–842. https://doi.org/10.1111/J.1699-0463.1997.TB05092.X

Moser C, Jensen PØ, Thomsen K et al (2021) Immune responses to *Pseudomonas aeruginosa* biofilm infections. Front Immunol 12:625597. https://doi.org/10.3389/fimmu.2021.625597

Nguyen LT, Haney EF, Vogel HJ (2011) The expanding scope of antimicrobial peptide structures and their modes of action. Trends Biotechnol 29:464–472. https://doi.org/10.1016/j.tibtech.2011.05.001

Nijnik A, Madera L, Ma S et al (2010) Synthetic cationic peptide IDR-1002 provides protection against bacterial infections through chemokine induction and enhanced leukocyte recruitment. J Immunol 184:2539–2550. https://doi.org/10.4049/jimmunol.0901813

Nippe N, Varga G, Holzinger D et al (2011) Subcutaneous infection with *S. aureus* in mice reveals association of resistance with influx of neutrophils and Th2 response. J Invest Dermatol 131: 125–132. https://doi.org/10.1038/jid.2010.282

Okuda KI, Zendo T, Sugimoto S et al (2013) Effects of bacteriocins on methicillin-resistant *Staphylococcus aureus* biofilm. Antimicrob Agents Chemother 57:5572–5579. https://doi.org/10.1128/AAC.00888-13

Oshiro KGN, Cândido ES, Chan LY et al (2019) Computer-aided design of Mastoparan-like peptides enables the generation of nontoxic variants with extended antibacterial properties. J Med Chem 62:8140–8151. https://doi.org/10.1021/acs.jmedchem.9b00915

Overhage J, Campisano A, Bains M et al (2008) Human host defense peptide LL-37 prevents bacterial biofilm formation. Infect Immun 76:4176–4182. https://doi.org/10.1128/IAI.00318-08

Paluch E, Rewak-Soroczyńska J, Jędrusik I et al (2020) Prevention of biofilm formation by quorum quenching. Appl Microbiol Biotechnol 104:1871–1881. https://doi.org/10.1007/s00253-020-10349-w

Park CB, Kim HS, Kim SC (1998) Mechanism of action of the antimicrobial peptide buforin II: buforin II kills microorganisms by penetrating the cell membrane and inhibiting cellular functions. Biochem Biophys Res Commun 244:253–257. https://doi.org/10.1006/bbrc.1998.8159

Passos da Silva D, Schofield M, Parsek M, Tseng B (2017) An update on the sociomicrobiology of quorum sensing in gram-negative biofilm development. Pathogens 6:51. https://doi.org/10.3390/pathogens6040051

Pati BA, Kurata WE, Horseman TS, Pierce LM (2021) Antibiofilm activity of chitosan/epsilon-poly-L-lysine hydrogels in a porcine *ex vivo* skin wound polymicrobial biofilm model. Wound Repair Regen 29:316–326. https://doi.org/10.1111/wrr.12890

Patil A, Hughes AL, Zhang G (2004) Rapid evolution and diversification of mammalian alpha-defensins as revealed by comparative analysis of rodent and primate genes. Physiol Genomics 20:1–11. https://doi.org/10.1152/physiolgenomics.00150.2004

Pena OM, Afacan N, Pistolic J et al (2013) Synthetic cationic peptide IDR-1018 modulates human macrophage differentiation. PLoS One 8:e52449. https://doi.org/10.1371/journal.pone.0052449

Pimentel-Filho NJ, de Freitas Martins MC, Nogueira GB et al (2014) Bovicin HC5 and nisin reduce *Staphylococcus aureus* adhesion to polystyrene and change the hydrophobicity profile and Gibbs free energy of adhesion. Int J Food Microbiol 190:1–8. https://doi.org/10.1016/j.ijfoodmicro.2014.08.004

Pletzer D, Hancock REW (2016) Antibiofilm peptides: potential as broad spectrum agents. J Bacteriol 198:2572–2578. https://doi.org/10.1128/JB.00017-16

Pletzer D, Hancock RE (2018) Is synergy the key to treating high-density infections? Future Microbiol 13:1629–1632. https://doi.org/10.2217/fmb-2018-0216

Pletzer D, Wolfmeier H, Bains M, Hancock REW (2017) Synthetic peptides to target stringent response-controlled virulence in a *Pseudomonas aeruginosa* murine cutaneous infection model. Front Microbiol 8:1867. https://doi.org/10.3389/fmicb.2017.01867

Potrykus K, Cashel M (2008) (p)ppGpp: still magical? Annu Rev Microbiol 62:35–51. https://doi.org/10.1146/annurev.micro.62.081307.162903

Prabhakara R, Harro JM, Leid JG et al (2011) Suppression of the inflammatory immune response prevents the development of chronic biofilm infection due to methicillin-resistant *Staphylococcus aureus*. Infect Immun 79:5010–5018. https://doi.org/10.1128/IAI.05571-11

Purcell EB, Tamayo R (2016) Cyclic diguanylate signaling in gram-positive bacteria. FEMS Microbiol Rev 40:753–773. https://doi.org/10.1093/femsre/fuw013

Qiu CC, Caricchio R, Gallucci S (2019) Triggers of autoimmunity: the role of bacterial infections in the extracellular exposure of lupus nuclear autoantigens. Front Immunol 10:2608. https://doi.org/10.3389/fimmu.2019.02608

Raheem N, Straus SK (2019) Mechanisms of action for antimicrobial peptides with antibacterial and antibiofilm functions. Front Microbiol 10:2866. https://doi.org/10.3389/fmicb.2019.02866

Ramamourthy G, Vogel HJ (2021) Antibiofilm activity of lactoferrin-derived synthetic peptides against *Pseudomonas aeruginosa* PAO1. Biochem Cell Biol 99:138–148. https://doi.org/10.1139/bcb-2020-0253

Römling U, Balsalobre C (2012) Biofilm infections, their resilience to therapy and innovative treatment strategies. J Intern Med 272:541–561. https://doi.org/10.1111/joim.12004

Roy R, Tiwari M, Donelli G, Tiwari V (2018) Strategies for combating bacterial biofilms: a focus on anti-biofilm agents and their mechanisms of action. Virulence 9:522–554. https://doi.org/10.1080/21505594.2017.1313372

Santana FL, Estrada K, Ortiz E, Corzo G (2021) Reptilian β-defensins: expanding the repertoire of known crocodylian peptides. Peptides 136:170473. https://doi.org/10.1016/j.peptides.2020.170473

Scheenstra MR, van den Belt M, Tjeerdsma-van Bokhoven JLM et al (2019) Cathelicidins PMAP-36, LL-37 and CATH-2 are similar peptides with different modes of action. Sci Rep 9:4780. https://doi.org/10.1038/s41598-019-41246-6

Scheper H, Wubbolts JM, Verhagen JAM et al (2021) SAAP-148 eradicates MRSA persisters within mature biofilm models simulating prosthetic joint infection. Front Microbiol 12:625952. https://doi.org/10.3389/fmicb.2021.625952

Scott MG, Yan H, Hancock REW (1999) Biological properties of structurally related α-helical cationic antimicrobial peptides. Infect Immun 67:2005–2009. https://doi.org/10.1128/.67.4.2005-2009.1999

Sengupta D, Leontiadou H, Mark AE, Marrink S-J (2008) Toroidal pores formed by antimicrobial peptides show significant disorder. Biochim Biophys Acta 1778:2308–2317. https://doi.org/10.1016/j.bbamem.2008.06.007

Shai Y (1999) Mechanism of the binding, insertion and destabilization of phospholipid bilayer membranes by alpha-helical antimicrobial and cell non-selective membrane-lytic peptides. Biochim Biophys Acta 1462:55–70. https://doi.org/10.1016/s0005-2736(99)00200-x

Shai Y (2002) Mode of action of membrane active antimicrobial peptides. Biopolymers 66:236–248. https://doi.org/10.1002/bip.10260

Steiner H, Hultmark D, Engström A et al (1981) Sequence and specificity of two antibacterial proteins involved in insect immunity. Nature 292:246–248. https://doi.org/10.1038/292246a0

Steinstraesser L, Hirsch T, Schulte M et al (2012) Innate defense regulator peptide 1018 in wound healing and wound infection. PLoS One 7:e39373. https://doi.org/10.1371/journal.pone.0039373

Sun E, Belanger CR, Haney EF, Hancock REW (2018) Host defense (antimicrobial) peptides. In: Peptide applications in biomedicine, biotechnology and bioengineering. Elsevier, Cambridge, pp 253–285

Traxler MF, Zacharia VM, Marquardt S et al (2011) Discretely calibrated regulatory loops controlled by ppGpp partition gene induction across the "feast to famine" gradient in *Escherichia coli*. Mol Microbiol 79:830–845. https://doi.org/10.1111/j.1365-2958.2010.07498.x

Ulrich LE, Koonin EV, Zhulin IB (2005) One-component systems dominate signal transduction in prokaryotes. Trends Microbiol 13:52–56. https://doi.org/10.1016/j.tim.2004.12.006

van Harten R, van Woudenbergh E, van Dijk A, Haagsman H (2018) Cathelicidins: immunomodulatory antimicrobials. Vaccines 6:63. https://doi.org/10.3390/vaccines6030063

Von Borowski RG, Chat S, Schneider R et al (2020) First-in-class matrix anti-assembly peptide prevents staphylococcal biofilm *in vitro* and *in vivo*. bioRxiv 2020. https://doi.org/10.1101/2020.04.03.022020

Wang Y, Ouyang J, Luo X et al (2021) Identification and characterization of novel bi-functional cathelicidins from the black-spotted frog (*Pelophylax nigromaculata*) with both anti-infective and antioxidant activities. Dev Comp Immunol 116:103928. https://doi.org/10.1016/j.dci.2020.103928

Wilkinson LV, Alford MA, Coleman SR et al (2021) Peptide 1018 inhibits swarming and influences Anr-regulated gene expression downstream of the stringent stress response in *Pseudomonas aeruginosa*. PLoS One 16:1–18. https://doi.org/10.1371/journal.pone.0250977

Wimley WC (2010) Describing the mechanism of antimicrobial peptide action with the interfacial activity model. ACS Chem Biol 5:905–917. https://doi.org/10.1021/cb1001558

Wu M, Maier E, Benz R, Hancock RE (1999) Mechanism of interaction of different classes of cationic antimicrobial peptides with planar bilayers and with the cytoplasmic membrane of *Escherichia coli*. Biochemistry 38:7235–7242. https://doi.org/10.1021/bi9826299

Wu H, Moser C, Wang H-Z et al (2015) Strategies for combating bacterial biofilm infections. Int J Oral Sci 7:1–7. https://doi.org/10.1038/ijos.2014.65

Wu BC, Lee AH-Y, Hancock REW (2017) Mechanisms of the innate defense regulator peptide-1002 anti-inflammatory activity in a sterile inflammation mouse model. J Immunol 199:3592–3603. https://doi.org/10.4049/jimmunol.1700985

Wu J, Yang J, Wang X et al (2018) A frog cathelicidin peptide effectively promotes cutaneous wound healing in mice. Biochem J 475:2785–2799. https://doi.org/10.1042/BCJ20180286

Wu BC, Haney EF, Akhoundsadegh N et al (2021) Human organoid biofilm model for assessing antibiofilm activity of novel agents. NPJ Biofilms Microbiomes 7:8. https://doi.org/10.1038/s41522-020-00182-4

Wuerth K, Lee AHY, Falsafi R et al (2019) Characterization of host responses during *Pseudomonas aeruginosa* acute infection in the lungs and blood and after treatment with the synthetic immunomodulatory peptide IDR-1002. Infect Immun 87:e00661–e00618. https://doi.org/10.1128/IAI.00661-18

Xie F, Zan Y, Zhang X et al (2020) Differential abilities of mammalian cathelicidins to inhibit bacterial biofilm formation and promote multifaceted immune functions of neutrophils. Int J Mol Sci 21:1871. https://doi.org/10.3390/ijms21051871

Yang L, Harroun TA, Weiss TM et al (2001) Barrel-stave model or toroidal model? A case study on melittin pores. Biophys J 81:1475–1485. https://doi.org/10.1016/S0006-3495(01)75802-X

Yang Y, Liu Z, He X et al (2019) A small mycobacteriophage-derived peptide and its improved isomer restrict mycobacterial infection via dual mycobactericidal-immunoregulatory activities. J Biol Chem 294:7615–7631. https://doi.org/10.1074/jbc.RA118.006968

Zasloff M (1987) Magainins, a class of antimicrobial peptides from *Xenopus* skin: isolation, characterization of two active forms, and partial cDNA sequence of a precursor. Proc Natl Acad Sci U S A 84:5449–5453. https://doi.org/10.1073/pnas.84.15.5449

Zasloff M (2002) Antimicrobial peptides of multicellular organisms. Nature 415:389–395. https://doi.org/10.1038/415389a

Zhu C, Tan H, Cheng T et al (2013) Human β-defensin 3 inhibits antibiotic-resistant *Staphylococcus* biofilm formation. J Surg Res 183:204–213. https://doi.org/10.1016/j.jss.2012.11.048

Learning from Nature: Naturally Derived Remedies

Masanobu Suzuki

Abstract The rise and spread of antimicrobial resistance among microbes represents one of today's most pressing global health issues. To combat the antimicrobial resistance of pathogens, the ongoing development of novel antibiotics is still the primary approach. However, due to the stalled antibiotic development pipeline, it is vital that novel, practical, and accessible methods, distinct from antibiotics, will be established to combat antimicrobial-resistant pathogens.

One promising strategy is the medical application of natural products exhibiting antimicrobial properties. Traditionally, many types of natural products were applied to the treatment of a variety of diseases or medical conditions as natural remedies. In particular, components—directly or indirectly derived from plants—have attracted interest among researchers and clinicians due to their antimicrobial and antibiofilm properties, good safety profiles, and less likelihood of resistance development.

In this section, we discuss natural products with potential clinical applications against biofilm-related infections, such as honey, essential oils, and teas. We present current research based on animal models and clinical research, as well as in vitro studies.

1 Introduction

Since 1928, a year made famous in medical history due to the discovery of penicillin by Alexander Fleming, numerous antibiotics have been developed and clinically applied. The extraordinary effectiveness of antibiotics against harmful microbes has made many infectious diseases controllable and has saved the lives of hundreds of millions of patients suffering from severe infections. Nowadays, antibiotics are applied not only for patients with infectious diseases but also for those who might have a future risk of infectious diseases; for example, for patients who will have

M. Suzuki (✉)
Department of Otolaryngology-Head and Neck Surgery, Faculty of Medicine and Graduate School of Medicine, Hokkaido University, Sapporo, Hokkaido, Japan
e-mail: masanobuwork@med.hokudai.ac.jp

K. Richter, K. Nørskov Kragh (eds.), *Antibiofilm Strategies*, Springer Series on Biofilms 11, https://doi.org/10.1007/978-3-031-10992-8_13

surgery and those with pancytopenia during chemotherapy. The prophylactic antibiotic contributes to a reduction in the rate of infection and, consequently, to the prevention of complications. On the other hand, the antibacterial effectiveness is inextricably associated with selective pressure on pathogens. Therefore, overuse and misuse of antibiotics also increases the incidence of antimicrobial resistance among microbes (Wu et al. 2019). Indeed, the rise and spread of antimicrobial resistance represents one of today's most pressing health issues (Barbosa et al. 2020). The bacterial group named ESKAPEE, an acronym for *Enterococcus faecium*, *Staphylococcus aureus*, *Klebsiella pneumoniae*, *Acinetobacter baumannii*, *Pseudomonas aeruginosa*, *Enterobacter spp.*, and *Escherichia coli*, show high to critical drug resistance (Yu et al. 2020). In 2019, the World Health Organization listed antimicrobial resistance as one of the top ten threats to global health (Scheres and Kuszewski 2019).

There are several mechanisms by which microbes gain antimicrobial resistance, including altered permeability in the cell membrane, over-expression of drug efflux pumps, inactivation of antibiotics, modification of target site within the microbes, and biofilm formation (Barbosa et al. 2020). Among them, biofilm formation significantly increases resistance to antimicrobial agents (Lu et al. 2019a, b). It was estimated that 100 to 1000 times the concentration of an antibiotic is required to eradicate a biofilm compared to the planktonic form of the same bacteria (Ceri et al. 1999). A recent study showed that vancomycin at a concentration of >1000 times the MIC is not enough to eradicate *S. aureus* biofilms (Post et al. 2017).

Against the widespread antimicrobial resistance of pathogens, continuous development of novel antibiotics remains the primary approach. However, this is becoming increasingly difficult because the continuous discovery and development of new antibacterial is not a straightforward task with the issue of commercial profitability (Wu et al. 2019; Butler and Paterson 2020). Indeed, only 30 new antibiotics have been approved for clinical use since 2000 (Butler et al. 2017; Yu et al. 2020). Therefore, the establishment of novel and practical strategies, other than antibiotics, against the antimicrobial resistance of pathogens has become an urgent matter (Lu et al. 2019a, b).

One of the promising strategies is the medical application of natural products with antibacterial properties (Alandejani et al. 2009). Traditionally, many natural products were used as remedies for the treatment of various diseases and medical conditions (Díaz-Nuñez et al. 2021). Components directly or indirectly derived from plants, in particular, have attracted interest among researchers and clinicians due to their antimicrobial and antibiofilm properties in vitro. Some plant-derived extracts affect biofilm formation not only by inhibiting microbial growth but also via the regulation of quorum sensing (QS) systems, which essentially mediates microbial communication and biofilm maturation (Zhang et al. 2020). Even in the modern era, natural products have remained a primary source of medicine. Indeed, more than half of the drugs approved by the US Food and Drug Administration (FDA) are derived or synthesized from natural products, with antimicrobial agents, in particular, heavily depending on natural products (Newman and Cragg 2016).

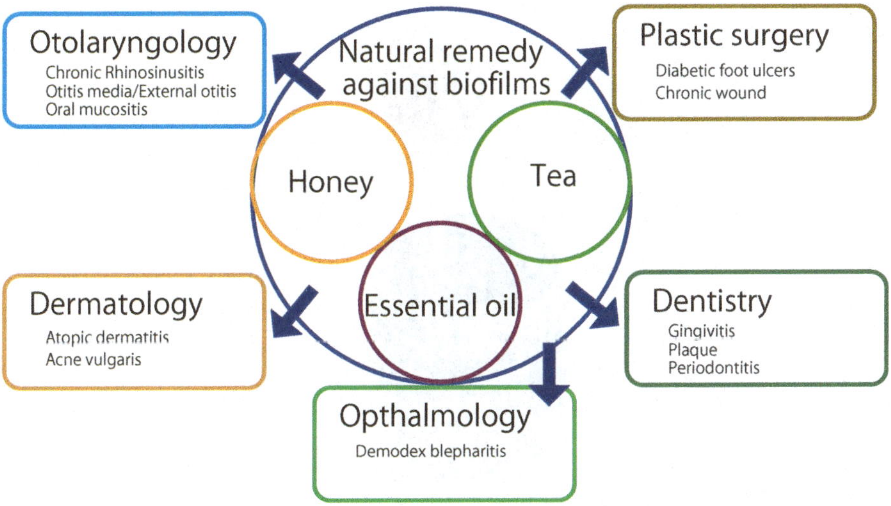

Fig. 1 Natural remedies against biofilms have been applied in several clinical fields. Natural products, such as honey, tea, and essential oils, have attracted the interest of researchers and clinicians due to their antibiofilm properties. Indeed, the products have been applied clinically in the fields of otolaryngology, plastic surgery, dermatology, dentistry, and ophthalmology

In this section, we discuss natural products with a potential for clinical applications against biofilm-related infections. These include honey, essential oils, and tea (Fig. 1). The research presented is based on animal and clinical research, as well as in vitro studies.

2 Honey

2.1 Antibiofilm Properties of Honey

Honey is a sweet and flavorful natural product produced by bees from plant nectars, plant secretions, and the secretion of plant-sucking insects (Alvarez-Suarez et al. 2014). Honey is classified into the following four categories: blossom honey, honeydew honey, monofloral honey, and multi-floral (or polyfloral) honey. Blossom honey is obtained predominantly from the nectar of flowers, while honeydew honey is produced by bees collecting the secretions of various insects. Monofloral honey and multi-floral honey are mainly produced from only one or many botanical sources (Alvarez-Suarez et al. 2014).

Because of its sweetness and nutritional richness, honey has long been used as a food substance around the world. Honey was also used for medicinal purposes in ancient India, Egypt, Greece, and the Islam world (Eteraf-Oskouei and Najafi 2013). In modern medicine, scientists have found evidence supporting the usefulness of

Fig. 2 Manuka honey. Manuka honey is a monofloral honey produced by Apis mellifera honey bees from the nectar of the flowers of the Manuka tree (Leptospermum scoparium) in New Zealand. Recent studies have shown that Manuka honey possesses an antibiofilm effect due to its higher phenolic acid level, hyperosmolarity, hydrogen peroxide level, and the active components of Manuka honey, bee defensin-1 and methylglyoxal (MGO). The photo was kindly provided by Yuki Suzuki

honey as a complementary therapy, based on its antimicrobial properties, wound-healing and anti-inflammatory effects, antioxidant activity, and anti-proliferative effects (Meo et al. 2017; Almasaudi 2021; Mărgăoan et al. 2021; Fernandez-Cabezudo et al. 2013; Erejuwa et al. 2012). There are about 320 different varieties of honey originating from various floral sources (Meo et al. 2017). As the components of honey vary depending on the weather and geographical conditions (Noori and Al Ghamdi 2013), each type of honey has different therapeutic effects (Lusby et al. 2005; Brady et al. 2004). The efficacy and safety profiles of individual honey should be cautiously assessed before clinical application.

To date, several kinds of honey have attracted the attention from the scientific community with regard to clinical application. One of the "front-runners" is Manuka honey (Fig. 2). Manuka honey is a monofloral honey produced by *Apis mellifera* honey bees from the nectar of the flowers of the Manuka tree (*Leptospermum scoparium*) in New Zealand (Johnston et al. 2018). Traditionally, the Manuka tree itself was also used for medical purposes, with its extracts used for sedation and wound healing (Alvarez-Suarez et al. 2014). Manuka honey is classified according to its Unique Manuka Factor (UMF), representing its antibacterial activity (Johnston et al. 2018).

Recent studies have shown that Manuka honey possesses an antibiofilm effect due to its higher phenolic acid level, hyperosmolarity, hydrogen peroxide level, and the active component of Manuka honey, methylglyoxal (MGO) (Ooi et al. 2019; Johnston et al. 2018). The honey exerts an antibiofilm property without any risk of inducing antibacterial resistance (Johnston et al. 2018) and also promotes the synthesis of various cytokines, including TNF-α, IL-1β, and IL-6, through TLR4 (Tonks et al. 2003).

In the following section, we provide greater detail on the clinical applications of honey, including Manuka honey.

2.2 *Clinical Application of Honey in Otolaryngology*

One of the most advanced areas for the medical application of honey is otolaryngology, with honey being investigated in detail as a potential complementary treatment for chronic rhinosinusitis, otitis media, and radiotherapy-induced oral mucositis.

2.2.1 Chronic Rhinosinusitis

Chronic rhinosinusitis (CRS) is an inflammatory disease of the nose and paranasal sinuses lasting over 12 weeks (Fokkens et al. 2020). CRS is a significant health problem with a high prevalence of 5–12% in the general population (Fokkens et al. 2020). CRS has been divided into two groups; chronic rhinosinusitis with nasal polyps (CRSwNP) and that without nasal polyps (CRSsNP) (Fokkens et al. 2012). Recent studies have shown that type-2 inflammation and eosinophilic inflammation are significantly associated with recalcitrant cases of CRSwNP. Therefore, classifications have recently been proposed to focus more on type-2 inflammation/eosinophilic inflammation (Fokkens et al. 2020; Fujieda et al. 2019).

The mainstay of treatment for CRSwNP is endoscopic sinus surgery (ESS). ESS allows the efficient resection of nasal polyps and reduces the inflammatory burden in the nasal mucosa without significant adverse risks. However, recurrence of inflammation after surgery is often observed, particularly in cases of CRSwNP. Several factors related to the recurrence have been identified, such as the presence of eosinophils, nasal polyposis, aspirin sensitivity, smoking, acid regurgitation, and biofilms.

The presence of bacterial biofilms is significantly associated with a worse preoperative radiological score, postoperative symptoms, endoscopic score, and mucosal outcomes (Psaltis et al. 2008; Singhal et al. 2010; Singhal et al. 2011). Interestingly, biofilms found in the nasal mucosa from CRS patients do not always consist of just one species. Rather, 60% of cases show polymicrobial biofilms, including *Staphylococcus aureus*, *Haemophilus influenza*, and those of fungal origin. Among them, *S. aureus* is significantly associated with the severity of disease (Singhal et al. 2011). The presence of *S. aureus* biofilms is also associated with eosinophilic inflammation, which has been identified as a prominent factor for recurrence (Foreman et al. 2011). Indeed, cytokines such as IL-5, IL-6, and eosinophil cationic protein (ECP) are significantly higher in nasal polyps with *S. aureus* biofilms than those without

S. aureus biofilms (Foreman et al. 2011). Thus, successful treatment of recalcitrant CRS may require the eradication of biofilms in the nasal mucosa.

Several studies have shown the substantial antibacterial effect of honey. Alandejani et al. examined the in vitro effect of Manuka and Sidr honey against biofilms produced by *Pseudomonas aeruginosa* and *S. aureus* (Alandejani et al. 2009), and found that both Manuka and Sidr honey eradicated the planktonic form of *P. aeruginosa* and *S. aureus*. In their study, the honey also showed antibacterial activity against biofilm forms of *S. aureus* and *P. aeruginosa*, although not as high as that against the planktonic forms. Notably, the antimicrobial activity of these types of honey was also effective against MRSA (Alandejani et al. 2009). In addition to Manuka and Sidr honey, other types of honey have been reported to inhibit *P. aeruginosa* biofilm formation (Lu et al. 2019a, b; Merckoll et al. 2009; Ansari et al. 2013). These antibacterial studies of honey have led researchers to take on the challenge of developing their clinical applications.

Prior to the clinical applications of honey, its safety and efficacy should be examined. In this regard, animal experiments have played a significant role in confirming both safety and effectiveness. Several studies have clarified the safety and efficacy of Manuka honey irrigation in animals. Kilty et al. undertook a histopathological investigation of the safety profile of honey on the respiratory epithelium in rabbits in which one side of the nasal cavity of rabbits was irrigated with a Manuka honey solution. The nasal mucosa was then harvested and examined for comparison with the control side. No histopathological evidence of inflammation, epithelial injury, or morphological changes was found for the side treated with Manuka honey (Kilty et al. 2010), supporting the safety of the honey for the respiratory epithelium.

Not only the honey itself, but the active component of Manuka honey, methylglyoxal (MGO), was also assessed for its safety and efficacy (Paramasivan et al. 2014). One side of the frontal sinus in sheep was irrigated twice daily for 14 days using several concentrations of MGO with/without Manuka honey. The other side was washed with a saline control. Histopathological examination showed cilia denudation, and squamous metaplasia was found in the mucosa from sinuses treated by honey, dependent on MGO concentration. However, when irrigation was performed at a concentration of less than 1.8 mg/ml of MGO, no abnormal structures in the epithelium or cilia were found. MGO in Manuka honey significantly reduced biofilm mass within a range from 0.9 mg/ml to 1.8 mg/ml. The group concluded that the optimized concentration of MGO in Manuka honey sinus irrigation is between 0.9 mg/dl and 1.8 mg/dl (Paramasivan et al. 2014).

Based on the safety profile and efficacy of honey applications to the nasal mucosa in in vitro and animal studies, several clinical trials have already been conducted. Ooi et al. reported the first randomized, single-blinded, placebo-controlled phase 1 clinical trial to investigate the safety and efficacy of Manuka honey sinus irrigation in CRS patients (Ooi et al. 2019). 25 CRS patients treated with ESS were randomly assigned to two groups: the Manuka honey and the oral-antibiotic groups. Patients in the Manuka honey group received 13 days of twice-daily Manuka honey and MGO sinonasal rinses and concurrent 10 days of placebo tablets after surgery. Patients in the oral-antibiotic group received 14 days of twice-daily saline irrigation with 10 days of oral antibiotics. Endoscopic score, smell function, microbiology results, subjective symptom score, and QOL score were then examined.

In the Manuka honey group, no serious adverse effects were reported, and all patients tolerated the honey irrigation. No smell dysfunction was reported in association with the treatment. In both groups, the endoscopic score was significantly improved after treatment, with no significant difference observed between the two groups. Further, there was no significant difference in the microbial clearance rate between the Manuka honey group and the oral-antibiotic group, and neither the subjective symptoms nor the QOL score showed significant differences between the two groups. Despite the short period of application (only 14 days), these data suggested Manuka honey could be as effective as postoperative oral antibiotics and provide a possible substitute for oral antibiotics. Longer term application of Manuka honey irrigation, which is infeasible in the case of antibiotics, might support the superiority of the Manuka honey irrigation over oral antibiotics during the postoperative period in cases undergoing ESS (Ooi et al. 2019).

Lee et al. investigated the efficacy of nasal irrigation with Manuka honey as an adjuvant therapy for CRS patients (Lee et al. 2017). 42 CRS patients were randomly assigned to two groups; one group received nasal irrigation with Manuka honey ($n = 20$) and the other underwent irrigation without honey ($n = 22$) for 30 days. There were no adverse effects observed for honey nasal irrigation. The subjective QOL scores significantly improved in both groups, with no significant difference between the groups. Further, the endoscopic scores improved in both groups, with no statistical significance between the two groups. Bacteria clearance was more likely in the honey irrigation group, but the difference was insignificant ($p = 0.11$). For patients who did not use oral antibiotics or steroids, bacterial clearance was significantly better in the honey irrigation group than in the control group, suggesting that honey irrigation alone may be effective in limiting the exacerbation of CRS and can reduce the frequency of oral antibiotic/steroid usage (Lee et al. 2017).

Lee et al. also examined the efficacy and safety of nasal irrigation with Manuka honey for CRS in cystic fibrosis patients. In a prospective, single-blinded randomized study, six cystic fibrosis patients received sinus irrigation with Manuka honey and seven patients underwent sinus irrigation with a saline control for 30 days. Due to the small sample size, no statistically significant difference in the subjective QOL score was observed between the two groups; however, the score significantly improved after irrigation for 30 days in the honey irrigation group but not in the control group. Manuka honey irrigation also considerably improved the endoscopic score, although complete clearance of bacteria in the nasal cavity was not achieved. As no adverse effects were reported in association with honey irrigation, Manuka honey also appears to be a promising therapy for improving QOL in CRS patients with cystic fibrosis (Lee et al. 2021).

Irrigation with Manuka honey not only exerts antibiofilm activity but also alters immunological status in the nasal mucosa. Immunological analysis was also performed for nasal irrigation using Manuka honey (Manji et al. 2019). In this study, 26 CRS patients treated with ESS received nasal irrigation with Manuka honey after surgery. Nasal specimens were collected during the surgery and during the postoperative period. 17 cytokines were examined through comparison to nasal specimens taken from CRS patients treated with nasal irrigation without Manuka honey. A multiplex analysis showed significantly increased expression levels of IL-6, IL-8, and MCP-1 compared to the controls at 5 weeks. The authors speculated this promotion of expression could be involved in the early phases of wound healing (Manji et al. 2019). On the contrary, Manuka honey irrigation significantly downregulated IL-13 expression at 3 months post-surgery in the controls (Manji et al. 2019). IL-13 is a type-2 inflammatory cytokines and is a promising therapeutic target molecule for CRSwNP. Nasal irrigation with honey might contribute to the regulation of type 2 inflammation.

Moreover, honey might contribute to CRS regulation through the production of nitrogen oxide (NO). A recent systematic review demonstrated that nasal NO was significantly decreased in CRSwNP patients in comparison with that in CRSsNP patients and healthy controls (Ambrosino et al. 2020). Also, the nasal NO level was significantly lower in eosinophilic CRS (eCRS) than in non-eCRS and healthy controls, and was negatively correlated to the eosinophil count in the nasal mucosa (Yoshida et al. 2019). As NO contributes to local host defenses against bacterial, viral, and fungal infections and helps maintain a bacteriostatic state in the paranasal sinus (Yoshida et al. 2019), decreased NO levels are also associated with the pathogenesis of CRS. Manuka honey can promote NO synthesis (Yang et al. 2020). Indeed, *P. aeruginosa* biofilms were significantly reduced by treatment with a combination of Manuka honey and nitrile in comparison with Manuka honey alone, implying that Manuka honey promoted the synthesis of NO from nitrile and the resultant NO enhanced the antibiofilm activity of Manuka honey (Yang et al. 2020).

In addition to nasal irrigation, other methods for applying honey as a local therapy for CRS have been reported. Hashemian et al. reported a double-blind, randomized controlled clinical trial to investigate the effect of honey spray on chronic rhinosinusitis (Hashemian et al. 2015). 64 CRS patients treated with ESS were randomly and blindly assigned to receive Iranian honey (Thyme honey) spray or placebo spray together with standard postoperative care. The endoscopic score was significantly improved more in the honey spray group. Synechiae formation and epistaxis were also less frequent in the honey group. Moreover, the honey spray did not increase the incidence of adverse effects in comparison with the placebo. The study group concluded that the thyme honey spray affords a low-priced potential adjuvant therapy, with an excellent safety profile, for postoperative care after ESS (Hashemian et al. 2015).

On the other hand, the application of honey to packing material after surgery appears not to be beneficial. Chang et al. examined the efficacy of Manuka honey-soaked packing in a study of 48 CRS patients undergoing bilateral endoscopic sinus surgery. After surgery, one nostril was packed with Manuka honey-soaked Merocel sponges, the other with nonmedicated Merocel sponges. Discomfort from nasal packing, endoscopic visualization, and histopathology were investigated, but there the honey-packed side was not found to have any superiority to the control side (Chang et al. 2011). Similarly, a Malaysian group examined 32 CRS patients undergoing ESS. After surgery, nasal dressing containing Tualang honey was applied, and postoperative nasal edema, crusting, secretions, scarring, and symptoms were examined. Compared to those fitted with nasal dressing containing a corticosteroid, no benefits were observed for the patients fitted with Tualang honey-treated dressing (Kumarasamy et al. 2020).

Recent advances in biology have led to the development of "bioengineered honey." Papadopoulou et al. examined the antimicrobial activity of engineered honey against *S. aureus* clinically isolated from CRS patients (Papadopoulou et al. 2020). The honey comprises gamma-irradiated honey and added glucose oxidase, which enhances the antimicrobial activity. The results showed that the engineered honey decreased biofilm viability, irrespective of the methicillin resistance of *S. aureus* (Papadopoulou et al. 2020). Although the engineered honey has not yet been applied clinically, such engineered honey could be used to treat sinusitis in the near future.

2.2.2 Otologic Diseases

In addition to the nasal cavity and paranasal sinuses, honey could be helpful for the treatment of ear diseases, such as otitis media and external otitis, through its antibacterial properties.

Aron et al. investigated the safety profile of the topical application of Manuka honey to the middle ear in an animal model, using female chinchillas (Aron et al. 2012, 2015). They injected Manuka honey unilaterally to the middle ear and investigated the auditory brainstem-evoked response (ABR), which reflects the auditory function in an objective manner. The application of 50% Manuka honey significantly impaired the ABR threshold, with scanning electron microscopy revealing the histologic evidence of hair cell toxicity. On the contrary, 4% Manuka honey did not change the ABR threshold and no pathological alteration or facial/vestibula nerve dysfunction was observed. They concluded that 4% Manuka honey is nontoxic to the middle ear of chinchillas and affords a promising therapeutic product for chronic refractory otitis due to biofilm in the middle ear (Aron et al. 2015). Kumar et al. reported a pilot study on Manuka honey for patients with otologic conditions, in which 20 patients with acute otitis media without perforation of the tympanic membrane were examined. The affected ear was packed with umbilical tape impregnated with 3–4 ml of Manuka honey gel and the pain score significantly decreased 1 day after the honey application without any adverse events reported. Although the contribution of Manuka honey to the treatment of acute otitis media is still unknown, as the authors themselves mentioned as a limitation, the application of honey to an external canal in patients without perforation of the tympanic membrane, at least, seems safe (Kumar et al. 2020).

Henatsch et al. reported the usefulness of honey eardrops for eczematous external otitis. Eczematous external otitis is a chronic inflammatory condition with associated discomfort and itchiness affecting the external canal (Henatsch et al. 2017). 15 patients with eczematous external otitis used honey eardrops three times daily for 2 weeks. Their discomfort and itching significantly decreased within a 2-week period. They also showed that the antibacterial activity of honey ear drops was irrelevant to bacterial resistance (Henatsch et al. 2017).

2.2.3 Radiation-induced Oral Mucositis

Honey was also applied for radiation-induced oral mucositis. Radiation is one of the major treatment strategies for the regulation of malignant head and neck lesions. The oral mucositis often occurs during and after the radiation period and is known to impair patient QOL. A systematic review showed honey could be useful as alternative therapy for radiation-induced oral mucositis (Münstedt et al. 2019). Interestingly, Manuka honey was not as effective as other types of honey, perhaps due to the higher level of MGO, which shows cytotoxicity at higher concentrations.

2.3 Diabetic Foot Ulcers and Chronic Wounds

Diabetic foot ulcers are one of the major severe complications of diabetes mellitus. In severe cases, lower-limb amputation is necessary, which not only significantly impairs the quality of life of patients but is also life-threatening. Diabetic foot ulcers represent an enormous burden from a socio-economic viewpoint (Hopkins et al. 2015).

Several studies have reported the efficacy of honey, particularly Manuka honey, for diabetic foot ulcers. Kamaratos et al. reported the results of a clinical trial in which 63 patients with diabetic foot ulcers were randomly assigned to two groups: one group was treated with Manuka honey dressings for 16 weeks, and the other was treated with conventional dressing (Kamaratos et al. 2014). The mean healing time was significantly shorter in the Manuka honey dressing group (Manuka honey group; 31 ± 4 days versus the conventional group; 43 ± 3 days). Also, the bacterial clearance ratio was significantly higher in the group treated with Manuka honey (Kamaratos et al. 2014).

In another report from a Malaysian group, 30 patients with diabetic foot ulcers were assigned to two groups receiving either honey dressings or conventional dressings. Although the healing time was not significantly shortened, they reported less pain, less edema, and less foul-smelling discharge in the honey dressing group when compared to the control group (Shukrimi et al. 2008).

A Pakistan group divided 100 patients with diabetic foot ulcers into two groups. One group was treated with honey dressings and the other with pyodine dressings. Although detailed information regarding the honey they used is absent, the honey dressing significantly shortened the wound healing time after the surgical debridement of the ulcer in the group treated with honey dressings. The healing rate and amputation ratio did not differ significantly (Jan et al. 2012).

One of the largest studies on the efficacy of honey treatment for diabetic foot ulcers is the report by Imran et al. (2015). The Pakistani group assigned 348 patients with diabetic foot ulcers into two groups: a honey dressing group ($n = 179$) using Beri (*Ziziphus jujuba*) honey and a saline dressing group ($n = 169$). The complete healing ratio was significantly higher in the honey dressing group than in the control group (75.97% versus 57.39%). Further, the time to achieve complete wound healing was significantly shorter in the honey dressing group than in the control group (Imran et al. 2015). Thus, a lot of research has shown the efficacy of honey for diabetic foot ulcers. A systematic review concluded that wound dressing using honey was effective for diabetic foot ulcers, supported by a shortened wound healing time and increased healing and bacterial clearance rates (Wang et al. 2019).

Royal jelly, a worker honeybee product, has also been applied to foot ulcers. Siavash et al. investigated the efficacy and safety of royal jelly on 60 diabetic foot ulcers from 25 patients. However, they failed to show any significant superiority of topical royal jelly to a placebo in terms of depth, length and width reduction rates, duration of the healing process, and ratio of complete healing (Siavash et al. 2015).

2.4 Honey and Skin Disease

Atopic dermatitis (AD) is a chronic, inflammatory, pruritic skin disorder with a high global prevalence. The symptoms of AD include rash and itching, as well as cataracts on occasion and retinal detachment in severe patients. Although these symptoms are usually not life-threatening, they can significantly impair patient QOL (Sugita and Akdis 2019; Katayama et al. 2017). The pathogenesis of atopic dermatitis consists of multiple pathogenic factors, including bacterial colonization. Recent studies have demonstrated that skin lesions and even non-lesions are often colonized with *S. aureus* and *Staphylococcus epidermidis* (Ogonowska et al. 2020; Gonzalez et al. 2017). Biofilms formed by these bacteria are associated with the severity of atopic dermatitis (Gonzalez et al. 2017, 2021; Allen et al. 2014). Several natural products have been tested for their activities on bacterial biofilms.

Alangari's group examined Manuka honey for its effect against atopic dermatitis (Alangari et al. 2017). Manuka honey was applied to one side of lesions resulting from atopic dermatitis in 14 patients. As a control, the other side of the lesions were untreated. Manuka honey significantly decreased the severity score; no reduction was reported for the untreated control. Interestingly, some patients reported that their eczema vanished entirely for 1 year after the study. On the contrary, honey treatment did not alter the bacterial skin flora of the patients recruited for the study. In vitro assays showed that the honey downregulated IL4-induced CCL26, one of the key molecules in the pathogenesis and severity of atopic dermatitis (Kagami et al. 2003; Owczarek et al. 2010). These results implied that not only the antibacterial effect but also the immune-modulatory effect of Manuka honey might contribute to the observed improvement in atopic dermatitis (Alangari et al. 2017).

Al-Waili et al. examined the therapeutic effect of a honey mixture (honey, beeswax, and olive oil) in the treatment of atopic dermatitis (Al-Waili 2003). In 11 patients with atopic dermatitis, the lesions on the right side of the body were treated with the honey mixture and those on the left side with soft paraffin. If patients had been using corticosteroids at the initiation of the study, the corresponding amounts of corticosteroids were also mixed in the honey mixture. The honey mixture significantly reduced the severity score of the atopic dermatitis in comparison to the control mixture in patients who had not been using corticosteroids. On the other hand, there was no significant difference of severity score between the two sides in cases that had already been treated with corticosteroids. The honey mixture significantly reduced the dose of corticosteroids required for the patients. The group concluded that the topical application of the honey mixture appeared to afford a significant improvement in symptoms in patients with atopic dermatitis (Al-Waili 2003).

Honey is also expected to prevent infection related to dialysis, where infection could directly threaten life prognosis. A randomized, controlled trial demonstrated that topical exit-site application of Medihoney prevented catheter-associated infection in hemodialysis (Johnson et al. 2005). On the contrary, a similar study (HONEYPOT trial) in patients with peritoneal dialysis failed to show effectiveness of topical honey in the prevention of infection (Johnson et al. 2014). In the HONEYPOT trial, one of the largest clinical trials on honey application, there were no significant differences in the incidence of serious adverse events or death between the topical honey-treated group and the control group. As both groups were additionally treated with topical nasal mupirocin, which is a standard prophylaxis measure for patients with peritoneal dialysis, the mupirocin usage might have affected the study results. Therefore, there is currently no evidence supporting the routine prophylactic usage of honey for catheter-associated infection in peritoneal dialysis and efficacy of Medihoney in preventing biofilm-associated transcutaneous infections (exit-site infections) is limited.

3 Essential Oils

Essential oils are concentrated plant extracts. Their desirable aroma and flavor make them widely used in food, drink, perfumes, cosmetics, and medicine (Raut and Karuppayil 2014). Numerous families of plants could be the source of essential oils; for example, *Alliaceae*, *Apiaceae*, *Asteraceae*, *Fabaceae*, *Lamiaceae*, *Myrtaceae*, *Poaceae*, and *Rutaceae* have often been used for medical and industrial grade essential oils (Raut and Karuppayil 2014; Karpiński et al. 2021). Essential oils have many unique effects on the human body, including sedative, anti-inflammatory, and local anesthetic properties (Bakkali et al. 2008). In addition, some essential oils are known to show in vitro antibiofilm activity against bacteria and fungi (Karpiński et al. 2021; Galvão et al. 2012). Due to the diversity of components and manufacturing methods, the properties of essential oils vary depending on type.

To date, several animal studies have been conducted to examine the efficacy and the safety of different essential oils against biofilms. First, an essential oil extracted from *Artemisia vestita* was evaluated using a mouse model of upper respiratory tract infections. Oral administration of the essential oil in mice infected with *S. pyogenes* significantly reduced the viable bacteria number in the lung in a dose-dependent manner. The group also confirmed that the administration did not impair hepatic or renal function at the effective concentration of the oil (Yang et al. 2015).

Another animal study examined the anti-Candida and anti-inflammatory activities of essential oil extracted from *Cymbopogon nardus* (Kandimalla et al. 2016). In the study, the essential oil was applied to *C. albicans*-infected wounds on diabetic mice. The microbial load was significantly reduced after the application of the essential oil. The essential oil treatment also significantly suppressed TNFα and IL-1β in the diabetic wounds. Histopathological examination revealed that complete wound healing was achieved in the wounds of mice treated with the essential oil, while fragments of degenerated tissue were observed in the wound of untreated mice. Instead, the wounds were also observed to spread, and necrosis was found in the mice not treated with the essential oil. The group concluded that essential oil extracted from *Cymbopogon nardus* promoted the wound healing process through its anti-Candida and anti-inflammatory activities (Kandimalla et al. 2016).

Oregano oil was also examined in in vitro and in vivo studies. A bacteriological survey by Lu et al. demonstrated its antibiofilm activity against 11 multidrug-resistant microbes, including *P. aeruginosa*, and four MRSA strains, without any potential risk of drug-resistance development. In their animal research, the group also showed that the topical application of oregano oil to infected burn wounds in mice significantly reduced the bacterial load without any significant adverse skin-related effects (Lu et al. 2018).

Further, essential oil extracted from *Rosmarinus officinalis* (Rosemary) has also been confirmed to promote wound healing in diabetic mice (Abu-Al-Basal 2010). A hexane extract from *Hesperozygis ringens* (HEHR) was shown to demonstrate antibacterial properties, with a significantly increased survival rate among juvenile silver catfish infected with *Aeromonas hydrophila* (Rosa et al. 2019).

Based on the promising results obtained from the preceding animal studies, essential oils have been evaluated for potential clinical applications. Tea tree oil (TTO), in particular, has attracted the interest of researcher. TTO consisting of 5% TTO in Tween 80 was found to be significantly more active against *E. coli* than a nonmedicated soft soap after 1 min of handwashing (Messager et al. 2005). Dryden et al. reported the results of a randomized controlled trial using a topical tea tree preparation for the clearance of MRSA colonization (Dryden et al. 2004). In their study, 110 hospital patients underwent a tea tree oil regimen consisting of a tea tree body wash and cream. One hundred and fourteen patients received standard treatment with mupirocin nasal ointment, chlorhexidine gluconate soap, and silver sulfadiazine cream. The tea tree regimen was more effective than the standard regimen at cleaning superficial skin sites and wounds (Dryden et al. 2004). William

A Bowler et al. reported a decolonization regimen against MRSA including TTO body wash. The regimen was applied for the residents of the five nursing homes, and the prevalence of MRSA was decreased by 67% (Bowler et al. 2010).

TTO is also expected to afford a useful treatment for acne vulgaris. The efficacy of TTO for acne vulgaris was examined in a randomized, double-blinded clinical trial (Enshaieh et al. 2007). 60 patients with mild to moderate acne vulgaris were randomly divided into TTO gel and placebo groups. The TTO gel improved total acne lesion count and the acne severity index without any severe side effects (Enshaieh et al. 2007). Malhi et al. conducted a phase II study to examine the efficacy and safety of TTO for facial acne (Malhi et al. 2017). 14 patients with mild to moderate facial acne received TTO products and, after TTO treatment for 12 weeks, the total number of facial acne lesions and the objective score were both significantly improved (Malhi et al. 2017).

Besides TTO, several essential oils were tested with the aim of treating acne vulgaris. Kwon et al. investigated the clinical efficacy and safety of extracts from *Chamaecyparis obtusa*, a tree found in East Asia (Kwon et al. 2014). 34 patients with acne vulgaris received *Lactobacillus*-fermented *C. obtusa* (LFCO) and TTO. Inflammatory acne lesions were significantly reduced by 65.3% on the LFCO side and by 38.2% on the TTO side without any severe adverse reactions. Histopathological and immunological investigation revealed the findings correlated well with the clinical acne grade, and treatment response with inflammatory cytokines subsequently decreased. The group concluded that LFCO treatment was a rapid and effective method for treating acne lesions in comparison to TTO (Kwon et al. 2014). Further, *Ocimum gratissimum* (Rosaceae) Linn Leaf Essential Oil was tested for acne vulgaris (Orafidiya et al. 2002). In the study, 126 subjects with acne vulgaris were randomly assigned to three groups: an Ocimum oil group, a reference product (benzoyl peroxide lotion) group, and a control group. The topical application was continued for 4 weeks, and the number of lesions and period to achieve a 50% reduction in lesion count were compared among the groups. The results showed that Ocimum oil was more effective and reduced lesions faster than the reference product with minimal adverse effects (Orafidiya et al. 2002). Copaiba essential oil also significantly decreased the area affected by acne in a double-blinded, placebo-controlled clinical trial (da Gomes et al. 2012).

In the field of dermatology, Chaijan et al. conducted a double-blinded, randomized clinical trial to evaluate the efficacy and safety of *Myrtus communis* L solution for the treatment of dandruff (Chaijan et al. 2018). 90 patients with dandruff received *M. communis* L. solution with a placebo shampoo or ketoconazole shampoo and a placebo solution in the clinical trial. The solution significantly improved pruritus, erythema, and severity of scaling in comparison with the control ketoconazole shampoo (Chaijan et al. 2018).

TTO was also tested for its potential clinical applications for Demodex blepharitis. One hundred and six patients with Demodex blepharitis received eyelid scrubbing with TTO and 54 patients underwent the same procedure without the oil (Koo et al. 2012). TTO eyelid scrubbing significantly reduced Demodex count with the subjective ocular symptom score also improved. Thus, TTO seems effective in improving blepharitis by reducing the bacterial burden.

Another medical field exploring the potential of essential oils is dentistry, with a number of essential oil-based mouth rinses clinically tested (Charles et al. 2004). One hundred and eight subjects were randomly assigned to receive an essential oil mouth rinse, chlorhexidine mouth rinse, or a hydro alcohol negative control mouth rinse. The gingivitis index was significantly improved in the patients using the essential oil and chlorhexidine mouth rinses. The plaque score also favored the essential oil and chlorhexidine mouth rinse groups, but the differences were not statistically significant (Charles et al. 2004).

In another study, 139 healthy adults with plaque and gingivitis were randomly assigned to use essential oil-containing antiseptic mouth rinse or a negative control placebo rinse. Gingivitis, plaque, and bleeding scores favored the patients using the essential oil-containing mouth rinse over the patients using a hydro alcohol negative control placebo rinsing (Charles et al. 2013).

Positive effects of essential oils on wound healing are also expected, and clinical trials have been conducted. 11 patients with wounds colonized with MRSA were treated with TTO solution at each dressing change. In eight of the 11 patients, the wound began to heal and reduced in size after the application of the tea tree oil solution, although the eradication of MRSA was not achieved (Edmondson et al. 2011).

4 Tea

Tea is an aromatic beverage made from *Camellia sinensis* (Fig. 3). Originating from China 2000 years ago, tea is the most commonly consumed drink in the world after water (Cabrera et al. 2006; Steinmann et al. 2013). In traditional Chinese medicine, tea was used for headaches, body aches and pains, digestion, depression, and also to prolong life (Cabrera et al. 2006).

There are three major types of tea; non-fermented "green tea," semi-fermented "oolong tea," and fermented "black tea" or "red tea" (Musial et al. 2020; Cabrera et al. 2006). Among the three types, green tea has attracted the most attention from clinicians and researchers as it contains more catechin than the other types of tea (Cabrera et al. 2006). Green tea catechins have been reported to have antibacterial and antiviral activities (Cabrera et al. 2006). There are four major catechins in green tea; epigallocatechin-3-gallate (EGCG), epigallocatechin (EGC), epicatechin-3-gallate (ECG), and epicatechin (EC) (McKay and Blumberg 2002). EGCG, an active polyphenolic catechin, accounts for 60% of catechins in green tea, followed by EGC (19%), ECG (13.6%), and EC (6.4%) (Steinmann et al. 2013; McKay and Blumberg 2002).

Among these catechins, EGCG most likely has clinical potentials due to its antioxidant and anti-allergic effects, as well as its antibacterial properties against *Staphylococci* (Shiota et al. 1999; Sudano Roccaro et al. 2004; Stapleton et al. 2004),

Fig. 3 Three major types of tea. Tea is an aromatic beverage made from *Camellia sinensis*. Tea is classified into three major types depending on fermentation: "green tea" (unfermented), "oolong tea" (semi-fermented), and "black tea" or "red tea" (fermented). Tea has been enjoyed for over 2000 years since it was first developed in ancient China and is the most commonly consumed drink in the world after water. The photo was kindly provided by Yuki Suzuki

E. coli (Cui et al. 2012), and *Helicobacter pylori* (Yanagawa et al. 2003). EGCG also inhibits biofilm formation by interfering with the extracellular polymeric materials (Sudano Roccaro et al. 2004). Several animal studies have elucidated the in vivo efficacy of EGCG in abnormal bone absorption. In an organ culture of mouse calvariae, EGCG significantly rescued LPS-induced bone absorption (Tominari et al. 2015). In addition, EGCG injection significantly inhibited LPS-induced mandibular alveolar bone loss in mice (Tominari et al. 2015).

In addition, the 3-O-methylated form of EGCG (EGCG3"Me) also shows a similar effect with more efficient absorption from the intestine. An animal study showed EGCG3"Me significantly suppressed LPS-induced bone resorption in organ cultures of alveolar bone from periodontitis model mice (Tominari et al. 2017). Moreover, EGCG3"Me was shown to attenuate alveolar bone loss in the lower gingiva. These results suggested that EGCG3"Me might contribute to protection against inflammatory bone loss in periodontitis (Tominari et al. 2017).

Another component of tea that is attracting the interest of researchers is Theaflavin. Theaflavin is one of the primary polyphenols in black tea. An animal study was conducted to elucidate the anti-inflammatory effect of Theaflavin on periodontitis (Wu et al. 2018). Topical application of the polyphenol showed a significant inhibition of alveolar bone loss and a reduction in inflammatory cell infiltration, with the expression levels of IL-6, MMP9, and RANKL all significantly decreased (Wu et al. 2018).

Several clinical trials targeting the clinical application of tea have been reported since the first study on the antibiotic properties of tea against *Bacillus typhosus* by McNaught in 1906 (McNaught 1906). Catechin, one of the major components in green tea, was applied as a mouthwash, and its antiplaque efficacy was examined (Kaur et al. 2014). Kaur et al. conducted a single-blinded, cross-over study on 30 patients (Kaur et al. 2014). They reported that the green tea catechin mouthwash had comparable antiplaque properties to a standard chlorhexidine gluconate mouthwash as there were no significant differences in the plaque score between the catechin group and the chlorhexidine group. They also mentioned the superiority of the green tea catechin mouthwash to chlorhexidine mouthwash in maintaining daily oral hygiene because the chlorhexidine gluconate mouthwash is not suitable for long-term usage due to adverse effects (Kaur et al. 2014). Thus, due to the absence of long-term adverse effects, the green tea mouthwash might be superior to the standard chlorhexidine gluconate mouthwash.

Green tea has also been applied as a toothpaste. Hirasawa et al. examined the usefulness of green tea catechin in the regulation of periodontal disease (Hirasawa et al. 2002). The group applied green tea catechin using a slow-release local delivery system in periodontal pockets in six patients with advanced periodontitis. After 8 weeks of application, the pocket depth and bacterial counts were significantly decreased without any adverse effects (Hirasawa et al. 2002).

5 Other Natural Products

Besides honey, essential oils, and tea, there are other natural products expected to have potentials for clinical applications, such as Aloe vera, and *Sasa senanensis Rehder*, Curcumin (Guimarães et al. 2011), Cinnamon (Yanakiev 2020), Ginseng (Wang et al. 2020), and European chestnut (Salam et al. 2021).

Syed et al. conducted a double-blinded, placebo-controlled study to examine the efficacy and safety of topical Aloe vera cream for psoriasis vulgaris (Syed et al. 1996). 60 patients with psoriasis vulgaris were randomly assigned to receive topical Aloe vera cream or a placebo. The cure ratio was significantly higher in patients receiving the Aloe vera cream (83.3%) than in the control group (6.6%), with the severity score significantly decreased in the Aloe vera group. Psoriatic plaque was also significantly reduced in the patients treated with the Aloe vera cream. There were no adverse effects observed and no dropouts in the Aloe vera group (Syed et al. 1996). The group concluded that Aloe vera cream could be an alternative treatment with high effectiveness and safety for patients suffering from psoriasis vulgaris.

Aloe vera is also expected to provide a first-aid treatment for burns, though the effect is limited (Cuttle et al. 2008). Cuttle et al. created burns on the skin of pigs and applied Aloe vera, essential oil, saliva from a human volunteer, or a control. Although the area of the burns did not significantly differ among the groups, Aloe vera significantly decreased the subdermal temperature more quickly than did the other treatments. However, no significant differences were found among the groups in the context of microbial inhibition, implying that Aloe vera did not affect the microbial status in the burns (Cuttle et al. 2008).

Fig. 4 *Sasa senanensis Rehder*. *Sasa senanensis Rehder*, a broad-leaf bamboo native to Japan, is expected to be effective in treatment of fatigue, low appetite, stomatitis, body odor, and halitosis. The photo was kindly provided by Dr. Koh Nakamura

Sasa senanensis Rehder, a broad-leaf bamboo, has been the focus of study as a potential remedy as the alkaline extract from its leaves has been shown to be effective against fatigue, low appetite, stomatitis, body odor, and halitosis (Fig. 4) (Sakagami et al. 2013). Sakagami et al. conducted a clinical test to examine if the alkaline extract from *Sasa Senanensis Rehder* had anti-halitosis activity (Sakagami et al. 2018). 12 volunteers brushed their teeth after each meal with toothpaste containing a *Sasa Senanensis Rehder* extract or a placebo toothpaste. The halitosis was evaluated by measuring the concentration of volatile sulfur compounds in the breath collected from participants. The toothpaste containing the *Sasa* extract significantly reduced halitosis, but the number of bacteria on the tongue did not differ (Sakagami et al. 2018). The same group also conducted multi-omics analysis and reported the alkaline extract from the *Sasa* leaves suppressed IL-1β-induced inflammation in human gingival fibroblasts (Sakagami et al. 2018). Thus, the alkaline extract from the *Sasa* leaves appears to have anti-inflammatory properties (Sakagami et al. 2018).

6 Conclusion

In China, there is an old saying, *wēn-gù-zhī-xīn*, which means "learning from the past." With the demand for new antibacterial strategies, modern scientists have explored a number of clues from ancient history and found various active components in natural products. The most successful example would be the discovery of penicillin as mentioned above in the introduction. Fleming found the antibiotic element, which was later named penicillin, in mold and applied it to clinical use, thereby changing medical history itself.

As discussed in this chapter, a great deal of effort has been devoted to identifying the effective components from within the natural products used in remedies. However, little attention has been paid to their combinations. It could be that some of the ingredients may not be effective alone while demonstrating excellent antimicrobial properties in combinations. In this regard, the study of Bald's eyesalve is noteworthy (Furner-Pardoe et al. 2020). Bald's eyesalve, described in a pre-modern European medical text written 1000 years ago, is made with garlic, onions, wine, and bile salts. Recently, the eyesalve was reconstructed and its efficacy against bacteria was scientifically examined. Bald's eyesalve showed antibacterial activity against the planktonic form of antimicrobial-resistant bacteria (ESKAPE) and antibiofilm activity (Furner-Pardoe et al. 2020). Interestingly, the individual components in Bald's eyesalve did not show any antimicrobial effect but did work in combination through a synergistic effect (Furner-Pardoe et al. 2020). This research suggests the significance of identifying individuals and combining multiple components from the various natural products.

The further exploration of the active ingredients in natural products is necessary toward the creation of alternatives to antibiotics against multidrug-resistant bacteria. Some clues might yet be found in history, traditional folk remedies, and natural products. Antibiotic resistance, regarded as one of the top ten threats to modern global health by the WHO, might be overcome by "learning from the past."

Acknowledgement The author is thankful to Dr. Koh Nakamura and Yuki Suzuki for providing the photos of *Sasa senanensis Rehder*, and Manuka honey and tea. The author is also grateful to Ayaka Kubota, Yuka Masuta, Shizuka Sugawara, and Yumiko Kimura for their help in preparing the manuscript. The authors declare that the chapter was written in the absence of any commercial or financial relationships that could be construed as a potential conflict of interest. This review was supported by JSPS KAKENHI Grant Number 17H06491, 18 K16871, and 18KK0444 for MS

References

Abu-Al-Basal MA (2010) Healing potential of Rosmarinus officinalis L. on full-thickness excision cutaneous wounds in alloxan-induced-diabetic BALB/c mice. J Ethnopharmacol 131:443–450

Alandejani T, Marsan J, Ferris W, Slinger R, Chan F (2009) Effectiveness of honey on Staphylococcus aureus and Pseudomonas aeruginosa biofilms. Otolaryngol Head Neck Surg 141:114–118

Alangari AA, Morris K, Lwaleed BA, Lau L, Jones K, Cooper R, Jenkins R (2017) Honey is potentially effective in the treatment of atopic dermatitis: clinical and mechanistic studies. Immun Inflamm Dis 5:190–199

Allen HB, Vaze ND, Choi C, Hailu T, Tulbert BH, Cusack CA, Joshi SG (2014) The presence and impact of biofilm-producing staphylococci in atopic dermatitis. JAMA Dermatol 150:260–265

Almasaudi S (2021) The antibacterial activities of honey. Saudi J Biol Sci 28:2188–2196

Alvarez-Suarez JM, Gasparrini M, Forbes-Hernández TY, Mazzoni L, Giampieri F (2014) The composition and biological activity of honey: a focus on Manuka honey. Foods 3:420–432

Al-Waili NS (2003) Topical application of natural honey, beeswax and olive oil mixture for atopic dermatitis or psoriasis: partially controlled, single-blinded study. Complement Ther Med 11: 226–234

Ambrosino P, Molino A, Spedicato GA, Parrella P, Formisano R, Motta A, Di Minno MND, Maniscalco M (2020) Nasal nitric oxide in chronic rhinosinusitis with or without nasal polyps: a systematic review with meta-analysis. J Clin Med 9:200

Ansari MJ, Al-Ghamdi A, Usmani S, Al-Waili NS, Sharma D, Nuru A, Al-Attal Y (2013) Effect of jujube honey on Candida albicans growth and biofilm formation. Arch Med Res 44:352–360

Aron M, Victoria Akinpelu O, Dorion D, Daniel S (2012) Otologic safety of manuka honey. J Otolaryngol Head Neck Surg 41(Suppl 1):S21–S30

Aron M, Akinpelu OV, Gasbarrino K, Daniel SJ (2015) Safety of transtympanic application of 4% manuka honey in a chinchilla animal model. Eur Arch Otorhinolaryngol 272:537–542

Bakkali F, Averbeck S, Averbeck D, Idaomar M (2008) Biological effects of essential oils–a review. Food Chem Toxicol 46:446–475

Barbosa F, Pinto E, Kijjoa A, Pinto M, Sousa E (2020) Targeting antimicrobial drug resistance with marine natural products. Int J Antimicrob Agents 56:106005

Bowler WA, Bresnahan J, Bradfish A, Fernandez C (2010) An integrated approach to methicillin-resistant Staphylococcus aureus control in a rural, regional-referral healthcare setting. Infect Control Hosp Epidemiol 31:269–275

Brady N, Molan P, Bang L (2004) A survey of non-manuka New Zealand honeys for antibacterial and antifungal activities. J Apic Res 43:47–52

Butler MS, Paterson DL (2020) Antibiotics in the clinical pipeline in October 2019. J Antibiot (Tokyo) 73:329–364

Butler MS, Blaskovich MAT, Cooper MA (2017) Antibiotics in the clinical pipeline at the end of 2015. J Antibiot 70:3–24

Cabrera C, Artacho R, Giménez R (2006) Beneficial effects of green tea--a review. J Am Coll Nutr 25:79–99

Ceri H, Olson ME, Stremick C, Read RR, Morck D, Buret A (1999) The Calgary biofilm device: new technology for rapid determination of antibiotic susceptibilities of bacterial biofilms. J Clin Microbiol 37:1771–1776

Chaijan MR, Handjani F, Zarshenas M, Rahimabadi MS, Tavakkoli A (2018) The myrtus communis L. solution versus ketoconazole shampoo in treatment of dandruff: a double blinded randomized clinical trial. J Pak Med Assoc 68:715–720

Chang EH, Alandejani T, Akbari E, Ostry A, Javer A (2011) Double-blinded, randomized, controlled trial of medicated versus nonmedicated merocel sponges for functional endoscopic sinus surgery. J Otolaryngol Head Neck Surg 40(Suppl 1):S14–S19

Charles CH, Mostler KM, Bartels LL, Mankodi SM (2004) Comparative antiplaque and antigingivitis effectiveness of a chlorhexidine and an essential oil mouthrinse: 6-month clinical trial. J Clin Periodontol 31:878–884

Charles CA, McGuire JA, Qaqish J, Amini P (2013) Increasing antiplaque/antigingivitis efficacy of an essential oil mouthrinse over time: an in vivo study. Gen Dent 61:23–28

Cui Y, Oh YJ, Lim J, Youn M, Lee I, Pak HK, Park W, Jo W, Park S (2012) AFM study of the differential inhibitory effects of the green tea polyphenol (−)-epigallocatechin-3-gallate (EGCG) against gram-positive and gram-negative bacteria. Food Microbiol 29:80–87

Cuttle L, Kempf M, Kravchuk O, George N, Liu PY, Chang HE, Mill J, Wang XQ, Kimble RM (2008) The efficacy of Aloe vera, tea tree oil and saliva as first aid treatment for partial thickness burn injuries. Burns 34:1176–1182

da Gomes S, Ary P, de Puziol F, Leitão RN, Gomes TR, Scherer R, Martins MLL, Cavalcanti ÁSS, Cavalcanti LC (2012) Application of the essential oil from copaiba (Copaifera langsdorffii Desf.) for acne vulgaris: a double-blind, placebo controlled clinical trial. Altern Med Rev 17:69–75

Díaz-Nuñez JL, García-Contreras R, Castillo-Juárez I (2021) The new antibacterial properties of the plants: Quo vadis studies of anti-virulence phytochemicals? Front Microbiol 12:667126

Dryden MS, Dailly S, Crouch M (2004) A randomized, controlled trial of tea tree topical preparations versus a standard topical regimen for the clearance of MRSA colonization. J Hosp Infect 56:283–286

Edmondson M, Newall N, Carville K, Smith J, Riley TV, Carson CF (2011) Uncontrolled, open-label, pilot study of tea tree (Melaleuca alternifolia) oil solution in the decolonisation of methicillin-resistant Staphylococcus aureus positive wounds and its influence on wound healing. Int Wound J 8:375–384

Enshaieh S, Jooya A, Siadat AH, Iraji F (2007) The efficacy of 5% topical tea tree oil gel in mild to moderate acne vulgaris: a randomized, double-blind placebo-controlled study. Indian J Dermatol Venereol Leprol 73:22

Erejuwa OO, Sulaiman SA, Ab Wahab MS (2012) Honey: a novel antioxidant. Molecules 17:4400–4423

Eteraf-Oskouei T, Najafi M (2013) Traditional and modern uses of natural honey in human diseases: a review. Iran J Basic Med Sci 16:731–742

Fernandez-Cabezudo MJ, El-Kharrag R, Torab F, Bashir G, George JA, El-Taji H, B. K. al-Ramadi. (2013) Intravenous administration of manuka honey inhibits tumor growth and improves host survival when used in combination with chemotherapy in a melanoma mouse model. PLoS One 8:e55993

Fokkens WJ, Lund VJ, Mullol J, Bachert C, Alobid I, Baroody F, Cohen N, Cervin A, Douglas R, Gevaert P, Georgalas C, Goossens H, Harvey R, Hellings P, Hopkins C, Jones N, Joos G, Kalogjera L, Kern B, Kowalski M, Price D, Riechelmann H, Schlosser R, Senior B, Thomas M, Toskala E, Voegels R, de Wang Y, Wormald PJ (2012) European position paper on rhinosinusitis and nasal polyps 2012. Rhinol Suppl 3. preceding table of contents:1–298

Fokkens WJ, Lund VJ, Hopkins C, Hellings PW, Kern R, Reitsma S, Toppila-Salmi S, Bernal-Sprekelsen M, Mullol J, Alobid I, Terezinha Anselmo-Lima W, Bachert C, Baroody F, von Buchwald C, Cervin A, Cohen N, Constantinidis J, De Gabory L, Desrosiers M, Diamant Z, Douglas RG, Gevaert PH, Hafner A, Harvey RJ, Joos GF, Kalogjera L, Knill A, Kocks JH, Landis BN, Limpens J, Lebeer S, Lourenco O, Meco C, Matricardi PM, O'Mahony L, Philpott CM, Ryan D, Schlosser R, Senior B, Smith TL, Teeling T, Tomazic PV, Wang DY, Wang D, Zhang L, Agius AM, Ahlstrom-Emanuelsson C, Alabri R, Albu S, Alhabash S, Aleksic A, Aloulah M, Al-Qudah M, Alsaleh S, Baban MA, Baudoin T, Balvers T, Battaglia P, Bedoya JD, Beule A, Bofares KM, Braverman I, Brozek-Madry E, Richard B, Callejas C, Carrie S, Caulley L, Chussi D, de Corso E, Coste A, El Hadi U, Elfarouk A, Eloy PH, Farrokhi S, Felisati G, Ferrari MD, Fishchuk R, Grayson W, Goncalves PM, Grdinic B, Grgic V, Hamizan AW, Heinichen JV, Husain S, Ping TI, Ivaska J, Jakimovska F, Jovancevic L, Kakande E, Kamel R, Karpischenko S, Kariyawasam HH, Kawauchi H, Kjeldsen A, Klimek L, Krzeski A, Kopacheva Barsova G, Kim SW, Lal D, Letort JJ, Lopatin A, Mahdjoubi A, Mesbahi A, Netkovski J, Nyenbue Tshipukane D, Obando-Valverde A, Okano M, Onerci M, Ong YK, Orlandi R, Otori N, Ouennoughy K, Ozkan M, Peric A, Plzak J, Prokopakis E, Prepageran N, Psaltis A, Pugin B, Raftopulos M, Rombaux P, Riechelmann H, Sahtout S, Sarafoleanu CC, Searyoh K, Rhee CS, Shi J, Shkoukani M, Shukuryan AK, Sicak M, Smyth D, Snidvongs K, Soklic Kosak T, Stjarne P, Sutikno B, Steinsvag S, Tantilipikorn P, Thanaviratananich S, Tran T, Urbancic J, Valiulius A, Vasquez de Aparicio C, Vicheva D, Virkkula PM, Vicente G, Voegels R, Wagenmann MM, Wardani RS, Welge-Lussen A, Witterick I,

Wright E, Zabolotniy D, Zsolt B, Zwetsloot CP (2020) European position paper on rhinosinusitis and nasal polyps 2020. Rhinology 58:1–464
Foreman A, Holtappels G, Psaltis AJ, Jervis-Bardy J, Field J, Wormald PJ, Bachert C (2011) Adaptive immune responses in *Staphylococcus aureus* biofilm-associated chronic rhinosinusitis. Allergy 66:1449–1456
Fujieda S, Imoto Y, Kato Y, Ninomiya T, Tokunaga T, Tsutsumiuchi T, Yoshida K, Kidoguchi M, Takabayashi T (2019) Eosinophilic chronic rhinosinusitis. Allergol Int 68:403–412
Furner-Pardoe J, Anonye BO, Cain R, Moat J, Ortori CA, Lee C, Barrett DA, Corre C, Harrison F (2020) Anti-biofilm efficacy of a medieval treatment for bacterial infection requires the combination of multiple ingredients. Sci Rep 10:12687
Galvão LC, de Carvalho V, Furletti F, Bersan SMF, Guilherme M, da Cunha A, Ruiz LTG, Ernesto J, de Carvalho A, Sartoratto VL, Rehder G, Figueira GM, Duarte MCT, Ikegaki M, de Alencar SM, Rosalen PL (2012) Antimicrobial activity of essential oils against *Streptococcus mutans* and their Antiproliferative effects. Evid Based Complement Alternat Med 2012:751435
Gonzalez T, Biagini Myers JM, Herr AB, Khurana Hershey GK (2017) Staphylococcal biofilms in atopic dermatitis. Curr Allergy Asthma Rep 17:81
Gonzalez T, Stevens ML, Baatyrbek Kyzy A, Alarcon R, He H, Kroner JW, Spagna D, Grashel B, Sidler E, Martin LJ, Biagini Myers JM, Khurana Hershey GK, Herr AB (2021) Biofilm propensity of Staphylococcus aureus skin isolates is associated with increased atopic dermatitis severity and barrier dysfunction in the MPAACH pediatric cohort. Allergy 76:302–313
Guimarães MR, Coimbra LS, de Aquino SG, Spolidorio LC, Kirkwood KL, Rossa C Jr (2011) Potent anti-inflammatory effects of systemically administered curcumin modulate periodontal disease in vivo. J Periodontal Res 46:269–279
Hashemian F, Baghbanian N, Majd Z, Rouini MR, Jahanshahi J, Hashemian F (2015) The effect of thyme honey nasal spray on chronic rhinosinusitis: a double-blind randomized controlled clinical trial. Eur Arch Otorhinolaryngol 272:1429–1435
Henatsch D, Nabuurs CH, van de Goor RM, Wolffs PF, Stokroos RJ (2017) Treatment of recurrent eczematous external otitis with honey eardrops: a proof-of-concept study. Otolaryngol Head Neck Surg 157:696–699
Hirasawa M, Takada K, Makimura M, Otake S (2002) Improvement of periodontal status by green tea catechin using a local delivery system: a clinical pilot study. J Periodontal Res 37:433–438
Hopkins RB, Burke N, Harlock J, Jegathisawaran J, Goeree R (2015) Economic burden of illness associated with diabetic foot ulcers in Canada. BMC Health Serv Res 15:13
Imran M, Hussain MB, Baig M (2015) A randomized, controlled clinical trial of honey-impregnated dressing for treating diabetic foot ulcer. J Coll Physicians Surg Pak 25:721–725
Jan WA, Shah H, Khan M, Fayaz M, Ullah N (2012) Comparison of conventional Pyodine dressing with honey dressing for the treatment of diabetic foot ulcers. J Postgrad Medical Inst (Peshawar - Pakistan) 26(4):402–407
Johnson DW, van Eps C, Mudge DW, Wiggins KJ, Armstrong K, Hawley CM, Campbell SB, Isbel NM, Nimmo GR, Gibbs H (2005) Randomized, controlled trial of topical exit-site application of honey (Medihoney) versus mupirocin for the prevention of catheter-associated infections in hemodialysis patients. J Am Soc Nephrol 16:1456–1462
Johnson DW, Badve SV, Pascoe EM, Beller E, Cass A, Clark C, de Zoysa J, Isbel NM, McTaggart S, Morrish AT, Geoffrey Playford E, Scaria A, Snelling P, Vergara LA, Hawley CM (2014) Antibacterial honey for the prevention of peritoneal-dialysis-related infections (HONEYPOT): a randomised trial. Lancet Infect Dis 14:23–30
Johnston M, McBride M, Dahiya D, Owusu-Apenten R, Nigam PS (2018) Antibacterial activity of Manuka honey and its components: an overview. AIMS Microbiol 4:655–664
Kagami S, Kakinuma T, Saeki H, Tsunemi Y, Fujita H, Nakamura K, Takekoshi T, Kishimoto M, Mitsui H, Torii H, Komine M, Asahina A, Tamaki K (2003) Significant elevation of serum levels of eotaxin-3/CCL26, but not of eotaxin-2/CCL24, in patients with atopic dermatitis: serum eotaxin-3/CCL26 levels reflect the disease activity of atopic dermatitis. Clin Exp Immunol 134:309–313

Kamaratos AV, Tzirogiannis KN, Iraklianou SA, Panoutsopoulos GI, Kanellos IE, Melidonis AI (2014) Manuka honey-impregnated dressings in the treatment of neuropathic diabetic foot ulcers. Int Wound J 11:259–263

Kandimalla R, Kalita S, Choudhury B, Dash S, Kalita K, Kotoky J (2016) Chemical composition and anti-candidiasis mediated wound healing property of Cymbopogon nardus essential oil on chronic diabetic wounds. Front Pharmacol 7:198

Karpiński TM, Ożarowski M, Seremak-Mrozikiewicz A, Wolski H, Adamczak A (2021) Plant preparations and compounds with activities against biofilms formed by Candida spp. J Fungi (Basel) 7(5):360

Katayama I, Aihara M, Ohya Y, Saeki H, Shimojo N, Shoji S, Taniguchi M, Yamada H (2017) Japanese guidelines for atopic dermatitis 2017. Allergol Int 66:230–247

Kaur H, Jain S, Kaur A (2014) Comparative evaluation of the antiplaque effectiveness of green tea catechin mouthwash with chlorhexidine gluconate. J Indian Soc Periodontol 18:178–182

Kilty SJ, AlMutairi D, Duval M, Groleau MA, De Nanassy J, Gomes MM (2010) Manuka honey: histological effect on respiratory mucosa. Am J Rhinol Allergy 24:e63–e66

Koo H, Kim TH, Kim KW, Wee SW, Chun YS, Kim JC (2012) Ocular surface discomfort and Demodex: effect of tea tree oil eyelid scrub in Demodex blepharitis. J Korean Med Sci 27:1574–1579

Kumar A, Mittal S, Tyagi A, Romesh H, Varshney S, Malhotra M (2020) Efficacy of medical grade manuka honey in acute otitis externa: a pilot study. Indian J Otol 26:151–154

Kumarasamy G, Ramli RR, Singh H, Abdullah B (2020) Tualang honey versus steroid impregnated nasal dressing following endoscopic sinus surgery: a randomized controlled trial. J Complement Integr Med 18:433–438

Kwon HH, Yoon JY, Park SY, Min S, Suh DH (2014) Comparison of clinical and histological effects between lactobacillus-fermented Chamaecyparis obtusa and tea tree oil for the treatment of acne: an eight-week double-blind randomized controlled split-face study. Dermatology 229: 102–109

Lee VS, Humphreys IM, Purcell PL, Davis GE (2017) Manuka honey sinus irrigation for the treatment of chronic rhinosinusitis: a randomized controlled trial. Int Forum Allergy Rhinol 7: 365–372

Lee VS, Humphreys IM, Purcell PL, Davis GE (2021) Manuka honey versus saline sinus irrigation in the treatment of cystic fibrosis-associated chronic rhinosinusitis: a randomised pilot trial. Clin Otolaryngol 46:168–174

Lu M, Dai T, Murray CK, Mei XW (2018) Bactericidal property of oregano oil against multidrug-resistant clinical isolates. Front Microbiol 9:2329

Lu J, Cokcetin NN, Burke CM, Turnbull L, Liu M, Carter DA, Whitchurch CB, Harry EJ (2019a) Honey can inhibit and eliminate biofilms produced by Pseudomonas aeruginosa. Sci Rep 9: 18160

Lu L, Hu W, Tian Z, Yuan D, Yi G, Zhou Y, Cheng Q, Zhu J, Li M (2019b) Developing natural products as potential anti-biofilm agents. Chin Med 14:11

Lusby PE, Coombes AL, Wilkinson JM (2005) Bactericidal activity of different honeys against pathogenic bacteria. Arch Med Res 36:464–467

Malhi HK, Jenny T, Riley TV, Kumarasinghe SP, Hammer KA (2017) Tea tree oil gel for mild to moderate acne; a 12 week uncontrolled, open-label phase II pilot study. Australas J Dermatol 58:205–210

Manji J, Thamboo A, Sunkaraneni V, Singh A, Tebbutt S, Garnis C, Javer A (2019) The association of leptospermum honey with cytokine expression in the sinonasal epithelium of chronic rhinosinusitis patients. World J Otorhinolaryngol Head Neck Surg 5:19–25

Mărgăoan R, Topal E, Balkanska R, Yücel B, Oravecz T, Cornea-Cipcigan M, Vodnar DC (2021) Monofloral honeys as a potential source of natural antioxidants, minerals and medicine. Antioxidants (Basel) 10:1023

McKay DL, Blumberg JB (2002) The role of tea in human health: an update. J Am Coll Nutr 21:1–13

McNaught JG (1906) On the action of cold or lukewarm tea on *Bacillus typhosus*. BMJ Mil Health 7:372–373

Meo SA, Al-Asiri SA, Mahesar AL, Ansari MJ (2017) Role of honey in modern medicine. Saudi J Biol Sci 24:975–978

Merckoll P, Jonassen TØ, Vad ME, Jeansson SL, Melby KK (2009) Bacteria, biofilm and honey: a study of the effects of honey on 'planktonic' and biofilm-embedded chronic wound bacteria. Scand J Infect Dis 41:341–347

Messager S, Hammer KA, Carson CF, Riley TV (2005) Effectiveness of hand-cleansing formulations containing tea tree oil assessed ex vivo on human skin and in vivo with volunteers using European standard EN 1499. J Hosp Infect 59:220–228

Münstedt K, Momm F, Hübner J (2019) Honey in the management of side effects of radiotherapy- or radio/chemotherapy-induced oral mucositis. A systematic review. Complement Ther Clin Pract 34:145–152

Musial C, Kuban-Jankowska A, Gorska-Ponikowska M (2020) Beneficial properties of green tea Catechins. Int J Mol Sci 21(5):1744

Newman DJ, Cragg GM (2016) Natural products as sources of new drugs from 1981 to 2014. J Nat Prod 79:629–661

Noori AL-W, Al Ghamdi MJA, Al-Attal Y, Al-Mubarak A, Salom K (2013) Differences in composition of honey samples and their impact on the antimicrobial activities against drug multiresistant bacteria and pathogenic fungi. Arch Med Res 44:307–316

Ogonowska P, Gilaberte Y, Barańska-Rybak W, Nakonieczna J (2020) Colonization with Staphylococcus aureus in atopic dermatitis patients: attempts to reveal the unknown. Front Microbiol 11:567090

Ooi ML, Jothin A, Bennett C, Ooi EH, Vreugde S, Psaltis AJ, Wormald PJ (2019) Manuka honey sinus irrigations in recalcitrant chronic rhinosinusitis: phase 1 randomized, single-blinded, placebo-controlled trial. Int Forum Allergy Rhinol 9:1470–1477

Orafidiya L, Agbani E, Oyedele A, Babalola O, Onayemi O (2002) Preliminary clinical tests on topical preparations of Ocimum gratissimum Linn leaf essential oil for the treatment of acne vulgaris. Clin Drug Investig 22:313–319

Owczarek W, Paplińska M, Targowski T, Jahnz-Różyk K, Paluchowska E, Kucharczyk A, Kasztalewicz B (2010) Analysis of eotaxin 1/CCL11, eotaxin 2/CCL24 and eotaxin 3/CCL26 expression in lesional and non-lesional skin of patients with atopic dermatitis. Cytokine 50:181–185

Papadopoulou D, Dabrowska A, Harries PG, Webb JS, Allan RN, Salib RJ (2020) Evaluation of a bioengineered honey and its synthetic equivalent as novel Staphylococcus aureus biofilm-targeted topical therapies in chronic rhinosinusitis. Am J Rhinol Allergy 34:80–86

Paramasivan S, Drilling AJ, Jardeleza C, Jervis-Bardy J, Vreugde S, Wormald PJ (2014) Methylglyoxal-augmented manuka honey as a topical anti-Staphylococcus aureus biofilm agent: safety and efficacy in an in vivo model. Int Forum Allergy Rhinol 4:187–195

Post V, Peter Wahl R, Richards G, Fintan Moriarty T (2017) Vancomycin displays time-dependent eradication of mature Staphylococcus aureus biofilms. J Orthop Res 35:381–388

Psaltis AJ, Weitzel EK, Ha KR, Wormald PJ (2008) The effect of bacterial biofilms on post-sinus surgical outcomes. Am J Rhinol 22:1–6

Raut JS, Karuppayil SM (2014) A status review on the medicinal properties of essential oils. Ind Crop Prod 62:250–264

Rosa IA, Rodrigues P, Bianchini AE, Silveira BP, Ferrari FT, Bandeira Junior G, Vargas APC, Baldisserotto B, Heinzmann BM (2019) Extracts of Hesperozygis ringens (Benth.) Epling: in vitro and in vivo antibacterial activity against fish pathogenic bacteria. J Appl Microbiol 126: 1353–1361

Sakagami H, Amano S, Yasui T, Satoh K, Shioda S, Kanamoto T, Terakubo S, Nakashima H, Watanabe K, Sugiura T, Kitajima M, Oizumi H, Oizumi T (2013) Biological interaction between *Sasa senanensis Rehder* leaf extract and toothpaste ingredients. In Vivo 27:275–284

Sakagami H, Watanabe T, Hoshino T, Suda N, Mori K, Yasui T, Yamauchi N, Kashiwagi H, Gomi T, Oizumi T, Nagai J, Uesawa Y, Takao K, Sugita Y (2018) Recent Progress of basic studies of natural products and their dental application. Medicines (Basel) 6(1):4

Salam AM, Porras G, Cho Y-SK, Brown MM, Risener CJ, Marquez L, Lyles JT, Bacsa J, Horswill AR, Quave CL (2021) Castaneroxy A from the leaves of Castanea sativa inhibits virulence in *Staphylococcus aureus*. Front Pharmacol 12:640179

Scheres J, Kuszewski K (2019) The ten threats to global health in 2018 and 2019. A welcome and informative communication of WHO to everybody. Zeszyty Naukowe Ochrony Zdrowia Zdrowie Publiczne i Zarzadzanie 17:2–8

Shiota S, Shimizu M, Mizushima T, Ito H, Hatano T, Yoshida T, Tsuchiya T (1999) Marked reduction in the minimum inhibitory concentration (MIC) of beta-lactams in methicillin-resistant Staphylococcus aureus produced by epicatechin gallate, an ingredient of green tea (Camellia sinensis). Biol Pharm Bull 22:1388–1390

Shukrimi A, Sulaiman AR, Halim AY, Azril A (2008) A comparative study between honey and povidone iodine as dressing solution for Wagner type II diabetic foot ulcers. Med J Malaysia 63: 44–46

Siavash M, Shokri S, Haghighi S, Shahtalebi MA, Farajzadehgan Z (2015) The efficacy of topical royal jelly on healing of diabetic foot ulcers: a double-blind placebo-controlled clinical trial. Int Wound J 12:137–142

Singhal D, Psaltis AJ, Foreman A, Wormald PJ (2010) The impact of biofilms on outcomes after endoscopic sinus surgery. Am J Rhinol Allergy 24:169–174

Singhal D, Foreman A, Jervis-Bardy J, Wormald PJ (2011) Staphylococcus aureus biofilms: nemesis of endoscopic sinus surgery. Laryngoscope 121:1578–1583

Stapleton PD, Shah S, Anderson JC, Hara Y, Hamilton-Miller JM, Taylor PW (2004) Modulation of beta-lactam resistance in Staphylococcus aureus by catechins and gallates. Int J Antimicrob Agents 23:462–467

Steinmann J, Buer J, Pietschmann T, Steinmann E (2013) Anti-infective properties of epigallocatechin-3-gallate (EGCG), a component of green tea. Br J Pharmacol 168:1059–1073

Sudano Roccaro A, Blanco AR, Giuliano F, Rusciano D, Enea V (2004) Epigallocatechin-gallate enhances the activity of tetracycline in staphylococci by inhibiting its efflux from bacterial cells. Antimicrob Agents Chemother 48:1968–1973

Sugita K, Akdis CA (2019) Recent developments and advances in atopic dermatitis and food allergy. Allergol Int 69(2):204–214

Syed TA, Ahmad SA, Holt AH, Ahmad SA, Ahmad SH, Afzal M (1996) Management of psoriasis with Aloe vera extract in a hydrophilic cream: a placebo-controlled, double-blind study. Tropical Med Int Health 1:505–509

Tominari T, Matsumoto C, Watanabe K, Hirata M, Grundler FMW, Miyaura C, Inada M (2015) Epigallocatechin gallate (EGCG) suppresses lipopolysaccharide-induced inflammatory bone resorption, and protects against alveolar bone loss in mice. FEBS Open Bio 5:522–527

Tominari T, Ichimaru R, Yoshinouchi S, Matsumoto C, Watanabe K, Hirata M, Grundler FMW, Inada M, Miyaura C (2017) Effects of O-methylated (−)-epigallocatechin gallate (EGCG) on LPS-induced osteoclastogenesis, bone resorption, and alveolar bone loss in mice. FEBS Open Bio 7:1972–1981

Tonks AJ, Cooper RA, Jones KP, Blair S, Parton J, Tonks A (2003) Honey stimulates inflammatory cytokine production from monocytes. Cytokine 21:242–247

Wang C, Guo M, Zhang N, Wang G (2019) Effectiveness of honey dressing in the treatment of diabetic foot ulcers: a systematic review and meta-analysis. Complement Ther Clin Pract 34: 123–131

Wang L, Huang Y, Yin G, Wang J, Wang P, Chen ZY, Wang T, Ren G (2020) Antimicrobial activities of Asian ginseng, American ginseng, and notoginseng. Phytother Res 34:1226–1236

Wu YH, Kuraji R, Taya Y, Ito H, Numabe Y (2018) Effects of theaflavins on tissue inflammation and bone resorption on experimental periodontitis in rats. J Periodontal Res 53:1009–1019

Wu SC, Liu F, Zhu K, Shen JZ (2019) Natural products that target virulence factors in antibiotic-resistant *Staphylococcus aureus*. J Agric Food Chem 67:13195–13211

Yanagawa Y, Yamamoto Y, Hara Y, Shimamura T (2003) A combination effect of epigallocatechin gallate, a major compound of green tea catechins, with antibiotics on *Helicobacter pylori* growth in vitro. Curr Microbiol 47:244–249

Yanakiev S (2020) Effects of Cinnamon (Cinnamomum spp.) in dentistry: a review. Molecules 25: 4184

Yang C, Dong-Hui H, Feng Y (2015) Essential oil of Artemisia vestita exhibits potent in vitro and in vivo antibacterial activity: investigation of the effect of oil on biofilm formation, leakage of potassium ions and survival curve measurement. Mol Med Rep 12:5762–5770

Yang C, Mavelli GV, Nacharaju P, Li K, Cleare LG, Nosanchuk JD, Friedman JM, Abuzeid WM (2020) Novel nitric oxide-generating platform using manuka honey as an anti-biofilm strategy in chronic rhinosinusitis. Int Forum Allergy Rhinol 10:223–232

Yoshida K, Takabayashi T, Imoto Y, Sakashita M, Narita N, Fujieda S (2019) Reduced nasal nitric oxide levels in patients with eosinophilic chronic rhinosinusitis. Allergol Int 68:225–232

Yu Z, Tang J, Khare T, Kumar V (2020) The alarming antimicrobial resistance in ESKAPEE pathogens: can essential oils come to the rescue? Fitoterapia 140:104433

Zhang L, Liang E, Cheng Y, Mahmood T, Ge F, Zhou K, Bao M, Lv L, Li L, Yi J, Lu C, Tan Y (2020) Is combined medication with natural medicine a promising therapy for bacterial biofilm infection? Biomed Pharmacother 128:110184

Historical Medical Remedies as Potential Anti-Biofilm Approaches: Can We Look Back to Move Forward?

Nijamuddin Shaikh and Karishma S. Kaushik

Abstract While the study of biofilms has grown in the last few decades, evidence of human infections per se has existed since ancient times. In this chapter, we review descriptions of biofilm infections across ancient compendia and treatises and summarize traditional and historical medical remedies suggested and used for these infections, classifying them into plant, animal, and inanimate sources. Finally, we discuss possible considerations towards developing these historical remedies as contemporary anti-biofilm approaches.

1 Introduction

Biofilms are structured communities of microorganisms, often bacteria and fungi, observed as aggregates of microbial cells adhered to each other or a substratum (Costerton et al. 1999; Bjarnsholt et al. 2013; Høiby 2017). Most often observed as polymicrobial microbial consortia, biofilms have significant implications on human health, particularly with respect to chronic and recalcitrant infection states (Costerton et al. 1999; Burmølle et al. 2010; Wolcott et al. 2013; Malone et al. 2017). Biofilm formation starts with the adherence of planktonic (free-floating) microbes or smaller microbial aggregates that adhere to each other to form larger aggregates or to a biotic or abiotic surface. Following initial attachment, bacteria and fungi produce a thick layer of extracellular matrix and, in the presence of favourable environmental conditions, proceed to establish mature, heterogeneous biofilm structures (Costerton 1999; Kokare et al. 2009; Høiby et al. 2011).

The coinage of the term 'biofilm', and recognition of biofilms as the predominant form of microbial life, was pioneered by J. William (Bill) Costerton in 1985 (Costerton et al. 1995). This work, along with other researchers in the decades to follow, elucidated the role and importance of adherent bacteria in health and infection (Høiby 1974; Mclean et al. 2012; Yan and Bassler 2019). Since then, the

N. Shaikh · K. S. Kaushik (✉)
Department of Biotechnology, Savitribai Phule Pune University, Pune, Maharashtra, India
e-mail: karishmaskaushik@unipune.ac.in

K. Richter, K. Nørskov Kragh (eds.), *Antibiofilm Strategies*, Springer Series on Biofilms 11, https://doi.org/10.1007/978-3-031-10992-8_14

field has grown significantly, with biofilms observed in a range of clinical infection states such as chronic wounds, lung and sinus infections, periodontitis, ocular and ear infections, as well as device-related infections on catheters, prosthetic valves, and joint replacements, to name a few (Donlan 2001; Hall-Stoodley and Stoodley 2009; Rudkjøbing et al. 2012; Bjarnsholt 2013; Ramos-Gallardo 2016). Clinical biofilms are notoriously recalcitrant to conventional antibiotics and prompt the need for prolonged and repeated antibiotic usage, which further fuels the emergence of antibiotic resistance (Stewart and Costerton 2001). This, together with a void in the discovery of new antibiotics, has exacerbated the challenges of treatment options for biofilm infections and prompts the urgent need to explore non-conventional therapeutic approaches (Overbye and Barrett 2005; Stanton 2013).

While the clinical relevance of biofilm infections has been increasingly recognized in the past few decades, human infections have existed since antiquity (Pappas et al. 2008a; Achtman 2016). Scriptures and historical texts across the world include accounts of protracted fevers and cachexia, as well as descriptions of suppurative conditions such as tonsillitis, pustules, abscesses, cellulitis, wound infections, and styes (hordeolum) (Zimmermann 1978; Cameron 1982; Kansupada and Sassani 1997; Cunha 2007; Pappas et al. 2008a; Kadam et al. 2020). Further, traditional and historical systems of medicine are replete with recipes of remedies and preparations, often from natural sources, and their applications in these conditions (Holland 1994; Raza 2006). Today, it is well established that several of these infection states are associated with microbial (often bacterial) biofilms, which often contribute to their chronic state, and their poor response to antibiotic therapy. Given this, exploring historical medical remedies as potential anti-biofilm therapeutics could provide hitherto unexplored approaches for the treatment of biofilm infections. In this chapter, we review historical descriptions of select biofilm infections, summarize available descriptions of potential remedies into those from plant, animal, and inanimate sources, and discuss considerations towards developing them as contemporary anti-biofilm approaches.

2 Descriptions in Historical Texts of Common Infections Known to Be Associated with Biofilms

While microscopic organisms were discovered in the late seventeenth century, descriptions that point to 'tiny organisms', and resultant signs of infection, can be traced across historical texts dating back several centuries (Gest 2004; Shulman 2004; Pappas et al. 2008a).

Wound Infections The earliest record of wound management is found on a clay tablet dating 2200 B.C., which describes the practice of washing wounds, making wound dressings, and bandaging the wound (Shah 2011). In historical texts, wounds are most often described in the context of warfare, physical injury with axes, spears, and arrow piercings, animal and snake bites, or then as 'internal' or 'body wounds', likely referring to wounds associated with co-morbid conditions. Historical descriptions also include evidence of infection in wounds, as well as chronic or non-healing

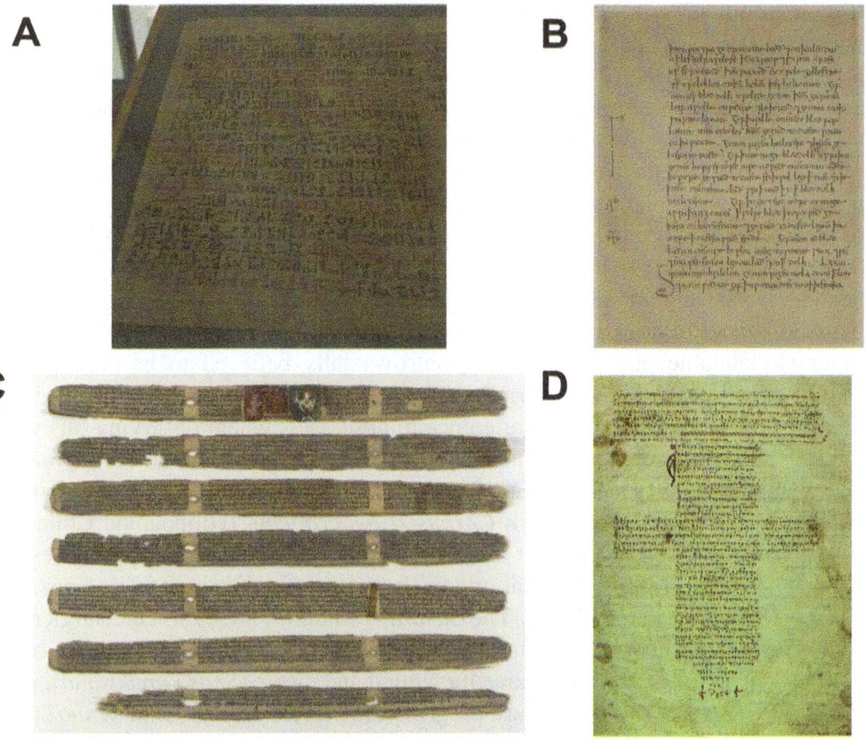

Fig. 1 Examples of historical texts known to have descriptions of infection states and remedies (**a**) Ebers' papyrus (1550 B.C.) from ancient Egypt (Reproduced from Ebers papyrus: Einsamer Schütze, CC BY-SA 3.0 https://creativecommons.org/licenses/by-sa/3.0,via Wikimedia Commons) (**b**) A facsimile page of Bald's Leechbook, an Anglo-Saxon medical text probably compiled in the ninth century (Original source: Cockayne, Oswald. 1865. Leechdoms, Wortcunning and Starcraft of Early England. London: Longman, Green, Longman, Roberts and Green. Frontispiece. Reproduced via Wikimedia Commons) (**c**) Sushruta Samhita, an Indian medical compendium written on palm leaves, displayed at the Los Angeles County Museum of Art (LACMA). Reproduced according to guidelines (https://www.lacma.org/).(**d**) Hippocratic Oath, an oath of ethics taken by physicians, written in Greek between the fifth and third centuries B.C. (Scan used in book: Foto de la Biblioteca Vaticanascan from book: User:Rmrfstar, Public domain, via Wikimedia Commons)

wounds (Donnelly 1998; Ovington 2002; Kadam et al. 2020). Today, it is well-known that the presence of microbial biofilms is a leading cause of impeded or arrested wound healing, contributing to the non-healing wound state. The Egyptian civilization left records of their practices in pictorial or written format, often through scribes, a particular group of people who would record and compile daily observations in the form of papyri (Fouly et al. 2012). Eight major medical papyri exist, among which Ebers' papyrus, dating back 1552 B.C., contains descriptions and treatments of 'sores and wounds' and 'burns' (Fig. 1) (von Klein 1905; Brandt-Rauf and Brandt-Rauf 1987). The medical text records of the renowned scribe and skilled physician Imhotep (around 2625 B.C.) were unearthed by Edwin Smith in 1862

(Edwin Smith papyrus) and describe several types of wound remedies using honey, willow leaves, acacia seeds, lint, and grease, as well as the use of raw meat to stop bleeding (Mukherjee 2010; Bhattacharya 2012; Gallin 2018). The Sumerian civilization (4500 B.C–1900 B.C.) of ancient Mesopotamia appears to have been familiar with managing infected wounds, with medical texts detailing a range of local prescriptions for wound care, including distilled compounds, washing agents, and preparations from plant sources (Gabriel 2011). In the series of books that form a part of Galenic Corpus, Galen of Pergamon (129–216 C.E.) details observations related to wound care, including the use of spice ointments and clay to promote healing. A vast amount of this knowledge was gleaned from his work as physician to the gladiators. Possibly the most well-known of these observations is related to the myth of 'laudable pus', which was controversially believed to be the sign of a healthy, healing wound (Freiberg 2017). However, a more nuanced interpretation likely indicates that this was in reference to abscesses, where observing pus would indicate the vital presence of drainage. A rational and evidence-based approach in Greek medicine is widely believed to have been introduced with the work of Hippocrates, in his seminal work *Corpus Hippocraticum* (Hippocratic Collection) (Pikoulis et al. 2004; Pappas et al. 2008b). In addition to other approaches, this compendium describes a surgical approach that uses a tin pipe to clear pus in an abscess (Broughton et al. 2006). The *Sushruta Samhita*, an Ayurvedic surgical text from the Indian subcontinent (treatise by Sushruta, fourth century A.D.), has an entire chapter describing procedures for wound and ulcers (*Shasti upakrama*), which mentions that certain wounds may progress to a 'bad' state (*Dushta-vrana*), if not appropriately cleaned. The cleansing of non-healing wounds using a decoction of medicinal herbs extracted in oil. The Ayurvedic Materia Medica (*Bhavaprakash Nighantu*) further details the benefits of these oil-based remedies, with specifications on types of plants and proportions to be used in the preparation (Shubhashree et al. 2020; Kadam et al. 2020). The medical health of the renowned Tudor, Henry VIII (1491–1547 C.E.) is recorded through official documents written to the government and contains descriptions of a purulent wound, following a jousting event, that turned into a foul-smelling, long-standing ulcer (Chalmers and Chaloner 2009). The wound was treated for several years, with Sloan MS 1047 containing records of medical prescriptions containing large amounts of lead (Charlton 2017).

Oral and Dental Infections Evidence of oral infections in ancient times can be observed via archaeological evidence; dental caries, for instance, have been confirmed in *Australopithecus*, the earliest hominid believed to have existed around 1.5 million years ago (Clark 1959; Napier and Napier 1985; Park and Yaacob 1991). Archaeological specimens also provide evidence of the role of diet in oral infections, with a major decline in carious lesions observed during 400–600 A.D. which was followed by an increase post 600 A.D., likely due to the production and consumption of refined flour and sugar (Park and Yaacob 1991). Palaeopathological examination of dental samples from ancient burial grounds provides insights into the consequences of serious forms of oral infections such as periodontitis. At St. Mary Graces cemetery, London, samples dating back 1350–1538 A.D. suggest an elevated risk of deaths among persons with oral infections, in comparison with unaffected counterparts during the late medieval period (DeWitte and Bekvalac 2010). In addition, genomic analysis of ancient dental pulps, teeth, and tartar has enabled the profiling of the oral microbiota among historical populations (Rascovan et al. 2016).

In seventeenth and eighteenth centuries, Anton Van Leeuwenhoek constructed simple microscopes and provided the first observational record of microorganisms in the form of 'animalcules' or 'small animals' in dental plaques (Toledo-Pereyra 2009; Seneviratne 2017). He described this as 'I saw an inconceivable number of exceedingly small animalcules and those of diverse sorts', likely referring to aggregated microorganisms in the sample from the 'scurf of teeth'. In more recent times, the association of biofilm aggregates, most often polymicrobial, in oral and dental pathologies such as caries, periodontitis, and endodontic infections is well established (Siqueira et al. 2010; Ricucci and Siqueira 2010; Colombo and Tanner 2019). Leeuwenhoek also observed that gargling with vinegar affected 'animalcules' present on the outside of the 'scurf' but did not pass through the entire substance of the scurf. These observations by Anton Van Leeuwenhoek are widely considered to be one of the oldest records of documented biofilms, and possibly of tolerance of microbes in biofilms to removal or treatments.

Eye Infections A range of ocular diseases have been reported since ancient times and are often clubbed under the broad description of 'ophthalmia' (Denholm and Hunt 2021). A well-known account of ophthalmia is that of General Hannibal's (217 B.C.) acute and painful eye condition that resulted following a long march through a swamp and led to a permanent loss of vision. From descriptions, the affliction appears to have been a form of keratitis, a corneal infection caused by a range of pathogens. Given that contact with marshy areas preceded the development of the infection, a water-borne ocular pathogen such as *Pseudomonas aeruginosa* remains a possibility (Denholm and Hunt 2021). Infections of the cornea (including contact-lens keratitis) are known to be associated with the presence of biofilms (Behlau and Gilmore 2008). Further down the ages, a medieval Anglo-Saxon treatise, Balds' Leechbook (tenth century), describes in detail the constituents of a potion to treat a 'wen', likely referring to a stye or infection of the eyelash follicle (hordeolum). Known as Balds' eyesalve, the recipe calls for a mixture of cropleac (Allium genus, including onion and leek), garlic, bile, and wine, among other constituents. The prescription also states that the active remedy is to be applied to the eye with a feather, likely referring to topical application at the site of the stye (Harrison et al. 2015).

Ear Infections Historical texts include several descriptions of ear diseases and a range of remedies, including from botanical and other natural sources. An archaeological examination of a skeleton from Spain, dating between the tenth and sixteenth century A.D., showed pathological changes due to otitis media and mastoiditis, presenting palaeopathological proof of ear infections (Olivé-Busom et al. 2021). King Francis II of France (1544–1560 A.D.) had been afflicted with adenoidal hypertrophy at a young age and later in life developed a brain abscess likely resulting from the spread of an ear infection (Srouji 2009).

In present times, ear infections such as acute and chronic otitis media (middle ear infection) are known to be biofilm-related infections (Bakaletz 2007; Akyıldız et al. 2013). In the tenth-century text Balds' Leechbook, 'sore ears' (possibly referring to the presence of an infection) are recommended to be treated with goat's gall (likely bile) mixed with cow's milk or equal quantities of buck, bull, or boar gall mixed with honey. Another remedy uses ram's bile mixed with the urine of the animal and butter (possibly for consistency) (Meaney 1984; Luft 2020). In parts of Northern Greece, acute otitis media has been reported to be treated by dropping human urine into the ear canal (Helidonis 1993). The antibacterial activity of urine is well documented and is likely due to biochemical components such as ammonia and urea, and these specific compounds have been characterized for their antimicrobial and anti-biofilm properties using modern-dat scientific methods (Kumar and Vadehra 1979; Richards et al. 2009; Kaushik et al. 2015; Torres et al. 2018; Jiang et al. 2020). The renowned French surgeon Ambroise Pare' (1510–1590 A.D.) suggested bull gall and iron ore boiled in strong vinegar followed by drying for the topical treatment of otorrhea (Srouji 2009) (Table 1).

3 Evaluation of Historical Remedies for Infections Associated with Biofilms

The therapeutic properties of medicinal plants and animal products has been leveraged for centuries; more recently, a range of compounds has been evaluated for antimicrobial properties, including potential anti-biofilm activity (Repetto and Llesuy 2002; Ahn et al. 2002; Pandey et al. 2016; Sánchez 2017; Lu et al. 2019; Süntar 2019; Valdivieso-Ugarte et al. 2019). While this has put forward several active ingredients, it is relevant to consider that historical remedies often focus on combinations of ingredients, as opposed to single constituents. This is important given that the activity of these remedies is based on several ingredients acting in combination with each other. Along these lines, in recent times, there has been a focus towards reconstituting 'whole' remedies, as opposed to evaluating single ingredients and chemical extracts (Harrison et al. 2015; Fuchs et al. 2018; Connelly et al. 2020; Kadam et al. 2020).

4 Plant-Based Remedies with Anti-Biofilm Potential

The tenth-century Anglo-Saxon text describes the preparation of a salve, recommended to treat a 'wen' or stye. The remedy consists of garlic, a second *Allium* plant (for example, onion or leek), wine, and ox gall, and the constituents

Table 1 Select historical remedies across different texts, time periods, and geographic regions with evidence of antimicrobial and anti-biofilm activity in the context of biofilm-relevant infections

Remedy	Text	Origin of text	Bioactivity
Bald's eyesalve (mixture of *Allium* spp., wine, oxgall, and other constituents). Stored in brass or bronze vessel.	Bald's Leechbook (tenth century A.D.)	Anglo-Saxon period	Antimicrobial (Harrison et al. 2015)
Turmeric (*Curcumae longa)*	Ebers papyrus (1500 B.C.), Sushruta Samhita (fourth century A.D.)	Ancient Egypt, ancient India	Wound healing, anti-inflammatory, antimicrobial (Aboelsoud 2010; Kandwal et al. 2015)
Saffron (*Crocus sativus*)	The canon of medicine/al-Qanun fi al-tibb (1025 A.D.)	Old Persia (modern day Iran)	Antimicrobial, antioxidant, anti-inflammatory (Hosseinzadeh and Nassiri-Asl 2013)
Topical ointment of egg yolk and turpentine	The Collected work of Ambriose Paré/Les Oeuvres d'Ambroise Paré (1575 A.D.)	Early modern France	Wound healing (Broughton et al. 2006)
Metals (copper, silver)	Sushruta Samhita (fourth century A.D.), Charaka Samhita (first century A.D.), Rasa Shastra (eighth century A.D.)	Ancient India	Antimicrobial (Ruddaraju et al. 2020)
Wine	Hippocratic collection/corpus hippocratum (between fifth and third centuries B.C.)	Ancient Greece	Antioxidant, anti-inflammatory (Guilford and Pezzuto 2011; Kleisiaris et al. 2014)
Maggot therapy	Research findings by William Baer based on Ambriose Paré and Dominique Jean Larrey's observation (1929)	Late modern France	Biological debridement, wound healing, anti-biofilm (Robinson 1940; Harris et al. 2009; Baer 2011; Manring and Calhoun 2011)
Holy basil or *tulsi (Ocimum tenuiflorum),* Bermuda grass or *durva (Cynodon dactylon), parnabeeja (Kalanchoe pinnata)* reconstituted in sesame oil	Sushruta Samhita (fourth century A.D.), Charaka Samhita (first century A.D.)	Ancient India	Antimicrobial, anti-biofilm (Kadam et al. 2020)

are mixed and stored in a brass or bronze container for 9 days. The remedy was reconstructed using modern ingredients and practices and evaluated against a range of bacterial pathogens, including *Staphylococcus aureus*, the bacterium it was likely designed to treat. The 1000-year-old remedy was capable of killing methicillin-resistant *S. aureus* (MRSA), *Acinetobacter baumannii*, *Staphylococcus epidermidis*, and *Streptococcus pyogenes*, and exhibited anti-biofilm activity in a synthetic biofilm model, soft-tissue wound biofilm model, and chronic wound mouse model. Notably, the effects were distinct from those observed with individual components, highlighting the importance of the combined activity of the ingredients (Harrison et al. 2015; Fuchs et al. 2018; Furner-Pardoe et al. 2020). More recently, the remedy is being evaluated in a clinical trial on healthy volunteers to study its safety profile on topical application to healthy human skin (www.isrctn.com/ISRCTN10773579).

In addition to specific remedy descriptions, the treatment of stubborn ulcers and suppurative lesions has been documented since ancient times (Forrest 1982). Hippocrates suggested 'sweet wine and much patience' for treating stubborn ulcers and recommended washing wounds with wine. Galen, the Roman philosopher and physician, reportedly noted a lack of suppuration in wounds treated with wine (Nikolova et al. 2018). The Nippur tablet from the Old Babylonian era also suggested the use of wine-containing ointments for skin diseases (Fehér et al. 2007). More recently, a fabricated co-polymer (TA561) containing tannic acid, inspired from the constituents of red wine, was evaluated for in vivo activity against methicillin-resistant *Staphylococcus aureus* (MRSA) infection (Chen et al. 2021), highlighting the possibility of distilling historical descriptions of treatments for contemporary evaluation.

Historical medical compendia from India, *Charaka Samhita* and *Sushruta Samhita,* have referenced plant-based treatments for chronic, non-healing wounds in chapters for wound management (*Vrana*) and chronic wounds (*Dushta-vrana*). *Sushruta Samhita* describes chronic wounds as 'pain, swelling and bleeding can be caused by germination of maggots or worms on the wound', alluding to a possible infected wound state (e-Samhita - National Institute of Indian Medical Heritage/ecaraka n.d.; e-Samhita - National Institute of Indian Medical Heritage/esushruta n. d.). The traditional remedies call for plant paste from leaves and grass of *Kalanchoe pinnata* ('Parnabeeja'), *Cynodon dactylon* (Bermuda grass), and *Ocimum tenuiflorum* (holy basil or 'Tulsi') to be reconstituted in sesame oil (KPE-SO, CDE-SO, and OTE-SP, respectively), to prepare the 'whole' remedy. The remedies were evaluated using standard biofilm assays, such as the crystal violet assay for biomass and the XTT assay for metabolic activity, against two common biofilm-forming wound pathogens *P. aeruginosa* and *S. aureus* (Kadam et al. 2020). The plant-based remedies had inhibitory effect on planktonic as well as biofilm modes of growth on *S. aureus* and *P. aeruginosa*, with differential effects on biomass formation and metabolic activity of biofilms. Evaluation of the effects of the remedies on biofilm formation and biofilm eradication indicated anti-biofilm potential, with KPE-SO, CDE-SO, and OTE-SE having an inhibitory effect on biomass and metabolic activity of *P. aeruginosa.* Although no effect was seen on *S. aureus* biomass, a reduction in metabolic activity was observed with KPE-SO and CDE-SO. An unexpected observation was the inherent antimicrobial activity of sesame oil, justifying its use as one of the key ingredients in the remedy. A reduction in the metabolic activity of pre-formed biofilms, in comparison to untreated biofilms of *P. aeruginosa* and *S. aureus*, was observed with KPE-SO, CDE-SO, and OTE-SO. The next step would be further evaluation of the remedies with mammalian cell toxicity assays and synthetic biofilm models that closely mimic the wound state. Taken together, the study presents a pipeline to evaluate traditional remedies for anti-biofilm potential, starting with mining historical treatise to evaluation with contemporary scientific assays and analyses (Kadam et al. 2020).

Another plant-based remedy, the maguey sap extracted from *Agave* species was traditionally used by the Aztecs for wound management. Laboratory analysis indicates that the sap is effective against pyogenic bacteria, and the bactericidal effect results in part from the saponins present in the sap. Further, the traditional addition of salt to the remedy was observed to enhance the efficacy of the remedy against *S. aureus* (Davidson and De Montellano 1983). Based on information in the public domain, the remedy does not seem to have been evaluated for specific anti-biofilm effects, which could be a valuable future line of investigation (Davidson and De Montellano 1983; Etkin 1986).

5 Animal-Based Remedies with Anti-Biofilm Potential

The role of historical animal-based remedies (zootherapy) as antimicrobial agents has been an understudied area for a range of reasons, including challenges with obtaining ingredients and ethical concerns with the use of animal products in research. While these are valid concerns, understanding the potential of animal products as antimicrobial and anti-biofilm approaches holds potential. Historically, a range of animal products such as tusk, bone, feathers, beak, eggshells, honey, wax, fat, flesh, skin, ear, tongue, teeth, insect cocoon, and blood have been used as ointments or protective layers (Alves and Alves 2011). In fact, in historical medicine, zootherapy is detailed rather extensively; with one study reporting that 15–20% of Ayurvedic (traditional Indian medicine) prescriptions are animal-based, nearly 42% of Ebers papyrus's prescriptions contain ingredients of animal origin, and over 1500 species of animals have been rendered as medicinally useful in Traditional Chinese medicine (TCM) (Sipos et al. 2004; Alves and Rosa 2005).

The use of maggots as a remedy for biological debridement and healing of wounds, also called larval therapy, has been considered since ancient times (Sherman et al. 2000; Sipos et al. 2004). The secretions of the maggot species *Phormia regina* was observed to contain high concentrations of ammonium bicarbonate, with therapeutic properties similar to other maggot products such as allantoin and urea (Robinson 1940). Maggots enable biological debridement of devitalized tissues and ingest bacteria in wounds, and thereby could promote wound infection clearance (Sipos et al. 2004; Kon and Rai 2014). Maggot therapy is divided into two types, depending on the application of larvae on the wounds. Free-range maggots are applied directly to the wounds. Bagged maggots, on the other hand, are enclosed in a polyvinyl netting, which is applied on the surface of the wound (Nigam 2021). A recent study of maggot debridement therapy on patients with diabetic foot ulcers reported complete debridement and subsequent healing (Siavash et al. 2020). The secretion system of maggots has been reported to contain enzymes capable of disrupting Staphylococcal biofilms (Harris et al. 2009). Along similar lines, sericin, a protein present in silk produced by silkworms, promotes fibroblast proliferation and collagen synthesis, and has antibacterial properties (Napavichayanun and Aramwit 2017). Silk was reportedly used by Galen, the famous Roman physician and philosopher, to stop the bleeding of vessels, and could possibly have also been used for its wound healing and antimicrobial effects (Broughton et al. 2006).

A recent study harvested cationic antimicrobial peptides (CAMPs) from the plasma of the American alligator, using a particle-assisted proteomics approach to extract peptides from small sample volumes. The newly identified peptides were synthesized and evaluated for antimicrobial activity against a range of bacterial pathogens. While additional work is warranted to evaluate their potential role as antimicrobial therapeutics, the approach provides 'proof of principle' for bioprospecting animal products for antimicrobial candidates (Bishop et al. 2015). In another study, a synthetic peptide was designed based on the histone-derived peptide identified in the *Komodo dragon* and evaluated for antimicrobial and anti-biofilm activity against *P. aeruginosa* and *S. aureus*. The synthetic peptide was observed to enhance wound healing in uninfected and biofilm infected wounds in a murine model, as well as promote clearance of infected wounds. In doing so, it displayed pathogen and host-directed activities through diverse mechanisms of action (Chung et al. 2017).

Honey is well-known for its antibacterial, antifungal, and anti-biofilm activities against a broad range of bacterial and fungal pathogens and is probably one of the best examples of the rediscovery of an ancient remedy (Anyanwu 2012; Roberts et al. 2015; Shirlaw et al. 2020). The human use of honey can be traced back nearly 8000 years, with depictions in Stone Age paintings (Eteraf-Oskouei and Najafi 2013). The activity of honey can be attributed to its low pH and osmolarity and the presence of peroxide and non-peroxide-based phytochemicals. It has been shown to prevent the attachment of planktonic bacterial cells, inhibit microbial processes such as quorum sensing, as well as possess tissue regenerative properties (Okhiria et al. 2009). Biocompatible wound dressings containing honey as a wound healing and antimicrobial agent, such as cellulose acetate (CA)—manuka honey (MH) nanofibrous mats and nanocellulose films coated with honey, are examples of modern-day approaches that incorporate an ancient remedy (Taher et al. 2020; Ullah et al. 2020).

6 Inanimate Material-Based Approaches with Anti-Biofilm Potential

Among inanimate materials used as remedies, metals as antimicrobial agents have been employed since historical times. Around 2000 B.C, copper was used as a topical wound disinfectant by the ancient Egyptians, and the Aztecs used it to treat skin conditions (Borkow and Gabbay 2012). In the 1800s, silver nitrate was topically used for ulcerations and infected wound dressings (Politano et al. 2013; Gao et al. 2018; Khansa et al. 2019) and continues to be used in wound care ointments and dressings. In recent times, the advent of nanotechnology has made targeted delivery of metals possible, with a reduction in systemic toxicity and improved local activity and efficacy for a range of applications (Lewis Oscar et al. 2015; Liu et al. 2019; Ruddaraju et al. 2020; Erkoc and Ulucan-Karnak 2021; Salah et al. 2021). Ancient records also demonstrate the use of natural clays as therapeutic agents (Carretero et al. 2013; de Gomes 2017; Davies and Behroozian 2020). The Egyptian physicians recommended the use of Nubian earth, and the Ebers papyrus (1600 BC) explains the use of medicinal clays for curative purposes. In present times, natural clays have been evaluated for sustained drug release and broad-spectrum antibacterial activity (de Rodrigues et al. 2013).

7 Conclusions

Historical medical treatises and remedies provide a vast set of descriptions of antimicrobial approaches, and could serve as untapped source of antimicrobial and anti-biofilm approaches. Prospecting these remedies involves several considerations, but holds substantial potential towards developing hitherto unknown treatment approaches (Kadam et al. 2020). Given this, it is both worthwhile and prudent to broaden our search for novel antimicrobial and anti-biofilm treatments by looking back at historical approaches, and bring them towards contemporary scientific evaluation.

Conflict of Interest The authors declare no professional or financial conflict of interest related to this work.

Funding The appointments of NS and KSK are funded by the Ramalingaswami Re-entry Fellowship, Department of Biotechnology, Government of India (BT/HRD/2005 to KSK).

References

Aboelsoud NH (2010) Herbal medicine in ancient Egypt. J Med Plants Res 4:082–086. https://doi.org/10.5897/JMPR09.013

Achtman M (2016) How old are bacterial pathogens? Proc R Soc B Biol Sci 283:1836. https://doi.org/10.1098/RSPB.2016.0990

Ahn J, Grün IU, Fernando LN (2002) Antioxidant properties of natural plant extracts containing polyphenolic compounds in cooked ground beef. J Food Sci 67:1364–1369. https://doi.org/10.1111/J.1365-2621.2002.TB10290.X

Akyıldız İ, Take G, Uygur K et al (2013) Bacterial biofilm formation in the middle-ear mucosa of chronic otitis media patients. Indian J Otolaryngol Head Neck Surg 65:557. https://doi.org/10.1007/S12070-012-0513-X

Alves R, Alves H (2011) The faunal drugstore: animal-based remedies used in traditional medicines in Latin America. J Ethnobiol Ethnomed 7:1–43. https://doi.org/10.1186/1746-4269-7-9

Alves R, Rosa I (2005) Why study the use of animal products in traditional medicines? J Ethnobiol Ethnomed 1:1–5. https://doi.org/10.1186/1746-4269-1-5
Anyanwu C (2012) Investigation of in vitro antifungal activity of honey. J Med Plants Res 6:3512–3516. https://doi.org/10.5897/JMPR12.577
Baer WS (2011) The classic: the treatment of chronic osteomyelitis with the maggot (larva of the blow fly). Clin Orthop Relat Res 469:920–944. https://doi.org/10.1007/s11999-010-1416-3
Bakaletz L (2007) Bacterial biofilms in otitis media: evidence and relevance. Pediatr Infect Dis J 26(10):17–19. https://doi.org/10.1097/INF.0B013E318154B273
Behlau I, Gilmore MS (2008) Microbial biofilms in ophthalmology and infectious disease. Arch Ophthalmol 126:1572–1581. https://doi.org/10.1001/ARCHOPHT.126.11.1572
Bhattacharya S (2012) Wound healing through the ages. Indian J Plast Surg 45:177. https://doi.org/10.4103/0970-0358.101255
Bishop BM, Juba ML, Devine MC et al (2015) Bioprospecting the American alligator (*Alligator mississippiensis*) host defense peptidome. PLoS One 10:e0117394. https://doi.org/10.1371/JOURNAL.PONE.0117394
Bjarnsholt T (2013) The role of bacterial biofilms in chronic infections. APMIS Suppl 1:1–51. https://doi.org/10.1111/apm.12099
Bjarnsholt T, Alhede M, Alhede M et al (2013) The in vivo biofilm. Trends Microbiol 21:466–474. https://doi.org/10.1016/j.tim.2013.06.002
Borkow G, Gabbay J (2012) Copper, an ancient remedy returning to fight microbial, fungal and viral infections. Curr Chem Biol 3:272–278. https://doi.org/10.2174/2212796810903030272
Brandt-Rauf PW, Brandt-Rauf SI (1987) History of occupational medicine: relevance of Imhotep and the Edwin Smith papyrus. Br J Ind Med 44:68. https://doi.org/10.1136/OEM.44.1.68
Broughton G, Janis JE, Attinger CE (2006) A brief history of wound care. Plast Reconstr Surg 117(17):6S–11S. https://doi.org/10.1097/01.PRS.0000225429.76355.DD
Burmølle M, Thomsen TR, Fazli M et al (2010) Biofilms in chronic infections - a matter of opportunity–monospecies biofilms in multispecies infections. FEMS Immunol Med Microbiol 59:324–336. https://doi.org/10.1111/j.1574-695X.2010.00714.x
Cameron ML (1982) The sources of medical knowledge in Anglo-Saxon England. Anglo Sax Engl 11:135–155. https://doi.org/10.1017/S0263675100002593
Carretero MI, Gomes CSF, Tateo F (2013) Clays, drugs, and human health. Dev Clay Sci 5:711–764. https://doi.org/10.1016/B978-0-08-098259-5.00025-1
Chalmers C, Chaloner E (2009) 500 years later: Henry VIII, leg ulcers and the course of history. J R Soc Med 102:513–517. https://doi.org/10.1258/JRSM.2009.090286
Charlton A (2017) A hypothesis: King Henry VIII's (1491–1547) personality change: a case of lead poisoning? J Med Biogr 25:72–80. https://doi.org/10.1177/0967772017694571
Chen C, Yang X, Li S-j et al (2021) Red wine-inspired tannic acid–KH561 copolymer: its adhesive properties and its application in wound healing. RSC Adv 11:5182–5191. https://doi.org/10.1039/D0RA07342C
Chung EMC, Dean SN, Propst CN et al (2017) Komodo dragon-inspired synthetic peptide DRGN-1 promotes wound-healing of a mixed-biofilm infected wound. npj Biofilms Microbiomes 3:1–13. https://doi.org/10.1038/s41522-017-0017-2
Clark W (1959) The antecedents of man; an introduction to the evolution of the primates. University Press, Edinburgh
Colombo APV, Tanner ACR (2019) The role of bacterial biofilms in dental caries and periodontal and Peri-implant diseases: a historical perspective. J Dent Res 98:373–385. https://doi.org/10.1177/0022034519830686
Connelly E, Del Genio CI, Harrison F (2020) Data mining a medieval medical text reveals patterns in ingredient choice that reflect biological activity against infectious agents. MBio 11(1):211. https://doi.org/10.1128/MBIO.03136-19
Costerton J (1999) Introduction to biofilm. Int J Antimicrob Agents 11:217–221. https://doi.org/10.1016/S0924-8579(99)00018-7

Costerton JW, Lewandowski Z, Caldwell DE et al (1995) Microbial biofilms. Annu Rev Microbiol 49:711–745. https://doi.org/10.1146/ANNUREV.MI.49.100195.003431
Costerton J, Steward P, Greenberg E (1999) Bacterial biofilms: a common cause of persistent infections. Science 284:1318–1322. https://doi.org/10.1126/SCIENCE.284.5418.1318
Cunha CB (2007) Prolonged and perplexing fevers in antiquity: malaria and typhoid fever. Infect Dis Clin N Am 21:857–866. https://doi.org/10.1016/J.IDC.2007.08.010
Davidson JR, De Montellano BRO (1983) The antibacterial properties of an Aztec wound remedy. J Ethnopharmacol 8:149–161. https://doi.org/10.1016/0378-8741(83)90051-X
Davies J, Behroozian S (2020) An ancient solution to a modern problem. Mol Microbiol 113:546–549. https://doi.org/10.1111/MMI.14481
de Gomes CSF (2017) Healing and edible clays: a review of basic concepts, benefits and risks. Environ Geochemistry Heal 405(40):1739–1765. https://doi.org/10.1007/S10653-016-9903-4
de Rodrigues LA, Figueiras A, Veiga F et al (2013) The systems containing clays and clay minerals from modified drug release: a review. Colloids Surfaces B Biointerfaces 103:642–651. https://doi.org/10.1016/J.COLSURFB.2012.10.068
Denholm JT, Hunt PN (2021) Hannibal's Ophthalmia–a new answer to an ancient question. Emerg Infect Dis 27:223–225. https://doi.org/10.3201/EID2701.191696
DeWitte SN, Bekvalac J (2010) Oral health and frailty in the medieval English cemetery of St Mary Graces. Am J Phys Anthropol 142:341. https://doi.org/10.1002/AJPA.21228
Donlan RM (2001) Biofilms and device-associated infections. Emerg Infect Dis 7:277. https://doi.org/10.3201/EID0702.010226
Donnelly J (1998) Wound healing–from poultices to maggots. (a short synopsis of wound healing throughout the ages). Ulster Med J 67(1):47–51
Erkoc P, Ulucan-Karnak F (2021) Nanotechnology-based antimicrobial and antiviral surface coating strategies. Prosthes 3:25–52. https://doi.org/10.3390/PROSTHESIS3010005
Eteraf-Oskouei T, Najafi M (2013) Traditional and modern uses of natural honey in human diseases: a review. Iran J Basic Med Sci 16:731
Etkin N (1986) Plants and indigenous medicine and diet: Biobehavioral approaches. Routledge, New York
Fehér J, Lengyel G, Lugasi A (2007) The cultural history of wine - theoretical background to wine therapy. Cent Eur J Med 24(2):379–391. https://doi.org/10.2478/S11536-007-0048-9
Forrest RD (1982) Early history of wound treatment. J R Soc Med 75:1
Fouly H, McCool WF, Koucoi J (2012) Ancient Egyptian Women's health care in relation to modern women's health care practices: an overview. Int J Childbirth 2:269–276. https://doi.org/10.1891/0886-6708.2.4.269
Freiberg JA (2017) The mythos of laudable pus along with an explanation for its origin. J Community Hosp Intern Med Perspect 7:196–198. https://doi.org/10.1080/20009666.2017.1343077
Fuchs AL, Weaver AJ Jr, Tripet BP et al (2018) Characterization of the antibacterial activity of Bald's eyesalve against drug resistant *Staphylococcus aureus* and *Pseudomonas aeruginosa*. PLoS One 13:e0208108. https://doi.org/10.1371/JOURNAL.PONE.0208108
Furner-Pardoe J, Anonye BO, Cain R et al (2020) Anti-biofilm efficacy of a medieval treatment for bacterial infection requires the combination of multiple ingredients. Sci Reports 101(10):1–14. https://doi.org/10.1038/s41598-020-69273-8
Gabriel RA (2011) Man and wound in the ancient world : a history of military medicine from Sumer to the fall of Constantinople. Potomac Books, Washington DC
Gallin JI (2018) A historical perspective on clinical research. Princ Pract Clin Res 1:1–15. https://doi.org/10.1016/B978-0-12-849905-4.00001-0
Gao SS, Zhao IS, Duffin S et al (2018) Revitalising silver nitrate for caries management. Int J Environ Res Public Heal 15:80. https://doi.org/10.3390/IJERPH15010080
Gest H (2004) The discovery of microorganisms by Robert Hooke and Antoni Van Leeuwenhoek, fellows of the Royal Society. Notes Rec R Soc Lond 58:187–201. https://doi.org/10.1098/RSNR.2004.0055

Guilford JM, Pezzuto JM (2011) Wine and health: a review. Am J Enol Vitic 4:471–486. https://doi.org/10.5344/ajev.2011.11013

Hall-Stoodley L, Stoodley P (2009) Evolving concepts in biofilm infections. Cell Microbiol 11: 1034–1043. https://doi.org/10.1111/J.1462-5822.2009.01323.X

Harris L, Bexfield A, Nigam Y et al (2009) Disruption of Staphylococcus epidermidis biofilms by medicinal maggot Lucilia sericata excretions/secretions. Int J Artif Organs 32:555–564. https://doi.org/10.1177/039139880903200904

Harrison F, Roberts AEL, Gabrilska R et al (2015) A 1,000-year-old antimicrobial remedy with antistaphylococcal activity. MBio 6:9. https://doi.org/10.1128/MBIO.01129-15

Helidonis ES (1993) The history of otolaryngology from ancient to modern times. Am J Otolaryngol 14(6):382–393

Heritage e-S-NI of I e-Samhita (n.d.) National Institute of Indian Medical Heritage

Heritage e-S-NI of IM e-Samhita (n.d.) National Institute of Indian Medical Heritage

Høiby N (1974) Pseudomonas Aeruginosa infection in cystic fibrosis. Relationship between mucoid strains of Pseudomonas aeruginosa and the humoral immune response. Acta Pathol Microbiol Scand Sect B Microbiol Immunol 82B:551–558. https://doi.org/10.1111/j.1699-0463.1974.tb02365.x

Høiby N (2017) A short history of microbial biofilms and biofilm infections. APMIS 125:272–275. https://doi.org/10.1111/apm.12686

Høiby N, Ciofu O, Johansen HK et al (2011) The clinical impact of bacterial biofilms. Int J Oral Sci 3:55–65. https://doi.org/10.4248/IJOS11026

Holland B (1994) Prospecting for drugs in ancient texts. Nature 369:702–702. https://doi.org/10.1038/369702A0

Hosseinzadeh H, Nassiri-Asl M (2013) Avicenna's (Ibn Sina) the canon of medicine and saffron (Crocus sativus): a review. Phyther Res 27:475–483. https://doi.org/10.1002/ptr.4784

Jiang Y, Geng M, Bai L (2020) Targeting biofilms therapy: current research strategies and development hurdles. Microorganisms 8:1–34. https://doi.org/10.3390/microorganisms8081222

Kadam S, Madhusoodhanan V, Bandgar A, Kaushik KS (2020) From treatise to test: evaluating traditional remedies for anti-biofilm potential. Front Pharmacol 11:1669. https://doi.org/10.3389/FPHAR.2020.566334

Kandwal A, Mamgain R, Mamgain P (2015) Comparative evaluation of turmeric gel with 2% chlorhexidine gluconate gel for treatment of plaque induced gingivitis: a randomized controlled clinical trial. AYU (An Int Q J Res Ayurveda) 36:145. https://doi.org/10.4103/0974-8520.175537

Kansupada KB, Sassani JW (1997) Sushruta: the father of Indian surgery and ophthalmology. Doc Ophthalmol 931(93):159–167. https://doi.org/10.1007/BF02569056

Kaushik KS, Ratnayeke N, Katira P, Gordon VD (2015) The spatial profiles and metabolic capabilities of microbial populations impact the growth of antibiotic-resistant mutants. J R Soc Interface 12(107):18. https://doi.org/10.1098/rsif.2015.0018

Khansa I, Schoenbrunner AR, Kraft CT, Janis JE (2019) Silver in wound care - friend or foe?: a comprehensive review. Plast Reconstr Surg - Glob Open 7:1–10. https://doi.org/10.1097/GOX.0000000000002390

Kleisiaris CF, Sfakianakis C, Papathanasiou IV (2014) Health care practices in ancient Greece: the hippocratic ideal. J Med Ethics Hist Med 7(1):6

Kokare CR, Chakraborty S, Khopade AN, Mahadik KR (2009) NOPR: biofilm: importance and applications

Kon K, Rai M (2014) Natural remedies for the treatment of wounds and wound infection. In: Kon KV, Rai M (eds) Microbiology for surgical infections: diagnosis, prognosis, and treatment. Elsevier, Amsterdam, pp 187–203. https://doi.org/10.1016/B978-0-12-411629-0.00011-8

Kumar R, Vadehra DV (1979) Antibacterial activity of human urine. Indian J Exp Biol 17:808–810

Lewis Oscar F, Mubarakali D, Nithya C et al (2015) One pot synthesis and anti-biofilm potential of copper nanoparticles (CuNPs) against clinical strains of Pseudomonas aeruginosa. Biofouling 31:379–391. https://doi.org/10.1080/08927014.2015.1048686

Liu Y, Shi L, Su L et al (2019) Nanotechnology-based antimicrobials and delivery systems for biofilm-infection control. Chem Soc Rev 48:428–446. https://doi.org/10.1039/C7CS00807D

Lu L, Hu W, Tian Z et al (2019) Developing natural products as potential anti-biofilm agents. Chinese Med 141(14):1–17. https://doi.org/10.1186/S13020-019-0232-2

Luft D (2020) Medieval welsh medical texts: Volume One: The Recipes. University of Wales Press, Cardiff

Malone M, Bjarnsholt T, McBain AJ et al (2017) The prevalence of biofilms in chronic wounds: a systematic review and meta-analysis of published data. J Wound Care 26:20–25. https://doi.org/10.12968/jowc.2017.26.1.20

Manring MM, Calhoun JH (2011) Biographical sketch: William S. Baer (1872-1931). Clin Orthop Relat Res 469:917–919. https://doi.org/10.1007/s11999-010-1415-4

Mclean RJC, Lam JS, Graham LL (2012) Training the biofilm generation–a tribute to J. W. Costerton. J Bacteriol 194:6706–6711. https://doi.org/10.1128/JB.01252-12

Meaney AL (1984) Variant versions of old English medical remedies and the compilation of Bald's Leechbook. Anglo Sax Engl 13:235–268. https://doi.org/10.1017/S0263675100003604

Mukherjee S (2010) The Emperor of all Maladies: a biography of cancer. Lions, London

Napavichayanun S, Aramwit P (2017) Effect of animal products and extracts on wound healing promotion in topical applications: a review. J Biomater Sci Polym Ed 28:703–729. https://doi.org/10.1080/09205063.2017.1301772

Napier JR, Napier PH (1985) The natural history of the primates. Zoo Biol 6:195–199. https://doi.org/10.1002/zoo.1430060210

Nigam Y (2021) The principles of maggot therapy and its role in contemporary wound care. Nursing Times

Nikolova P, Stoyanov Z, Doncheva D, Trendafilova S (2018) Wine as a medicine in ancient times. Scr Sci Pharm 5:14. https://doi.org/10.14748/SSP.V5I2.5610

Okhiria OA, Henriques AFM, Burton NF et al (2009) Honey modulates biofilms of Pseudomonas aeruginosa in a time and dose dependent manner. J ApiProduct ApiMedical Sci 1:6–10. https://doi.org/10.3896/IBRA.4.01.1.03

Olivé-Busom J, López-Costas O, Quer-Agustí M et al (2021) Evidence of otitis media and mastoiditis in a medieval Islamic skeleton from Spain and possible implications for ancient surgical treatment of the condition. Int J Paleopathol 32:17–22. https://doi.org/10.1016/J.IJPP.2020.10.009

Overbye KM, Barrett JF (2005) Antibiotics: where did we go wrong? Drug Discov Today 10:45–52. https://doi.org/10.1016/S1359-6446(04)03285-4

Ovington L (2002) The evolution of wound management: ancient origins and advances of the past 20 years. Home Healthc Nurse 20:652–656. https://doi.org/10.1097/00004045-200210000-00009

Pandey S, Cabot PJ, Shaw PN, Hewavitharana AK (2016) Anti-inflammatory and immunomodulatory properties of Carica papaya. J Immunotoxicol 13:590–602. https://doi.org/10.3109/1547691X.2016.1149528

Pappas G, Kiriaze IJ, Falagas ME (2008a) Insights into infectious disease in the era of Hippocrates. Int J Infect Dis 12:347–350. https://doi.org/10.1016/J.IJID.2007.11.003

Pappas G, Kiriaze IJ, Falagas ME (2008b) Insights into infectious disease in the era of Hippocrates. Int J Infect Dis 12:347–350. https://doi.org/10.1016/J.IJID.2007.11.003

Park AW, Yaacob HB (1991) The ancient origins of oral pathology. J Nihon Univ Sch Dent 33: 211–243. https://doi.org/10.2334/JOSNUSD1959.33.211

Pikoulis EA, Petropoulos J, Tsigris C, Pikoulis N et al (2004) Trauma management in ancient Greece: value of surgical principles through the years. World J Surg 28:425–430. https://doi.org/10.1007/S00268-003-6931-X

Politano AD, Campbell KT, Rosenberger LH, Sawyer RG (2013) Use of silver in the prevention and treatment of infections: silver review. Surg Infect 14:8. https://doi.org/10.1089/SUR.2011.097

Ramos-Gallardo G (2016) Chronic wounds in burn injury: a case report on importance of biofilms. World J Plast Surg 5:175

Rascovan N, Huynh H, Chouin G et al (2016) Tracing back ancient oral microbiomes and oral pathogens using dental pulps from ancient teeth. npj Biofilms Microbiomes 2:1–4. https://doi.org/10.1038/s41522-016-0008-8

Raza M (2006) A role for physicians in ethnopharmacology and drug discovery. J Ethnopharmacol 104:297–301. https://doi.org/10.1016/J.JEP.2006.01.007

Repetto MG, Llesuy SF (2002) Antioxidant properties of natural compounds used in popular medicine for gastric ulcers. Braz J Med Biol Res 35:2002

Richards JJ, Reyes S, Stowe SD et al (2009) Amide isosteres of oroidin: assessment of antibiofilm activity and C. elegans toxicity. J Med Chem 52:4582–4585. https://doi.org/10.1021/JM900378S/SUPPL_FILE/JM900378S_SI_001.PDF

Ricucci D, Siqueira JF (2010) Biofilms and apical periodontitis: study of prevalence and association with clinical and histopathologic findings. J Endod 36:1277–1288. https://doi.org/10.1016/J.JOEN.2010.04.007

Roberts AEL, Brown HL, Jenkins RE (2015) On the antibacterial effects of manuka honey: mechanistic insights. Res Rep Biol 6:215–224. https://doi.org/10.2147/RRB.S75754

Robinson W (1940) Ammonium bicarbonate secreted by surgical maggots stimulates healing in purulent wounds. Am J Surg 47:111–115. https://doi.org/10.1016/S0002-9610(40)90125-8

Ruddaraju L, Pammi S, Guntuku G et al (2020) A review on anti-bacterials to combat resistance: from ancient era of plants and metals to present and future perspectives of green nano technological combinations. Asian J Pharm Sci 15:42–59. https://doi.org/10.1016/J.AJPS.2019.03.002

Rudkjøbing VB, Thomsen TR, Alhede M et al (2012) The microorganisms in chronically infected end-stage and non-end-stage cystic fibrosis patients. FEMS Immunol Med Microbiol 65:236–244. https://doi.org/10.1111/j.1574-695X.2011.00925.x

Salah I, Parkin IP, Allan E (2021) Copper as an antimicrobial agent: recent advances. RSC Adv 11: 18179–18186. https://doi.org/10.1039/D1RA02149D

Sánchez C (2017) Reactive oxygen species and antioxidant properties from mushrooms. Synth Syst Biotechnol 2:13–22. https://doi.org/10.1016/J.SYNBIO.2016.12.001

Seneviratne CJ (2017) Microbial biofilms: omics biology, antimicrobials and clinical implications. CRC Press, Boca Raton

Shah JB (2011) The history of wound care. J Am Col Certif Wound Spec 3:65. https://doi.org/10.1016/J.JCWS.2012.04.002

Sherman RA, Hall M, Thomas S (2000) Medicinal maggots: an ancient remedy for some contemporary afflictions. Annu Rev Entomol 45:55–81. https://doi.org/10.1146/ANNUREV.ENTO.45.1.55

Shirlaw O, Billah Z, Attar B et al (2020) Antibiofilm activity of heather and Manuka honeys and Antivirulence potential of some of their constituents on the DsbA1 enzyme of Pseudomonas aeruginosa. Antibiotics 9:911. https://doi.org/10.3390/ANTIBIOTICS9120911

Shubhashree M, Naik R, Doddamani SH, Bhavya B (2020) Preventive strategies to combat infections–a review of traditional practices and Ayurveda concepts. Int J Altern Complement Med 13(3):125–129. https://doi.org/10.15406/ijcam.2020.13.00505

Shulman ST (2004) The history of pediatric infectious diseases. Pediatr Res 551(55):163–176. https://doi.org/10.1203/01.pdr.0000101756.93542.09

Siavash M, Najjarnezhad A, Mohseni N et al (2020) Efficacy of maggot debridement therapy on refractory atypical diabetic foot ulcers: an open-label study. Int J Low Extrem Wounds 20(4):315–320. https://doi.org/10.1177/1534734620920403

Sipos P, Győry H, Hagymási K et al (2004) Special wound healing methods used in ancient Egypt and the mythological background. World J Surg 282(28):211–216. https://doi.org/10.1007/S00268-003-7073-X

Siqueira JF, Rôças IN, Ricucci D (2010) Biofilms in endodontic infection. Endod Top 22:33–49. https://doi.org/10.1111/J.1601-1546.2012.00279.X

Srouji I (2009) Did one ear infection in France change the history of Britain? The illness and death of Francis II (1544-60). J Med Biogr 17:231–234. https://doi.org/10.1258/JMB.2009.009028

Stanton TB (2013) A call for antibiotic alternatives research. Trends Microbiol 21:111–113. https://doi.org/10.1016/J.TIM.2012.11.002

Stewart PS, Costerton JW (2001) Antibiotic resistance of bacteria in biofilms. Lancet 358:135–138. https://doi.org/10.1016/S0140-6736(01)05321-1

Süntar I (2019) Importance of ethnopharmacological studies in drug discovery: role of medicinal plants. Phytochem Rev 195(19):1199–1209. https://doi.org/10.1007/S11101-019-09629-9

Taher MA, Zahan KA, Rajaie MA et al (2020) Nanocellulose as drug delivery system for honey as antimicrobial wound dressing. Mater Today Proc 31:14–17. https://doi.org/10.1016/J.MATPR.2020.01.076

Toledo-Pereyra LH (2009) The strange little animals of Antony van Leeuwenhoek surgical revolution. J Investig Surg 22:4–8. https://doi.org/10.1080/08941930802634439

Torres NS, Montelongo-Jauregui D, Abercrombie JJ et al (2018) Antimicrobial and antibiofilm activity of synergistic combinations of a commercially available small compound library with colistin against Pseudomonas aeruginosa. Front Microbiol 9:2541. https://doi.org/10.3389/fmicb.2018.02541

Ullah A, Ullah S, Khan MQ et al (2020) Manuka honey incorporated cellulose acetate nanofibrous mats: fabrication and in vitro evaluation as a potential wound dressing. Int J Biol Macromol 155:479–489. https://doi.org/10.1016/J.IJBIOMAC.2020.03.237

Valdivieso-Ugarte M, Gomez-Llorente C, Plaza-Díaz J, Gil Á (2019) Antimicrobial, antioxidant, and immunomodulatory properties of essential oils: a systematic review. Nutrients 11:2786. https://doi.org/10.3390/NU11112786

von Klein CH (1905) The medical features of the Papyrus Ebers. J Am Med Assoc XLV:1928–1935. https://doi.org/10.1001/JAMA.1905.52510260014001E

Wolcott R, Costerton JW, Raoult D, Cutler SJ (2013) The polymicrobial nature of biofilm infection. Clin Microbiol Infect 19:107–112. https://doi.org/10.1111/J.1469-0691.2012.04001.X

www.isrctn.com/ISRCTN10773579 ISRCTN - ISRCTN10773579: Assessment of the safety of a "medieval antibiotic" used to treat infected diabetic ulcers on the skin of healthy non-diabetic volunteers

Yan J, Bassler BL (2019) Surviving as a community: antibiotic tolerance and persistence in bacterial biofilms. Cell Host Microbe 26:15–21. https://doi.org/10.1016/j.chom.2019.06.002

Zimmermann F (1978) From classic texts to learned practice: methodological remarks on the study of Indian medicine. Soc Sci Med Part B Med Anthropol 12:97–103. https://doi.org/10.1016/0160-7987(78)90016-9

Part III
Biofilm Control

The Action of Phytochemicals in the Control of Pathogenic Biofilms

Mariana Sousa, Inês B. Gomes, Lúcia C. Simões, Manuel Simões, and Marta Ribeiro

Abstract The treatment of bacterial biofilms has been progressively troubled due to increasing antibiotic resistance. Biofilms exacerbate the fight against infections as they provide a protective environment for microbial cells, hindering the penetration of antimicrobial agents and favoring the uptake of elements necessary for cell survival like water, oxygen, and nutrients. Indeed, many first-line antimicrobial agents have become ineffective in treating biofilm-related infections, instigating the search for new antimicrobial agents. Natural products, particularly plant-derived compounds known as phytochemicals, have been shown to be effective, with an excellent broad-spectrum antibacterial profile, even against drug-resistant bacteria. This chapter addresses the main concepts of biofilm-associated mechanisms that promote bacterial survival. Moreover, different phytochemicals are described for biofilm prevention and control, correlating the mode of action of phytochemicals to the inhibition/alteration of virulence factors and other biofilm mechanisms. The scientific evidence of a wide variety of phytochemicals being described in this chapter should support future efforts to fast-track in vitro research to clinical applications to fight biofilm-related infectious diseases.

M. Sousa · I. B. Gomes · M. Simões (✉)
LEPABE—Laboratory for Process Engineering, Environment, Biotechnology and Energy, Faculty of Engineering, University of Porto, Porto, Portugal
e-mail: mvs@fe.up.pt

L. C. Simões
CEB—Centre of Biological Engineering, University of Minho, Braga, Portugal

LABBELS—Associate Laboratory, Guimarães, Braga, Portugal

M. Ribeiro
LEPABE—Laboratory for Process Engineering, Environment, Biotechnology and Energy, Faculty of Engineering, University of Porto, Porto, Portugal

CIQUP/Department of Chemistry and Biochemistry, Faculty of Sciences, University of Porto, Porto, Portugal

K. Richter, K. Nørskov Kragh (eds.), *Antibiofilm Strategies*, Springer Series on Biofilms 11, https://doi.org/10.1007/978-3-031-10992-8_15

1 Introduction

Since their discovery, antibiotics have been essential for the treatment and prevention of diseases, allowing many previously fatal infections to be treated or controlled (Nigam et al. 2014; Allen et al. 2019; Micoli et al. 2021). Nevertheless, the improper use and over-prescription of antibiotics, poor hygiene and sanitation habits, poor infection control, and the absence of new antibiotics led to a rapid emergence and spread of resistant microbial strains (Allen et al. 2019; Liu et al. 2021; Hawkings et al. 2007; Ferri et al. 2017; Barbieri et al. 2017). In fact, antibiotic resistance is one of the greatest threats to public health, limiting the prevention and treatment of a large range of previously treatable infectious diseases (Micoli et al. 2021; Jindal et al. 2015; Monte et al. 2014; Levy and Marshall 2004). Antimicrobial resistance (AMR) is a natural result of the adaptation of infectious pathogens to antibiotics used in a variety of fields. Moreover, the economic burden associated with these multidrug-resistant bacteria is massive (Monte et al. 2014; Prestinaci et al. 2015). Since antibiotic resistance is multifactorial, the WHO recommends a series of coordinated actions, from preventing the spread of the infection to promoting the research and development of new treatment strategies (Barbieri et al. 2017; Neu 1992; Ventola 2015). In addition, microorganisms may live in biofilms, which consist of aggregates of microbial cells attached to biotic or abiotic surfaces and embedded in a self-produced matrix of extracellular polymeric substances that protect cells from environmental stressors. It is well-known that the acquisition of resistance to antibiotics by bacteria is facilitated when they are present in biofilms (de Carvalho 2018; Yin et al. 2019; Kostakioti et al. 2013; Bridier et al. 2011). Biofilms are linked to serious and difficult-to-treat infections and combating them can be unmanageable or requires high doses of antibiotics (Llor and Bjerrum 2014; Frieri et al. 2017). Bacteria in biofilms can be up to 1000 times more tolerant to antibiotics than free-floating bacteria, which severely limits treatment options (Potera 2010; Sharma et al. 2019; Venkatesan et al. 2015).

To find novel antimicrobial agents distinct from antibiotics, plants have been explored as a source of compounds that show promising effectiveness against a variety of organisms, including fungi, yeasts, bacteria, and viruses (Monte et al. 2014; Wintola and Afolayan 2015; Cowan 1999). Phytochemicals are non-nutritive plant secondary metabolites with biological activity (Huang et al. 2016; Jimenez-Garcia et al. 2018; Liu 2013; Arendt and Zannini 2013; Diep et al. 2015). Different parts of plants from several species, such as leaves, roots, fruits, seeds, barks, stem bark, and flowers, are rich in different classes of phytochemicals, which may be a source of effective, inexpensive, and safe antimicrobial agents (Barbieri et al. 2017; Aung et al. 2020; Chouhan et al. 2017). They have been shown to inhibit peptidoglycan synthesis, destroy the structure of microbial membranes, change the hydrophobicity of bacterial membranes, and interfere with quorum sensing (QS) (Monte et al. 2014; Nazzaro et al. 2013). Plant extracts and their phytochemicals have been highlighted as promising antimicrobial agents due to their cost-effectiveness, eco-friendliness, large structural diversity, and the possibility of reducing the resistance to antimicrobial drugs (Sakarikou et al. 2020; Giaouris and Simões 2018).

2 Biofilm-Related Infections

Biofilms are surface-related microbial structures that can be present in a large variety of environments, either abiotic or biotic (Moshynets and Spiers 2016; Balcázar et al. 2015; Lebeaux et al. 2014). One of the most important characteristics of biofilms is their capacity to ensure greater protection of bacterial cells and survive in the presence of external stressors, such as high concentrations of antimicrobials. Biofilms recalcitrance may lead to treatment failure and infection recurrence (Lebeaux et al. 2014; Aggarwal et al. 2015; Giulia and O'Toole 2021). Therefore, even in unfavorable environments, biofilms can promote the survival of pathogenic microorganisms and facilitate their spread and the recolonization of new niches (Muhammad et al. 2020; Del Pozo 2018; Roy et al. 2018; Wu et al. 2015; Kostakioti et al. 2013). Biofilms are also associated with device-related infectious diseases, which can be responsible for the development of malfunctions of devices and even chronic infections (Del Pozo 2018; Nandakumar et al. 2013; Khatoon et al. 2018). In fact, biofilms are responsible for up to 80% of human infections, such as cystic fibrosis (CFs), endocarditis, osteomyelitis, and sepsis. Thus, given the relevance of infections caused by biofilms, it is essential to highlight the pathogens considered most critical, as well as the stages of development of a biofilm, their main specific tolerance mechanisms, and their respective virulence factors.

2.1 Critical Multidrug-Resistant Pathogens

According to the WHO, certain microorganisms are particularly problematic for human health, including Gram-negative bacteria, such as *Enterobacteriaceae*, *Acinetobacter baumannii,* and *Pseudomonas aeruginosa* and ESKAPE pathogens, i.e. a group of six highly virulent, pathogenic, and antibiotic-resistant pathogens, including *Enterococcus faecium*, *Staphylococcus aureus*, *Klebsiella pneumoniae*, *Acinetobacter baumannii*, *Pseudomonas aeruginosa,* and *Enterobacter* spp. These Gram-positive and Gram-negative bacteria can evade or "escape" commonly used antibiotics due to the spread of multidrug resistance, being linked to increasing AMR infections worldwide (Santajit and Indrawattana 2016; Schultz et al. 2020; Mulani et al. 2019; Skariyachan and Garka 2018; Julian and Blumberg 2017; Karlowsky et al. 2017). Table 1 shows the main virulence factors associated with some of the most concerning bacteria, as defined by the WHO. These virulence factors are correlated with their tolerance to antimicrobial agents, as well as the development of infection in the host system.

Table 1 Main virulence factors of selected pathogens

Bacterium	Main virulence factors	References
A. baumannii	Porins; capsular polysaccharides; phospholipases; lipopolysaccharides; outer membrane vesicles; metal acquisition systems; protein secretion systems.	(Lee et al. 2017; Sarshar et al. 2021; Aliramezani et al. 2019)
P. aeruginosa	Pyocyanin; pyochelin; other pigments; elastase; phospholipase C; protease A; exotoxins and cytotoxins; flagella and pili; QS regulatory system proteins.	(Alonso et al. 2020; Britigan et al. 1992; Cox 1986)
S. aureus	Membrane-damaging toxins; toxic shock syndrome toxin; epidermolytic toxin; staphylococcal enterotoxins; pyrogenic exotoxin; exoenzymes; enterotoxin; panton-valentine Leukocidin (PVL) toxin; thermostable nuclease (TNase); adhesins; phenol-soluble modulins; surface proteins; capsular polysaccharides; invasins; surface inhibitors of phagocytosis; membrane-damaging enzymes.	(Harvey et al. 1999; Tam and Torres 2019; Dinges et al. 2000; Etter et al. 2020; Fisher et al. 2018; Balasubramanian et al. 2017; Watkins et al. 2012; Pidwill et al. 2021; Flannagan et al. 2015)
E. faecium	Proteins and carbohydrates; enzymes; enterococcal surface protein (Esp); cytolysin; cell wall adhesins efaAfm; gelatinase (gelE); aggregation substances (agg).	(Huycke et al. 1998; Mascini et al. 2006; Ike 2017; Mundy et al. 2000; Barbosa et al. 2010)

2.2 *Biofilm Formation and Specific Mechanisms*

Biofilm formation has been recognized as a multistage process. It is believed that the biofilm development process occurs through several stages, as schematized in Fig. 1.

For the successful development of a biofilm, specific mechanisms and associated virulence factors are crucial. Virulence factors are molecules produced by bacteria, which are needed to enable bacteria to attack eukaryotic cells and cause infections. They are usually encoded on mobile and integrative genetic elements, namely plasmids, bacteriophages, conjugative transposons, integrative and conjugative elements, and pathogenicity islands (Abedon et al. 2009; Sharma et al. 2017; Lami 2019; Boyd 2012). Virulence factors allow the pathogens to evade the host immune response and facilitate the establishment and long-term survival of biofilms in tissues (Zhao and Ma 2015; Phillips and Schultz 2012). Besides virulence factors, biofilms also developed several mechanisms, which are extremely important for their survival and proliferation in host tissues (summary in Table 2).

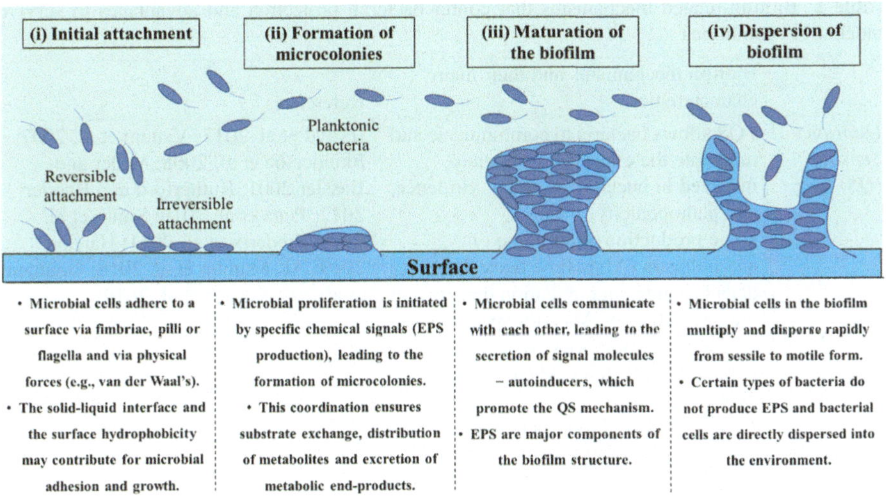

Fig. 1 Representation of the biofilm development stages (based on Jamal et al. 2018; López and Soto 2020)

3 Antibacterial Strategies Based on Natural Products to Target Biofilms

Bacterial resistance mechanisms are specific strategies of microbial evolution (Simões et al. 2009; Upmanyu et al. 2020; Ray et al. 2017). Thus, it is not surprising that new antibiotic resistance mechanisms will emerge (Barbieri et al. 2017; Diggle and Whiteley 2020; Munita and Arias 2016; Sekyere and Asante 2018). Some phenomena can explain the resistance of microorganisms to antibiotics: (i) cells can present an intrinsic and natural resistance to drugs, being "genetically programmed" to resist antibiotics (Holmes et al. 2016; Beceiro et al. 2013); (ii) microorganisms may suffer some genetic modifications like mutations, which make them insensitive to antibiotics (Fair and Tor 2014; Beceiro et al. 2013; Meredith et al. 2015); (iii) cells may become resistant to antimicrobial agents through the horizontal acquisition of genes that confer resistance to certain antibiotics from other microorganisms, which is a very extensive phenomenon and increases the probability of survival under the selective pressure caused by antibiotics (Beceiro et al. 2013; Meredith et al. 2015; Bello-López et al. 2019). Moreover, bringing a new antibiotic to the market may take up to 10 years and it is very expensive and time consuming (Barbieri et al. 2017; Spellberg 2014; Power 2006). Thus, new strategies aiming to fight bacterial infections, especially those caused by biofilms, are highly necessary (Bi et al. 2021; Srinivasan et al. 2021; Zhang et al. 2020a; Simões et al. 2009). Indeed, natural compounds, especially those obtained from plants, have become promising candidates for these much-needed treatments due to excellent antibacterial activity (Barbieri et al. 2017; Ćirić et al. 2019; Melander et al. 2020; Chassagne et al. 2021; Lahlou 2013).

Table 2 Biofilm-related mechanisms that confer bacterial protection and advantages to survive under stress conditions

	Biofilm mechanisms and their main characteristics	References
Quorum Sensing (QS)	• QS allows bacteria to communicate and modulate the expression of genes involved in bacterial survival, virulence, and pathogenicity. • The production and release of autoinducers (AIs) allow bacteria to detect changes in population density and the interaction of AIs with specific transcriptional regulators leads to population-dependent changes in gene expression. • QS systems in Gram-negative bacteria are related to a LuxI/R-like system. LuxI-type proteins are synthase proteins that catalyze the synthesis of Acyl-Homoserine Lactones (AHL) while LuxR type proteins are transcriptional activator proteins to which AHL binds, triggering transcriptional modifications. • In Gram-positive bacteria, peptides are secreted and detected by the two-component systems but can also be actively transported via ABC transporters to bind a cytoplasmic regulator.	(Floyd et al. 2017; Vattem et al. 2007; Skindersoe et al. 2008; Miller and Bassler 2001; Rutherford and Bassler 2012; Pena et al. 2019; Smith et al. 2020; Frederix et al. 2011; Harriott 2019; AL-Mamun et al. 2018; Goulden et al. 2013; Clark et al. 2019)
Adhesion	• Adhesion is the first essential step for biofilm development and it is controlled by a complex set of chemical and physical interactions. • The first adhesion process of a planktonic microorganism is reversible, and it is dictated by several physicochemical interactions, being mostly associated with van der Waals forces and electrical double layer forces (DLVO theory). • The adhesion will depend on the net sum of attractive and repulsive forces between the surfaces and the microorganism. • The second stage implies a stronger and irreversible binding of the microorganism to the surface through bacterial adhesins.	(Garrett et al. 2008; Almaguer-Flores 2013; Stones and Krachler 2016; Abraham et al. 2015; Berne et al. 2018; Meireles et al. 2015)
Motility	• Motility is one of the most dynamic bacterial phenomena, contributing to virulence through adhesion and formation of biofilms. • Six different categories of bacterial motility have been identified:	(Monte et al. 2014; Xu et al. 2014; Harshey 2003; Pollitt and Diggle 2017; Eberl et al. 1999; Tomada et al. 2016)

(continued)

Table 2 (continued)

	Biofilm mechanisms and their main characteristics	References
	Swimming, swarming, gliding, twitching, sliding, and darting. • Swimming and swarming are dependent on flagella. • Twitching and some ways of gliding require type IV pili. • Sliding/spreading are passive translocation types of motility. Information about darting is still scarce.	
EPS production	• EPS are organic polymers, namely exopolysaccharides, proteins, lipids, and DNA, crucial for the interaction between bacteria and the environment. • The extracellular matrix, composed by water and EPS, is essential for the establishment and maintenance of the structure and properties of biofilms. • The matrix stability is guaranteed by non-covalent binding between EPS involving weak physicochemical forces. • The EPS network confers cohesion and viscoelasticity to the structure.	(Costerton et al. 1999; Di Martino 2018; Jiao et al. 2010; Sheng et al. 2010; Gao et al. 2019; Flemming and Wingender 2010; Costa et al. 2018; Caro-Astorga et al. 2020; Chiba et al. 2015; Jachlewski et al. 2015)

Medicinal plants are the most abundant biological resources in traditional and modern medicine, nutraceuticals, food supplements, folk medicines, pharmaceutical intermediates, and chemical components of synthetic medicines (Desai et al. 2015; Saranraj et al. 2016; Pan et al. 2013). The intake of phytochemicals in the diet can promote health and protect the human body from diseases, being also able to reduce the risk of some chronic diseases (Jimenez-Garcia et al. 2018; Liu 2013; Yoo et al. 2018; Septembre-Malaterre et al. 2017). Phytochemicals with nutritious and health-care properties in foods are of great significance due to their beneficial effects on human health. They are able to prevent and assist in the combat of a large variety of diseases, such as cancer, coronary heart disease, diabetes, hypertension, inflammation, microbial, viral, and parasitic infections, psychosis, spastic diseases, ulcers, osteoporosis and related diseases, among others (Thakur et al. 2020; Zhang et al. 2015; Forni et al. 2019; Guan et al. 2021; Kibe et al. 2017; Visioli et al. 2000; Howes and Simmonds 2014).

A large variety of well-known phytochemicals have been identified over the years, such as lycopene present in tomatoes, isoflavones in soybeans, and flavonoids found in fruits (Zhang et al. 2015; Ghoshal 2018; Waheed Janabi et al. 2020). An interesting feature of these compounds is their powerful antioxidant potential. In fact, the regular consumption of fruits, vegetables, and whole grains has shown to reduce the risk of various diseases related to oxidative damage, acting as free radical scavengers like hydrogen donors, electron donors, peroxide decomposers, singlet oxygen quenchers, enzyme inhibitors, synergists, and metal chelators (Thatoi et al. 2014; Yu et al. 2021; Chen et al. 2014; Asaduzzaman 2018; Manganaris et al. 2018; Martínez et al. 2014). Phytochemicals that have significant health benefits belong to the categories of phenolic compounds, alkaloids, organosulfur compounds, phytoestrogens, terpenoids, carotenoids, limonoids, and phytosterols. All of these have shown to be highly effective in preventing and combating a wide variety of diseases (Huang et al. 2016; Jimenez-Garcia et al. 2018; Thakur et al. 2020; Singh et al. 2020; Bayir et al. 2019; Andrade et al. 2020; Koche et al. 2018; Patra 2012). Thus, the exciting properties of phytochemicals, combined with the growing need for new antimicrobial agents highlight these molecules as potential antibiofilm strategies of the future.

3.1 Targeting Bacterial Biofilms with Phytochemicals

Phytochemicals can present various mechanisms of action that have the capacity to harm the establishment and development of biofilms (Simões et al. 2009). These mechanisms include the inhibition of peptidoglycan synthesis (Monte et al. 2014; Nazzaro et al. 2013), cytoplasmatic membranes and cell wall damage and disruption (Suarez et al. 2005; Oussalah et al. 2006; Haraguchi et al. 1996), modification of hydrophobicity and permeabilization of the bacterial membranes (Cox et al. 2000; Carson et al. 2002; Trombetta et al. 2005; Melzig et al. 2001; Ultee et al. 1999), efflux pump inhibition (Aeschlimann et al. 1999; Schmitz et al. 1998; Khan et al. 2006; Micol et al. 2001), inhibition of RNA or DNA synthesis (Mori et al. 1987; Cushnie and Lamb 2005; Feldberg et al. 1988; Sundar and Chang 1992), inhibition/interference of enzyme activity and of the electron transport chain (Lin et al. 2005; Mirzoeva et al. 1997; Sinha Babu et al. 1997; Mandal et al. 2005). Despite the wide variety of phytochemical molecules that have been described as bioactive and of potential interest as food supplement or drug, only few were approved by the US Food and Drug Administration (FDA) (Kongkham et al. 2020).

Phytochemicals with antibiofilm properties include, for example, quercetin, which can inhibit the production of alginate, leading to a decrease in the adhesion during the biofilm development (Górniak et al. 2019; Memariani et al. 2019; Lee et al. 2013). Another example is (+)-usnic acid, which has been used to modify polyurethane surfaces to evaluate its influence on the development of *S. aureus* and *P. aeruginosa* biofilms (Francolini et al. 2004). The results showed that the formation of *S. aureus* biofilms for a period of up to 6 days was inhibited on polyurethane surfaces with (+)-usnic acid. On the contrary, *P. aeruginosa* biofilms were able to develop on surfaces of (+)-usnic acid-treated polymer. In fact, the (+)-usnic acid has higher antimicrobial activity against *S. aureus* than against *P. aeruginosa,* as translated by the minimum inhibitory concentration (MIC), which was 32 μg/mL for *S. aureus* and 256 μg/mL for *P. aeruginosa.* (+)-usnic acid was also able to affect the morphology of *P. aeruginosa* biofilms, since the biofilm was thinner and flat in the control polymer and in the (+)-usnic acid-treated polymer the biofilm was substantially thicker and rougher (Francolini et al. 2004). Another phytochemical that exhibited very interesting results is emodin, which inhibited the development of biofilms by *P. aeruginosa, E. coli, and S. aureus* through the decrease of expression of key genes involved in biofilm formation. Moreover, several studies also evaluated the effect of plant extracts on biofilm control, instead of pure and isolated phytochemicals. The use of extracts containing different molecules may result in synergistic interactions, which may explain the positive results obtained with low doses of active compounds in herbal products. This highlights the hypothesis that the use of plant extracts may be advantageous in comparison with the use of single phytochemicals (Borges et al. 2016).

The effectiveness of phytochemicals and plant extracts in combating infections may be justified by alterations caused in specific biofilm mechanisms, as described in the following sections.

3.1.1 Effects of Phytochemicals in QS Mechanisms

QS is a crucial regulatory mechanism involved in biofilm formation and differentiation (Borges et al. 2014a). It allows bacteria to communicate through the delivery and sensing of small signal molecules, ensuring relevant advantages for bacteria in regard to colonization, biofilm development, defense, adaptation, virulence, and pathogenicity (Rutherford and Bassler 2012; Pena et al. 2019; Li and Tian 2012; Grainha et al. 2020). Therefore, studying natural molecules with the ability to interfere with QS and understanding their effects may be an important way to fight bacterial tolerance and biofilms (Rutherford and Bassler 2012; Borges et al. 2014a; Lade et al. 2014). In fact, several studies have shown the action of specific phytochemicals on QS and consequently the possibility to retard or avoid biofilm development. For instance, the synthetic halogenated furanone, a secondary metabolite compound derived from furanone present in the Australian macroalgae *Delisea pulchra*, has shown great potential in affecting the bacterial signaling QS, as well as the motility of cells. Based on the structural similarity between *D. pulchra* furanone and Acyl-homoserine lactone (AHL) molecules, it is hypothesized that this furanone may be responsible for interfering with the interaction of the putative regulatory protein and AHL molecule through its competitive binding to the receptor. Moreover, furanones can also inhibit surface aggregation traits, when in an ecologically significant concentration (Roy et al. 2018). Hentzer et al. suggested that rhl system, which is a QS mechanism in *P. aeruginosa*, is a target to furanone (Hentzer et al. 2002). This compound can penetrate the biofilm matrix of *P. aeruginosa*, interfering with genes related to QS biofilm maturity expression. Hence, this compound is capable of changing the biofilm structure, which promotes the detachment of bacteria and leads to the loss of bacterial biomass in the matrix, also mediating the displacement of AHL molecules from LuxR receptor sites (Roy et al. 2018; Hentzer et al. 2002; Hentzer and Givskov 2003; Asfour 2018; Alasil et al. 2015).

Certain polyphenol compounds, such as epigallocatechin gallate (EGCG), tannic acid, ellagic acid, gallic acid, and ferulic acid are also promising molecules for inhibiting biofilm development. Some studies linked the antibiofilm activity of these polyphenolic molecules to the interference in QS. For instance, Huber et al. demonstrated that polyphenolic compounds, such as gallic acid, (−)-epigallocatechin gallate, (+)-catechin and tannic acid were capable of interfering with QS mechanisms of *E. coli* and *P. putida* through the blockade of the AHL system (Quave et al. 2012; Huber et al. 2003). Quercetin, as well as some others flavonoids, have also shown to be efficient in affecting the QS mechanisms of several microorganisms, such as *P. aeruginosa*, *S. aureus*, *Escherichia coli*, *Enterococcus faecalis,* and *Streptococcus mutans*.

Curcumin, from the rhizome of *Curcuma longa*, showed strong and effective biofilm inhibitory activity related to the regulation of the expression of genes involved in QS and associated with the production of alginate and exopolysaccharides. Curcumin has also shown effects in inhibiting the swimming and swarming motilities and increasing biofilm susceptibility to antibiotics (Borges et al. 2016; Packiavathy et al. 2013; Packiavathy et al. 2014). A study performed by Kali et al. demonstrated that curcumin in combination with ciprofloxacin was effective against biofilms of Gram-positive bacteria. Moreover, this compound was also effective against Gram-negative biofilms when combined with the antibiotics amikacin, gentamicin (GEN), and cefepime. These results revealed the potential of using curcumin in a combination therapy (Kali et al. 2016). In addition, emodin induced the proteolysis of *E. coli* QS signal receptor TraR and enhanced the activity of ampicillin against *P. aeruginosa. This compound* has also been associated with the downregulation of the *cidA* gene, which is involved in cell lysis and eDNA release (Borges et al. 2016; Yan et al. 2017; Harapanahalli et al. 2015; Zhang et al. 2020b).

Regarding the use of plant extracts, Zhang et al. evaluated the antibiofilm and QS inhibition activity of an extract from *Rosa rugosa* tea, whose main components are polyphenols and flavonoids (Zhang et al. 2014). This extract inhibited the production of violacein, controlled by QS in *C. violaceum*. Furthermore, this extract also inhibited the biofilm formation by *E. coli* and *P. aeruginosa, which may be related not only to its quorum quenching activity, but also to the inhibition of bacteria swarming motility*. Glycosyl flavonoids, such as chlorogenic acid, isoorientin, orientin, isovitexin, vitexin, and rutin were also capable of inhibiting the QS in *C. violaceum* and *E. coli*. Moreover, there are more plant components with QS inhibition properties, such as extracts from liverwort *Lepidozia chordulifera, where it is possible to find* sesquiterpenoid viridiflorol, triterpenoids, ursolic and betulinic acids (Zhang et al. 2014). Other relevant and efficient phytochemicals with antiquorum sensing effects include N-(heptylsulfanylacetyl)-L-homoserine lactone, extracted from garlic (able to interrupt QS signaling by competitively inhibiting transcriptional regulators LuxR and LasR), limonoids, and hordenine (Klančnik et al. 2021). Furthermore, baicalin hydrate, cinnamaldehyde, and hamamelitannin have been shown to significantly improve the susceptibility of *P. aeruginosa*, *B. cenocepacia,* and *S. aureus* biofilms, including MRSA, in combination with conventional antibiotics. In fact, hamamelitannin can act as a QS inhibitor by increasing both the in vitro and in vivo susceptibility of MRSA biofilms, which can be associated with its ability to interact with the TraP receptor and affect the release of eDNA (Brackman et al. 2011; Yuan et al. 2020; Jiang et al. 2019).

There are more examples of studies that equally show promising results of phytochemicals regarding antiquorum sensing activity. It was demonstrated by Niu et al. that *Cinnamomum cassia* extract, containing cinnamaldehyde and eugenol, significantly inhibited QS of *E. coli* and *Vibrio harveyi* (Niu et al. 2006). In addition, clove, cinnamon, peppermint, and lavender also exhibited antiquorum sensing activity, according to Khan et al. (Khan et al. 2009). Iberin, which is an isothiocyanate and sulfoxide, interfered with the QS mechanism of *P. aeruginosa* and induced apoptosis. Organosulfur compounds from garlic, such as allicin and ajoene, were able to inhibit QS of *P. aeruginosa* and *E. coli* (Borges et al. 2013). In another study by Vikram et al. flavonoids were evaluated for their capacity to interfere with QS of *V. harveyi* and inhibited the development of *E. coli* O157:H7 and *V. harveyi* biofilms (Vikram et al. 2010). The results showed that kaempferol, naringenin, quercetin, and apigenin were able to act as nonspecific inhibitors of autoinducer-mediated cell–cell signaling processes. In addition, these molecules inhibited the formation of *V. harveyi* and *E. coli* O157:H7 biofilms (Vikram et al. 2010).

3.1.2 Effects of Phytochemicals on Motility

Motility, which generally depends on flagella and pili, is one of the crucial factors that contribute to biofilm formation and development, especially at an earlier stage, since these mechanisms require a multicellular movement and dispersion on a surface (Monte et al. 2014; Lemon et al. 2007; Cai et al. 2020; Kearns 2010; Köhler et al. 2000). It has been already shown that many natural compounds, extracts or pure products, have the capacity to interfere with the bacterial motility of several microorganisms (Vattem et al. 2007; Liaqat et al. 2018), which may be reflected on the reduction of biofilm formation ability. Swimming and swarming motilities of *P. aeruginosa*, *Proteus mirabilis,* and *Serratia marcescens* were inhibited by methanolic extracts of *Cuminum cyminum*, especially methyl eugenol, a compound with a well-studied antibiofilm activity. In addition, cinnamaldehyde and eugenol, from *Cinnamomum cassia*, were able to compromise the development of *E. coli* biofilms by interfering with their swimming motility (Monte et al. 2014). Gallic acid and ferulic acid displayed potential to inhibit the motility and adhesion of four pathogenic bacteria, including *E. coli*, *P. aeruginosa*, *S. aureus,* and *Listeria monocytogenes*. Identical results were obtained for isothiocyanates allyl isothiocyanate (AITC) and 2-phenylethyl isothiocyanate (PEITC) (Borges et al. 2012). Ferulic acid and salicylic acid (SA) were reported to inhibit swimming motility of *Bacillus cereus* and *Pseudomonas fluorescens* (Lemos et al. 2014).

Camellia nitidissima Chi is a well-known edible plant from China with diverse biological and medicinal properties, especially antibacterial activity. A study performed by Yang et al. assessed the inhibitory activity of the dichloromethane fraction (DF) of *C. nitidissima* Chi flowers on the pyocyanin production, as well as on the swarming and swimming motility of *P. aeruginosa* PAO1, at sub-minimum inhibitory concentrations (Yang et al. 2018). DF was associated with a concentration-dependent inhibition activity in relation to swarming and swimming motility. Moreover, the half maximal inhibitory concentrations (IC50) were determined and they found values of 0.158 ± 0.009 mg/mL for pyocyanin production, 0.139 ± 0.004 mg/mL for swarming motility, and 0.334 ± 0.049 mg/mL for swimming motility. In addition, the results of a real-time polymerase chain reaction (RT-PCR) showed that DF significantly downregulated the expression of rhlR and lasR (Yang et al. 2018). In fact, *las* and *rhl* systems are the two principal QS mechanism elements of *P. aeruginosa.* Finally, through the high-performance liquid chromatography (HPLC)-triple-time of flight (TOF)-MS/MS analysis, it was found that DF is mainly composed of gallic acid, catechin, ellagic acid, chlorogenic acid, quercetin, and kaempferol. These six identified compounds displayed inhibitory effects on pyocyanin production, swarming, and swimming motilities. Ellagic acid showed the strongest effect with IC50 values of 0.067 ± 0.002 mg/mL for pyocyanin production, 0.024 ± 0.008 mg/mL for swarming motility, and 0.020 ± 0.003 mg/mL for swimming motility (Yang et al. 2018). Therefore, it is possible to infer that all these alterations caused by DF may cause significant effects on the ability of *P. aeruginosa* to form biofilms.

In another study by Borges et al., the effect of AITC and PEITC on planktonic cell susceptibility, bacterial motility and adhesion, and biofilms of *E. coli, P. aeruginosa, S. aureus,* and *L. monocytogenes* was evaluated (Borges et al. 2014b). AITC caused complete inhibition of swimming motility of *P. aeruginosa* and total inhibition of swarming motility of *E. coli*, while PEITC caused complete inhibition of swimming motility of *E. coli, P. aeruginosa,* and *L. monocytogenes* and swarming motility of *E. coli* and *P. aeruginosa*. Moreover, the spreading motility of *S. aureus* was completely inhibited by PEITC. AITC and PEITC presented a preventive effect on biofilm formation, which may be related to the effects on the inhibition of bacterial motility. Moreover, AITC and PEITC showed a high potential to reduce the mass of biofilm formed by the Gram-negative bacteria (Borges et al. 2014b).

3.1.3 Effects of Phytochemicals on Adhesion

Bacterial adhesion to the surface of materials is the first step that leads to the formation of a biofilm and, from the moment the adhesion occurs, it is possible for the bacteria to communicate with each other and establish a community (Klančnik et al. 2021; Dunne 2002; Flemming et al. 2016). Therefore, since bacterial adhesion to a surface is an initial prerequisite for biofilm formation, it is essential to identify natural compounds capable of interfering with this phenomenon, which may constitute an important target when developing strategies for controlling pathogenicity (Kostakioti et al. 2013; Verderosa et al. 2019). Within this context, phenolic compounds have been shown to exhibit important roles in preventing bacterial adhesion, including the adhesion of *Campylobacter jejuni* (Klančnik et al. 2021). In a study of Pogačar et al., thyme ethanolic extract (TE), thyme post-hydrodistillation residue (TE-R), and olive leaf extract (OE) were evaluated for their phytochemical composition using high-performance liquid chromatography (HPLC) with photodiode array, and antimicrobial activity against *C. jejuni* (Šikić Pogačar et al. 2016). The analysis showed that the main components present in TE and TE-R were flavone glucuronides and rosmarinic acid derivatives, while in OE verbascoside, luteolin 7-O-glucoside and oleuroside were the major compounds. The compounds TE and TE-R were able to decrease *C. jejuni* adhesion to the abiotic surface by up to 35%, at a concentration range between 50–200 μg/mL, and up to 30%, at a concentration range between 0.2–12.5 μg/mL. TE-R was more effective than TE and displayed a meaningful inhibition of *C. jejuni* adhesion (higher than 30%), reaching up to 40%. When considering a concentration range of OE between 3.125–12.5 μg/mL, biofilm formation was affected and the adhesion inhibition reached 10% to 23%, respectively. In addition, the anti-adhesion effect of *C. jejuni* to cell cultures was also assessed and it was observed that the *C. jejuni* adhesion toward pig small intestine epithelial cell line, PSI cl1 cells, was remarkably decreased up to 30% in the presence of TE, TE-R, and OE at a concentration range between 0.78–200 μg/mL (Šikić Pogačar et al. 2016).

The North American cranberry is also recognized for its high levels of phytochemicals, including phenolic acids, flavonoids, and ellagic acid (Walton 2014). The antimicrobial properties of cranberry species have been associated with high levels of polyphenol compounds, particularly proanthocyanidin (PACs). So far, several data have revealed that PACs have unique characteristics that allow the inhibition of bacterial adhesion to epithelial cells. Indeed, the studies developed by various authors have shown that there is a direct correlation between the ingestion of cranberries and the prevention of urinary tract infections (UTIs) in females (Walton 2014; Kontiokari et al. 2005; Carson and Riley 2003; Tempera et al. 2010). Several in vitro studies have already been performed and demonstrated that the anti-adhesion effects of PACs are responsible for a reduction in the bacterial adherence to urinary tract cells such as uroepithelial cells, but also to other biological materials (Walton 2014; Kontiokari et al. 2005). Furthermore, cranberry extracts also seem to have similar inhibitory effects on tissue adhesion with regard to Gram-negative bacteria, including the ones of the genera *Proteus*, *Klebsiella*, *Enterobacter,* and *Pseudomonas*. The intake of cranberry has been associated with a decrease in the incidence of gastric ulcer and gastric cancer (Walton 2014). Burger et al. showed that for cultured gastric epithelial cells, the phytochemicals found in cranberry juice can specifically interfere with a sialyllactose-specific adhesion mechanism that allows *Helicobacter pylori* to adhere to the gastric mucosa. Consequently, these authors inferred that the cranberry juice could also inhibit the adhesion of bacteria to the stomach in vivo and acting in the prevention of stomach ulcers that are caused by *H. pylori*. (Walton 2014; Burger et al. 2002).

Phenolic and polyphenolic compounds have also been shown to have substantial anti-adhesion effects. A work performed by Toivanen et al. evaluated the anti-adhesion activity of different molecular size fractions of wild cranberry *Vaccinium oxycoccos* polyphenols against *Streptococcus pneumoniae* and *Streptococcus agalactiae* (Toivanen et al. 2010). Moreover, according to Borges et al., there are other phytochemicals with very impressive anti-adhesion activity, namely cinnamaldehyde and eugenol, from *Cinnamomum cassia*, which interfered with adhesion and biofilm formation of *E. coli* (Borges et al. 2013). Polyanacardic acid, polysalicylic acid, catechin, epigallocatechin, and tannic acid also exhibited anti-adhesion activity and inhibited the formation of *Streptococcus mutans* and *P. aeruginosa* biofilms (Borges et al. 2013). In another work of Lee et al., the potential effects of phloretin, which can be found, for instance, in apples, were investigated (Lee et al. 2011). Phloretin reduced the attachment of *E. coli* O157:H7 to human colon epithelial cells and decreased the formation of *E. coli* O157:H7 biofilms. It was also demonstrated that phloretin was able to repress genes associated with toxicity, hlyE and stx(2); the autoinducer-2 importer genes, lsrACDBF; curli genes csgA and csgB; and also prophage genes of *E. coli* O157:H7 biofilms (Lee et al. 2011).

3.1.4 Effects of Phytochemicals on the EPS Production

EPS have a wide range of biological functions, being part of the carbon and energy reserves, preventing desiccation, protecting against environmental stresses, providing protection against toxins and antibiotics, as well as playing a crucial role in pathogenicity, symbiosis phenomena, and adhesion to surfaces (Singha 2012; Whitfield et al. 1993; Roberts 1996; Khan and Iqbal 2017). Hence, it is fundamental to discover and understand the role of natural compounds in inhibiting the production of EPS with regard to controlling and fighting biofilms (Mishra et al. 2020; Koo et al. 2017). Several studies described significant action of extracts and specific phytochemicals on EPS inhibition and removal. For instance, Borges et al showed that methyl eugenol interfered with EPS production (Borges et al. 2013). In a study developed by Packiavathy et al., the inhibitory properties of spices and vegetables, commonly found in South India, on the QS and EPS production of the bacteria *C. violaceum* were evaluated (Packiavathy et al. 2012). In the 22 samples tested, the QS inhibitory compound present in the methanolic extract of *Cuminum cyminum*, at a concentration of 2 mg/mL, was able to inhibit the production of violacein by *C. violaceum*. In addition, it was demonstrated that the *C. cyminum* extract highly interfered with the physiological functions regulated by AHL and the formation of biofilms, including flagellar movement and EPS production. The QS inhibitory effects shown by this plant were assigned to methyl eugenol, considering the results of molecular docking analysis. At a sub-MIC level, it promoted damage to the biofilm structure and strongly inhibited the biofilm formation of *P. aeruginosa* PAO1, *P. mirabilis,* and *Serratia marcescens* (Packiavathy et al. 2012). Abraham et al. assessed the inhibition of EPS production by methanolic fraction of the dried fruits of *Caesalpinia spinosa.* Also, the ability to inhibit biofilm formation by *E. coli*, *P. mirabilis*, *S. marcescens,* and *P. aeruginosa* PAO1 was assessed (Issac Abraham et al. 2011). In addition, its effect in inhibiting QS-dependent phenomena, such as violacein production in *C. violaceum* and swimming and swarming motility was also investigated. It was observed that the extract of *C. spinosa* interfered with swimming and swarming motility, EPS production and biofilm formation in *E. coli, P. mirabilis, S. marcescens,* and *P. aeruginosa* PAO1. Here, the extract revealed a strong QS inhibition effect, in a concentration-dependent way, however, without interfering with the microbial growth. At a concentration of 2 mg/mL, this extract significantly restricted the EPS production to 58, 46, 66, and 67%, as well as limited biofilm formation to 79, 75, 73, 70%, in *S. marcescens, P. aeruginosa* PAO1, *E. coli,* and *P. mirabilis*, respectively (Issac Abraham et al. 2011).

There are other phytochemicals with a proven ability to inhibit EPS production. A well-known compound is (−)-epigallocatechin gallate (EGCg), which has several antimicrobial action modes, including the ability to decrease EPS production (Borges et al. 2013). EGCg is the major component of tea catechins and displayed an inhibitory effect, EPS degradation capacity, and a destructive action against *E. coli* biofilms (Maeyama et al. 2005). Taking into account the EGCg effects, Maeyama et al. developed a new bactericidal surface based on a catechin polymer (Maeyama et al. 2005). The surfaces tested, containing EGCg, were prepared through photopolymerization of liquid biodegradable polyesters and the rate of releasing was potentiated by increasing the rate of surface-erosion of the polymers. Interestingly, polymers exhibiting a higher releasing ability showed a lower biofilm development on the surfaces. Moreover, EGCg induced a biofilm-destructing effect, such as EPS degradation, damage of bacterial membrane, and cell detachment (Maeyama et al. 2005). Allicin is another phytochemical with remarkable proven evidence in relation to EPS production. Lihua et al. investigated the activity of allicin against *P. aeruginosa* biofilm development, specifically their effect on the production of virulence factors controlling the QS mechanism and EPS (Lihua et al. 2013). The authors found that the EPS production increased over time, but decreased with the concentration of allicin used in the treatment. The study also demonstrated that although the dry weight of bacteria between the different groups was almost equal to the one registered at the beginning of the experiments, the total amount of EPS was significantly reduced when allicin was used. Moreover, bacterial adhesion was significantly reduced compared to the saline control group, when 128 μg/mL of allicin was used. Interestingly, when the biofilm was treated with 10 μg/mL of allicin, it became thinner, and its structure was compromised. Similar results were obtained for the group treated with 128 μg/mL of allicin, in which the biofilm thickness was reduced from 28.83 μm to 16.50 μm. The effect of allicin on the virulence factors exotoxin A, elastase, pyoverdine, and rhamnolipid was also assessed, being demonstrated that allicin significantly downregulated the expression of exotoxin A and elastase. In addition, there was a complete inhibition of rhamnolipid and pyoverdine production. Thus, allicin can affect the development and maturation of *P. aeruginosa* biofilms, suggesting that this compound may be a promising therapy for treating bacterial biofilms (Lihua et al. 2013).

4 Conclusions and Future Perspectives

Antibiotic resistance is a serious problem in modern society. It has been claimed for many years that several natural plant products have medicinal effects, which make them promising for the treatment of various diseases. The development of natural, safe, and effective therapies helps control and minimize antibiotic resistance. Within the broad therapeutic and nutritional effects of phytochemicals, the ability to impair biofilm mechanisms, damage the bacterial membrane, restrict biofilm formation and silence virulence factors make them highly sought-after, effective antimicrobial agents. Besides the interference of phytochemicals with many planktonic bacterial processes, these natural compounds also have promising antimicrobial effects on biofilm mechanisms, such as adhesion, motility, QS and EPS production, displaying a strong activity against virulence factors of both planktonic and sessile cells. Nevertheless, further research is needed to develop and tailor specific applications of phytochemicals for therapeutic and clinical purposes, to assess their safety profile, to confirm their efficacy in vivo and in humans, and to ensure appropriate and selected use to prevent the future emergence of resistances.

Acknowledgments This work was financially supported by: Base Funding—UIDB/00511/2020 of the Laboratory for Process Engineering, Environment, Biotechnology and Energy—LEPABE—funded by national funds through the FCT/MCTES (PIDDAC); Projects Biocide_for_Biofilm—TDC/BII-BTI/30219/2017—POCI-01-0145-FEDER-030219; Germirrad—POCI-01-0247-FEDER-072237 funded by FEDER funds through COMPETE2020—Programa Operacional Competitividade e Internacionalização (POCI) and by national funds (PIDDAC) through FCT/MCTES; Project "HealthyWaters—Identification, Elimination, Social Awareness and Education of Water Chemical and Biological Micropollutants with Health and Environmental Implications," with reference NORTE-01-0145-FEDER-000069, supported by Norte Portugal Regional Operational Programme (NORTE 2020), under the PORTUGAL 2020 Partnership Agreement, through the European Regional Development Fund (ERDF).

The authors declare no conflict of interest.

References

Abedon S, Duffy S, Turner PE (2009) Bacteriophage ecology. Encycl Microbiol 1:42–57

Abraham SN, Sharon N, Ofek I, Schwartzman JD (2015) In: Tang Y-W, Sussman M, Liu D, Poxton I, J. B. T.-M. M. M. (Second E. Schwartzman) (eds) Chapter 24: Adhesion and colonization. Academic Press, Boston, pp 409–421

Aeschlimann JR, Dresser LD, Kaatz GW, Rybak MJ (1999) Effects of nor a inhibitors on in vitro antibacterial activities and postantibiotic effects of levofloxacin, ciprofloxacin, and norfloxacin in genetically related strains of Staphylococcus aureus. Antimicrob Agents Chemother 43(2): 335–340. https://doi.org/10.1128/AAC.43.2.335

Aggarwal S, Stewart PS, Hozalski RM (2015) Biofilm cohesive strength as a basis for biofilm recalcitrance: are bacterial biofilms overdesigned? Microbiol Insights 82:MBI.S31444. https://doi.org/10.4137/MBI.S31444

Alasil SM, Omar R, Ismail S, Yusof MY (2015) Inhibition of quorum sensing-controlled virulence factors and biofilm formation in *Pseudomonas aeruginosa* by culture extract from novel bacterial species of *Paenibacillus* using a rat model of chronic lung infection. Int J Bacteriol 2015:671562. https://doi.org/10.1155/2015/671562

Aliramezani A, Soleimani M, Fard RMN, Nojoomi F (2019) Virulence determinants and biofilm formation of Acinetobacter baumannii isolated from hospitalized patients. Germs 9(3):148–153. https://doi.org/10.18683/germs.2019.1171

Allen N, Mulvihill DP, Hiscock JR (2019) Novel antimicrobial agents for clinical applications. Access Microbiol 1(1A):0301. https://doi.org/10.1099/acmi.ac2019.po0250

Almaguer-Flores A (2013) 8: Biofilms in the oral environment. In: Yan MI (ed) Woodhead publishing series in biomaterials. Woodhead Publishing, pp 169–186

AL-Mamun M, Chowdhury T, Biswas B, Absar N (2018) In: Grumezescu AM, A. M. B. T.-F. S, Holban P (eds) Chapter 11: Food poisoning and intoxication: a global leading concern for human health. Academic Press, pp 307–352

Alonso B et al (2020) Characterization of the virulence of Pseudomonas aeruginosa strains causing ventilator-associated pneumonia. BMC Infect Dis 20(1):909. https://doi.org/10.1186/s12879-020-05534-1

Andrade M, Malheiro J, Borges F, Saavedra MJ, Simões M (2020) In: Simoes M, Borges A, L. B. T.-R. T. in B. S, Simoes TC (eds) Chapter 12: The potential of phytochemical products in biofilm control. Academic Press, pp 273–293

Arendt EK, Zannini E (2013) 4–Barley. In: Arendt EK, E. B. T.-C. G. for the F, Zannini BI (eds) Woodhead publishing series in food science, technology and nutrition. Woodhead Publishing, pp 155–201e

Asaduzzaman GAEE-TAE-M (2018) Free radicals and the role of plant phytochemicals as antioxidants against oxidative stress-related diseases. Intech Open, Rijeka. Ch. 4

Asfour HZ (2018) Anti-quorum sensing natural compounds. J Microsc Ultrastruct 6(1):1–10. https://doi.org/10.4103/JMAU.JMAU_10_18

Aung EE, Kristanti AN, Aminah NS, Takaya Y, Ramadhan R (2020) Plant description, phytochemical constituents and bioactivities of Syzygium genus: a review. Open Chem 18(1): 1256–1281. https://doi.org/10.1515/chem-2020-0175

Balasubramanian D, Harper L, Shopsin B, Torres VJ (2017) Staphylococcus aureus pathogenesis in diverse host environments. Pathog Dis 75(1):ftx 005. https://doi.org/10.1093/femspd/ftx005

Balcázar JL, Subirats J, Borrego CM (2015) The role of biofilms as environmental reservoirs of antibiotic resistance. Front Microbiol 6:1216. https://doi.org/10.3389/fmicb.2015.01216

Barbieri R et al (2017) Phytochemicals for human disease: an update on plant-derived compounds antibacterial activity. Microbiol Res 196:44–68. https://doi.org/10.1016/j.micres.2016.12.003

Barbosa J, Gibbs PA, Teixeira P (2010) Virulence factors among enterococci isolated from traditional fermented meat products produced in the north of Portugal. Food Control 21(5): 651–656. https://doi.org/10.1016/j.foodcont.2009.10.002

Bayir AG, Kiziltan HS, Kocyigit A (2019) Chapter 1: Plant family, carvacrol, and putative protection in gastric cancer. In: Watson RR, Preedy GD (eds) Dietary interventions in gastrointestinal diseases. Academic Press, London, pp 3–18

Beceiro A, Tomás M, Bou G (2013) Antimicrobial resistance and virulence: a successful or deleterious association in the bacterial world? Clin Microbiol Rev 26(2):185–230. https://doi.org/10.1128/CMR.00059-12

Bello-López JM et al (2019) Horizontal gene transfer and its association with antibiotic resistance in the genus Aeromonas spp. Microorganisms 7(9):363. https://doi.org/10.3390/microorganisms7090363

Berne C, Ellison CK, Ducret A, Brun YV (2018) Bacterial adhesion at the single-cell level. Nat Rev Microbiol 16(10):616–627. https://doi.org/10.1038/s41579-018-0057-5

Bi Y et al (2021) Therapeutic strategies against bacterial biofilms. Fundam Res 1(2):193–212. https://doi.org/10.1016/j.fmre.2021.02.003

Borges A, Saavedra MJ, Simões M (2012) The activity of ferulic and gallic acids in biofilm prevention and control of pathogenic bacteria. Biofouling 28(7):755–767. https://doi.org/10.1080/08927014.2012.706751

Borges A, Abreu A, Malheiro JF, Saavedra M, Simões M (2013) Biofilm prevention and control by dietary phytochemicals. In: Méndez-Vilas A (ed) Microbial pathogens and strategies for combating them: science, technology and education. Formatex Research Center, Badajoz, pp 32–41

Borges A, Serra S, Cristina Abreu A, Saavedra MJ, Salgado A, Simões M (2014a) Evaluation of the effects of selected phytochemicals on quorum sensing inhibition and in vitro cytotoxicity. Biofouling 30(2):183–195. https://doi.org/10.1080/08927014.2013.852542

Borges A, Simões L, Saavedra M, Simões M (2014b) The action of selected isothiocyanates on bacterial biofilm prevention and control. Int Biodeterior Biodegradation 86:25–33. https://doi.org/10.1016/j.ibiod.2013.01.015

Borges A, Abreu AC, Dias C, Saavedra MJ, Borges F, Simões M (2016) New perspectives on the use of phytochemicals as an emergent strategy to control bacterial infections including biofilms. Molecules 21(7):877. https://doi.org/10.3390/molecules21070877

E. F. Boyd, "Chapter 4: Bacteriophage-encoded bacterial virulence factors and phage–Pathogenicity Island interactions," in Bacteriophages, Part A, vol. 82, M. Łobocka and W. T. B. T.-A. in V. R. Szybalski, Eds. Academic Press, 2012, pp. 91–118

Brackman G, Cos P, Maes L, Nelis H, Coenye T (2011) Quorum sensing inhibitors increase the susceptibility of bacterial biofilms to antibiotics in vitro and in vivo. Antimicrob Agents Chemother 55:2655–2661. https://doi.org/10.1128/AAC.00045-11

Bridier A, Briandet R, Thomas V, Dubois-Brissonnet F (2011) Resistance of bacterial biofilms to disinfectants: a review. Biofouling 27(9):1017–1032. https://doi.org/10.1080/08927014.2011.626899

Britigan BE, Roeder TL, Rasmussen GT, Shasby DM, McCormick ML, Cox CD (1992) Interaction of the Pseudomonas aeruginosa secretory products pyocyanin and pyochelin generates hydroxyl radical and causes synergistic damage to endothelial cells. Implications for pseudomonas-associated tissue injury. J Clin Invest 90(6):2187–2196. https://doi.org/10.1172/JCI116104

Burger O, Weiss E, Sharon N, Tabak M, Neeman I, Ofek I (2002) Inhibition of Helicobacter pylori adhesion to human gastric mucus by a high-molecular-weight constituent of cranberry juice. Crit Rev Food Sci Nutr 42(3 Suppl):279–284. https://doi.org/10.1080/10408390209351916

Cai Y, Hutchin A, Craddock J, Walsh MA, Webb JS, Tews I (2020) Differential impact on motility and biofilm dispersal of closely related phosphodiesterases in Pseudomonas aeruginosa. Sci Rep 10(1):6232. https://doi.org/10.1038/s41598-020-63008-5

Caro-Astorga J et al (2020) Two genomic regions encoding exopolysaccharide production systems have complementary functions in B. cereus multicellularity and host interaction. Sci Rep 10(1): 1000. https://doi.org/10.1038/s41598-020-57970-3

Carson CF, Riley TV (2003) Non-antibiotic therapies for infectious diseases. Commun Dis Intell Q Rep 27(Suppl):S143–S146

Carson CF, Mee BJ, Riley TV (2002) Mechanism of action of Melaleuca alternifolia (tea tree) oil on Staphylococcus aureus determined by time-kill, lysis, leakage, and salt tolerance assays and electron microscopy. Antimicrob Agents Chemother 46(6):1914–1920. https://doi.org/10.1128/AAC.46.6.1914-1920.2002

Chassagne F et al (2021) A systematic review of plants with antibacterial activities: a taxonomic and phylogenetic perspective. Front Pharmacol 11:2069. https://doi.org/10.3389/fphar.2020.586548

Chen Y et al (2014) Phytochemical profiles and antioxidant activities in six species of ramie leaves. PLoS One 9(9):e108140. https://doi.org/10.1371/journal.pone.0108140

Chiba A, Sugimoto S, Sato F, Hori S, Mizunoe Y (2015) A refined technique for extraction of extracellular matrices from bacterial biofilms and its applicability. Microb Biotechnol 8(3): 392–403. https://doi.org/10.1111/1751-7915.12155

Chouhan S, Sharma K, Guleria S (2017) Antimicrobial activity of some essential oils-present status and future perspectives. Medicines (Basel, Switzerland) 4(3):58. https://doi.org/10.3390/medicines4030058

Ćirić AD et al (2019) Natural products as biofilm formation antagonists and regulators of quorum sensing functions: a comprehensive review update and future trends. South African J Bot 120: 65–80. https://doi.org/10.1016/j.sajb.2018.09.010
Clark DP, Pazdernik NJ, McGehee MR (2019) Chapter 16: Regulation of transcription in prokaryotes. In: Clark DP, Pazdernik NJ, McGehee MR (eds) Molecular biology. Academic Press, London, pp 522–559
Costa OYA, Raaijmakers JM, Kuramae EE (2018) Microbial extracellular polymeric substances: ecological function and impact on soil aggregation. Front Microbiol 9:1636. https://doi.org/10.3389/fmicb.2018.01636
Costerton JW, Stewart PS, Greenberg EP (1999) Bacterial biofilms: a common cause of persistent infections. Science 284(5418):1318–1322. https://doi.org/10.1126/science.284.5418.1318
Cowan MM (1999) Plant products as antimicrobial agents. Clin Microbiol Rev 12(4):564–582. https://doi.org/10.1128/CMR.12.4.564
Cox CD (1986) Role of pyocyanin in the acquisition of iron from transferrin. Infect Immun 52(1): 263–270. https://doi.org/10.1128/iai.52.1.263-270.1986
Cox SD et al (2000) The mode of antimicrobial action of the essential oil of Melaleuca alternifolia (tea tree oil). J Appl Microbiol 88(1):170–175. https://doi.org/10.1046/j.1365-2672.2000.00943.x
Cushnie TPT, Lamb AJ (2005) Antimicrobial activity of flavonoids. Int J Antimicrob Agents 26(5): 343–356. https://doi.org/10.1016/j.ijantimicag.2005.09.002
de Carvalho CCCR (2018) Marine biofilms: a successful microbial strategy with economic implications. Front Mar Sci 5:126. https://doi.org/10.3389/fmars.2018.00126
Del Pozo JL (2018) Biofilm-related disease. Expert Rev Anti-Infect Ther 16(1):51–65. https://doi.org/10.1080/14787210.2018.1417036
Desai SD, Saheb S, Das K, Haseena S (2015) Phytochemical analysis of Nigella sativa and it's antidiabetic effect. J Pharm Sci Res 7:527–532
Di Martino P (2018) Extracellular polymeric substances, a key element in understanding biofilm phenotype. AIMS Microbiol 4(2):274–288. https://doi.org/10.3934/microbiol.2018.2.274
Diep CS, Baranowski J, Baranowski T (2015) 4–The impact of fruit and vegetable intake on weight management. In: T. B. T.-M, Gill PO (eds) Woodhead publishing series in food science, technology and nutrition. Woodhead Publishing, pp 59–78
Diggle SP, Whiteley M (2020) Microbe profile: Pseudomonas aeruginosa: opportunistic pathogen and lab rat. Microbiology 166(1):30–33. https://doi.org/10.1099/mic.0.000860
Dinges MM, Orwin PM, Schlievert PM (2000) Exotoxins of Staphylococcus aureus. Clin Microbiol Rev 13(1):16–34. https://doi.org/10.1128/CMR.13.1.16
Dunne WM (2002) Bacterial adhesion: seen any good biofilms lately? Clin Microbiol Rev 15(2): 155–166. https://doi.org/10.1128/CMR.15.2.155-166.2002
Eberl L, Molin S, Givskov M (1999) Surface motility of serratia liquefaciens MG1. J Bacteriol 181(6):1703–1712. https://doi.org/10.1128/JB.181.6.1703-1712.1999
Etter D, Schelin J, Schuppler M, Johler S (2020) Staphylococcal enterotoxin C—an update on SEC variants, their structure and properties, and their role in foodborne intoxications. Toxins 12(9): 584. https://doi.org/10.3390/toxins12090584
Fair RJ, Tor Y (2014) Antibiotics and bacterial resistance in the 21st century. Perspect Medicin Chem 6:25–64. https://doi.org/10.4137/PMC.S14459
Feldberg RS et al (1988) In vitro mechanism of inhibition of bacterial cell growth by allicin. Antimicrob Agents Chemother 32(12):1763–1768. https://doi.org/10.1128/AAC.32.12.1763
Ferri M, Ranucci E, Romagnoli P, Giaccone V (2017) Antimicrobial resistance: a global emerging threat to public health systems. Crit Rev Food Sci Nutr 57(13):2857–2876. https://doi.org/10.1080/10408398.2015.1077192
Fisher EL, Otto M, Cheung GYC (2018) Basis of virulence in enterotoxin-mediated staphylococcal food poisoning. Front Microbiol 9:436. https://doi.org/10.3389/fmicb.2018.00436

Flannagan RS, Heit B, Heinrichs DE (2015) Antimicrobial mechanisms of macrophages and the immune evasion strategies of Staphylococcus aureus. Pathogens 4(4):826–868. https://doi.org/10.3390/pathogens4040826

Flemming H-C, Wingender J (2010) The biofilm matrix. Nat Rev Microbiol 8(9):623–633. https://doi.org/10.1038/nrmicro2415

Flemming H-C, Wingender J, Szewzyk U, Steinberg P, Rice SA, Kjelleberg S (2016) Biofilms: an emergent form of bacterial life. Nat Rev Microbiol 14(9):563–575. https://doi.org/10.1038/nrmicro.2016.94

Floyd KA, Eberly AR, Hadjifrangiskou M (2017) In: Deng Y, W. B. T.-B. and I. M. D. Lv (eds) 3-Adhesion of bacteria to surfaces and biofilm formation on medical devices. Woodhead Publishing, pp 47–95

Forni C et al (2019) Beneficial role of phytochemicals on oxidative stress and age-related diseases. Biomed Res Int 2019:8748253. https://doi.org/10.1155/2019/8748253

Francolini I, Norris P, Piozzi A, Donelli G, Stoodley P (2004) Usnic acid, a natural antimicrobial agent able to inhibit bacterial biofilm formation on polymer surfaces. Antimicrob Agents Chemother 48(11):4360–4365. https://doi.org/10.1128/AAC.48.11.4360-4365.2004

M. Frederix and J. A. Downie, "Chapter 2: Quorum sensing: regulating the regulators," vol. 58, R. K. B. T.-A. in M. P. Poole, Ed. Academic Press, 2011, pp. 23–80

Frieri M, Kumar K, Boutin A (2017) Antibiotic resistance. J Infect Public Health 10(4):369–378. https://doi.org/10.1016/j.jiph.2016.08.007

Gao L et al (2019) Mediation of extracellular polymeric substances in microbial reduction of hematite by Shewanella oneidensis MR-1. Front Microbiol 10:575. https://doi.org/10.3389/fmicb.2019.00575

Garrett TR, Bhakoo M, Zhang Z (2008) Bacterial adhesion and biofilms on surfaces. Prog Nat Sci 18(9):1049–1056. https://doi.org/10.1016/j.pnsc.2008.04.001

Ghoshal G (2018) Chapter 2: Biotechnology in food processing and preservation: an overview. In: Holban AM, A. M. B. T.-A. in B. for F. I. Grumezescu (eds) Handbook of Food Bioengineering. Academic Press, pp 27–54

Giaouris E, Simões M (2018) Pathogenic biofilm formation in the food industry and alternative control strategies. In: Dodd CER (ed) Foodborne diseases. Elsevier, London, pp 309–377

Giulia O, O'Toole GA (2021) 'It takes a village': mechanisms underlying antimicrobial recalcitrance of polymicrobial biofilms. J Bacteriol 202(1):e00530–e00519. https://doi.org/10.1128/JB.00530-19

Górniak I, Bartoszewski R, Króliczewski J (2019) Comprehensive review of antimicrobial activities of plant flavonoids. Phytochem Rev 18(1):241–272. https://doi.org/10.1007/s11101-018-9591-z

E. F. Goulden, L. Høj, and M. R. Hall, "8–Microbial management for bacterial pathogen control in invertebrate aquaculture hatcheries," in Woodhead publishing series in food science, technology and nutrition, G. Allan and G. B. T.-A. in A. H. T. Burnell, Eds. Woodhead Publishing, 2013, pp. 246–285

Grainha T, Jorge P, Alves D, Lopes SP, Pereira MO (2020) Unraveling Pseudomonas aeruginosa and Candida albicans communication in coinfection scenarios: insights through network analysis. Front Cell Infect Microbiol 10:676. https://doi.org/10.3389/fcimb.2020.550505

Guan R et al (2021) A review of dietary phytochemicals and their relation to oxidative stress and human diseases. Chemosphere 271:129499. https://doi.org/10.1016/j.chemosphere.2020.129499

Haraguchi H, Oike S, Muroi H, Kubo I (1996) Mode of antibacterial action of totarol, a diterpene from Podocarpus nagi. Planta Med 62(2):122–125. https://doi.org/10.1055/s-2006-957832

Harapanahalli AK, Chen Y, Li J, Busscher HJ, van der Mei HC (2015) Influence of adhesion force on Ica a and cid a gene expression and production of matrix components in Staphylococcus aureus biofilms. Appl Environ Microbiol 81(10):3369–3378. https://doi.org/10.1128/AEM.04178-14

Harriott BS (2019) Biofilms and antibiotics. In: Caplan MJ (ed) Reference module in biomedical sciences. Elsevier, Amsterdam

Harshey RM (2003) Bacterial motility on a surface: many ways to a common goal. Annu Rev Microbiol 57:249–273. https://doi.org/10.1146/annurev.micro.57.030502.091014

Harvey J, Gilmour A (1999) Staphylococcus/Staphylococcus Aureus. In: Robinson FM (ed) Encyclopedia of food microbiology. Elsevier, Oxford, pp 2066–2071

Hawkings NJ, Wood F, Butler CC (2007) Public attitudes towards bacterial resistance: a qualitative study. J Antimicrob Chemother 59(6):1155–1160. https://doi.org/10.1093/jac/dkm103

Hentzer M, Givskov M (2003) Pharmacological inhibition of quorum sensing for the treatment of chronic bacterial infections. J Clin Invest 112(9):1300–1307. https://doi.org/10.1172/JCI20074

Hentzer M et al (2002) Inhibition of quorum sensing in Pseudomonas aeruginosa biofilm bacteria by a halogenated furanone compound. Microbiology 148(Pt 1):87–102. https://doi.org/10.1099/00221287-148-1-87

Holmes AH et al (2016) Understanding the mechanisms and drivers of antimicrobial resistance. Lancet (London, England) 387(10014):176–187. https://doi.org/10.1016/S0140-6736(15)00473-0

Howes M-JR, Simmonds MSJ (2014) The role of phytochemicals as micronutrients in health and disease. Curr Opin Clin Nutr Metab Care 17(6):558–566. [Online]. Available: https://journals.lww.com/co-clinicalnutrition/Fulltext/2014/11000/The_role_of_phytochemicals_as_micronutrients_in.12.aspx

Huang Y, Xiao D, Burton-Freeman B, Edirisinghe I (2016) Chemical changes of bioactive phytochemicals during thermal processing. In: Reference module in food science

Huber B, Eberl L, Feucht W, Polster J (2003) Influence of polyphenols on bacterial biofilm formation and quorum-sensing. Zeitschrift für Naturforsch C 58(11–12):879–884. https://doi.org/10.1515/znc-2003-11-1224

Huycke MM, Sahm DF, Gilmore MS (1998) Multiple-drug resistant enterococci: the nature of the problem and an agenda for the future. Emerg Infect Dis J 4(2):239. https://doi.org/10.3201/eid0402.980211

Ike Y (2017) Pathogenicity of enterococci. Nihon Saikingaku Zasshi 72(2):189–211. https://doi.org/10.3412/jsb.72.189

Issac Abraham SVP, Palani A, Ramaswamy BR, Shunmugiah KP, Arumugam VR (2011) Antiquorum sensing and antibiofilm potential of Capparis spinosa. Arch Med Res 42(8): 658–668. https://doi.org/10.1016/j.arcmed.2011.12.002

Jachlewski S, Jachlewski WD, Linne U, Bräsen C, Wingender J, Siebers B (2015) Isolation of extracellular polymeric substances from biofilms of the Thermoacidophilic archaeon Sulfolobus acidocaldarius. Front Bioeng Biotechnol 3:123. https://doi.org/10.3389/fbioe.2015.00123

Jamal M et al (2018) Bacterial biofilm and associated infections. J Chinese Med Assoc 81(1):7–11. https://doi.org/10.1016/j.jcma.2017.07.012

Jiang Q, Chen J, Yang C, Yin Y, Yao K (2019) Quorum sensing: a prospective therapeutic target for bacterial diseases. Biomed Res Int 2019:2015978. https://doi.org/10.1155/2019/2015978

Jiao Y et al (2010) Characterization of extracellular polymeric substances from acidophilic microbial biofilms. Appl Environ Microbiol 76(9):2916–2922. https://doi.org/10.1128/AEM.02289-09

Jimenez-Garcia SN et al (2018) Chapter 13: Phytochemical and pharmacological properties of secondary metabolites in berries. In: Holban AM, Grumezescu AMBT-TF (eds) Handbook of food bioengineering. Academic Press, pp 397–427

Jindal AK, Pandya K, Khan ID (2015) Antimicrobial resistance: a public health challenge. Med Journal, Armed Forces India 71(2):178–181. https://doi.org/10.1016/j.mjafi.2014.04.011

Julian KG, Blumberg EA (2017) In: Cohen J, Powderly WG, S. M. B. T.-I. D. Fourth E. Opal (eds) Practice point 19–urinary tract infections in kidney transplant recipients. Elsevier, pp 557–558.e1

Kali A, Bhuvaneshwar D, Charles P, Seetha K (2016) Antibacterial synergy of curcumin with antibiotics against biofilm producing clinical bacterial isolates. J Basic Clin Pharm 7:93. https://doi.org/10.4103/0976-0105.183265

Karlowsky JA, Hoban DJ, Hackel MA, Lob SH, Sahm DF (2017) Antimicrobial susceptibility of gram-negative ESKAPE pathogens isolated from hospitalized patients with intra-abdominal and urinary tract infections in Asia-Pacific countries: SMART 2013-2015. J Med Microbiol 66(1): 61–69. https://doi.org/10.1099/jmm.0.000421

Kearns DB (2010) A field guide to bacterial swarming motility. Nat Rev Microbiol 8(9):634–644. https://doi.org/10.1038/nrmicro2405

Khan MSA, Iqbal MMA (2017) Chemical nature of biofilm matrix and its significance. In: Ahmad I (ed) Biofilms in plant and soil health. Wiley, Chichester, pp 151–177

Khan IA, Mirza ZM, Kumar A, Verma V, Qazi GN (2006) Piperine, a phytochemical potentiator of ciprofloxacin against Staphylococcus aureus. Antimicrob Agents Chemother 50(2):810–812. https://doi.org/10.1128/AAC.50.2.810-812.2006

Khan MSA, Zahin M, Hasan S, Husain FM, Ahmad I (2009) Inhibition of quorum sensing regulated bacterial functions by plant essential oils with special reference to clove oil. Lett Appl Microbiol 49(3):354–360. https://doi.org/10.1111/j.1472-765X.2009.02666.x

Khatoon Z, McTiernan CD, Suuronen EJ, Mah T-F, Alarcon EI (2018) Bacterial biofilm formation on implantable devices and approaches to its treatment and prevention. Heliyon 4(12):e01067–e01067. https://doi.org/10.1016/j.heliyon.2018.e01067

Kibe M, Konyole S, Oloo M, Ochieng N, Kathure D (2017) The role of phytochemicals in prevention and control of chronic diseases. Int J Curr Res 9:62540–62543

Klančnik A, Šimunović K, Sterniša M, Ramić D, Smole Možina S, Bucar F (2021) Anti-adhesion activity of phytochemicals to prevent campylobacter jejuni biofilm formation on abiotic surfaces. Phytochem Rev 20(1):55–84. https://doi.org/10.1007/s11101-020-09669-6

Koche D, Shirsat R, Kawale M (2018) An overerview of major classes of phytochemicals: their types and role in disease prevention. Hislopia 9:2016

Köhler T, Curty LK, Barja F, van Delden C, Pechère J-C (2000) Swarming of *Pseudomonas aeruginosa* is dependent on cell-to-cell signaling and requires flagella and pili. J Bacteriol 182(21):5990–5996. https://doi.org/10.1128/JB.182.21.5990-5996.2000

Kongkham B, Prabakaran D, Puttaswamy H (2020) Opportunities and challenges in managing antibiotic resistance in bacteria using plant secondary metabolites. Fitoterapia 147:104762. https://doi.org/10.1016/J.FITOTE.2020.104762

Kontiokari T, Salo J, Eerola E, Uhari M (2005) Cranberry juice and bacterial colonization in children--a placebo-controlled randomized trial. Clin Nutr 24(6):1065–1072. https://doi.org/10.1016/j.clnu.2005.08.009

Koo H, Allan RN, Howlin RP, Stoodley P, Hall-Stoodley L (2017) Targeting microbial biofilms: current and prospective therapeutic strategies. Nat Rev Microbiol 15(12):740–755. https://doi.org/10.1038/nrmicro.2017.99

Kostakioti M, Hadjifrangiskou M, Hultgren SJ (2013) Bacterial biofilms: development, dispersal, and therapeutic strategies in the dawn of the postantibiotic era. Cold Spring Harb Perspect Med 3(4):a010306. https://doi.org/10.1101/cshperspect.a010306

Lade H, Paul D, Kweon JH (2014) N-acyl homoserine lactone-mediated quorum sensing with special reference to use of quorum quenching bacteria in membrane biofouling control. Biomed Res Int 2014:162584. https://doi.org/10.1155/2014/162584

Lahlou M (2013) The success of natural products in drug discovery. Pharmacol Pharm 04:17–31. https://doi.org/10.4236/pp.2013.43A003

Lami R (2019) In: Tommonaro GBT-QS (ed) Chapter 3: Quorum sensing in marine biofilms and environments. Academic Press, pp 55–96

Lebeaux D, Ghigo J-M, Beloin C (2014) Biofilm-related infections: bridging the gap between clinical management and fundamental aspects of recalcitrance toward antibiotics. Microbiol Mol Biol Rev 78(3):510–543. https://doi.org/10.1128/MMBR.00013-14

Lee J-H et al (2011) Apple flavonoid phloretin inhibits Escherichia coli O157:H7 biofilm formation and ameliorates colon inflammation in rats. Infect Immun 79(12):4819–4827. https://doi.org/10.1128/IAI.05580-11

Lee J-H, Park J-H, Cho HS, Joo SW, Cho MH, Lee J (2013) Anti-biofilm activities of quercetin and tannic acid against Staphylococcus aureus. Biofouling 29(5):491–499. https://doi.org/10.1080/08927014.2013.788692

Lee C-R et al (2017) Biology of Acinetobacter baumannii: pathogenesis, antibiotic resistance mechanisms, and prospective treatment options. Front Cell Infect Microbiol 7:55. https://doi.org/10.3389/fcimb.2017.00055

Lemon KP, Higgins DE, Kolter R (2007) Flagellar motility is critical for *listeria monocytogenes* biofilm formation. J Bacteriol 189(12):4418–4424. https://doi.org/10.1128/JB.01967-06

Lemos M et al (2014) The effects of ferulic and salicylic acids on Bacillus cereus and Pseudomonas fluorescens single-and dual-species biofilms. Int Biodeterior Biodegradation 86:42–51. https://doi.org/10.1016/j.ibiod.2013.06.011

Levy SB, Marshall B (2004) Antibacterial resistance worldwide: causes, challenges and responses. Nat Med 10(12):S122–S129. https://doi.org/10.1038/nm1145

Li Y-H, Tian X (2012) Quorum sensing and bacterial social interactions in biofilms. Sensors (Basel) 12(3):2519–2538. https://doi.org/10.3390/s120302519

Liaqat I et al (2018) Flagellar motility plays important role in biofilm formation of Bacillus cereus and Yersinia enterocolitica. Pak J Pharm Sci 31(5):2047–2052

Lihua L, Jianhuit W, Jialini Y, Yayin L, Guanxin L (2013) Effects of allicin on the formation of Pseudomonas aeruginosa biofinm and the production of quorum-sensing controlled virulence factors. Polish J Microbiol 62(3):243–251

Lin YT, Kwon YI, Labbe RG, Shetty K (2005) Inhibition of Helicobacter pylori and associated urease by oregano and cranberry phytochemical synergies. Appl Environ Microbiol 71(12): 8558–8564. https://doi.org/10.1128/AEM.71.12.8558-8564.2005

Liu RH (2013) Health-promoting components of fruits and vegetables in the diet. Adv Nutr 4(3): 384S–392S. https://doi.org/10.3945/an.112.003517

Liu Y, Tong Z, Shi J, Li R, Upton M, Wang Z (2021) Drug repurposing for next-generation combination therapies against multidrug-resistant bacteria. Theranostics 11(10):4910–4928. https://doi.org/10.7150/thno.56205

Llor C, Bjerrum L (2014) Antimicrobial resistance: risk associated with antibiotic overuse and initiatives to reduce the problem. Ther Adv Drug Saf 5(6):229–241. https://doi.org/10.1177/2042098614554919

López Y, Soto SM (2020) The usefulness of microalgae compounds for preventing biofilm infections. Antibiotics 9(1):9. https://doi.org/10.3390/antibiotics9010009

Maeyama R, Kwon IK, Mizunoe Y, Anderson JM, Tanaka M, Matsuda T (2005) Novel bactericidal surface: Catechin-loaded surface-erodible polymer prevents biofilm formation. J Biomed Mater Res A 75(1):146–155. https://doi.org/10.1002/jbm.a.30346

Mandal P, Sinha Babu SP, Mandal NC (2005) Antimicrobial activity of saponins from Acacia auriculiformis. Fitoterapia 76(5):462–465. https://doi.org/10.1016/j.fitote.2005.03.004

Manganaris GA, Goulas V, Mellidou I, Drogoudi P (2018) Antioxidant phytochemicals in fresh produce: exploitation of genotype variation and advancements in analytical protocols. Front Chem 5:95. https://doi.org/10.3389/fchem.2017.00095

Martínez A, Stinco CM, Meléndez-Martínez AJ (2014) Free radical scavenging properties of phytofluene and phytoene isomers as compared to lycopene: a combined experimental and theoretical study. J Phys Chem B 118(33):9819–9825. https://doi.org/10.1021/jp503227j

Mascini EM et al (2006) Genotyping and preemptive isolation to control an outbreak of vancomycin-resistant enterococcus faecium. Clin Infect Dis an Off Publ Infect Dis Soc Am 42(6):739–746. https://doi.org/10.1086/500322

Meireles A, Gonçalves AL, Gomes IB, Simões LC, Manuel S (2015) Methods to study microbial adhesion on abiotic surfaces. AIMS Bioeng 2(4):297–309. https://doi.org/10.3934/bioeng.2015.4.297

Melander R, Basak A, Melander C (2020) Natural products as inspiration for the development of bacterial antibiofilm agents. Nat Prod Rep 37:1454–1477. https://doi.org/10.1039/D0NP00022A
Melzig MF, Bader G, Loose R (2001) Investigations of the mechanism of membrane activity of selected triterpenoid saponins. Planta Med 67(1):43–48. https://doi.org/10.1055/s-2001-10632
Memariani H, Memariani M, Ghasemian A (2019) An overview on anti-biofilm properties of quercetin against bacterial pathogens. World J Microbiol Biotechnol 35(9):143. https://doi.org/10.1007/s11274-019-2719-5
Meredith HR, Srimani JK, Lee AJ, Lopatkin AJ, You L (2015) Collective antibiotic tolerance: mechanisms, dynamics and intervention. Nat Chem Biol 11(3):182–188. https://doi.org/10.1038/nchembio.1754
Micol V, Mateo CR, Shapiro S, Aranda FJ, Villalaín J (2001) Effects of (+)-totarol, a diterpenoid antibacterial agent, on phospholipid model membranes. Biochim Biophys Acta-Biomembr 1511(2):281–290. https://doi.org/10.1016/S0005-2736(01)00284-X
Micoli F, Bagnoli F, Rappuoli R, Serruto D (2021) The role of vaccines in combatting antimicrobial resistance. Nat Rev Microbiol 19(5):287–302. https://doi.org/10.1038/s41579-020-00506-3
Miller MB, Bassler BL (2001) Quorum sensing in bacteria. Annu Rev Microbiol 55:165–199. https://doi.org/10.1146/annurev.micro.55.1.165
Mirzoeva OK, Grishanin RN, Calder PC (1997) Antimicrobial action of propolis and some of its components: the effects on growth, membrane potential and motility of bacteria. Microbiol Res 152(3):239–246. https://doi.org/10.1016/S0944-5013(97)80034-1
Mishra R, Panda AK, De Mandal S, Shakeel M, Bisht SS, Khan J (2020) Natural anti-biofilm agents: strategies to control biofilm-forming pathogens. Front Microbiol 11:2640. https://doi.org/10.3389/fmicb.2020.566325
Monte J, Abreu AC, Borges A, Simões LC, Simões M (2014) Antimicrobial activity of selected phytochemicals against Escherichia coli and Staphylococcus aureus and their biofilms. Pathogens (Basel, Switzerland) 3(2):473–498. https://doi.org/10.3390/pathogens3020473
Mori A, Nishino C, Enoki N, Tawata S (1987) Antibacterial activity and mode of action of plant flavonoids against Proteus vulgaris and Staphylococcus aureus. Phytochemistry 26(8): 2231–2234. https://doi.org/10.1016/S0031-9422(00)84689-0
Moshynets O, Spiers A (2016) Viewing biofilms within the larger context of bacterial aggregations. IntechOpen, London
Muhammad MH et al (2020) Beyond risk: bacterial biofilms and their regulating approaches. Front Microbiol 11:928. https://doi.org/10.3389/fmicb.2020.00928
Mulani MS, Kamble EE, Kumkar SN, Tawre MS, Pardesi KR (2019) Emerging strategies to combat ESKAPE pathogens in the era of antimicrobial resistance: a review. Front Microbiol 10:539. https://doi.org/10.3389/fmicb.2019.00539
Mundy LM, Sahm DF, Gilmore M (2000) Relationships between enterococcal virulence and antimicrobial resistance. Clin Microbiol Rev 13(4):513–522. https://doi.org/10.1128/CMR.13.4.513
Munita JM, Arias CA (2016) Mechanisms of antibiotic resistance. Microbiol Spectr 4(2):11. https://doi.org/10.1128/microbiolspec.VMBF-0016-2015
Nandakumar V, Chittaranjan S, Kurian VM, Doble M (2013) Characteristics of bacterial biofilm associated with implant material in clinical practice. Polym J 45(2):137–152. https://doi.org/10.1038/pj.2012.130
Nazzaro F, Fratianni F, De Martino L, Coppola R, De Feo V (2013) Effect of essential oils on pathogenic bacteria. Pharmaceuticals (Basel) 6(12):1451–1474. https://doi.org/10.3390/ph6121451
Neu HC (1992) The crisis in antibiotic resistance. Science 257(5073):1064–1073. https://doi.org/10.1126/science.257.5073.1064
Nigam A, Gupta D, Sharma A (2014) Treatment of infectious disease: beyond antibiotics. Microbiol Res 169(9):643–651. https://doi.org/10.1016/j.micres.2014.02.009

Niu C, Afre S, Gilbert ES (2006) Subinhibitory concentrations of cinnamaldehyde interfere with quorum sensing. Lett Appl Microbiol 43(5):489–494. https://doi.org/10.1111/j.1472-765X.2006.02001.x

Oussalah M, Caillet S, Lacroix M (2006) Mechanism of action of Spanish oregano, Chinese cinnamon, and savory essential oils against cell membranes and walls of Escherichia coli O157:H7 and listeria monocytogenes. J Food Prot 69(5):1046–1055. https://doi.org/10.4315/0362-028x-69.5.1046

Packiavathy IASV, Agilandeswari P, Musthafa KS, Pandian SK, Ravi AV (2012) Antibiofilm and quorum sensing inhibitory potential of Cuminum cyminum and its secondary metabolite methyl eugenol against gram negative bacterial pathogens. Food Res Int 45(1):85–92. https://doi.org/10.1016/j.foodres.2011.10.022

Packiavathy IASV, Sasikumar P, Pandian SK, Veera Ravi A (2013) Prevention of quorum-sensing-mediated biofilm development and virulence factors production in vibrio spp. by curcumin. Appl Microbiol Biotechnol 97(23):10177–10187. https://doi.org/10.1007/s00253-013-4704-5

Packiavathy IASV, Priya S, Pandian SK, Ravi AV (2014) Inhibition of biofilm development of uropathogens by curcumin–an anti-quorum sensing agent from Curcuma longa. Food Chem 148:453–460. https://doi.org/10.1016/j.foodchem.2012.08.002

Pan S-Y et al (2013) New perspectives on how to discover drugs from herbal medicines: CAM's outstanding contribution to modern therapeutics. Evid Based Complement Alternat Med 2013: 627375. https://doi.org/10.1155/2013/627375

Patra AK (2012) An overview of antimicrobial properties of different classes of phytochemicals. Diet Phytochem Microbes 1:1–32. https://doi.org/10.1007/978-94-007-3926-0_1

Pena RT et al (2019) Relationship between quorum sensing and secretion systems. Front Microbiol 10:1100. https://doi.org/10.3389/fmicb.2019.01100

Phillips PL, Schultz GS (2012) Molecular mechanisms of biofilm infection: biofilm virulence factors. Adv Wound Care 1(3):109–114. https://doi.org/10.1089/wound.2011.0301

Pidwill GR, Gibson JF, Cole J, Renshaw SA, Foster SJ (2021) The role of macrophages in Staphylococcus aureus infection. Front Immunol 11:3506. https://doi.org/10.3389/fimmu.2020.620339

Pollitt EJG, Diggle SP (2017) Defining motility in the Staphylococci. Cell Mol Life Sci 74(16): 2943–2958. https://doi.org/10.1007/s00018-017-2507-z

Potera C (2010) Antibiotic Resistance: biofilm dispersing agent rejuvenates older antibiotics. Environ Health Perspect 118(7):A288. [Online]. Available: https://www.ncbi.nlm.nih.gov/pmc/articles/PMC2920928/

Power E (2006) Impact of antibiotic restrictions: the pharmaceutical perspective. Clin Microbiol Infect 12:25–34. https://doi.org/10.1111/j.1469-0691.2006.01528.x

Prestinaci F, Pezzotti P, Pantosti A (2015) Antimicrobial resistance: a global multifaceted phenomenon. Pathog Glob Health 109(7):309–318. https://doi.org/10.1179/2047773215Y.0000000030

Quave CL et al (2012) Ellagic acid derivatives from Rubus ulmifolius inhibit Staphylococcus aureus biofilm formation and improve response to antibiotics. PLoS One 7(1):e28737. https://doi.org/10.1371/journal.pone.0028737

Ray S, Das S, Suar M (2017) In: Arora G, Sajid A, Kalia VC (eds) Molecular mechanism of drug resistance BT–drug resistance in bacteria, fungi, malaria, and cancer. Springer International Publishing, Cham, pp 47–110

Roberts IS (1996) The biochemistry and genetics of capsular polysaccharide production in bacteria. Annu Rev Microbiol 50:285–315. https://doi.org/10.1146/annurev.micro.50.1.285

Roy R, Tiwari M, Donelli G, Tiwari V (2018) Strategies for combating bacterial biofilms: a focus on anti-biofilm agents and their mechanisms of action. Virulence 9(1):522–554. https://doi.org/10.1080/21505594.2017.1313372

Rutherford ST, Bassler BL (2012) Bacterial quorum sensing: its role in virulence and possibilities for its control. Cold Spring Harb Perspect Med 2(11):a012427. https://doi.org/10.1101/cshperspect.a012427

Sakarikou C, Kostoglou D, Simões M, Giaouris E (2020) Exploitation of plant extracts and phytochemicals against resistant *Salmonella* spp. in biofilms. Food Res Int 128:108806. https://doi.org/10.1016/j.foodres.2019.108806
Santajit S, Indrawattana N (2016) Mechanisms of antimicrobial resistance in ESKAPE pathogens. Biomed Res Int 2016:2475067. https://doi.org/10.1155/2016/2475067
Saranraj P, Sivasakthi S, Deepa M (2016) Phytochemistry of pharmacologically important medicinal plants–a review. Int J Curr Res Chem Pharm Sci 3:56–66. https://doi.org/10.22192/ijcrcps.2016.03.11.009
Sarshar M, Behzadi P, Scribano D, Palamara AT, Ambrosi C (2021) Acinetobacter baumannii: an ancient commensal with weapons of a pathogen. Pathogens 10(4):387. https://doi.org/10.3390/pathogens10040387
Schmitz FJ et al (1998) The effect of reserpine, an inhibitor of multidrug efflux pumps, on the in-vitro activities of ciprofloxacin, sparfloxacin and moxifloxacin against clinical isolates of Staphylococcus aureus. J Antimicrob Chemother 42(6):807–810. https://doi.org/10.1093/jac/42.6.807
Schultz F et al (2020) Targeting ESKAPE pathogens with anti-infective medicinal plants from the greater Mpigi region in Uganda. Sci Rep 10(1):11935. https://doi.org/10.1038/s41598-020-67572-8
Sekyere JO, Asante J (2018) Emerging mechanisms of antimicrobial resistance in bacteria and fungi: advances in the era of genomics. Future Microbiol 13(2):241–262. https://doi.org/10.2217/fmb-2017-0172
Septembre-Malaterre A, Remize F, Poucheret P (2017) Fruits and vegetables, as a source of nutritional compounds and phytochemicals: changes in bioactive compounds during lactic fermentation. Food Res Int 104:86–99. https://doi.org/10.1016/j.foodres.2017.09.031
Sharma AK et al (2017) Bacterial virulence factors: secreted for survival. Indian J Microbiol 57(1):1–10. https://doi.org/10.1007/s12088-016-0625-1
Sharma D, Misba L, Khan AU (2019) Antibiotics versus biofilm: an emerging battleground in microbial communities. Antimicrob Resist Infect Control 8(1):76. https://doi.org/10.1186/s13756-019-0533-3
Sheng G-P, Yu H-Q, Li X-Y (2010) Extracellular polymeric substances (EPS) of microbial aggregates in biological wastewater treatment systems: a review. Biotechnol Adv 28(6):882–894. https://doi.org/10.1016/j.biotechadv.2010.08.001
Šikić Pogačar M, Klančnik A, Bucar F, Langerholc T, Smole Možina S (2016) Anti-adhesion activity of thyme (Thymus vulgaris L.) extract, thyme post-distillation waste, and olive (Olea europea L.) leaf extract against Campylobacter jejuni on polystyrene and intestine epithelial cells. J Sci Food Agric 96(8):2723–2730. https://doi.org/10.1002/jsfa.7391
Simões M, Bennett R, Rosa E (2009) Understanding antimicrobial activities of phytochemicals against multidrug resistant bacteria and biofilms. Nat Prod Rep 26:746–757. https://doi.org/10.1039/b821648g
Singh MK, Singh SK, Singh AV, Verma H, Singh PP, Kumar A (2020) In: B. B. T.-F. and P. P. of P. Prakash (ed) 12–Phytochemicals: intellectual property rights. Academic Press, pp 363–375
Singha T (2012) Microbial extracellular polymeric substances: production, isolation and applications. IOSR J Pharm 2:276–281. https://doi.org/10.9790/3013-0220276281
Sinha Babu SP, Sarkar D, Ghosh NK, Saha A, Sukul NC, Bhattacharya S (Dec. 1997) Enhancement of membrane damage by saponins isolated from Acacia auriculiformis. Jpn J Pharmacol 75(4):451–454. https://doi.org/10.1254/jjp.75.451
Skariyachan S, Garka S (2018) In: Grumezescu Graphenes AMBT-F, Nanotubes (eds) Chapter 1: Exploring the binding potential of carbon nanotubes and fullerene towards major drug targets of multidrug resistant bacterial pathogens and their utility as novel therapeutic agents. William Andrew Publishing, pp 1–29
Skindersoe ME et al (2008) Effects of antibiotics on quorum sensing in Pseudomonas aeruginosa. Antimicrob Agents Chemother 52(10):3648–3663. https://doi.org/10.1128/AAC.01230-07

Smith RP, Barraza I, Quinn RJ, Fortoul MC (2020) Chapter One: The mechanisms and cell signaling pathways of programmed cell death in the bacterial world. In: Spetz JKE, L. B. T.-I. R. of C. and M. B. Galluzzi (eds) Cell death regulation in health and disease-part B, vol 352. Academic Press, pp 1–53

Spellberg B (2014) The future of antibiotics. Crit Care 18(3):228. https://doi.org/10.1186/cc13948

Srinivasan R, Santhakumari S, Poonguzhali P, Geetha M, Dyavaiah M, Xiangmin L (2021) Bacterial biofilm inhibition: a focused review on recent therapeutic strategies for combating the biofilm mediated infections. Front Microbiol 12:1106. [Online]. Available: https://www.frontiersin.org/article/10.3389/fmicb.2021.676458

Stones DH, Krachler AM (2016) Against the tide: the role of bacterial adhesion in host colonization. Biochem Soc Trans 44(6):1571–1580. https://doi.org/10.1042/BST20160186

Suarez M et al (2005) Structure-function characterization and optimization of a plant-derived antibacterial peptide. Antimicrob Agents Chemother 49(9):3847–3857. https://doi.org/10.1128/AAC.49.9.3847-3857.2005

Sundar L, Chang FN (1992) The role of guanosine-3′,5′-bis-pyrophosphate in mediating antimicrobial activity of the antibiotic 3, 5-dihydroxy-4-ethyl-trans-stilbene. Antimicrob Agents Chemother 36(12):2645–2651. https://doi.org/10.1128/AAC.36.12.2645

Tam K, Torres VJ (2019) Staphylococcus aureus secreted toxins and extracellular enzymes. Microbiol Spectr 7(2):315. https://doi.org/10.1128/microbiolspec.GPP3-0039-2018

Tempera G, Corsello S, Genovese C, Caruso FE, Nicolosi D (2010) Inhibitory activity of cranberry extract on the bacterial adhesiveness in the urine of women: an ex-vivo study. Int J Immunopathol Pharmacol 23(2):611–618. https://doi.org/10.1177/039463201002300223

Thakur M, Singh K, Khedkar R (2020) In: B. B. T.-F. and P. P. of P. Prakash (ed) 11–Phytochemicals: extraction process, safety assessment, toxicological evaluations, and regulatory issues. Academic Press, pp 341–361

Thatoi HN, Patra JK, Das SK (2014) Free radical scavenging and antioxidant potential of mangrove plants: a review. Acta Physiol Plant 36(3):561–579. https://doi.org/10.1007/s11738-013-1438-z

Toivanen M et al (2010) Screening of binding activity of Streptococcus pneumoniae, Streptococcus agalactiae and Streptococcus suis to berries and juices. Phytother Res 24(Suppl 1):S95–S101. https://doi.org/10.1002/ptr.2939

Tomada S, Puopolo G, Perazzolli M, Musetti R, Loi N, Pertot I (2016) Pea broth enhances the biocontrol efficacy of Lysobacter capsici AZ78 by triggering cell motility associated with biogenesis of type IV pilus. Front Microbiol 7:1136. https://doi.org/10.3389/fmicb.2016.01136

Trombetta D et al (2005) Mechanisms of antibacterial action of three monoterpenes. Antimicrob Agents Chemother 49(6):2474–2478. https://doi.org/10.1128/AAC.49.6.2474-2478.2005

Ultee A, Kets EP, Smid EJ (1999) Mechanisms of action of carvacrol on the food-borne pathogen Bacillus cereus. Appl Environ Microbiol 65(10):4606–4610. https://doi.org/10.1128/AEM.65.10.4606-4610.1999

N. Upmanyu and V. N. Malviya, “16–Multidrug resistance in pathogenic microorganisms,” P. Chowdhary, A. Raj, D. Verma, Y. B. T.-M. for S. E. H. Akhter, Eds. Elsevier, 2020, pp. 329–341

Vattem DA, Mihalik K, Crixell SH, McLean RJC (2007) Dietary phytochemicals as quorum sensing inhibitors. Fitoterapia 78(4):302–310. https://doi.org/10.1016/j.fitote.2007.03.009

Venkatesan N, Perumal G, Doble M (2015) Bacterial resistance in biofilm-associated bacteria. Future Microbiol 10(11):1743–1750. https://doi.org/10.2217/fmb.15.69

Ventola CL (2015) The antibiotic resistance crisis: part 1: causes and threats. P T 40(4):277–283. [Online]. Available: https://pubmed.ncbi.nlm.nih.gov/25859123

Verderosa AD, Totsika M, Fairfull-Smith KE (2019) Bacterial biofilm eradication agents: a current review. Front Chem 7:824. https://doi.org/10.3389/fchem.2019.00824

Vikram A, Jayaprakasha GK, Jesudhasan PR, Pillai SD, Patil BS (2010) Suppression of bacterial cell-cell signalling, biofilm formation and type III secretion system by citrus flavonoids. J Appl Microbiol 109(2):515–527. https://doi.org/10.1111/j.1365-2672.2010.04677.x

Visioli F, Borsani L, Galli C (2000) Diet and prevention of coronary heart disease: the potential role of phytochemicals. Cardiovasc Res 47(3):419–425. https://doi.org/10.1016/S0008-6363(00)00053-5

Waheed Janabi AH et al (2020) Flavonoid-rich foods (FRF): a promising nutraceutical approach against lifespan-shortening diseases. Iran J Basic Med Sci 23(2):140–153. https://doi.org/10.22038/IJBMS.2019.35125.8353

Walton EW (2014) Topical phytochemicals: applications for wound healing. Adv Skin Wound Care 27(7):324–328. https://doi.org/10.1097/01.ASW.0000450101.97743.0f

Watkins RR, David MZ, Salata RA (2012) Current concepts on the virulence mechanisms of meticillin-resistant Staphylococcus aureus. J Med Microbiol 61(Pt 9):1179–1193. https://doi.org/10.1099/jmm.0.043513-0

C. Whitfield and M. A. Valvano, "Biosynthesis and expression of cell-surface polysaccharides in gram-negative bacteria," vol. 35, A. H. B. T.-A. in M. P. Rose, Ed. Academic Press, 1993, pp. 135–246

Wintola OA, Afolayan AJ (2015) The antibacterial, phytochemicals and antioxidants evaluation of the root extracts of Hydnora africanaThunb. Used as antidysenteric in Eastern Cape Province, South Africa. BMC Complement Altern Med 15(1):307. https://doi.org/10.1186/s12906-015-0835-9

Wu H, Moser C, Wang H-Z, Høiby N, Song Z-J (2015) Strategies for combating bacterial biofilm infections. Int J Oral Sci 7(1):1–7. https://doi.org/10.1038/ijos.2014.65

Xu T et al (2014) Mutations of flagellar genes fli C12, fli A and flh DC of Edwardsiella tarda attenuated bacterial motility, biofilm formation and virulence to fish. J Appl Microbiol 116(2): 236–244. https://doi.org/10.1111/jam.12357

Yan X, Gu S, Shi Y, Cui X, Wen S, Ge J (2017) The effect of emodin on Staphylococcus aureus strains in planktonic form and biofilm formation in vitro. Arch Microbiol 199(9):1267–1275. https://doi.org/10.1007/s00203-017-1396-8

Yang R et al (2018) Phytochemicals from Camellia nitidissima chi flowers reduce the Pyocyanin production and motility of Pseudomonas aeruginosa PAO1. Front Microbiol 8:2640. https://doi.org/10.3389/fmicb.2017.02640

Yin W, Wang Y, Liu L, He J (2019) Biofilms: the microbial 'protective clothing' in extreme environments. Int J Mol Sci 20(14):3423. https://doi.org/10.3390/ijms20143423

Yoo S, Kim K, Nam H, Lee D (2018) Discovering health benefits of phytochemicals with integrated analysis of the molecular network, chemical properties and ethnopharmacological evidence. Nutrients 10(8):1042. https://doi.org/10.3390/nu10081042

Yu M, Gouvinhas I, Rocha J, Barros AIRNA (2021) Phytochemical and antioxidant analysis of medicinal and food plants towards bioactive food and pharmaceutical resources. Sci Rep 11(1): 10041. https://doi.org/10.1038/s41598-021-89437-4

Yuan Z et al (2020) Thymol inhibits biofilm formation, eliminates pre-existing biofilms, and enhances clearance of methicillin-resistant Staphylococcus aureus (MRSA) in a mouse peritoneal implant infection model. Microorganisms 8(1):99. https://doi.org/10.3390/microorganisms8010099

Zhang J, Rui X, Wang L, Guan Y, Sun X, Dong M (2014) Polyphenolic extract from Rosa rugosa tea inhibits bacterial quorum sensing and biofilm formation. Food Control 42:125–131. https://doi.org/10.1016/j.foodcont.2014.02.001

Zhang Y-J et al (2015) Antioxidant phytochemicals for the prevention and treatment of chronic diseases. Molecules 20(12):21138–21156. https://doi.org/10.3390/molecules201219753

Zhang K, Li X, Yu C, Wang Y (2020a) Promising therapeutic strategies against microbial biofilm challenges. Front Cell Infect Microbiol 10:359. https://doi.org/10.3389/fcimb.2020.00359

Zhang L et al (2020b) Is combined medication with natural medicine a promising therapy for bacterial biofilm infection? Biomed Pharmacother 128:110184. https://doi.org/10.1016/j.biopha.2020.110184

Zhao Z, Ma L (2015) In: Tang Y-W, Sussman M, Liu D, Poxton I, J. B. T.-M. M. M. Second E. Schwartzman (eds) Chapter 45: Skin and soft tissue infections: a clinical overview. Academic Press, Boston, pp 825–835

Current and Future Applications to Control Polymicrobial Biofilms Associated with Oral Disease

Peter Zilm, Laura S. Weyrich, Richard Bright, Simona Gatej, Giampiero Rossi-Fedele, Sabine Selbach, Don Hashan Kevin Ketagoda, Aws Alani, Dimitra Lekkas, and Krasimir Vasilev

Abstract In 1683, Antony van Leeuwenhoek made drawings of "little animals" taken from scrapings from the teeth observed under his newly invented microscope. As such, oral biofilms were the first studied polymicrobial biofilms.

The oral environment is a complex bio-system that consists of up to 800 different bacterial species. Collectively referred to as the oral microbiota, the ecosystem has distinct environments that are affected by *inter alia*, redox potential, pH, and saliva flow.

Oral diseases are primarily the result of a shift in oral biofilm homeostasis to dysbiosis as the microbial composition of the biofilm reacts to changes in the local environment. These changes lead to an imbalance in biofilm composition as low-level commensal organisms proliferate in response to exposure to dietary sugars, acids, poor saliva flow, or an abnormal host inflammatory response. Here, we describe current clinical treatments for dental caries, periodontal and endodontic diseases, and denture stomatitis. We also discuss the advantage of fluoride and the use of oral antiseptics as adjuncts to clinical intervention. The use of new innovative treatments such as D-amino acids, quorum sensing inhibitors, probiotics, and oral

P. Zilm (✉) · S. Gatej · G. Rossi-Fedele · S. Selbach · D. H. K. Ketagoda · D. Lekkas
The Adelaide Dental School, The University of Adelaide, Adelaide, SA, Australia
e-mail: peter.zilm@adelaide.edu.au

L. S. Weyrich
Department of Anthropology and Huck Institutes of the Life Sciences, Pennsylvania State University and School of Biological Sciences, University of Adelaide, Adelaide, SA, Australia

R. Bright
The University of South Australia, STEM, Adelaide, SA, Australia

A. Alani
Department of Restorative Dentistry and Traumatology, Kings College Hospital, London, UK

K. Vasilev
The University of South Australia, STEM, Adelaide, SA, Australia

College of Medicine and Public Health, Flinders, University, Adelaide, SA, Australia

K. Richter, K. Nørskov Kragh (eds.), *Antibiofilm Strategies*, Springer Series on Biofilms 11, https://doi.org/10.1007/978-3-031-10992-8_16

microbiome transplantation is also discussed along with nanotechnology and advances in endodontics that will influence modern dentistry.

1 The Oral Microbiome: Understanding the Complexity

The human body contains 1.3 times more microorganisms than human cells, suggesting that humans are more microbial than human (Sender et al. 2016). In fact, over 1,800 species of bacteria, fungi, parasites, and viruses (collectively referred to as a microbiome) have been described in the human body, accounting for 1.4 kg of human body weight—equal in mass to the human brain (Turnbaugh et al. 2007). These microbes also maintain over 99% of the genetic diversity within the human body, as each human contains 2–5 million microbial genes, compared to <30,000 human genes (Turnbaugh et al. 2006; Venter et al. 2001). Nearly every surface of the body, inside and out, has been colonized by unique microbial communities, including the gastrointestinal, urogenital, oral and respiratory tracts, skin, and the surface of the eye (Costello et al. 2009; Shin et al. 2016).

Of these, the oral microbiome is one of the most stable, yet complex, microbiomes in the whole body. While more than 800 bacterial microbes have been identifed in the mouths of different human populations, each person maintains roughly 200 bacterial species. The oral microbiome within a single person also consists of several distinct oral microbial assemblages associated with different locations and tissues in the mouth, including the teeth, gingival sulcus, attached gingiva, tongue, cheek, lip, and hard and soft palates (Dewhirst et al. 2010; Kilian et al. 2016).

The microbiomes on hard surface of the mouth are typically characterised as biofilms (i.e. dental plaque or tartar) (Kilian et al. 2016). Here, microorganisms in these communities form close, interdependent relationships that have been nutured through thousands of years of co-evolution (Fellows Yates et al. 2021; Weyrich et al. 2017). Plaque biofilms build and develop by leveraging interactions between micro-organisms, host factors (salivary flow rate and buffering capacity), and dietary inputs (e.g. fermentable carbohydrates) (Marsh and Martin 2009) and are notoriously difficult to remove or treat because the microbes are enmeshed in a protective extracellular matrix, making them recalcitrant to antibiotics. Plaque biofilms on teeth can also be highly heterogeneous, even within a single person's mouth, as the microbial composition can differ according to which tooth and tooth surface they are bound (Simon-Soro et al. 2013). These micro-ecosystems may be deteremined by access to salivary proteins, oxygen, and other factors, but further research is needed to better understand this variation (Mark Welch et al. 2020).

Our understanding of the local environmental factors that shape the quantity and diversity of microbes that live in the mouth is growing almost daily. Many of the oral microbes within these microbiomes are shared within a family or population, as they are vertically inherited or transferred from a primary care giver (*e.g.* mother) early in life (Caufield et al. 1988). In contrast, there is evidence of unique microorganisms found in the mouth that can be introduced by diet or distinct environmental exposures (Handsley-Davis et al. 2020). There is also evidence that the environment, smoking, salivary gland function, oral hygiene, and diet also play secondary roles in shaping oral microbiota (Shaw et al. 2017). Dietary influences are currently poorly described, but impacts of coffee and tea, as well as broad scale shifts in fat and protein consumption all help contribute to the types of microorganisms in the mouth (Weyrich 2021). These shifts in oral microbiota composition can be harmless or can be linked to disease (e.g. a dysbiosis or shift from a comparative, healthy community), as oral microbiome composition has been linked to almost all oral diseases examined to date, including caries, periodontal disease, gingivitis, halitosis, and oral cancers (Kilian et al. 2016).

2 Oral Diseases Caused by Bacterial Biofilms

2.1 Dental Caries

Dental caries is a common and costly, preventable multifactorial biofilm disease affecting about 60–90% of people and is characterised by the demineralisation of the tooth's enamel (Zhu et al. 2013). The annual treatment cost of oral biofilm-related diseases is approximately $81 billion in the USA alone and according to reports from the National Institute of Health (NIH), dental biofilm infections are suggested to increase by 60% in the developed world alone (Chrisopoulos et al. 2016).

The oral microbiota consists of several distinct microbial communities (Dewhirst et al. 2010; Kilian et al. 2016). Although 800 different oral bacterial species have been described in different human populations, each individual maintains ~200 bacterial species with about 50% thought to be unculturable (Aas et al. 2005; Kolenbrander et al. 2002; Marsh 2006). Typically, the proportions of bacteria in a healthy oral microbiota are distinct from those found in disease states (Xiao et al. 2016). Regular exposure to dietary sucrose and acids, poor saliva flow and quantity, and failure to reduce plaque build-up cause the plaque pH to become more acidic and this gives a selective advantage to aciduric (acid tolerant) and acidogenic (acid producing) bacteria that exist in small numbers as commensal bacteria (Marsh 2003, 2006; Takahashi and Nyvad 2008). The bacterial composition of the biofilm therefore reacts to changes in the local environment and can lead to an imbalance in composition, changes in the overall metabolic activity of the biofilm, which subsequently leads to disease (dental caries).

The main modulators of the development of cariogenic biofilms are collectively known as mutans streptococci, which includes *Streptococcus mutans* and *Streptococcus sobrinus*. *S. mutans* is an opportunistic pathogen due to its ability to liberate lactic acid and to proliferate in acidic environments (Pitts et al. 2017; Salli et al. 2016).

Biofilms (dental plaque) are notoriously difficult to remove or treat as the bacteria are enmeshed in a protective extracellular matrix. Bacterial biofilms attach to acquired pellicle consisting of glycoproteins and macromolecules found in saliva and range in thickness from 1 μm to 10 μm depending on the site (Fejerskov and Kidd 2008). Acquired pellicle also allows for specific short-ranged interactions which act as receptors for adhesins on the microbial cell surface of early colonisers including Streptococci and Actinomyces. When left undisturbed, late colonisers adhere through cellular co-adhesion producing a complex structure consisting of different bacterial species involving complex environmental and metabolic interactions which form mature dental plaque (Marsh and Martin 2009; Kolenbrander et al. 2002).

The inefficiency of common chemotherapeutics against cariogenic biofilms is due to poor retention of topically introduced agents on dental surfaces, reduced drug availability within the biofilm matrix, and drug recalcitrance of biofilm bacteria.

2.2 *Periodontal Disease*

Periodontitis is a bacterial biofilm-induced chronic inflammatory disease characterised by the destruction of the periodontium (supporting structures of teeth) that leads to tooth exfoliation in severe cases. Present in all age groups, the disease has world-wide financial implications with an estimated $US39 billion treatment costs in 2015 alone (Righolt et al. 2018).

The complex interactions between the host and the subgingival (below the gum line) biofilm have been studied intensively over the last 40 years yet the initial cause of the disease remains equivocal (Dongari-Bagtzoglou 2008; Van Dyke et al. 2020). The microbial profile of the healthy periodontium consists of commensal Gram-positive organisms, which include Streptococci, Corynebacteria, Rothia spp. which co-exist in homeostasis with the host (Kirst et al. 2015). Periodontal disease is initiated by the host response to bacterial infection caused by the accumulation of dental plaque (a polymicrobial biofilm) around the gingival margin which can lead to a reversible inflammatory condition known as gingivitis. The biofilm consists mainly of commensal bacteria and leads to localised inflammation, pocket formation below the gingival margin, and an increase in microbial diversity (Lamont et al. 2018). Removal of the biofilm may lead to a resolution of inflammation and the gingivitis; however, in certain susceptible individuals, the disease can progress to chronic periodontitis (Van Dyke et al. 2020).

The mechanism or trigger for the progression to periodontitis remains equivocal. The disease is characterised by changes in the diversity, richness, and relative proportions of species in the subgingival biofilm. This is driven by the transition to an anaerobic environment and the influx of nutrients (hemin, amino acids) as a result of the increased flow of gingival crevicular fluid into the developing pocket (Lamont et al. 2018; Naginyte et al. 2019). Transcriptomic studies also indicate that the altered environment induces the upregulation of virulence factors in the developing polymicrobial biofilm (Lamont et al. 2018; Yost et al. 2015).

Significantly, there is a shift to a predominance of Gram-negative, anaerobic species (Eloe-Fadrosh and Rasko 2013) with the key periodontal pathogens termed the "red complex" (How et al. 2016; Socransky et al. 1998) which includes *Porphyromonas gingivalis*, *Tannerella forsythia,* and *Treponema denticola*. The most significant of these is *P. gingivalis* which is considered a "keystone" pathogen in subgingival biofilms (Honda 2011) as it is detected in approximately 80% of periodontitis patients (Bao et al. 2014; Griffen et al. 1998; Yang et al. 2004). It plays a major role in the dysbiotic ratio of beneficial and pathogenic microbial communities within the oral cavity, resulting in complete disruption of the homeostatic environment required to maintain oral health. It is unclear whether the host inflammatory response or the biofilm dysbiosis is the cause of the progression from gingivitis to periodontitis, but it is recognised that both are necessary (Van Dyke et al. 2020). The reduction or removal of the bacterial biofilm is an important clinical intervention that reduces inflammation (Chen et al. 2018) and leads to a resolution of the disease in many cases. Taken together, current understanding is that periodontal disease is a complex interplay between the host inflammatory response and the oral microbiome.

Further insights and a review of current thinking can be found in the recent review by Van Dyke et al. 2020.

2.3 *Endodontic Infections*

Endodontic diseases are associated with the loss of integrity of the external surface of the tooth due to caries, traumatic dental injuries, tooth wear, and/or operative procedures related to dentistry. These allow the exposure of the root canal space to the microbiota normally residing in the oral cavity leading to the demise of the dental pulp and colonisation of the root canal space. The root canal walls are non-shedding surfaces with routine oral hygiene procedures being unable to alter the environment, or the intracanal infection as inaccessible. The internal tooth surface presents with an irregular surface that allows for biofilm attachment, biofilm formation, and subsequent possible penetration of the channels present in the root canal wall (dentinal tubuli) (Sundqvist and Figdor 2003), as illustrated in Fig. 1. Overall, the root canal space has complex and highly variable anatomy that depends on multiple factors including tooth type and ethnic group (Ahmed and Rossi-Fedele 2020). An example of the morphological complexities in a mandibular premolar tooth is illustrated in Fig. 2.

The endodontic milieu is particular in regard to its environment, including the source of nutrients, redox potential, and availability of oxygen, amongst others (Sundqvist 1994). This, together with antagonists and synergistic relationships between different microorganisms, leads to a selection of specific species in the colonisation, establishment and affects the composition of the intracanal infection and biofilms (Sundqvist 1994). Endodontic microbiological research has extensively focused on the identification of species associated with endodontic diseases using

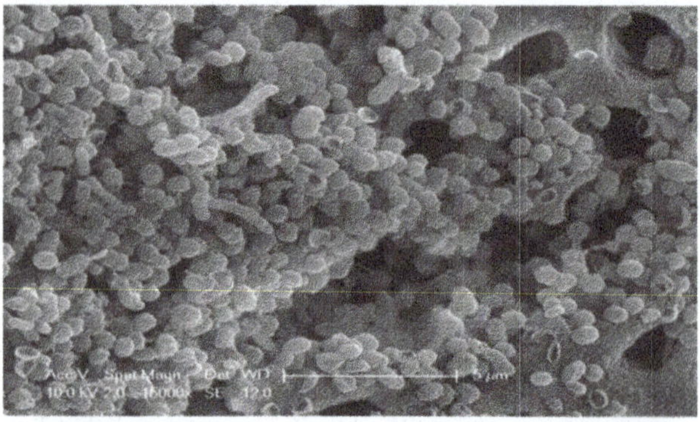

Fig. 1 Multispecies engineered biofilms on root canal surface. Image from Chan et al. (2020)

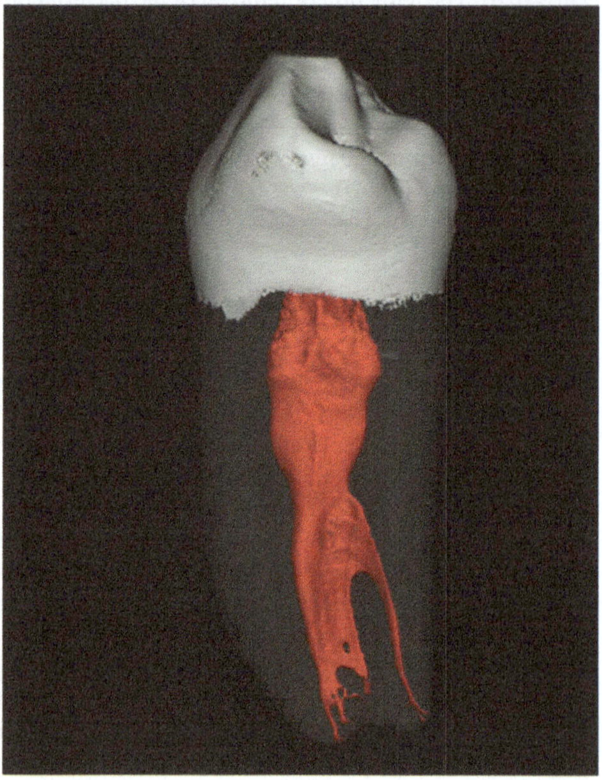

Fig. 2 Micro-computed tomography image illustrating the complexity of the root canal morphology of a mandibular premolar tooth (image courtesy of Dr Hany Ali Mohammed, University of Malaya, Kuala Lumpur, Malaysia)

cultural and molecular techniques, without a definitive conclusion reached (Sundqvist 1994; Sundqvist and Figdor 2003). What is evident is that multispecies biofilms, which may include fungi, are present in this field of dentistry (Sundqvist 1994; Yoo et al. 2020). From the clinical standpoint, an important issue is to differentiate between biofilms present in previously untreated canals and root filled canals. In the latter clinical presentation, the microbiota often includes species that are purportedly more able to survive current biocidal agents commonly used in clinical practice (Sundqvist and Figdor 2003). Furthermore, microorganisms in previously filled root canals are assumed to be in a physiological state, which allows survival to the multiple stresses of a root canal space following root canal filling, though still often viable. The latter leads to a further limitation in the available nutrients, among others, in an already unfavourable environment (Sundqvist and Figdor 2003).

Root canal infections are the causative agent of apical periodontitis, which is a chronic and/or acute inflammatory disease caused by the host response to the presence of intracanal biofilms (Peddis et al. 2019). Apical periodontitis is highly prevalent globally, and it has been calculated that half of the world population presents with this disease (Tiburcio-Machado et al. 2021). Furthermore, apical periodontitis, being a chronic inflammatory disease can participate in the inflammatory burden and has been associated with systemic conditions such as diabetes, cardiovascular, and hepatic diseases (Cintra et al. 2021). This highlights the significance of endodontic infections in general health.

2.4 Denture Stomatitis

Denture stomatitis is a chronic inflammatory condition of the oral mucosal tissues which supports either a full or partial removable prosthesis (Gendreau and Loewy 2011). The condition is largely asymptomatic and almost always present in the maxillary as opposed to mandibular arch.

Any prosthesis introduced into the oral cavity represents an opportunity for oral biofilm to develop. This is more acute if the prosthesis is not optimally provided or maintained by the patient or the dental professional. The surface of dentures provides ample opportunity for debris to develop, otherwise known as "denture plaque", resulting in the development of a mature biofilm. Unfortunately, the elderly patient group are more likely to be denture wearers and also exhibit poorer oral hygiene habits (Frenkel et al. 2000). This is also set on a backdrop of the patients presenting with medical co-morbidities, such as arthritis, physically hampering self-care, or neurological conditions, such as dementia, making compliance challenging (Martori et al. 2014). Polypharmacy is likely to make xerostomia more likely (Martori et al. 2014).

Candida albicans is heavily implicated in the development of denture stomatitis and accounts for the majority of cases (Dagistan et al. 2009). Candida itself is a commensal organism but due to change in oral environmental and systemic factors its role migrates to that of a parasite. Candida co-aggregates with bacteria commonly associated with other conditions such as *Porphyromonas gingivalis* (periodontal) or *Streptococcus mutans* (caries) resulting in a general proliferation within the denture substrate (Sachdeo et al. 2008). The relative corrugated, rough, porous, and large surface area provides an excellent environment for biofilm to develop. The fit surface of dentures although seemingly smooth to the naked eye contains numerous areas and voids that can be inhabited by microorganisms. If patients wear their dentures at night, a common presenting sign, candida can colonise the denture and flourish undisturbed in an anaerobic, low pH environment created by the low salivary flow during sleep. Other risk factors include sugar consumption where glucose encourages the growth of fungal species whilst also improving biofilm adhesion to the denture substrate (Martori et al. 2014). Immunocompromised individuals and those with dietary deficiencies are also more likely to exhibit denture stomatitis (Nikawa et al. 1998).

2.5 Current Management of Denture Stomatitis

Traditional management of the condition is largely instigated through improved hygiene measures in debriding the surface of the denture through mechanical brushing to disrupt biofilm formation coupled with refraining from wearing the denture at night (Lombardi and Budtz-Jorgensen 1993). Where the clinical denture fit is suboptimal, improving marginal fit through either a reline or a new denture will also reduce the ingress of plaque between the denture and the palate.

2.6 Future Applications to Control Denture Stomatitis

With the increased advent of "smart" materials, novel approaches to managing denture stomatitis may emerge (Alani et al. 2009). These may broadly either prevent the biofilm from adhering initially or target specific organisms within the microbial scaffold to eradicate once established (Wang et al. 2015). Incorporating curcumin, a naturally occurring substance that has broad antimicrobial properties, into the denture base, has been investigated. Curcumin demonstrated anti-adhesive properties against Candida albicans biofilms resulting in decreased biofilm formation (Shahzad et al. 2014; Alalwan et al. 2017). Utilising nanoparticles incorporated into denture bases has also been investigated to create surfaces that can prevent biofilm formation or a topography less lucrative for adhesion (Gad et al. 2017).

3 Current Strategies for Preventing Disease

3.1 Clinical Intervention for the Management of Dental Caries

Scientific evidence and improved understanding about the aetiology and complex nature of the dental caries process has seen a monumental shift in the clinical management of caries away from just "drilling and filling holes in teeth". Caries is a biofilm-mediated, sugar-driven, multifactorial and dynamic disease that alters the balance between pathological and protective factors and results in phasic demineralisation-remineralisation of the dental hard tissues (Philip et al. 2018; Pitts et al. 2017). The current philosophy underpinning the clinical management of dental caries is a Minimal Intervention (MI) approach that includes preventing new caries formation, identifying and arresting (remineralisation) early non-cavitated carious lesions, and preserving tooth structure once the tooth has irreversibly cavitated. Key preventive strategies at an individual level include identifying and modifying the caries risk (pathological) factors and promoting protective factors. Our current understanding of the caries process is thus important to enable an evidence-based and holistic approach to the prevention of dental caries over the lifespan of a person, from birth to old age.

Fluoride has a predominantly physico-chemical preventive role in caries management. Daily tooth brushing with fluoride containing toothpastes (for example, 1100 ppm) provides a low and constant source of fluoride ion close to the tooth surface to provide resistance against and repair any early demineralisation. Professionally applied fluoride products such as varnishes and gels can arrest early non-cavitated lesions (Featherstone et al. 2018). Products containing calcium in different forms (e.g. calcium stabilised by casein derivatives) can also be used to aid remineralisation, and these products are useful for patients who have reduction in salivary protective factors (including calcium and phosphate ions that are needed as part of remineralization) (Reynolds 2009).

With our improved understanding of the role of the oral microbiome in the caries process, preventive strategies should be aimed to not just physically prevent or repair demineralised tooth structure, but to manipulate the oral environment to encourage an ecologically balanced and diverse microbiome that favours remineralisation over demineralisation (Featherstone et al. 2018).

Fermentable sugars are the drivers of biofilm dysbiosis and cariogenic activity and common patient advice is to reduce or replace dietary fermentable sugars. The use of non-fermentative sugar polyols, such as xylitol, sorbitol, and erythritol has also been used to try to inhibit virulent cariogenic bacteria and allow healthy bacteria to proliferate in the plaque biofilm. Recent studies indicate that erythritol may be more effective in preventing caries compared with xylitol (de Cock et al. 2016; Falony et al. 2016). Diet modification does require behaviour change and a motivated patient and hence diet control can be difficult to achieve.

Mechanical disruption and reduction of the plaque biofilm through the use of tooth brushing and interdental cleaning is the traditional approach to biofilm control, and also requires sustained patient behaviour change. More novel approaches that aim to enhance the growth of health-promoting microbial communities within the plaque biofilm have been proposed. Two of these will be discussed briefly here, and in more detail later in this chapter. Prebiotics such as arginine can alkalinise the biofilm and reduce the growth of acidophilic and aciduric bacteria (Liu et al. 2012). Recent studies have shown promising anti-caries effects from arginine containing toothpaste (Li et al. 2015). Further research is needed to confirm initial findings to wider population groups. There is also emerging evidence of the ecological role of casein phosphopeptide-amorphous calcium phosphate (CPP-ACP) by anti-adhesion, buffering, and biofilm disrupting actions (Philip and Walsh 2019; Sionov et al. 2021).

3.2 Clinical Intervention for the Management of Periodontal Disease

The keystone to prevent and control biofilm-associated periodontal diseases is effective biofilm management undertaken in collaboration between the patient and dental professional. Non-specific biofilm mass reduction can be achieved by regular oral hygiene measures at home, namely thorough tooth brushing with supplemental interdental cleaning and the use of adjunctive antiseptic agents like dentifrices or mouth rinses. The space below the gum level remains mostly inaccessible by any home care product and individuals are unable to accomplish a complete biofilm removal.

Conventionally dental professionals use micro bladed hand instruments (scalers, curettes) or powered instruments (sonic, ultrasonic) to gain access into subgingival areas. Following an overlapping stroke pattern, the tooth and root surface is denuded from biofilm and calcified deposits. Regular professional periodontal care visits should be undertaken 3 to 12 monthly depending on the patient's risk profile (Sanz et al. 2020). Abrasion of the tooth's hard substance might become substantial during repeated instrumentation (Flemmig et al. 2007). The new technology of air polishing with low abrasive powders (see Fig. 3) is more effective and efficient in disrupting and removing biofilm and causes less impact on hard and soft tissues than traditional instrumentation does (Bastendorf et al. 2021). Nevertheless, a long-term elimination of periodontopathogenic bacteria is not achieved by either technique (Reinhardt et al. 2019). Adjunctive laser application has gained popularity, but currently there is insufficient evidence for its recommendation (Sanz et al. 2020).

Systemic antibiotics in combination with instrumentation improve the clinical treatment outcome significantly (Teughels et al. 2020). However, microbiological shifts after root surface instrumentation occur regardless of the use of adjunctive systematic antibiotics (Dilber et al. 2020). Routine use is not recommended due to a high frequency of side effects as well as resistance formation in pathogens (Sanz et al. 2020).

Besides biofilm management, host and environmental factors modifying the inflammatory immune response such as diabetes and smoking must be recognised.

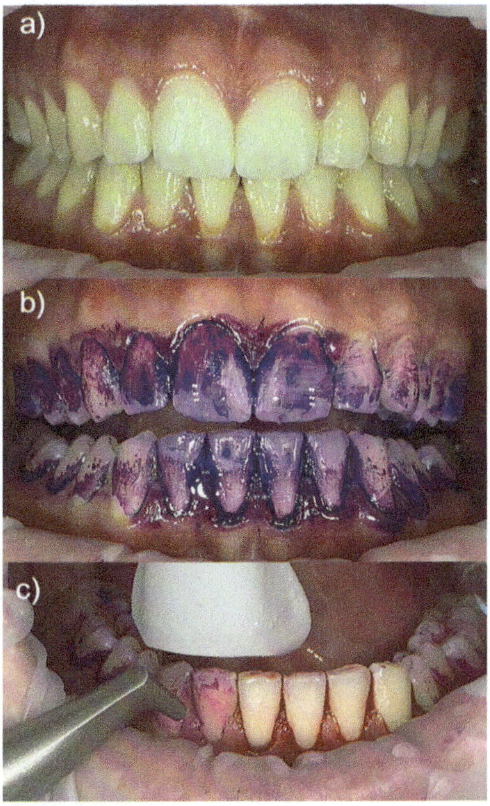

Fig. 3 Supragingival biofilm before (**a**) and after disclosing (**b**). Interdental spaces and gingival margins present an increased accumulation of matured biofilm (blue/dark purple colour) due to insufficient oral hygiene practice. (**c**) An air polishing device with a low abrasive powder is used to disrupt and remove the biofilm in a minimally invasive targeted approach. Bleeding from the gingival margins is indicative of a pre-existing tissue inflammation. Photos: Courtesy, Tabitha Acret BOH, 2021

3.3 Clinical Intervention for the Management of Endodontic Disease

Intracanal biofilms are the causative agent of apical periodontitis, a chronic inflammatory disease affecting the tissues surrounding the teeth (Fig. 4). Current strategies to prevent and/or treat apical periodontitis include root canal treatment, apical surgery, or tooth extraction. Tooth retention is considered desirable as tooth loss has a negative effect on mastication, aesthetics, and overall quality of life (Tan et al. 2016). Root canal treatment encompasses various steps. The cause of loss of integrity of the external surface of the tooth needs to be identified and addressed to prevent the recurrence of microorganism penetration into the root canal space. Subsequently, different endodontic files are used to clean and shape the root canal and eliminate its contents. Irrigants solutions with biocidal activity, more commonly sodium hypochlorite, are used with the aim to alter biofilms that remained undisturbed by the instrumentation process (Savani et al. 2014). The latter should also address the various anatomical variations and irregularities highlighted previously considering that a large percent of the root canal surface remains untouched by the endodontic

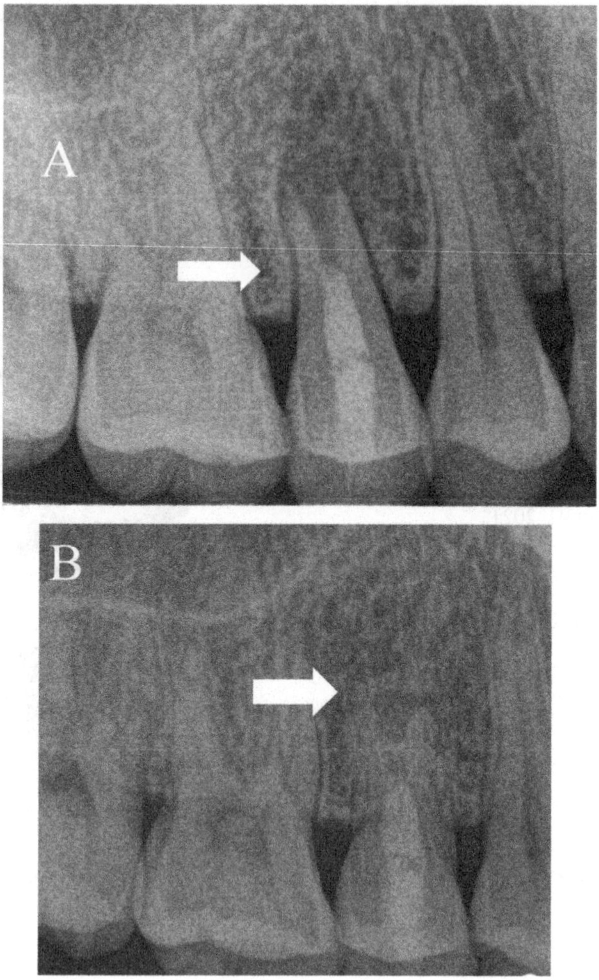

Fig. 4 Apical periodontitis on second maxillary premolar tooth treated using regenerative endodontics. Preoperative periapical radiograph showing a radiolucent area (arrow) in the periapical area suggestive of apical periodontitis (A). One year recall showing the formation of hard tissue intracanal and the disappearance of the previously present radiolucency (arrow), suggestive of healing (B). Case courtesy of Dr Bill Kahler, Queensland.

files (Paque et al. 2009). Some clinicians also include an interim intracanal medicament, such as calcium hydroxide, to reduce further the microbial load (Sharma et al. 2018). Finally, root canals are filled, commonly using a core material made of zinc oxide and gutta-percha (gutta-percha points) and a sealer that sets after completion of the procedure. The previously described steps aim towards the elimination of intracanal biofilms and to create a local environment where any residual microorganisms are unable to elicit a host response. Apical surgery aims to remove those anatomical complexities likely harboring intracanal infection and to disinfect and fill the apical part of the endodontic space.

3.4 Non-clinical Intervention to Prevent Oral Diseases

Clinical intervention is one of the several strategies used to limit the development of oral biofilms that cause oral diseases. Issues related to dental anxiety associated with fear of pain and cost are among several factors that reduce the chance of people visiting the dentist (Armfield 2012). As such, public health strategies such as the fluoridation of drinking water has seen a dramatic decrease in the incidence of dental caries. The development of oral antiseptics has also proven to be a useful adjunct to maintaining good oral health by limiting and reducing the development of dysbiotic oral biofilms. Being freely available, a number of oral care products are marketed by large oral care companies and contain antiseptics that have a proven record based on scientific research. Here we discuss two of the most commonly used oral antiseptics and discuss how fluoride has played a significant role in the prevention of dental caries.

4 Mouthwash Antiseptics

4.1 Cetylpyridinium Chloride

The use of mouthwashes or dentifrices containing antiplaque or antimicrobial agents has been the predominant method of controlling oral biofilms and their associated diseases over almost 80 years (Huyck 1945). They are widely available as over the counter products, so their use has been uncontrolled, and little is known about the potential risks associated with antiseptic resistance. Here we describe the mechanism of action of two most commonly used oral antiseptics and discuss their effect on oral biofilms.

First used in the 1930s as an antiseptic, cetylpyridinium chloride (CPC) is a mono-cationic quaternary ammonium compound which consists of quaternary nitrogen connected with one or more hydrophobic (alkyl) side chains (Gilbert and Moore 2005). The length of the alkyl side chain determines its antimicrobial efficacy with the greatest effect against Gram-positive bacteria being 12 to 14 carbons and 14 to 16 carbons for Gram-negatives (Maris 1995). The positively charged pyridine ion is thought to displace Mg^{2+} and Ca^{2+} ions that bind to the negative charge of the bacterial cell surface and the alkyl tail integrates into the lipid membrane which interferes with cellular osmoregulation at low concentrations or causes disintegration of membranes at high concentrations (Gilbert and Moore 2005). More detail on the effect of CPC on biofilms can be found in the review by Mao et al (2020).

The vast majority of research into oral biofilms has relied on dual or the so-called polymicrobial systems that may consist of only 3 species (Cieplik et al. 2019). More recently, the use of confocal microscopy and live/dead staining has enabled the study of much more complex biofilm systems derived from whole saliva or dental plaque. In a dual species system, the cariogenic organism *Streptococcus mutans* showed a decreased susceptibility to CPC when co-cultured with *Veillonella parvula* (Luppens et al. 2008) compared to axenic cultures of *S. mutans*. Cieplik et al. (2019) used a biofilm system comprised of *S. mutans*, *Actinomyces naeslundii*, and *Actinomyces odontolyticus* and found that 0.1% CPC reduced the levels of *S. mutans* by 4 log10, of *A. naeslundii* by 6 log10, and of *A. odontolyticus* by 5.2 log10 after 10 min exposure (Cieplik et al. 2019).

In a more sophisticated system, biofilms derived from saliva were grown on hydroxyapatite (the inorganic mineral in tooth enamel) using a disc biofilm reactor. Biofilms treated twice daily with 0.075% CPC for four days showed a 3 log10 fall in numbers compared to controls (Latimer et al. 2015).

4.2 *Chlorhexidine*

Chlorhexidine is a bisbiguanide base, cationic broad-spectrum antiseptic that has been used as an antimicrobial agent (chlorhexidine gluconate) for almost 40 years for the control of oral biofilms associated with caries, gingivitis, and periodontal disease (Marsh 2010; Nagle and Turnbull 1978). It is commonly used as an adjunct to brushing and flossing and is favoured due to its substantivity as about 30% is retained in the mouth following rinsing (Cummins and Creeth 1992; Varoni et al. 2012). This property is due to chlorhexidine binding to pellicle—the salivary glycoproteins that cover all hard and mucosal surfaces in the mouth. On rinsing, its bactericidal action can be maintained over an extended period thus inhibiting biofilm formation for up to12 hours (Bonesvoll 1977) before being slowly displaced by calcium ions (Greenstein et al. 1986).

Its effect is concentration dependent as high concentrations dissolve cell membranes causing the precipitation of cellular proteins (Jenkins et al. 1988). At sub-lethal doses, chlorhexidine affects sugar transport and acid production in cariogenic streptococci; perturbs osmotic balance through leakage of potassium and phosphorus, and a major protease (gingipain) in *Porphyromonas gingivalis* (Marsh 2010; Varoni et al. 2012). Although the bactericidal action is more effective against Gram-positive cocci (Jenkins et al. 1988), chlorhexidine has been reported to cause resistance in the oral bacteria species, i.e. *Streptococcus gordonii*, *Enterococcus faecalis*, *Fusobacterium nucleatum,* and *P. gingivalis* (Wang and Ren 2017).

Chlorhexidine is commonly added a final concentration of 0.2% which is effective in controlling gingival inflammation and plaque index, but can lead to dysgeusia and tooth staining, the latter can be exacerbated by the consumption of tea and coffee (Varoni et al. 2012). However, despite these minor adverse effects chlorhexidine remains the gold standard in oral antiseptics for the control of oral biofilms.

4.3 Fluoride

The common theme with the treatment of dysbiotic oral biofilms is not complete eradication as therapies should be targeted towards the manipulation to allow a transition to health.

Fluoride has been one of the largest public health interventions that has seen significant decreases in dental caries (Jones et al. 2005). Used in combination, fluoride in drinking water and oral health care products increases the resistance of enamel to acidic pH by promoting remineralisation of fluorapatite crystallites (ten Cate 2013). Fluoride also inhibits the glycolytic enzyme enolase, enhances oxidative damage to bacteria, inhibits the proton-translocating F-ATPases, and acidifies the cytoplasm by increasing membrane permeability to protons (Lopez-Lopez and Mira 2020; Van Loveren 2001). However, saliva flow and the transit of food/water means that the concentration of any agent can be reduced by 3-fold (Naumova et al. 2012); therefore, the metabolic effects and potential changes in the biofilm composition caused by fluoride exposure remain equivocal.

Lopez-Lopez and Mira (2020) investigated fluoride-induced changes in the bacterial composition of oral biofilms containing hundreds of species grown in vitro under controlled conditions. Using human saliva as a source of bacteria, biofilms were exposed to 500 ppm fluoride (1500 ppm which is the standard concentration in commercial toothpastes). DNA- and an RNA-based bacterial composition analyses showed that although the biofilm formation was not affected by exposure to fluoride, *Streptococcus oralis* decreased from 25% to around 2% while *Streptococcus salivarius* and *Rothia mucilaginosa* increased by 4–5-fold and 8-fold, respectively. Significantly, the two latter species are associated with good oral health (Rosier et al. 2018; Wescombe et al. 2012). It was noted that the changes in composition could be attributed to either resistance to fluoride (Liao et al. 2017) or that the allogenic changes (fluoride exposure) promoted the growth of some species by the repression of genes involved in the transport of carbohydrates, and the overexpression of pyruvate oxidase (Lopez-Lopez and Mira 2020).

Fluoride exposure appears to significantly affect bacterial proportions and gene expression profiles which combine to limit the pH fall in oral biofilms. This highlights the important application of fluoride to oral health.

5 In Vitro Oral Biofilm Systems

The oral milieu is a complex ecosystem made up of microorganisms that have developed both synergistic and antagonistic mechanisms that enable all species to survive and maintain homeostasis with the host. Given the complexity of these interactions, numerous in vitro biofilm systems have become popular as the long-term study of oral biofilms in vivo has several ethical implications and can be confounded by subject compliance (Salli and Ouwehand 2015). In vitro models enable the study of the plaque biofilm and growth under controlled conditions (Sanchez et al. 2014). Here, we review some of the in vitro oral biofilm models presently used and discuss factors to consider when selecting or applying an in vitro system.

5.1 Batch Culture Models

Several batch culture models have been developed using microtitre plates (Salli and Ouwehand 2015; Walker and Sedlacek 2007). Biofilms develop on the walls or the bottom of plate as a closed system and thus the population dynamics change over time as nutrients become depleted and waste products/signalling molecules accumulate. Batch culture systems may consist of axenic or multiple bacterial species and are often used to study early stage biofilm formation by monitoring microbial attachment to an abiotic surface (O'Toole 2011). They are also widely used to study the effect of various antimicrobials on biofilm development (Furiga et al. 2014; Giertsen et al. 2011; Guggenheim and Meier 2011), oral probiotics (Marttinen et al. 2013), and dental caries (Arthur et al. 2013; Guggenheim et al. 2004).

5.1.1 Zurich Biofilm Model

The so-called Zurich biofilm model was developed over twenty years ago to study cariology and harbours a specified inoculum comprising between five and ten oral species (Guggenheim et al. 2001a). Thurnheer et al. (2004) used a mixed microbial consortia consisting of *Actinomyces naeslundii, Candida albicans*, *Fusobacterium nucleatum, Streptococcus oralis, Streptococcus sobrinus, and Veillonella dispar* (Thurnheer et al. 2004). Their results highlighted the spatial arrangement and synergistic interactions that occur during biofilm development and that the presence of certain strains facilitated adherence to pellicle and early colonisation due to intraspecies aggregation, interspecies co-aggregation, and co-adhesion (Guggenheim et al. 2001b; Thurnheer et al. 2004).

Despite the low cost, reliability, and high-throughput screening of batch culture models, the method has limited application in dental research. Notably, detailed microscopy of the biofilm and quantitative analyses are not possible (Peeters et al. 2008). Additionally, the low species number often used provides limited insight as supragingival plaque consists of several hundred bacterial species (Xiao et al. 2016).

To address this issue, researchers have resorted to using saliva or dental plaque (both subgingival and supragingival) as an inoculum (Segata et al. 2012). Walker and Sedlacek (2007) were the first to use subgingival dental plaque (Walker and Sedlacek 2007) and grew biofilms for 10 days on hydroxyapatite discs, coated with sterilised human saliva to mimic the tooth surface. The experiment produced a heterogenous biofilm consisting of 15–20 cultivable species that were identified using traditional culture methods and checkerboard DNA-DNA hybridisation.

Velsko and Shaddox also used the same apparatus (Velsko and Shaddox 2018) to grow biofilms from both orally diseased and healthy individuals for 8 days with the growth medium changed every 48 h. After 8 days, the disc was removed and used as a fresh inoculum to inoculate a new hydroxyapatite disc. This was repeated 10 times (total of 80 days), they reported a reduction in the number of species from 116 to 64 after 80 days which although significant, was within the range reported for other in vitro systems (Edlund et al. 2013; Fernandez et al. 2017; Kistler et al. 2015; Walker and Sedlacek 2007). Velsko and Shaddox showed that the biofilm retained a core group of species that were consistently identified. Edlund et al. (2013) used pooled human saliva and maintained 60–80% of species in the original inoculum in the biofilm over 48 h which were grown on a pre-saliva coated 24-well plate (Edlund et al. 2013). Next-Generation Sequencing identified 250 species in the biofilms grown for 48 h compared to 410 species in inoculated saliva. Biofilms showed high reproducibility and diversity between samples and although biofilms did not harbour all species in saliva, the study recorded the highest diversity in an in vitro model for the time.

5.1.2 The Calgary Biofilm Device

The Calgary Biofilm Device (CBD) consists of a 96-well microtitre tray with a modified lid with pegs projecting into the wells on which, biofilms form. Developed in 1999, the system has been commonly used to study axenic or multispecies biofilms (Ali et al. 2006; Ceri et al. 1999). Using this device, Kistler et al. (2015) used pooled saliva to grow an oral biofilm for 7 to 14 days (Kistler et al. 2015) and maintained 250 species (78%) compared to an average of 320 species in the inoculum. Fernandez et al. (2017) used a custom-made lid fixed with nylon clamps, to which glass discs were attached (Fernandez et al. 2017). Subgingival plaque was harvested from 7 donors with active periodontitis and compared with 2 healthy donors, biofilms were cultured for 2 and 4 weeks. DNA sequencing identified an average of 61 species (21%) from all biofilms across both time-points in all donors compared to 284 species in the initial inoculum. It is important to note that the large difference in species retention is a by-product of the different inoculum source. Subgingival plaque contains significant numbers of obligate anaerobic and asaccharolytic species which are difficult to maintain.

5.2 The Artificial Mouth Model

Artificial biofilm model (AMM) systems are defined as an oral biofilm model that is intermittently or continuously provided with growth medium simulating saliva flowing over the surface (McBain 2009). The first AMM appeared in the literature in 1975, Russell and Coulter (1975) placed caries-free molars inside a glass incubation chamber and delivered a continuous flow of synthetic saliva over the tooth surface (Russell and Coulter 1975). The tooth was inoculated using human saliva and species retention was monitored microscopically by traditional bacterial culture. This early work discovered that in the presence of sucrose, the pH decreased and favoured predominantly streptococcal species whereas in the absence of sucrose, the pH increased and a mixed bacterial community was maintained.

Studies similar were performed by others in the 1970s using 1–4 species (Russell and Tagg 1981; Ahmed and Russell 1978). However, Dibdin et al. (1976) was the first to make significant changes to the design by adding six separate chambers allowing each plaque biofilm to be independently controlled (Dibdin et al. 1976). Growth medium was supplied to two halves of pre-cleaned human pre-molars wired together to mimic two adjacent teeth. Dibdin et al. (1976) discovered that a plaque biofilm successfully formed on dentine and enamel for both single and mixed oral bacteria, but it was notably thicker in crevices between the halves of the pre-molars where there were minimal shear forces. It was noted that nutrient media supplemented with sucrose led to the formation of thicker biofilm (plaque) and that the majority of the bacteria were Gram-positive streptococci.

Sissons et al. (1991) established a modified version of the AMM. This allowed the assessment of various gases and temperature on the long-term growth of oral biofilms derived from supra gingival plaque and saliva samples (Sissons et al. 1991). Supra gingival plaque was grown for 6 weeks on saliva coated cover slips with simulated oral fluid (Basal Medium Mucin) as the nutrient medium. The chamber was periodically supplemented with sucrose. Traditional culture methods and microscopy were used to analyse biofilms. It was reported that although *Haemophilus* species were less abundant, aerobic Gram-negative rods increased and a significant number of presumptive *Veillonella* and *Fusobacteria* were observed.

The development of different AMMs involves specialised equipment, which is commercially unavailable. Additionally, studies have reported significant disparities in biofilm mass, underlining the obstacles in creating a standardised testing environment without utilising identical equipment (Aas et al. 2005; Segata et al. 2012)

5.3 Flow Cells

Flow cells systems are the most utilised tool for biofilm research and frequently used to study the development of oral biofilm formation and morphology (McBain 2009). More information on the design and components of flow cell systems can be found in the paper published by Crusz et al. (2012). Generally, the system consists of a medium reservoir, waste container, a bubble trap, and the main chamber containing a surface for biofilm formation. Flow cells can also be designed so that a surface can be removed without terminating the experiment (Caiazza and O'Toole 2004; Davey and O'Toole 2000; Sauer et al. 2002). As with the other systems, flow cells can be inoculated with single or multiple bacterial species and colonisation can be examined in real time by Confocal Laser Scanning Microscopy (CLSM) (Pamp et al. 2009).

Although frequently used, there are only few studies studying human dental plaque. All published studies have used either single (Dashper et al. 2016; Foster et al. 2004; Rath et al. 2017) or pre-selected multispecies bacteria (Blanc et al. 2014; Cook et al. 1998; Foster and Kolenbrander 2004) which do not fully represent the species diversity in the oral cavity. Dashper et al. (2016) utilised a modified commercially available flow cell system (Stovall Life Science, USA) to investigate the biofilm disruptive capability of casein phosphopeptide-amorphous calcium phosphate (CPP-ACP). *S. mutans* biofilms were cultured on glass ionomer cement for 16 hours and then challenged with CPP-ACP. CLSM revealed that biofilms exposed to CPP-ACP had a significant reduction in biovolume and thickness when compared to the negative control (Dashper et al. 2016).

6 Future Applications for the Control of Oral Biofilms to Prevent Disease

6.1 Silver Nanoparticles

For over a century, silver has been utilised as an antimicrobial agent due to its broad spectrum, low toxicity, and absence of bacterial resistance (Fakhruddin et al. 2020; Gao et al. 2018). Primarily, it was used as silver nitrate ($AgNO_3$), and later in association with fluorine (AgF) (Fernandez et al. 2021). However, its use as a prophylactic in preventing dental caries is limited due to the black pigmentation resulting in tooth discolouration (Moazami et al. 2018; Hamdy et al. 2021; Sayed et al. 2018, 2019). Silver nanoparticles (AgNPs) are nano-sized particles of silver of between 1 nm and 100 nm in size.

The development of non-staining AgNP (Yin et al. 2020) with potent antimicrobial, antifungal, and antiviral properties and reduced toxicity has emerged as a promising compound to prevent dental caries (Abadi et al. 2013), candidiasis (Lara et al. 2015) and as an endodontic irrigant (Bhandi et al. 2021). Furthermore, there is potential for AgNP to be attached to the surface of dental implants, composite resins, porcelain restorations, and orthodontic and endodontic materials (Weerasinghe et al. 2020; Vasilev et al. 2011; Vasilev 2019).

AgNPs are synthesised using a precursor, usually silver nitrate, a reducing agent (e.g. sodium citrate, ammonium formate, sodium borohydride, and natural green sources (Rodríguez-León et al. 2013)), and a stabilising agent (e.g. thiols, chitosan, polyvinyl pyrrolidone, and sodium carboxyl methyl cellulose) which ensures stabilisation and prevents self-assembly and aggregation (Fernandez et al. 2021). AgNPs display excellent antibacterial properties due to the dissolution of free Ag^+ ion which then binds to the bacterial cell membrane, proteins, lipids, and DNA disrupting cellular metabolism, replication and causing lysis. AgNPs also have a large surface area, providing high contact with microorganisms. An important mechanism of action of AgNP is the induction of reactive oxygen species (ROS) production (Abdal Dayem et al. 2017), the hydroxyl radicals being the main species responsible for the oxidative damage (Quinteros et al. 2018). Furthermore, AgNPs also damage the cell membrane, nucleic acids, interfere in the cellular metabolic process thus exhausting the levels of intracellular ATP (Porenczuk et al. 2019).

Many researchers have reported AgNP to be an effective antibacterial agent against a plethora of dental bacteria (Ahmad et al. 2019; Martínez-Robles et al. 2016; Panpaliya et al. 2019; Choi et al. 2019). Haidari et al. reported polycationic silver nanoclusters not only exhibited excellent antibacterial activity against aerobic pathogens, but also killed *Fusobacterium nucleatum* and *Streptococcus sanguinis*, two dental pathogens implicated in dental caries, periodontal disease, and peri-implantitis. Additionally, AgNP showed strong capacity to prevent the development of bacterial resistance (Haidari et al. 2021a).

Particle size and shape can also influence the mechanistic action of AgNP. Recently, Haidari et al. (2019) reported that ultra-small AgNP of approximately 3.0 nm showed high bacterial killing potency and low cytotoxicity to fibroblasts. This work concluded, that ultra-small AgNPs with high concentration of silver ions in the extremely reactive +1 oxidation state were highly antibacterial to Gram-negative and Gram-positive bacteria. Additionally, AgNPs with a sustained release of silver ions for long-term antibacterial activity, would be of great interest. The use of biocompatible hydrogels, chitosan, and pH responsive polymers for a controlled delivery has shown promising results and could be translatable to dental treatments (Haidari et al. 2020, 2021b; Divakar et al. 2018).

The potential for AgNP is immense due to their antimicrobial, antiviral, antifungal actions, and the capacity to overcome bacterial resistance. Silver enhances the mechanical properties of materials used in restorative dentistry. Additionally, AgNP could also be used as a prophylactic mouth wash or incorporated into toothpaste to prevent dental carries. Their potential to be incorporated into dental composites and dental implants makes AgNP a promising antibacterial/antibiofilm therapy. Finally, combinational treatment of photodynamic therapy and AgNP may be proposed as a substitute for use against intracanal infections due to their ability to disrupt biofilm and prevent bacterial adhesion to dentin (Aydın et al. 2020). To date, most dentistry related studies are *in vitro,* using limited microorganism and cell lines to study cytotoxicity. To move towards clinical applications, long-term in vivo studies to validate antibacterial effectiveness, biocompatibility, and safety are required.

6.2 Probiotics

Dental caries and periodontal disease are two of the most prevalent oral health burdens world-wide (Petersen 2003). Traditional biofilm control interventions result in a temporary reduction of the bacterial load and associated side effects which, in some cases, has proven not sufficient to control the disease (Gatej et al. 2017a). The administration of probiotics, beneficial bacteria with antimicrobial and anti-inflammatory properties, is one of the several contemporary strategies being considered to address bacterial imbalances and prevent, control, and eradicate biofilms (Gatej et al. 2020).

The 2002 definition of probiotics by the World Health Organization as “live micro-organisms which, when administered in adequate amounts, confer a health benefit on the host” has been recently reinforced (Hill et al. 2014; World Heath Organization 2002). In the oral tract, the majority of probiotics are ubiquitous and belong to the Firmicutes, Actinobacteria, and Proteobacteria phyla, playing essential roles in immunity, physiology, and metabolism (Lee and Salminen 2009; Reid 2016). The most studied probiotic microorganisms species are *Lactobacillus* and *Bifidobacterium* spp. (Reid 2016).

6.2.1 Probiotics in the Treatment of Dental Caries

Several studies have investigated the effect of probiotic supplementation in caries prevention. A recent systematic review and meta-analysis of nineteen studies investigated the effect of probiotics on the incidence of caries or the levels of mutans streptococci and/or Lactobacilli species. When probiotic and control groups were compared after treatment, significantly more patients in the probiotic group had low mutans streptococci counts ($<10^5$ CFU/ml) and significantly less patients had high mutans streptococci counts ($>10^6$ CFU/ml) (Laleman et al. 2014). Additionally, at the end of the probiotic treatment, the use of probiotics did not result in significant differences in the Lactobacilli counts between the probiotic and control groups (Laleman et al. 2014).

Probiotics intervention in caries treatment has been conducted in different age patients. In a 7-month randomised, double-blind, placebo-controlled intervention study of 594, 1–6 years old children, *Lactobacillus rhamnosus GG* in milk significantly reduced the risk of caries ($p = 0.01$) and lowered mutans streptococci counts in saliva and dental plaque when compared with milk alone (Näse et al. 2001). A double-blind, placebo-controlled clinical trial of 160 healthy subjects, 58–84 years of age, found that daily intake of milk supplemented with fluoride and/or probiotic bacteria (*L. rhamnosus* LB21) may reverse soft primary root caries lesions in older adults (Petersson et al. 2011). Future studies are needed to determine specific probiotic strains, doses, delivery methods, and treatment schedules.

6.2.2 Probiotics in the Treatment of Periodontal Disease

Traditional treatments for periodontitis include subgingival debridement and, in some cases, the use of antimicrobials and systemic broad-spectrum antibiotics such as doxycycline, amoxicillin, or metronidazole, all of which contribute to significantly reduce the pathogenic biofilm and associated inflammation (Gatej et al. 2017a). However, the reduction is only temporarily, recolonisation re-occurring within months (Rhemrev et al. 2006). Additionally, treatment compliance is compromised due to adverse gastrointestinal side effects including antibiotic-associated diarrhoea, nausea, vomiting, and intestinal dysbiosis (Lynne 2006). Current results of using probiotics in the prevention and/or treatment of periodontal disease are encouraging. A literature review of eight clinical studies using probiotics in patients with chronic periodontitis indicated decreased clinical parameters such as plaque index, calculus, gingival inflammation, and bleeding on probing, together with decreased pro-inflammatory markers PGE2, MMP and INF-γ levels in saliva or gingival crevicular fluid in treated periodontitis patients when compared with controls or placebo (Gatej et al. 2017a). Other results included decreased periodontal pocket depth and clinical attachment loss and significant attachment gain for scaling and root planing plus probiotic treatment versus scaling and root planing alone. A reduction in *Porphyromonas gingivalis* total viable count and the proportion of obligate anaerobic bacteria in the probiotic groups when compared with controls was also reported (Gatej et al. 2017a). More importantly, in the light of increased bacterial resistance, the use of probiotics for the prevention and treatment of periodontal disease presents as a low-risk option in controlling the subgingival biofilm and achieving periodontal health (Gatej et al. 2017b).

6.2.3 Mechanisms of Action

The mechanism of probiotic treatment is currently under study and may depend on a variety of factors such as the condition being treated, the strain and concentration of probiotics used, the mode of delivery, and the stage when they are introduced (Gatej et al. 2017a). Generally, probiotics produce short chain fatty acids with roles in the secretion of mucin, the sticky layer protecting intestinal epithelium from contact with pathogenic microorganisms (Fig. 5) (Gatej 2018). Specifically, in caries, the role of probiotics is associated with the inhibition of pathogenic bacteria either indirectly by competing with pathogens for essential nutrients or changing the environmental pH and consequently restricting the pathogens' adhesion capabilities, or directly by secreting antimicrobial substances such as bacteriocins, lactic acid, and hydrogen peroxide. In periodontal disease, the probiotic mechanisms may include modulating the host's innate and adaptive immune system by decreasing the production of pro-inflammatory cytokines IL-6, IL-1β, TNFα, IFN-α, IL-8, IL-17, IL-1β and increasing the production of anti-inflammatory cytokines IL-10, IL-12, TGF-β (Gatej et al. 2017a).

Probiotics exert their effects without necessarily colonising the mouth. In a parallel randomised clinical trial of 39 subjects consuming a product containing one of the probiotics *Enterococcus faecalis*, *Lactobacillus casei*, or *Lactobacillus rhamnosus GG* for 28 days, the probiotics did not integrate or persist in the dental biofilm; however, the proportion of Streptococci spp. was significantly reduced ($p < 0.05$) (Arweiler et al. 2020).

Investigations into the role of probiotics in biofilm-induced oral conditions demonstrate probiotics can influence and control tooth-colonising and subgingival

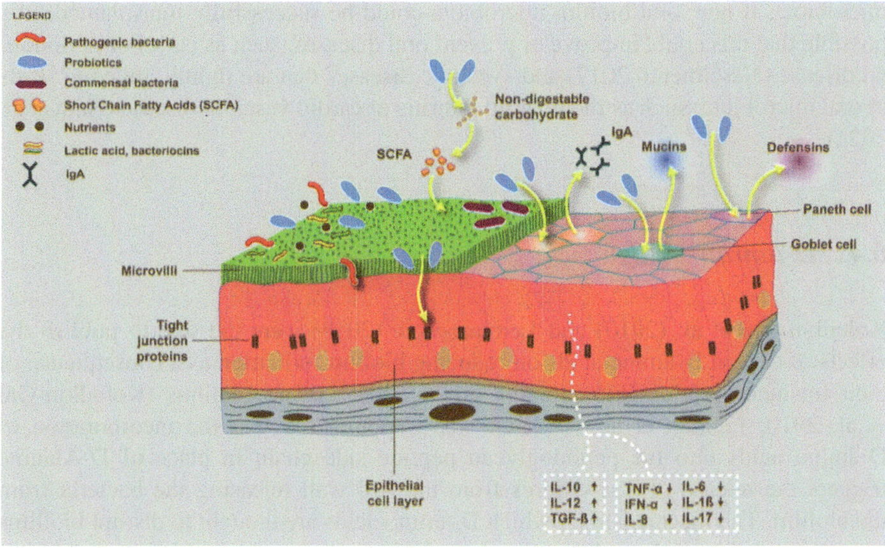

Fig. 5 Proposed mechanisms of probiotic action include restriction of the pathogens' adhesion capabilities, competition with pathogens for nutrients and secretion of antimicrobial substances (in caries control) and modulating the host's immune system (in periodontal disease) (Gatej 2018)

biofilms offering an inexpensive and easy to use adjunctive option for the treatment or prevention of caries and periodontal disease. Additionally, in the light of increased bacterial resistance, the use of probiotics may present as a low-risk option in controlling the subgingival biofilm. More randomised clinical control trials are needed to provide health professionals with accurate information regarding probiotic doses, delivery methods, treatment schedule, and how to maintain the results of the probiotic interventions (Kilian et al. 2016).

6.3 Oral Microbiome Transplant

As oral biofilms are notoriously hard to treat using conventional antimicrobial methods (Socransky and Haffajee 2002), new techniques need to be developed. A promising new area of research works to overturn existing microbial communities in the body and replace them with donor microbiota. This procedure, entitled microbial transplantation, has been incredibly successful in the gut in treating infectious disease, such as infection by *Clostridium difficile*, and has had mild success in treating chronic, non-communicable disease, such as inflammatory bowel disease (Smits et al. 2013). It is possible that microbial transplantation may be a new tool to shift oral biofilms. Existing oral biofilms would need to be first removed or significantly disrupted, followed by the transplantation or introduction of donor plaque microbiota. If new oral biofilm microbiota could be successfully transplanted, it is possible that this could improve or prevent oral diseases, such as caries or periodontal disease (Nascimento 2017) and systemic diseases that are tightly linked to shifts in oral microbiota, such as rheumatoid arthritis or cardio vascular disease (Nath et al. 2021).

6.4 D-amino Acids

Kolodkin-Gal et al. (2010) and Leiman et al. (2013) were the first to publish the effects of certain D-amino acids on reducing biofilm formation as a consequence of their misincorporation into proteins produced in mature biofilms (Kolodkin-Gal et al. 2010; Leiman et al. 2013). It was hypothesised that the incorporation of D-amino acids into the peptidoglycan peptide side chain in place of D-Alanine triggers the release of TasA fibres from the cell wall releasing the bacteria from the biofilm. The mechanism by which D-amino acids are thought to disrupt biofilms however remains equivocal, but they remain an important area of interest and have been termed biofilm breakers.

Enterococcus faecalis is a significant nosocomial pathogen that is the most frequent species isolated from obturated root canals, with prevalence values reaching up to 90% (Molander et al. 1998; Peciuliene et al. 2001; Pinheiro et al. 2003). Root canal therapy relies on hydroxyl ions from medicaments such as calcium hydroxide diffusing through the dentinal tubules, which are part of the complex anatomy of the tooth root (Fig. 2). The ability to tolerate pH extremes (pH 12.2) and the ability of *E. faecalis* to penetrate deep into the dentinal tubules allow the organism to survive while other bacteria are destroyed (Haapasalo and Orstavik 1987). Another important feature is that hydroxyl ions may take up to seven days to reach the outer root dentine and up to four weeks to reach peak pH levels (Athanassiadis et al. 2007). A slow rise in pH allows bacteria such as *E. faecalis* to respond by eliciting a stress response, one such example is the formation of biofilms (Zilm et al. 2017). Wilson et al. (2015) reported that six clinical strains of *E. faecalis*, which included one oral isolate produced significantly increased biofilms in response to sub-minimum inhibitory concentrations of antimicrobials used in endodontic medicaments (Wilson et al. 2015).

Zilm et al. (2017) examined whether D-amino acids could inhibit initial biofilm formation and whether they could disrupt established biofilms in 10 clinical isolates of *E. faecalis* that were shown to produce biofilms. Experiments were also performed in the presence of sub-inhibitory levels of the commonly used endodontic medicaments, sodium hypochlorite, Odontopaste™ (5% w/w clindamycin hydrochloride; 1% w/w triamcinolone acetonide) and Odontocide™ (calcium hydroxide 20% w/w; ibuprofen 7% w/w).

They found that the D-amino acids added to the growth medium (25 mM D-Leu, D-Met, and D-Trp and 0.25 mM D-Tyr) did not affect the generation time but did significantly inhibit biofilm formation as well as disrupt established biofilms of *E. faecalis* isolates cultured over 6 days.

E. faecalis biofilms were also unaffected by the presence of sub-MIC levels of the medicaments over 6 days; however, the addition of D-amino acids to the medicaments significantly reduced biofilm formation compared to the addition of cognate L-amino acids. It was concluded that D-amino acids reduce biofilm formation in the presence of agents that stimulate biofilm formation. Xu et al. (2012) who reported that antimicrobial stress improved biofilm disruption using D-amino acids support this conclusion (Xu et al. 2012).

D-amino acids have potential to be biofilm breakers that may improve the efficacy of antimicrobials against pathogens; however, more research is required to investigate their clinical application.

6.5 *Quorum Sensing Inhibitors*

Quorum Sensing is an elegant cell-to-cell signalling mechanism that allows inter- and intra-species communication to regulate gene expression of numerous virulence factors including biofilm formation, in response to a given population size or a "quorum" (Ng and Bassler 2009). Quorum sensing pathways are an alternative target for control of biofilms which may not promote biological resistance as the intervention does not affect cellular function (Rasmussen and Givskov 2006a; Kang et al. 2019). There are three main targets for artificial manipulation of these systems: the gene that codes for autoinducer(s), (i.e. small diffusible signalling molecules) the chemical alteration or magnification of the autoinducer, and the chemical alteration of or interaction with the receptor. In regard to the latter, many studies have investigated quorum sensing inhibitors that act as an antagonist analogue of the autoinducer and consequently block quorum sensing receptors (Rasmussen and Givskov 2006b).

There are currently several natural and synthetic quorum sensing inhibitors which show potential for dental treatment. Both in vitro and in vivo studies have demonstrated that quorum sensing inhibitors inhibit biofilm formation by several cariogenic bacterial species (Table 1) and periodontal pathogens (Slobodníkova et al. 2016; Ben Amara et al. 2018).

Streptococcus mutans has been shown to use the ComCDE quorum sensing pathway, which is linked to the regulation of genetic competence, mutacin and bacteriocin production, stress response, sucrose-dependent adherence, and biofilm formation (Smith and Spatafora 2012; Leung et al. 2015; Shanker and Federle 2017). A detailed description of this pathway has been reported (Dong et al. 2016). The pathway uses competence stimulating peptide (CSP) as the signalling molecule which influences specific genes for biofilm formation, namely *gbcC* that promotes a glucan-binding protein and *dblB* increases dextran-dependent aggregation (Lee et al. 2015).

Table 1 Published research on quorum sensing inhibitors and their impact on biofilm formation of oral bacteria

References	Bacteria	QS inhibitor	Result
Muras et al. (2018)	*S. mutans*, *S. oralis*, *S. dentisani*	FuranoneC30, *T.* sp. 20J	Furanone C30 no effect and effective *T.* sp. 20J.
Tsuchiya et al. (1994)	*S. mutans* *S sobrinus*	Purified phytoalexins from *Sophora exiguna*	Effective *S. mutans*, effective *S. sobrinus*.
Sato et al. (1996)	*S. mutans* *S sobrinus*	Artocarpin and Artocarpesin	Effective. S. *mutans* and *S. sobrinus*.
Uzel et al. (2005)	*S. mutans* *S. sobrinus*	Four Anatolian propolis samples: mainly flavonoids (naringenin and quercetin included)	Effective *S. mutans* and *S. sobrinus*.
Wu (2009)	*S. mutans* *P. gingivalis*	Thompson seedless raisins, raisins	Certain compounds effective on each bacterium
Koo et al. (2010)	*S. mutans* *S. anginosus*	Proanthocyanidins	Decreased biofilm accumulation and acidogenicity of *S. mutans*.
He et al. (2012)	*S. mutans* *luxS* mutant	Furanone C30	Decreased biofilm and change in biofilm structure.
Wittpahl et al. (2015)	*S. mutans*	Cistus Incanus Herbal tea	Effective against *S. mutans* biofilms.
Janus et al. (2016)	*S. mutans* *S. mitis* *S. gordonii* *S. salivarius* *S. sanguinis* *S. oralis* *S. sobrinus*	Furanone C30 Homoserine lactone	Furanone C30 minimised lactic acid synthesis.
Vijayakumar and Ramanathan (2018)	*S. pyogenes* *S. mutans*, *S. aureus*, *S. epidermidis*	5-hydroxymethylfurfural	Effective *S. mutans* and *S epidermis* at 100 µg/ml, *S. pyogenes* and *S. aureus* at 125 µg/ml.
Senpuku et al. (2019)	*S. mutans*, *sunL* mutant	Adzuki extract (7S globulin)	Effective at inhibiting biofilm formation, impact CSP low pH conditions but not at neutral conditions.
Kang et al. (2019)	*S. mutans*	Furanone C30	Furanone C30 increased time for biofilm inhibition.

Another QS pathway in *S. mutans* is LuxS (Yoshida et al. 2005; Merritt et al. 2003; Wen and Burne 2004; Vendeville et al. 2005; Smith and Spatafora 2012; Sztajer et al. 2008). The auto inducer LuxS controls several specific quorum sensing genes including the glycosyltransferase genes (*gtfA, gtfB, gtfC, gtfD*) and a number of glucan-binding genes (Sztajer et al. 2008). Both the glucan-binding region of the Gtf enzyme and glucan-binding proteins (Gbp) have been associated with *S. mutans* initial and specific adherence to acquired pellicle (Conrads et al. 2014). However, in the case of *S. mutans* and other bacteria, the deletion of the luxS gene caused an increase in aggregation rather than a decrease in aggregation.

There are several studies that investigate the effect of QSIs on *S. mutans*, which have produced a few notable discrepancies (Table 1). Muras et al. (2018) investigated furanone C30 and *T.* sp. 20J against a mixed species biofilm of *S. mutans, Streptococcus oralis,* and *Streptococcus dentisani* and found furanone C30 to be ineffective at controlling biofilm formation, yet *T.* sp. 20J was not (Muras et al. 2018). These results were gained using impedance technology so measured biofilm formation in real time and was supported with more traditional biofilm measurement techniques. However, this contradicts results found by He et al. (2012) and Janus et al. (2016), which found furanone C30 to be effective against *S. mutans* single species biofilm and on bacterial species taken from saliva of volunteers (Table 1).

The potential application of quorum sensing inhibitors for the treatment of oral diseases such as caries and periodontal disease remains equivocal. The oral environment poses significant challenges as flavonoid compounds, a major class of quorum sensing inhibitors, are sensitive to changes in temperature, light, moisture, pH, and oxygen, and therefore may have reduced effectiveness (Tsirigotis-Maniecka et al. 2017; Senpuku et al. 2019). To be effective, quorum sensing inhibitors must also be maintained at correct concentrations where there is continuous flushing by saliva (Humphrey and Williamson 2001). It is also important to note that there is potential for synergism of different therapies using quorum sensing inhibitors in combination with antimicrobials and other caries control methods.

6.6 *Endodontic Irrigant/Medicaments*

Despite extensive laboratory research with ensuing publications on the field, commonly used endodontic irrigant solutions and medicaments have remained almost unchanged throughout the recent decades. The focus of current clinical investigations assessing the disinfection of the root canal space is the use of technologies to enhance the dynamics of irrigation, contact time, and volume using previously established irrigant solutions (Sigurdsson et al. 2018). This aims to address the difficulties for biocidal agents to reach the less accessible areas of the root canal space.

The use of different antibiotic mixtures for the disinfection of the intracanal space in regenerative Endodontics has gained interest similarly (Fig. 4), however, some are associated with tooth discolouration, which is an undesirable treatment complication (Kahler and Rossi-Fedele 2016). Further understanding of the possible disinfection protocols for these procedures based on clinical trials, including their association with patient and clinician-reported outcome measures, is required (Kahler et al. 2017).

New concepts focussing on biological strategies to alter the abilities of microbes to form biofilms have been suggested (Nair 2014). In addition, the ability to elicit a host response by the biofilm matrix has been shown (Ramirez et al. 2019); therefore, apart from biocidal activity, biofilm matrix disaggregation should be considered desirable. One example of a biological strategy in endodontics is the use of D-Amino acids, which has been shown in vitro to prevent the formation and cause the disaggregation of biofilms in bacterial species commonly present in previously filled root canals (Zilm et al. 2017). Other authors have suggested other nanoparticles, including silver nanoparticles as well as intracanal activation of novel materials (Abusrewil et al. 2020); however, their clinical application remains unknown together with their proven efficacy in regard to patient and clinician-reported outcomes. Clinical studies for endodontic irrigants, medicaments, and root filling materials that have demonstrated the ability to alter intracanal biofilms in laboratory studies, including those previously discussed in the present chapter, are required in this field of dentistry.

7 Conclusion

The study of oral biofilms has led the way in the understanding of the complex interactions that occur within and between bacteria as they interact with changes in the environment. The complexity of the oral microbiome has developed in conjunction with advances made in high-throughput sequencing and presently is believed to be around 800 species, consisting of a core microbiome of about 200 bacterial species.

In recent years, the ability to monitor changes in the oral microbiome has modified our understanding and treatment of oral diseases. This has become increasingly important as we discover the importance at restoring and maintaining a healthy oral microbiome that is in homeostasis with the host. Non-specific clinical treatments directed towards periodontal, endodontic infections are aimed at decreasing total microbial numbers in the biofilm in order to remediate the disease. These include removal of the biofilm by mechanical and chemical treatments with oral antiseptics or by the addition of fluoride in oral care products and in drinking water as a major public health measure.

New "cutting edge" treatment modalities for dental diseases are now focussed on moving away from the use of systemic antibiotics that promote antibacterial resistance and generalised destruction of the oral biofilm. Nanotechnology, probiotics, targeted quorum sensing inhibitors, biofilm disruptors, and oral microbiome transplants are a few of the new technologies that aim to manipulate a dysbiotic biofilm to shift the balance back to homeostasis with the host. These more targeted therapies pave the way for more specialised precision-based dentistry that may help to reduce the global epidemic of oral diseases.

References

Aas JA, Paster BJ, Stokes LN, Olsen I, Dewhirst FE (2005) Defining the normal bacterial flora of the oral cavity. J Clin Microbiol 43(11):5721–5732. https://doi.org/10.1128/JCM.43.11.5721-5732.2005

Abadi MF, Mehrabian S, Asghari B, Namvar AE, Ezzatifar F, Lari AR (2013) Silver nanoparticles as active ingredient used for alcohol-free mouthwash. GMS Hyg. Infect Control 8(1):Doc05. https://doi.org/10.3205/dgkh000205

Abdal Dayem A, Hossain MK, Lee SB, Kim K, Saha SK, Yang GM, Choi HY, Cho SG (2017) The role of reactive oxygen species (ROS) in the biological activities of metallic nanoparticles. Int J Mol Sci 18(1). https://doi.org/10.3390/ijms18010120

Abusrewil S, Alshanta OA, Albashaireh K, Alqahtani S, Nile CJ, Scott JA, McLean W (2020) Detection, treatment and prevention of endodontic biofilm infections: what's new in 2020? Crit Rev Microbiol 46(2):194–212. https://doi.org/10.1080/1040841X.2020.1739622

Ahmad S, Munir S, Zeb N, Ullah A, Khan B, Ali J, Bilal M, Omer M, Alamzeb M, Salman SM, Ali S (2019) Green nanotechnology: a review on green synthesis of silver nanoparticles - an ecofriendly approach. Int J Nanomed 14:5087–5107. https://doi.org/10.2147/ijn.S200254

Ahmed HMA, Rossi-Fedele G (2020) Preferred reporting items for root and canal anatomy in the human dentition (PROUD 2020) - a systematic review and a proposal for a standardized protocol. Eur Endod J 5(3):159–176. https://doi.org/10.14744/eej.2020.88942

Ahmed FI, Russell C (1978) Plaque formation in vitro by Actinomyces viscosus in the presence of Streptococcus sanguis or Streptococcus mutans. Microbios 23(92):93–98

Alalwan H, Rajendran R, Lappin DF, Combet E, Shahzad M, Robertson D, Nile CJ, Williams C, Ramage G (2017) The anti-adhesive effect of curcumin on Candida albicans biofilms on denture materials. Front Microbiol 20(8):659

Alani A, Knowles JC, Chranzowski W, Ng YL, Gulabivala K (2009) Ion release characteristics, precipitate formation and sealing ability of a phosphate glass-polycaprolactone-based composite for use as a root canal obturation material. Dent Mater 25(3):400–410

Ali L, Khambaty F, Diachenko G (2006) Investigating the suitability of the calgary biofilm device for assessing the antimicrobial efficacy of new agents. Bioresour Technol 97(15):1887–1893. https://doi.org/10.1016/j.biortech.2005.08.025

Armfield J (2012) The avoidance and delaying of dental visits in Australia. Aust Dent J 57(2): 243–247. https://doi.org/10.1111/j.1834-7819.2012.01697.x

Arthur RA, Waeiss RA, Hara AT, Lippert F, Eckert GJ, Zero DT (2013) A defined-multispecies microbial model for studying enamel caries development. Caries Res 47(4):318–324. https://doi.org/10.1159/000347050

Arweiler NB, Auschill TM, Heumann C, Hellwig E, Ahmad AA (2020) Influence of probiotics on the salivary microflora oral streptococci and their integration into oral biofilm. Antibiotics (Basel) 9(11):1–16. https://doi.org/10.3390/antibiotics9110803

Athanassiadis B, Abbott PV, Walsh LJ (2007) The use of calcium hydroxide, antibiotics and biocides as antimicrobial medicaments in endodontics. Aust Dent J 52(1 Suppl):S64–S82. https://doi.org/10.1111/j.1834-7819.2007.tb00527.x

Aydın H, Er K, Kuştarcı A, Akarsu M, Gençer GM, Er H, Felek R (2020) Antibacterial activity of silver nanoparticles activated by photodynamic therapy in infected root canals. Dent Med Probl 57(4):393–400. https://doi.org/10.17219/dmp/123615

Bao K, Belibasakis GN, Thurnheer T, Aduse-Opoku J, Curtis MA, Bostanci N (2014) Role of Porphyromonas gingivalis gingipains in multi-species biofilm formation. BMC Microbiol 14: 258. https://doi.org/10.1186/s12866-014-0258-7

Bastendorf KD, Strafela-Bastendorf N, Lussi A (2021) Mechanical removal of the biofilm: is the curette still the gold standard? Monogr Oral Sci 29:105–118. https://doi.org/10.1159/000510187

Ben Amara H, Song HY, Ryu E, Park JS, Schwarz F, Kim BM, Choi BK, Koo KT (2018) Effects of quorum-sensing inhibition on experimental periodontitis induced by mixed infection in mice. Eur J Oral Sci 126(6):449–457. https://doi.org/10.1111/eos.12570

Bhandi S, Mehta D, Mashyakhy M, Chohan H, Testarelli L, Thomas J, Dhillon H, Raj AT, Madapusi Balaji T, Varadarajan S, Patil S (2021) Antimicrobial efficacy of silver nanoparticles as root canal irrigant's: a systematic review. J Clin Med 10(6). https://doi.org/10.3390/jcm10061152

Blanc V, Isabal S, Sánchez MC, Llama-Palacios A, Herrera D, Sanz M, León R (2014) Characterization and application of a flow system for in vitro multispecies oral biofilm formation. J Periodontal Res 49(3):323–332. https://doi.org/10.1111/jre.12110

Bonesvoll P (1977) Oral pharmacology of chlorhexidine. J Clin Periodontol 4(5):49–65. https://doi.org/10.1111/j.1600-051x.1977.tb00051.x

Caiazza NC, O'Toole GA (2004) SadB is required for the transition from reversible to irreversible attachment during biofilm formation by Pseudomonas aeruginosa PA14. J Bacteriol 186(14):4476–4485. https://doi.org/10.1128/JB.186.14.4476-4485.2004

Caufield PW, Ratanapridakul K, Allen DN, Cutter GR (1988) Plasmid-containing strains of Streptococcus mutans cluster within family and racial cohorts: implications for natural transmission. Infect Immun 56(12):3216–3220. https://doi.org/10.1128/iai.56.12.3216-3220.1988

Ceri H, Olson ME, Stremick C, Read RR, Morck D, Buret A (1999) The Calgary biofilm device: new technology for rapid determination of antibiotic susceptibilities of bacterial biofilms. J Clin Microbiol 37(6):1771–1776. https://doi.org/10.1128/JCM.37.6.1771-1776.1999

Chan W, Chowdhury NR, Sharma G, Zilm P, Rossi-Fedele G (2020) Comparison of the biocidal efficacy of sodium dichloroisocyanurate and calcium hydroxide as intracanal medicaments over a 7-day contact time: an ex vivo study. J Endod 46(9):1273–1278. https://doi.org/10.1016/j.joen.2020.05.011

Chen C, Hemme C, Beleno J, Shi ZJ, Ning D, Qin Y, Tu Q, Jorgensen M, He Z, Wu L, Zhou J (2018) Oral microbiota of periodontal health and disease and their changes after nonsurgical periodontal therapy. ISME J 12(5):1210–1224. https://doi.org/10.1038/s41396-017-0037-1

Choi S-H, Jang Y-S, Jang J-H, Bae T-S, Lee S-J, Lee M-H (2019) Enhanced antibacterial activity of titanium by surface modification with polydopamine and silver for dental implant application. J Appl Biomater Funct Mater 17(3):2280800019847067. https://doi.org/10.1177/2280800019847067

Chrisopoulos S, Harford JE, Ellershaw A (2016) Oral health and dental care in Australia: key facts and figures vol Cat. no. DEN 229. Australian Institute of Health and Welfare, Australian Hospital Statistics, Canberra

Cieplik F, Kara E, Muehler D, Enax J, Hiller KA, Maisch T, Buchalla W (2019) Antimicrobial efficacy of alternative compounds for use in oral care toward biofilms from caries-associated bacteria in vitro. Microbiologyopen 8(4):e00695. https://doi.org/10.1002/mbo3.695

Cintra LTA, Gomes MS, da Silva CC, Faria FD, Benetti F, Cosme-Silva L, Samuel RO, Pinheiro TN, Estrela C, Gonzalez AC, Segura-Egea JJ (2021) Evolution of endodontic medicine: a critical narrative review of the interrelationship between endodontics and systemic pathological conditions. Odontology 109(4):741–769. https://doi.org/10.1007/s10266-021-00636-x

Conrads G, de Soet H, Song L, Henne K, Sztajer H, Wagner-Döbler I, Zeng A-P (2014) Comparing the cariogenic species Streptococcus sobrinus and S. mutans on whole genome level. J Oral Microbiol. https://doi.org/10.3402/jom.v6.26189

Cook GS, Costerton JW, Lamont RJ (1998) Biofilm formation by Porphyromonas gingivalis and Streptococcus gordonii. J Periodontal Res 33(6):323–327. https://doi.org/10.1111/j.1600-0765.1998.tb02206.x

Costello EK, Lauber CL, Hamady M, Fierer N, Gordon JI, Knight R (2009) Bacterial community variation in human body habitats across space and time. Science 326(5960):1694–1697. https://doi.org/10.1126/science.1177486

Crusz SA, Popat R, Rybtke MT, Camara M, Givskov M, Tolker-Nielsen T, Diggle SP, Williams P (2012) Bursting the bubble on bacterial biofilms: a flow cell methodology. Biofouling 28(8): 835–842. https://doi.org/10.1080/08927014.2012.716044

Cummins D, Creeth JE (1992) Delivery of antiplaque agents from dentifrices, gels, and mouthwashes. J Dent Res 71(7):1439–1449. https://doi.org/10.1177/00220345920710071601

Dagistan S, Aktas AE, Caglayan F, Ayyildiz A, Bilge M (2009) Differential diagnosis of denture-induced stomatitis, Candida, and their variations in patients using complete denture: a clinical and mycological study. Mycoses 52(3):266–271. https://doi.org/10.1111/j.1439-0507.2008.01592.x

Dashper SG, Catmull DV, Liu SW, Myroforidis H, Zalizniak I, Palamara JE, Huq NL, Reynolds EC (2016) Casein phosphopeptide-amorphous calcium phosphate reduces streptococcus mutans biofilm development on glass ionomer cement and disrupts established biofilms. PLoS One 11(9):e0162322. https://doi.org/10.1371/journal.pone.0162322

Davey ME, O'Toole GA (2000) Microbial biofilms: from ecology to molecular genetics. Microbiol Mol Biol Rev 64(4):847–867. https://doi.org/10.1128/MMBR.64.4.847-867.2000

de Cock P, Makinen K, Honkala E, Saag M, Kennepohl E, Eapen A (2016) Erythritol is more effective than xylitol and sorbitol in managing oral health endpoints. Int J Dent 2016:9868421. https://doi.org/10.1155/2016/9868421

Dewhirst FE, Chen T, Izard J, Paster BJ, Tanner AC, Yu WH, Lakshmanan A, Wade WG (2010) The human oral microbiome. J Bacteriol 192(19):5002–5017. https://doi.org/10.1128/JB.00542-10

Dibdin GH, Shellis RP, Wilson CM (1976) An apparatus for the continuous culture of micro-organisms on solid surfaces with special reference to dental plaque. J Appl Bacteriol 40(3): 261–268. https://doi.org/10.1111/j.1365-2672.1976.tb04173.x

Dilber E, Hagenfeld D, Ehmke B, Faggion CM Jr (2020) A systematic review on bacterial community changes after periodontal therapy with and without systemic antibiotics: an analysis with a wider lens. J Periodontal Res 55(6):785–800. https://doi.org/10.1111/jre.12803

Divakar DD, Jastaniyah NT, Altamimi HG, Alnakhli YO, Muzaheed AAA, Haleem S (2018) Enhanced antimicrobial activity of naturally derived bioactive molecule chitosan conjugated silver nanoparticle against dental implant pathogens. Int J Biol Macromol 108:790–797. https://doi.org/10.1016/j.ijbiomac.2017.10.166

Dong G, Tian XL, Cyr K, Liu T, Lin W, Tziolas G, Li YH (2016) Membrane topology and structural insights into the peptide pheromone receptor ComD, a quorum-sensing histidine protein kinase of Streptococcus mutans. Sci Rep 6:26502. https://doi.org/10.1038/srep26502

Dongari-Bagtzoglou A (2008) Mucosal biofilms: challenges and future directions. Expert Rev Anti Infect Ther 6(2):141–144. https://doi.org/10.1586/14787210.6.2.141

Edlund A, Yang Y, Hall AP, Guo L, Lux R, He X, Nelson KE, Nealson KH, Yooseph S, Shi W, McLean JS (2013) An in vitro biofilm model system maintaining a highly reproducible species and metabolic diversity approaching that of the human oral microbiome. Microbiome 1(1):25. https://doi.org/10.1186/2049-2618-1-25

Eloe-Fadrosh EA, Rasko DA (2013) The human microbiome: from symbiosis to pathogenesis. Annu Rev Med 64:145–163. https://doi.org/10.1146/annurev-med-010312-133513

Fakhruddin KS, Egusa H, Ngo HC, Panduwawala C, Pesee S, Samaranayake LP (2020) Clinical efficacy and the antimicrobial potential of silver formulations in arresting dental caries: a systematic review. BMC Oral Health 20(1):160. https://doi.org/10.1186/s12903-020-01133-3

Falony G, Honkala S, Runnel R, Olak J, Nommela R, Russak S, Saag M, Makinen PL, Makinen K, Vahlberg T, Honkala E (2016) Long-term effect of erythritol on dental caries development during childhood: a posttreatment survival analysis. Caries Res 50(6):579–588. https://doi.org/10.1159/000450762

Featherstone JD, Fontana M, Wolff M (2018) Novel anticaries and remineralization agents: future research needs. J Dent Res 97(2):125–127. https://doi.org/10.1177/0022034517746371

Fejerskov O, Kidd E (2008) Dental caries: the disease and its clinical management, 2nd edn. Blackwell Munksgaard Ltd, Carlton, VIC

Fellows Yates JA, Velsko IM, Aron F, Posth C, Hofman CA, Austin RM, Parker CE, Mann AE, Nagele K, Arthur KW, Arthur JW, Bauer CC, Crevecoeur I, Cupillard C, Curtis MC, Dalen L, Diaz-Zorita Bonilla M, Diez Fernandez-Lomana JC, Drucker DG, Escribano Escriva E, Francken M, Gibbon VE, Gonzalez Morales MR, Grande Mateu A, Harvati K, Henry AG, Humphrey L, Menendez M, Mihailovic D, Peresani M, Rodriguez Moroder S, Roksandic M, Rougier H, Sazelova S, Stock JT, Straus LG, Svoboda J, Tessmann B, Walker MJ, Power RC, Lewis CM, Sankaranarayanan K, Guschanski K, Wrangham RW, Dewhirst FE, Salazar-Garcia DC, Krause J, Herbig A, Warinner C (2021) The evolution and changing ecology of the African hominid oral microbiome. Proc Natl Acad Sci U S A 118(20). https://doi.org/10.1073/pnas.2021655118

Fernandez YMM, Exterkate RAM, Buijs MJ, Beertsen W, van der Weijden GA, Zaura E, Crielaard W (2017) A reproducible microcosm biofilm model of subgingival microbial communities. J Periodontal Res 52(6):1021–1031. https://doi.org/10.1111/jre.12473

Fernandez CC, Sokolonski AR, Fonseca MS, Stanisic D, Araújo DB, Azevedo V, Portela RD, Tasic L (2021) Applications of silver nanoparticles in dentistry: advances and technological innovation. Int J Mol Sci 22(5). https://doi.org/10.3390/ijms22052485

Flemmig TF, Hetzel M, Topoll H, Gerss J, Haeberlein I, Petersilka G (2007) Subgingival debridement efficacy of glycine powder air polishing. J Periodontol 78(6):1002–1010. https://doi.org/10.1902/jop.2007.060420

Foster JS, Kolenbrander PE (2004) Development of a multispecies oral bacterial community in a saliva-conditioned flow cell. Appl Environ Microbiol 70(7):4340–4348. https://doi.org/10.1128/AEM.70.7.4340-4348.2004

Foster J, Pan P, Kolenbrander P, Kolenbrander D (2004) Effects of antimicrobial agents on oral biofilms in saliva-conditioned flowcell. Biofilms 1:5–12. https://doi.org/10.1017/S1479050503001017

Frenkel H, Harvey I, Newcombe RG (2000) Oral health care among nursing home residents in Avon. Gerodontology 17(1):33–38. https://doi.org/10.1111/j.1741-2358.2000.00033.x

Furiga A, Roques C, Badet C (2014) Preventive effects of an original combination of grape seed polyphenols with amine fluoride on dental biofilm formation and oxidative damage by oral bacteria. J Appl Microbiol 116(4):761–771. https://doi.org/10.1111/jam.12395

Gad MM, Fouda SM, Al-Harbi FA, Näpänkangas R, Raustia A (2017) PMMA denture base material enhancement: a review of fiber, filler, and nanofiller addition. Int J Nanomedicine 17(12):3801–3812

Gao SS, Zhao IS, Duffin S, Duangthip D, Lo ECM, Chu CH (2018) Revitalising silver nitrate for caries management. Int J Environ Res Public Health 15(1). https://doi.org/10.3390/ijerph15010080

Gatej SM (2018) Clinical and microbiological effects of probiotics in experimental induced periodontitis. Thesis, The University of Adelaide

Gatej S, Gully N, Gibson R, Bartold PM (2017a) Probiotics and periodontitis – a literature review. J Int Acad Periodontol 19(2):42–50

Gatej S, Marino V, Richard B, Fitzsimmons TR, Gully N, Zilm PS, Gibson RJ, Edwards S, Bartold PM (2017b) Probiotic Lactobacillus rhamnosus GG prevents alveolar bone loss in a mouse model of experimental periodontitis. J Clin Periodontol:1–9. https://doi.org/10.1111/jcpe.12838

Gatej SM, Bright R, Weyrich LS, Marino V, Christophersen CT, Gibson RJ, Gully N, Zilm P, Bartold PM (2020) Probiotic Lactobacillus Rhamnosus GG protects against P. Gingivalis And F. Nucleatum gut dysbiosis. J Int Acad Periodontol 22:18–27

Gendreau L, Loewy ZG (2011) Epidemiology and etiology of denture stomatitis. J Prosthodont 20(4):251–260. https://doi.org/10.1111/j.1532-849X.2011.00698.x

Giertsen E, Arthur RA, Guggenheim B (2011) Effects of xylitol on survival of mutans streptococci in mixed-six-species in vitro biofilms modelling supragingival plaque. Caries Res 45(1):31–39. https://doi.org/10.1159/000322646

Gilbert P, Moore LE (2005) Cationic antiseptics: diversity of action under a common epithet. J Appl Microbiol 99(4):703–715. https://doi.org/10.1111/j.1365-2672.2005.02664.x

Greenstein G, Berman C, Jaffin R (1986) Chlorhexidine. An adjunct to periodontal therapy. J Periodontol 57(6):370–377. https://doi.org/10.1902/jop.1986.57.6.370
Griffen AL, Becker MR, Lyons SR, Moeschberger ML, Leys EJ (1998) Prevalence of Porphyromonas gingivalis and periodontal health status. J Clin Microbiol 36(11):3239–3242. https://doi.org/10.1128/JCM.36.11.3239-3242.1998
Guggenheim B, Meier A (2011) In vitro effect of chlorhexidine mouth rinses on polyspecies biofilms. Schweiz Monatsschr Zahnmed 121(5):432–441
Guggenheim B, Giertsen E, Schupbach P, Shapiro S (2001a) Validation of an in vitro biofilm model of supragingival plaque. J Dent Res 80(1):363–370. https://doi.org/10.1177/00220345010800011201
Guggenheim M, Shapiro S, Gmur R, Guggenheim B (2001b) Spatial arrangements and associative behavior of species in an in vitro oral biofilm model. Appl Environ Microbiol 67(3):1343–1350. https://doi.org/10.1128/AEM.67.3.1343-1350.2001
Guggenheim B, Guggenheim M, Gmur R, Giertsen E, Thurnheer T (2004) Application of the Zurich biofilm model to problems of cariology. Caries Res 38(3):212–222. https://doi.org/10.1159/000077757
Haapasalo M, Orstavik D (1987) In vitro infection and disinfection of dentinal tubules. J Dent Res 66(8):1375–1379. https://doi.org/10.1177/00220345870660081801
Haidari H, Goswami N, Bright R, Kopecki Z, Cowin AJ, Garg S, Vasilev K (2019) The interplay between size and valence state on the antibacterial activity of sub-10 nm silver nanoparticles. Nanoscale Advances 1(6):2365–2371. https://doi.org/10.1039/C9NA00017H
Haidari H, Kopecki Z, Bright R, Cowin AJ, Garg S, Goswami N, Vasilev K (2020) Ultrasmall AgNP-impregnated biocompatible hydrogel with highly effective biofilm elimination properties. ACS Appl Mater Interfaces 12(37):41011–41025. https://doi.org/10.1021/acsami.0c09414
Haidari H, Bright R, Kopecki Z, Garg S, Cowin AJ, Vasilev K, Goswami N (2021a) Polycationic silver nanoclusters comprising nanoreservoirs of Ag^{+}-ions with high antimicrobial and antibiofilm activity. ACS Appl Mater Interfaces 14(1):390–403
Haidari H, Kopecki Z, Sutton AT, Garg S, Cowin AJ, Vasilev K (2021b) pH-responsive "smart" hydrogel for controlled delivery of silver nanoparticles to infected wounds. Antibiotics (Basel) 10(1). https://doi.org/10.3390/antibiotics10010049
Hamdy D, Giraki M, Abd Elaziz A, Badran A, Allam G, Ruettermann S (2021) Laboratory evaluation of the potential masking of color changes produced by silver diamine fluoride in primary molars. BMC Oral Health 21(1):337. https://doi.org/10.1186/s12903-021-01697-8
Handsley-Davis M, Jamieson L, Kapellas K, Hedges J, Weyrich LS (2020) The role of the oral microbiota in chronic non-communicable disease and its relevance to the indigenous health gap in Australia. BMC Oral Health 20(1):327. https://doi.org/10.1186/s12903-020-01308-y
He Z, Wang Q, Hu Y, Liang J, Jiang Y, Ma R, Tang Z, Huang Z (2012) Use of the quorum sensing inhibitor furanone C-30 to interfere with biofilm formation by Streptococcus mutans and its luxS mutant strain. Int J Antimicrob Agents 40(1):30–35. https://doi.org/10.1016/j.ijantimicag.2012.03.016
Hill C, Guarner F, Reid G, Gibson G, Merenstein D, Pot B, Morelli L, Canani R, Flint H, Salminen S, Calder PC, Sanders ME (2014) The International Scientific Association for Probiotics and Prebiotics consensus statement on the scope and appropriate use of the term probiotic. Nat Rev Gastroenterol Hepatol 11(8):506–514. https://doi.org/10.1038/nrgastro.2014.66
Honda K (2011) Porphyromonas gingivalis sinks teeth into the oral microbiota and periodontal disease. Cell Host Microbe 10(5):423–425. https://doi.org/10.1016/j.chom.2011.10.008
How KY, Song KP, Chan KG (2016) Porphyromonas gingivalis: an overview of periodontopathic pathogen below the gum line. Front Microbiol 7:53. https://doi.org/10.3389/fmicb.2016.00053
Humphrey SP, Williamson RT (2001) A review of saliva: normal composition, flow, and function. J Prosthet Dent 85(2):162–169. https://doi.org/10.1067/mpr.2001.113778
Huyck CL (1945) The effect of cetylpyridinium chloride on the bacterial growth in the oral cavity. J Am Pharm Assoc Am Pharm Assoc (Baltim) 34:5–11

Janus MM, Crielaard W, Zaura E, Keijser BJ, Brandt BW, Krom BP (2016) A novel compound to maintain a healthy oral plaque ecology in vitro. J Oral Microbiol 8:32513. https://doi.org/10.3402/jom.v8.32513

Jenkins S, Addy M, Wade W (1988) The mechanism of action of chlorhexidine. A study of plaque growth on enamel inserts in vivo. J Clin Periodontol 15(7):415–424. https://doi.org/10.1111/j.1600-051x.1988.tb01595.x

Jones S, Burt BA, Petersen PE, Lennon MA (2005) The effective use of fluorides in public health. Bull World Health Org 83(9):670–676

Kahler B, Rossi-Fedele G (2016) A review of tooth discoloration after regenerative endodontic therapy. J Endod 42(4):563–569. https://doi.org/10.1016/j.joen.2015.12.022

Kahler B, Rossi-Fedele G, Chugal N, Lin LM (2017) An evidence-based review of the efficacy of treatment approaches for immature permanent teeth with pulp necrosis. J Endod 43(7): 1052–1057. https://doi.org/10.1016/j.joen.2017.03.003

Kang M, Kim S, Kim H, Song Y, Jung D, Kang S, Seo JH, Nam S, Lee Y (2019) Calcium-binding polymer-coated poly(lactide-co-glycolide) microparticles for sustained release of quorum sensing inhibitors to prevent biofilm formation on hydroxyapatite surfaces. ACS Appl Mater Interfaces 11(8):7686–7694. https://doi.org/10.1021/acsami.8b18301

Kilian M, Chapple IL, Hannig M, Marsh PD, Meuric V, Pedersen AM, Tonetti MS, Wade WG, Zaura E (2016) The oral microbiome - an update for oral healthcare professionals. Br Dent J 221(10):657–666. https://doi.org/10.1038/sj.bdj.2016.865

Kirst ME, Li EC, Alfant B, Chi YY, Walker C, Magnusson I, Wang GP (2015) Dysbiosis and alterations in predicted functions of the subgingival microbiome in chronic periodontitis. Appl Environ Microbiol 81(2):783–793. https://doi.org/10.1128/AEM.02712-14

Kistler JO, Pesaro M, Wade WG (2015) Development and pyrosequencing analysis of an in-vitro oral biofilm model. BMC Microbiol 15:24. https://doi.org/10.1186/s12866-015-0364-1

Kolenbrander PE, Andersen RN, Blehert DS, Egland PG, Foster JS, Palmer RJ Jr (2002) Communication among oral bacteria. Microbiol Mol Biol Rev 66(3):486. https://doi.org/10.1128/MMBR.66.3.486-505.2002

Kolodkin-Gal I, Romero D, Cao S, Clardy J, Kolter R, Losick R (2010) D-amino acids trigger biofilm disassembly. Science 328(5978):627–629. https://doi.org/10.1126/science.1188628

Koo H, Duarte S, Murata RM, Scott-Anne K, Gregoire S, Watson GE, Singh AP, Vorsa N (2010) Influence of cranberry proanthocyanidins on formation of biofilms by Streptococcus mutans on saliva-coated apatitic surface and on dental caries development in vivo. Caries Res 44(2): 116–126. https://doi.org/10.1159/000296306

Laleman I, Detailleur V, Slot DE, Slomka V, Quirynen M, Teughels W (2014) Probiotics reduce mutans streptococci counts in humans: a systematic review and meta-analysis. Clin Oral Investig 18(6):1539–1552. https://doi.org/10.1007/s00784-014-1228-z

Lamont RJ, Koo H, Hajishengallis G (2018) The oral microbiota: dynamic communities and host interactions. Nat Rev Microbiol 16(12):745–759. https://doi.org/10.1038/s41579-018-0089-x

Lara HH, Romero-Urbina DG, Pierce C, Lopez-Ribot JL, Arellano-Jiménez MJ, Jose-Yacaman M (2015) Effect of silver nanoparticles on Candida albicans biofilms: an ultrastructural study. J Nanobiotechnol 13:91–91. https://doi.org/10.1186/s12951-015-0147-8

Latimer J, Munday JL, Buzza KM, Forbes S, Sreenivasan PK, McBain AJ (2015) Antibacterial and anti-biofilm activity of mouthrinses containing cetylpyridinium chloride and sodium fluoride. BMC Microbiol 15:169. https://doi.org/10.1186/s12866-015-0501-x

Lee YK, Salminen S (2009) Handbook of probiotics and prebiotics, 2nd edn. Wiley, Hoboken, NJ

Lee HJ, Kim SC, Kim J, Do A, Han SY, Lee BD, Lee HH, Lee MC, Lee SH, Oh T, Park S, Hong SH (2015) Synergistic inhibition of Streptococcal biofilm by ribose and xylitol. Arch Oral Biol 60(2):304–312. https://doi.org/10.1016/j.archoralbio.2014.11.004

Leiman SA, May JM, Lebar MD, Kahne D, Kolter R, Losick R (2013) D-amino acids indirectly inhibit biofilm formation in Bacillus subtilis by interfering with protein synthesis. J Bacteriol 195(23):5391–5395. https://doi.org/10.1128/JB.00975-13

Leung V, Dufour D, Levesque CM (2015) Death and survival in Streptococcus mutans: differing outcomes of a quorum-sensing signaling peptide. Front Microbiol 6:1176. https://doi.org/10.3389/fmicb.2015.01176

Li J, Huang Z, Mei L, Li G, Li H (2015) Anti-caries effect of arginine-containing formulations in vivo: a systematic review and meta-analysis. Caries Res 49(6):606–617. https://doi.org/10.1159/000435894

Liao Y, Brandt BW, Li J, Crielaard W, Van Loveren C, Deng DM (2017) Fluoride resistance in Streptococcus mutans: a mini review. J Oral Microbiol 9(1):1344509. https://doi.org/10.1080/20002297.2017.1344509

Liu YL, Nascimento M, Burne RA (2012) Progress toward understanding the contribution of alkali generation in dental biofilms to inhibition of dental caries. Int J Oral Sci 4(3):135–140. https://doi.org/10.1038/ijos.2012.54

Lombardi T, Budtz-Jorgensen E (1993) Treatment of denture-induced stomatitis: a review. Eur J Prosthodont Restor Dent 2(1):17–22

Lopez-Lopez A, Mira A (2020) Shifts in composition and activity of oral biofilms after fluoride exposure. Microb Ecol 80(3):729–738. https://doi.org/10.1007/s00248-020-01531-8

Luppens SB, Kara D, Bandounas L, Jonker MJ, Wittink FR, Bruning O, Breit TM, Ten Cate JM, Crielaard W (2008) Effect of Veillonella parvula on the antimicrobial resistance and gene expression of Streptococcus mutans grown in a dual-species biofilm. Oral Microbiol Immunol 23(3):183–189. https://doi.org/10.1111/j.1399-302X.2007.00409.x

Lynne VM (2006) Meta-analysis of probiotics for the prevention of antibiotic associated diarrhea and the treatment of clostridium difficile disease. Am J Gastroenterol 101(4):812–822. https://doi.org/10.1111/j.1572-0241.2006.00465.x

Mao X, Auer DL, Buchalla W, Hiller KA, Maisch T, Hellwig E, Al-Ahmad A, Cieplik F (2020) Cetylpyridinium chloride: mechanism of action, antimicrobial efficacy in biofilms, and potential risks of resistance. Antimicrob Agents Chemother 64(8). https://doi.org/10.1128/AAC.00576-20

Maris P (1995) Modes of action of disinfectants. Rev Sci Tech 14(1):47–55. https://doi.org/10.20506/rst.14.1.829

Mark Welch JL, Ramirez-Puebla ST, Borisy GG (2020) Oral microbiome geography: micron-scale habitat and niche. Cell Host Microbe 28(2):160–168. https://doi.org/10.1016/j.chom.2020.07.009

Marsh PD (2003) Are dental diseases examples of ecological catastrophes? Microbiology (Reading, England) 149(2):279. https://doi.org/10.1099/mic.0.26082-0

Marsh PD (2006) Dental plaque as a biofilm and a microbial community - implications for health and disease. BMC Oral Health 6(Suppl 1):S14. https://doi.org/10.1186/1472-6831-6-S1-S14

Marsh PD (2010) Controlling the oral biofilm with antimicrobials. J Dent 38(Suppl 1):S11–S15. https://doi.org/10.1016/S0300-5712(10)70005-1

Marsh PD, Martin MV (2009) Oral microbiology. Elsevier Churchill Livingstone, New York

Martínez-Robles ÁM, Loyola-Rodríguez JP, Zavala-Alonso NV, Martinez-Martinez RE, Ruiz F, Lara-Castro RH, Donohué-Cornejo A, Reyes-López SY, Espinosa-Cristóbal LF (2016) Antimicrobial properties of biofunctionalized silver nanoparticles on clinical isolates of streptococcus mutans and its serotypes. Nanomaterials (Basel, Switzerland). 6(7):136. https://doi.org/10.3390/nano6070136

Martori E, Ayuso-Montero R, Martinez-Gomis J, Vinas M, Peraire M (2014) Risk factors for denture-related oral mucosal lesions in a geriatric population. J Prosthet Dent 111(4):273–279. https://doi.org/10.1016/j.prosdent.2013.07.015

Marttinen AM, Haukioja AL, Keskin M, Soderling EM (2013) Effects of Lactobacillus reuteri PTA 5289 and L. paracasei DSMZ16671 on the adhesion and biofilm formation of Streptococcus mutans. Curr Microbiol 67(2):193–199. https://doi.org/10.1007/s00284-013-0352-3

McBain AJ (2009) Chapter 4: In vitro biofilm models: an overview. Adv Appl Microbiol 69:99–132. https://doi.org/10.1016/S0065-2164(09)69004-3

Merritt J, Qi F, Goodman SD, Anderson MH, Shi W (2003) Mutation of luxS affects biofilm formation in streptococcus mutans. Infect Immun 71(4):1972. https://doi.org/10.1128/IAI.71.4.1972-1979.2003

Moazami F, Sahebi S, Ahzan S (2018) Tooth discoloration induced by imidazolium based silver nanoparticles as an intracanal irrigant. J Dent (Shiraz) 19(4):280–286

Molander A, Reit C, Dahlen G, Kvist T (1998) Microbiological status of root-filled teeth with apical periodontitis. Int Endod J 31(1):1–7

Muras A, Mayer C, Romero M, Camino T, Ferrer MD, Mira A, Otero A (2018) Inhibition of Steptococcus mutans biofilm formation by extracts of Tenacibaculum sp. 20J, a bacterium with wide-spectrum quorum quenching activity. J Oral Microbiol 10(1). https://doi.org/10.1080/20002297.2018.1429788

Naginyte M, Do T, Meade J, Devine DA, Marsh PD (2019) Enrichment of periodontal pathogens from the biofilms of healthy adults. Sci Rep 9(1):5491. https://doi.org/10.1038/s41598-019-41882-y

Nagle PJ, Turnbull RS (1978) Chlorhexidine: an ideal plaque inhibiting agent? Dent J 44(2):73–77

Nair PN (2014) Endodontic biofilm, technology and pulpal regenerative therapy: where do we go from here? Int Endod J 47(11):1003–1011. https://doi.org/10.1111/iej.12287

Nascimento MM (2017) Oral microbiota transplant: a potential new therapy for oral diseases. J Calif Dent Assoc 45(10):565–568

Näse L, Hatakka K, Savilahti E, Saxelin M, Pönkä A, Poussa T, Korpela R, Meurman JH (2001) Effect of long–term consumption of a probiotic bacterium, Lactobacillus rhamnosus GG, in milk on dental caries and caries risk in children. Caries Res 35(6):412–420. https://doi.org/10.1159/000047484

Nath S, Zilm, P., Jamieson, L. Kapellas, K. Goswami, N., Ketagoda, K, and Laura S. Weyrich. (2021) Development and characterization of an oral microbiome transplant among australians for the treatment of dental caries and periodontal disease: a study protocol. PLoS One 16(11): e0260433

Naumova EA, Kuehnl P, Hertenstein P, Markovic L, Jordan RA, Gaengler P, Arnold WH (2012) Fluoride bioavailability in saliva and plaque. BMC Oral Health 12:3. https://doi.org/10.1186/1472-6831-12-3

Ng WL, Bassler BL (2009) Bacterial quorum-sensing network architectures. Annu Rev Genet 43: 197–222. https://doi.org/10.1146/annurev-genet-102108-134304

Nikawa H, Hamada T, Yamamoto T (1998) Denture plaque--past and recent concerns. J Dent 26(4): 299–304. https://doi.org/10.1016/s0300-5712(97)00026-2

O'Toole GA (2011) Microtiter dish biofilm formation assay. J Vis Exp 47. https://doi.org/10.3791/2437

Pamp SJ, Sternberg C, Tolker-Nielsen T (2009) Insight into the microbial multicellular lifestyle via flow-cell technology and confocal microscopy. Cytometry A 75(2):90–103. https://doi.org/10.1002/cyto.a.20685

Panpaliya NP, Dahake PT, Kale YJ, Dadpe MV, Kendre SB, Siddiqi AG, Maggavi UR (2019) In vitro evaluation of antimicrobial property of silver nanoparticles and chlorhexidine against five different oral pathogenic bacteria. Saudi Dent J 31(1):76–83. https://doi.org/10.1016/j.sdentj.2018.10.004

Paque F, Ganahl D, Peters OA (2009) Effects of root canal preparation on apical geometry assessed by micro-computed tomography. J Endod 35(7):1056–1059. https://doi.org/10.1016/j.joen.2009.04.020

Peciuliene V, Reynaud AH, Balciuniene I, Haapasalo M (2001) Isolation of yeasts and enteric bacteria in root-filled teeth with chronic apical periodontitis. Int Endod J 34(6):429–434. https://doi.org/10.1046/j.1365-2591.2001.00411.x

Peddis N, Musu D, Ideo F, Rossi-Fedele G, Cotti E (2019) Interaction of biologic therapy with apical periodontitis and periodontitis: a systematic review. Aust Dent J 64(2):122–134. https://doi.org/10.1111/adj.12684

Peeters E, Nelis HJ, Coenye T (2008) Comparison of multiple methods for quantification of microbial biofilms grown in microtiter plates. J Microbiol Methods 72(2):157–165. https://doi.org/10.1016/j.mimet.2007.11.010

Petersen P (2003) The World Oral Health Report 2003: continuous improvement of oral health in the 21st century - the approach of the WHO Global Oral Health Programme. Community Dent Oral Epidemiol 31:3–23

Petersson LG, Magnusson K, Hakestam U, Baigi A, Twetman S (2011) Reversal of primary root caries lesions after daily intake of milk supplemented with fluoride and probiotic lactobacilli in older adults. Acta Odontol Scand 69(6):321–327. https://doi.org/10.3109/00016357.2011.568962

Philip N, Walsh L (2019) The potential ecological effects of casein phosphopeptide-amorphous calcium phosphate in dental caries prevention. Aust Dent J 64(1):66–71. https://doi.org/10.1111/adj.12661

Philip N, Suneja B, Walsh L (2018) Beyond Streptococcus mutans: clinical implications of the evolving dental caries aetiological paradigms and its associated microbiome. Br Dent J 224(4): 219–225. https://doi.org/10.1038/sj.bdj.2018.81

Pinheiro ET, Gomes BP, Ferraz CC, Sousa EL, Teixeira FB, Souza-Filho FJ (2003) Microorganisms from canals of root-filled teeth with periapical lesions. Int Endod J 36(1):1–11. https://doi.org/10.1046/j.1365-2591.2003.00603.x

Pitts NB, Zero DT, Marsh PD, Ekstrand K, Weintraub JA, Ramos-Gomez F, Tagami J, Twetman S, Tsakos G, Ismail A (2017) Dental caries. Nat Rev Dis Primers 3:17030. https://doi.org/10.1038/nrdp.2017.30

Porenczuk A, Grzeczkowicz A, Maciejewska I, Gołaś M, Piskorska K, Kolenda A, Gozdowski D, Kopeć-Swoboda E, Granicka L, Olczak-Kowalczyk D (2019) An initial evaluation of cytotoxicity, genotoxicity and antibacterial effectiveness of a disinfection liquid containing silver nanoparticles alone and combined with a glass-ionomer cement and dentin bonding systems. Adv Clin Exp Med. 28(1):75–83. https://doi.org/10.17219/acem/76160

Quinteros MA, Viviana CA, Onnainty R, Mary VS, Theumer MG, Granero GE, Paraje MG, Páez PL (2018) Biosynthesized silver nanoparticles: decoding their mechanism of action in Staphylococcus aureus and Escherichia coli. Int J Biochem Cell Biol 104:87–93. https://doi.org/10.1016/j.biocel.2018.09.006

Ramirez T, Shrestha A, Kishen A (2019) Inflammatory potential of monospecies biofilm matrix components. Int Endod J 52(7):1020–1027. https://doi.org/10.1111/iej.13093

Rasmussen TB, Givskov M (2006a) Quorum sensing inhibitors: a bargain of effects. Microbiology 152(4):895–904

Rasmussen TB, Givskov M (2006b) Quorum sensing inhibitors: a bargain of effects. Microbiology (Reading) 152(Pt 4):895–904. https://doi.org/10.1099/mic.0.28601-0

Rath H, Stumpp SN, Stiesch M (2017) Development of a flow chamber system for the reproducible in vitro analysis of biofilm formation on implant materials. PLoS One 12(2):e0172095. https://doi.org/10.1371/journal.pone.0172095

Reid G (2016) Probiotics: definition, scope and mechanisms of action. Best Pract Res Clin Gastroenterol 30(1):17–25. https://doi.org/10.1016/j.bpg.2015.12.001

Reinhardt B, Klocke A, Neering SH, Selbach S, Peters U, Flemmig TF, Beikler T (2019) Microbiological dynamics of red complex bacteria following full-mouth air polishing in periodontally healthy subjects-a randomized clinical pilot study. Clin Oral Investig 23(10): 3905–3914. https://doi.org/10.1007/s00784-019-02821-3

Reynolds EC (2009) Casein phosphopeptide-amorphous calcium phosphate: the scientific evidence. Adv Dent Res 21(1):25–29. https://doi.org/10.1177/0895937409335619

Rhemrev GE, Timmerman MF, Veldkamp I, Van Winkelhoff AJ, Van Der Velden U (2006) Immediate effect of instrumentation on the subgingival microflora in deep inflamed pockets under strict plaque control. J Clin Periodontol 33(1):42–48. https://doi.org/10.1111/j.1600-051X.2005.00871.x

Righolt AJ, Jevdjevic M, Marcenes W, Listl S (2018) Global-, regional-, and country-level economic impacts of dental diseases in 2015. J Dent Res 97(5):501–507. https://doi.org/10.1177/0022034517750572

Rodríguez-León E, Iñiguez-Palomares R, Navarro RE, Herrera-Urbina R, Tánori J, Iñiguez-Palomares C, Maldonado A (2013) Synthesis of silver nanoparticles using reducing agents obtained from natural sources (Rumex hymenosepalus extracts). Nanoscale Res Lett 8(1):318. https://doi.org/10.1186/1556-276X-8-318

Rosier BT, Marsh PD, Mira A (2018) Resilience of the oral microbiota in health: mechanisms that prevent dysbiosis. J Dent Res 97(4):371–380. https://doi.org/10.1177/0022034517742139

Russell C, Coulter WA (1975) Continuous monitoring of pH and Eh in bacterial plaque grown on a tooth in an artificial mouth. Appl Microbiol 29(2):141–144. https://doi.org/10.1128/am.29.2.141-144.1975

Russell C, Tagg JR (1981) Role of bacteriocin during plaque formation by Streptococcus salivarius and Streptococcus sanguis on a tooth in an artificial mouth. J Appl Bacteriol 50(2):305–313. https://doi.org/10.1111/j.1365-2672.1981.tb00895.x

Sachdeo A, Haffajee AD, Socransky SS (2008) Biofilms in the edentulous oral cavity. J Prosthodont 17(5):348–356. https://doi.org/10.1111/j.1532-849X.2008.00301.x

Salli KM, Ouwehand AC (2015) The use of in vitro model systems to study dental biofilms associated with caries: a short review. J Oral Microbiol 7:26149. https://doi.org/10.3402/jom.v7.26149

Salli KM, Forssten SD, Lahtinen SJ, Ouwehand AC (2016) Influence of sucrose and xylitol on an early Streptococcus mutans biofilm in a dental simulator. Arch Oral Biol 70:39–46. https://doi.org/10.1016/j.archoralbio.2016.05.020

Sanchez MC, Llama-Palacios A, Fernandez E, Figuero E, Marin MJ, Leon R, Blanc V, Herrera D, Sanz M (2014) An in vitro biofilm model associated to dental implants: structural and quantitative analysis of in vitro biofilm formation on different dental implant surfaces. Dent Mater 30(10):1161–1171. https://doi.org/10.1016/j.dental.2014.07.008

Sanz M, Herrera D, Kebschull M, Chapple I, Jepsen S, Beglundh T, Sculean A, Tonetti MS, Participants EFPW, Methodological C (2020) Treatment of stage I-III periodontitis-The EFP S3 level clinical practice guideline. J Clin Periodontol 47(Suppl 22):4–60. https://doi.org/10.1111/jcpe.13290

Sato M, Fujiwara S, Tsuchiya H, Fujii T, Iinuma M, Tosa H, Ohkawa Y (1996) Flavones with antibacterial activity against cariogenic bacteria. J Ethnopharmacol 54(2–3):171–176. https://doi.org/10.1016/S0378-8741(96)01464-X

Sauer K, Camper AK, Ehrlich GD, Costerton JW, Davies DG (2002) Pseudomonas aeruginosa displays multiple phenotypes during development as a biofilm. J Bacteriol 184(4):1140–1154. https://doi.org/10.1128/jb.184.4.1140-1154.2002

Savani GM, Sabbah W, Sedgley CM, Whitten B (2014) Current trends in endodontic treatment by general dental practitioners: report of a United States National Survey. J Endod 40(5):618–624. https://doi.org/10.1016/j.joen.2014.01.029

Sayed M, Matsui N, Hiraishi N, Nikaido T, Burrow MF, Tagami J (2018) Effect of glutathione bio-molecule on tooth discoloration associated with silver diammine fluoride. Int J Mol Sci 19(5). https://doi.org/10.3390/ijms19051322

Sayed M, Matsui N, Hiraishi N, Inoue G, Nikaido T, Burrow MF, Tagami J (2019) Evaluation of discoloration of sound/demineralized root dentin with silver diamine fluoride: in-vitro study. Dent Mater J 38(1):143–149. https://doi.org/10.4012/dmj.2018-008

Segata N, Haake SK, Mannon P, Lemon KP, Waldron L, Gevers D, Huttenhower C, Izard J (2012) Composition of the adult digestive tract bacterial microbiome based on seven mouth surfaces, tonsils, throat and stool samples. Genome Biol 13(6):R42. https://doi.org/10.1186/gb-2012-13-6-r42

Sender R, Fuchs S, Milo R (2016) Revised estimates for the number of human and bacteria cells in the body. PLoS Biol 14(8):e1002533. https://doi.org/10.1371/journal.pbio.1002533

Senpuku H, Mohri S, Mihara M, Arai T, Suzuki Y, Saeki Y (2019) Effects of 7S globulin 3 derived from the adzuki bean [Vigna angularis] on the CSP- and eDNA- dependent biofilm formation of Streptococcus mutans. Arch Oral Biol 102:256–265. https://doi.org/10.1016/j.archoralbio.2019.04.010

Shahzad M, Sherry L, Rajendran R, Edwards CA, Combet E, Ramage G (2014) Utilising polyphenols for the clinical management of Candida albicans biofilms. Int J Antimicrob Agents 44(3):269–273

Shanker E, Federle MJ (2017) Quorum sensing regulation of competence and bacteriocins in Streptococcus pneumoniae and mutans. Genes (Basel) 8(1). https://doi.org/10.3390/genes8010015

Sharma G, Ahmed HMA, Zilm PS, Rossi-Fedele G (2018) Antimicrobial properties of calcium hydroxide dressing when used for long-term application: A systematic review. Aust Endod J 44(1):60–65. https://doi.org/10.1111/aej.12216

Shaw L, Ribeiro ALR, Levine AP, Pontikos N, Balloux F, Segal AW, Roberts AP, Smith AM (2017) The human salivary microbiome is shaped by shared environment rather than genetics: evidence from a large family of closely related individuals. mBio 8(5). https://doi.org/10.1128/mBio.01237-17

Shin H, Price K, Albert L, Dodick J, Park L, Dominguez-Bello MG (2016) Changes in the eye microbiota associated with contact lens wearing. mBio 7(2):e00198. https://doi.org/10.1128/mBio.00198-16

Sigurdsson A, Garland RW, Le KT, Rassoulian SA (2018) Healing of periapical lesions after endodontic treatment with the gentlewave procedure: a prospective multicenter clinical study. J Endod 44(3):510–517. https://doi.org/10.1016/j.joen.2017.12.004

Simon-Soro A, Tomas I, Cabrera-Rubio R, Catalan MD, Nyvad B, Mira A (2013) Microbial geography of the oral cavity. J Dent Res 92(7):616–621. https://doi.org/10.1177/0022034513488119

Sionov RV, Tsavdaridou D, Aqawi M, Zaks B, Steinberg D, Shalish M (2021) Tooth mousse containing casein phosphopeptide-amorphous calcium phosphate prevents biofilm formation of Streptococcus mutans. BMC Oral Health 21(1):136. https://doi.org/10.1186/s12903-021-01502-6

Sissons CH, Cutress TW, Hoffman MP, Wakefield JS (1991) A multi-station dental plaque microcosm (artificial mouth) for the study of plaque growth, metabolism, pH, and mineralization. J Dent Res 70(11):1409–1416. https://doi.org/10.1177/00220345910700110301

Slobodníkova L, Fialova S, Rendekova K, Kovac J, Mucaji P (2016) Antibiofilm activity of plant polyphenols. Molecules 21(12):1717. https://doi.org/10.3390/molecules21121717

Smith EG, Spatafora GA (2012) Gene regulation in S. mutans: complex control in a complex environment. J Dent Res 91(2):133–141. https://doi.org/10.1177/0022034511415415

Smits LP, Bouter KE, de Vos WM, Borody TJ, Nieuwdorp M (2013) Therapeutic potential of fecal microbiota transplantation. Gastroenterology 145(5):946–953. https://doi.org/10.1053/j.gastro.2013.08.058

Socransky SS, Haffajee AD (2002) Dental biofilms: difficult therapeutic targets. Periodontol 2000 28:12–55. https://doi.org/10.1034/j.1600-0757.2002.280102.x

Socransky SS, Haffajee AD, Cugini MA, Smith C, Kent RL Jr (1998) Microbial complexes in subgingival plaque. J Clin Periodontol 25(2):134–144

Sundqvist G (1994) Taxonomy, ecology, and pathogenicity of the root canal flora. Oral Surg Oral Med Oral Pathol 78(4):522–530. https://doi.org/10.1016/0030-4220(94)90047-7

Sundqvist G, Figdor D (2003) Life as an endodontic pathogen. Endodontic Topics 6(1):3–28. https://doi.org/10.1111/j.1601-1546.2003.00054.x

Sztajer H, Lemme A, Vilchez R, Schulz S, Geffers R, Yip CY, Levesque CM, Cvitkovitch DG, Wagner-Dobler I (2008) Autoinducer-2-regulated genes in Streptococcus mutans UA159 and global metabolic effect of the luxS mutation. J Bacteriol 190(1):401–415. https://doi.org/10.1128/jb.01086-07

Takahashi N, Nyvad B (2008) Caries ecology revisited: microbial dynamics and the caries process. Caries Res 42(6):409–418. https://doi.org/10.1159/000159604

Tan H, Peres KG, Peres MA (2016) Retention of teeth and oral health-related quality of life. J Dent Res 95(12):1350–1357. https://doi.org/10.1177/0022034516657992

ten Cate JM (2013) Contemporary perspective on the use of fluoride products in caries prevention. Br Dent J 214(4):161–167. https://doi.org/10.1038/sj.bdj.2013.162

Teughels W, Feres M, Oud V, Martin C, Matesanz P, Herrera D (2020) Adjunctive effect of systemic antimicrobials in periodontitis therapy: a systematic review and meta-analysis. J Clin Periodontol 47(Suppl 22):257–281. https://doi.org/10.1111/jcpe.13264

Thurnheer T, Gmur R, Guggenheim B (2004) Multiplex FISH analysis of a six-species bacterial biofilm. J Microbiol Methods 56(1):37–47. https://doi.org/10.1016/j.mimet.2003.09.003

Tiburcio-Machado CS, Michelon C, Zanatta FB, Gomes MS, Marin JA, Bier CA (2021) The global prevalence of apical periodontitis: a systematic review and meta-analysis. Int Endod J 54(5): 712–735. https://doi.org/10.1111/iej.13467

Tsirigotis-Maniecka M, Gancarz R, Wilk KA (2017) Polysaccharide hydrogel particles for enhanced delivery of hesperidin: fabrication, characterization and in vitro evaluation. Colloids Surf A Physicochem Eng Asp 532:48–56. https://doi.org/10.1016/j.colsurfa.2017.07.001

Tsuchiya H, Sato M, Iinuma M, Yokoyama J, Ohyama M, Tanaka T, Takase I, Namikawa I (1994) Inhibition of the growth of cariogenic bacteria in vitro by plant flavanones. Experientia 50(9): 846–849. https://doi.org/10.1007/bf01956469

Turnbaugh PJ, Ley RE, Mahowald MA, Magrini V, Mardis ER, Gordon JI (2006) An obesity-associated gut microbiome with increased capacity for energy harvest. Nature 444(7122): 1027–1031. https://doi.org/10.1038/nature05414

Turnbaugh PJ, Ley RE, Hamady M, Fraser-Liggett CM, Knight R, Gordon JI (2007) The human microbiome project. Nature 449(7164):804–810. https://doi.org/10.1038/nature06244

Uzel A, Sorkun K, Oncag O, Cogulu D, Gencay O, Salih B (2005) Chemical compositions and antimicrobial activities of four different Anatolian propolis samples. Microbiol Res 160(2): 189–195

Van Dyke TE, Bartold PM, Reynolds EC (2020) The nexus between periodontal inflammation and dysbiosis. Front Immunol 11:511. https://doi.org/10.3389/fimmu.2020.00511

Van Loveren C (2001) Antimicrobial activity of fluoride and its in vivo importance: identification of research questions. Caries Res 35(Suppl 1):65–70. https://doi.org/10.1159/000049114

Varoni E, Tarce M, Lodi G, Carrassi A (2012) Chlorhexidine (CHX) in dentistry: state of the art. Minerva Stomatol 61(9):399–419

Vasilev K (2019) Nanoengineered antibacterial coatings and materials: a perspective. Coatings 9(10):654

Vasilev K, Griesser SS, Griesser HJ (2011) Antibacterial surfaces and coatings produced by plasma techniques. Plasma Process Polymers. 8(11):1010–1023. https://doi.org/10.1002/ppap.201100097

Velsko IM, Shaddox LM (2018) Consistent and reproducible long-term in vitro growth of health and disease-associated oral subgingival biofilms. BMC Microbiol 18(1):70. https://doi.org/10.1186/s12866-018-1212-x

Vendeville A, Winzer K, Heurlier K, Tang CM, Hardie KR (2005) Making 'sense' of metabolism: autoinducer-2, LuxS and pathogenic bacteria. Nat Rev Microbiol 3(5):383. https://doi.org/10.1038/nrmicro1146

Venter JC, Adams MD, Myers EW, Li PW, Mural RJ, Sutton GG, Smith HO, Yandell M, Evans CA, Holt RA, Gocayne JD, Amanatides P, Ballew RM, Huson DH, Wortman JR, Zhang Q, Kodira CD, Zheng XH, Chen L, Skupski M, Subramanian G, Thomas PD, Zhang J, Gabor Miklos GL, Nelson C, Broder S, Clark AG, Nadeau J, McKusick VA, Zinder N, Levine AJ, Roberts RJ, Simon M, Slayman C, Hunkapiller M, Bolanos R, Delcher A, Dew I, Fasulo D, Flanigan M, Florea L, Halpern A, Hannenhalli S, Kravitz S, Levy S, Mobarry C, Reinert K, Remington K, Abu-Threideh J, Beasley E, Biddick K, Bonazzi V, Brandon R, Cargill M, Chandramouliswaran I, Charlab R, Chaturvedi K, Deng Z, Di Francesco V, Dunn P, Eilbeck K,

Evangelista C, Gabrielian AE, Gan W, Ge W, Gong F, Gu Z, Guan P, Heiman TJ, Higgins ME, Ji RR, Ke Z, Ketchum KA, Lai Z, Lei Y, Li Z, Li J, Liang Y, Lin X, Lu F, Merkulov GV, Milshina N, Moore HM, Naik AK, Narayan VA, Neelam B, Nusskern D, Rusch DB, Salzberg S, Shao W, Shue B, Sun J, Wang Z, Wang A, Wang X, Wang J, Wei M, Wides R, Xiao C, Yan C, Yao A, Ye J, Zhan M, Zhang W, Zhang H, Zhao Q, Zheng L, Zhong F, Zhong W, Zhu S, Zhao S, Gilbert D, Baumhueter S, Spier G, Carter C, Cravchik A, Woodage T, Ali F, An H, Awe A, Baldwin D, Baden H, Barnstead M, Barrow I, Beeson K, Busam D, Carver A, Center A, Cheng ML, Curry L, Danaher S, Davenport L, Desilets R, Dietz S, Dodson K, Doup L, Ferriera S, Garg N, Gluecksmann A, Hart B, Haynes J, Haynes C, Heiner C, Hladun S, Hostin D, Houck J, Howland T, Ibegwam C, Johnson J, Kalush F, Kline L, Koduru S, Love A, Mann F, May D, McCawley S, McIntosh T, McMullen I, Moy M, Moy L, Murphy B, Nelson K, Pfannkoch C, Pratts E, Puri V, Qureshi H, Reardon M, Rodriguez R, Rogers YH, Romblad D, Ruhfel B, Scott R, Sitter C, Smallwood M, Stewart E, Strong R, Suh E, Thomas R, Tint NN, Tse S, Vech C, Wang G, Wetter J, Williams S, Williams M, Windsor S, Winn-Deen E, Wolfe K, Zaveri J, Zaveri K, Abril JF, Guigo R, Campbell MJ, Sjolander KV, Karlak B, Kejariwal A, Mi H, Lazareva B, Hatton T, Narechania A, Diemer K, Muruganujan A, Guo N, Sato S, Bafna V, Istrail S, Lippert R, Schwartz R, Walenz B, Yooseph S, Allen D, Basu A, Baxendale J, Blick L, Caminha M, Carnes-Stine J, Caulk P, Chiang YH, Coyne M, Dahlke C, Mays A, Dombroski M, Donnelly M, Ely D, Esparham S, Fosler C, Gire H, Glanowski S, Glasser K, Glodek A, Gorokhov M, Graham K, Gropman B, Harris M, Heil J, Henderson S, Hoover J, Jennings D, Jordan C, Jordan J, Kasha J, Kagan L, Kraft C, Levitsky A, Lewis M, Liu X, Lopez J, Ma D, Majoros W, McDaniel J, Murphy S, Newman M, Nguyen T, Nguyen N, Nodell M, Pan S, Peck J, Peterson M, Rowe W, Sanders R, Scott J, Simpson M, Smith T, Sprague A, Stockwell T, Turner R, Venter E, Wang M, Wen M, Wu D, Wu M, Xia A, Zandieh A, Zhu X (2001) The sequence of the human genome. Science 291(5507):1304–1351. https://doi.org/10.1126/science.1058040

Vijayakumar K, Ramanathan T (2018) Antiquorum sensing and biofilm potential of 5-hydroxymethylfurfural against Gram positive pathogens. Microb Pathog 125:48–50. https://doi.org/10.1016/j.micpath.2018.09.008

Walker C, Sedlacek MJ (2007) An in vitro biofilm model of subgingival plaque. Oral Microbiol Immunol 22(3):152–161. https://doi.org/10.1111/j.1399-302X.2007.00336.x

Wang H, Ren D (2017) Controlling Streptococcus mutans and Staphylococcus aureus biofilms with direct current and chlorhexidine. AMB Express 7(1):204. https://doi.org/10.1186/s13568-017-0505-z

Wang W, Liao S, Zhu Y, Liu M, Zhao Q, Fu Y (2015) Recent applications of nanomaterials in prosthodontics. J Nanomater 2015; Article ID 408643

Weerasinghe J, Li W, Zhou R, Zhou R, Gissibl A, Sonar P, Speight R, Vasilev K, Ostrikov KK (2020) Bactericidal silver nanoparticles by atmospheric pressure solution plasma processing. Nanomaterials (Basel) 10(5). https://doi.org/10.3390/nano10050874

Wen ZT, Burne RA (2004) LuxS-mediated signaling in Streptococcus mutans is involved in regulation of acid and oxidative stress tolerance and biofilm formation. J Bacteriol 186(9): 2682–2691. https://doi.org/10.1128/jb.186.9.2682-2691.2004

Wescombe PA, Hale JD, Heng NC, Tagg JR (2012) Developing oral probiotics from Streptococcus salivarius. Future Microbiol 7(12):1355–1371. https://doi.org/10.2217/fmb.12.113

Weyrich LS (2021) The evolutionary history of the human oral microbiota and its implications for modern health. Periodontol 2000 85(1):90–100. https://doi.org/10.1111/prd.12353

Weyrich LS, Duchene S, Soubrier J, Arriola L, Llamas B, Breen J, Morris AG, Alt KW, Caramelli D, Dresely V, Farrell M, Farrer AG, Francken M, Gully N, Haak W, Hardy K, Harvati K, Held P, Holmes EC, Kaidonis J, Lalueza-Fox C, de la Rasilla M, Rosas A, Semal P, Soltysiak A, Townsend G, Usai D, Wahl J, Huson DH, Dobney K, Cooper A (2017) Neanderthal behaviour, diet, and disease inferred from ancient DNA in dental calculus. Nature 544(7650):357–361. https://doi.org/10.1038/nature21674

Wilson CE, Cathro PC, Rogers AH, Briggs N, Zilm PS (2015) Clonal diversity in biofilm formation by Enterococcus faecalis in response to environmental stress associated with endodontic irrigants and medicaments. Int Endod J 48(3):210–219. https://doi.org/10.1111/iej.12301

Wittpahl G, Kolling-Speer I, Basche S, Herrmann E, Hannig M, Speer K, Hannig C (2015) The polyphenolic composition of cistus incanus herbal tea and its antibacterial and anti-adherent activity against Streptococcus mutans. Planta Med 81(18):1727–1735. https://doi.org/10.1055/s-0035-1557822

World Heath Organization (2002) Joint FAO/WHO working group report on drafting guidelines for the evaluation of probiotics in food. WHO Media Centre. ftp://ftp.fao.org/es/esn/food/wgreport2.pdf. Accessed 15 November 2015

Wu CD (2009) Grape products and oral health. J Nutr 139(9):1818s–1823s. https://doi.org/10.3945/jn.109.107854

Xiao C, Ran S, Huang Z, Liang J (2016) Bacterial diversity and community structure of supragingival plaques in adults with dental health or caries revealed by 16S pyrosequencing. Front Microbiol 7:1145. https://doi.org/10.3389/fmicb.2016.01145

Xu D, Wen J, Fu W, Gu T, Raad I (2012) D-amino acids for the enhancement of a binary biocide cocktail consisting of THPS and EDDS against an SRB biofilm. World J Microbiol Biotechnol 28(4):1641–1646. https://doi.org/10.1007/s11274-011-0970-5

Yang HW, Huang YF, Chou MY (2004) Occurrence of Porphyromonas gingivalis and Tannerella forsythensis in periodontally diseased and healthy subjects. J Periodontol 75(8):1077–1083. https://doi.org/10.1902/jop.2004.75.8.1077

Yin IX, Zhao IS, Mei ML, Lo ECM, Tang J, Li Q, So LY, Chu CH (2020) Synthesis and characterization of fluoridated silver nanoparticles and their potential as a non-staining anti-caries agent. Int J Nanomed 15:3207–3215. https://doi.org/10.2147/ijn.S243202

Yoo YJ, Kim AR, Perinpanayagam H, Han SH, Kum KY (2020) Candida albicans virulence factors and pathogenicity for endodontic infections. Microorganisms 8(9). https://doi.org/10.3390/microorganisms8091300

Yoshida A, Ansai T, Takehara T, Kuramitsu H (2005) LuxS-based signaling affects Streptococcus mutans biofilm formation. Appl Environ Microbiol 71(5):2372–2380. https://doi.org/10.1128/AEM.71.5.2372-2380.2005

Yost S, Duran-Pinedo AE, Teles R, Krishnan K, Frias-Lopez J (2015) Erratum to: functional signatures of oral dysbiosis during periodontitis progression revealed by microbial metatranscriptome analysis. Genome Med 7(1):111. https://doi.org/10.1186/s13073-015-0231-6

Zhu Y, Dashper SG, Chen YY, Crawford S, Slakeski N, Reynolds EC (2013) Porphyromonas gingivalis and Treponema denticola synergistic polymicrobial biofilm development. PLoS One 8(8):e71727. https://doi.org/10.1371/journal.pone.0071727

Zilm PS, Butnejski V, Rossi-Fedele G, Kidd SP, Edwards S, Vasilev K (2017) D-amino acids reduce Enterococcus faecalis biofilms in vitro and in the presence of antimicrobials used for root canal treatment. PLoS One 12(2):e0170670. https://doi.org/10.1371/journal.pone.0170670

Clinical Management of Fungal Biofilm Infections

Gordon Ramage, Bryn Short, Emily McKloud, Om Alkhir Alshanta, Mark Butcher, William McLean, and Jason L. Brown

Abstract Fungal biofilms are highly resilient to antifungal therapies, of which there are relatively few licensed options available in clinical medicine. Nonetheless, there is a vibrant research culture aimed at enhancing and expanding the arsenal of antifungals capable of inhibiting, killing, and disrupting fungal biofilms. This chapter aims to explore the wide variety of fungal biofilms affecting human health and to discuss the clinical options for existing and novel chemotherapeutics.

1 Introduction

Fungal biofilms have gained notoriety over the past two decades, with studies and reviews of *Candida* biofilms alone amassing approximately 4000 publications. Whilst relatively slower to generate general interest within the biofilm community, it is increasingly clear that they are a clinically challenging issue due to the difficulties in treating these infections (Ramage et al. 2009). Microbiologically, fungi are amongst the most important clinical infections, globally accounting for 300 million infections, and mortality rates of up to 50% in some diseases (Brown et al. 2012). Many of these fungal infections predominantly affect immunocompromised individuals and may be impacted by a diverse range of risk factors (Table 1), which can result in a wide range of fungal infections (Table 2). Management of these patient groups, or lack thereof, is exacerbated by a limited arsenal of antifungal agents capable of killing or disrupting these tenacious structures.

So, what antifungals are commonly used to manage fungal infections, and how do they fare against biofilms? Classically, azole antifungals are the mainstay of treatment for many pathogenic fungi, albeit with a few exceptions; e.g. *C. glabrata* and

G. Ramage (✉) · B. Short · E. McKloud · O. A. Alshanta · M. Butcher · W. McLean · J. L. Brown
Glasgow Biofilm Research Network, Oral Sciences Research Group, School of Medicine, Dentistry and Nursing, University of Glasgow, Glasgow, UK
e-mail: gordon.ramage@glasgow.ac.uk

K. Richter, K. Nørskov Kragh (eds.), *Antibiofilm Strategies*, Springer Series on Biofilms 11, https://doi.org/10.1007/978-3-031-10992-8_17

Table 1 Predisposing factors for fungal infections

Systemic	Local
Human immunodeficiency virus (HIV)	Dentures
Broad-spectrum antibiotic treatment	Xerostomia
Cancers	Indwelling prosthetic devises
Immunosuppressive therapy or condition, e.g. organ transplant	Burns
Hyperglycemia	Trauma
Cytotoxic chemotherapy	Vaginal douching
Radiotherapy	Contraceptive pills
Nutrition, e.g. iron, folate, and vitamin C, B, A	
Infections such as tuberculosis	
Chronic renal failure	
Pregnancy	
Impaired liver function	
Genetic susceptibility	

C. krusei are intrinsically insensitive. Azoles function as fungistatic antimicrobials, targeting ergosterol biosynthesis. Specifically, they act upon the 14-lanosterol demethylase enzyme pathway, which results in depletion of ergosterol molecules in the cell membrane, and at the same time an accumulation of sterol precursors, such as 14-alpha methylase (Pristov and Ghannoum 2019). This leads to membrane instability of growing cells and a static impact on fungal growth. The triazoles are the most widely used azoles, which are heterocyclic compounds that include fluconazole, itraconazole, and voriconazole, amongst some others. These are not without their problems though, as resistance is common through alterations in the ergosterol biosynthesis pathway, upregulated efflux pumps, activation of heat shock proteins, and of course biofilm-mediated tolerance.

Polyenes, such as amphotericin B (AMB), nystatin, and their liposomal formulations, offer a fungicidal option. These antifungal drugs are thought to insert into the lipid membrane adjacent to ergosterol, which in turn destabilises the cell membrane by forming pores and enabling cellular lysis (Carolus et al. 2020). Moreover, oxidative stress induction is also thought to additionally contribute to its effectiveness as a fungicidal agent. Resistance is rare due to its membrane-based target, but in some cases alterations to sterols and anti-oxidative stress mechanisms can protect the cell. Cell wall changes in the form of enhanced 1,3-alpha- and 1,3-beta-glucans have also been shown to correlate with AMB resistance.

Echinocandins, which include caspofungin, micafungin, and anidulafungin, act by inhibiting 1,3-beta-glucan synthase, which results in cell wall destabilisation and fungicidal activity. They can be considered analogous to penicillin interfering with peptidoglycan in bacteria. They have a wide spectrum of activity, though are limited against septate fungi such as A*spergillus fumigatus*, as they require actively growing cells to be fungicidal. Notably, these compounds are effective, albeit paradoxically, against *Candida albicans* biofilms (Pristov and Ghannoum 2019). Nevertheless, as a

Table 2 Clinical manifestations of fungal infections. A non-exhaustive overview of the most associated fungal pathogens, their predominant affected sites, and alternative names for several clinically relevant fungal infections

Clinical manifestation	Alias	Affected site	Common pathogen
Seborrheic dermatitis	Dandruff	Scalp, hair follicles	*Malassezia* species (Grimshaw et al. 2019)
Dermatophytosis	Ringworm	Skin (Scalp)	*Trichophyton rubrum* (Wang et al. 2006)
Fungal Meningitis	None	CSF/Brain	*Cryptococcus neoformans* (Charalambous et al. 2018)
Mycotic keratitis	Corneal inflammation	Eye	*Candida albicans* or filamentous *Fusarium solani* and *Aspergillus* spp. (Thomas and Kaliamurthy 2013)
Endophthalmitis	Intraocular inflammation	Eye	*Candida albicans* (Regan et al. 2020)
Otomycosis	Ear infection	Ear	*Candida* species and *Aspergillus niger* (Anwar and Gohar 2014)
Fungal sinusitis	Rhinosinusitis	Mucosa/ paranasal sinuses	*Aspergillus fumigatus* (Chakrabarti et al. 2009)
Angular cheilitis	Lip disease	Exterior corners of mouth	*Candida albicans* (Lugovic-Mihic et al. 2018)
Oral candidiasis	Oral thrush	Interior mouth mucosa/ submucosa	*Candida albicans* (Singh et al. 2014)
Onychomycosis	Nail infection	Nail bed	*Trichophyton rubrum* (Ghannoum and Isham 2014)
Vulvovaginal candidiasis	Yeast infection	Vagina/Vulva	*Candida albicans* (McKloud et al. 2021)
Blastomycosis	Gilchrist's/ Chicago disease	Lungs	*Blastomyces dermatitidis* (McBride et al. 2017)
Aspergillosis	Brooder's pneumonia	Bronchioles	*Aspergillus* species (Sherif and Segal 2010)
Mycetoma	Madura foot	Skin/subcutaneous tissue	*Madurella mycetomatis* (Emmanuel et al. 2018)
Fungal septicaemia	Bloodstream infection	Systemic	*Candida albicans* (Delaloye and Calandra 2014)

second option for azole-insensitive yeasts and moulds, their rise in use has led to echinocandin resistance through alteration of the glucan synthase enzymes (Fks1-Fks2 complex), changes in chitin composition, and stimulation of stress pathways.

Other antifungals do exist, but these are used more infrequently. This emphasises why there is a need to improve our pipeline of antifungal drugs (Perfect 2017). These are taking the form of oral formulations, nanoparticles, pathway inhibitors, and augmentative strategies. This chapter outlines how conventional and novel

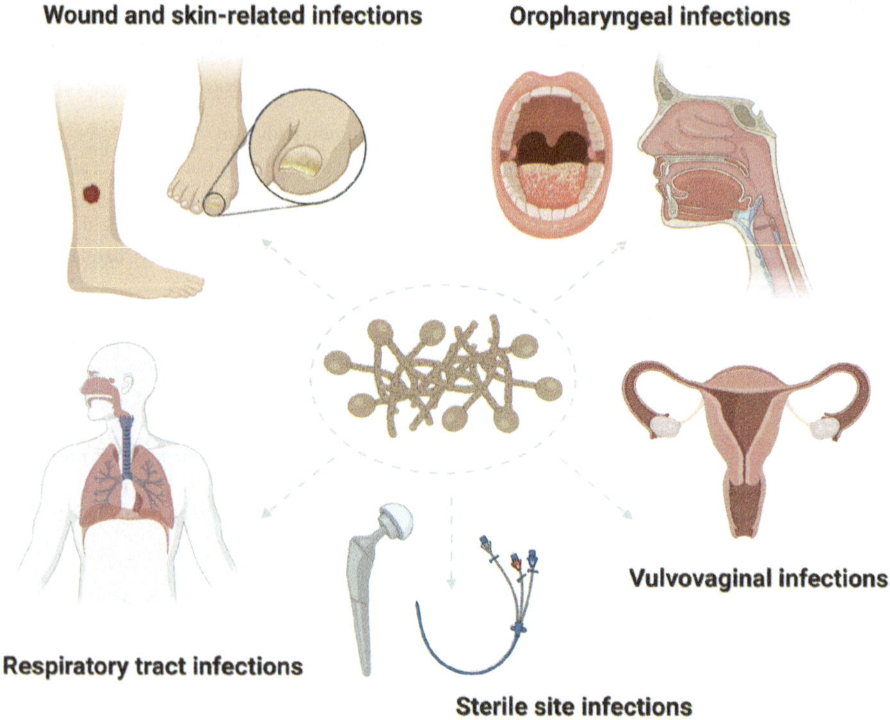

Fig. 1 Examples of sources of fungal biofilms on the human body (images produced in BioRender [biorender.com])

antifungals are being used for consideration in the management of different diseases associated with fungal biofilms, as illustrated in Fig. 1.

2 Oropharyngeal Infections

Fungal infections of the oral cavity are mainly opportunistic in nature due to host immunity impairment, which is a result of local, or more importantly, systemic factors. The frequency of these infections is increasing globally due to rise in the use of immunosuppressive drugs, broad-spectrum antibiotics, malignancies, diabetes, human immunodeficiency virus (HIV), and increased life expectancy (Richardson and Lass-Flörl 2008). Importantly, within the oral cavity these fungal species have the capacity to co-aggregate with microbial species in the form of biofilms on biological and inert substrates, or as aggregates within saliva. This has a profound effect on our ability to manage these infections with antifungal agents.

Oral fungal infections, or "oral mycoses", are broadly categorised as candidal and non-candidal fungal infections, or as superficial and deep mycoses (Santosh et al.

2021). Oral candidiasis (candidosis) is the most frequently reported oral fungal infection. This form of superficial mycoses is a result of the overgrowth of *Candida* species, mainly *Candida albicans.* Other non-*albicans* species, *Candida parapsilosis, Candida krusei, Candida stellatoidea, Candida tropicalis, Candida glabrata, Candida guilliermondii,* and *Candida dubliniensis*, have also contributed to oral candidiasis to a lesser extent. The diagnosis of oral candidiasis is usually based on the cytological/histopathological examinations and clinical presentation of the infection (Rautemaa and Ramage 2011). As *C. albicans* is a natural habitant of the oral microbiome in the majority of healthy individuals, oral samples with a positive culture for *Candida* species with absence of clinical manifestation are diagnostically inconclusive. Non-candidal oral mycoses (deep mycoses) are less common, and most infections are exotic with specific geographic distribution (Iatta et al. 2009). These infections are usually deep mycoses of other body areas other than the oral cavity, mainly pulmonary infections, which present with oral manifestations as an indication of systemic and disseminated condition. However, isolated oral lesions without systemic involvement are also reported (Scully 1913). Examples of oral non-candida infections are aspergillosis, cryptococcosis, histoplasmosis, blastomycosis, zygomycosis, mucormycosis, geotrichosis, and rhinosporidiosis (Williams and Ramage 2015)

Dental caries and periodontal diseases are amongst the most prevalent chronic diseases in humans, and both are biofilm related (Casamassimo et al. 2009; Nazir et al. 2020). Whilst the role of bacteria in such conditions is well recognised, the role of fungi is largely unknown or unexplored due to the bias towards bacterial biofilm communities. There is growing evidence that fungi have a significant influence on oral microbiome composition and pathogenicity (Delaney et al. 2018). Clearly, interkingdom interactions play a critical role in promoting tolerance to biocides and antifungals through their extra-polymeric substrates acting like "drug sponges" (Kim et al. 2018). High levels of *Candida* species were shown to be found in children with caries (Raja et al. 2010). Though, there is limited evidence for its direct association with dental caries (Sridhar et al. 2020). Indeed, it was recently shown that *C. albicans* was not more abundant in children with caries, although those children showed less fungal abundance and diversity compared with caries-free children (Fechney et al. 2019). Likewise, higher fungal abundance was reported in patients with periodontal diseases compared with healthy individuals and shown to be associated with disease severity (Canabarro et al. 2013; Urzúa et al. 2008; Peters et al. 2017). Indeed, a recent systematic review from 21 available studies showed a positive association between periodontal disease and *Candida* species (Suresh Unniachan et al. 2020). Again, it is unclear whether this higher abundance has a direct role in disease causation, or it is simply a consequence of bacterial dysbiosis and environmental change that favours fungal growth. There is a significant growth of studies investigating synergism and antagonism amongst these interkingdom diseases to define the importance of *Candida* in the oral cavity (Delaney et al. 2018), so the increasing use of newer molecular tools begins to define a clearer idea of the role that fungi play in these seemingly bacterial centric infections. The advent of mycobiome research has helped, but also hindered our progress of understanding. The first ever mycobiome study by Ghannoum (2010) and colleagues reported over 101 separate species from 21 individuals, though whether these fungi play defined roles remains to be ascertained (Ghannoum et al. 2010). The same authors contend that changes to the basal mycobiome can be observed in patients admitted to intensive care units (ICU) (Watkins et al. 2017).

The traditional clinical management of oral fungal infections involves the use of topical and systemic antifungals. In oropharyngeal candidiasis, identification and removal of local and systemic predisposing factors is paramount if feasible. Oral candidiasis is usually successfully managed with topical azoles and polyenes in the form of oral suspensions, lozenges, gels, creams, and ointment. Nystatin is usually effective for treatment of oral candidiasis. Amphotericin and miconazole can also be used, both of which elicit positive fungicidal effects (Farah et al. 2010). Refractory and recurrent infections usually require the use of systemic antifungals such as ketoconazole, fluconazole, and itraconazole and amphotericin in conjunction with topical agents to control the infection (Epstein and Polsky 1998). Despite treatment with antifungal, recurrence of oral candidiasis is not uncommon (Rautemaa and Ramage 2011). Recurrent infection can be due to incorrect diagnosis, inability to identify or treat underlying factors or inappropriate drug selection, or simply that the infection is biofilm-based and therefore intrinsically tolerant (Darwazeh and Darwazeh 2014). Systemic antifungals are mainly used for the treatment of deep mycoses, and drug selection depends on the severity of the infection and may require surgical debridement of necrotic tissue in some infections such mucormycosis and invasive aspergillosis. Amphotericin B, caspofungin, voriconazole, itraconazole, miconazole, ketoconazole, and fluconazole are the most commonly used systemic agents for deep fungal infections (Santosh et al. 2021).

Given that the currently available antifungal agents have many limitations, mainly drug resistance and cytotoxicity, efforts have been directed towards discovering novel antifungals. Repurposing drugs that have been previously approved for human use, plant derivatives and high-throughput screening are an appealing approach. Many anticancer, antimalarial, and antibacterial compounds have shown antifungal activity either alone or in combination with classical antifungals with less potential to develop drug resistance. However, most novel compounds are not translated into clinical trials for many reasons, mainly cost and lack of sufficient evidence. Antifungal resistance is a serious concern with classical antifungal especially with azoles. Therefore, drug combination was proposed to overcome drug resistance. With β-1,3-D-glucan of fungi being an ideal drug target, combining drugs that act on this essential cell wall component will potentially help in resolving antifungal resistance. MK-3118 (SCY-078) is a semisynthetic potent β-1,3-D-glucan synthases inhibitor. It showed a powerful *in vitro* effect on wild-type and resistant strains of *Aspergillus* species (Pfaller et al. 2013). It was also effective against *C. albicans*, C. *parapsilosis,* and *C. tropicalis* (Scorneaux et al. 2017). MK-3118 was also tested in a murine model of invasive candidiasis, and its effect was similar to that of intravenous echinocandin (Lepak et al. 2015).

3 Respiratory Tract Infections

Typically, fungi in the respiratory tract are often thought of as an innocent by-stander, but there are increasing amounts of evidence suggesting they play a major role in respiratory biofilm-associated diseases such as cystic fibrosis (CF), bronchiectasis, and chronic obstructive pulmonary disease (COPD) (Pendleton et al. 2017; Garczewska et al. 2016). However, as for fungal biofilms, our understanding of their role in COPD and bronchiectasis is in its infancy.

CF is an autosomal recessive condition, and although it is a multi-organ disease, mortality is most often determined by excess mucus production plugging the airways, chronic inflammation, and infection (Mall and Hartl 2014). Bacteria remain the most common causative agent of CF infections, but the isolation of fungi is becoming more common (Delfino et al. 2019). Lower respiratory tract infections of the CF lung are often associated with biofilms due to a decrease in volumes of periciliary fluid and diminished mucus detachment (Singh et al. 2000; Høiby et al. 2010). Indeed, there are reliable studies reporting the importance of fungal cells as a critical component of these biofilm aggregates (Mowat et al. 2007; Muller et al. 2011).

The loss of this innate clearance mechanism in the airways provides prime conditions for long-term colonisation. Spores formed by *Aspergillus,* one of the most abundant fungal species found throughout the environment (Paulussen et al. 2017), are readily inhaled, with *Aspergillus fumigatus* being the most common species identified in sputum cultures (Sabino et al. 2015). We were the first to demonstrate the biofilm forming capacity of this organism (Mowat et al. 2007), which has been followed up and expanded upon by us and others (Boisvert et al. 2016; Ramage et al. 2011). However, many CF airway infections are polymicrobial where microbe–microbe interactions can influence patient outcome. *Pseudomonas aeruginosa* is the most frequently isolated bacterial pathogen from CF sputum (Williams and Davies 2012). In vitro, *A. fumigatus* and *P. aeruginosa* exhibit a mutually antagonistic relationship (Reece et al. 2018; Shirazi et al. 2016). However, *in vivo* co-isolation of these organisms is associated with poorer patient outcomes, which may be explained by an increase in *P. aeruginosa* elastase production in the presence of *A. fumigatus* (Reece et al. 2017; Smith et al. 2015).

In cases where *A. fumigatus* colonises CF patients, the most commonly prescribed antifungals are triazoles such as voriconazole and itraconazole (Boyle et al. 2019). The consensus regarding the efficacy of azoles in CF is obscured with several smaller, non-controlled, open-label studies reporting a beneficial role for azole treatments (Hilliard et al. 2005; Shoseyov et al. 2006; Coughlan et al. 2012), whereas the only published randomised control trial (RCT) study investigating the efficacy of azoles in CF therapy reported no significant benefit in patients who were chronically colonised by *A. fumigatus* (Aaron et al. 2012). The chronic colonisation and ineffectiveness of azole interventions in this RCT points towards the presence of fungal biofilms which display high levels of azole resistance (Rajendran et al. 2011).

Aspergillus is not the only fungal species identified from the lungs of CF patients. *Candida albicans* is the most commonly isolated yeast and can be isolated from up to 75% of patients (Williams et al. 2016). As with *A. fumigatus,* co-isolation of *C. albicans* and *P. aeruginosa* from the same patient is associated with worse clinical outcomes (Dhamgaye et al. 2016). *P. aeruginosa* within these dual-species biofilms has displayed increased tolerance to meropenem when compared to a mono-species biofilm (Alam et al. 2020). This has been attributed to fungal mannans and glucans within the extracellular matrix of the biofilm. This trend of *C. albicans* protecting bacteria within a biofilm has been observed with other bacteria such as *Staphylococcus aureus,* another commonly found bacteria in CF, through production of ECM components such as glucans and extracellular DNA (Kean et al. 2017; Harriott and Noverr 2009; Kong et al. 2016).

Unlike *A. fumigatus,* there is some controversy surrounding the treatment of *C. albicans* in the airways due to clouded lines distinguishing colonisation and active infection (Pendleton et al. 2017). However, several studies have identified an association between *Candida* colonisation and worsening FEV1 in CF patients (Gileles-Hillel et al. 2015; Dhamgaye et al. 2016; Williams et al. 2016). Although *C. albicans* may not be actively causing infection, its interactions with certain bacteria indicate an indirect role in disease through protecting pathogenic bacteria and increasing their virulence (Dhamgaye et al. 2016; Kean et al. 2017; Todd et al. 2019). This raises the question: should *Candida* colonisation be addressed in order to improve patient outcomes? Although there is not a generally accepted treatment option for *C. albicans* in CF, azoles will likely prove to be ineffective due to high levels of azole tolerance in *C. albicans* biofilms, and the ease in which the organism develops resistance (El-Azizi et al. 2015; Kean et al. 2017; Rajendran et al. 2016b). Therefore, the use of echinocandins or polyenes may yield more promising clinical outcomes, particularly if nebulised to enhance delivery (Liao and Lam 2021).

Although things may appear bleak concerning our current ability to treat fungal and interkingdom biofilms in CF, there are novel drugs that show promising results. For example, a recent study by Miesel and colleagues reported on the efficacy of rezafungin, an echinocandin for use in respiratory therapy (Miesel et al. 2021). Their findings showed that prophylactic rezafungin of 10 and 20mg/kg resulted in 100% survival of mice who were subsequently challenged with *A. fumigatus.* Rezafungin has also shown promising effects against *Candida* and *Pneumocystis* (Miesel et al. 2021). Other studies have aimed to identify novel therapeutics or repurpose old ones that can be recruited in the fight against fungal biofilms. One 2016 study aimed to identify compounds with anti-*Candida* activities (Vila and Lopez-Ribot 2017), which identified several compounds that were capable of inhibiting *C. albicans* biofilm formation by more than 50%. However, only one was effective against pre-formed biofilms (MMV688768).

Currently, there are a few other potential agents with antifungal activities in preclinical stages, one of which is aureobasidin A. Although it was discovered in 1989, more recently it has shown promise as an agent to inhibit inositol phosphorylceramide synthase, an enzyme involved in sphingolipid synthesis, with activity against both planktonic and biofilm *Candida* species (Tan and Tay 2013). An alternative compound whose exact mechanism of action remains unknown but causes mitochondrial membrane collapse is T-2307. T-2307 is a novel arylamidine that has been tested against a panel of relevant fungi such as *Aspergillus, Candida*, and *Cryptococcus* species and was more effective than fluconazole, micafungin, and amphotericin B (Yamashita et al. 2019; Mitsuyama et al. 2008). At the time of writing, T-2307 is in phase 1 clinical trials and no clinical efficacy data are available.

4 Vulvovaginal Infections

It is estimated that up to three-quarters of women will suffer from at least one episode of vulvovaginal candidiasis (VVC) during their child-bearing years (Sobel 1992). Up to 8% of these women are expected to develop recurrent VVC (RVVC) (Sobel et al. 1998), defined as three or more episodes within one year (Sobel 2016). Although not associated with mortality, the symptoms of RVVC are debilitating, impact quality of life, and can result in psychological implications (Yano et al. 2019). *C. albicans* is reported as the causative organism in up to 90% of VVC episodes (Sobel et al. 1998). Uncomplicated (sporadic) VVC is associated with mild symptoms caused by *C. albicans* and can be treated with a single dose of oral or topical fluconazole in 80% of cases (Dovnik et al. 2015; Pappas et al. 2009). Fluconazole has remained the frontline treatment for VVC owing to its high cure rates and availability at clinics as well as over the counter. It is unclear whether any azole is more effective and whether oral or topical agents can impact clinical outcome (Dharmik et al. 2013; Whaley et al. 2016).

Treatment for complicated VVC (RVVC) requires prolonged maintenance azole therapies which are often unsuccessful. Fluconazole treatments are ineffective against *C. albicans* biofilms, suggesting their formation could contribute to failed clinical treatment. Treatment for RVVC caused by azole-resistant *C. glabrata* involves daily treatment with boric acid or nystatin pessaries for 14 days (Sobel et al. 2003; Fan et al. 2015). Alternative treatments include topical 17% flucytosine administered alone or in combination with 3% amphotericin B, daily for 14 days (Phillips 2005). Failed treatment of RVVC is suppressed with 10–14-day maintenance therapy using topical or oral fluconazole followed by a weekly dose of fluconazole for the next 6 months (Donders et al. 2008). Although suppressive therapies are often sufficient to relieve symptoms and recurrence during treatment, RVVC remains uncured and subsequently patients are prescribed these treatments for years (Sobel et al. 2004). Long-term use of azoles can drive antifungal resistance in *Candida*; however, if treatment options remain limited for women with persistent RVVC, this is inevitable.

The presence of *Candida* biofilms on vaginal mucosa during VVC is an area of controversy. Some authors dispute the presence of these biofilms, suggesting VVC is a result of polymicrobial invasion of vaginal tissues (Swidsinski et al. 2019; Sobel 2015). Conversely, *C. albicans* biofilm formation on vaginal mucosa in a murine model of VVC has been visualised using confocal and electron microscopy (Harriott et al. 2010). At present, there are no large-scale studies which aim to visualise *Candida* biofilm formation on mucosa from vaginal biopsies from women with VVC/RVVC, such as those carried out to investigate biofilm formation of *Gardnerella vaginalis* in bacterial vaginosis (Machado et al. 2015). Further, there are no characterised biofilm models representative of the vaginal environment during VVC to study potential *Candida* biofilm formation. Studies such as these would provide important knowledge of pathogenesis and resistance of RVVC which could improve future diagnosis and improve clinical treatment.

A novel potential avenue for the treatment of RVVC in the presence or absence of biofilms is the novel drug, ibrexafungerp (Ghannoum et al. 2019). It is the first of a new class of triterpenoid glucan synthase inhibitor antifungals. Orally administered ibrexafungerp has been shown to destabilise the fungal cell wall through the reduction of (1,3)-β-D-glucan polymers. This novel drug has shown efficacy against a range of *Candida* species, including azole and echinocandin-resistant isolates (Jimenez-Ortigosa et al. 2017). Importantly, it has been shown to inhibit *C. albicans* and *C. glabrata* biofilms *in vitro*, displaying lower MICs than fluconazole (Marcos-Zambrano et al. 2017). The recent FDA approval of ibrexafungerp and its broad range of activity against *Candida* species mean it could be used for the treatment of RVVC soon. Additionally, it displayed tolerability and low toxicity in phase 1 and 2 clinical trials (Akizawa et al. 2018). Although the presence of biofilms may not be universally accepted or currently diagnosed in RVVC, this drug could provide an exciting alternative treatment for women with azole-resistant, potential biofilm-associated VVC/RVVC. Moreover, topical echinocandin CD101 has also displayed significant promise against azole-resistant fungal species in the context of RVVC (Boikov et al. 2017). There is now even scope for consideration of probiotics for the management of RVVC (Pendharkar et al. 2015; Vladareanu et al. 2018).

5 Wound and Skin-Related Infections

There has been a growing recognition that complex biofilm communities of the mucosa and skin contain cross-kingdom biofilm communities of fungi and bacteria (Dowd et al. 2011; Kalan and Grice 2018; Kalan et al. 2016). Intriguingly, these relationships elicit reciprocal antimicrobial tolerance (Kong et al. 2016), meaning that we need to carefully consider and target fungi within complex infections until we more fully understand the consequences of broad-spectrum antimicrobial therapies. Regarding the local fungal microbiota, a publication by Oh et al. (2014) investigated the biogeography of the human skin, suggesting that the mycobiome consists of $<10\%$ of the total microbial population (Oh et al. 2014). The authors demonstrated that such multi-kingdom diversity is strongly shaped by the local skin microenvironment, with levels of fungal organisms varying from site to site. *Malassezia* species have been identified as the most prevalent fungal species at the skin barrier (Findley et al. 2013; Oh et al. 2014), comprising of up to 80% of the total skin population (Gao et al. 2010). Such organisms are also well represented within chronic wounds, alongside several opportunistic fungal pathogens such as *Trichosporon*, *Rhodotorula*, *Candida,* and *Cladosporidium* species (Chellan et al. 2010; Dowd et al. 2011; Kalan and Grice 2018). Alongside the common knowledge that bacterial accumulation in the infected wound can interfere with sufficient healing and repair, it is clear from the above studies that polymicrobial biofilm formation in the wound bed needs careful consideration during clinical management of chronic wounds such as diabetic foot ulcers (DFUs).

For most infected wounds, physical debridement of the tissue is recommended which largely results in removal of the biofilm and appropriate dressing of the infected area. Empirical antibiotic therapy is then often utilised as the first-line treatment for patients, to primarily target a wide range of Gram-positive cocci, including *Staphylococcus* species such as *S. aureus* which is often the most common pathogen in infected wounds (Lipsky et al. 2016b). However, as outlined by guidance published by the International Working Group on the Diabetic Foot (IWGDF), initial treatment is often administered based on "likely or proven causative pathogens", which can obviously be problematic to determine in polymicrobial infections (Lipsky et al. 2016a). Nevertheless, depending on severity, in antibiotic naïve patients, early therapy for mild infections often consists of flucloxacillin treatment and used in combination with metronidazole for more moderate to severe infections, whilst ciprofloxacin use is recommended in severe cases particularly when DFUs are accompanied by osteomyelitis (Barwell et al. 2017). Follow-up definitive therapies may be required based on culture and susceptibility results, and the patients' response to the empirical therapy (Lipsky et al. 2016b). At this juncture, it is important to note that antifungal therapeutics are not commonly recommended for DFU and other related chronic wound therapy.

With other skin infections, antifungal treatments are only utilised for nail infections such as onychomycosis, tinea pedis (athlete's foot), and tinea corporis (ringworm), whereby the sole causative agent is fungi. For example, in onychomycosis, dermatophytes such as *Trichophyton* species can account for almost 90% of all cases (Thomas et al. 2010). Typical onychomycosis or tinea-related antifungal therapy includes oral administration of terbinafine, with azoles such as itraconazole as the first line of treatment, with fluconazole used in some cases as an alternative therapy (Thomas et al. 2010; Hay 2018). For such infections, early management involves application of topical ointments, which can contain a range of antifungals, including azoles (Kawa et al. 2019). Common disinfectants and antiseptics such as chlorhexidine and povidone iodine have also been used to treat cases of onychomycosis (Capriotti and Capriotti 2015; Silva-Neves et al. 2021), whilst such treatments are also commonplace in wound care to prevent regrowth of the microbial population (Atiyeh et al. 2009). However, careful consideration for use of such antiseptics is required given concerns over their cytotoxic effects to the host, affecting wound healing and skin regeneration.

As discussed above, although antifungal therapies are uncommon for infected wounds, they have been shown to be effective in a small number of clinical studies and *in vitro* model systems. For example, a study by Heald et al. (2001) demonstrated that antifungal treatment (a mixture of flucytosine, fluconazole, itraconazole, and terbinafine) resulted in an improvement in wound healing in 17 DFU patients who were non-responsive to antibacterial therapy (Heald et al. 2001). Similar results were demonstrated elsewhere, where diabetic foot wounds in 38 patients given standard care alongside oral administration of fluconazole healed faster than those that received standard care alone (Chellan et al. 2012). Moving forward, research into biomaterials for "local delivery" of antibiotics has peaked in the last few years, providing a potential alternative for drug administration to infected wounds (Saghazadeh et al. 2018). Indeed, recent publications have highlighted that antifungals can be incorporated into materials such as calcium sulphate beads or polymer microparticles that can be used to effectively control fungal growth, both *in vitro* using a wide range of fungal isolates (Butcher et al. 2021) and *in vivo* in a murine model of cutaneous aspergillosis (Tatara et al. 2019). Nanotechnology is also beginning to be considered as prospective therapeutic avenue to revolutionise wound treatment using nano-drug delivery systems, although at the time of writing such research is still well in its infancy (Wang et al. 2019).

6 Medical Device-Related Infections

Nosocomial infection is a wide-scale health concern across the world, with approximately 60–70% of all hospital-based infections being accounted for by direct contact with implanted medical devices (Bryers 2008). Infection management can be successfully maintained through the removal, sterilisation, and replacement of implanted biomaterials (Khatoon et al. 2018), though biofilms remain an issue. Indeed, there is a vast range of indwelling biomaterials that have been associated with fungal biofilm infection, which have been reviewed extensively elsewhere (Ramage et al. 2006; Williams and Ramage 2015).

Prosthetic joint infection (PJI) is a significant complication to an otherwise ordinarily successful procedure. Despite strict surgical hygiene protocols, as well as carefully administered antimicrobial regimens before and after the procedure, PJI is still a prevalent outcome of joint replacement (van de Belt et al. 2001). This is not surprising given that prostheses can provide an optimal substrate and surrounding environment for the growth and development of polymicrobial biofilms (Tande and Patel 2014). PJI presents a unique issue for post-clinical management, which presents very significant and fundamental therapeutic challenges, including sepsis, amputation, and even death,

Colonisation of PJI implants is predominantly driven by biofilm adhesion and growth, which have the propensity to haematologically spread and impact distal sites (Ramage et al. 2006; Kaplan 2010). Typically there is a panel of usual suspects associated with PJIs, with the prime candidate being *Staphylococcus* species (Hall and Mah 2017; Perry and Hanssen 2017). Notably though, pathogenic fungi have been shown to exist in these biofilm-centric infections. The ability for interkingdom relationships within the biofilm can present extreme difficulty for establishing a therapeutic regimen. For example, fungal organisms account for only approximately

1% of PJI cases (Brown et al. 2018), but this small percentage can result in drastic oversight and clinical repercussions if not addressed with rapidity.

In recent years, research surrounding the management of PJI has highlighted the need for void-filling and moisture stability within the surgical site (Jones et al. 2016). This has, in turn, presented potential for novel therapeutics and the release of antimicrobial agents in a localised manner to areas of compromised vasculature using conventional filler material such as drug-loaded calcium sulphate (Butcher et al. 2021). This allows exposure to much higher effective doses of antimicrobials than would normally only be possible through a systemic route. While evidence has shown these methods to be effective, there remains the threat of infectious organisms developing antimicrobial resistance. Indeed, prevention of biofilm formation at the point of contact through alteration of the biomaterial surface proposes an interesting alternative to conventional therapy. Through alteration of variables which may significantly impact microbial adhesion, such as surface roughness and electrostatic charge (Gallo et al. 2014), the possibility for colonisation may also significantly diminish (Rzhepishevska et al. 2013; Yoda et al. 2014). Indeed, work conducted by Mayahara et al. in 2014 highlighted that adhesion of *Candida* yeast cells was up to two times greater on rougher surfaces when comparing roughness across the same substrate (Mayahara et al. 2014). Additional work has also taken place in investigating the efficacy of antimicrobial peptides against fungal pathogens (Delattin et al. 2017). More recently, it was shown that nanotopographical alterations to surface structure could significantly decrease yeast adhesion, paving out a promising strategy for implanted biomaterials (Alalwan et al. 2018).

Indwelling medical devices, such as intravascular catheters, voice prostheses, and ventricular-assisted devices (VADs), are commonly colonised with *Candida spp.* (Aslam et al. 2010; Elving et al. 2002). These infections are highly problematic due to the difficulty in effectively diagnosing them due to similar clinical symptoms to bacterial biofilm infections. Moreover, *Candida*, as in colonisation of the vocal prosthesis, is often discovered in a polymicrobial community, further exacerbating clinical identification (Leonhard and Schneider-Stickler 2015). Clinically, unless swiftly diagnosed the failure to treat a *Candida* infection in the ICU in the first 24 h can lead to a 30-fold increased likelihood of mortality (Kollef et al. 2012). Therefore, diagnosis, speed, and the choice of antifungal are all critically important in managing *Candida* spp. biofilms.

Within critical care there are a plethora of indwelling lines where adherent biofilm communities can thrive, from which cells can detach and cause a fungemia by spreading throughout the human body. These detached cells have been shown to be pathogenically primed and are associated with higher mortality rates in a murine model (Uppuluri et al. 2010). Literature has reported that biofilm formation correlates with clinical outcome, with those biofilms forming the greatest levels of bioburden being independent predictors of mortality (Rajendran et al. 2016a, b; Tumbarello et al. 2007). Moreover, it is clear from these studies that the choice of antifungal agent used is an important driver of clinical outcome, where the use of azoles is negatively correlated with clinical outcomes, whereas the use of echinocandins and liposomal formulations of amphotericin B leads to positive patient outcomes (Tumbarello et al. 2007).

Innovations in biofilm prevention, as well as novel methods of drug delivery, indicate a hopeful outlook for future therapy. A range of studies, such as those conducted by Free et al., in 2001, and Rodrigues et al., in 2004, have indicated the production of biosurfactants produced by probiotic bacteria as having significant potential for treatment of *Candida*-driven infection of prostheses (Free et al. 2001; Rodrigues et al. 2004). Additionally, there has been a series of studies that positively highlight the use of antifungals, such as liposomal amphotericin B, as antifungal line locks in the prevention and management of fungal line infections (McGhee et al. 2016; Paul DiMondi et al. 2014). Though, the wide acceptance of such approaches is still limited due to clinical apprehensiveness of fungal line infection management, apart from line removal. Caspofungin remains an important antifungal in the management of fungal line infection and has even been shown in formulations to be useful against *Candida auris* (Sumiyoshi et al. 2020).

7 *Candida auris*: The New Superfungus on the Block

Since its discovery in 2009 (Satoh et al. 2009), the nosocomial pathogen *Candida auris* has received global attention in the world of medical mycology. This is largely due to the organisms' ability to persist within the environment (Welsh et al. 2017), exhibit unique biofilm forming capabilities (Borman et al. 2016; Sherry et al. 2017), and demonstrate an unusually high resistance to three common classes of antifungals (Kean and Ramage 2019). The simultaneous emergence of four independent geographical clades (Lockhart et al. 2017), which have been linked with the rising temperature around the globe (Casadevall et al. 2019), has added further complexity to this enigmatic panfungal pathogen. In addition, a certain biofilm heterogeneity exists within this species, which is determined by an aggregative and non-aggregative morphology: whereby some isolates can form relatively large clusters of cells vs. some that remain as single-celled entities (Borman et al. 2016; Sherry et al. 2017; Brown et al. 2020). The geographical clade and these unique morphological phenotypes can influence antifungal susceptibility. It is widely accepted that most isolates are resistant to fluconazole, likely a result of a mutation in *ERG11* gene (which encodes for lanosterol 14-alpha-demethylase) and/or overexpression in drug efflux pumps. Varying rates in susceptibility to echinocandin and polyenes have been reported in the literature too, resistance to the former likely owing to genetic mutation in the *FKS1* gene, responsible for synthesis of β-glucan, a key component of the fungal cell wall (reviewed in Chaabane et al. 2019; Kean and Ramage 2019). Due to these multidrug resistance mechanisms, the search is now on for innovative therapeutics to combat *C. auris* pathogenicity. To date, several *in vitro* biofilm studies, *in vivo* model systems, and clinical trials are in the process of investigating the novel antifungal agents. Of note, fosmanogepix (Berkow et al. 2017; Larkin et al. 2017; Ghannoum et al. 2020; Arendrup et al. 2020) and ibrexafungerp (Hager et al. 2018; Zhao et al. 2018; Wiederhold et al. 2019), which target β-glucan synthase pathways, have been reported as effective alternatives in controlling *C. auris* biofilm formation *in vitro* and/or associated candidiasis in animal models. At the time of writing, both drugs are in phase 2 and phase 3 of clinical trials, respectively. This yeast is certainly one that will continue to be a global threat, and its propensity to form biofilms in healthcare environments, coupled with panfungal resistance, is a significant issue in this COVID era.

8 Concluding Comments

Fungal biofilms are an important clinical entity, and increasingly they are recognised as interkingdom structures alongside a myriad of bacterial species throughout the body. These difficult-to-treat infections are poorly responsive to antifungal agents routinely used in the clincial environment. However, this book chapter has demonstrated that above and beyond azoles, polyenes, and echinocandins, there are options that are in clinical trials and many other in preclinical development. The future is promising for management of fungal biofilms, but only if the wider clinical and scientific community recognise their importance.

References

Aaron SD, Vandemheen KL, Freitag A, Pedder L, Cameron W, Lavoie A, Paterson N, Wilcox P, Rabin H, Tullis E, Morrison N, Ratjen F (2012) Treatment of Aspergillus fumigatus in patients with cystic fibrosis: a randomized, placebo-controlled pilot study. PloS One 7(4):e36077–e36077. https://doi.org/10.1371/journal.pone.0036077

Akizawa T, Shimazaki R, Fukagawa M, Evocalcet Study Group (2018) Phase 2b study of evocalcet (KHK7580), a novel calcimimetic, in Japanese patients with secondary hyperparathyroidism undergoing hemodialysis: a randomized, double-blind, placebo-controlled, dose-finding study. PLoS One 13(10):e0204896. https://doi.org/10.1371/journal.pone.0204896

Alalwan H, Nile CJ, Rajendran R, McKerlie R, Reynolds P, Gadegaard N, Ramage G (2018) Nanoimprinting of biomedical polymers reduces candidal physical adhesion. Nanomedicine 14(3):1045–1049. https://doi.org/10.1016/j.nano.2018.01.011

Alam F, Catlow D, Di Maio A, Blair JMA, Hall RA (2020) Candida albicans enhances meropenem tolerance of Pseudomonas aeruginosa in a dual-species biofilm. J Antimicrobial Chemother 75(4):925–935. https://doi.org/10.1093/jac/dkz514

Anwar K, Gohar MS (2014) Otomycosis; clinical features, predisposing factors and treatment implications. Pak J Med Sci 30(3):564–567. https://doi.org/10.12669/pjms.303.4106

Arendrup MC, Jorgensen KM, Hare RK, Chowdhary A (2020) In vitro activity of Ibrexafungerp (SCY-078) against Candida auris isolates as determined by EUCAST methodology and comparison with activity against C. albicans and C. glabrata and with the activities of six comparator agents. Antimicrob Agents Chemother 64(3). https://doi.org/10.1128/AAC.02136-19

Aslam S, Hernandez M, Thornby J, Zeluff B, Darouiche RO (2010) Risk factors and outcomes of fungal ventricular-assist device infections. Clin Infect Dis 50(5):664–671. https://doi.org/10.1086/650454

Atiyeh BS, Dibo SA, Hayek SN (2009) Wound cleansing, topical antiseptics and wound healing. Int Wound J 6(6):420–430. https://doi.org/10.1111/j.1742-481X.2009.00639.x

Barwell ND, Devers MC, Kennon B, Hopkinson HE, McDougall C, Young MJ, Robertson HMA, Stang D, Dancer SJ, Seaton A, Leese GP, Scottish Diabetes Foot Action Group (2017) Diabetic foot infection: antibiotic therapy and good practice recommendations. Int J Clin Pract 71(10). https://doi.org/10.1111/ijcp.13006

Berkow EL, Angulo D, Lockhart SR (2017) In vitro activity of a novel glucan synthase inhibitor, SCY-078, against clinical isolates of Candida auris. Antimicrob Agents Chemother 61(7). https://doi.org/10.1128/AAC.00435-17

Boikov DA, Locke JB, James KD, Bartizal K, Sobel JD (2017) In vitro activity of the novel echinocandin CD101 at pH 7 and 4 against Candida spp. isolates from patients with vulvovaginal candidiasis. J Antimicrob Chemother 72(5):1355–1358. https://doi.org/10.1093/jac/dkx008

Boisvert AA, Cheng MP, Sheppard DC, Nguyen D (2016) Microbial biofilms in pulmonary and critical care diseases. Ann Am Thorac Soc 13(9):1615–1623. https://doi.org/10.1513/AnnalsATS.201603-194FR

Borman AM, Szekely A, Johnson EM (2016) Comparative pathogenicity of United Kingdom Isolates of the emerging pathogen candida auris and other key pathogenic Candida Species. mSphere 1(4). https://doi.org/10.1128/mSphere.00189-16

Boyle M, Moore JE, Whitehouse JL, Bilton D, Downey DG (2019) The diagnosis and management of respiratory tract fungal infection in cystic fibrosis: a UK survey of current practice. Med Mycol 57(2):155–160. https://doi.org/10.1093/mmy/myy014

Brown GD, Denning DW, Gow NA, Levitz SM, Netea MG, White TC (2012) Hidden killers: human fungal infections. Sci Transl Med 4(165):165rv113. https://doi.org/10.1126/scitranslmed.3004404

Brown TS, Petis SM, Osmon DR, Mabry TM, Berry DJ, Hanssen AD, Abdel MP (2018) Periprosthetic joint infection with fungal pathogens. J Arthroplasty 33(8):2605–2612. https://doi.org/10.1016/j.arth.2018.03.003

Brown JL, Delaney C, Short B, Butcher MC, McKloud E, Williams C, Kean R, Ramage G (2020) Candida auris phenotypic heterogeneity determines pathogenicity in vitro. mSphere 5(3). https://doi.org/10.1128/mSphere.00371-20

Bryers JD (2008) Medical biofilms. Biotechnol Bioeng 100(1):1–18. https://doi.org/10.1002/bit.21838

Butcher MC, Brown JL, Hansom D, Wilson-van Os R, Delury C, Laycock PA, Ramage G (2021) Assessing the bioactive profile of antifungal-loaded calcium sulfate against fungal biofilms. Antimicrob Agents Chemother 65(6). https://doi.org/10.1128/AAC.02551-20

Canabarro A, Valle C, Farias M, Santos F, Lazera M, Wanke B (2013) Association of subgingival colonization of C andida albicans and other yeasts with severity of chronic periodontitis. J Periodontal Res 48(4):428–432

Capriotti K, Capriotti JA (2015) Onychomycosis treated with a dilute povidone-iodine/dimethyl sulfoxide preparation. Int Med Case Rep J 8:231–233. https://doi.org/10.2147/IMCRJ.S90775

Carolus H, Pierson S, Lagrou K, Van Dijck P (2020) Amphotericin B and other polyenes-discovery, clinical use, mode of action and drug resistance. J Fungi (Basel) 6(4). https://doi.org/10.3390/jof6040321

Casadevall A, Kontoyiannis DP, Robert V (2019) On the emergence of Candida auris: climate change, azoles, swamps, and birds. mBio 10(4). https://doi.org/10.1128/mBio.01397-19

Casamassimo PS, Thikkurissy S, Edelstein BL, Maiorini E (2009) Beyond the dmft: the human and economic cost of early childhood caries. J Am Dental Assoc 140(6):650–657

Chaabane F, Graf A, Jequier L, Coste AT (2019) Review on antifungal resistance mechanisms in the emerging pathogen Candida auris. Front Microbiol 10:2788. https://doi.org/10.3389/fmicb.2019.02788

Chakrabarti A, Denning DW, Ferguson BJ, Ponikau J, Buzina W, Kita H, Marple B, Panda N, Vlaminck S, Kauffmann-Lacroix C, Das A, Singh P, Taj-Aldeen SJ, Kantarcioglu AS, Handa

KK, Gupta A, Thungabathra M, Shivaprakash MR, Bal A, Fothergill A, Radotra BD (2009) Fungal rhinosinusitis: a categorization and definitional schema addressing current controversies. Laryngoscope 119(9):1809–1818. https://doi.org/10.1002/lary.20520
Charalambous LT, Premji A, Tybout C, Hunt A, Cutshaw D, Elsamadicy AA, Yang S, Xie J, Giamberardino C, Pagadala P, Perfect JR, Lad SP (2018) Prevalence, healthcare resource utilization and overall burden of fungal meningitis in the United States. J Med Microbiol 67(2):215–227. https://doi.org/10.1099/jmm.0.000656
Chellan G, Shivaprakash S, Karimassery Ramaiyar S, Varma AK, Varma N, Thekkeparambil Sukumaran M, Rohinivilasam Vasukutty J, Bal A, Kumar H (2010) Spectrum and prevalence of fungi infecting deep tissues of lower-limb wounds in patients with type 2 diabetes. J Clin Microbiol 48(6):2097–2102. https://doi.org/10.1128/JCM.02035-09
Chellan G, Neethu K, Varma AK, Mangalanandan TS, Shashikala S, Dinesh KR, Sundaram KR, Varma N, Jayakumar RV, Bal A, Kumar H (2012) Targeted treatment of invasive fungal infections accelerates healing of foot wounds in patients with Type 2 diabetes. Diabet Med 29(9):e255–e262. https://doi.org/10.1111/j.1464-5491.2012.03574.x
Coughlan CA, Chotirmall SH, Renwick J, Hassan T, Low TB, Bergsson G, Eshwika A, Bennett K, Dunne K, Greene CM, Gunaratnam C, Kavanagh K, Logan PM, Murphy P, Reeves EP, McElvaney NG (2012) The effect of Aspergillus fumigatus infection on vitamin D receptor expression in cystic fibrosis. Am J Respir Crit Care Med 186(10):999–1007. https://doi.org/10.1164/rccm.201203-0478OC
Darwazeh AM, Darwazeh TA (2014) What makes oral candidiasis recurrent infection? A clinical view. J Mycol 2014:758394
Delaloye J, Calandra T (2014) Invasive candidiasis as a cause of sepsis in the critically ill patient. Virulence 5(1):161–169. https://doi.org/10.4161/viru.26187
Delaney C, Kean R, Short B, Tumelty M, McLean W, Nile CJ, Ramage G (2018) Fungi at the scene of the crime: innocent bystanders or accomplices in oral infections? Curr Clin Microbiol Rep 5(3):190–200
Delattin N, Brucker K, Cremer K, Cammue BP, Thevissen K (2017) Antimicrobial peptides as a strategy to combat fungal biofilms. Curr Top Med Chem 17(5):604–612. https://doi.org/10.2174/1568026616666160713142228
Delfino E, Del Puente F, Briano F, Sepulcri C, Giacobbe DR (2019) Respiratory fungal diseases in adult patients with cystic fibrosis. Clin Med Insights Circ Respir Pulm Med 13: 1179548419849939–1179548419849939. https://doi.org/10.1177/1179548419849939
Dhamgaye S, Qu Y, Peleg AY (2016) Polymicrobial infections involving clinically relevant Gram-negative bacteria and fungi. Cell Microbiol 18(12):1716–1722
Dharmik PG, Gomashe AV, Upadhyay VG (2013) Susceptibility pattern of various azoles against Candida species causing vulvovaginal candidiasis. J Obstet Gynaecol India 63(2):135–137. https://doi.org/10.1007/s13224-012-0280-3
Donders G, Bellen G, Byttebier G, Verguts L, Hinoul P, Walckiers R, Stalpaert M, Vereecken A, Van Eldere J (2008) Individualized decreasing-dose maintenance fluconazole regimen for recurrent vulvovaginal candidiasis (ReCiDiF trial). Am J Obstet Gynecol 199(6):613. https://doi.org/10.1016/j.ajog.2008.06.029
Dovnik A, Golle A, Novak D, Arko D, Takac I (2015) Treatment of vulvovaginal candidiasis: a review of the literature. Acta Dermatovenerol Alp Pannonica Adriat 24(1):5–7. https://doi.org/10.15570/actaapa.2015.2
Dowd SE, Delton Hanson J, Rees E, Wolcott RD, Zischau AM, Sun Y, White J, Smith DM, Kennedy J, Jones CE (2011) Survey of fungi and yeast in polymicrobial infections in chronic wounds. J Wound Care 20(1):40–47. https://doi.org/10.12968/jowc.2011.20.1.40
El-Azizi M, Farag N, Khardori N (2015) Antifungal activity of amphotericin B and voriconazole against the biofilms and biofilm-dispersed cells of Candida albicans employing a newly developed in vitro pharmacokinetic model. Ann Clin Microbiol Antimicrobials 14(1):21. https://doi.org/10.1186/s12941-015-0083-3

Elving GJ, van der Mei HC, Busscher HJ, van Weissenbruch R, Albers FW (2002) Comparison of the microbial composition of voice prosthesis biofilms from patients requiring frequent versus infrequent replacement. Ann Otol Rhinol Laryngol 111(3 Pt 1):200–203. https://doi.org/10.1177/000348940211100302

Emmanuel P, Dumre SP, John S, Karbwang J, Hirayama K (2018) Mycetoma: a clinical dilemma in resource limited settings. Ann Clin Microbiol Antimicrob 17(1):35. https://doi.org/10.1186/s12941-018-0287-4

Epstein JB, Polsky B (1998) Oropharyngeal candidiasis: a review of its clinical spectrum and current therapies. Clin Therap 20(1):40–57

Fan S, Liu X, Wu C, Xu L, Li J (2015) Vaginal nystatin versus oral fluconazole for the treatment for recurrent vulvovaginal candidiasis. Mycopathologia 179(1–2):95–101. https://doi.org/10.1007/s11046-014-9827-4

Farah C, Lynch N, McCullough M (2010) Oral fungal infections: an update for the general practitioner. Austr Dent J 55:48–54

Fechney JM, Browne GV, Prabhu N, Irinyi L, Meyer W, Hughes T, Bockmann M, Townsend G, Salehi H, Adler CJ (2019) Preliminary study of the oral mycobiome of children with and without dental caries. J Oral Microbiol 11(1):1536182

Findley K, Oh J, Yang J, Conlan S, Deming C, Meyer JA, Schoenfeld D, Nomicos E, Park M, Program NIHISCCS, Kong HH, Segre JA (2013) Topographic diversity of fungal and bacterial communities in human skin. Nature 498(7454):367–370. https://doi.org/10.1038/nature12171

Free RH, Busscher HJ, Elving GJ, van der Mei HC, van Weissenbruch R, Albers FW (2001) Biofilm formation on voice prostheses: in vitro influence of probiotics. Ann Otol Rhinol Laryngol 110(10):946–951. https://doi.org/10.1177/000348940111001010

Gallo J, Holinka M, Moucha CS (2014) Antibacterial surface treatment for orthopaedic implants. Int J Mol Sci 15(8):13849–13880. https://doi.org/10.3390/ijms150813849

Gao Z, Perez-Perez GI, Chen Y, Blaser MJ (2010) Quantitation of major human cutaneous bacterial and fungal populations. J Clin Microbiol 48(10):3575–3581. https://doi.org/10.1128/JCM.00597-10

Garczewska B, Jarzynka S, Kuś J, Skorupa W, Augustynowicz-Kopeć E (2016) Fungal infection of cystic fibrosis patients - single center experience. Pneumonol Alergol 84(3):151–159. https://doi.org/10.5603/PiAP.2016.0017

Ghannoum M, Isham N (2014) Fungal nail infections (onychomycosis): a never-ending story? PLoS Pathog 10(6):e1004105. https://doi.org/10.1371/journal.ppat.1004105

Ghannoum MA, Jurevic RJ, Mukherjee PK, Cui F, Sikaroodi M, Naqvi A, Gillevet PM (2010) Characterization of the oral fungal microbiome (mycobiome) in healthy individuals. PLoS Pathog 6(1):e1000713. https://doi.org/10.1371/journal.ppat.1000713

Ghannoum M, Long L, Isham N, Hager C, Wilson R, Borroto-Esoda K, Barat S, Angulo D (2019) Activity of a novel 1,3-beta-D-glucan synthase inhibitor, Ibrexafungerp (formerly SCY-078), against Candida glabrata. Antimicrob Agents Chemother. https://doi.org/10.1128/AAC.01510-19

Ghannoum M, Isham N, Angulo D, Borroto-Esoda K, Barat S, Long L (2020) Efficacy of Ibrexafungerp (SCY-078) against Candida auris in an in vivo guinea pig cutaneous infection model. Antimicrob Agents Chemother 64(10). https://doi.org/10.1128/AAC.00854-20

Gileles-Hillel A, Shoseyov D, Polacheck I, Korem M, Kerem E, Cohen-Cymberknoh M (2015) Association of chronic Candida albicans respiratory infection with a more severe lung disease in patients with cystic fibrosis. Pediatr Pulmonol 50(11):1082–1089

Grimshaw SG, Smith AM, Arnold DS, Xu E, Hoptroff M, Murphy B (2019) The diversity and abundance of fungi and bacteria on the healthy and dandruff affected human scalp. PLoS One 14(12):e0225796. https://doi.org/10.1371/journal.pone.0225796

Hager CL, Larkin EL, Long L, Zohra Abidi F, Shaw KJ, Ghannoum MA (2018) In vitro and in vivo evaluation of the antifungal activity of APX001A/APX001 against Candida auris. Antimicrob Agents Chemother 62(3). https://doi.org/10.1128/AAC.02319-17

Hall CW, Mah TF (2017) Molecular mechanisms of biofilm-based antibiotic resistance and tolerance in pathogenic bacteria. FEMS Microbiol Rev 41(3):276–301. https://doi.org/10.1093/femsre/fux010

Harriott MM, Noverr MC (2009) Candida albicans and Staphylococcus aureus form polymicrobial biofilms: effects on antimicrobial resistance. Antimicrobial Agents Chemother 53(9): 3914–3922

Harriott MM, Lilly EA, Rodriguez TE, Fidel PL, Noverr MC (2010) Candida albicans forms biofilms on the vaginal mucosa. Microbiology 156(Pt 12):3635–3644. https://doi.org/10.1099/mic.0.039354-0

Hay R (2018) Therapy of skin, hair and nail fungal infections. J Fungi (Basel) 4(3). https://doi.org/10.3390/jof4030099

Heald AH, O'Halloran DJ, Richards K, Webb F, Jenkins S, Hollis S, Denning DW, Young RJ (2001) Fungal infection of the diabetic foot: two distinct syndromes. Diabet Med 18(7): 567–572. https://doi.org/10.1046/j.1464-5491.2001.00523.x

Hilliard T, Edwards S, Buchdahl R, Francis J, Rosenthal M, Balfour-Lynn I, Bush A, Davies J (2005) Voriconazole therapy in children with cystic fibrosis. J Cyst Fibros 4(4):215–220. https://doi.org/10.1016/j.jcf.2005.05.019

Høiby N, Ciofu O, Bjarnsholt T (2010) Pseudomonas aeruginosa biofilms in cystic fibrosis. Future Microbiol 5(11):1663–1674. https://doi.org/10.2217/fmb.10.125

Iatta R, Napoli C, Borghi E, Montagna MT (2009) Rare mycoses of the oral cavity: a literature epidemiologic review. Oral Surg Oral Med Oral Pathol Oral Radiol Endod 108(5):647–655

Jimenez-Ortigosa C, Perez WB, Angulo D, Borroto-Esoda K, Perlin DS (2017) De novo acquisition of resistance to SCY-078 in Candida glabrata involves FKS mutations that both overlap and are distinct from those conferring echinocandin resistance. Antimicrob Agents Chemother 61(9). https://doi.org/10.1128/AAC.00833-17

Jones Z, Brooks AE, Ferrell Z, Grainger DW, Sinclair KD (2016) A resorbable antibiotic eluting bone void filler for periprosthetic joint infection prevention. J Biomed Mater Res B Appl Biomater 104(8):1632–1642. https://doi.org/10.1002/jbm.b.33513

Kalan L, Grice EA (2018) Fungi in the wound microbiome. Adv Wound Care (New Rochelle) 7(7): 247–255. https://doi.org/10.1089/wound.2017.0756

Kalan L, Loesche M, Hodkinson BP, Heilmann K, Ruthel G, Gardner SE, Grice EA (2016) Redefining the chronic-wound microbiome: fungal communities are prevalent, dynamic, and associated with delayed healing. mBio 7(5). https://doi.org/10.1128/mBio.01058-16

Kaplan JB (2010) Biofilm dispersal: mechanisms, clinical implications, and potential therapeutic uses. J Dent Res 89(3):205–218. https://doi.org/10.1177/0022034509359403

Kawa N, Lee KC, Anderson RR, Garibyan L (2019) Onychomycosis: a review of new and emerging topical and device-based treatments. J Clin Aesthet Dermatol 12(10):29–34

Kean R, Ramage G (2019) Combined antifungal resistance and biofilm tolerance: the global threat of candida auris. mSphere 4(4). https://doi.org/10.1128/mSphere.00458-19

Kean R, Rajendran R, Haggarty J, Townsend EM, Short B, Burgess KE, Lang S, Millington O, Mackay WG, Williams C, Ramage G (2017) Candida albicans mycofilms support Staphylococcus aureus colonization and enhances miconazole resistance in dual-species interactions. Front Microbiol 8(258). https://doi.org/10.3389/fmicb.2017.00258

Khatoon Z, McTiernan CD, Suuronen EJ, Mah TF, Alarcon EI (2018) Bacterial biofilm formation on implantable devices and approaches to its treatment and prevention. Heliyon 4(12):e01067. https://doi.org/10.1016/j.heliyon.2018.e01067

Kim D, Liu Y, Benhamou RI, Sanchez H, Simon-Soro A, Li Y, Hwang G, Fridman M, Andes DR, Koo H (2018) Bacterial-derived exopolysaccharides enhance antifungal drug tolerance in a cross-kingdom oral biofilm. ISME J 12(6):1427–1442. https://doi.org/10.1038/s41396-018-0113-1

Kollef M, Micek S, Hampton N, Doherty JA, Kumar A (2012) Septic shock attributed to Candida infection: importance of empiric therapy and source control. Clin Infect Dis 54(12):1739–1746. https://doi.org/10.1093/cid/cis305

Kong EF, Tsui C, Kucharíková S, Andes D, Van Dijck P, Jabra-Rizk MA (2016) Commensal protection of Staphylococcus aureus against antimicrobials by Candida albicans biofilm matrix. mBio 7(5):e01365. https://doi.org/10.1128/mBio.01365-16

Larkin E, Hager C, Chandra J, Mukherjee PK, Retuerto M, Salem I, Long L, Isham N, Kovanda L, Borroto-Esoda K, Wring S, Angulo D, Ghannoum M (2017) The emerging pathogen Candida auris: growth phenotype, virulence factors, activity of antifungals, and effect of SCY-078, a novel glucan synthesis inhibitor, on growth morphology and biofilm formation. Antimicrob Agents Chemother 61(5). https://doi.org/10.1128/AAC.02396-16

Leonhard M, Schneider-Stickler B (2015) Voice prostheses, microbial colonization and biofilm formation. Adv Exp Med Biol 830:123–136. https://doi.org/10.1007/978-3-319-11038-7_8

Lepak AJ, Marchillo K, Andes DR (2015) Pharmacodynamic target evaluation of a novel oral glucan synthase inhibitor, SCY-078 (MK-3118), using an in vivo murine invasive candidiasis model. Antimicrobial Agents Chemother 59(2):1265–1272

Liao Q, Lam JKW (2021) Inhaled antifungal agents for the treatment and prophylaxis of pulmonary mycoses. Curr Pharm Des 27(12):1453–1468. https://doi.org/10.2174/1381612826666210101153547

Lipsky BA, Aragon-Sanchez J, Diggle M, Embil J, Kono S, Lavery L, Senneville E, Urbancic-Rovan V, Van Asten S, International Working Group on the Diabetic Foot, Peters EJ (2016a) IWGDF guidance on the diagnosis and management of foot infections in persons with diabetes. Diabetes Metab Res Rev 32(Suppl 1):45–74. https://doi.org/10.1002/dmrr.2699

Lipsky BA, Dryden M, Gottrup F, Nathwani D, Seaton RA, Stryja J (2016b) Antimicrobial stewardship in wound care: a position paper from the British Society for Antimicrobial Chemotherapy and European Wound Management Association. J Antimicrob Chemother 71(11):3026–3035. https://doi.org/10.1093/jac/dkw287

Lockhart SR, Etienne KA, Vallabhaneni S, Farooqi J, Chowdhary A, Govender NP, Colombo AL, Calvo B, Cuomo CA, Desjardins CA, Berkow EL, Castanheira M, Magobo RE, Jabeen K, Asghar RJ, Meis JF, Jackson B, Chiller T, Litvintseva AP (2017) Simultaneous emergence of multidrug-resistant Candida auris on 3 continents confirmed by whole-genome sequencing and epidemiological analyses. Clin Infect Dis 64(2):134–140. https://doi.org/10.1093/cid/ciw691

Lugovic-Mihic L, Pilipovic K, Crnaric I, Situm M, Duvancic T (2018) Differential diagnosis of cheilitis - how to classify cheilitis? Acta Clin Croat 57(2):342–351. https://doi.org/10.20471/acc.2018.57.02.16

Machado D, Castro J, Palmeira-de-Oliveira A, Martinez-de-Oliveira J, Cerca N (2015) Bacterial vaginosis biofilms: challenges to current therapies and emerging solutions. Front Microbiol 6: 1528. https://doi.org/10.3389/fmicb.2015.01528

Mall MA, Hartl D (2014) CFTR: cystic fibrosis and beyond. Eur Respir J 44(4):1042–1054. https://doi.org/10.1183/09031936.00228013

Marcos-Zambrano LJ, Gomez-Perosanz M, Escribano P, Bouza E, Guinea J (2017) The novel oral glucan synthase inhibitor SCY-078 shows in vitro activity against sessile and planktonic Candida spp. J Antimicrob Chemother 72(7):1969–1976. https://doi.org/10.1093/jac/dkx010

Mayahara M, Kataoka R, Arimoto T, Tamaki Y, Yamaguchi N, Watanabe Y, Yamasaki Y, Miyazaki T (2014) Effects of surface roughness and dimorphism on the adhesion of Candida albicans to the surface of resins: scanning electron microscope analyses of mode and number of adhesions. J Investig Clin Dent 5(4):307–312. https://doi.org/10.1111/jicd.12055

McBride JA, Gauthier GM, Klein BS (2017) Clinical manifestations and treatment of blastomycosis. Clin Chest Med 38(3):435–449. https://doi.org/10.1016/j.ccm.2017.04.006

McGhee W, Michaels MG, Martin JM, Mazariegos GV, Green M (2016) Antifungal lock therapy with liposomal amphotericin B: a prospective trial. J Pediatric Infect Dis Soc 5(1):80–84. https://doi.org/10.1093/jpids/piu083

McKloud E, Delaney C, Sherry L, Kean R, Williams S, Metcalfe R, Thomas R, Richardson R, Gerasimidis K, Nile CJ, Williams C, Ramage G (2021) Recurrent vulvovaginal Candidiasis: a dynamic interkingdom biofilm disease of Candida and Lactobacillus. mSystems. 6(4): e0062221. https://doi.org/10.1128/mSystems.00622-21

Miesel L, Cushion Melanie T, Ashbaugh A, Lopez Santiago R, Ong V (2021) Efficacy of rezafungin in prophylactic mouse models of invasive Candidiasis, Aspergillosis, and Pneumocystis Pneumonia. Antimicrobial Agents Chemother 65(3):e01992–e01920. https://doi.org/10.1128/AAC.01992-20

Mitsuyama J, Nomura N, Hashimoto K, Yamada E, Nishikawa H, Kaeriyama M, Kimura A, Todo Y, Narita H (2008) In vitro and in vivo antifungal activities of T-2307, a novel arylamidine. Antimicrobial Agents Chemotherapy 52(4):1318–1324. https://doi.org/10.1128/AAC.01159-07

Mowat E, Butcher J, Lang S, Williams C, Ramage G (2007) Development of a simple model for studying the effects of antifungal agents on multicellular communities of Aspergillus fumigatus. J Med Microbiol 56(Pt 9):1205–1212. https://doi.org/10.1099/jmm.0.47247-0

Muller FM, Seidler M, Beauvais A (2011) Aspergillus fumigatus biofilms in the clinical setting. Med Mycol 49(Suppl 1):S96–S100. https://doi.org/10.3109/13693786.2010.502190

Nazir M, Al-Ansari A, Al-Khalifa K, Alhareky M, Gaffar B, Almas K (2020) Global prevalence of periodontal disease and lack of its surveillance. Sci World J 2020:2146160

Oh J, Byrd AL, Deming C, Conlan S, Program NCS, Kong HH, Segre JA (2014) Biogeography and individuality shape function in the human skin metagenome. Nature 514(7520):59–64. https://doi.org/10.1038/nature13786

Pappas PG, Kauffman CA, Andes D, Benjamin DK Jr, Calandra TF, Edwards JE Jr, Filler SG, Fisher JF, Kullberg BJ, Ostrosky-Zeichner L, Reboli AC, Rex JH, Walsh TJ, Sobel JD, Infectious Diseases Society of America (2009) Clinical practice guidelines for the management of candidiasis: 2009 update by the Infectious Diseases Society of America. Clin Infect Dis 48(5):503–535. https://doi.org/10.1086/596757

Paul DiMondi V, Townsend ML, Johnson M, Durkin M (2014) Antifungal catheter lock therapy for the management of a persistent Candida albicans bloodstream infection in an adult receiving hemodialysis. Pharmacotherapy 34(7):e120–e127. https://doi.org/10.1002/phar.1433

Paulussen C, Hallsworth JE, Álvarez-Pérez S, Nierman WC, Hamill PG, Blain D, Rediers H, Lievens B (2017) Ecology of aspergillosis: insights into the pathogenic potency of Aspergillus fumigatus and some other Aspergillus species. Microbial Biotechnol 10(2):296–322. https://doi.org/10.1111/1751-7915.12367

Pendharkar S, Brandsborg E, Hammarstrom L, Marcotte H, Larsson PG (2015) Vaginal colonisation by probiotic lactobacilli and clinical outcome in women conventionally treated for bacterial vaginosis and yeast infection. BMC Infect Dis 15:255. https://doi.org/10.1186/s12879-015-0971-3

Pendleton KM, Huffnagle GB, Dickson RP (2017) The significance of Candida in the human respiratory tract: our evolving understanding. Pathogens Dis 75(3). https://doi.org/10.1093/femspd/ftx029

Perfect JR (2017) The antifungal pipeline: a reality check. Nat Rev Drug Discov 16(9):603–616. https://doi.org/10.1038/nrd.2017.46

Perry KI, Hanssen AD (2017) Orthopaedic infection: prevention and diagnosis. J Am Acad Orthop Surg 25(Suppl 1):S4–S6. https://doi.org/10.5435/JAAOS-D-16-00634

Peters BA, Wu J, Hayes RB, Ahn J (2017) The oral fungal mycobiome: characteristics and relation to periodontitis in a pilot study. BMC Microbiol 17(1):157. https://doi.org/10.1186/s12866-017-1064-9

Pfaller MA, Messer SA, Motyl MR, Jones RN, Castanheira M (2013) In vitro activity of a new oral glucan synthase inhibitor (MK-3118) tested against Aspergillus spp. by CLSI and EUCAST broth microdilution methods. Antimicrobial Agents Chemother 57(2):1065–1068

Phillips AJ (2005) Treatment of non-albicans Candida vaginitis with amphotericin B vaginal suppositories. Am J Obstet Gynecol 192(6):2009–2012. https://doi.org/10.1016/j.ajog.2005.03.034

Pristov KE, Ghannoum MA (2019) Resistance of Candida to azoles and echinocandins worldwide. Clin Microbiol Infect 25(7):792–798. https://doi.org/10.1016/j.cmi.2019.03.028

Raja M, Hannan A, Ali K (2010) Association of oral candidal carriage with dental caries in children. Caries Res 44(3):272–276

Rajendran R, Mowat E, McCulloch E, Lappin DF, Jones B, Lang S, Majithiya JB, Warn P, Williams C, Ramage G (2011) Azole resistance of Aspergillus fumigatus biofilms is partly associated with efflux pump activity. Antimicrobial Agents Chemother 55(5):2092–2097. https://doi.org/10.1128/AAC.01189-10

Rajendran R, Sherry L, Deshpande A, Johnson EM, Hanson MF, Williams C, Munro CA, Jones BL, Ramage G (2016a) A prospective surveillance study of candidaemia: epidemiology, risk factors, antifungal treatment and outcome in hospitalized patients. Front Microbiol 7:915. https://doi.org/10.3389/fmicb.2016.00915

Rajendran R, Sherry L, Nile CJ, Sherriff A, Johnson EM, Hanson MF, Williams C, Munro CA, Jones BJ, Ramage G (2016b) Biofilm formation is a risk factor for mortality in patients with Candida albicans bloodstream infection-Scotland, 2012-2013. Clin Microbiol Infect 22(1): 87–93. https://doi.org/10.1016/j.cmi.2015.09.018

Ramage G, Martinez JP, Lopez-Ribot JL (2006) Candida biofilms on implanted biomaterials: a clinically significant problem. FEMS Yeast Res 6(7):979–986. https://doi.org/10.1111/j.1567-1364.2006.00117.x

Ramage G, Mowat E, Jones B, Williams C, Lopez-Ribot J (2009) Our current understanding of fungal biofilms. Crit Rev Microbiol 35(4):340–355. https://doi.org/10.3109/10408410903241436

Ramage G, Rajendran R, Gutierrez-Correa M, Jones B, Williams C (2011) Aspergillus biofilms: clinical and industrial significance. FEMS Microbiol Lett 324(2):89–97. https://doi.org/10.1111/j.1574-6968.2011.02381.x

Rautemaa R, Ramage G (2011) Oral candidosis--clinical challenges of a biofilm disease. Crit Rev Microbiol 37(4):328–336. https://doi.org/10.3109/1040841X.2011.585606

Reece E, Segurado R, Jackson A, McClean S, Renwick J, Greally P (2017) Co-colonisation with Aspergillus fumigatus and Pseudomonas aeruginosa is associated with poorer health in cystic fibrosis patients: an Irish registry analysis. BMC Pulmonary Medicine 17(1):70. https://doi.org/10.1186/s12890-017-0416-4

Reece E, Doyle S, Greally P, Renwick J, McClean S (2018) Aspergillus fumigatus inhibits Pseudomonas aeruginosa in co-culture: implications of a mutually antagonistic relationship on virulence and inflammation in the CF airway. Front Microbiol 9:1205–1205. https://doi.org/10.3389/fmicb.2018.01205

Regan KA, Radhakrishnan NS, Hammer JD, Wilson BD, Gadkowski LB, Iyer SSR (2020) Endogenous endophthalmitis: yield of the diagnostic evaluation. BMC Ophthalmol 20(1):138. https://doi.org/10.1186/s12886-020-01418-9

Richardson M, Lass-Flörl C (2008) Changing epidemiology of systemic fungal infections. Clin Microbiol Infect 14(Suppl 4):5–24. https://doi.org/10.1111/j.1469-0691.2008.01978.x

Rodrigues L, van der Mei HC, Teixeira J, Oliveira R (2004) Influence of biosurfactants from probiotic bacteria on formation of biofilms on voice prostheses. Appl Environ Microbiol 70(7): 4408–4410. https://doi.org/10.1128/AEM.70.7.4408-4410.2004

Rzhepishevska O, Hakobyan S, Ruhal R, Gautrot J, Barbero D, Ramstedt M (2013) The surface charge of anti-bacterial coatings alters motility and biofilm architecture. Biomater Sci 1(6): 589–602. https://doi.org/10.1039/c3bm00197k

Sabino R, Ferreira JA, Moss RB, Valente J, Veríssimo C, Carolino E, Clemons KV, Everson C, Banaei N, Penner J (2015) Molecular epidemiology of Aspergillus collected from cystic fibrosis patients. J Cyst Fibros 14(4):474–481

Saghazadeh S, Rinoldi C, Schot M, Kashaf SS, Sharifi F, Jalilian E, Nuutila K, Giatsidis G, Mostafalu P, Derakhshandeh H, Yue K, Swieszkowski W, Memic A, Tamayol A, Khademhosseini A (2018) Drug delivery systems and materials for wound healing applications. Adv Drug Deliv Rev 127:138–166. https://doi.org/10.1016/j.addr.2018.04.008

Santosh ABR, Muddana K, Bakki SR (2021) Fungal infections of oral cavity: diagnosis, management, and association with COVID-19. SN Compr Clin Med 2021:1–12

Satoh K, Makimura K, Hasumi Y, Nishiyama Y, Uchida K, Yamaguchi H (2009) Candida auris sp. nov., a novel ascomycetous yeast isolated from the external ear canal of an inpatient in a Japanese hospital. Microbiol Immunol 53(1):41–44. https://doi.org/10.1111/j.1348-0421.2008.00083.x

Scorneaux B, Angulo D, Borroto-Esoda K, Ghannoum M, Peel M, Wring S (2017) SCY-078 is fungicidal against Candida species in time-kill studies. Antimicrobial Agents Chemother 61(3): e01961–e01916

Scully C (1913) Noncandidal fungal infections of the mouth. Culture 11:02

Sherif R, Segal BH (2010) Pulmonary aspergillosis: clinical presentation, diagnostic tests, management and complications. Curr Opin Pulm Med 16(3):242–250. https://doi.org/10.1097/MCP.0b013e328337d6de

Sherry L, Ramage G, Kean R, Borman A, Johnson EM, Richardson MD, Rautemaa-Richardson R (2017) Biofilm-forming capability of highly virulent, multidrug-resistant candida auris. Emerg Infect Dis 23(2):328–331. https://doi.org/10.3201/eid2302.161320

Shirazi F, Ferreira JAG, Stevens DA, Clemons KV, Kontoyiannis DP (2016) Biofilm filtrates of pseudomonas aeruginosa strains isolated from cystic fibrosis patients inhibit preformed aspergillus fumigatus biofilms via apoptosis. PLoS One 11(3):e0150155. https://doi.org/10.1371/journal.pone.0150155

Shoseyov D, Brownlee KG, Conway SP, Kerem E (2006) Aspergillus bronchitis in cystic fibrosis. Chest 130(1):222–226. https://doi.org/10.1378/chest.130.1.222

Silva-Neves V, Hugo V, Alves P, Amado JC, Pais-Vieira C, Sousa F, Cerqueira F, Pinto E, Pais-Vieira M (2021) Quality of life and therapeutic regimen management in onychomycosis patients and in vitro study of antiseptic solutions. Sci Rep 11(1):12789. https://doi.org/10.1038/s41598-021-92111-4

Singh PK, Schaefer AL, Parsek MR, Moninger TO, Welsh MJ, Greenberg E (2000) Quorum-sensing signals indicate that cystic fibrosis lungs are infected with bacterial biofilms. Nature 407(6805):762–764

Singh A, Verma R, Murari A, Agrawal A (2014) Oral candidiasis: an overview. J Oral Maxillofac Pathol 18(Suppl 1):S81–S85. https://doi.org/10.4103/0973-029X.141325

Smith K, Rajendran R, Kerr S, Lappin D, Mackay W, Williams C, Ramage G (2015) Aspergillus fumigatus enhances elastase production in Pseudomonas aeruginosa co-cultures. Med Mycol 53. https://doi.org/10.1093/mmy/myv048

Sobel JD (1992) Pathogenesis and treatment of recurrent vulvovaginal candidiasis. Clin Infect Dis 14(Suppl 1):S148–S153. https://doi.org/10.1093/clinids/14.supplement_1.s148

Sobel JD (2015) Editorial commentary: vaginal biofilm: much ado about nothing, or a new therapeutic challenge? Clin Infect Dis 61(4):607–608. https://doi.org/10.1093/cid/civ358

Sobel JD (2016) Recurrent vulvovaginal candidiasis. Am J Obstet Gynecol 214(1):15–21. https://doi.org/10.1016/j.ajog.2015.06.067

Sobel JD, Faro S, Force RW, Foxman B, Ledger WJ, Nyirjesy PR, Reed BD, Summers PR (1998) Vulvovaginal candidiasis: epidemiologic, diagnostic, and therapeutic considerations. Am J Obstet Gynecol 178(2):203–211. https://doi.org/10.1016/s0002-9378(98)80001-x

Sobel JD, Chaim W, Nagappan V, Leaman D (2003) Treatment of vaginitis caused by Candida glabrata: use of topical boric acid and flucytosine. Am J Obstet Gynecol 189(5):1297–1300. https://doi.org/10.1067/s0002-9378(03)00726-9

Sobel JD, Wiesenfeld HC, Martens M, Danna P, Hooton TM, Rompalo A, Sperling M, Livengood C 3rd, Horowitz B, Von Thron J, Edwards L, Panzer H, Chu TC (2004) Maintenance fluconazole therapy for recurrent vulvovaginal candidiasis. N Engl J Med 351(9):876–883. https://doi.org/10.1056/NEJMoa033114

Sridhar S, Suprabha BS, Shenoy R, Suman E, Rao A (2020) Association of Streptococcus Mutans, Candida Albicans and oral health practices with activity status of caries lesions among 5-year-old children with early childhood caries. Oral Health Prev Dent 18(1):911–919. https://doi.org/10.3290/j.ohpd.a45411

Sumiyoshi M, Miyazaki T, Makau JN, Mizuta S, Tanaka Y, Ishikawa T, Makimura K, Hirayama T, Takazono T, Saijo T, Yamaguchi H, Shimamura S, Yamamoto K, Imamura Y, Sakamoto N, Obase Y, Izumikawa K, Yanagihara K, Kohno S, Mukae H (2020) Novel and potent antimicrobial effects of caspofungin on drug-resistant Candida and bacteria. Sci Rep 10(1):17745. https://doi.org/10.1038/s41598-020-74749-8

Suresh Unniachan A, Krishnavilasom Jayakumari N, Sethuraman S (2020) Association between Candida species and periodontal disease: a systematic review. Curr Med Mycol 6:63–68

Swidsinski A, Guschin A, Tang Q, Dorffel Y, Verstraelen H, Tertychnyy A, Khayrullina G, Luo X, Sobel JD, Jiang X (2019) Vulvovaginal candidiasis: histologic lesions are primarily polymicrobial and invasive and do not contain biofilms. Am J Obstet Gynecol 220(1):91. https://doi.org/10.1016/j.ajog.2018.10.023

Tan H, Tay S (2013) The inhibitory effects of aureobasidin A on Candida planktonic and biofilm cells. Mycoses 56(2):150–156

Tande AJ, Patel R (2014) Prosthetic joint infection. Clin Microbiol Rev 27(2):302–345. https://doi.org/10.1128/CMR.00111-13

Tatara AM, Watson E, Albert ND, Kontoyiannis PD, Kontoyiannis DP, Mikos AG (2019) A murine model of cutaneous aspergillosis for evaluation of biomaterials-based local delivery therapies. J Biomed Mater Res A 107(9):1867–1874. https://doi.org/10.1002/jbm.a.36671

Thomas PA, Kaliamurthy J (2013) Mycotic keratitis: epidemiology, diagnosis and management. Clin Microbiol Infect 19(3):210–220. https://doi.org/10.1111/1469-0691.12126

Thomas J, Jacobson GA, Narkowicz CK, Peterson GM, Burnet H, Sharpe C (2010) Toenail onychomycosis: an important global disease burden. J Clin Pharm Ther 35(5):497–519. https://doi.org/10.1111/j.1365-2710.2009.01107.x

Todd OA, Fidel PL, Harro JM, Hilliard JJ, Tkaczyk C, Sellman BR, Noverr MC, Peters BM (2019) Candida albicans augments Staphyloccus aureus virulence by engaging the Staphylococcal quorum sensing system. mBio 10(3):e00910. https://doi.org/10.1128/mBio.00910-19

Tumbarello M, Posteraro B, Trecarichi EM, Fiori B, Rossi M, Porta R, de Gaetano DK, La Sorda M, Spanu T, Fadda G, Cauda R, Sanguinetti M (2007) Biofilm production by Candida species and inadequate antifungal therapy as predictors of mortality for patients with candidemia. J Clin Microbiol 45(6):1843–1850. https://doi.org/10.1128/JCM.00131-07

Uppuluri P, Chaturvedi AK, Srinivasan A, Banerjee M, Ramasubramaniam AK, Kohler JR, Kadosh D, Lopez-Ribot JL (2010) Dispersion as an important step in the Candida albicans biofilm developmental cycle. PLoS Pathog 6(3):e1000828. https://doi.org/10.1371/journal.ppat.1000828

Urzúa B, Hermosilla G, Gamonal J, Morales-Bozo I, Canals M, Barahona S, Cóccola C, Cifuentes V (2008) Yeast diversity in the oral microbiota of subjects with periodontitis: Candida albicans and Candida dubliniensis colonize the periodontal pockets. Sabouraudia 46(8):783–793

van de Belt H, Neut D, Schenk W, van Horn JR, van der Mei HC, Busscher HJ (2001) Infection of orthopedic implants and the use of antibiotic-loaded bone cements. A review. Acta Orthop Scand 72(6):557–571. https://doi.org/10.1080/000164701317268978

Vila T, Lopez-Ribot JL (2017) Screening the pathogen box for identification of Candida albicans biofilm inhibitors. Antimicrobial Agents Chemother 61(1):e02006–e02016. https://doi.org/10.1128/AAC.02006-16

Vladareanu R, Mihu D, Mitran M, Mehedintu C, Boiangiu A, Manolache M, Vladareanu S (2018) New evidence on oral L. plantarum P17630 product in women with history of recurrent vulvovaginal candidiasis (RVVC): a randomized double-blind placebo-controlled study. Eur Rev Med Pharmacol Sci 22(1):262–267

Wang L, Ma L, Leng W, Liu T, Yu L, Yang J, Yang L, Zhang W, Zhang Q, Dong J, Xue Y, Zhu Y, Xu X, Wan Z, Ding G, Yu F, Tu K, Li Y, Li R, Shen Y, Jin Q (2006) Analysis of the dermatophyte Trichophyton rubrum expressed sequence tags. BMC Genomics 7:255. https://doi.org/10.1186/1471-2164-7-255

Wang W, Lu KJ, Yu CH, Huang QL, Du YZ (2019) Nano-drug delivery systems in wound treatment and skin regeneration. J Nanobiotechnol 17(1):82. https://doi.org/10.1186/s12951-019-0514-y

Watkins RR, Mukherjee PK, Chandra J, Retuerto MA, Guidry C, Haller N, Paranjape C, Ghannoum MA (2017) Admission to the intensive care unit is associated with changes in the oral mycobiome. J Intensive Care Med 32(4):278–282. https://doi.org/10.1177/0885066615627757

Welsh RM, Bentz ML, Shams A, Houston H, Lyons A, Rose LJ, Litvintseva AP (2017) Survival, persistence, and isolation of the emerging multidrug-resistant pathogenic yeast Candida auris on a plastic health care surface. J Clin Microbiol 55(10):2996–3005. https://doi.org/10.1128/JCM.00921-17

Whaley SG, Berkow EL, Rybak JM, Nishimoto AT, Barker KS, Rogers PD (2016) Azole antifungal resistance in candida albicans and emerging non-albicans Candida species. Front Microbiol 7:2173. https://doi.org/10.3389/fmicb.2016.02173

Wiederhold NP, Najvar LK, Shaw KJ, Jaramillo R, Patterson H, Olivo M, Catano G, Patterson TF (2019) Efficacy of delayed therapy with fosmanogepix (APX001) in a murine model of Candida auris invasive candidiasis. Antimicrob Agents Chemother 63(11). https://doi.org/10.1128/AAC.01120-19

Williams HD, Davies JC (2012) Basic science for the chest physician: Pseudomonas aeruginosa and the cystic fibrosis airway. Thorax 67(5):465–467

Williams C, Ramage G (2015) Fungal biofilms in human disease. Adv Exp Med Biol 831:11–27. https://doi.org/10.1007/978-3-319-09782-4_2

Williams C, Ranjendran R, Ramage G (2016) Pathogenesis of fungal infections in cystic fibrosis. Curr Fungal Infect Rep 10(4):163–169

Yamashita K, Miyazaki T, Fukuda Y, Mitsuyama J, Saijo T, Shimamura S, Yamamoto K, Imamura Y, Izumikawa K, Yanagihara K, Kohno S, Mukae H (2019) The novel arylamidine T-2307 selectively disrupts yeast mitochondrial function by inhibiting respiratory chain complexes. Antimicrobial Agents and Chemotherapy 63(8):e00374–e00319. https://doi.org/10.1128/AAC.00374-19

Yano J, Sobel JD, Nyirjesy P, Sobel R, Williams VL, Yu Q, Noverr MC, Fidel PL Jr (2019) Current patient perspectives of vulvovaginal candidiasis: incidence, symptoms, management and post-treatment outcomes. BMC Womens Health 19(1):48. https://doi.org/10.1186/s12905-019-0748-8

Yoda I, Koseki H, Tomita M, Shida T, Horiuchi H, Sakoda H, Osaki M (2014) Effect of surface roughness of biomaterials on Staphylococcus epidermidis adhesion. BMC Microbiol 14:234. https://doi.org/10.1186/s12866-014-0234-2

Zhao M, Lepak AJ, VanScoy B, Bader JC, Marchillo K, Vanhecker J, Ambrose PG, Andes DR (2018) In vivo pharmacokinetics and pharmacodynamics of APX001 against Candida spp. in a neutropenic disseminated Candidiasis mouse model. Antimicrob Agents Chemother 62(4). https://doi.org/10.1128/AAC.02542-17

Metal-Based Nanoparticles for Biofilm Treatment and Infection Control: From Basic Research to Clinical Translation

Paula Facal Marina, Laurine Kaul, Nils Mischer, and Katharina Richter

Abstract Multidrug-resistant bacteria represent a growing threat to human health worldwide and are caused by the overuse and misuse of broad-spectrum antibiotics. Therefore, the need for alternative, innovative intervention strategies arose and metallic nanoparticles, e.g. based on silver, gold, zinc, copper, iron, titanium, and selenium, stand out as novel agents to effectively control the proliferation of multidrug-resistant bacteria. In recent years, numerous studies have proposed mechanisms by which metallic nanoparticles target and kill bacteria, mostly based on their small size, which renders them highly reactive and effective in penetrating biofilms. However, research has shown that monotherapy is often not sufficient to eradicate biofilm-forming pathogens and annihilate established biofilms. Therefore, synergistic formulations combining multiple antibacterial agents are emerging as one of the most promising approaches. Recent advances in in vitro and in vivo experiments are paving the way for clinical trials, already showing promising results of metallic nanoparticles as wound dressing, implant coating, or rinse solution against bacteria causing oral, nasal, or skin infections. More clinical trials are expected to occur in the future, promoting metallic nanoparticles as innovative antimicrobial therapy. This

P. Facal Marina · K. Richter (✉)
Richter Lab, Surgery Department, Basil Hetzel Institute for Translational Health Research, The University of Adelaide, Adelaide, SA, Australia

Institute for Photonics and Advanced Sensing, The University of Adelaide, Adelaide, SA, Australia
e-mail: Katharina.Richter@adelaide.edu.au

L. Kaul
Richter Lab, Surgery Department, Basil Hetzel Institute for Translational Health Research, The University of Adelaide, Adelaide, SA, Australia

Department of Pharmaceutical Technology and Biopharmacy, Institute of Pharmaceutical Sciences, University of Freiburg, Freiburg, Germany

N. Mischer
Richter Lab, Surgery Department, Basil Hetzel Institute for Translational Health Research, The University of Adelaide, Adelaide, SA, Australia

K. Richter, K. Nørskov Kragh (eds.), *Antibiofilm Strategies*, Springer Series on Biofilms 11, https://doi.org/10.1007/978-3-031-10992-8_18

chapter reviews recent advances in the use of metallic nanoparticles as a promising strategy to control biofilm infections.

1 Introduction

Biofilms are colonies of bacteria, fungi, or yeasts that form organised and heterogeneous 3D structures on biotic or abiotic surfaces protected by a matrix of extracellular polymeric substances (EPS), also known as "slime". The EPS matrix is a dynamic compartment that has multiple functions; e.g. it resembles a defensive barrier that protects individual cells from the immune system and antibacterial agents. The matrix provides shielding of the embedded cells from antibiotics and nanoparticles by acting as a diffusion hindrance, provides protection from toxic radicals, and allows for intra- and inter-species interactions and exchange of information. These properties can ultimately lead to microbial persistence, increased tolerance, and resistance to antimicrobials, including antibiotics and nanoparticles. For further reading on the biofilm matrix and interactions with nanoparticles, please refer to the literature (Fulaz et al. 2019; Joo and Aggarwal 2018; Flemming et al. 2016).

Once the biofilm is established, a survival mechanism is activated that impedes the diffusion of antimicrobials and simultaneously induces multidrug efflux pumps to remove antimicrobials from the individual cells sticking to the biofilm (Jolivet-Gougeon and Bonnaure-Mallet 2019; Geddes-McAlister et al. 2019; Uruén et al. 2021). The tolerance to antimicrobials and, therefore, the survival of biofilm is also conditioned by the production of persister cells (Carvalho et al. 2018). Biofilms are associated with over 60% of infections in humans, and there are a number of excellent reviews on protective mechanisms against antimicrobial agents and host innate immunity (Flemming et al. 2016; Yan and Bassler 2019). There is a consensus that treatment strategies with elevated efficacy against biofilms are urgently needed.

The emergence of nanotechnology in 1959 and the advances in the design, synthesis, and manipulation of particles led to noteworthy applications of nanomaterials in biotechnology. Nano refers to any material in the size range of 0.1–100 nm (1 nm = 10^{-9} m), in which basic biological functions operate. As physico-chemical properties can be tailored for targeted drug delivery to any site in the body, there has been a boom in research of nanoparticles for biomedical applications over the past decade.

Nanomaterials have been successfully adopted for diagnosis and sensing, novel treatments against a wide range of diseases, and as antibacterial agents (Abdel-Karim et al. 2020). In the latter case, nanomaterials have been extensively used for antibacterial coatings for implants, prostheses, and medical devices or locally delivered to infection sites as single-agent treatments or as components in multidrug delivery systems (Rezaei et al. 2019; Kirtane et al. 2021). The main advantage compared to conventional treatments is that nanomaterials have a higher reactivity conferred by their large surface-area-to-volume ratio and that it is possible to control their physico-chemical properties, thus being able to customise their interaction with biological entities (Navya and Daima 2016). A wide variety of nanomaterials have been tested for biomedical applications including metals (e.g. silver, gold, zinc, copper, iron, titanium, selenium) and the corresponding metal oxides, polymers (e.g. nanoporous polymers, hydrogels) and polymer composites, and technologies for the encapsulation of antibacterial agents comprising liposomes and responsive smart nanomaterials, amongst others (Kirtane et al. 2021).

This chapter comprehensively summarises current knowledge and breakthroughs in metal-based nanoparticles employed as antimicrobial agents, with a special emphasis on the treatment of biofilms from concept to real-world healthcare applications.

2 Silver Nanoparticles

Extensive interrelated research has been carried out to utilise metal-based nanoparticles for biomedical applications (Sánchez-López et al. 2020). In particular, silver nanoparticles (Ag-NPs) stand out for their antibacterial properties and have found multiple applications in the health field, including antibacterial treatments, antibacterial coatings for medical devices, and textile fibres, amongst others (Richter et al. 2017; Ooi et al. 2018; Lee and Jun 2019; Dong et al. 2019; Talapko et al. 2020; Sánchez-López et al. 2020). Although there are some concerns about the safety of using Ag-NPs for biomedical applications, in low concentrations (i.e. 0.001–0.1% weight/weight, based on 0.01–1 mg/mL concentration) Ag-NPs are not considered toxic (Liu et al. 2018a). Unfortunately, there is a lack of in vivo data elucidating the mechanism of Ag-NP toxicity; the majority of knowledge is based on in vitro studies, which calls for more investigations to determine safety of Ag-NP products in animals and humans. Literature has reported oxidative stress, DNA damage, and cytokine induction to be the three main mechanisms of Ag-NP toxicity (Ferdous and Nemmar 2020; Stensberg et al. 2011). Some animal studies showed Ag-NP effects on skin (Samberg et al. 2010), circulatory (Kim et al. 2008; Sung et al. 2009), respiratory (Sung et al. 2008), central nervous (Tang et al. 2008), and hepatic systems, depending on the route of Ag-NP delivery. An accumulation of Ag^+ and Ag-NPs in the liver has been observed in animals (Sung et al. 2008, 2009; Kim et al. 2008). However, there is an ongoing debate on dose-dependent antimicrobial efficacy and safety of Ag-NPs and their associated effects on intracellular functions, which are attributed to the shape, size, and mobility of Ag-NPs (Richter et al. 2017; Cheon et al. 2019; Dong et al. 2019; Anees Ahmad et al. 2020).

2.1 Mechanism of Antibacterial Action

Although it is well known that Ag-NPs possess antibacterial properties, the mechanisms by which the nanoparticles target and destroy bacteria have not yet been fully

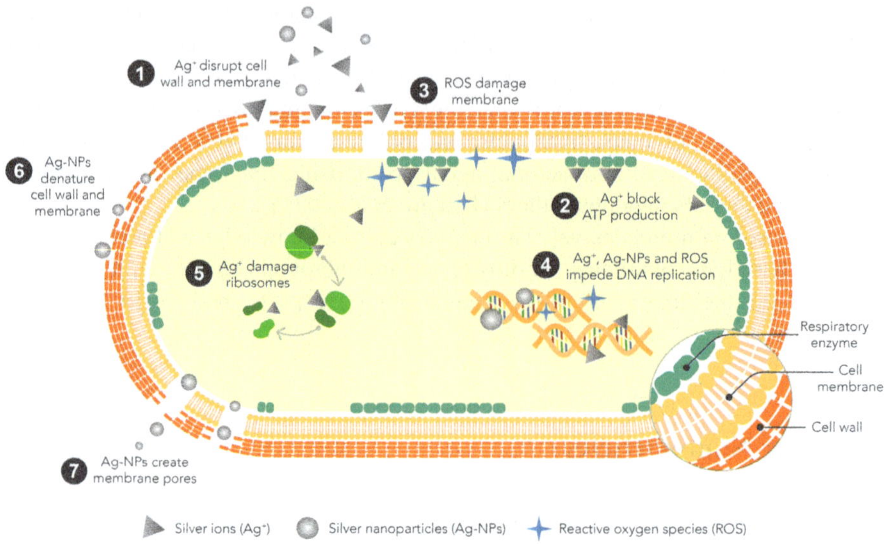

Fig. 1 Mechanisms of action that collectively contribute to antibacterial effects of silver nanoparticles (Ag-NPs). (1) Release of silver ions (Ag^+) from Ag-NPs disrupts the bacterial cell wall and membrane; (2) Ag^+ block ATP production and inhibit respiration; (3) Reactive oxygen species (ROS, generated as a downstream result of Ag^+/Ag-NPs inhibiting bacterial respiration and metabolic pathways) damage the bacterial membrane; (4) Ag^+, Ag-NPs, and ROS can impede DNA replication; (5) Ag^+ can damage ribosomes; (6) Denaturation of the bacterial cell wall and membrane by Ag-NPs; (7) Creation of membrane pores by Ag-NPs inducing cytoplasmic leakage

elucidated. Figure 1 summarises the mechanisms that potentially contribute to the antibacterial properties of Ag-NPs.

There is an ongoing discussion on how physico-chemical properties affect the antibacterial activity of Ag-NPs and how they enter cells (Slavin et al. 2017; Shaikh et al. 2019; Lin et al. 2020).

1. It has been suggested that the antibacterial properties of Ag-NPs are predominantly dependent on the release of silver ions (Ag^+) from Ag-NPs surfaces, which produce pores and react with the peptidoglycan component, and then penetrate through the bacterial cell wall (Fig. 1.1; Jung et al. 2008; Singh et al. 2021). Once inside the bacterial cell, Ag^+ can block adenosine triphosphate (ATP) production and inhibit respiration by deactivating respiratory enzymes on the bacterial cell membrane (Fig. 1.2), resulting in the generation of reactive oxygen species (ROS) that can further damage the bacterial membrane (Fig. 1.3; Morones-Ramirez et al. 2013). Moreover, Ag^+ (as well as Ag-NPs and ROS) can impede DNA replication (Fig. 1.4) by preventing the hybridisation of single-stranded DNA to double-stranded DNA, damage ribosomes by inhibiting protein synthesis (Fig. 1.5), and disrupt intracellular metabolic pathways, creating more ROS (Fig. 1.3; Yakabe et al. 1980; Bao et al. 2015; Sadoon et al. 2020). An excellent review on the bactericidal mechanisms of silver is available elsewhere (Möhler et al. 2018).

2. Some studies linked the bactericidal activity of Ag^+ ions and Ag-NPs to the morphological and structural changes produced in the bacterial cell (Markowska et al. 2013). Ag^+ ions and Ag-NPs have the capacity to alter the permeability of bacterial membranes/cell wall through attachment to the surface (interfering with membrane proteins, such as sulfur proteins), changing the membrane/cell wall structure and modifying the cell potential (Liao et al. 2019, Lok et al. 2007). This can contribute to the denaturation or disruption of the membrane/cell wall (Fig. 1.6), the creation of membrane pores, and cytoplasmic leakage (Fig. 1.7), leading to bacterial cell death (Yin et al. 2020).
3. Physico-chemical properties such as particle size, shape, and zeta potential (i.e. the potential difference at the interface between the surface of the particle and the bulk solution) affect the nanoparticles' ability to overcome biological barriers to reach the site of action. The mechanisms of action of the antibacterial properties of Ag-NPs can be classified as:

   - *Influence of size:* Due to their small size and large surface area, Ag-NPs are capable of penetrating into bacterial cells (Lin et al. 2020). The surface area increases with smaller particle size, which is associated with greater interactions between NPs and the surrounding area (Markowska et al. 2013). When 10–100 nm sized Ag-NPs were tested against *Pseudomonas aeruginosa* biofilms, the results showed that smaller Ag-NPs (10 and 20 nm) were more effective in reducing the bacterial viability and showed a lower minimum biofilm eradication concentration than 100 nm particles (Habash et al. 2014; Thieme et al. 2019). Similar findings have been reported in *Vibrio natriegens,* where 10 nm sized Ag-NPs completely inhibited bacterial and biofilm growth, and caused more cell damage and oxidative stress response by inducing more ROS than larger particles (Dong et al. 2019).
   - *Influence of shape:* It has been reported that 9 nm-sized spherical and 21 nm-sized quasi-spherical Ag-NPs showed selective size- and shape-dependent antibacterial activity against various pathogenic fungi and Gram-positive and Gram-negative bacteria. However, cubic Ag-NPs were not effective (Osonga et al. 2020). Another study showed significant activity of triangular and spherical Ag-NPs against *P. aeruginosa* and *Escherichia coli.* Spherical Ag-NPs at 50 nm were most effective, followed by triangular Ag-NPs at 150 nm, whereas spherical Ag-NPs at 90 nm were least effective (Raza et al. 2016). There is an understanding that the shape of Ag-NPs determines the surface-area-to-volume ratio. The larger the specific surface, the faster the Ag-NPs dissolve. Recent data suggest that Ag-NPs dissolve in aqueous media if oxygen is present, leading to the release of Ag^+ ions, and that the rate of dissolution is highly dependent on the surface area and the shape of the Ag-NPs (Cheon et al. 2019). Considering that antibacterial assays are typically performed in aqueous media in vitro and in vivo (culture medium, body fluids, blood), it is key to further investigate the relationship between the shape of Ag-NPs, Ag^+ release, and its influence on the antibacterial activity.

- *Influence of zeta potential:* A positive zeta potential was reported to promote interactions of Ag-NPs with negatively charged bacterial membranes causing membrane disruption, which led to bacterial flocculation and a reduction in viability (Seil and Webster 2012). In contrast, Ag-NPs with a negative zeta potential could also exhibit antibacterial effects, though due to reduced electrostatic interactions with bacterial membranes, higher Ag-NPs concentrations or a small particle size is required for an extended antibacterial effect (Seil and Webster 2012). As an example, spherical Ag-NPs with a zeta potential of −30.8 mV effectively interacted with *E. coli* and *S. aureus*, disrupted bacterial respiration and cell permeability, and induced reactive nitrogen and oxygen species, thus generating oxidative stress on DNA, which in turn caused cell death. This was, however, mainly relying on the small Ag-NP size of 7 nm (Riaz et al. 2021). These examples show that the zeta potential has an effect on antibiofilm activity, but there are multiple factors that need to be taken into account to fully understand the antibacterial properties of Ag-NPs. If NPs have a similar size, the ones with a positive zeta potential will show greater antibacterial activity than the ones with a negative zeta potential (Seil and Webster 2012).

Although most studies investigated effects of Ag-NPs against planktonic bacteria, some studies carried out biofilm assays. Ramalingam et al. (2014) showed that Ag-NPs of 35 nm were effective against *P. aeruginosa* and *S. aureus* biofilms. Martinez-Gutierrez et al. (2013) evaluated the antibiofilm activity of Ag-NPs in *P. aeruginosa* biofilms, which were generated under static conditions and high fluid shear using a bioreactor. They concluded that Ag-NPs not only prevented biofilm formation, but also killed bacteria in established biofilms. Interestingly, the antibiofilm activity of Ag-NPs in this study showed less efficacy against Gram-positive bacteria (Martinez-Gutierrez et al. 2013), which is in contrast to other literature that found elevated antibiofilm effects of spherical Ag-NPs against MRSA compared to *P. aeruginosa* (Richter et al. 2017).

2.2 Silver Nanoparticles in Combination with Other Antimicrobial Agents

Multi-pronged treatment strategies have been investigated to maximise antibacterial efficacy, overcome biofilm tolerance to monotherapy, and circumvent potential drug resistance development.

As an example, Ag-NPs have been combined with antibiotics. Habash et al. (2014) have tested Ag-NPs in combination with aztreonam, which is an antibiotic used for the treatment of infections with Gram-negative bacteria. The sole administration of aztreonam was not effective in eradicating *P. aeruginosa* PAO1 biofilm, but instead stimulated its growth by 250% compared to untreated biofilms. Notably, the combination of 10 nm-sized Ag-NPs and the antibiotic achieved a 98% reduction in biofilm biomass and a 50% reduction in biofilm thickness (Habash et al. 2014). Another study confirmed substantial damage to *E. coli, Salmonella typhimurium, and S aureus* bacterial cell walls, resulting in 95% growth inhibition when treated with Ag-NPs and kanamycin. The mechanism of the synergistic effect was explained as a process in stages. Firstly, Ag-NPs destabilised the bacterial cell membrane, thus facilitating the penetration of the antibiotic. Secondly, increased uptake of kanamycin led to elevated antibiotic efficacy, inducing bacterial cell death (Vazquez-Muñoz et al. 2019). In contrast, when Ag-NPs were combined with ß-lactam antibiotics (e.g. ampicillin, aztreonam, and biapenem) which target the bacterial cell wall, no synergistic effect was observed. An excellent review on the synergistic effect of Ag-NPs with antibiotics is available elsewhere (Prasher et al. 2018).

Another approach is the combination of antimicrobial peptides (AMP) and Ag-NPs, which showed enhanced antimicrobial activity against several pathogens, including *Klebsiella pneumoniae* (Pal et al. 2019). This literature concluded that Ag-NPs were responsible for the localised delivery of AMP close to the bacterial cell membrane. This in turn facilitated the interaction between AMPs and the outer membrane of the bacterial wall, inducing its destabilisation. Moreover, Li et al. (2020) developed a multifunctional peptide (MFP)-coated Ag-NPs (MFP@Ag-NPs), which exhibited excellent antibacterial activity against Gram-positive bacteria *S. aureus* and methicillin-resistant *S. aureus* (MRSA), and Gram-negative bacteria *E. coli* and multidrug-resistant *Acinetobacter baumannii* in vitro. Furthermore, outcomes of a pneumonia model in mice infected with multidrug-resistant *A. baumannii* confirmed excellent antibacterial activity of MFP@Ag-NPs in vivo (Li et al. 2020).

Xu et al. (2021) synthesised a nanocomposite consisting of modified Ag-NPs with AMP using polydopamine (PDA). The incorporation of PDA served as green reducing agent and also contributed to enhance antibacterial properties. The nanocomposite inhibited the growth of *E. coli, P. aeruginosa*, and *S. aureus* by inhibiting the expression of biofilm-related genes in vitro (Xu et al. 2021).

GO-PEG-Ag (graphene oxide, polyethylene glycol, and silver) nanocomposites were synthesised by Zhao et al. (2017). The nanocomposite was tested in vitro against *E. coli* and *S. aureus*. The results revealed that the incorporation of PEG increased the compatibility and stability of the nanocomposite. Furthermore, GO-PEG-Ag nanocomposites were capable of damaging bacterial cells, inducing leakage of cytoplasm, producing ROS, and reducing bacterial metabolism, ultimately leading to bacterial cell death (Zhao et al. 2017).

2.3 *Clinical Trials and Applications of Silver Nanoparticles*

There are a growing number of clinical trials underway or recently completed evaluating the therapeutic potential of Ag-NPs in various conditions, including COVID-19 and biofilm co-infections, chronic sinus infections, wound care, dentistry, and implant- and medical device-related infections, amongst others (Table 1). Different routes of Ag-NP administration are being studied, such as local

Table 1 Selection of clinical trials using silver nanoparticles (source: https://clinicaltrials.gov/ct2/search)

Application	Status	Trial number	Time frame	Condition	Location
Ag-coated endotracheal tube	Completed (Pirrone et al. 2019)	NCT02120001	2014/06–2015/08	Ventilator-associated pneumonia	Massachusetts General Hospital, USA
Ag-Au-Pd alloy-coated endotracheal tube	Completed (Thorarinsdottir et al. 2020)	NCT02284438	2014/04–2020/02	Ventilator-associated pneumonia, COVID-19	Skåne University Hospital, Sweden
Ag-NPs as topical gel dished in oral mucosa	Completed (no results posted)	NCT02761525	2014/10–2015/07	Oral pathogens in mechanically ventilated patients	University of de San Luis Potosí, Mexico
Ag-NPs as solution taken orally or inhaled	Recruiting	NCT04978025	2020/11–2021/11	SARS (severe acute respiratory syndrome)	University Hospital Sahloul, Tunisia
Ag-NPs as nasal rinse and/or mouthwash	Completed (no results posted)	NCT04894409	2020/04–2020/09	COVID-19	Tijuana General Hospital, Mexico
	Completed (Al-sharani et al. 2018)	ACTRN12618001265268	2017/01–2017/12	Plaque-induced gingivitis	Yemen University, Yemen
	Completed (Ooi et al. 2018)	ACTRN12616001558415	2016/11–2017/06	Chronic rhinosinusitis	University of Adelaide, Australia
	Completed (Scott et al. 2017)	NCT02403479	2015/12–2017/04	Chronic rhinosinusitis	Saint Joseph's Health Care, Canada
Ag-NPs as wound dressing	Not yet recruiting	NCT04684121	2021/12–2024/04	Burn wounds	First String Research, Inc., USA

	Not yet recruiting	ACTRN12618002003257	2019/01	Burn wounds	Royal Adelaide Hospital, Australia
	Recruiting	NCT03621592	2018/08–2022/01	Venous leg ulcer	University of Miami, USA
	Recruiting	ACTRN12614001234606	2014/12	Diabetic foot ulcers	The Royal Melbourne Hospital, Australia
	Recruiting	NCT02652169	2014/06–2022/04	Chronic skin ulcer	Medical University of Vienna, Austria
	Completed (Nherera et al. 2017)	NCT02852148	2016/11–2017/12	Wounds and injuries	Northumbria Healthcare, UK
	Completed (Tsang et al. 2017)	NCT02577900	2013/02–2015/08	Diabetic foot	Queen Elizabeth Hospital, Hong Kong
	Completed (no results posted)	NCT02288884	2011/05–2016/10	Surgical site infection	Regina General Hospital, Canada
Ag as wound dressing	Not yet recruiting	NCT05045430	2021/10–2022/02	Wound infections	Advanced Medical Solutions Ltd., UK
	Recruiting	NCT04297215	2020/03–2023/01	Atopic dermatitis	Erasmus University Medical Center, Netherlands
	Completed (Moreira et al. 2021)	NCT02108535	2014/04–2021/05	Burn wounds	University of Sorocaba, Brazil
		NCT04582045	2019/09–2021/04	Surgical site infections	Regions Hospital Saint Paul, USA

(continued)

Table 1 (continued)

Application	Status	Trial number	Time frame	Condition	Location
	Completed (no results posted)				
	Completed (results posted)	NCT03136159	2017/06–2019/06	Post-caesarean section wound infections	Loyola University Medical Center, USA
	Completed (no results posted)	NCT03277131	2017/12–2019/06	Wounds and leg ulcer	Multi-centre trial, K
	Completed (no results posted)	NCT02660541	2016/06–2017/05	Burn wounds	Hangang Sacred Heart Hospital, Korea
	Completed (no results posted)	NCT04410497	2015/05–2018/05	Burn wounds	University Hospital Linkoeping, Sweden
	Completed (results posted)	NCT02581488	2015/10–2017/01	Diabetic foot ulcer	Multi-centre trial, USA and Canada
	Completed (results posted)	NCT02210208	2014/09–2016/06	Burn wounds	The Arizona Burn Center, USA
	Completed (results posted)	NCT01697748	2013/09–2018/05	Surgical site infections	University of South Florida, USA
	Completed (Verbelen et al. 2014)	NCT00343824	2006/08–2015/09	Burn wounds	Ghent University Hospital, Belgium

Ag-doped hydroxyapatite coated implants	Completed (no results posted)	NCT03343288	2017/03–2020/03	Implant infection	Eskisehir Osmangazi University, Turkey
Ag graft	Completed (no results posted)	NCT01970605	2013/10–2018/12	Peripheral artery occlusive disease	Hospital Magdeburg, Germany
Ag-modified glass ionomer	Not yet recruiting	NCT04329858	2021/07	Dental caries	Cairo University, Egypt
Ag-NPs contained in bioceramic sealer	Not yet recruiting	NCT04481945	2021/06–2022/01	Endodontic disease	Misr International University, Egypt
Nano-silver fluoride as solution or varnish (Ag-NPs or mixture of Ag- and Au-NPs)	Not yet recruiting	NCT04887389	2021/07–2023/01	Carious dentine	Mansoura University, Egypt
	Not yet recruiting	NCT04731766	2021/11–2022/01	Dentin hypersensitivity	Cairo University, Egypt
	Recruiting	NCT04930458	2021/08–2022/03	Carious dentine	University of Baghdad, Iraq
	Recruiting	NCT04929509	2020/12–2022/01	Enamel white spot lesions	Alexandria University, Egypt
	Completed (no results posted)	NCT03669224	2019/11–2021/03	Caries class II	Cairo University, Egypt
	Completed (Abuhashema et al. 2020)	NCT03193606	2018/07–2019/03	Carious dentine	Cairo University, Egypt
	Completed (results posted)	NCT03186261	2018/04–2018/12	Carious dentine	Cairo University, Egypt
	Completed (Al-Nerabieah et al. 2020)	ACTRN12618001865202	2018/11–2019/03	Carious dentine	Damascus University, Syria

(continued)

Table 1 (continued)

Application	Status	Trial number	Time frame	Condition	Location
	Completed (dos Santos Jr et al. 2014)	NCT01950546	2014/09–2015/01	Carious dentine	University of Pernambuco, Brazil
Ag-NPs as intracanal medication with or without calcium hydroxide	Recruiting	NCT04338633	2020/04	Necrotic pulp	Minia University, Egypt
	Completed (Hassan et al. 2021)	NCT03692286	2019/06–2020/09	Necrotic pulp	Cairo University, Egypt
	Completed (no results posted)	NCT04213716	2015/06–2016/12	Symptomatic root treatment failure	Cairo University, Egypt
Ag-NPs against multidrug-resistant bacteria	Completed (Elsawy et al. 2021)	NCT04431440	2020/06–2020/10	Methicillin- and vancomycin-resistant *Staphylococcus aureus* (MRSA, VRSA) infections in critically ill patients	Assiut University, Egypt
Mixture of Ag- and Cu-NPs against nosocomial bacteria	Recruiting	NCT04775238	2021/02–2021/07	Nosocomial *Staphylococcus aureus* and *Pseudomonas aeruginosa* infection in critically ill patients	Sohag University, Egypt
Ag-NPs against fungi	Completed (no results posted)	NCT04431804	2020/06–2020/09	Invasive aspergillosis with *Aspergillus fumigatus*	Assiut University, Egypt

applications, including aerosols, injections, or wound dressings, and systemic applications through oral, intravenous, or intraperitoneal administration. Additionally, multiple clinical trials tested different types of pharmaceutical formulations, coatings, and nanocomposites for controlled and localised release of Ag-NPs, which resulted in a number of patented Ag-NP products (Sim et al. 2018). A recent review by Dhiman et al. (2019) thoroughly summarises antibacterial applications of Ag-NPs.

- *COVID-19 and Biofilm Co-infections*

Whilst COVID-19 is a virus infection, multiple cases of biofilm co-infections have been reported around the world, contributing to exacerbation of disease and high mortality rates (Thorarinsdottir et al. 2020; Qu et al. 2021; Hughes et al. 2020; Lansbury et al. 2020; Rawson et al. 2020; Vaughn et al. 2021; Zhu et al. 2020). Bacterial co-infections can easily emerge during a viral lung infection due to a weakened immune system and disrupted respiratory microbiota of the host, creating a favourable environment for bacterial colonisation (Hanada et al. 2018). Most common co-infections have been associated with *Haemophilus influenzae*, *S. aureus*, *K. pneumoniae*, *Mycoplasma pneumoniae*, *Streptococcus pneumoniae*, and *P. aeruginosa* (Lansbury et al. 2020; Zhu et al. 2020).

Specifically, bacterial co-infection in severe cases of COVID-19 can be lethal when patients require mechanical ventilation for life support and rely on endotracheal tubes to maintain an open airway. Unfortunately, the tubes can be rapidly and frequently colonised by biofilms, which are difficult to eradicate and serve as a pathogen reservoir, putting patients at risk of relapsing ventilator-associated pneumonia (Thorarinsdottir et al. 2020). In a clinical trial (NCT02284438), endotracheal tubes coated with an alloy of silver, gold, and palladium reduced high-grade biofilm formation and, therefore, the risk for ventilator-associated pneumonia, which could be a life-saving intervention for mechanically ventilated COVID-19 patients.

Another clinical trial (NCT04894409) investigated Ag-NPs for the prevention of COVID-19, which was based on in vitro findings of Ag-NPs inhibiting SARS-CoV-2 infection in cultured cells. The clinical trial tested Ag-NPs in a mouthwash and nose rinse to prevent the viral spread amongst health workers. Findings are not published yet.

- *Chronic Sinus Infections*

Several clinical trials have focused on the evaluation of Ag-NPs for the treatment of chronic sinus infections. For example, Richter et al. (2017) developed a nasal rinse to treat recalcitrant chronic rhinosinusitis, which could also be used as an antibiotic-free lavage in wound care (Richter et al. 2017). After confirming substantial antibiofilm activity in vitro and in vivo with 96%, 97%, and 98% biofilm reduction of *S. aureus*, MRSA, and *P. aeruginosa*, respectively, the Ag-NPs nasal rinse was translated into human clinical trials (ACTRN12616001558415). The nasal rinse contained 0.015 mg/mL Ag-NPs and was used twice daily for 10 days. The outcomes confirmed safety of the Ag-NPs, improvement in symptoms and endoscopic scores, and similar antibiofilm efficacy of Ag-NPs compared to culture-directed oral antibiotics (Ooi et al. 2018).

Another study (NCT02403479) compared the efficacy of a commercially available nasal spray containing colloidal Ag-NPs to saline spray (Scott et al. 2017). Set up as a crossover study, chronic rhinosinusitis patients used one spray twice daily for 6 weeks and continued with the other spray twice daily for 6 weeks. Outcomes indicated no negative health effects of the Ag-NP spray, though no clinical improvements after Ag-NP treatment have been observed.

- *Wound Care*

Wound dressings containing Ag-NPs, such as Ag-NP-loaded hydrogels made of biocompatible polymers (Zhong et al. 2020), have been the focus of numerous clinical trials. For example, one study (NCT02210208) investigated the wound healing capacity of a dressing containing Ag-NPs in the treatment of skin grafts in surgical burn wounds. The dressing supported adequate fixation of the skin graft and significantly improved epithelialisation and wound healing.

Two examples for successful translation from bench to bedside and market approval are the silver dressings Aquacel and Acticoat. Aquacel is a liquid absorbing hydrofibre dressing incorporating silver ions. Acticoat is a three-layered antimicrobial barrier dressing containing an absorbent core surrounded by a nanocrystalline silver-coated mesh. In a clinical study (NCT00343824) the treatment efficacy and cost-effectiveness of both wound dressings was evaluated in burn wounds. Whilst wound healing time and antibacterial efficacy were similar for both dressings, study outcomes confirmed significant benefits of Aquacel regarding ease of use and application, patient comfort, and higher cost-effectiveness than Acticoat (Verbelen et al. 2014).

In another burn wound clinical trial (NCT02108535) Acticoat was compared to silver sulfadiazine cream. Both treatments showed comparable antimicrobial efficacy and safety, whilst the use of Acticoat required fewer dressing changes/less medical material and fewer hospital visits/less professional and administrative labour and was therefore more cost-effective than silver sulfadiazine (Moreira et al. 2021).

For further reading, please refer to current systematic reviews on wound management (Nherera et al. 2017; Zhang et al. 2019) and clinical trials (Table 1).

- *Dentistry*

Silver-based products have found their way into multiple dental applications, including as Ag-NPs to prevent and treat biofilm infection in the oral cavities. As an example, nano-silver fluoride was shown to prevent biofilm growth and effectively halted dentine caries without staining teeth (NCT01950546, dos Santos et al. 2014; ACTRN12618001865202; Al-Nerabieah et al. 2020). Moreover, nano-silver fluoride increased remineralisation of carious dentin (NCT03193606, Abuhashema et al. 2020). Another study (NCT03669224) used a mixture of silver and gold nanoparticles as pre-treatment for caries class II interventions and assessed the effect on marginal integrity of resin composite restoration. Findings are not published yet.

- *Implants and Medical Devices*

Nanocomposite formulations consisting of biocompatible polymers and Ag-NPs (e.g. spherical Ag-NPs, nanowires, and nanolayers) have been evaluated as antimicrobial coatings for implants and medical devices, such as endotracheal tubes, catheters, tissue scaffolds, surgical meshes, and other polymeric fibres (Polívková et al. 2017). One example is an antibacterial titanium implant for the prevention of periprosthetic infection. Ag-NPs were immobilised on the implant surface via plasma immersion ion implantation. The antibiofilm activity of the implant was assessed in vivo using a model of implant-related tibia osteomyelitis in rats, which showed effective prevention of *Staphylococcus epidermidis* biofilms (Qin et al. 2014). Unfortunately, this innovation has not been translated to human studies. There is a lack of ongoing/upcoming clinical trials testing the antibiofilm activity of Ag-NP containing implants and medical devices.

3 Gold Nanoparticles

In contrast to Ag-NPs, the antimicrobial activity of gold nanoparticles (Au-NPs) is less pronounced due to metallic gold being inert and nontoxic in nature (Boda et al. 2015). However, Au-NPs are also considered the most biocompatible metal nanoparticles with relatively low toxicity because of their high chemical and physical stability (Aminabad et al. 2019; Hu et al. 2020b). Therefore, Au-NPs are found in various medical fields such as imaging, diagnostics, or as drug delivery carrier and therapeutic agent in cancer and infectious diseases (Hu et al. 2020b).

The antimicrobial activity of Au-NPs has been determined for both the planktonic and the biofilm form in a wide range of Gram-positive and Gram-negative bacteria (Qayyum and Khan 2016; Abdalla et al. 2020). Whilst the mechanism for antibiofilm activity is not yet determined, it has been suggested that Au-NPs induce ROS production (Sathyanarayanan et al. 2013), create electrostatic interactions between NPs and bacterial cell membrane or cell membrane proteins (Sathyanarayanan et al. 2013; Giri et al. 2015), inhibit the exopolysaccharide synthesis (Manju et al. 2016; Ali et al. 2020), and inhibit quorum-sensing-controlled virulence factors (Qais et al. 2021).

Overall, Au-NPs were studied as monotherapy, in combination with other antimicrobial agents, and in combination with photothermal therapy. Whilst in vitro studies of new nanoparticle structures are still ongoing, first in vivo results are available for specific Au-NPs, including therapeutic approaches such as band aids (Wang et al. 2017), dental aligners (Zhang et al. 2020), and ureteral stents (Gao et al. 2020). These outcomes hold promise to progress Au-NP therapy to first human clinical trials in the future.

3.1 Gold Nanoparticle Effects on Biofilms

Similar to Ag-NPs, the shape and size of Au-NPs plays an important role in the antimicrobial activity and is influenced by the synthetic procedure. Ultrasmall nanoparticles with sizes below or around 10 nm have been chemically synthesised and showed in vitro antibiofilm activity against *S. aureus*, *P. aeruginosa*, and *Shigella* (Boda et al. 2015; Giri et al. 2015; Wu et al. 2021). Wu et al. (2021) described the antibiofilm effect of ultrasmall Au-nanoclusters against *Shigella* based on the induction of ROS. Furthermore, in a mouse model of Shigella-induced colitis the treatment with Au-nanoclusters resulted in reduced bacterial load in the faeces and an overall recovering of mice health. Furthermore, the intestinal mucosa and epithelial cell damage caused by *Shigella* gradually improved after treatment with Au-nanoclusters (Wu et al. 2021).

Chatterjee et al. (2021) investigated the antibiofilm activity of spherical and rod-shaped Au-NPs against *Vibrio cholerae*. Whilst all tested Au-NPs showed no toxicity against human colonic epithelial cells, only spherical Au-NPs destroyed biofilms. After oral treatment of *V. cholerae*-infected mice with spherical Au-NPs, disruption of *V. cholerae* attachment onto the epithelial surface and biofilm inhibition was observed. The antibiofilm activity of spherical Au-NPs is based on the disruption and crossing of the *V. cholerae* membrane, inhibition of ATP synthesis, and DNA damage, consequently resulting in bacterial death (Chatterjee et al. 2021).

Chemical synthesis allows us to control the specific shape and size of Au-NPs. However, as chemical procedures are linked to a range of problems, including energy loss, environmental effects, and safety challenges, the focus has shifted towards green synthesis of Au-NPs in recent years (Sánchez-López et al. 2020). Table 2 highlights examples of biogenic procedures using extracts of plants, bacteria, and fungi.

Some Au-NPs that were synthesised with different green methods were investigated for their in vivo activity. Au-NPs synthesised with *Woodfordia fruticosa* flowers have shown antibiofilm activity in vitro against the fungus strains *Candida albicans* and *Cryptococcus neoformans,* whilst a topical application of the Au-NPs in a rat model promoted wound healing and prevented scar formation (Raghuwanshi et al. 2017). Furthermore, treatment with Au-NPs from the hyper-thermophilic bacterial strain *Caldicellulosiruptor changbaiensis* was investigated on catheters causing periprosthetic infection in mice. These small and moderate sized Au-NPs coupled with H_2O_2 exhibited high peroxidase activity, therefore increasing the ROS production causing the destruction of the lipid membrane of bacteria and biofilms. After 7 days, no sign of infection or *S. aureus* colonisation remained, the catheters were unstained, and low side effects on the mice were observed (Bing et al. 2018).

Table 2 *Examples of Au-NPs developed with green methods and their* in vitro antibiofilm activity

Synthesis involving	Specifics	Antibiofilm activity against	References
Plants	Stem of *Tinospora cordifolia*	*P. aeruginosa*	Ali et al. (2020)
	Cannabis sativa	*P. aeruginosa* *E. coli*	Singh et al. (2018)
	Capsicum annuum aqueous extract	*P. aeruginosa* *Serratia marcescens*	Qais et al. (2021)
	Various plant extracts of the Philippines	*C. albicans*	Judan Cruz et al. (2021)
	Essential oil extracted from *Nigella sativa* seed	*S. aureus* *V. harveyi*	Manju et al. (2016)
Bacteria	*Acinetobacter baumannii* collected from an infected apple sample	*P. aeruginosa* *V. cholera* *Brevibacterium linens*	Rajput and Bankar (2017)
	Lytic phages of *Salmonella serovar*	*P. aeruginosa*	Ahiwale et al. (2017)
Fungus	Ectomycorrhizal strain *Lacaria fraternal*	*P. aeruginosa*	Samanta et al. (2017)

3.2 Gold Nanoparticles in Combination with Other Antimicrobial Agents

Dual treatment with other antimicrobial agents has been investigated to reduce Au-NP dosages and increase antibiofilm activity. Antimicrobials can either be directly coated onto the surface of Au-NPs or be attached through an intermediate coating. As an example, when cinnamaldehyde was directly coated onto the Au-NPs surface, the nanostructure showed in vitro antibiofilm activity against *E. coli*, *P. aeruginosa*, *S. aureus*, and MRSA. In addition, the bacterial virulence was reduced and viability of *S. aureus*-infected *C. elegans* was prolonged, indicating strong antimicrobial activity in vivo (Ramasamy et al. 2017). Similarly, chlorhexidine-coated Au-NPs inhibited the formation and eradicated preformed biofilms of *Klebsiella pneumoniae* (Ahmed et al. 2016). In another example, glucosamine-based Au-NPs (Au-GluN) synthesised by Yang et al. (2019) showed antibiofilm activity against MRSA. After exhibiting no haemolytic activity in vitro, cytotoxicity was assessed by intraperitonetal injection in vivo. No side effects were observed, no serious immune response over a 7-day period, and no histopathological changes in the heart, liver, spleen, lung, and kidney, indicating low pathological toxicity of Au-GluN in mice. After intraperitonetal MRSA injection, treatment with Au-GluN increased the survival of the mice. Furthermore, Au-GluN treatment of MRSA-infected wounds in mice resulted in smaller wound size, lower inflammation, and re-epithelialisation, suggesting Au-GluN to be a promising treatment for MRSA wound infections (Yang et al. 2019).

Antimicrobial agents can also be attached to the surface of Au-NPs by an intermediate coating, for example in a lipid double layer of phosphatidylcholine, loading gentamicin onto Au-NPs. This complex damaged established biofilm and inhibited *S. aureus*, *P. aeruginosa*, *E. coli*, and *Listeria monocytogenes* biofilm formation. In addition, the gentamicin-loaded Au-NPs killed intracellular bacteria in infected macrophages (Mu et al. 2016). Another combination of Au-NP with imidazole and chitosan showed antibiofilm properties against *S. aureus* and *E. coli* biofilms. The cationic amine on the surface interacts with the negatively charged sites of bacterial cells. The disruption of mature biofilms occurred through synergistic effects of the Au-NPs and imidazole by interrupting bacteria–bacteria interactions at the bacterial surface. The combination showed no haemolytic activity nor significant cell toxicity and, in a rabbit wound model with *S. aureus* biofilm infection, treatment with modified Au-NPs led to complete animal recovery after 14 days. In addition, wound healing was accelerated and the local bacterial count was significantly decreased (Lu et al. 2018).

However, a major component limiting treatment efficacy is the biofilm matrix, which provides a protective barrier that can hinder the penetration of antimicrobial agents, shield bacteria, and facilitate tolerance and resistance to nanoparticles (Qayyum and Khan 2016). An excellent summary on nanoparticle–biofilm matrix interactions can be found elsewhere (Fulaz et al. 2019). Multiple strategies have been proposed to disrupt the biofilm matrix, e.g. by photothermal treatment combined with amoxicillin-coated Au-NPs. Benefiting from the photosensitising properties of amoxicillin, the coated Au-NPs showed synergistic effects with phototherapy, damaging *P. aeruginosa* and *S. aureus* biofilm matrixes, whilst requiring a short irradiation time and a low Au-NP concentration (Rocca et al. 2020). In another study, daptomycin was incorporated in polydopamine-coated gold nanocages and conjugated to staphylococcal specific surface proteins. This structure showed activity against *S. aureus* and *S. epidermidis* biofilms after photothermal activation (Meeker et al. 2016). Similar results were observed against *P. aeruginosa* biofilms with gentamicin Au-NPs conjugated to an adapted outer membrane protein (Meeker et al. 2018).

3.3 Gold Nanoparticle Delivery Systems and Applications

Au-NP delivery systems with controlled release, as well as Au-NP-embedded medical devices, such as catheters, band aids, or aligners, have been developed for medical applications.

A photothermal nanoswimmer with spherically embedded Au-NPs and enclosed vancomycin was developed by Cui et al. (2020). When exposed to near-infrared laser irradiation, the nanoswimmers quickly penetrated and diffused into *S. aureus* biofilm, where the encapsulated vancomycin was released. In vitro more than 90% of the biofilm mass was removed after treatment, without spreading live bacteria in the environment. Treatment of a mouse implant-related periprosthetic infection showed very few bacteria and no biofilm remaining on the catheters. In addition, no toxicity to major organs was observed and wound healing was accelerated. This new technology provides the opportunity of eliminating deep layered biofilms, without causing damage to healthy tissue (Cui et al. 2020).

Wang et al. (2017) investigated the antimicrobial activity of Au-NPs with graphitic carbon nitride in a mouse wound model and an in vivo acute lung infected model. Application of an Au-NP-band aid with H_2O_2 on a drug-resistant *S. aureus*-infected wound resulted in prevention of wound infection, reduced presence of bacteria, and significantly accelerated the wound healing process. After intranasal instillation of MRSA into mice, the Au-NPs were given intravenously and led to a reduced number of bacteria in the lung compared to vancomycin treatment. These novel nanostructures with versatile functionalities show promising advances in different fields of nanomedical application (Wang et al. 2017).

A nanoparticle consisting of a gold core with a silver shell (Ag-Au-NP) was synthesised by Gao et al. (2020). The nanoparticle was embedded in a PGA/PLGA ureteral stent and has been investigated for cytotoxic and antibacterial properties against *E. coli* and *S. aureus* in a farm pig model. The antibiofilm activity of the stent was based on the biodegradable properties of PGA/PLGA, resulting in a constant exfoliating and removal of attached bacteria. In addition, through the self-cleaning process the embedded Ag-Au-NPs were exposed, which further eliminated adherent bacteria. The stent inhibited bacterial adhesion, contributed to biofilm removal, and reduced inflammatory response in vivo, displaying excellent characteristics as novel treatment for stent-induced urinary tract infections (Gao et al. 2020).

Au-NPs can also have an impact on oral health. Zhang et al. (2020) investigated the activity of 4,6-diamino-2pyrimidinethiol-modified Au-NPs coated on aligners against *Porphyromonas gingivalis* biofilms, the main cause of periodontitis. In the presence of the coated aligner, the biofilm formation was prevented, and no haemolysis induction was observed. As no irritation of the mucosa was observed after a mucosal contact test on golden hamsters, the Au-NP-coated aligner exhibited favourable biocompatible characteristics and is a promising new dental device for oral bacterial related diseases (Zhang et al. 2020).

4 Other Metal-Based Nanoparticles

In contrast to silver and gold nanoparticles, other metallic NPs typically contain the metal oxide form. These NPs are less common but have also been investigated for antibiofilm properties on their own or as part of mixed metal NPs (Shkodenko et al. 2020). Zinc oxide, copper-based, iron, titanium dioxide, and selenium NPs showed in vitro antibiofilm activity against Gram-positive and Gram-negative bacteria, which is partly based on the release of free metal ions (Kaul et al. 2022; Qayyum and Khan 2016; Sánchez-López et al. 2020; Shkodenko et al. 2020). An overview of recent in vivo biofilm studies and their application in human clinical trials is presented in Table 3.

Table 3 Selection of clinical trials using metal-based nanoparticles, not including silver nanoparticles (source: https://clinicaltrials.gov/ct2/search)

Application	Status	Trial number	Time frame	Condition	Location
ZnO-NPs coated on socks with/without chlorhexidine scrub	Not yet recruiting	NCT04332796	2021/09–2022/10	Treatment of foot dermatoses	Mahidol University, Thailand
	Not yet recruiting	NCT04337749	2021/11–2022/10	Prevention of foot dermatoses	Mahidol University, Thailand
	Completed (Ongsri et al. 2021)	NCT04000386	2019/06–2021/02	Foot dermatoses	Mahidol University, Thailand
ZnO as topical ointment	Recruiting	NCT03561376	2021/09–2022/06	Surgical wounds	UPMC Dermatology St. Margaret, USA
	Recruiting	NCT04505670	2020/08–2023/02	Wounds	University of Maryland, USA
	Recruiting	NCT02673229	2014/01–2021/12	Burn wounds	University of Kansas Medical Center, USA
TiO_2-NPs with chitosan glass ionomer	Completed (no results posted)	NCT04365270	2019/01–2021/01	Carious dentine	Ain Shams University, Egypt
Dentures modified with TiO_2-NPs	Completed (no results posted)	NCT04991064	2017/01–2017/06	Denture retention time	Cairo University, Egypt
Ti meshes for surgery	Completed (Cucchi et al. 2021)	NCT04332679	2013/09–2019/09	Mandibular bone atrophies	University of Bologna, Italy
CuZn nano-hydrocolloid dressing	Recruiting	NCT04634838	2021/02–2021/09	Diabetic foot ulcer	Loewenstein Rehabilitation Center, Israel
	Completed (no results posted)	NCT04963998	2019/10–2020/10	Wound infection	Rambam Health Care Campus, Israel
	Completed (Arendsen et al. 2021)	NCT03284749	2016/01–2017/12	Wound infection	Croydon Health Services, United Kingdom
Cu-based surfaces plus Cu-enriched linen	Recruiting	NCT04873557	2020/11–2021/10	Healthcare-associated infections	Hospital Regional de Iquique, Chile
Cu- or Ti-coated surfaces	Completed (no results posted)	NCT02463214	2015/09–2018/03	Transplantation infection	Vancouver General Hospital, Canada
Ferumoxytol as topical antibiofilm application	Completed (no results posted)	NCT03678012	2018/06–2018/08	Dental caries	University of Pennsylvania, USA

4.1 *Zinc Oxide Nanoparticles*

Zinc oxide nanoparticles (ZnO-NPs) exhibit beneficial properties, including high toxicity against drug-resistant microorganisms, whilst being recognised as safe material (Mahamuni-Badiger et al. 2020). ZnO-NPs alone or as nanocomposite have shown high antibacterial and antibiofilm activity in vitro and in vivo against a range of Gram-positive and Gram-negative bacteria (Yazhiniprabha et al. 2019; Saravanakumar et al. 2020). In addition, in vivo experiments resulted in effective *S. aureus* biofilm prevention of a wound dressing coated with ZnO-NPs and orange oil (Rădulescu et al. 2016) and of an interconnected porous granules of nanostructured hydroxyapatite incorporated with ZnO-NPs, simulating a potential use in orthopaedic and dental application (Grenho et al. 2015).

As medical device, ZnO-NPs coated socks are investigated in multiple clinical trials against foot dermatoses caused by Gram-positive bacteria, including *Corynebacterium* species, *Kytococcus sedentarius*, *Dermatophilus congolensis*, and *Actinomyces* species. In a randomised placebo-controlled clinical trial, the effect of wearing ZnO-NP-coated socks on unpleasant foot odour and fungal feet infection was investigated over 2 weeks (NCT04000386, Table 3). ZnO-NP-coated socks inhibited the development of pitted keratolysis and reduced the level of foot odour (Ongsri et al. 2021), leading to a follow-up study that started in 2021. The effect of chlorhexidine scrubs, ZnO-NP-coated socks, and the combination of both will be examined as treatment and prevention of pitted keratolysis. In these studies, the improvement (NCT04332796) and occurrence (NCT04337749) of pitted lesion after treatment and side effects will be assessed.

4.2 *Copper-Based Nanoparticles*

Copper-based nanoparticles (Cu-NPs) were investigated for their antibacterial properties in healthcare-associated infections (Table 3). The addition of copper zinc nano-hydrocolloid dressing to sanitary towels significantly reduced the infection rate after vaginal delivery (Arendsen et al. 2021), and the impact of copper-enriched linen on healthcare-associated infections is ongoing (NCT04873557). Another current clinical trial (NCT04775238) investigates the antimicrobial and antibiofilm activity of Cu-NPs against *S. aureus* and *P. aeruginosa* in nosocomial infections. In addition, synergistic effects of Cu-NPs with a range of antibiotics will be examined (Kaul et al. 2022; Table 1). Previously, the combination of both Cu- and Ag-NPs has been investigated in vivo. Copper oxide microspheres decorated with Ag-NPs were constructed onto an implant used for bone defect repair. The structure promoted bone regeneration and osseointegration in rabbit tibia. The mechanism of action relied on the liberation of Cu^{2+} and Ag^{+} ions, which induced cell membrane damage in *S. aureus* and *E. coli* biofilms, ROS production, and lipid peroxidation (Yan et al. 2020). In another study, a silver-copper-boron nanocomposite showed extensive elimination of *S. aureus* bone infection in mice after intravenous or intramuscular administration without displaying adverse immune response (Qadri et al. 2017).

Copper sulfide NPs were investigated for their antibiofilm activity in combination with near-infrared (NIR) laser irradiation on different surfaces. On glass, a monolayer of CuS-NPs showed in vitro antibiofilm activity against *S. aureus* and *E. coli* (Gargioni et al. 2020).

4.3 Iron Nanoparticles

Iron nanoparticles (Fe-NPs) are nontoxic in nature and the most common iron oxide polymorphs are Fe_2O_3, Fe_3O_4, and FeO (Fahmy et al. 2018). Fe-NPs have been shown to inhibit biofilm formation and eradicate existing biofilms in various Gram-positive and Gram-negative bacteria in vitro (Sathyanarayanan et al. 2013). Their antimicrobial activity typically relies on superparamagnetic properties by magnetically immobilising bacterial cells and making them sensitive to an external magnetic field, resulting in the inhibition of biofilm formation (Taylor et al. 2012; Ranmadugala et al. 2018), or on the synergistic combination with other agents, such as oleic acid (Velusamy et al. 2021), poly(acrylic acid), and antibiotics (Armijo et al. 2020; Yang et al. 2020). Specifically, the dual treatment with antibiotics has revealed promising results.

- Ficai et al. (2018) observed a significant antibiofilm activity of cefepime-functionalised Fe_3O_4 nanoparticles dispersed in a PLGA-based matrix in vitro against *E. coli* and *S. aureus* whilst being biocompatible and safe in vivo, making it a suitable material for the treatment of implantable devices surfaces (Ficai et al. 2018).
- Chitosan/poly(acrylic acid) particles co-loaded with superparamagnetic Fe-NPs and amoxicillin were used for *Helicobacter pylori*-associated biofilm infections. This system showed multiple advantages: (i) the presence of chitosan allowed the nanoparticles to adhere onto and penetrate the gastric mucus layer upon exposure to a magnetic field; (ii) the poly(acrylic acid) allowed for a continuous release of amoxicillin; and (iii) the nanocarrier protected the amoxicillin from being dissociated by gastric acid. In vitro and in vivo results showed a prolonged residual time in the stomach by applying a magnetic field, thereby increasing the eradication rate of *Helicobacter pylori* and its biofilm. This new approach allowed for a reduction in dosage and treatment frequency, leading to a better tolerability and efficacy of *Helicobacter pylori* therapy (Yang et al. 2020).
- Another approach used magnetically propelling Fe-NPs that dig artificial channels into biofilms to promote deep penetration of antibiotics. This concept was tested in vitro and in vivo (Quan et al. (2019). The combination of magnetically propelled Fe-NPs with gentamicin resulted in significant *S. aureus* colony

forming unit reduction without adverse effect on body weight and blood chemistry of mice (Quan et al. 2019).

Other innovative treatment approaches are antibiotic-independent. Some Fe-NPs displayed peroxidase-like activity in the presence of H_2O_2, as the iron oxide can catalyse H_2O_2 leading to extracellular matrix degeneration, cell membrane damage, and bacterial death within the biofilm (Gao et al. 2016). This process was pH related and especially interesting in the treatment of dental biofilms. The successive treatment of different Fe-NPs and H_2O_2 was therefore investigated in an in vivo *Streptococcus mutans* orally infected rat model. Fe-NPs only containing Fe_3O_4 (Gao et al. 2016), the nanoparticle formulation ferumoxytol (Liu et al. 2018b), and dextran-coated Fe-NPs (Naha et al. 2019) showed a disruption of the development of biofilms, a suppression of caries development, and an attenuation of severe caries lesions in the presence of H_2O_2. These results are highly promising, as another dextran-coated Fe-NP formulation (Feridex) and ferumoxytol already received FDA approval, making these Fe-NPs excellent candidates for the treatment of dental caries (Liu et al. 2018b; Naha et al. 2019). Ferumoxytol was investigated as topical antibiofilm treatment on dental carries in a clinical trial (NCT03678012, Table 3). Results are not available yet. To overcome the problem of a two-component application, Huang et al. (2021) developed dextran-coated Fe-NPs conjugated with glucose oxidase. This nanohybrid does not require additional H_2O_2, as the glucose oxidase catalyses the oxidation of glucose to H_2O_2. Since glucose accumulates in oral biofilms, the nanohybrid enhanced in vitro *S. mutans* killing without affecting other commensal bacteria. Similar results were observed in vivo; when compared to chlorhexidine, the nanohybrid had significantly increased efficacy, without affecting the gastrointestinal microbiota and host tissue (Huang et al. 2021).

4.4 Titanium Dioxide Nanoparticles

Medical devices are frequently made of a titanium base with an antimicrobial coating, such as Ag-NPs, which is mainly responsible for antibiofilm effects (Wang et al. 2016; Hu et al. 2020a). An example has been described by Guan et al. (2019), who developed a titanium plate featuring polydopamine-coated TiO_2 nanorods with attached Ag-NPs. This new surface coating displayed adequate long-term antibacterial and antibiofilm activity against MRSA and *E. coli* through selective physical puncture of bacteria and the controlled release of Ag^+. In addition, satisfying biocompatibility was observed in vitro and in vivo, making it a promising application for orthopaedics and dental implants (Guan et al. 2019).

Furthermore, the antibiofilm activity of titanium dioxide nanoparticles (TiO_2-NPs) was observed independently of the presence of silver, mainly through production of ROS and resulting damage of cell membranes and cellular constituents (Ismail et al. 2019; Alexpandi et al. 2020). TiO_2 nanocomposites synthesised with the extract of *Diplocyclos palmatus* showed antibiofilm activity against *Vibrio harveyi*, making this a potential anti-infection candidate in aquaculture (Alexpandi et al. 2020). In a medical setting, TiO_2-NPs incorporated into a gellan gum biofilm were applied as a wound dressing material on an open excision wound model in rats. The combination showed antimicrobial activity against *S. aureus* and *E. coli* and promoted cell growth, leading to a better re-epithelialisation and no scar formation (Ismail et al. 2019).

According to clinicaltrials.gov, a phase 3 clinical trial (NCT04365270) investigated the effect of TiO_2-NPs with chitosan glass ionomer on the bacterial count of children's dentin samples. Furthermore, a clinical trial (NCT04991064) examined the effect of TiO_2-NPs' addition to denture base material on their retention (Table 3). Results for both clinical trials have not yet been published.

4.5 Selenium Nanoparticles

Selenium-NPs (Se-NPs) previously showed activity against leishmania, fungus, planktonic bacteria, and biofilms (Tan et al. 2018). Se-NPs on titanium implants inhibited MRSA and MRSE biofilm formation and reduced the viable bacteria count in surrounding tissue of an infected femur rat. Implants coated with Se-NPs have the potential to reduce the risk for antibiotic-resistant orthopaedic implant infections (Tran et al. 2019). Prateeksha et al. (2017) determined that the combination of Se-NPs with polyphenols of honey induced wound healing and prevented fatal *P. aeruginosa* infection in mice. The antibiofilm activity was linked to the downregulation of quorum sensing and of quorum-sensing-related virulence genes (Prateeksha et al. 2017).

5 Conclusions and Future Perspectives

Metallic nanoparticles have opened new avenues for the treatment of biofilm-related infections and present an alternative strategy to antibiotics to combat multidrug-resistant bacteria. Numerous studies have been conducted, particularly using silver and gold nanoparticles, which showed a strong antibiofilm effect against a variety of bacterial species in vitro and in vivo. Notably, synergistic combinations with other antimicrobial agents seem to be most promising. Ongoing and upcoming clinical trials will validate the potential of metallic nanoparticles as effective antimicrobial treatment in clinical applications. However, there are still considerable gaps in our knowledge about the mechanism of action (especially regarding the role of the EPS in the regulation of nanoparticle–biofilm interactions), safety, and targeted drug delivery platforms. Therefore, more research and pharmaceutical formulation development is warranted.

Acknowledgments This review was supported by the National Health and Medical Research Council (GNT1163634, GNT2004036), the University of Adelaide (Joint PhD Scholarship held by LK), and The Hospital Research Foundation, Australia.

The authors kindly acknowledge Animate Your Science (www.animateyour.science) for professional graphics.

References

Abdalla SSI, Katas H, Azmi F, Busra MFM (2020) Antibacterial and anti-biofilm biosynthesised silver and gold nanoparticles for medical applications: mechanism of action, toxicity and current status. Curr Drug Deliv 17(2):88–100. https://doi.org/10.2174/1567201817666191227094334

Abdel-Karim R, Reda Y, Abdel-Fattah A (2020) Review—nanostructured materials-based nanosensors. J Electrochem Soc 167(3):037554. https://doi.org/10.1149/1945-7111/ab67aa

Abuhashema RA, Akah MM, Abu-Taleb NS, Safwat OM (2020) Radiographic assessment of glass ionomer restorations with and without prior application of nano silver fluoride in occlusal carious molars treated with partial caries removal technique: a randomized clinical trial. Egypt Dent J Res 4(1):293–302

Ahiwale SS, Bankar AV, Tagunde S, Kapadnis BP (2017) A bacteriophage mediated gold nanoparticles synthesis and their anti-biofilm activity. Indian J Microbiol 57(2):188–194. https://doi.org/10.1007/s12088-017-0640-x

Ahmed A, Khan AK, Anwar A, Ali SA, Shah MR (2016) Biofilm inhibitory effect of chlorhexidine conjugated gold nanoparticles against Klebsiella pneumoniae. Microb Pathog 98:50–56. https://doi.org/10.1016/j.micpath.2016.06.016

Alexpandi R, Gopi C, Durgadevi R, Kim HJ, Pandian SK, Ravi AV (2020) Metal sensing-carbon dots loaded TiO(2)-nanocomposite for photocatalytic bacterial deactivation and application in aquaculture. Sci Rep 10(1):12883. https://doi.org/10.1038/s41598-020-69888-x

Ali SG, Ansari MA, Alzohairy MA, Alomary MN, AlYahya S, Jalal M, Khan HM, Asiri SMM, Ahmad W, Mahdi AA, El-Sherbeeny AM, El-Meligy MA (2020) Biogenic gold nanoparticles as potent antibacterial and antibiofilm nano-antibiotics against Pseudomonas aeruginosa. Antibiotics (Basel) 9(3):100. https://doi.org/10.3390/antibiotics9030100

Al-Nerabieah Z, Arrag EA, Comisi JC, Rajab A (2020) Effectiveness of a novel nano-silver fluoride with green tea extract compared with silver diamine fluoride: a randomized, controlled, non-inferiority trial. Int J Dentistry Oral Sci 7(6):753–761

Al-sharani A, Al-Hajj W, Madfa A (2018) Clinical efficacy of nanosilver and chlorhexidine in the treatment of plaque-induced gingivitis: a randomized controlled clinical trial. J Oral Res 7(7): 238–244

Aminabad NS, Farshbaf M, Akbarzadeh A (2019) Recent advances of gold nanoparticles in biomedical applications: state of the art. Cell Biochem Biophys 77(2):123–137. https://doi.org/10.1007/s12013-018-0863-4

Anees Ahmad S, Sachi Das S, Khatoon A, Tahir Ansari M, Afzal M, Saquib Hasnain M, Kumar Nayak A (2020) Bactericidal activity of silver nanoparticles: a mechanistic review. Mater Sci Energy Technol 3:756–769. https://doi.org/10.1016/j.mset.2020.09.002

Arendsen LP, Thakar R, Bassett P, Sultan AH (2021) A double blind randomized controlled trial using copper impregnated maternity sanitary towels to reduce perineal wound infection. Midwifery 92:102858

Armijo LM, Wawrzyniec SJ, Kopciuch M, Brandt YI, Rivera AC, Withers NJ, Cook NC, Huber DL, Monson TC, Smyth HDC, Osiński M (2020) Antibacterial activity of iron oxide, iron nitride, and tobramycin conjugated nanoparticles against Pseudomonas aeruginosa biofilms. J Nanobiotechnol 18(1):35. https://doi.org/10.1186/s12951-020-0588-6

Bao H, Yu X, Xu C, Li X, Li Z, Wei D, Liu Y (2015) New toxicity mechanism of silver nanoparticles: promoting apoptosis and inhibiting proliferation. PLoS One 10(3):e0122535–e0122535. https://doi.org/10.1371/journal.pone.0122535

Bing W, Sun H, Wang F, Song Y, Ren J (2018) Hydrogen-producing hyperthermophilic bacteria synthesized size-controllable fine gold nanoparticles with excellence for eradicating biofilm and antibacterial applications. J Mater Chem B 6(28):4602–4609. https://doi.org/10.1039/c8tb00549d

Boda SK, Broda J, Schiefer F, Weber-Heynemann J, Hoss M, Simon U, Basu B, Jahnen-Dechent W (2015) Cytotoxicity of ultrasmall gold nanoparticles on planktonic and biofilm encapsulated gram-positive staphylococci. Small 11(26):3183–3193. https://doi.org/10.1002/smll.201403014

Carvalho G, Balestrino D, Forestier C, Mathias J-D (2018) How do environment-dependent switching rates between susceptible and persister cells affect the dynamics of biofilms faced with antibiotics? Biofilms Microbiomes 4(1):6. https://doi.org/10.1038/s41522-018-0049-2

Chatterjee T, Saha T, Sarkar P, Hoque KM, Chatterjee BK, Chakrabarti P (2021) The gold nanoparticle reduces Vibrio cholerae pathogenesis by inhibition of biofilm formation and disruption of the production and structure of cholera toxin. Colloids Surf B Biointerfaces 204: 111811. https://doi.org/10.1016/j.colsurfb.2021.111811

Cheon JY, Kim SJ, Rhee YH, Kwon OH, Park WH (2019) Shape-dependent antimicrobial activities of silver nanoparticles. Int J Nanomed 14:2773–2780. https://doi.org/10.2147/IJN.S196472

Cucchi A, Vignudelli E, Fiorino A, Pellegrino G, Corinaldesi G (2021) Vertical ridge augmentation (VRA) with Ti-reinforced d-PTFE membranes or Ti meshes and collagen membranes: 1-year results of a randomized clinical trial. Clin Oral Implants Res 32(1):1–14

Cui T, Wu S, Sun Y, Ren J, Qu X (2020) Self-propelled active photothermal nanoswimmer for deep-layered elimination of biofilm in vivo. Nano Lett 20(10):7350–7358. https://doi.org/10.1021/acs.nanolett.0c02767

Dhiman NK, Agnihotri S, Shukla R (2019) Silver-based polymeric nanocomposites as antimicrobial coatings for biomedical applications. In: Singh S, Maurya PK (eds) Nanotechnology in modern animal biotechnology: recent trends and future perspectives. Springer Singapore, Singapore, pp 115–171. https://doi.org/10.1007/978-981-13-6004-6_4

Dong Y, Zhu H, Shen Y, Zhang W, Zhang L (2019) Antibacterial activity of silver nanoparticles of different particle size against Vibrio natriegens. PLoS One 14(9):e0222322–e0222322. https://doi.org/10.1371/journal.pone.0222322

dos Santos Jr VE, Vasconcelos Filho A, Targino AGR, Flores MAP, Galembeck A, Caldas AF Jr, Rosenblatt A (2014) A new "silver-bullet" to treat caries in children–nano silver fluoride: a randomised clinical trial. J Dent 42(8):945–951

Elsawy S, Elsherif WM, Hamed R (2021) Effect of silver nanoparticles on vancomycin resistant Staphylococcus aureus infection in critically ill patients. Pathogens Global Health 115(5): 315–324. https://doi.org/10.1080/20477724.2021.1914412

Fahmy HM, Mohamed FM, Marzouq MH, Mustafa ABE-D, Alsoudi AM, Ali OA, Mohamed MA, Mahmoud FA (2018) Review of green methods of iron nanoparticles synthesis and applications. BioNanoScience 8(2):491–503. https://doi.org/10.1007/s12668-018-0516-5

Ferdous Z, Nemmar A (2020) Health impact of silver nanoparticles: a review of the biodistribution and toxicity following various routes of exposure. Int J Mol Sci 21(7):2375. https://doi.org/10.3390/ijms21072375

Ficai D, Grumezescu V, Fufă OM, Popescu RC, Holban AM, Ficai A, Grumezescu AM, Mogoanta L, Mogosanu GD, Andronescu E (2018) Antibiofilm coatings based on PLGA and nanostructured cefepime-functionalized magnetite. Nanomaterials (Basel) 8(9):doi:10.3390/nano8090633

Flemming H-C, Wingender J, Szewzyk U, Steinberg P, Rice SA, Kjelleberg S (2016) Biofilms: an emergent form of bacterial life. Nat Rev Microbiol 14(9):563–575. https://doi.org/10.1038/nrmicro.2016.94

Fulaz S, Vitale S, Quinn L, Casey E (2019) Nanoparticle–biofilm interactions: the role of the EPS matrix. Trends Microbiol 27(11):915–926. https://doi.org/10.1016/j.tim.2019.07.004

Gao L, Liu Y, Kim D, Li Y, Hwang G, Naha PC, Cormode DP, Koo H (2016) Nanocatalysts promote Streptococcus mutans biofilm matrix degradation and enhance bacterial killing to suppress dental caries in vivo. Biomaterials 101:272–284. https://doi.org/10.1016/j.biomaterials.2016.05.051

Gao L, Wang Y, Li Y, Xu M, Sun G, Zou T, Wang F, Xu S, Da J, Wang L (2020) Biomimetic biodegradable Ag@Au nanoparticle-embedded ureteral stent with a constantly renewable contact-killing antimicrobial surface and antibiofilm and extraction-free properties. Acta Biomater 114:117–132. https://doi.org/10.1016/j.actbio.2020.07.025

Gargioni C, Borzenkov M, D'Alfonso L, Sperandeo P, Polissi A, Cucca L, Dacarro G, Grisoli P, Pallavicini P, D'Agostino A, Taglietti A (2020) Self-assembled monolayers of copper sulfide nanoparticles on glass as antibacterial coatings. Nanomaterials (Basel) 10(2). https://doi.org/10.3390/nano10020352

Geddes-McAlister J, Kugadas A, Gadjeva M (2019) Tasked with a challenging objective: why do neutrophils fail to battle Pseudomonas aeruginosa biofilms. Pathogens 8(4):283

Giri K, Yepes LR, Duncan B, Parameswaran PK, Yan B, Jiang Y, Bilska M, Moyano DF, Thompson M, Rotello VM, Prakash YS (2015) Targeting bacterial biofilms via surface engineering of gold nanoparticles. RSC Adv 5(128):105551–105559. https://doi.org/10.1039/C5RA16305F

Grenho L, Salgado CL, Fernandes MH, Monteiro FJ, Ferraz MP (2015) Antibacterial activity and biocompatibility of three-dimensional nanostructured porous granules of hydroxyapatite and zinc oxide nanoparticles: an in vitro and in vivo study. Nanotechnology 26(31):315101. https://doi.org/10.1088/0957-4484/26/31/315101

Guan M, Chen Y, Wei Y, Song H, Gao C, Cheng H, Li Y, Huo K, Fu J, Xiong W (2019) Long-lasting bactericidal activity through selective physical puncture and controlled ions release of polydopamine and silver nanoparticles-loaded TiO(2) nanorods in vitro and in vivo. Int J Nanomedicine 14:2903–2914. https://doi.org/10.2147/ijn.S202625

Habash MB, Park AJ, Vis EC, Harris RJ, Khursigara CM (2014) Synergy of silver nanoparticles and aztreonam against Pseudomonas aeruginosa PAO1 biofilms. Antimicrob Agents Chemother 58(10):5818–5830. https://doi.org/10.1128/AAC.03170-14

Hanada S, Pirzadeh M, Carver KY, Deng JC (2018) Respiratory viral infection-induced microbiome alterations and secondary bacterial pneumonia. Front Immunol 9:2640. https://doi.org/10.3389/fimmu.2018.02640. PMID: 30505304; PMCID: PMC6250824

Hassan N, Diab A, Ahmed G (2021) Post-operative pain and antibacterial efficacy of silver nanoparticles formulations intracanal medication: a randomized controlled clinical study. Open Access Maced J Med Sci 9(D):248–256

Hu CC, Chang CH, Chang Y, Hsieh JH, Ueng SW (2020a) Beneficial effect of TaON-Ag nanocomposite titanium on antibacterial capacity in orthopedic application. Int J Nanomed 15:7889–7900. https://doi.org/10.2147/ijn.S264303

Hu X, Zhang Y, Ding T, Liu J, Zhao H (2020b) Multifunctional gold nanoparticles: a novel nanomaterial for various medical applications and biological activities. Front Bioeng Biotechnol 8(990). https://doi.org/10.3389/fbioe.2020.00990

Huang Y, Liu Y, Shah S, Kim D, Simon-Soro A, Ito T, Hajfathalian M, Li Y, Hsu JC, Nieves LM, Alawi F, Naha PC, Cormode DP, Koo H (2021) Precision targeting of bacterial pathogen via bi-functional nanozyme activated by biofilm microenvironment. Biomaterials 268:120581. https://doi.org/10.1016/j.biomaterials.2020.120581

Hughes S, Troise O, Donaldson H, Mughal N, Moore LSP (2020) Bacterial and fungal coinfection among hospitalized patients with COVID-19: a retrospective cohort study in a UK secondary-care setting. Clin Microbiol Infect 26(10):1395–1399. https://doi.org/10.1016/j.cmi.2020.06.025

Ismail NA, Amin KAM, Majid FAA, Razali MH (2019) Gellan gum incorporating titanium dioxide nanoparticles biofilm as wound dressing: physicochemical, mechanical, antibacterial properties and wound healing studies. Mater Sci Eng C Mater Biol Appl 103:109770. https://doi.org/10.1016/j.msec.2019.109770

Jolivet-Gougeon A, Bonnaure-Mallet M (2019) Bacterial persistence in biofilms and antibiotics. In: Antibiotic drug resistance. Wiley, New York, pp 181–210. https://doi.org/10.1002/9781119282549.ch9

Joo SH, Aggarwal S (2018) Factors impacting the interactions of engineered nanoparticles with bacterial cells and biofilms: mechanistic insights and state of knowledge. J Environ Manag 225: 62–74. https://doi.org/10.1016/j.jenvman.2018.07.084

Judan Cruz KG, Alfonso ED, Fernando SID, Watanabe K (2021) Candida albicans biofilm inhibition by ethnobotanicals and ethnobotanically-synthesized gold nanoparticles. Front Microbiol 12:665113. https://doi.org/10.3389/fmicb.2021.665113

Jung WK, Koo HC, Kim KW, Shin S, Kim SH, Park YH (2008) Antibacterial activity and mechanism of action of the silver ion in *Staphylococcus aureus* and *Escherichia coli*. Appl Environ Microbiol 74(7):2171–2178. https://doi.org/10.1128/AEM.02001-07

Kaul L, Abdo AI, Coenye T, Krom BP, Hoogenkamp MA, Zannettino ACW, Süss R, Richter K (2022) The combination of diethyldithiocarbamate and copper ions is active against Staphylococcus aureus and Staphylococcus epidermidis biofilms in vitro and in vivo. Front Microbiol 13:999893. https://doi.org/10.3389/fmicb.2022.999893

Kim J, Pitts B, Stewart PS, Camper A, Yoon J (2008) Comparison of the antimicrobial effects of chlorine, silver ion, and tobramycin on biofilm. Antimicrob Agents Chemother 52(4): 1446–1453. https://doi.org/10.1128/AAC.00054-07

Kirtane AR, Verma M, Karandikar P, Furin J, Langer R, Traverso G (2021) Nanotechnology approaches for global infectious diseases. Nat Nanotechnol 16(4):369–384. https://doi.org/10.1038/s41565-021-00866-8

Lansbury L, Lim B, Baskaran V, Lim WS (2020) Co-infections in people with COVID-19: a systematic review and meta-analysis. J Infect 81(2):266–275. https://doi.org/10.1016/j.jinf.2020.05.046

Lee SH, Jun B-H (2019) Silver nanoparticles: synthesis and application for nanomedicine. Int J Mol Sci 20(4):865

Li W, Li Y, Sun P, Zhang N, Zhao Y, Qin S, Zhao Y (2020) Antimicrobial peptide-modified silver nanoparticles for enhancing the antibacterial efficacy. RSC Adv 10(64):38746–38754. https://doi.org/10.1039/D0RA05640E

Liao C, Li Y, Tjong SC (2019) Bactericidal and cytotoxic properties of silver nanoparticles. Int J Mol Sci 20(2):449. https://doi.org/10.3390/ijms20020449

Lin J, Miao L, Zhong G, Lin C-H, Dargazangy R, Alexander-Katz A (2020) Understanding the synergistic effect of physicochemical properties of nanoparticles and their cellular entry pathways. Commun Biol 3(1):205. https://doi.org/10.1038/s42003-020-0917-1

Liu M, Zhang H, Song X, Wei C, Xiong Z, Yu F, Li C, Ai F, Guo G, Wang X (2018a) NaCl: for the safer in vivo use of antibacterial silver based nanoparticles. Int J Nanomed 13:1737–1748. https://doi.org/10.2147/IJN.S153168

Liu Y, Naha PC, Hwang G, Kim D, Huang Y, Simon-Soro A, Jung HI, Ren Z, Li Y, Gubara S, Alawi F, Zero D, Hara AT, Cormode DP, Koo H (2018b) Topical ferumoxytol nanoparticles disrupt biofilms and prevent tooth decay in vivo via intrinsic catalytic activity. Nat Commun 9(1):2920. https://doi.org/10.1038/s41467-018-05342-x

Lok CN, Ho CM, Chen R, He QY, Yu WY, Sun H, Tam PK, Chiu JF, Che CM (2007) Silver nanoparticles: partial oxidation and antibacterial activities. J Biol Inorg Chem 12(4):527–534. https://doi.org/10.1007/s00775-007-0208-z

Lu B, Lu F, Ran L, Yu K, Xiao Y, Li Z, Dai F, Wu D, Lan G (2018) Imidazole-molecule-capped chitosan–gold nanocomposites with enhanced antimicrobial activity for treating biofilm-related infections. J Colloid Interface Sci 531:269–281

Mahamuni-Badiger PP, Patil PM, Badiger MV, Patel PR, Thorat-Gadgil BS, Pandit A, Bohara RA (2020) Biofilm formation to inhibition: role of zinc oxide-based nanoparticles. Mater Sci Eng C Mater Biol Appl 108:110319. https://doi.org/10.1016/j.msec.2019.110319

Manju S, Malaikozhundan B, Vijayakumar S, Shanthi S, Jaishabanu A, Ekambaram P, Vaseeharan B (2016) Antibacterial, antibiofilm and cytotoxic effects of Nigella sativa essential oil coated gold nanoparticles. Microb Pathog 91:129–135. https://doi.org/10.1016/j.micpath.2015.11.021
Markowska K, Grudniak AM, Wolska KI (2013) Silver nanoparticles as an alternative strategy against bacterial biofilms. Acta Biochim Pol 60(4):523–530
Martinez-Gutierrez F, Boegli L, Agostinho A, Sánchez EM, Bach H, Ruiz F, James G (2013) Anti-biofilm activity of silver nanoparticles against different microorganisms. Biofouling 29(6): 651–660. https://doi.org/10.1080/08927014.2013.794225
Meeker DG, Jenkins SV, Miller EK, Beenken KE, Loughran AJ, Powless A, Muldoon TJ, Galanzha EI, Zharov VP, Smeltzer MS, Chen J (2016) Synergistic photothermal and antibiotic killing of biofilm-associated Staphylococcus aureus using targeted antibiotic-loaded gold nanoconstructs. ACS Infect Dis 2(4):241–250. https://doi.org/10.1021/acsinfecdis.5b00117
Meeker DG, Wang T, Harrington WN, Zharov VP, Johnson SA, Jenkins SV, Oyibo SE, Walker CM, Mills WB, Shirtliff ME, Beenken KE, Chen J, Smeltzer MS (2018) Versatility of targeted antibiotic-loaded gold nanoconstructs for the treatment of biofilm-associated bacterial infections. Int J Hyperth 34(2):209–219. https://doi.org/10.1080/02656736.2017.1392047
Möhler JS, Sim W, Blaskovich MAT, Cooper MA, Ziora ZM (2018) Silver bullets: a new lustre on an old antimicrobial agent. Biotechnol Adv 36(5):1391–1411. https://doi.org/10.1016/j.biotechadv.2018.05.004
Moreira SS, de Camargo MC, Caetano R, Alves MR, Itria A, Pereira TV, Lopes LC (2021) Efficacy and costs of nanocrystalline silver dressings versus 1% silver sulfadiazine dressings to treat burns in adults in the outpatient setting: a randomized clinical trial. Burns. https://doi.org/10.1016/j.burns.2021.05.014
Morones-Ramirez JR, Winkler JA, Spina CS, Collins JJ (2013) Silver enhances antibiotic activity against gram-negative bacteria. Sci Transl Med 5(190):190ra181. https://doi.org/10.1126/scitranslmed.3006276
Mu H, Tang J, Liu Q, Sun C, Wang T, Duan J (2016) Potent antibacterial nanoparticles against biofilm and intracellular bacteria. Sci Rep 6:18877. https://doi.org/10.1038/srep18877
Naha PC, Liu Y, Hwang G, Huang Y, Gubara S, Jonnakuti V, Simon-Soro A, Kim D, Gao L, Koo H, Cormode DP (2019) Dextran-coated iron oxide nanoparticles as biomimetic catalysts for localized and pH-activated biofilm disruption. ACS Nano 13(5):4960–4971. https://doi.org/10.1021/acsnano.8b08702
Navya PN, Daima HK (2016) Rational engineering of physicochemical properties of nanomaterials for biomedical applications with nanotoxicological perspectives. Nano Convergence 3(1):1. https://doi.org/10.1186/s40580-016-0064-z
Nherera L, Trueman P, Roberts C, Berg L (2017) Silver delivery approaches in the management of partial thickness burns: a systematic review and indirect treatment comparison. Wound Repair Regen 25(4):707–721
Ongsri P, Leeyaphan C, Limphoka P, Kiratiwongwan R, Bunyaratavej S (2021) Effectiveness and safety of zinc oxide nanoparticle-coated socks compared to uncoated socks for the prevention of pitted keratolysis: a double-blinded, randomized, controlled trial study. Int J Dermatol 60(7): 864–867
Ooi ML, Richter K, Bennett C, Macias-Valle L, Vreugde S, Psaltis AJ, Wormald PJ (2018) Topical colloidal silver for the treatment of recalcitrant chronic rhinosinusitis. Front Microbiol 9:720. https://doi.org/10.3389/fmicb.2018.00720
Osonga FJ, Akgul A, Yazgan I, Akgul A, Eshun GB, Sakhaee L, Sadik OA (2020) Size and shape-dependent antimicrobial activities of silver and gold nanoparticles: a model study as potential fungicides. Molecules 25(11). https://doi.org/10.3390/molecules25112682
Pal I, Bhattacharyya D, Kar RK, Zarena D, Bhunia A, Atreya HS (2019) A peptide-nanoparticle system with improved efficacy against multidrug resistant bacteria. Sci Rep 9(1):4485. https://doi.org/10.1038/s41598-019-41005-7
Pirrone M, Imber DA, Marrazzo F, Pinciroli R, Zhang C, Bry L, Delaney ML, Dubois AM, Thomas JG, Nistico L, Melton-Kreft R, Bittner EA, Kacmarek RM, Berra L (2019) Silver-coated

endotracheal tubes cleaned with a mechanism for secretion removal. Respir Care 64(1):1–9. https://doi.org/10.4187/respcare.06222

Polívková M, Hubáček T, Staszek M, Švorčík V, Siegel J (2017) Antimicrobial treatment of polymeric medical devices by silver nanomaterials and related technology. Int J Mol Sci 18(2):419. https://doi.org/10.3390/ijms18020419

Prasher P, Singh M, Mudila H (2018) Silver nanoparticles as antimicrobial therapeutics: current perspectives and future challenges. 3 Biotech 8(10):411. https://doi.org/10.1007/s13205-018-1436-3

Prateeksha SBR, Shoeb M, Sharma S, Naqvi AH, Gupta VK, Singh BN (2017) Scaffold of selenium nanovectors and honey phytochemicals for inhibition of Pseudomonas aeruginosa quorum sensing and biofilm formation. Front Cell Infect Microbiol 7:93. https://doi.org/10.3389/fcimb.2017.00093

Qadri S, Haik Y, Mensah-Brown E, Bashir G, Fernandez-Cabezudo MJ, Al-Ramadi BK (2017) Metallic nanoparticles to eradicate bacterial bone infection. Nanomedicine 13(7):2241–2250. https://doi.org/10.1016/j.nano.2017.05.013

Qais FA, Ahmad I, Altaf M, Alotaibi SH (2021) Biofabrication of gold nanoparticles using capsicum annuum extract and its antiquorum sensing and antibiofilm activity against bacterial pathogens. ACS Omega 6(25):16670–16682. https://doi.org/10.1021/acsomega.1c02297

Qayyum S, Khan AU (2016) Nanoparticles vs. biofilms: a battle against another paradigm of antibiotic resistance. MedChemComm 7(8):1479–1498. https://doi.org/10.1039/C6MD00124F

Qin H, Cao H, Zhao Y, Zhu C, Cheng T, Wang Q, Peng X, Cheng M, Wang J, Jin G, Jiang Y, Zhang X, Liu X, Chu PK (2014) In vitro and in vivo anti-biofilm effects of silver nanoparticles immobilized on titanium. Biomaterials 35(33):9114–9125. https://doi.org/10.1016/j.biomaterials.2014.07.040

Qu J, Cai Z, Liu Y, Duan X, Han S, Liu J, Zhu Y, Jiang Z, Zhang Y, Zhou C, Liu Y, Liu Y, Liu L, Yang L (2021) Persistent bacterial coinfection of a COVID-19 patient caused by a genetically adapted Pseudomonas aeruginosa chronic colonizer. Front Cell Infect Microbiol 11:129. https://doi.org/10.3389/fcimb.2021.641920

Quan K, Zhang Z, Chen H, Ren X, Ren Y, Peterson BW, van der Mei HC, Busscher HJ (2019) Artificial channels in an infectious biofilm created by magnetic nanoparticles enhanced bacterial killing by antibiotics. Small 15(39):e1902313. https://doi.org/10.1002/smll.201902313

Rădulescu M, Andronescu E, Cirja A, Holban AM, Mogoantă L, Bălşeanu TA, Cătălin B, Neagu TP, Lascăr I, Florea DA, Grumezescu AM, Ciubuca B, Lazăr V, Chifiriuc MC, Bolocan A (2016) Antimicrobial coatings based on zinc oxide and orange oil for improved bioactive wound dressings and other applications. Romanian J Morphol Embryol 57(1):107–114

Raghuwanshi N, Kumari P, Srivastava AK, Vashisth P, Yadav TC, Prasad R, Pruthi V (2017) Synergistic effects of Woodfordia fruticosa gold nanoparticles in preventing microbial adhesion and accelerating wound healing in Wistar albino rats in vivo. Mater Sci Eng C Mater Biol Appl 80:252–262. https://doi.org/10.1016/j.msec.2017.05.134

Rajput N, Bankar A (2017) Bio-inspired gold nanoparticles synthesis and their anti-biofilm efficacy. J Pharm Investig 47(6):521–530. https://doi.org/10.1007/s40005-016-0280-x

Ramalingam V, Rajaram R, PremKumar C, Santhanam P, Dhinesh P, Vinothkumar S, Kaleshkumar K (2014) Biosynthesis of silver nanoparticles from deep sea bacterium Pseudomonas aeruginosa JQ989348 for antimicrobial, antibiofilm, and cytotoxic activity. J Basic Microbiol 54(9): 928–936. https://doi.org/10.1002/jobm.201300514

Ramasamy M, Lee JH, Lee J (2017) Development of gold nanoparticles coated with silica containing the antibiofilm drug cinnamaldehyde and their effects on pathogenic bacteria. Int J Nanomed 12:2813–2828. https://doi.org/10.2147/ijn.S132784

Ranmadugala D, Ebrahiminezhad A, Manley-Harris M, Ghasemi Y, Berenjian A (2018) Magnetic immobilization of bacteria using iron oxide nanoparticles. Biotechnol Lett 40(2):237–248. https://doi.org/10.1007/s10529-017-2477-0

Rawson TM, Moore LSP, Zhu N, Ranganathan N, Skolimowska K, Gilchrist M, Satta G, Cooke G, Holmes A (2020) Bacterial and fungal co-infection in individuals with coronavirus: a rapid

review to support COVID-19 antimicrobial prescribing. Clin Infect Dis 71(9):2459–2468. https://doi.org/10.1093/cid/ciaa530
Raza MA, Kanwal Z, Rauf A, Sabri AN, Riaz S, Naseem S (2016) Size- and shape-dependent antibacterial studies of silver nanoparticles synthesized by wet chemical routes. Nanomaterials (Basel) 6(4):74. https://doi.org/10.3390/nano6040074
Rezaei R, Safaei M, Mozaffari HR, Moradpoor H, Karami S, Golshah A, Salimi B, Karami H (2019) The role of nanomaterials in the treatment of diseases and their effects on the immune system. Open Access Maced J Med Sci 7(11):1884–1890. https://doi.org/10.3889/oamjms.2019.486
Riaz M, Mutreja V, Sareen S, Ahmad B, Faheem M, Zahid N, Jabbour G, Park J (2021) Exceptional antibacterial and cytotoxic potency of monodisperse greener AgNPs prepared under optimized pH and temperature. Sci Rep 11(1):2866. https://doi.org/10.1038/s41598-021-82555-z
Richter K, Facal P, Thomas N, Vandecandelaere I, Ramezanpour M, Cooksley C, Prestidge CA, Coenye T, Wormald P-J, Vreugde S (2017) Taking the silver bullet-colloidal silver particles for the topical treatment of biofilm-related infections. ACS Appl Mater Interfaces 9(26): 21631–21638. https://doi.org/10.1021/acsami.7b03672
Rocca DM, Silvero CMJ, Aiassa V, Cecilia Becerra M (2020) Rapid and effective photodynamic treatment of biofilm infections using low doses of amoxicillin-coated gold nanoparticles. Photodiagn Photodyn Ther 31:101811. https://doi.org/10.1016/j.pdpdt.2020.101811
Sadoon AA, Khadka P, Freeland J, Gundampati RK, Manso RH, Ruiz M, Krishnamurthi VR, Thallapuranam SK, Chen J, Wang Y, Liu S-J (2020) Silver ions caused faster diffusive dynamics of histone-like nucleoid-structuring proteins in live bacteria. Appl Environ Microbiol 86(6):e02479–e02419. https://doi.org/10.1128/AEM.02479-19
Samanta S, Singh BR, Adholeya A (2017) Intracellular synthesis of gold nanoparticles using an ectomycorrhizal strain EM-1083 of Laccaria fraterna and its nanoanti-quorum sensing potential against Pseudomonas aeruginosa. Indian J Microbiol 57(4):448–460. https://doi.org/10.1007/s12088-017-0662-4
Samberg ME, Oldenburg SJ, Monteiro-Riviere NA (2010) Evaluation of silver nanoparticle toxicity in skin in vivo and keratinocytes in vitro. Environ Health Perspect 118(3):407–413. https://doi.org/10.1289/ehp.0901398
Sánchez-López E, Gomes D, Esteruelas G, Bonilla L, Lopez-Machado AL, Galindo R, Cano A, Espina M, Ettcheto M, Camins A, Silva AM, Durazzo A, Santini A, Garcia ML, Souto EB (2020) Metal-based nanoparticles as antimicrobial agents: an overview. Nanomaterials 10(2): 292. https://doi.org/10.3390/nano10020292
Saravanakumar K, Jeevithan E, Hu X, Chelliah R, Oh DH, Wang MH (2020) Enhanced anti-lung carcinoma and anti-biofilm activity of fungal molecules mediated biogenic zinc oxide nanoparticles conjugated with β-D-glucan from barley. J Photochem Photobiol B 203:111728. https://doi.org/10.1016/j.jphotobiol.2019.111728
Sathyanarayanan MB, Balachandranath R, Genji Srinivasulu Y, Kannaiyan SK, Subbiahdoss G (2013) The effect of gold and iron-oxide nanoparticles on biofilm-forming pathogens. ISRN Microbiol 2013:272086. https://doi.org/10.1155/2013/272086
Scott JR, Krishnan R, Rotenberg BW, Sowerby LJ (2017) The effectiveness of topical colloidal silver in recalcitrant chronic rhinosinusitis: a randomized crossover control trial. J Otolaryngol Head Neck Surg 46(1):64. https://doi.org/10.1186/s40463-017-0241-z
Seil JT, Webster TJ (2012) Antimicrobial applications of nanotechnology: methods and literature. Int J Nanomed 7:2767–2781. https://doi.org/10.2147/ijn.S24805
Shaikh S, Nazam N, Rizvi SMD, Ahmad K, Baig MH, Lee EJ, Choi I (2019) Mechanistic insights into the antimicrobial actions of metallic nanoparticles and their implications for multidrug resistance. Int J Mol Sci 20(10):2468. https://doi.org/10.3390/ijms20102468
Shkodenko L, Kassirov I, Koshel E (2020) Metal oxide nanoparticles against bacterial biofilms: perspectives and limitations. Microorganisms 8(10):doi:10.3390/microorganisms8101545

Sim W, Barnard RT, Blaskovich MAT, Ziora ZM (2018) Antimicrobial silver in medicinal and consumer applications: a patent review of the past decade (2007-2017). Antibiotics (Basel) 7(4): 93. https://doi.org/10.3390/antibiotics7040093

Singh P, Pandit S, Garnæs J, Tunjic S, Mokkapati VR, Sultan A, Thygesen A, Mackevica A, Mateiu RV, Daugaard AE, Baun A, Mijakovic I (2018) Green synthesis of gold and silver nanoparticles from Cannabis sativa (industrial hemp) and their capacity for biofilm inhibition. Int J Nanomed 13:3571–3591. https://doi.org/10.2147/ijn.S157958

Singh P, Pandit S, Jers C, Joshi AS, Garnæs J, Mijakovic I (2021) Silver nanoparticles produced from Cedecea sp. exhibit antibiofilm activity and remarkable stability. Sci Rep 11(1):12619. https://doi.org/10.1038/s41598-021-92006-4

Slavin YN, Asnis J, Häfeli UO, Bach H (2017) Metal nanoparticles: understanding the mechanisms behind antibacterial activity. J Nanobiotechnol 15(1):65. https://doi.org/10.1186/s12951-017-0308-z

Stensberg MC, Wei Q, McLamore ES, Porterfield DM, Wei A, Sepúlveda MS (2011) Toxicological studies on silver nanoparticles: challenges and opportunities in assessment, monitoring and imaging. Nanomedicine 6(5):879–898. https://doi.org/10.2217/nnm.11.78

Sung JH, Ji JH, Yoon JU, Kim DS, Song MY, Jeong J, Han BS, Han JH, Chung YH, Kim J, Kim TS (2008) Lung function changes in Sprague-Dawley rats after prolonged inhalation exposure to silver nanoparticles. Inhal Toxicol 20(6):567–574. https://doi.org/10.1080/08958370701874671

Sung JH, Ji JH, Park JD, Yoon JU, Kim DS, Jeon KS, Song MY, Jeong J, Han BS, Han JH, Chung YH (2009) Subchronic inhalation toxicity of silver nanoparticles. Toxicol Sci 108(2):452–461. https://doi.org/10.1093/toxsci/kfn246

Talapko J, Matijević T, Juzbašić M, Antolović-Požgain A, Škrlec I (2020) Antibacterial activity of silver and its application in dentistry, cardiology and dermatology. Microorganisms 8(9):1400. https://doi.org/10.3390/microorganisms8091400

Tan VLC, Hinchman A, Williams R, Tran PA, Fox K (2018) Nanostructured biomedical selenium at the biological interface (review). Biointerphases 13(6):06d301. https://doi.org/10.1116/1.5042693

Tang J, Xiong L, Wang S, Wang J, Liu L, Li J, Wan Z, Xi T (2008) Influence of silver nanoparticles on neurons and blood-brain barrier via subcutaneous injection in rats. Appl Surf Sci 255(2): 502–504. https://doi.org/10.1016/j.apsusc.2008.06.058

Taylor EN, Kummer KM, Durmus NG, Leuba K, Tarquinio KM, Webster TJ (2012) Superparamagnetic iron oxide nanoparticles (SPION) for the treatment of antibiotic-resistant biofilms. Small 8(19):3016–3027. https://doi.org/10.1002/smll.201200575

Thieme L, Hartung A, Tramm K, Klinger-Strobel M, Jandt KD, Makarewicz O, Pletz MW (2019) MBEC versus MBIC: the lack of differentiation between biofilm reducing and inhibitory effects as a current problem in biofilm methodology. Biol Proced Online 21(1):18. https://doi.org/10.1186/s12575-019-0106-0

Thorarinsdottir HR, Kander T, Holmberg A, Petronis S, Klarin B (2020) Biofilm formation on three different endotracheal tubes: a prospective clinical trial. Crit Care 24(1):1–12

Tran PA, O'Brien-Simpson N, Palmer JA, Bock N, Reynolds EC, Webster TJ, Deva A, Morrison WA, O'Connor AJ (2019) Selenium nanoparticles as anti-infective implant coatings for trauma orthopedics against methicillin-resistant Staphylococcus aureus and epidermidis: in vitro and in vivo assessment. Int J Nanomed 14:4613–4624. https://doi.org/10.2147/ijn.S197737

Tsang KK, Kwong EWY, To TSS, Chung JWY, Wong TKS (2017) A pilot randomized, controlled study of nanocrystalline silver, Manuka honey, and conventional dressing in healing diabetic foot ulcer. Evid Based Complement Alternat Med 2017. https://doi.org/10.1155/2017/5294890

Uruén C, Chopo-Escuin G, Tommassen J, Mainar-Jaime RC, Arenas J (2021) Biofilms as promoters of bacterial antibiotic resistance and tolerance. Antibiotics 10(1):3

Vaughn VM, Gandhi TN, Petty LA, Patel PK, Prescott HC, Malani AN, Ratz D, McLaughlin E, Chopra V, Flanders SA (2021) Empiric antibacterial therapy and community-onset bacterial coinfection in patients hospitalized with coronavirus disease 2019 (COVID-19): a multi-hospital cohort study. Clin Infect Dis 72(10):e533–e541. https://doi.org/10.1093/cid/ciaa1239

Vazquez-Muñoz R, Meza-Villezcas A, Fournier PGJ, Soria-Castro E, Juarez-Moreno K, Gallego-Hernández AL, Bogdanchikova N, Vazquez-Duhalt R, Huerta-Saquero A (2019) Enhancement of antibiotics antimicrobial activity due to the silver nanoparticles impact on the cell membrane. PLoS One 14(11):e0224904–e0224904. https://doi.org/10.1371/journal.pone.0224904

Velusamy P, Su CH, Kannan K, Kumar GV, Anbu P, Gopinath SCB (2021) Surface engineered iron oxide nanoparticles as efficient materials for antibiofilm application. Biotechnol Appl Biochem. https://doi.org/10.1002/bab.2146

Verbelen J, Hoeksema H, Heyneman A, Pirayesh A, Monstrey S (2014) Aquacel® Ag dressing versus Acticoat™ dressing in partial thickness burns: a prospective, randomized, controlled study in 100 patients. Part 1: burn wound healing. Burns 40(3):416–427

Wang J, Li J, Qian S, Guo G, Wang Q, Tang J, Shen H, Liu X, Zhang X, Chu PK (2016) Antibacterial surface design of titanium-based biomaterials for enhanced bacteria-killing and cell-assisting functions against periprosthetic joint infection. ACS Appl Mater Interfaces 8(17): 11162–11178. https://doi.org/10.1021/acsami.6b02803

Wang Z, Dong K, Liu Z, Zhang Y, Chen Z, Sun H, Ren J, Qu X (2017) Activation of biologically relevant levels of reactive oxygen species by Au/g-C(3)N(4) hybrid nanozyme for bacteria killing and wound disinfection. Biomaterials 113:145–157. https://doi.org/10.1016/j.biomaterials.2016.10.041

Wu X, Chen Y, Zhang Y, Shan Y, Peng Z, Gu B, Yang H (2021) Au nanoclusters ameliorate Shigella infectious colitis by inducing oxidative stress. Int J Nanomedicine 16:4545–4557. https://doi.org/10.2147/ijn.S315481

Xu J, Li Y, Wang H, Zhu M, Feng W, Liang G (2021) Enhanced antibacterial and anti-biofilm activities of antimicrobial peptides modified silver nanoparticles. Int J Nanomed 16:4831–4846. https://doi.org/10.2147/IJN.S315839

Yakabe Y, Sano T, Ushio H, Yasunaga T (1980) Kinetic studies of the interaction between silver ion and deoxyribonucleic acid. Chem Lett 9(4):373–376. https://doi.org/10.1246/cl.1980.373

Yan J, Bassler BL (2019) Surviving as a community: antibiotic tolerance and persistence in bacterial biofilms. Cell Host Microbe 26(1):15–21. https://doi.org/10.1016/j.chom.2019.06.002

Yan J, Xia D, Zhou W, Li Y, Xiong P, Li Q, Wang P, Li M, Zheng Y, Cheng Y (2020) pH-responsive silk fibroin-based CuO/Ag micro/nano coating endows polyetheretherketone with synergistic antibacterial ability, osteogenesis, and angiogenesis. Acta Biomater 115:220–234. https://doi.org/10.1016/j.actbio.2020.07.062

Yang X, Wei Q, Shao H, Jiang X (2019) Multivalent aminosaccharide-based gold nanoparticles as narrow-spectrum antibiotics in vivo. ACS Appl Mater Interfaces 11(8):7725–7730. https://doi.org/10.1021/acsami.8b19658

Yang SJ, Huang CH, Yang JC, Wang CH, Shieh MJ (2020) Residence time-extended nanoparticles by magnetic field improve the eradication efficiency of Helicobacter pylori. ACS Appl Mater Interfaces 12(49):54316–54327. https://doi.org/10.1021/acsami.0c13101

Yazhiniprabha M, Vaseeharan B, Sonawane A, Behera A (2019) In vitro and in vivo toxicity assessment of phytofabricated ZnO nanoparticles showing bacteriostatic effect and larvicidal efficacy against Culex quinquefasciatus. J Photochem Photobiol B 192:158–169. https://doi.org/10.1016/j.jphotobiol.2019.01.014

Yin IX, Zhang J, Zhao IS, Mei ML, Li Q, Chu CH (2020) The antibacterial mechanism of silver nanoparticles and its application in dentistry. Int J Nanomed 15:2555–2562. https://doi.org/10.2147/IJN.S246764

Zhang L, Yin H, Lei X, Lau JN, Yuan M, Wang X, Zhang F, Zhou F, Qi S, Shu B, Wu J (2019) A systematic review and meta-analysis of clinical effectiveness and safety of hydrogel dressings in the management of skin wounds. Front Bioeng Biotechnol 7:342. https://doi.org/10.3389/fbioe.2019.00342

Zhang M, Liu X, Xie Y, Zhang Q, Zhang W, Jiang X, Lin J (2020) Biological safe gold nanoparticle-modified dental aligner prevents the Porphyromonas gingivalis biofilm formation. ACS Omega 5(30):18685–18692. https://doi.org/10.1021/acsomega.0c01532

Zhao R, Lv M, Li Y, Sun M, Kong W, Wang L, Song S, Fan C, Jia L, Qiu S, Sun Y, Song H, Hao R (2017) Stable nanocomposite based on pegylated and silver nanoparticles loaded graphene oxide for long-term antibacterial activity. ACS Appl Mater Interfaces 9(18):15328–15341. https://doi.org/10.1021/acsami.7b03987

Zhong Y, Xiao H, Seidi F, Jin Y (2020) Natural polymer-based antimicrobial hydrogels without synthetic antibiotics as wound dressings. Biomacromolecules 21(8):2983–3006. https://doi.org/10.1021/acs.biomac.0c00760

Zhu X, Ge Y, Wu T, Zhao K, Chen Y, Wu B, Zhu F, Zhu B, Cui L (2020) Co-infection with respiratory pathogens among COVID-2019 cases. Virus Res 285:198005. https://doi.org/10.1016/j.virusres.2020.198005

Nitric Oxide-Mediated Dispersal as an Adjunctive Strategy for the Control of Biofilm-Associated Infection

Declan Power and Jeremy Webb

Abstract Structured biofilm aggregates offer an increased tolerance to antimicrobials, providing either a physical barrier to antimicrobial penetration or leading to physiological adaptations amongst biofilm bacteria that may impact on antibiotic efficacy. Consequently, biofilms are causative of a range of chronic infections where the use of antimicrobials rarely eradicates the underlying infection. One feature of the development of microbial biofilm communities is that they often undergo lifecycle changes between aggregated and planktonic modes of growth. This transition between sessile and motile growth modes is referred to as biofilm dispersal. Understanding and controlling the dispersal process is leading to novel adjunctive strategies to disrupt clinically important biofilms. Exogenous nitric oxide (NO) has been shown to regulate, in a dose-dependent manner, biofilm lifecycle dynamics and can induce the disruption of biofilm aggregates. The use of NO therefore offers a potential therapeutic approach to address the challenge of biofilm-associated antimicrobial tolerance. This chapter explores the current and prospective NO therapies and the underpinning mechanisms of NO-mediated biofilm regulation. It also reviews NO-releasing chemistries, prodrugs, and materials for clinical use.

D. Power
School of Biological Sciences, University of Southampton, Southampton, UK

National Biofilms Innovation Centre, Centre for Biological Sciences, University of Southampton, Southampton, UK

Cystic Fibrosis Trust Strategic Research Centre, Imperial College London, London, UK

J. Webb (✉)
School of Biological Sciences, University of Southampton, Southampton, UK

National Biofilms Innovation Centre, Centre for Biological Sciences, University of Southampton, Southampton, UK
e-mail: J.S.Webb@soton.ac.uk

K. Richter, K. Nørskov Kragh (eds.), *Antibiofilm Strategies*, Springer Series on Biofilms 11, https://doi.org/10.1007/978-3-031-10992-8_19

1 Introduction

Biofilms are structured aggregates of bacterial cells, encased in a self-produced matrix of extracellular polymeric substances (EPS). They offer increased tolerance to antimicrobials and are currently recognised as one of the major contributing factors responsible for chronic infections, with 80% of bacterial infections treated in the developed world caused by organisms growing within biofilms (NIH 2002).

The ESKAPE pathogens (*Enterococcus faecium, Staphylococcus aureus, Klebsiella pneumoniae, Acinetobacter baumannii, Pseudomonas aeruginosa,* and *Enterobacter* species) are commonly associated with multidrug resistance. These pathogens are frequently responsible for hospital-acquired and community-acquired infections and are able to colonise and develop biofilms on natural tissues, medical devices, and implants leading to chronic infection, e.g. diabetic ulcers, burns, surgical wounds, and ventilator-associated pneumonia (Lin et al. 2018; Mulani et al. 2019; Lazar et al. 2021). Notably, *P. aeruginosa* results in increased morbidity and mortality in cystic fibrosis (CF), with chronic *P. aeruginosa* biofilm infection affecting up to 60% of adult patients on the UK CF registry (Cystic Fibrosis Trust 2018). This trend is generally observed on most CF registries worldwide. A multitude of features give biofilms an intrinsically high tolerance to antibiotics, up to 1000-fold, compared to their planktonic counterparts (Lin et al. 2018; Ciofu and Tolker-Nielsen 2019). With the challenges of antimicrobial resistance (AMR) and the high antimicrobial tolerance of biofilms (Ciofu et al. 2022), the treatment options for controlling such chronic infections are limited; thus, there is a significant unmet need for novel therapeutics and alternative anti-biofilm strategies.

The development and lifecycle of biofilm communities is a complex and dynamic process but can broadly be categorised into five distinct stages: (i) transient polar attachment, (ii) irreversible attachment, (iii) proliferation into microcolonies, (iv) a mature biofilm structure formation, and (v) detachment and dispersal (Sauer et al. 2002; Høiby et al. 2010; Lappin-Scott et al. 2014; Hughes and Webber 2017). Detachment can be either active or passive, the latter through external forces such as increased fluid shear or human intervention and debridement (Stoodley et al. 2002; Santajit and Indrawattana 2016). Active detachment (or dispersal) is a result of internal biofilm processes, quorum sensing, intracellular signalling, organised cell death, and endogenous enzymatic degradation (Webb et al. 2003; Hall-Stoodley et al. 2004; Sharma et al. 2019). This end stage of the biofilm lifecycle provides an intriguing target for novel therapeutic interventions. By exploiting the biofilm lifecycle, particularly the detachment and dispersal phases of development, the biofilm biomass can be disrupted, with the released cells reverting to their more antimicrobial susceptible planktonic form.

2 Nitric Oxide-Mediated Biofilm Dispersal

Nitric oxide (NO) is a small diatomic free radical that has an extremely short half-life (<1 s) (Bryan and Grisham, 2007). In eukaryotic cells, it is produced endogenously from L-arginine by a family of enzymes named nitric oxide synthases (NOSs) which exist in 3 isoforms and play important roles in vascular endothelial cells, neuron function, and host immune defence mechanisms (Bredt 1999; Bryan and Grisham 2007). Production of NO by the NOSs of neutrophils may be activated during chronic biofilm infection (Kolpen et al. 2014), and NO has been studied extensively as a signalling molecule in mammalian systems; however, the role of NO signalling in the regulation of bacterial developmental processes within biofilms is a relatively new field of study.

Nitric oxide has relatively recently been discovered as a potentiator of biofilm dispersal. The *P. aeruginosa* biofilm dispersal effect induced by low-dose NO was first described by Barraud et al. (2006). Investigating the role of endogenous NO in biofilm dispersal, reductase-deficient mutants (Δ*nirS*/Δ*norCB*) were shown to have significant differences in dispersal compared to the wild type. Δ*nirS* mutants (unable to produce NO) did not show dispersal after 6 days and formed thicker and more confluent biofilms. On the other hand, Δ*norCB* mutants (unable to reduce NO) showed an increased number of dispersed cells, with multiple hollow voids in biofilm structures, thus implicating the role of NO in biofilm dispersal.

Using an optimised sub-lethal concentration (500 nM) of the NO donor sodium nitroprusside (SNP) resulted in a dispersal response of preestablished biofilms and a tenfold reduction in the biofilm to planktonic ratio. Further to dispersal, the authors combined NO with antimicrobial treatment with tobramycin. Combination treatment saw an enhanced efficacy of the antimicrobials, with the removal of 80% of cells from the biofilm surface. The dispersal effect was linked to increases of cellular motility, with 25% increase of swimming, and 77% swarming motilities with SNP treatment at the same concentration (Barraud et al. 2006).

Following the discovery of the NO-induced dispersal effect in *P. aeruginosa*, Barraud et al. looked at a broader range of clinically relevant, biofilm forming bacteria. *Serratia marcescens*, *Vibrio cholerae*, *Escherichia coli*, *Staphylococcus epidermidis*, *Fusobacterium nucleatum*, and *Candida albicans* were the clinically relevant microbial strains utilised in the study, all of which (to varying degrees) demonstrated a reduction in biofilm surface coverage following treatment with low concentrations of the NO donor, SNP (Barraud et al. 2009). These findings suggested a novel therapeutic use for NO in the control of clinical biofilms.

3 The NO Response Is Dose-Dependent

The effect that NO has on biofilm architecture, development, and dispersal is influenced by the concentration at which NO is delivered. Biofilm dispersal occurs in the nanomolar/low micromolar range dependent on species, and whether anaerobic or aerobically grown (Arora et al. 2015). Even within the same species, if the concentration of NO delivered to the biofilm is not optimised, it can be seen to promote biofilm growth. For example, in *P. aeruginosa*, with sub-lethal concentrations of >25 mM donor SNP (~>2500 nM NO), biofilm growth is enhanced (Barraud et al. 2006). Similarly, different clinical isolates can have varied responses to NO when given the same concentration (Soren et al. 2020). Before low-dose NO therapy is employed, screening of the bacterial response may be beneficial to prevent adverse outcomes and worsening infection.

At higher doses within the micromolar range and above, NO can lead to broad-spectrum antibacterial activity through multiple mechanisms. Toxic derivative NO molecules are created after reacting with oxygen or superoxide. Reactive nitrogen oxide species S-nitrosothiols (RNOS), peroxynitrite ($OONO^-$), nitrogen dioxide (NO_2), and dinitrogen trioxide (N_2O_3) cause oxidative and nitrosative damage leading to DNA cleavage, thiol nitrosation, and destruction of bacterial membranes through lipid peroxidation (Hetrick et al. 2008; Schairer et al. 2012; Rumbaugh and Sauer 2020). Many studies (including some that we refer to in this chapter) have evaluated the use of NO as a therapeutic, but do not necessarily consider whether the reduction in bacterial load is attributed to biofilm-specific activity and dispersal, and/or the bactericidal effects of NO. This understanding and distinction is important because it will impact on the response to treatment and on additional therapeutic requirements and patient outcomes. Furthermore, high-dose NO may have issues with cytotoxicity and generation of reactive nitrogen species in host cells (Vieira et al. 2021). At the time of writing there are examples of ongoing clinical trials evaluating the safety and efficacy of high-dose NO (e.g. ClinicalTrails.gov Identifier: NCT04685720).

4 Mechanism of NO-Mediated Dispersal

Bis-(3′-5′)-cyclic dimeric guanosine monophosphate (c-di-GMP) is an intracellular secondary messenger that regulates many bacterial behaviours such as motility and virulence through multiple complex signalling pathways. Notably, c-di-GMP is key to the phenotypic switch from a planktonic, free-swimming lifestyle to the sessile biofilm phenotype and vice versa with dispersal (Hengge 2009; Romling et al. 2013; Valentini and Filloux 2016; Rinaldo et al. 2018).

The c-di-GMP molecule was first described in 1987 as an allosteric factor to control the activation of cellulose biosynthesis in *Gluconacetobacter xylinus* and has since been implicated in many other bacterial functions across species (Romling et al. 2013). The effects of c-di-GMP are dependent on the intracellular concentration. This is a tightly regulated dynamic equilibrium controlled by the synthesis and degradation by diguanylate cyclases (DGCs) and phosphodiesterases (PDEs), respectively. High intracellular concentrations are favourable for biofilm formation with upregulation of adhesins, exopolysaccharides, and inhibition of motility (Fig. 1), whereas low concentrations have the opposite effect and favour the planktonic lifestyle with increased production of acute virulence factors (Hengge 2009; Wolska et al. 2016; Jenal et al. 2017).

The c-di-GMP is synthesised from two GTP molecules by regulatory proteins: DGCs containing catalytic GGDEF domains. Likewise, the degradation of c-di-GMP is by metal-dependent PDE proteins carrying either the EAL or HD-GYP catalytic domains (Ha and O'Toole 2015). The degradation of c-di-GMP is a two-step process, it is first split into 5′-phosphoguanylyl-(3′-5′)-guanosine (pGpG) by either EAL or HD-GYP catalytic domains; the pGpG is subsequently split into two separate GMP molecules by HD-GYP itself or other non-specific cellular PDEs (Hengge 2009).

The intracellular c-di-GMP concentrations are dependent on DGC and PDE activity as previously discussed. Most of the GGDEF, EAL, and HD-GYP domains are linked to sensory input domains; sensory input to these domains can include oxygen and redox conditions, light, nitric oxide, nutrients, cell membrane stress, surface contact, and antibiotics. Thus, c-di-GMP signalling can be controlled by a multitude of external environmental stimuli (Hengge 2009; Wolska et al. 2016; Jenal et al. 2017).

Several bacterial species, including *Vibrio harveyi*, *Pseudoalteromonas atlantica*, and *Shewanella oneidensis* (Arora and Boon 2012; Plate and Marletta 2012; Henares et al. 2012), have been shown to possess proteins containing known NO-sensing H-NOX (heme-nitric oxide/oxygen) domains. These H-NOX domains have been shown to directly bind NO and regulate c-di-GMP levels and thus biofilm formation (Liu et al. 2012; Plate and Marletta 2012). However, *P. aeruginosa* does not encode for H-NOX proteins, yet still displays a biofilm dispersal response to NO, suggesting the existence of a different NO-sensing protein/s.

More recently, a novel NO-sensing protein (NosP), comparable to that of H-NOX, has been characterised in *P. aeruginosa*, and found widely conversed (more so than H-NOX) across multiple bacterial species (Hossain and Boon 2017; Hossain et al. 2017). NosP is encoded within the same gene cluster as the NosP-associated histidine kinase (NahK), which has been previously implicated in biofilm regulation (Hossain and Boon 2017). With NO bound, NosP has a slow (NO) dissociation rate and is able to control the phosphorelay activity of NahK (Hossain and Boon 2017). *P. aeruginosa* histidine-containing phosphotransfer protein (HptB) is known to accept phosphate from NahK, modulating flagella-related gene expression, and thus biofilm dispersal regulation. It is proposed that in the presence of NO, NosP is able to suppress NahK-mediated phosphorelay to HptB, which ultimately leads to modulation of biofilm in *P. aeruginosa* (Hossain and Boon 2017). The exact mechanisms underlying NO/NosP biofilm dispersal in *P. aeruginosa* are yet to be elucidated.

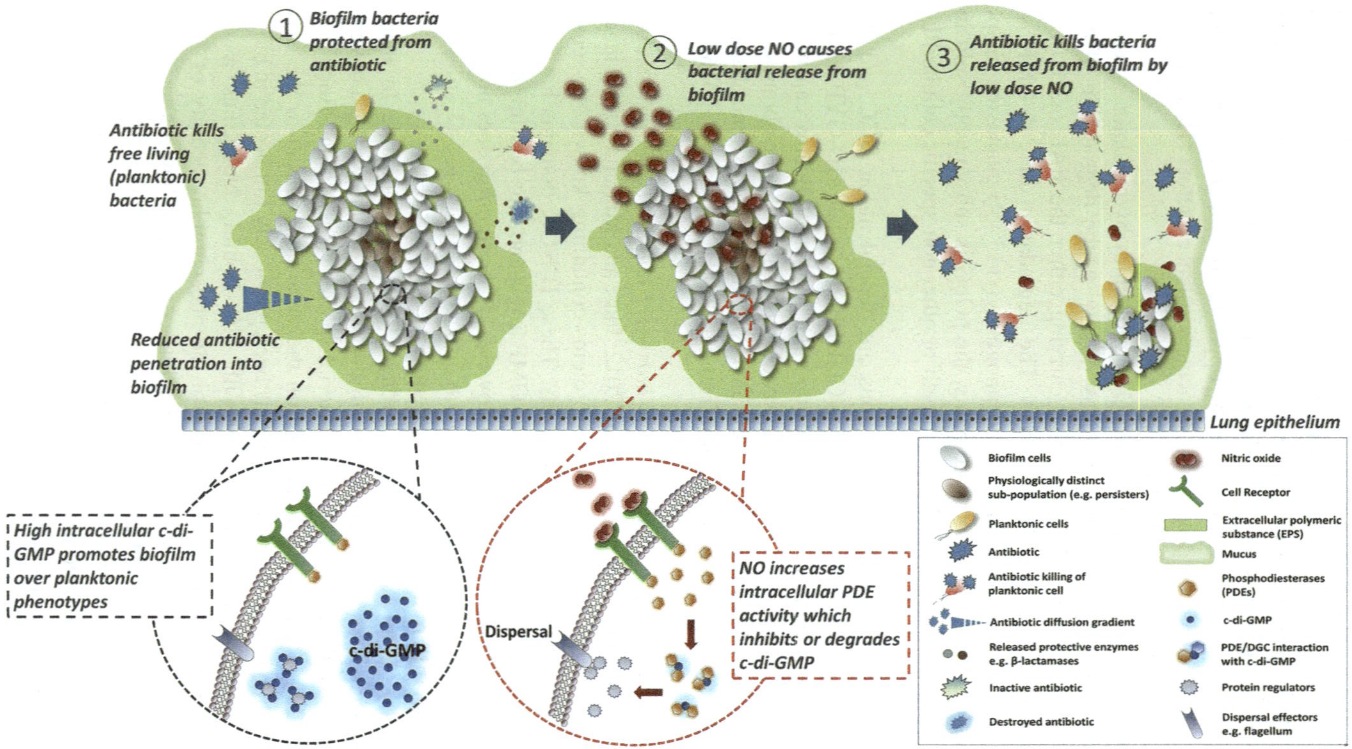

Fig. 1 Role of NO in disrupting antibiotic tolerance mechanisms associated with the biofilm structure. Low-dose NO diffuses into the biofilm. Cell receptors upregulate cellular phosphodiesterases (PDEs) accelerating c-di-GMP degradation. This prevents c-di-GMP from interacting with proteins and leads to cell surface and physiological changes associated with dispersal and motility. Reversion of the bacteria to a planktonic phenotype renders them more susceptible to antibiotic-mediated killing. Reproduced from (Howlin et al. 2017) with permission

5 Direct Use of Gaseous NO

Following the discovery of the NO biofilm dispersal effects, the potential of NO as a therapeutic for *P. aeruginosa* biofilms has been investigated. A proof-of-concept clinical trial involving ex vivo CF sputum samples and a cohort of 12 patients in a randomised double-blind clinical trial was undertaken (Howlin et al. 2017). Ex vivo experiments with fluorescence in situ hybridisation (FISH) of sputum samples showed significant reduction in mean *P. aeruginosa* biofilm thickness when treated with SNP, and again in adjunctive therapy with antibiotics when compared to untreated controls. Twelve patients were randomised to receive either low-dose gaseous NO (gNO) inhalation or a placebo in conjunction with antibiotics. Analysis proved a significant reduction of *P. aeruginosa* biofilm aggregates (10 and 20 cells in size). Unfortunately, complete eradication was not observed, and *P. aeruginosa* biofilm aggregates were again observed in the sputum following cessation of NO treatment.

More recently, gNO has been used in the CF clinic for the treatment of *Burkholderia multivorans* for the first time (Bartley et al. 2020). Following the successes of the above study and others for the safety and efficacy of gNO for bacterial lung infections against *P. aeruginosa* (Miller et al. 2012, 2013), gNO was utilised in a young CF patient with frequent pulmonary exacerbations, and primary *B. multivorans* infection that would only briefly respond to single antibiotic therapy. Intermittent inhalations of 160 ppm NO over a 28-day period increased the patient's pulmonary function and decreased bacterial load, with a notable change in the antibiotic resistance profile of *B. multivorans* (Bartley et al. 2020). This displays promising results for the use of gNO for the management of recalcitrant infection.

Subsequent studies into the use of gNO for bacterial infection have been undertaken (Bentur et al. 2020; Wiegand et al. 2021) and are expected to continue to assess the viability and efficacy of the anti-infective potential in adults with cystic fibrosis. Unfortunately, an ongoing trial (ClinicalTrials.gov Identifier: NCT02498535) has been terminated due to patient safety concerns surrounding the COVID-19 pandemic.

The antimicrobial properties of gNO are not limited to inhalation therapy alone; it can also be applied in the context of chronic wounds. Ghaffari et al. (2007) utilised a rabbit model in which full-thickness wounds were infected with *S. aureus* on the dorsal skin surface. Wounds were treated with 200 ppm gNO for 8 h/day for 3 consecutive days and demonstrated a significant reduction in viable bacterial load, highlighting the potential therapy for non-healing chronic ulcers and burns. The same group had previously demonstrated the potential application of gNO in a wide range of wound pathogens including methicillin-resistant *S. aureus* (MRSA), Group B Streptococcus, *C. albicans*, *E. coli*, and *P. aeruginosa* (Ghaffari et al. 2006).

Whilst exogenous, inhaled NO has demonstrated its potential for clinical anti-biofilm treatment in respiratory disease, the use of gaseous NO has a number of limitations including its expense and additional cost for complex delivery equipment, and is not currently a viable option for routine patient care within a clinical setting or for therapy at home (Yang et al. 2015; Yu et al. 2019). To overcome these issues, some companies have developed technologies to create NO on demand from ambient air. One such company, Beyond Air, has the LungFit® platform, which removes the need for high-pressure cylinders, and devices are able to be used for home care outside of the clinic (Beyond Air Inc. 2022). Clinical trials evaluating this system are ongoing and the technology is awaiting FDA approval. If approved, this may remove a number of the logistical challenges associated with delivery of NO and lead to an increase in the clinical use of NO.

6 Nitric Oxide Donor Chemistries

Many NO donor compounds and chemistries exist that have potential application in the disruption of microbial biofilms. The properties of a number of the key NO-donating materials are summarised here:

6.1 Sodium Nitroprusside

The NO donor first used to study biofilm dispersal effects was sodium nitroprusside (Barraud et al. 2006). The release of NO from SNP is derived from the nitroso or nitrosyl functional groups without the need of enzymatic oxidation (Gresele et al. 2019). The donor SNP has been used extensively in laboratory studies for biofilm dispersal although to our knowledge has not yet been used clinically in this context. Nevertheless, there is a precedent for the human therapeutic use of SNP as it is safely tolerated in the treatment of acute hypertension and controlled hypotension during surgery (Simsic et al. 2003; Drover et al. 2015) which may suggest that clinical studies of anti-biofilm efficacy may be feasible. Cyanide or thiocyanate metabolites of SNP can accumulate however; therefore, blood-cyanide parameters must be continually monitored if used clinically (Shishido and De Oliveira 2001; Cai and Webb 2020).

6.2 Furoxans

Furoxans are heterocyclic compounds characterised by a 1,2,5-oxadiazole 2-oxide ring and two substituent groups at positions 3 and 4 of the furoxan ring and were first synthesised over 100 years ago (Cerecetto and Porcal 2012). The substituents in the furoxan ring determine solubility, rate of NO release, and mechanism of NO release, either spontaneously or thiol activated. Hence hybrid furoxans with the most desirable NO release properties can be designed (Feelisch et al. 1992; Poh et al. 2017). It was recently demonstrated that these modifiable compounds in low concentrations (200 μM) can induce *P. aeruginosa* biofilm dispersal and reduce biofilm biomass by >70% compared to the untreated control after 1 h of treatment. Other furoxans included in the study did not show immediate dispersal effects; however, they prevented biofilm formation over a more prolonged period, and also repressed the production of pyoverdine, an essential siderophore and virulence factor of *P. aeruginosa* (Poh et al. 2017).

6.3 Acidified Nitrite

Sodium nitrite has long been used as a food preservative due to its antimicrobial properties. The antimicrobial properties are, in part, due to the generation of NO under acidic conditions and are known to prevent biofilm formation through interference with respiratory pathways, and generation of nitrosative stress affecting proteins and DNA as previously discussed. It has also been shown to induce biofilm dispersal (Schlag et al. 2007; Zemke et al. 2014, 2015, 2020).

A bactericidal tandem drug AB569, which kills most pathogenic bacteria including antibiotic-resistant *P. aeruginosa* in multiple disease settings, has been discovered by the research group led by Daniel Hassett (2021). AB569 comprises two component parts: acidified nitrite ($A\text{-}NO_2-$) and Na_2-EDTA (disodium ethylenediaminetetraacetic acid). The group reports the synergistic killing of many Gram-negative and Gram-positive organisms by widely available and inexpensive compounds that do not affect mammalian cell viability (McDaniel et al. 2020). The exact mechanism of the synergistic killing remains unclear, but the compound was able to effectively inhibit biofilm formation of *A. baumannii* and induce dispersal to reduce biofilm biomass by 38% (Bogue et al. 2021).

6.4 Diazeniumdiolates (NONOates)

Diazeniumdiolates have been recognised and synthesised since 1960 (Drago and Paulik 1960). However, attention has intensified since the 1990s, when their NO donor properties were considered in biological settings (Maragos et al. 1991).

NONOates are a series of compounds containing an X-[N(O)NO] − functional parent nuclear structure where X is a secondary amine. NONOates are stable in solid state. However, NONOate compounds can rapidly self-decompose when dissolved in aqueous solution and generate up to two molecules of NO per [N(O)NO] − unit under physiological conditions. The half-life and release of NO from NONOates can vary from seconds to hours, but this release rate can be accurately calculated (Davies et al. 2001; Li et al. 2020).

Recently, DETA-NONOate has been investigated as a donor to restore NO to the hyperoxia-compromised innate immunity often seen in ventilator-associated pneumonia and increase the clearance of *P. aeruginosa*. DETA-NONOate treatment significantly increased the survival of mice inoculated with *P. aeruginosa* under hyperoxic conditions, and significantly decreased bacterial loads in the lung, attenuating lung injury (Gore et al. 2020). Commercially available spermine NONOate was illustrated to significantly disperse the biofilms in a panel of CF *P. aeruginosa* isolates within 2 h of low dose [250 μM] treatment in vitro, without the release of cyanide as with SNP (Cai and Webb 2020).

6.5 Cephalosporin-3′-Diazeniumdiolates

Recently produced NO donor prodrugs, cephalosporin-3′-diazeniumdiolates (C3Ds), release NO locally following cleavage of the β-lactam ring by bacterial β-lactamases (Fig. 2), thus triggering biofilm dispersion at the site of chronic infection with maximised NO delivery. The prototype compound used in this study, DEA-C3D ('DiEthylAmin-Cephalosporin-3'-Diazeniumdiolate'), synthesised by Kelso et al. has been shown to disperse PA01 laboratory isolate biofilms of *P. aeruginosa* (Yepuri et al. 2013). Soren et al. (2020) demonstrated the efficacy of C3Ds and investigated the dispersal effect on various clinical *P. aeruginosa* isolates. It was found that DEA-C3D alone was able to significantly reduce the biofilm biomass by 50.2% compared with the untreated control but had no direct antibacterial activity. The authors also demonstrated that when DEA-C3D was administered in conjunction with colistin, the mean average biofilm biomass was reduced 97.8% compared to the untreated control. The mean biofilm thickness was also reduced with a colistin combination, with a decrease in the number of large living cell clusters observed. Thus, it was concluded that DEA-C3D was an effective biofilm dispersal agent particularly when given in combination with colistin and would be favourable over antibiotic treatment alone, albeit with differing responses dependent on clinical isolate (Soren et al. 2020).

The research group led by Kelso subsequently looked to synthesise a next generation "all-in-one" C3D compound with direct anti-biofilm effects and antibacterial killing of dispensed planktonic cells (Rineh et al. 2020). They identified

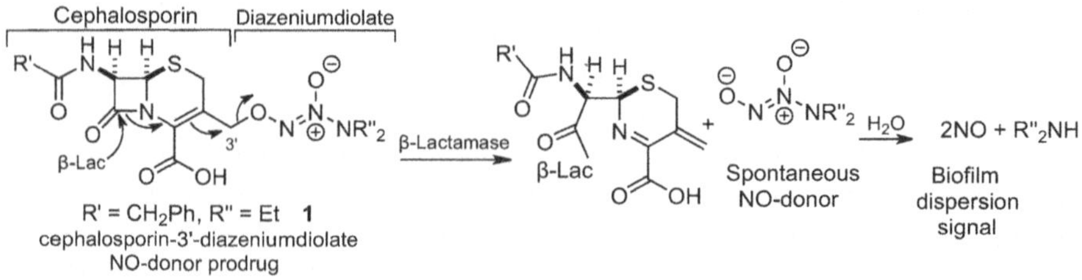

Fig. 2 Proposed mechanism of first-generation cephalosporin-3′-diazeniumdiolates (C3Ds). β-lactamase cleavage of the β-lactam moiety releases nitric oxide locally at the site of infection to mediate *P. aeruginosa* biofilm dispersal. Reproduced from (Yepuri et al. 2013) with permission

and tested a highly active C3D AMINOPIP2-ceftazidime which produced reduction in biofilm biomass suggesting NO release was responsible for the observed effect, supporting the "all-in-one" hypothesis. However, further experimentation utilising the NO scavenger 2-phenyl-4,4,5,5-tetramethylimidazoline-1-oxyl-3-oxide (PTIO) to confirm the involvement of NO in biofilm dispersal as previously described (Barraud et al. 2009) would be beneficial. These preliminary screening experiments of AMINOPIP2-ceftazidime against established biofilms were performed with clinical isolates that were known to undergo low-dose NO-mediated biofilm dispersal. A better representation of the efficacy AMINOPIP2-ceftazidime against biofilms can be achieved with a larger panel of clinical isolates that display differing NO responses as previously shown (Soren et al. 2020).

7 Photodynamic Therapy

Photodynamic therapies (PDT) are precise, controllable, non-invasive therapies that use a photosensitiser (PS) applied or injected into the required areas. The PS is irradiated with specific wavelengths of light and in conjunction with O_2 can produce highly reactive oxygen species in a spatially confined manner (de Melo et al. 2013; Hu et al. 2018). Photodynamic therapies have been used in the past for tumour destruction and have shown bactericidal properties; however, due to increased tolerances of biofilms to antimicrobial agents, the PS concentrations required to treat chronic biofilm infection could lead to complications with cytotoxicity. Rather than using PS alone, coupling the PS with an NO donor would first disperse cells from the biofilm, and subsequently kill the new planktonic population, allowing concentrations to be kept low enough so not to cause cytotoxicity (Zhao et al. 2021).

Zhao et al. (2021) demonstrated such effect with NO/PS, with a novel ruthenium ii-based compound with boric acid and light responsive NO release properties. The authors showed that the positively charged boric acid within the PS structure allowed for selective, preferential binding to the negatively charged outer membrane of bacteria within the biofilms through boron-polyol based boronolectin chemistry (Yang et al. 2017). The strict photosensitive nature of the PS/NO resulted in accurate control of NO release with 425 nm light irradiation. These PS compounds were evaluated against established *P. aeruginosa* biofilms and were able to induce dispersal, reduce bacterial viability by 90%, and reduce the biofilm biomass by 80% under irradiation, giving better efficacy than the NONOate donor also tested. Biofilms formed by *S. aureus* were also included, whilst PS/NO treatment was not as effective in *P. aeruginosa* biofilm biomass; it was nevertheless reduced by 50%. The dose-dependent response of the PS/NO compound was comparable to that of colistin, highlighting the potential preferential treatment over an antibiotic drug of last resort.

Similarly, a group of researchers in China have reported a nitric oxide (NO)-releasing gold nanocage (AuNC@NO) that is stimulated by near-infrared (NIR) irradiation to deliver NO and generate hyperthermia for biofilm elimination. The

AuNCs contain a void space in which NO can be immobilised with cupferron (Cup), an ammonium salt of N-nitroso-N-phenylhydroxylamine, which provides temperature-sensitive NO donation. At physiological conditions, slow release of sustained NO promoted dispersal of MRSA. Upon NIR irradiation, AuNC quickly heated accelerating the release of NO for bacterial killing with ROS, achieving 4 orders of magnitude bacterial reduction and 85.4% biofilm elimination (Tang et al. 2021). PDT coupled with NO donors offers a highly effective, localised treatment for clinical biofilms for a multitude of organisms.

8 NO-Releasing Dressings

In order to control and prolong the release of NO, film-type delivery systems embedded into wound dressings have been explored. These film-type dressings are of particular interest as they are able to maintain a moist environment that facilitates wound healing (Matica et al. 2019). Multiple studies have explored the intrinsic antimicrobial properties of chitosan alone, and coupled with NO-releasing properties for the treatment of chronic wound infection (Kim et al. 2015; Yang et al. 2018; Alven and Aderibigbe 2020; Hasan et al. 2022).

In particular, Choi et al. (2020) recently demonstrated that S-nitrosoglutathione (GSNO) can be incorporated into a film of the natural biopolymer, chitosan (CS), and can be applied as a topical wound dressing and deliver a sustained NO release. Their study focused on the synergistic effects of CS and GNSO as a CS/NO film for the treatment of MRSA biofilm-infected wounds. A 20% GNSO content in the CS film saw a >3 log reduction in bacterial viability in vitro (~99.9% killing), and a 59% reduction in biofilm biomass after 48 hrs of CS/NO treatment compared to a 38.5% reduction with CS alone. Murine models of MRSA biofilm infection in diabetic and non-diabetic wounds also demonstrated the efficacy of CS/NO films with reduction in bacterial viability by >6 logs (99.9999% bacterial viability) by day 15 post-injury highlighting the enhanced antibacterial and anti-biofilm effect, and promotion of wound healing with NO incorporation (Choi et al. 2020).

Maloney et al. (2021) have created an NO-releasing hyaluronic acid (HA) for the treatment of chronic wounds. The modified HA contains two N-diazeniumdiolate donors and is capable of decreasing bacterial viability in common wound pathogens and exerts bactericidal properties against both susceptible and multidrug-resistant *P. aeruginosa* in vitro. The in vivo murine modelling of *P. aeruginosa*-infected wounds showed that the modified HA reduced the relative abundance of bacterial genomic material; however, no further experiments were performed to address the viability of the remaining bacterial cells in the wound.

9 Safety Considerations and NO-Mediated Dispersal as an Adjunctive Strategy

The nature of NO-mediated biofilm disruption and dispersal introduces considerations for its use therapeutically. Biofilm dispersal (without co-administration of antibiotic treatments) has previously been implicated in the spread of systemic infection in mice. Following the induction of dispersal with glycoside hydrolases, fatal systemic septicaemia with *P. aeruginosa* was observed (Fleming and Rumbaugh 2018). The authors also identified that this effect was aided by bacterial motility. Highly motile progeny cells dispersed from the biofilm were able to disseminate quickly from the initial localised infection and colonise elsewhere. Dispersed cells were also shown to have a distinct phenotype from the initial planktonic colonisers or within the biofilm, with large-scale changes in the transcriptome, resulting in more virulent cells, more effective in penetrating and killing macrophages, which has the potential to worsen clinical outcomes (Chua et al. 2014; Guilhen et al. 2017; Wille and Coenye 2020). Therefore, utilising NO in low doses to disrupt biofilms should be considered as an adjunctive strategy, in combination with antimicrobial treatments and with appropriate safety studies to ensure that the dispersed population of biofilms cells can be removed.

Despite the promising anti-biofilm nature of low-dose NO therapies, the issue of AMR remains. Dispersed cell populations may still have intrinsic resistance to antimicrobials, or may have acquired antimicrobial resistance genes whilst in the biofilm structure via conjugation by plasmids, transduction by bacteriophages, or natural transformation by extracellular DNA (Schroeder et al. 2017; Lerminiaux and Cameron 2019; Flores-Vargas et al. 2021). This may contribute to the issues of dissemination of biofilm infection if the dispersed population cannot be eradicated effectively. However, the use of NO causes damage to the bacterial membrane as discussed, and has been shown to potentiate the action of antimicrobials, even those classified as multidrug resistant (Rouillard et al. 2021). Some of the therapies discussed in this chapter address alternative methods of the killing of dispersed cell populations, namely AuNC PDT that utilise NO as the bactericidal factor. It has been previously demonstrated that spontaneous and serial passage mutagenesis assays with repeated NO exposure resulted in no increase in minimum inhibitory concentration of NO in vitro, supporting the use of NO as a clinical antimicrobial therapy (Privett et al. 2012).

10 Conclusions/Future Perspectives

The mechanisms that underpin biofilm dispersal represent novel therapeutic targets to address clinical challenges associated with biofilms. NO-mediated control of biofilm dispersal is one such promising strategy and a range of laboratory, pre-clinical, and clinical studies have unravelled the underpinning mechanisms and have demonstrated the potential for NO to disrupt biofilms and enhance their susceptibility to antimicrobial compounds. Clinical NO delivery through traditional high-pressure gaseous cylinders may soon be complemented by new devices that can generate NO from air in a controlled way and which may allow for far more convenient delivery of NO in therapeutic applications. Novel NO donor chemistries can provide targeted NO delivery at the site of infection and offer more precise and controllable spatiotemporal NO release limiting effects in undesirable locations. This is particularly applicable to PDT; unfortunately, these are limited for topical use by the need for light irradiation. Therapeutics that utilise NO in low doses are not to be viewed as a sole treatment strategy, but as an adjunctive in combination with agents that are able to limit the dissemination of the dispersed population. Further clinical trials will be needed to assess safety and tolerance of these therapies in vivo, and considerations surrounding the effects of biofilm dispersal on patient outcomes are required for such methods to be implemented in the clinic.

References

Alven S, Aderibigbe BA (2020) Chitosan and cellulose-based hydrogels for wound management. Int J Mol Sci 21:1–30. https://doi.org/10.3390/IJMS21249656

Arora DP, Boon EM (2012) Nitric oxide regulated two-component signaling in Pseudoalteromonas atlantica. Biochem Biophys Res Commun 421:521–526. https://doi.org/10.1016/j.bbrc.2012.04.037

Arora DP, Hossain S, Xu Y, Boon EM (2015) Nitric oxide regulation of bacterial biofilms. Biochemistry 54:3717–3728. https://doi.org/10.1021/bi501476n

Barraud N, Hassett DJ, Hwang SH et al (2006) Involvement of nitric oxide in biofilm dispersal of Pseudomonas aeruginosa. J Bacteriol 188:7344–7353. https://doi.org/10.1128/JB.00779-06

Barraud N, Storey MV, Moore ZP et al (2009) Nitric oxide-mediated dispersal in single- and multi-species biofilms of clinically and industrially relevant microorganisms. Microb Biotechnol 2: 370–378. https://doi.org/10.1111/j.1751-7915.2009.00098.x

Bartley BL, Gardner KJ, Spina S et al (2020) High-dose inhaled nitric oxide as adjunct therapy in cystic fibrosis targeting Burkholderia multivorans. Case Rep Pediatr 2020:1–6. https://doi.org/10.1155/2020/1536714

Bentur L, Gur M, Ashkenazi M et al (2020) Pilot study to test inhaled nitric oxide in cystic fibrosis patients with refractory Mycobacterium abscessus lung infection. J Cyst Fibros 19:225–231. https://doi.org/10.1016/j.jcf.2019.05.002

Beyond Air Inc (2022) Differentiated nitric oxide technology, generated from ambient air. https://www.beyondair.net/technology

Bogue AL, Panmanee W, McDaniel CT et al (2021) AB569, a non-toxic combination of acidified nitrite and EDTA, is effective at killing the notorious Iraq/Afghanistan combat wound pathogens, multi-drug resistant Acinetobacter baumannii and Acinetobacter spp. PLoS One 16: e0247513. https://doi.org/10.1371/JOURNAL.PONE.0247513

Bredt DS (1999) Endogenous nitric oxide synthesis: biological functions and pathophysiology. Free Radic Res 31:577–596. https://doi.org/10.1080/10715769900301161

Bryan NS, Grisham MB (2007) Methods to detect nitric oxide and its metabolites in biological samples. Free Radic Biol Med 43:645–657

Cai YM, Webb JS (2020) Optimization of nitric oxide donors for investigating biofilm dispersal response in Pseudomonas aeruginosa clinical isolates. Appl Microbiol Biotechnol 104:8859. https://doi.org/10.1007/S00253-020-10859-7

Cerecetto H, Porcal W (2012) Pharmacological properties of furoxans and benzofuroxans: recent developments. Mini Rev Med Chem 5:57–71. https://doi.org/10.2174/1389557053402864

Choi M, Hasan N, Cao J et al (2020) Chitosan-based nitric oxide-releasing dressing for anti-biofilm and in vivo healing activities in MRSA biofilm-infected wounds. Int J Biol Macromol 142:680–692. https://doi.org/10.1016/j.ijbiomac.2019.10.009

Chua SL, Liu Y, Yam JKH et al (2014) Dispersed cells represent a distinct stage in the transition from bacterial biofilm to planktonic lifestyles. Nat Commun 5:1–12. https://doi.org/10.1038/ncomms5462

Ciofu O, Tolker-Nielsen T (2019) Tolerance and resistance of Pseudomonas aeruginosa biofilms to antimicrobial agents—how P. aeruginosa can escape antibiotics. Front Microbiol 10(913). https://doi.org/10.3389/fmicb.2019.00913

Ciofu O, Moser C, Jensen PØ, Høiby N (2022) Tolerance and resistance of microbial biofilms. Nat Rev Microbiol. https://doi.org/10.1038/s41579-022-00682-4

ClinicalTrials.gov (2015) Efficacy and safety of inhaled nitric oxide (NO) in cystic fibrosis (CF) patients. Identifier NCT02498535. https://clinicaltrials.gov/ct2/show/NCT02498535. Accessed 19 Jan 2022

ClinicalTrials.gov (2020) A pilot study to assess the effect of intermittent iNO on the treatment of NTM lung infection in CF and non-CF patients. Identifier NCT04685720. https://clinicaltrials.gov/ct2/show/NCT04685720. Accessed 19 Jan 2022

Cystic Fibrosis Trust (2018) UK cystic fibrosis registry annual data report 2018

Davies KM, Wink DA, Saavedra JE, Keefer LK (2001) Chemistry of the diazeniumdiolates. 2. Kinetics and mechanism of dissociation to nitric oxide in aqueous solution. J Am Chem Soc 123:5473–5481. https://doi.org/10.1021/ja002899q

de Melo WCMA, Avci P, de Oliveira MN et al (2013) Photodynamic inactivation of biofilm: taking a lightly colored approach to stubborn infection. Expert Rev Anti Infect Ther 11:669–693. https://doi.org/10.1586/14787210.2013.811861

Drago RS, Paulik FE (1960) The reaction of nitrogen(II) oxide with diethylamine. J Am Chem Soc 82:96–98. https://doi.org/10.1021/ja01486a021

Drover DR, Hammer GB, Barrett JS et al (2015) Evaluation of sodium nitroprusside for controlled hypotension in children during surgery. Front Pharmacol 6:136. https://doi.org/10.3389/fphar.2015.00136

Feelisch M, Schönafingeri K, Noack H (1992) Thiol-mediated generation of nitric oxide accounts for the vasodilator action of furoxans. Biochem Pharmacol 44:1149–1157. https://doi.org/10.1016/0006-2952(92)90379-W

Fleming D, Rumbaugh K (2018) The consequences of biofilm dispersal on the host. Sci Rep 8. https://doi.org/10.1038/S41598-018-29121-2

Flores-Vargas G, Bergsveinson J, Lawrence JR, Korber DR (2021) Environmental biofilms as reservoirs for antimicrobial resistance. Front Microbiol 12:3880

Ghaffari A, Miller CC, McMullin B, Ghahary A (2006) Potential application of gaseous nitric oxide as a topical antimicrobial agent. Nitric Oxide 14:21–29. https://doi.org/10.1016/J.NIOX.2005.08.003

Ghaffari A, Jalili R, Ghaffari M et al (2007) Efficacy of gaseous nitric oxide in the treatment of skin and soft tissue infections. Wound Repair Regen 15:368–377. https://doi.org/10.1111/j.1524-475X.2007.00239.x

Gore A, Gauthier AG, Lin M et al (2020) The nitric oxide donor, (Z)-1-[N-(2-aminoethyl)-N-(2-ammonioethyl)amino]diazen-1-ium-1,2-diolate (DETA-NONOate/D-NO), increases survival by attenuating hyperoxia-compromised innate immunity in bacterial clearance in a mouse model of ventilator-associated. Biochem Pharmacol 176:113817. https://doi.org/10.1016/j.bcp.2020.113817

Gresele P, Momi S, Guglielmini G (2019) Nitric oxide-enhancing or -releasing agents as antithrombotic drugs. Biochem Pharmacol 166:300–312. https://doi.org/10.1016/J.BCP.2019.05.030

Guilhen C, Forestier C, Balestrino D (2017) Biofilm dispersal: multiple elaborate strategies for dissemination of bacteria with unique properties. Mol Microbiol 105:188–210. https://doi.org/10.1111/MMI.13698

Ha D-G, O'Toole GA (2015) c-di-GMP and its effects on biofilm formation and dispersion: a pseudomonas aeruginosa review. Microbiol Spectr 3. https://doi.org/10.1128/microbiolspec.mb-0003-2014

Hall-Stoodley L, Costerton JW, Stoodley P (2004) Bacterial biofilms: from the natural environment to infectious diseases. Nat Rev Microbiol 2:95–108

Hasan N, Lee J, Ahn HJ et al (2022) Nitric oxide-releasing bacterial cellulose/chitosan crosslinked hydrogels for the treatment of polymicrobial wound infections. Pharmaceutics 14:22. https://doi.org/10.3390/PHARMACEUTICS14010022

Hassett DJ, Kovall RA, Schurr MJ et al (2021) The bactericidal tandem drug, AB569: how to eradicate antibiotic-resistant biofilm Pseudomonas aeruginosa in multiple disease settings including cystic fibrosis, burns/wounds and urinary tract infections. Front Microbiol 12:1003. https://doi.org/10.3389/fmicb.2021.639362

Henares BM, Higgins KE, Boon EM (2012) Discovery of a nitric oxide responsive quorum sensing circuit in Vibrio harveyi. ACS Chem Biol 7:1331–1336. https://doi.org/10.1021/cb300215t

Hengge R (2009) Principles of c-di-GMP signalling in bacteria. Nat Rev Microbiol 7:263–273

Hetrick EM, Shin JH, Stasko NA et al (2008) Bactericidal efficacy of nitric oxide-releasing silica nanoparticles. ACS Nano 2:235–246. https://doi.org/10.1021/nn700191f

Høiby N, Bjarnsholt T, Givskov M et al (2010) Antibiotic resistance of bacterial biofilms. Int J Antimicrob Agents 35:322–332. https://doi.org/10.1016/j.ijantimicag.2009.12.011

Hossain S, Boon EM (2017) Discovery of a novel nitric oxide binding protein and nitric-oxide-responsive signaling pathway in Pseudomonas aeruginosa. ACS Infect Dis 3:454–461. https://doi.org/10.1021/acsinfecdis.7b00027

Hossain S, Nisbett LM, Boon EM (2017) Discovery of two bacterial nitric oxide-responsive proteins and their roles in bacterial biofilm regulation. Acc Chem Res 50:1633–1639. https://doi.org/10.1021/acs.accounts.7b00095

Howlin RP, Cathie K, Hall-Stoodley L et al (2017) Low-dose nitric oxide as targeted anti-biofilm adjunctive therapy to treat chronic Pseudomonas aeruginosa infection in cystic fibrosis. Mol Ther 25:2104–2116. https://doi.org/10.1016/j.ymthe.2017.06.021

Hu X, Huang YY, Wang Y et al (2018) Antimicrobial photodynamic therapy to control clinically relevant biofilm infections. Front Microbiol 9:1299

Hughes G, Webber MA (2017) Novel approaches to the treatment of bacterial biofilm infections. Br J Pharmacol 174:2237. https://doi.org/10.1111/BPH.13706

Jenal U, Reinders A, Lori C (2017) Cyclic di-GMP: second messenger extraordinaire. Nat Rev Microbiol 15:271–284

Kim JO, Noh JK, Thapa RK et al (2015) Nitric oxide-releasing chitosan film for enhanced antibacterial and in vivo wound-healing efficacy. Int J Biol Macromol 79:217–225. https://doi.org/10.1016/j.ijbiomac.2015.04.073

Kolpen M, Kühl M, Bjarnsholt T et al (2014) Nitrous oxide production in sputum from cystic fibrosis patients with chronic Pseudomonas aeruginosa lung infection. PLoS One 9. https://doi.org/10.1371/journal.pone.0084353

Lappin-Scott H, Burton S, Stoodley P (2014) Revealing a world of biofilms-the pioneering research of Bill Costerton. Nat Rev Microbiol 12:781–787

Lazar V, Holban AM, Curutiu C, Chifiriuc MC (2021) Modulation of quorum sensing and biofilms in less investigated gram-negative ESKAPE pathogens. Front Microbiol 12:2072

Lerminiaux NA, Cameron ADS (2019) Horizontal transfer of antibiotic resistance genes in clinical environments. Can J Microbiol 65:34–44. https://doi.org/10.1139/cjm-2018-0275

Li B, Ming Y, Liu Y et al (2020) Recent developments in pharmacological effect, mechanism and application prospect of diazeniumdiolates. Front Pharmacol 11. https://doi.org/10.3389/fphar.2020.00923

Lin Q, Deslouches B, Montelaro RC, Di YP (2018) Prevention of ESKAPE pathogen biofilm formation by antimicrobial peptides WLBU2 and LL37. Int J Antimicrob Agents 52:667–672. https://doi.org/10.1016/j.ijantimicag.2018.04.019

Liu N, Xu Y, Hossain S et al (2012) Nitric oxide regulation of cyclic di-GMP synthesis and hydrolysis in Shewanella woodyi. Biochemistry 51:2087–2099. https://doi.org/10.1021/bi201753f

Maloney SE, McGrath KV, Ahonen MJR et al (2021) Nitric oxide-releasing hyaluronic acid as an antibacterial agent for wound therapy. Biomacromolecules 22:867–879. https://doi.org/10.1021/acs.biomac.0c01583

Maragos CM, Morley D, Wink DA et al (1991) Complexes of NO with nucleophiles as agents for the controlled biological release of nitric oxide. Vasorelaxant effects. J Med Chem 34:3242–3247. https://doi.org/10.1021/jm00115a013

Matica MA, Aachmann FL, Tøndervik A et al (2019) Chitosan as a wound dressing starting material: antimicrobial properties and mode of action. Int J Mol Sci 20. https://doi.org/10.3390/IJMS20235889

McDaniel CT, Panmanee W, Winsor GL et al (2020) AB569, a nontoxic chemical tandem that kills major human pathogenic bacteria. Proc Natl Acad Sci U S A 117:4921–4930. https://doi.org/10.1073/pnas.1911927117

Miller C, Miller M, McMullin B et al (2012) A phase I clinical study of inhaled nitric oxide in healthy adults. J Cyst Fibros 11:324–331. https://doi.org/10.1016/j.jcf.2012.01.003

Miller CC, Hergott CA, Rohan M et al (2013) Inhaled nitric oxide decreases the bacterial load in a rat model of Pseudomonas aeruginosa pneumonia. J Cyst Fibros 12:817–820. https://doi.org/10.1016/J.JCF.2013.01.008

Mulani MS, Kamble EE, Kumkar SN et al (2019) Emerging strategies to combat ESKAPE pathogens in the era of antimicrobial resistance: a review. Front Microbiol 10:539. https://doi.org/10.3389/FMICB.2019.00539

NIH (2002) Research on microbial biofilms: PA NUMBER: PA-03-047. National Institutes of Health

Plate L, Marletta MA (2012) Nitric oxide modulates bacterial biofilm formation through a multicomponent cyclic-di-GMP signaling network. Mol Cell 46:449–460. https://doi.org/10.1016/j.molcel.2012.03.023

Poh WH, Barraud N, Guglielmo S et al (2017) Furoxan nitric oxide donors disperse Pseudomonas aeruginosa biofilms, accelerate growth, and repress Pyoverdine production. ACS Chem Biol 12: 2097–2106. https://doi.org/10.1021/acschembio.7b00256

Privett BJ, Broadnax AD, Bauman SJ et al (2012) Examination of bacterial resistance to exogenous nitric oxide. Nitric Oxide 26:169–173. https://doi.org/10.1016/j.niox.2012.02.002

Rinaldo S, Giardina G, Mantoni F et al (2018) Beyond nitrogen metabolism: nitric oxide, cyclic-di-GMP and bacterial biofilms. FEMS Microbiol, Lett, p 365

Rineh A, Soren O, McEwan T et al (2020) Discovery of Cephalosporin-3′-Diazeniumdiolates that show dual antibacterial and antibiofilm effects against Pseudomonas aeruginosa clinical cystic fibrosis isolates and efficacy in a murine respiratory infection model. ACS Infect Dis 6:1460–1479. https://doi.org/10.1021/acsinfecdis.0c00070

Romling U, Galperin MY, Gomelsky M (2013) Cyclic di-GMP: the first 25 years of a universal bacterial second messenger. Microbiol Mol Biol Rev 77:1–52. https://doi.org/10.1128/mmbr.00043-12

Rouillard KR, Novak OP, Pistiolis AM et al (2021) Exogenous nitric oxide improves antibiotic susceptibility in resistant bacteria. ACS Infect Dis 7:23–33. https://doi.org/10.1021/acsinfecdis.0c00337

Rumbaugh KP, Sauer K (2020) Biofilm dispersion. Nat Rev Microbiol 18:571–586

Santajit S, Indrawattana N (2016) Mechanisms of antimicrobial resistance in ESKAPE pathogens. Biomed Res Int 2016:1–8. https://doi.org/10.1155/2016/2475067

Sauer K, Camper AK, Ehrlich GD et al (2002) Pseudomonas aeruginosa displays multiple phenotypes during development as a biofilm. J Bacteriol 184:1140–1154. https://doi.org/10.1128/jb.184.4.1140-1154.2002

Schairer DO, Chouake JS, Nosanchuk JD, Friedman AJ (2012) The potential of nitric oxide releasing therapies as antimicrobial agents. Virulence 3:271–279

Schlag S, Nerz C, Birkenstock TA et al (2007) Inhibition of staphylococcal biofilm formation by nitrite. J Bacteriol 189:7911–7919. https://doi.org/10.1128/JB.00598-07

Schroeder M, Brooks BD, Brooks AE (2017) The complex relationship between virulence and antibiotic resistance. Genes (Basel) 8:39

Sharma D, Misba L, Khan AU (2019) Antibiotics versus biofilm: an emerging battleground in microbial communities. Antimicrob Resist Infect Control 8:76. https://doi.org/10.1186/s13756-019-0533-3

Shishido SM, De Oliveira MG (2001) Photosensitivity of aqueous sodium nitroprusside solutions: nitric oxide release versus cyanide toxicity. Prog React Kinet Mech 26:239–261. https://doi.org/10.3184/007967401103165271

Simsic JM, Bradley SM, Mulvihill DM (2003) Sodium nitroprusside infusion after bidirectional superior cavopulmonary connection: preserved cerebral blood flow velocity and systemic oxygenation. J Thorac Cardiovasc Surg 126:186–190. https://doi.org/10.1016/S0022-5223(03)00582-8

Soren O, Rineh A, Silva DG et al (2020) Cephalosporin nitric oxide-donor prodrug DEA-C3D disperses biofilms formed by clinical cystic fibrosis isolates of Pseudomonas aeruginosa. J Antimicrob Chemother 75:117–125. https://doi.org/10.1093/jac/dkz378

Stoodley P, Cargo R, Rupp CJ et al (2002) Biofilm material properties as related to shear-induced deformation and detachment phenomena. J Ind Microbiol Biotechnol 29:361–367

Tang Y, Wang T, Feng J et al (2021) Photoactivatable nitric oxide-releasing gold nanocages for enhanced hyperthermia treatment of biofilm-associated infections. ACS Appl Mater Interfaces 13:50668–50681. https://doi.org/10.1021/acsami.1c12483

Valentini M, Filloux A (2016) Biofilms and cyclic di-GMP (c-di-GMP) signaling: lessons from Pseudomonas aeruginosa and other bacteria. J Biol Chem 291:12547–12555

Vieira F, Makoni M, Szyld E, Sekar K (2021) The controversy persists: is there a qualification criterion to utilize inhaled nitric oxide in pre-term newborns? Front Pediatr 9:212

Webb JS, Thompson LS, James S et al (2003) Cell death in Pseudomonas aeruginosa biofilm development. J Bacteriol 185:4585–4592. https://doi.org/10.1128/JB.185.15.4585-4592.2003

Wiegand SB, Traeger L, Nguyen HK et al (2021) Antimicrobial effects of nitric oxide in murine models of Klebsiella pneumonia. Redox Biol 39. https://doi.org/10.1016/j.redox.2020.101826

Wille J, Coenye T (2020) Biofilm dispersion: the key to biofilm eradication or opening Pandora's box? Biofilms 2:100027. https://doi.org/10.1016/J.BIOFLM.2020.100027

Wolska KI, Grudniak AM, Rudnicka Z, Markowska K (2016) Genetic control of bacterial biofilms. J Appl Genet 57:225–238

Yang Y, Qi PK, Yang ZL, Huang N (2015) Nitric oxide based strategies for applications of biomedical devices. Biosurface Biotribol 1:177–201. https://doi.org/10.1016/j.bsbt.2015.08.003

Yang P, Bam M, Pageni P et al (2017) Trio act of boronolectin with antibiotic-metal complexed macromolecules toward broad-spectrum antimicrobial efficacy. ACS Infect Dis 3:845–853. https://doi.org/10.1021/acsinfecdis.7b00132

Yang L, Feura ES, Ahonen MJR, Schoenfisch MH (2018) Nitric oxide-releasing macromolecular scaffolds for antibacterial applications. Adv Healthc Mater 7:e1800155. https://doi.org/10.1002/ADHM.201800155

Yepuri NR, Barraud N, Mohammadi NS et al (2013) Synthesis of cephalosporin-3-′-diazeniumdiolates: biofilm dispersing NO-donor prodrugs activated by β-lactamase. Chem Commun 49:4791–4793. https://doi.org/10.1039/c3cc40869h

Yu B, Ichinose F, Bloch DB, Zapol WM (2019) Inhaled nitric oxide. Br J Pharmacol 176:246–255

Zemke AC, Shiva S, Burns JL et al (2014) Nitrite modulates bacterial antibiotic susceptibility and biofilm formation in association with airway epithelial cells. Free Radic Biol Med 77:307. https://doi.org/10.1016/J.FREERADBIOMED.2014.08.011

Zemke AC, Gladwin MT, Bomberger JM (2015) Sodium nitrite blocks the activity of aminoglycosides against Pseudomonas aeruginosa biofilms. Antimicrob Agents Chemother 59:3329. https://doi.org/10.1128/AAC.00546-15

Zemke AC, D'Amico EJ, Snell EC et al (2020) Dispersal of epithelium-associated Pseudomonas aeruginosa biofilms. mSphere 5:10.1128/mSphere.00630-20

Zhao Z, Li H, Tao X et al (2021) Light-triggered nitric oxide release by a photosensitizer to combat bacterial biofilm infections. Chemistry 27:5453–5460. https://doi.org/10.1002/chem.202004698

Biofouling Control in Water Filtration Systems

Andreas Møllebjerg and Rikke Louise Meyer

Abstract Biofouling is one of the biggest impediments to the production of clean and safe water by membrane filtration. As contaminated water is filtered through the membrane, organic matter and microorganisms accumulate on the surface to form a biofilm which obstructs the water flow. A range of strategies have been developed to tackle this issue. Microorganisms and their nutrients can be removed from the feed water by different pretreatment methods to limit colonization and proliferation of microorganisms on the membrane. Furthermore, the membrane surface can be modified to achieve anti-adhesive, biocidal, or quorum quenching properties. These fouling prevention strategies can only delay the development of a biofilm, and the membranes must eventually be cleaned. Physical cleaning methods include air bubbles or ultrasound, but the most commonly used methods use chemical cleaning by applying biocides, acids, caustics, chelators, and surfactants. Chemical cleaning agents come with environmental and economic drawbacks, and biological cleaning solutions are therefore under development. These solutions include treatment with bacteriophages, quorum sensing inhibitors, and matrix-degrading enzymes. Despite extensive research on combatting biofouling, there is still no anti-fouling panacea. However, the additive effect of feedwater pretreatment, anti-adhesive membrane surfaces and the use of gentle biological membrane cleaning methods is bringing the industry closer to long-lasting mitigation of biofouling.

A. Møllebjerg
Interdisciplinary Nanoscience Center (iNANO), Aarhus University, Aarhus C, Denmark

Novozymes A/S, Kongens Lyngby, Denmark

R. L. Meyer (✉)
Interdisciplinary Nanoscience Center (iNANO), Aarhus University, Aarhus C, Denmark

Department of Biology, Aarhus University, Aarhus C, Denmark
e-mail: rikke.meyer@inano.au.dk

K. Richter, K. Nørskov Kragh (eds.), *Antibiofilm Strategies*, Springer Series on Biofilms 11, https://doi.org/10.1007/978-3-031-10992-8_20

1 Introduction

The global demand for clean water is rapidly increasing due to rising demands from emerging economies. Meanwhile, freshwater availability is decreasing due to pollution and climate change. The disparity between supply and demand of clean water is one of the major challenges to global sustainable growth and ensuring access to clean water is one of UN's sustainable development goals (Lee et al. 2011). Efficient and environmentally friendly methods of generating clean water are sorely needed. In many regions, ground water or surface freshwater can be used directly with minimal treatment. In regions with scarce freshwater sources, seawater can be desalinated, or wastewater can even be cleaned sufficiently to reclaim potable water. Water from all these sources can be cleaned by membrane technology. Among the various filtration technologies, nanofiltration (NF) and reverse osmosis (RO) membranes are the predominant and least energy-consuming methods for potable water production (Shannon et al. 2008; Fritzmann et al. 2007).

The Achilles heel of membrane filtration is the clogging of membranes due to biofouling (Flemming et al. 1997). Combatting fouling requires many resources, including membrane cleaning, pretreatment, manual labor, and chemical purchase and disposal. As fouling accumulates, the hydraulic resistance increases, which in turn increases energy consumption. Membranes must therefore be cleaned regularly and at great cost in the form of manpower and shorter lifetime of membranes due to oxidative damage by cleaning chemicals. The cumulative cost of these burdens contribute ~30% of the operating cost of membrane filtration systems (Flemming et al. 1994; Flemming 1997). In this chapter, we will provide an overview of industrial anti-fouling strategies through the lens of tackling fouling of water filtration membranes, and we discuss innovations developed to mitigate biofouling in water filtration. Novel technologies that could decrease the economic and environmental footprint of membrane biofouling will have a significant global impact.

2 Fouling of RO Membranes

Fouling can be divided into colloidal fouling, inorganic fouling, organic fouling, and biofouling. Colloidal fouling is the deposition of suspended, non-biological particles on the surface; inorganic fouling is the precipitation of inorganic compounds (also known as scaling); organic fouling is deposition of organic, non-cellular matter; and biofouling is the adhesion and proliferation of microorganisms, algae, and spores or larvae of higher organisms. These four phenomena have been studied separately, and their respective mechanisms are well understood. However, in water filtration systems all four fouling categories are interdependent and can happen simultaneously. This results in biofilms with incorporated inorganic solids, colloidal particles, and recalcitrant organic chemicals. In most fouled membrane modules, the majority of the dry mass is organic and biological, with a smaller inorganic fraction

(Baker and Dudley 1998). This complexity makes the chance of finding a silver bullet for fouling prevention unlikely. However, as the fouling process becomes better understood, tailored anti-fouling strategies may be designed for individual filtration systems.

Reverse osmosis (RO) filtration uses a high pressure to drive water through a membrane against the osmotic pressure, leaving the solutes behind. To ease the mechanical stress on the membrane, RO filtration is applied in a crossflow setup, where the water flows parallel to the membrane surface, and only a fraction is filtered through. This design minimizes the accumulation of solutes near the membrane surface, as most are flushed out with the unfiltered water. Biofilms that form in these systems must withstand high shear forces and are thus mechanically robust. Biofilms on RO membranes decrease the flow of water by two mechanisms. Firstly, the biofilm exerts hydraulic resistance to water flow, and secondly, it influences the movement of solutes at the membrane surface, which increases the local solute concentration and osmotic pressure (Herzberg and Elimelech 2007).

The factors that lead to biofouling of RO membranes involve characteristics of the membrane, the feedwater, and the system design, enabling adhesion and subsequent proliferation of bacteria on the membrane surface.

2.1 *Adhesion to Membranes*

As the properties of the membrane influence foulant adhesion, it is worthwhile to understand the membrane surface chemistry. Reverse osmosis (RO) (pore size of 0.1–0.01 nm) and nanofiltration (NF) (pore size of 1–0.1 nm) are the two principal membrane classes in RO filtration (Tchobanoglous et al. 2003). The most common membrane type used for RO and NF is the multi-layered thin-film composite (TFC) membrane with a polyamide surface (Kang and Cao 2012). Although cellulose acetate surface layers are occasionally used, the primary focus here is on membranes with polyamide surface chemistry.

Membrane fouling is primarily governed by the electrostatic and hydrophilic/hydrophobic interactions between the bacteria and the membrane (Liu et al. 2001). These interactions can be partially described by XDLVO theory (van Oss and Jan. 2008), where bacteria are modeled as a colloidal particles, and the forces between the particle and the surface depend on the surface properties of both and the composition of the solution. The most relevant parameters are surface charge and hydrophobicity. The polyamide membrane surface contains amine, amide, and carboxyl groups, and have a net negative charge resulting in electrostatic repulsion of bacteria (Hurwitz et al. 2010). It is therefore hydrophobicity that drives bacterial adhesion to the polyamide membrane (Hurwitz et al. 2010; Bayoudh et al. 2009), and hydrophobic bacteria have a larger propensity for adhering to this type of membrane (Kochkodan et al. 2008).

The charge and hydrophobicity of surfaces depend on the chemistry of the solution. The protonation of carboxyl groups will render the membrane surface less negatively charged and more hydrophobic at low pH, facilitating bacterial adhesion (Hurwitz et al. 2010). Increasing the ionic strength also facilitates adhesion by shielding surface charges and reducing electrostatic repulsion. Finally, divalent cations promote adhesion by screening negative surface charges (Hurwitz et al. 2010) or neutralizing electrostatic repulsion by the formation of salt bridges (Liu et al. 2001). Both calcium and magnesium ions facilitate adhesion and aggregation of microorganisms through these mechanisms (He et al. 2016b) and thereby cause high levels of fouling (Liu et al. 2001).

2.2 Bacterial Proliferation and Biofilm Formation

After adhesion, a continuous supply of nutrients to the membrane surface sustains bacterial growth and biofilm formation. In water filtration systems, complex natural organic matter (NOM) in the water sources adsorbs to the membrane and fuels biofilm growth. Although the precise NOM composition is undefined, it is known to contain proteins, polysaccharides, and other biodegradable compounds that can serve as a nutrient source for adhered bacteria (Leenheer et al. 2000; Kwon et al. 2005). The fouling behavior of NOM constituents has been studied individually. A large fraction of NOM is made up by humic substances including humic acids (Aiken et al. 1986). These are complex macromolecules of low to moderate molecular weight containing aromatic and aliphatic components and mainly carboxylic and phenolic groups, rendering them negatively charged at neutral pH (Aiken et al. 1986). Humic substances often include many recalcitrant compounds, but a significant fraction is biodegradable (Camper 2004). NOM also contains proteins, DNA, and polysaccharides. DNA and polysaccharides readily adsorb to RO membranes (Herzberg et al. 2009) although adsorptive properties of polysaccharides depend on structure and charge. Branched-chain polysaccharides appear to foul membranes more severely than straight-chain polysaccharides, and acidic polysaccharides that are negatively charged at neutral pH are repelled (Tong et al. 2021). Proteins adsorb less well to RO membranes (Herzberg et al. 2009), but much remains to be learned about protein adsorption from NOM, as protein adsorption to RO membranes has mainly been studied using model proteins like bovine serum albumin (BSA), under the assumption that it represents general protein behavior.

Similarly to its effect on bacterial adhesion, the chemistry of the solution also affects nutrient availability via its effect on NOM adsorption. For example, the ionic strength does not affect the overall adsorption of NOM, but it does affect the composition of what is adsorbed (Li and Elimelech 2004; Ang et al. 2006). Negatively charged components adsorb better, and positively charged components adsorb less as the ionic strength increases (Ang and Elimelech 2007). Likewise, pH influences NOM adsorption by protonation of carboxyl groups, promoting adsorption at low pH (Clark and Lucas 1998). Divalent cations, primarily calcium, strongly promote nutrient adsorption by forming complexes with humic substances, proteins, polysaccharides, and DNA, inducing charge neutralization, and decreasing steric repulsion (Ang et al. 2006; Li and Elimelech 2004). Calcium ions can bridge the anionic groups of proteins, polysaccharides, humic acids and DNA and the membrane surface, increasing adsorption (Mo et al. 2008; Herzberg et al. 2009; Li and Elimelech 2004). Furthermore, calcium ions can induce conformational changes in humic acids, making them smaller and coiled, which allows them to more readily diffuse through the biofilm (Clark and Lucas 1998). Hence, nutrient availability depends not only on the concentration of nutrients in the feed water, but to a large degree also on the feed water chemistry through its role in nutrient adsorption.

2.3 *The Biofilm Composition*

The microbial composition of membrane biofouling is very diverse and depends on the origin of the feed water (wastewater, freshwater, seawater), the season and geographical location, and the amount of time the fouling layer has had to develop (Zhang et al. 2011; Bereschenko et al. 2008). Common to all membrane filtration systems is the dominance of proteobacteria. Systems used to filter wastewater tend to be dominated by α-proteobacteria, followed by γ-proteobacteria and β-proteobacteria (Pang and Liu 2007; Chen et al. 2004). In seawater, α- and γ-proteobacteria are also dominating (Zhang et al. 2011).

Among α-proteobacteria, the genus *Sphingomonas* appears to be important fouling organisms in all three applications and are therefore well characterized. These organisms are oligotrophic and non-fastidious, making them well adapted to compete for nutrients on the membrane surface (Eguchi et al. 2001; Schut et al. 1997). Sphingomonads adhere to the membrane using surface-associated glycosphingolipids that are more hydrophobic than conventional lipopolysaccharides (Gutman et al. 2014). Sphingomonads also produce a range of adhesive polysaccharides, called sphingans, that both facilitate adhesion and stabilize the resulting biofilm (Gutman et al. 2014). *Pseudomonas* species also constitute a significant fraction of biofilms on membranes used in both freshwater and wastewater filtration (Ridgway et al. 1983; Bereschenko et al. 2008, 2010). Pseudomonads adhere well to polyamide surfaces (Ghayeni et al. 1998) and produce a range of extracellular polysaccharides (Pel, Psl, and alginate) that facilitate adhesion and stabilize the biofilm (Barnes et al. 2014; Friedman and Kolter 2004; Jackson et al. 2004). Although *Pseudomonas* is often used to study anti-fouling strategies, *Sphingomonas* is rarely applied, a crucial omission considering its apparent importance.

Polysaccharides in general play an important role in the biofilm matrix, and the bacterial composition indicates that in addition to Psl, Pel, and alginate, significant polysaccharides include xanthan and sphingans (Li et al. 2021b; Barnes et al. 2014). Similar to its effect on bacterial adhesion and nutrient adsorption, the chemistry of the feed water affects intermolecular interactions in the biofilm matrix and thereby the matrix properties at large. Again, calcium has a prominent role. Calcium ions increase the viscosity of straight-chain acidic polysaccharides (Tong et al. 2021) by binding the carboxyl groups of, e.g., alginate in the egg-box model, bridging adjacent alginate molecules and form a gel (Grant et al. 1973). Calcium also crosslinks DNA (Das et al. 2014), and its ability to crosslink humic acids and induce conformational changes that compact the biofilm will increase its resistance to hydrodynamic shear (Li and Elimelech 2004; Hong and Elimelech 1997). Due to the outsized effect of calcium on many mechanisms in biofouling, some fouling control strategies specifically target divalent cations.

3 Fouling Prevention

3.1 Feed Water Treatment

One way to prevent fouling is to limit the number of microorganisms that come in contact with the membrane, and to limit nutrient availability. It is inevitable that the feed water contains dissolved or suspended contaminants, since these are what the membranes are designed to remove. However, removing the bulk of the contaminants before it arrives to the membrane significantly reduces fouling in the membrane modules that are hard to clean. This can be done by pre-treating the feed water. Pretreatment complicates the water treatment system and only moves the fouling problem to pretreatment systems, but it can provide an overall energy saving (Shahalam et al. 2002). Pretreatment strategies can be divided into two major categories: (1) Addition of anti-fouling compounds to the feed water, or (2) removal of nutrients and bacteria before the membrane filtration step. The latter will be discussed here. Whereas RO and NF membranes retain bacteria on the membrane surface, pretreatment filtration binds the bacteria within the filter medium by adsorption to, e.g., an activated carbon filter, or straining by cartridge filtration, microfiltration, ultrafiltration, or slow sand filtration. In contrast to other treatment methods, like photocatalysis and chemical treatment, the physical removal mechanisms do not produce toxic by-products (Gutiérrez-Pulpeiro et al. 2021). These pretreatment filters are less discriminatory than the RO and NF membranes and eventually end up allowing some bacteria to pass through, but if the nutrient concentration of the feed water is kept low, these bacteria will not form a biofilm (Griebe and Flemming 1998). Nutrient limitation is therefore the next line of defense.

Nutrient limitation has great potential to limit biofouling and is performed mainly by limiting the supply of biodegradable carbon to the membrane. Microbial growth can also be limited by removing other macro-nutrients, such as phosphorus or nitrogen (Miqueleto et al. 2010; Vrouwenvelder et al. 2010) although this strategy only works under conditions where phosphorous and nitrogen cannot be scavenged from biomolecules in NOM and carbon availability is limited (Javier et al. 2020). If carbon remains readily available, the excess carbon is directed toward EPS production (Huang et al. 2019), which still causes hydraulic resistance during filtration. It is therefore important to balance the C:P and C:N ratios (Javier et al. 2020; Miqueleto et al. 2010).

Nutrients are usually removed by the same filters that remove bacteria. Microfiltration (MF) and ultrafiltration (UF) membranes have larger pores than RO and NF, with >50 nm and 2–50 nm, respectively (Tchobanoglous et al. 2003; Wolf et al. 2005). Of these, UF is most commonly used for feed water pretreatment, and it removes bacteria and larger inorganic colloids by straining. UF also removes some dissolved nutrients (Gao et al. 2011). Over time, the ultrafiltration membrane also accumulates a fouling layer, but they are easier to clean than RO and NF membranes, simply by backwashing and air scouring (Gao et al. 2011; Wolf et al. 2005). MF and UF can effectively remove phosphate from the feed water, when combined with coagulants (Jacobson et al. 2009), and their use as pretreatment for RO resulted in minimal biofouling (Corral et al. 2014).

Membrane bioreactors (MBRs) combine the physical removal mechanism of MF and UF membranes with the biological removal of nutrients through bacterial metabolic processes (Shannon et al. 2008). In addition to physically filtering the water, MBRs also support growth of a planktonic microbial phase that can be optimized to remove any biodegradable nutrients passing through. In contrast to other biological pretreatment strategies, the active biological phase is not a biofilm within a filter, but suspended as flocs or aggregates. MBR is an attractive RO pretreatment method to remove both bacteria and nutrients from wastewater and seawater (Hosseinzadeh et al. 2013; Visvanathan et al. 2003).

Rapid sand filtration (RSF) removes contaminants both by adsorption and straining by moving contaminated water through a sand filter (Hoslett et al. 2018). RSF removes significant fraction of microorganisms, suspended solids, and dissolved organics (Rajala et al. 2003; Hamoda et al. 2004). Slow sand filtration (SSF) employs a similar setup but at ~300 times slower flow rates (Bar-Zeev et al. 2012). Due to the slow flow, a biofilm referred to as the "schmutzdecke" develops in the upper layer of the sand and removes organic matter (Verma et al. 2017), nitrate (Aslan and Cakici 2007), and bacteria (Weber-Shirk and Dick 1997). Compounds that are not degraded in the schmutzdecke are often removed by physical adsorption to the sand grains (Escolà Casas 2015). Although sand filters clog over time, they are easily cleaned by backwashing or removal of the top layer, which carries the schmutzdecke (Rodgers et al. 2004). These maintenance strategies make sand filtration easy and cost-effective, and although it is less effective than, e.g., membrane bioreactors (Visscher 1987; Verma et al. 2017), both RSF and SSF of feed water have been shown to reduce fouling of RO filtration membranes (de Oliveira and Schneider 2019; Bar-Zeev et al. 2012).

Activated carbon filtration is primarily based on physical adsorption and biological degradation. Due to the high porosity, activated carbon has a large specific surface area (Cao et al. 2006), increasing both the number of available binding sites and the probability of encountering a binding site. Activated carbon contains a lot of different functional surface groups, like carboxyl, carbonyl, and alcohol, which allows binding of many different ligands (Fuente et al. 2003). Due to the high surface area and chemical versatility, activated carbon efficiently adsorbs dissolved organic matter and adheres suspended microorganisms (Monnot et al. 2016). This eventually leads to biofilm formation inside the filter, referred to as biologically activated carbon (Gutiérrez-Pulpeiro et al. 2021), which can further remove biodegradable compounds. The synergy between physical adsorption and biological degradation increases the range of contaminants that can be removed as well as increases the lifetime of the activated carbon. When activated carbon was combined with SSF and ozonation as RO pretreatment, very little membrane fouling was observed (Van der Hoek et al. 2000).

The complimentary mechanisms of physical and biological removal are key to achieving high-quality water for RO treatment. The biological phase can adapt to target the biodegradable compounds that are specific to the water source while physical removal can remove the more recalcitrant compounds. Due to the varying interplay between physical and biological removal in the different pretreatment methods, the best suited pretreatment depends on the local water quality as well as the properties and cleaning protocol of the downstream membrane.

3.2 *Surface Modification*

Although sophisticated pretreatment methods will reduce the amount of nutrients and microorganisms in the feed water, they are not 100% effective. It is therefore still relevant to consider other strategies to control membrane fouling, such as altering the properties of the membrane surface itself (Fig. 1). The surface properties are modified by one of the three strategies: modified membrane synthesis, grafting, and physical coating. In modified synthesis, the membrane recipe is changed to introduce components that convey anti-fouling properties. Grafting works to functionalize the original membrane surface with covalently bound surface groups. Finally, the membranes can be non-covalently coated in a range of materials to alter the surface properties. Much attention has been paid to novel ways of engineering modified surfaces. However, for the purpose of this chapter, we focus on the resulting surface properties rather than the method of synthesis. The surface of RO membranes is optimized to be highly water-permeable and efficiently reject colloids and solutes. Introducing changes to the membrane surface to prevent biofouling must therefore not compromise these essential properties.

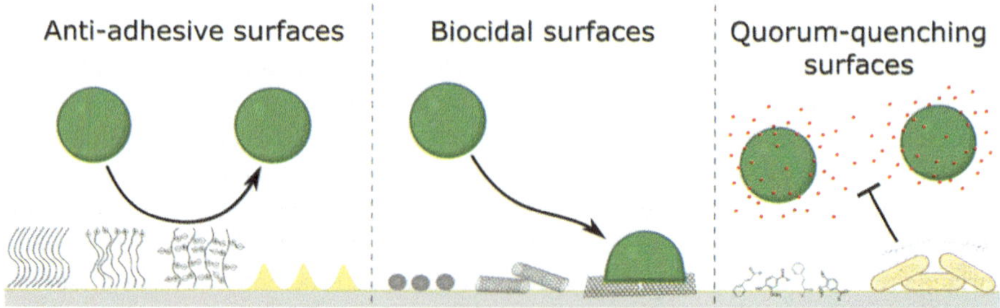

Fig. 1 Fouling prevention strategies by surface modification

3.2.1 Anti-Adhesive Surfaces

Bacterial adhesion is a well-studied phenomenon that partially operates according to known physico-chemical principles, and guidelines have been proposed for how to minimize adhesion. Surfaces that best resist bacterial adhesion are generally hydrophilic, H-bond acceptors, non-H-bond donors, and neutrally charged (Ostuni et al. 2001). Another parameter relevant to bacterial adhesion is surface topography. It is generally understood that smoother membranes adsorb fewer biomolecules, colloidal particles, and bacteria (Li et al. 2007; Vrijenhoek et al. 2001; Subramani and Hoek 2008). Conventional RO and NF membranes have a rather rough ridge and valley structure, which provides favorable conditions for bacterial adhesion and prevents subsequent detachment (Shang et al. 2020; Vrijenhoek et al. 2001). Most surface modifications have thus sought to smoothen the membrane to prevent fouling.

A popular approach to optimize surface hydrophilicity, charge, and smoothness is to use polymer coatings such as polyethylene glycol (PEG) and its derivatives. PEG prevents fouling by increasing hydrophilicity, neutralizing surface charge, and sterically repulsing approaching foulants (Sagle et al. 2009; Kang et al. 2007b). PEG coatings were shown to prevent fouling of a common water contaminant, a cationic surfactant (Kang et al. 2007b) and BSA (Gol and Jewrajka 2014), and PEG-based hydrogel coatings have also increased fouling resistance to oil/water emulsions (Sagle et al. 2009). Various PEG derivatives that have shown anti-adhesive properties for RO membranes include PEG diglycidyl ether (Van Wagner et al. 2011), Triglyme (Zou et al. 2011), PEG methacrylate (Belfer et al. 2001; Gilron et al. 2001), and block copolymer of PEG and Nylon-6 (Louie et al. 2006).

Although PEG-based coatings can be effective, PEG polymers are susceptible to oxidative damage (Ostuni et al. 2001), making these coatings unsuited for long-term applications. More stable polymers with similar properties include the hydrophilic and neutrally charged polyvinyl alcohol (PVA), which is applied in many commercial membranes. PVA-coated membranes are smoother, more hydrophilic, and neutrally charged (Tang et al. 2007a). However, the PVA coating does not mitigate fouling in all membrane types, and it simultaneously reduces permeability (Tang et al. 2009).

Instead of neutralizing the surface charge, some researchers have tried reversing it by applying cationic polyethyleneimine (PEI), while increasing hydrophilicity and smoothness (Chiao et al. 2020). Due to the altered surface charge, the PEI coatings improve fouling resistance toward cationic proteins and surfactants (Xu et al. 2015). It is doubtful that the positively charged membranes will be resistant to negatively charged foulants like bacteria and many EPS molecules. However, post-treatment with alkaline can increase the hydrophilicity, negative charge, and fouling resistance (Shen et al. 2017). PEI can also be combined with other polymers in layer-by-layer polyelectrolyte coatings, where fouling is mitigated due to the formation of hydrophilic ionic crosslinks (Wang et al. 2009).

Neutralizing or reversing the surface charge with hydrophilic polymers shows good rejection of certain foulants, but the same can be obtained with zwitterionic polymers made of monomers containing both anionic and cationic groups, but with a net neutral charge. The ionic groups in zwitterionic polymers can bind the water molecules more densely than polar groups in uncharged hydrophilic polymers like PEG (Chen et al. 2005). Hence, zwitterionic polymers form hydration shells that are thermodynamically harder to displace for any approaching material, rendering them more resistant to adsorption of foulants (He et al. 2008). Furthermore, when zwitterionic polymers are compressed by an approaching foulant, their motility and excluded volume decrease, which increases Gibbs free energy. This drives the polymers back into the swelling state, rejecting the foulant (He et al. 2016a). The ionic strength of the liquid is an important factor to zwitterionic performance. At low salinity, the strong repulsion between anionic and cationic groups causes polymer aggregation which exposes hydrophobic sites to foulants and decreases steric hindrance and the hydration shell (He et al. 2016a). In the presence of sufficient salt, the charged groups are screened from each other and hydrated. This makes zwitterionic polymers a particularly potent strategy for avoiding biofouling in desalination systems where the high salt concentrations may decrease performance of other polymers. Some of the conventional anti-adhesive polymers can be chemically modified to become zwitterionic, such as by carboxylation of PEI coatings (Choi et al. 2015). Several different zwitterionic polymer coatings have been shown to decrease fouling on RO and NF membranes (Shafi et al. 2015; Ji et al. 2012; Abdelhamid et al. 2015). One of the more interesting zwitterionic polymers to be applied in commercial membranes is the dopamine derivative L-DOPA, the adhesive protein found in mussels. When commercial membranes were coated with poly L-DOPA, the anti-adhesive properties improved (Azari and Zou 2012).

A non-polymeric strategy for improving anti-adhesive properties is to coat the membrane with inorganic particles. Zeolite nanocrystals incorporated into the membrane surface greatly improve hydrophilicity and smoothness (Lind et al. 2009) and improve resistance to protein fouling (Safarpour et al. 2017). However, zeolite functionalization also increases the negative surface charge, which may make the membrane more prone to fouling in the presence of divalent cations (Lind et al. 2009).

Although many anti-adhesive membranes also reduce surface roughness, the majority of the engineering focus has been directed at optimizing surface chemistry. However, seeing as the surface topography also affects fouling without negatively influencing surface chemistry, it is a good investment to also optimize this parameter. Although improved smoothness decreases fouling, having a rough surface with a specific surface pattern can also be a potent anti-adhesive strategy. The geometry of the surface pattern greatly influences the fouling behavior (Choi et al. 2020) and it may be difficult to predict which specific surface topographies best prevent foulant adhesion. Instead of attempting to identify optimal anti-adhesive patterns from first principles, many researchers look to nature for inspiration. Aquatic organisms have had millions of years to evolve surfaces that are not excessively colonized by marine microorganisms. Interesting examples of biological inspiration include a membrane with a sharkskin-mimetic surface pattern. When these membranes were exposed to *E. coli* and a natural feed source under flow, fouling resistance was increased remarkably. The decreased adhesion was attributed to the direction and velocity of the flow generated at the membrane surface, as a result of the engineered topography (Choi et al. 2018).

Anti-adhesive surface modifications also have a range of problems and limitations. Although superhydrophilic surfaces may prevent the majority of bacterial adhesion, certain species adhere better to hydrophilic surfaces (Boks et al. 2008). Likewise, abiotic fouling is still a problem at highly hydrophilic surfaces (Koc et al. 2019). Many of the polymer coatings have limited long-term physical and chemical stability. Stability may be enhanced by covalent attachment but this method is expensive and require techniques that are not easily scaled to industrial production. Surface smoothening is not a silver bullet either, as different bacteria are selected for (Shang et al. 2020), and some biopolymers, including alginate, even adsorb better to smoother surfaces (Lin et al. 2010). It is therefore likely that improving hydrophilicity and decreasing roughness will lead to the formation of biofilms of different species. Due to the difference in surface charge, hydrophilicity, and roughness, different surface coatings are prone to adsorption by different foulants. Choosing a membrane coating that is vulnerable to the foulants that are most effectively removed by the used pretreatment methods may be a good way to complement these anti-fouling strategies and ensure that the net fouling is minimal.

3.2.2 Biocidal Surfaces

Anti-adhesive surfaces can postpose but not prevent biofilm formation, and anti-adhesive properties may therefore be combined with antibacterial properties. Contact-killing surfaces can be achieved by incorporating biocidal chemicals, e.g. chlorhexidine (Kim et al. 2019) or triclosan (Park et al. 2018), into the membrane, or immobilizing antibacterial enzymes, e.g. lysozyme (Saeki et al. 2013), on the membrane surface.

Membranes have also been functionalized with antibacterial nanomaterials (Ong et al. 2016). Photocatalytic titanium dioxide nanoparticles can kill adhered bacteria when exposed to UV light (Kim et al. 2003), but the requirement to deliver UV light to the membrane makes the practical implementation of this approach difficult. Copper and silver nanoparticles, however, exert their bactericidal effect by releasing ions that cause membrane disruption, enzyme deactivation, ROS production, and DNA binding (Yin et al. 2020). Membranes functionalized with copper (Ben-Sasson et al. 2014b) and silver nanoparticles (Yin et al. 2013) effectively kill adhered bacteria short-term, but the biocidal activity weans over time (Ben-Sasson et al. 2014a; Cervantes and Gutierrez-Corona 1994). Carbon-based nanomaterials, such as carbon nanotubes, do not need to be released from the surface and will damage the bacterial membrane by direct contact (Kang et al. 2008). Membranes have been functionalized with both single-walled (Tiraferri et al. 2011) and multi-walled carbon nanotubes (Kim et al. 2014), which improve antibacterial properties. The antibacterial effects of silver nanoparticles and carbon nanotubes have also been combined to yield superior bactericidal activity (Kim et al. 2012; Gunawan et al. 2011). The two-dimensional equivalent of carbon nanotubes, graphene, has also been used as an antibacterial material. Graphene oxide inactivates bacteria by damaging the membrane through physical disruption, formation of ROS, charge transfer, and extraction of lipids from the membrane (Misdan et al. 2016). Graphene-functionalized membranes can thus inactivate bacteria and reduce fouling (Perreault et al. 2014). Antibacterial graphene has also been combined with silver nanoparticles on RO membranes, where they worked synergistically to inactivate bacteria (Soroush et al. 2016). Although nanomaterial-modified membranes show promising properties, they have yet to be commercialized, and the challenge to do so is the cost (Ong et al. 2016).

All biocidal surfaces have intrinsic problems that prevent them from becoming the long-sought easy fix for biofouling. If not combined with anti-adhesive properties, abiotic fouling and dead cells will build up and shield incoming bacteria from the biocidal properties. They may also select for colonization by species that are resistant to the biocidal agents. It is therefore important to conduct long-term evaluation of fouling build-up to determine the true potential of new antibacterial technologies.

3.2.3 Quorum-Quenching Surfaces

Most anti-fouling strategies seek to either prevent adhesion or kill bacteria upon adhesion. However, biofouling only becomes problematic after the biofilm has grown to a certain size. Another strategy for fouling control is therefore to mitigate biofilm formation after the bacteria have attached by repressing the biofilm phenotype. To prevent the switch to the biofilm phenotype, quorum sensing can be disrupted by quorum sensing inhibitors (QSIs), in a phenomenon known as quorum quenching. QSIs, like vanillin and cinnamaldehyde, have been used to functionalize membranes to prevent any adhered bacteria from expressing the biofilm behavior. Vanillin inhibits AHL-based quorum sensing, probably by binding AHLs directly and preventing their detection (Ponnusamy et al. 2009). Cinnamaldehyde interferes with Al-2-based quorum sensing by preventing binding of a quorum sensing regulator to its target promoter (Brackman et al. 2008). When deposited on the membrane, vanillin and cinnamaldehyde could reduce the amount of bacteria and matrix-associated polysaccharide (Katebian et al. 2018), indicating that biofilm behavior was repressed.

QS inhibitors can only work by diffusing from the surface and are therefore depleted over time. However, continuous QS inhibition can be achieved enzymatically by degrading QS molecules. The enzyme acylase cleaves AHL (Yeon et al. 2009), and immobilizing acylase on membranes has resulted in a smaller biofilm thickness (Kim et al. 2011). Continuous synthesis of QSIs directly on the surface can also be achieved by coating with genetically engineered quorum-quenching bacteria that also target AHL (Wood et al. 2016). However, far from all species employ AHL quorum sensing, and this narrow strategy will merely select for strains that use other QS molecules.

Biofilm dispersal can also be induced in a more generic way by nitric oxide, which causes degradation of the secondary messenger cyclic diguanylate (cyclic-di-GMP), a global regulator of biofilm formation (Barraud et al. 2009). NO is short lived and can be synthesized by genetically modified *E. coli* at the membrane surface, which has shown potential for preventing *P. aeruginosa* biofilms (Wood et al. 2016).

The incorporation of different biofilm-disruption mechanisms into the coating gives the membrane broader resistance toward different species of colonizing bacteria. The beneficial biofilm coating could also be used to confer unrelated but desirable features to the membrane, like expression of contaminant-degrading enzymes (Wood et al. 2016). However, coating the surface with anti-fouling bacteria or enzymes is vulnerable to chemical cleaning and colonization by organisms insensitive to the targeted quorum sensing pathway, that may replace them over time.

3.2.4 Combination Strategies

The three strategies presented above each have their advantages and shortcomings. Anti-adhesive surfaces delay adhesion but have no adverse effect on already adhered bacteria. Once the surface is coated with bacteria, the anti-adhesive properties disappear. Biocidal surfaces may kill bacteria upon contact, preventing initial growth, but become ineffective when coated with organic matter and dead bacteria. The merging of anti-adhesive and biocidal properties may provide additional benefits. For example, silver nanoparticles have been combined with either PEG or zwitterionic polymers to synergistically prevent biofouling, while also inactivating any adhered bacteria (Zhang et al. 2013; Yi et al. 2019). Furthermore, the negative zwitterionic groups can effectively bind silver to increase the total silver content and possibly delay depletion (Yi et al. 2019). Some hydrophilic coatings are ideally suited to incorporate biocides. For example, dendrimers with graphene nanosheets have mitigated fouling by industrial wastewater (Vatanpour and Sanadgol 2020), and polydopamine with biocidal nanoparticles has shown excellent anti-fouling properties while simultaneously protecting the polyamide membrane against oxidation by hydrogen peroxide during cleaning (Guha et al. 2017).

Although it is a potentially powerful strategy, the combination of anti-adhesive properties with quorum-quenching activity is less explored. One example is a micro-patterned surface of carbon nanotubes doped with the quorum-quenching enzyme acylase, which showed good anti-fouling behavior (Lee et al. 2019). Biocidal and quorum quenching properties have also been combined. For example, graphene-functionalized membranes dosed with a *Pseudomonas* quinolone signal inhibitor effectively prevented fouling by *P. aeruginosa* (Li et al. 2021a). In another example, the QSI cinnamaldehyde was encapsulated with the antibiotic kanamycin in biodegradable particles to prevent biofilms (Zodrow et al. 2014). However, due to the release-based platform, the anti-fouling chemicals eventually deplete.

Anti-fouling properties can also be combined with other desirable features. For example, a coating with hydantoin monomers can render the membrane bactericidal while preventing membrane damage by chlorine (Wei et al. 2010). Once exposed to chlorine, the chemical structure of the monomers is modified and becomes biocidal in a circular reaction scheme which restores the original structure. The hydantoin-modified membranes showed improved antibacterial properties when treated with chlorine (Wei et al. 2010). Another interesting strategy is to functionalize polydopamine coatings with copper- and manganese oxide nanoparticles which convert hydrogen peroxide into oxygen bubbles. The oxygen bubbles formed at the membrane surface beneath the foulants will mechanically dislodge the fouling layer, inhibiting biofilm formation (Guha et al. 2017). Thus, membrane coatings can also be used to improve efficacy and decrease adverse effects of membrane cleaning. This is especially potent as none of the described fouling prevention strategies can completely prevent fouling, making cleaning a necessity regardless of the anti-fouling technology used.

4 Membrane Cleaning

No anti-fouling technology can circumvent the need for cleaning, and many of the cleaning methods can also be used preventatively, by continuously applying them during membrane filtration. The distinction between fouling prevention and membrane cleaning is therefore blurry. Fouling prevention may be defined as a continuous process, whereas membrane cleaning refers to the periodic, high-intensity strategies applied once the fouling severity has passed a certain threshold. However, the prevention and cleaning mechanisms are similar and will be discussed together. Membrane cleaning may be divided into three categories that occasionally overlap: physical, chemical, and biological cleaning (Fig. 2).

4.1 Physical Cleaning

As the fouling layer is not covalently attached to the membrane, it can be dislodged or disintegrated by applying a sufficiently high shear force. One way to introduce shear force is to increase the flow rate, which also makes it more difficult for bacteria to adhere (Shang et al. 2020). However, high shear forces select for species that adhere better and form a more robust matrix, resulting in a compact biofilm that is harder to remove (Vrouwenvelder 2011). Although simulations indicate that high flow rates still lead to less biofouling, this is counterbalanced by a higher energy cost and lower water recovery (Radu et al. 2012). Thus, a constant high flow rate is not necessarily a good anti-fouling strategy.

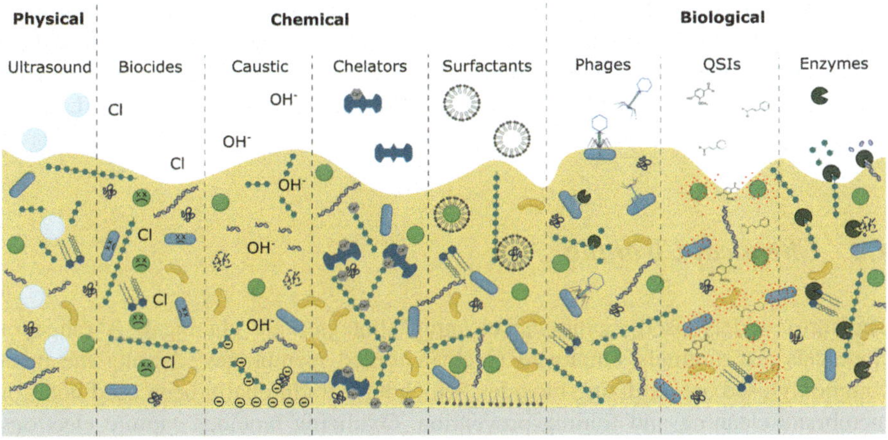

Fig. 2 Membrane fouling cleaning strategies

Increasing turbulence near the membrane has also been suggested to decrease fouling. Turbulence can be induced by modifying the geometry of the feed spacer that separates the two opposing membranes, which controls the flow pattern above the membrane surface (Williams and Wakeman 2000). However, increasing the turbulence and crossflow velocity also increases the transport of nutrients to the membrane surface, stimulating bacterial growth (Nguyen et al. 2012; Radu et al. 2012). It is therefore doubtful how effective increased turbulence is as an anti-fouling strategy. Instead, a viable strategy might be hydraulic cleaning by periodically increasing the flow rate. However, hydraulic cleaning has not shown consistent performance (Javier et al. 2021; Gwon et al. 2003). Simulations indicate that a temporary increase in crossflow velocity leads to sloughing of the biofouling layer, followed by a rapid regrowth once initial crossflow conditions are restored (Gwon et al. 2003).

Higher shear forces can be induced by applying air. When air bubbles travel over the membrane surface, the foulant layer passes through an air–water interface which exerts high shear and detaches the fouling layer from the membrane. This strategy, referred to as air scouring or air sparging, is widely used across industries, but few have implemented it in RO and NF membranes although daily air scouring has proven to reduce long-term fouling of NF membranes manyfold (Cornelissen et al. 2007).

The most intensive physical cleaning strategy used for membranes is ultrasound. Ultrasound waves cause cavitation in the liquid in which they propagate. This involves formation, expansion, and implosion of air bubbles, exerting large forces on the fouling layer. Ultrasound could remove inorganic and organic foulants (Feng et al. 2006) and effectively detach a model bacterial monolayer from an RO membrane (Zips et al. 1990). Despite the intense mechanical stress exerted by ultrasonic waves, they do not cause membrane damage (Feng et al. 2006; Wang et al. 2016). However the ultrasound effect is difficult to distribute throughout a whole membrane module, and the treatment is problematic to apply at industrial scale. Despite some promising research, physical cleaning of membranes at best serves an auxiliary role in membrane cleaning. However, when combined with chemical cleaning, like acids, the removal efficiency is often synergistic (Liu and Craig 2009).

4.2 *Chemical Cleaning*

The most widely used membrane cleaning strategies are chemical. A range of chemicals are routinely employed to clean fouled membranes. Among the most commonly used are oxidizing biocides, which have traditionally been used both for membrane cleaning and fouling prevention. Oxidizing biocides capture electrons from biomolecules, leading to the oxidation of the biofilm matrix and damage to the bacterial cells (Bajpai 2015). Oxidizing chemicals also increase the number of polar functional groups of the foulants, which increases electrostatic and hydrophobic repulsion (Liu et al. 2001). The biofilm is therefore both degraded and repelled from the membrane simultaneously. As oxidizing chemicals are consumed when reacting with the foulants, they need time to penetrate the fouling layer completely. This may not occur within a normal cleaning cycle if the fouling layer is thick. Oxidizing biocides therefore work best when the fouling layer is relatively thin.

Chlorine, either as a gas or hypochlorite (ClO^-), is the most widely used oxidizing biocide, principally due to the low cost. However, chlorine has a limited biocidal activity against many species (Shannon et al. 2008), limiting its use as a disinfectant. Additionally, chlorine disrupts hydrogen bonds within polyamide chains on the membrane surface, decreasing water permeability (Kang et al. 2007a). Therefore, caution has to be applied when using chlorine in polyamide membranes over longer time periods. Although some chlorine resistant surface coatings have been developed (Kang et al. 2007a; Wei et al. 2010), their long-term viability has not been confirmed. Chlorine has the further disadvantage of reacting with water solutes and producing toxic and carcinogenic by-products, which need further deactivation downstream (Richardson 2003; Shannon et al. 2008; Al-Abri et al. 2019). The oxidizing chemicals have the further disadvantage of degrading recalcitrant humic acids into more biodegradable components for the biofilm, which further stimulates microbial growth (Świetlik et al. 2004). The alternative oxidizing disinfectant ozone is able to kill certain pathogens that chlorine cannot (Shannon et al. 2008). However, ozone also produces toxic by-products (Zhao et al. 2012) and is more expensive than other disinfectants.

To prevent the indiscriminate oxidation of chlorine and ozone, non-oxidizing biocides have been developed as alternative cleaning agents. Non-oxidizing biocides have specific and narrow modes of action and are therefore compatible with the polyamide membrane chemistry. These include dibromonitrilopropionamide (DBNPA), glutaraldehyde, isothiazoline, and quaternary ammonium cations (Tan et al. 2017; Kim and Park 2015). DBNPA inactivates microbial proteins by reacting with sulfur-containing nucleophiles, like cysteine and methionine (Paulus 2005). DBNPA can be continuously dosed or shock dosed to the feed water to minimize biofouling (Majamaa et al. 2012; Siddiqui et al. 2017). Continuous dosing of commercial non-oxidizing biocides can likewise greatly decrease the active fouled biomass (Tan et al. 2017). However, DBNPA could not disperse an already established biofilm (Siddiqui et al. 2017).

This demonstrates that even if efficient disinfection of the fouling layer succeeds, cleaning by biocidal action does not restore flow through the membrane as the dead biomass remains on the membrane surface. Biocides have even been observed to worsen biofouling (Hamida 1996). This has been attributed to several mechanisms. The fraction of bacteria that survive the biocide treatment may have increased EPS production as a defense mechanism (Baker and Dudley 1998). Meanwhile, dead bacteria become a nutrient source for any surviving cells and fuel regrowth of the biofilm. Biofouling control therefore needs to move away from the medical paradigm of disinfection and instead focus on removing the biomass. This can be done by disrupting both the internal biofilm cohesion and the adhesion between the biofilm and the membrane, while effectively transporting loosened biomass away from the surface. This is well demonstrated by another non-oxidizing biocide: the quaternary ammonium analog tributyl tetradecyl phosphonium chloride (TTPC). TTPC is a surface-active compound that kills bacteria by disrupting the cell membrane (Xue and Xiao 2016), but its surface-active properties also solubilize the fouling layer (Kim and Park 2015). The other commonly applied cleaning agents also work to disrupt biofilm cohesion and adhesion.

The most common cleaning agents are caustics and acids, which are usually applied in sequence. Common caustic compounds include sodium hydroxide, potassium hydroxide, and ammonium hydroxide (Nguyen et al. 2012). Although caustic treatment causes reversible swelling of the polyamide membrane, resulting in decreased solute rejection (Kallioinen et al. 2016), it is thought to be relatively safe for the membrane integrity. Caustic agents, like sodium hydroxide, clean by hydrolysis and solubilization (Liu et al. 2001). Polysaccharides and proteins are hydrolyzed while fats and oils are saponified by reacting with NaOH and form water-soluble soap micelles. The caustic agents simultaneously increase the pH and induce negative charges on humic acids and the membrane, which in turn decreases adsorption and improves solubility of foulants (Tang et al. 2007b).

Acidic cleaning agents include hydrochloric acid, nitric acid, sulfuric acid, phosphoric acid, citric acid, and oxalic acid (Nguyen et al. 2012). Although strong acids weaken the stability of polyamide membranes (Jun et al. 2018), weak acids are thought to be relatively safe. Acidic cleaning agents are primarily used to remove scaling and metal oxides and not biofouling, although they can also hydrolyze polysaccharides and proteins (Liu et al. 2001). Of the acidic cleaning agents hydrochloric acid is most effective, partially due to the high solubility of its salt (Madaeni and Samieirad 2010). Although caustic cleaning is more effective (Madaeni and Samieirad 2010), the acids target different components in the fouling layer. Cleaning with caustic followed by acid therefore improves performance (Madaeni and Samieirad 2010). However, caustic treatment is largely inefficient at cleaning some fouling components, including calcium-complexed NOM, and needs to be combined with other cleaning agents like chelators or surfactants to be effective (Li and Elimelech 2004). When applied simultaneously with other cleaning agents, the caustic agents increase the available surface area, thus directly increasing their efficiency (Nguyen et al. 2012).

Cleaning solutions often include metal chelators like ethylenediaminetetraacetic acid (EDTA). Like caustics, EDTA can cause membrane swelling and decrease solute rejection (Simon et al. 2012), but are still considered relatively safe cleaning agents. Chelators clean by sequestering both free and complexed iron and calcium ions through ligand-exchange interactions (Ang et al. 2006). Sequestering calcium prevents formation of salt bridges between the negatively charged bacteria, EPS, and membrane, reducing adhesion forces (Ang et al. 2006). Continuously dosing the feed water with EDTA can therefore prevent adsorption and fouling by NOM (Hong and Elimelech 1997). However, EDTA can also be used to remove an already established fouling layer (Al Ashhab et al. 2017). As EDTA removes calcium ions from humic-calcium complexes, the electrostatic repulsion between NOM and the membrane increases, favoring desorption (Li and Elimelech 2004). Meanwhile, chelators also disrupt the gelling effect of some calcium-complexed polysaccharides, like alginate, in the presence of calcium ions (Ang et al. 2006; Hong and Elimelech 1997). The cleaning efficiency of EDTA increases significantly when combined with other cleaning agents like caustic. Caustic agents are especially potent with EDTA as the carboxylic groups of EDTA deprotonate at increased pH and enhance their affinity for calcium ions, improving cleaning (Ang et al. 2006). On the other hand, EDTA can become inhibited by certain biocides to decrease the cleaning efficiency (Liu et al. 2018), highlighting the importance of careful combination of cleaning agents.

As demonstrated by the non-oxidizing biocide TTPC, surfactants are potent cleaning agents. Although surfactants can adsorb on the membrane surface, they do not exert permanent damage and are generally considered safe for polyamide membranes (Huang et al. 2020). The most studied surfactant in biofouling is the anionic sodium dodecyl sulfate (SDS). Surfactants can form micelles with fats, oils, and proteins and solubilize macromolecules within the micelles (Li and Elimelech 2004; Liu et al. 2001). Surfactants can also disrupt the lipid membrane of bacteria, exerting a biocidal effect (Liu et al. 2001). The amphiphilic surfactants can disrupt the hydrophobic interaction between bacteria, foulants, and membrane, effectively eliminating the adhesion forces between the fouling layer and the membrane surface (Ang et al. 2006). SDS binds to NOM, reducing hydrophobicity and relaxing the tight, coiled conformation (Li and Elimelech 2004). Compared with EDTA, SDS is more effective at removing less dense fouling layers (Ang et al. 2006). SDS cleaning efficiency improves with increasing concentration and markedly improves when above the critical micelle concentration (Li and Elimelech 2004). Though some studies observe little effect (Ang et al. 2006), caustics generally improves SDS performance, most likely by loosening up the biofilm structure (Madaeni and Samieirad 2010). SDS also has improved effectiveness when combined with EDTA due to the complimentary cleaning strategies (Salahi et al. 2010). Surfactants also exert an anti-adhesive effect after cleaning by forming a hydrophilic coating on the hydrophobic membranes, decreasing adhesion during subsequent operation.

Chemical cleaning is the most applied cleaning method in the industry, but since chemicals cannot be discharged into the water supply, they require special treatment after use. This complicates the cleaning protocols and increases the economic cost of membrane cleaning. Application of environmentally friendly cleaning agents that could be discharged to the environment after use would benefit the industry as well as the environment. Chemical cleaning agents also damage the membrane and shorten its lifespan as they attack organic matter through generic mechanisms that do not only target the fouling layer. Cleaning methods that target the fouling layer more selectively are therefore needed.

4.3 Biological Cleaning

Biological cleaning solutions target the fouling layer with higher specificity than chemical cleaning, and they are of less environmental concern. Despite these desirable characteristics, most biological cleaning solutions are still in the experimental phase and are not yet widely applied in the industry.

One of the more promising strategies for fouling mitigation is the use of bacteriophages, which can be added to the feed water and lead to lysis of infected bacteria. As an additional mode of action, the phage genome can be designed to express EPS-degrading enzymes through the bacterial machinery, which stimulates the removal of biofilm (Harper et al. 2014). Although promising, the use of phages to remove membrane fouling has not been widely demonstrated. Some have used phages to disperse biofilms from ultrafiltration membranes (Goldman et al. 2009), while others could use phages to disperse biofilms from species that were known to foul RO membranes (Belgini et al. 2014). Although the specificity of the phages prevents membrane damage, it may also be the biggest limitation of phages. Because phages are highly host-specific (Milho et al. 2021), application of phages may quickly drive the fouling population toward dominance of non-targeted species. Meanwhile, the targeted species may develop immunity via the CRISPR-cas system, making subsequent phage therapies ineffective.

Biofilms can also be removed by activating their own mechanisms of dispersal. This can be achieved with nitric oxide or with QSIs in a similar manner as described for fouling prevention. For cleaning purposes, QSIs are dosed to the feed water (Kim et al. 2015; Kappachery et al. 2010), but continuous dosing is not practical due to the high cost of QSIs. An alternative to dosing QSIs to the feed water is the use of quorum-quenching bacteria that are genetically modified to express AHL-degrading enzymes, but successful implementation requires that these organisms can be established in the system to provide a continuous source of quorum-quenching enzymes (Oh et al. 2017). In addition to these challenges, the application of QSI and quorum-quenching strategies faces the same fundamental challenge as phage therapy: Their action is directed at a narrow range of microorganisms.

A more broad-spectrum approach to biological fouling removal is through enzymatic degradation of the biofilm matrix. The matrix confers the biofilm with mechanical stability and resistance to shear forces. By degrading the matrix polymers, the fouling layer is more easily removed by hydrodynamic forces (Brindle et al. 2011). Although individual enzymes are also highly specific, formulations can include many different enzymes that collectively provide broad-spectrum activity. Thus, enzymes are also the only biological cleaning agents that are routinely applied commercially. To improve broad-spectrum activity, most studies test complex enzyme mixtures with many different enzyme classes, usually containing proteases as a minimum and sometimes in combination with other cleaning agents. For example, Whittaker et al. tested a combination of enzymes and other cleaning agents against a fouled cellulose acetate membrane (Whittaker et al. 1984). Enzymes did not remove the fouling layer when applied without other cleaning agents, however, trypsin, papain, protease, esterase, and pancreatin effectively removed the fouling when combined with a surfactant (Triton X-100) and/or a chelator (EDTA) (Whittaker et al. 1984). Chelators typically improve enzyme activity by preventing inhibition by toxic metal ions, and the surfactant may have improved fouling removal by aiding the penetration of enzymes into the fouling layer. Such adjuvants can therefore increase the efficacy of enzymatic fouling removal considerably.

Although proteases are the most widely applied enzymes, other biofilm components, like lipids and polysaccharides, are worth targeting. Lipids both coat the bacterial surfaces and form a fraction of the extracellular matrix, and lipid degradation is a good supplement to protease treatment (Li et al. 2021b; Conrad et al. 2003). Degradation of surface-associated lipid structures by a carboxylic ester hydrolase could effectively remove mycobacteria from cellulose acetate membranes (Ridgway et al. 1984). Polysaccharides also constitute a significant fraction of the matrix. Removal of fully established fouling layers by carbohydrases is poorly documented, but some potential has been demonstrated for this enzyme class. Amyloglucosidases targeting α-1,4- and α-1,6-glucan bonds could detach mycobacteria from cellulose acetate membranes (Ridgway et al. 1984). However, the polysaccharide composition in multi-species biofilms is often diverse, necessitating the need for adding several different carbohydrases to the cleaning agents.

Identifying the right enzyme composition to combat biofouling is challenging, as the matrix composition is so poorly understood and highly variable from system to system. Membranes fouled under different conditions most likely require different enzyme formulations for optimal fouling control. A shortcoming of enzyme-based fouling control is the inability of enzymes to target inorganic foulants, and supplementary chemical cleaning can therefore be necessary. Such cleaning may also be necessary to remove enzymes that adsorb to the membrane where they can serve as an additional bacterial nutrient source (Leroy et al. 2008). Despite these challenges, enzymatic cleaning is one of the most promising and versatile anti-fouling strategies available.

5 Conclusion

Despite the many years of research, the hunt for an anti-fouling panacea is still ongoing. Although many cutting-edge anti-fouling methods are promising, they have only been stress-tested in model systems which cannot accurately represent the complexity of biofouling in filtration systems. Most functionalized membrane surfaces have only been exposed to a few model organic foulants, often with similar charge or hydrophobicity (Kang et al. 2007b; Sagle et al. 2009; Choi et al. 2015; Ji et al. 2012). Likewise, it is common to screen only a few bacterial species against the membranes (Choi et al. 2018; Matin et al. 2014; Park et al. 2018; Lee et al. 2019; Yi et al. 2019). The tested species are often common laboratory strains, such as *E. coli*, *P. aeruginosa,* and *S. aureus,* although they may not be the most relevant bacteria for membrane fouling. Thus, there is no guarantee that the anti-fouling properties are retained when the membrane is exposed to the diverse array of nutrients and organisms that are found in conventional feed water. Furthermore, most anti-fouling surfaces are only tested over short time frames, from a few hours to a week (Zodrow et al. 2014; Kim et al. 2019; Wang et al. 2018; Kang et al. 2007b; Chiao et al. 2020). As most membrane modules are used for several years, the short-term fouling behavior does not necessarily predict long-term fouling resistance.

Similar shortcomings apply to novel and experimental cleaning methods. These methods are often tested against model biofilms, some of which are monospecies biofilms of the same common lab strains (Kim and Park 2015; Kim et al. 2015). These model biofilms are often grown in nutrient-rich laboratory media and without the flow or pressure conditions that normal membrane fouling is exposed to (Kim and Park 2015; Belgini et al. 2014). These growth conditions result in a fouling layer that is very different and less resilient than the ones that the cleaning solutions would encounter in situ. As the membranes become fouled again after cleaning, novel cleaning methods also need to be verified by periodic cleaning over longer time scales. However, the cleaning methods are often only tested once, and their efficacy against slower irreversible fouling is not documented (Kim and Park 2015; Wibisono et al. 2015; Whittaker et al. 1984). It is therefore difficult to assess whether any fraction of the fouling composition is resistant to the cleaning method. Although testing experimental anti-fouling strategies in simpler preliminary studies may be necessary for initial screening and evaluation, they need to be verified over a longer time spans in pilot-scale or full-scale systems.

Despite the immaturity of many novel anti-fouling strategies, they show great potential to improve the sustainability and ease the economic burden of potable water production in the future. The ideal setup would include effective pretreatment, anti-fouling membrane and cleaning. However, no fixed setup works as a global optimum for water treatment. Depending on organic and bacterial composition and load of the feed water, combinations of different anti-fouling strategies may constitute local optima, and fouling management in the future will likely be tailored specifically to certain sources of feed water. There remains no silver bullet for fouling control but combining and developing novel anti-fouling strategies, industrial fouling is becoming an ever more manageable problem.

References

Abdelhamid AE, Elawady MM, El-Ghaffar MAA, Rabie AM, Larsen P, Christensen ML (2015) Surface modification of reverse osmosis membranes with zwitterionic polymer to reduce biofouling. Water Sci Technol Water Supply 15:999–1010

Aiken GR, McKnight DM, Wershaw RL, Maccarthy P (1986) Humic substances in soil, sediment, and water. 1985. Soil Sci 142:323

Al-Abri M, Al-Ghafri B, Bora T, Dobretsov S, Dutta J, Castelletto S, Rosa L, Boretti A (2019) Chlorination disadvantages and alternative routes for biofouling control in reverse osmosis desalination. NPJ Clean Water 2:1–16

Al Ashhab A, Sweity A, Bayramoglu B, Herzberg M, Gillor O (2017) Biofouling of reverse osmosis membranes: effects of cleaning on biofilm microbial communities, membrane performance, and adherence of extracellular polymeric substances. Biofouling 33:397–409

Ang WS, Elimelech M (2007) Protein (BSA) fouling of reverse osmosis membranes: implications for wastewater reclamation. J Membr Sci 296:83–92

Ang WS, Lee S, Elimelech M (2006) Chemical and physical aspects of cleaning of organic-fouled reverse osmosis membranes. J Membr Sci 272:198–210

Aslan S, Cakici H (2007) Biological denitrification of drinking water in a slow sand filter. J Hazard Mater 148:253–258

Azari S, Zou L (2012) Using zwitterionic amino acid l-DOPA to modify the surface of thin film composite polyamide reverse osmosis membranes to increase their fouling resistance. J Membr Sci 401:68–75

Bajpai P (2015) The control of microbiological problems. Pulp Paper Industry 2015:103

Baker JS, Dudley LY (1998) Biofouling in membrane systems—a review. Desalination 118:81–89

Barnes RJ, Bandi RR, Chua F, Low JH, Aung T, Barraud N, Fane AG, Kjelleberg S, Rice SA (2014) The roles of Pseudomonas aeruginosa extracellular polysaccharides in biofouling of reverse osmosis membranes and nitric oxide induced dispersal. J Membr Sci 466:161–172

Barraud N, Schleheck D, Klebensberger J, Webb JS, Hassett DJ, Rice SA, Kjelleberg S (2009) Nitric oxide signaling in Pseudomonas aeruginosa biofilms mediates phosphodiesterase activity, decreased cyclic di-GMP levels, and enhanced dispersal. J Bacteriol 191:7333–7342

Bar-Zeev E, Belkin N, Liberman B, Berman T, Berman-Frank I (2012) Rapid sand filtration pretreatment for SWRO: microbial maturation dynamics and filtration efficiency of organic matter. Desalination 286:120–130

Bayoudh S, Othmane A, Mora L, Ouada HB (2009) Assessing bacterial adhesion using DLVO and XDLVO theories and the jet impingement technique. Colloids Surf B: Biointerfaces 73:1–9

Belfer S, Gilron J, Purinson Y, Fainshtain R, Daltrophe N, Priel M, Tenzer B, Toma A (2001) Effect of surface modification in preventing fouling of commercial SWRO membranes at the Eilat seawater desalination pilot plant. Desalination 139:169–176

Belgini DRB, Dias RS, Siqueira VM, Albanese JM, Souza RS, Torres APR, Sousa MP, Silva CC, De Paula SO, Oliveira VM (2014) Culturable bacterial diversity from a feed water of a reverse osmosis system, evaluation of biofilm formation and biocontrol using phages. World J Microbiol Biotechnol 30:2689–2700

Ben-Sasson M, Lu X, Bar-Zeev E, Zodrow KR, Nejati S, Qi G, Giannelis EP, Elimelech M (2014a) In situ formation of silver nanoparticles on thin-film composite reverse osmosis membranes for biofouling mitigation. Water Res 62:260–270

Ben-Sasson M, Zodrow KR, Genggeng Q, Kang Y, Giannelis EP, Elimelech M (2014b) Surface functionalization of thin-film composite membranes with copper nanoparticles for antimicrobial surface properties. Environ Sci Technol 48:384–393

Bereschenko LA, Heilig GHJ, Nederlof MM, Van Loosdrecht MCM, Stams AJM, Euverink GJW (2008) Molecular characterization of the bacterial communities in the different compartments of a full-scale reverse-osmosis water purification plant. Appl Environ Microbiol 74:5297

Bereschenko LA, Stams AJM, Euverink GJW, Van Loosdrecht MCM (2010) Biofilm formation on reverse osmosis membranes is initiated and dominated by Sphingomonas spp. Appl Environ Microbiol 76:2623–2632

Boks NP, Norde W, van der Mei HC, Busscher HJ (2008) Forces involved in bacterial adhesion to hydrophilic and hydrophobic surfaces. Microbiology 154:3122–3133

Brackman G, Defoirdt T, Miyamoto C, Bossier P, Van Calenbergh S, Nelis H, Coenye T (2008) Cinnamaldehyde and cinnamaldehyde derivatives reduce virulence in vibrio spp. by decreasing the DNA-binding activity of the quorum sensing response regulator LuxR. BMC Microbiol 8:1–14

Brindle ER, Miller DA, Stewart PS (2011) Hydrodynamic deformation and removal of Staphylococcus epidermidis biofilms treated with urea, chlorhexidine, iron chloride, or DispersinB. Biotechnol Bioeng 108:2968–2977

Camper AK (2004) Involvement of humic substances in regrowth. Int J Food Microbiol 92:355–364

Cao Q, Xie K-C, Lv Y-K, Bao W-R (2006) Process effects on activated carbon with large specific surface area from corn cob. Bioresour Technol 97:110–115

Cervantes C, Gutierrez-Corona F (1994) Copper resistance mechanisms in bacteria and fungi. FEMS Microbiol Rev 14:121–137

Chen C-L, Liu W-T, Chong M-L, Wong M-T, Ong SL, Seah H, Ng WJ (2004) Community structure of microbial biofilms associated with membrane-based water purification processes as revealed using a polyphasic approach. Appl Microbiol Biotechnol 63:466–473

Chen S, Zheng J, Li L, Jiang S (2005) Strong resistance of phosphorylcholine self-assembled monolayers to protein adsorption: insights into nonfouling properties of zwitterionic materials. J Am Chem Soc 127:14473–14478

Chiao Y-H, Sengupta A, Chen S-T, Hung W-S, Lai J-Y, Upadhyaya L, Qian X, Ranil S, Wickramasinghe (2020) Novel thin-film composite forward osmosis membrane using polyethylenimine and its impact on membrane performance. Sep Sci Technol 55:590–600

Choi H, Jung Y, Han S, Tak T, Kwon Y-N (2015) Surface modification of SWRO membranes using hydroxyl poly(oxyethylene) methacrylate and zwitterionic carboxylated polyethyleneimine. J Membr Sci 486:97–105

Choi W, Lee C, Lee D, Won YJ, Lee GW, Shin MG, Chun B, Kim T-S, Park H-D, Jung HW (2018) Sharkskin-mimetic desalination membranes with ultralow biofouling. J Mater Chem A 6: 23034–23045

Choi W, Lee C, Yoo CH, Shin MG, Lee GW, Kim T-S, Jung HW, Lee JS, Lee J-H (2020) Structural tailoring of sharkskin-mimetic patterned reverse osmosis membranes for optimizing biofouling resistance. J Membr Sci 595:117602

Clark MM, Lucas P (1998) Diffusion and partitioning of humic acid in a porous ultrafiltration membrane. J Membr Sci 143:13–25

Conrad A, Kontro M, Keinänen MM, Cadoret A, Faure P, Mansuy-Huault L, Block J-C (2003) Fatty acids of lipid fractions in extracellular polymeric substances of activated sludge flocs. Lipids 38:1093–1105

Cornelissen ER, Vrouwenvelder JS, Heijman SGJ, Viallefont XD, Van Der Kooij D, Wessels LP (2007) Periodic air/water cleaning for control of biofouling in spiral wound membrane elements. J Membr Sci 287:94–101

Corral AF, Yenal U, Strickle R, Yan D, Holler E, Hill C, Ela WP, Arnold RG (2014) Comparison of slow sand filtration and microfiltration as pretreatments for inland desalination via reverse osmosis. Desalination 334:1–9

Das T, Sehar S, Koop L, Wong YK, Ahmed S, Siddiqui KS, Manefield M (2014) Influence of calcium in extracellular DNA mediated bacterial aggregation and biofilm formation. PLoS One 9:e91935
de Oliveira FF, Schneider RP (2019) Slow sand filtration for biofouling reduction in seawater desalination by reverse osmosis. Water Res 155:474–486
Eguchi M, Ostrowski M, Fegatella F, Bowman J, Nichols D, Nishino T, Cavicchioli R (2001) Sphingomonas alaskensis strain AFO1, an abundant oligotrophic ultramicrobacterium from the North Pacific. Appl Environ Microbiol 67:4945–4954
Escolà Casas M, Bester K (2015) Can those organic micro-pollutants that are recalcitrant in activated sludge treatment be removed from wastewater by biofilm reactors (slow sand filters)? Sci Total Environ 506–507:315–322
Feng D, van Deventer JSJ, Aldrich C (2006) Ultrasonic defouling of reverse osmosis membranes used to treat wastewater effluents. Sep Purif Technol 50:318–323
Flemming H-C (1997) Reverse osmosis membrane biofouling. Exp Thermal Fluid Sci 14:382–391
Flemming HC, Schaule G, McDonogh R, Ridgway HF (1994) Mechanism and extent of membrane biofouling. In: Biofouling and biocorrosion in industrial water systems. Lewis, Chelsea, pp 63–89
Flemming H-C, Schaule G, Griebe T, Schmitt J, Tamachkiarowa A (1997) Biofouling—the Achilles heel of membrane processes. Desalination 113:215–225
Friedman L, Kolter R (2004) Genes involved in matrix formation in Pseudomonas aeruginosa PA14 biofilms. Mol Microbiol 51:675–690
Fritzmann C, Löwenberg J, Wintgens T, Melin T (2007) State-of-the-art of reverse osmosis desalination. Desalination 216:1–76
Fuente E, Menéndez JA, Suárez D, Montes-Morán MA (2003) Basic surface oxides on carbon materials: a global view. Langmuir 19:3505–3511
Gao W, Liang H, Ma J, Han M, Chen Z-l, Han Z-s, Li G-b (2011) Membrane fouling control in ultrafiltration technology for drinking water production: a review. Desalination 272:1–8
Ghayeni SBS, Beatson PJ, Schneider RP, Fane AG (1998) Adhesion of waste water bacteria to reverse osmosis membranes. J Membr Sci 138:29–42
Gilron J, Belfer S, Väisänen P, Nyström M (2001) Effects of surface modification on antifouling and performance properties of reverse osmosis membranes. Desalination 140:167–179
Gol RM, Jewrajka SK (2014) Facile in situ PEGylation of polyamide thin film composite membranes for improving fouling resistance. J Membr Sci 455:271–282
Goldman G, Starosvetsky J, Armon R (2009) Inhibition of biofilm formation on UF membrane by use of specific bacteriophages. J Membr Sci 342:145–152
Grant GT, Morris ER, Rees DA, Smith PJC, Thom D (1973) Biological interactions between polysaccharides and divalent cations: the egg-box model. FEBS Lett 32:195–198
Griebe T, Flemming H-C (1998) Biocide-free antifouling strategy to protect RO membranes from biofouling. Desalination 118:153-IN9
Guha R, Xiong B, Geitner M, Moore T, Wood TK, Velegol D, Kumar M (2017) Reactive micromixing eliminates fouling and concentration polarization in reverse osmosis membranes. J Membr Sci 542:8–17
Gunawan P, Guan C, Song X, Zhang Q, Jan SS, Leong CT, Chen Y, Chan-Park MB, Chang MW, Wang K, Rong X (2011) Hollow fiber membrane decorated with Ag/MWNTs: toward effective water disinfection and biofouling control. ACS Nano 5:10033–10040
Gutiérrez-Pulpeiro M, Grillini V, Pavlović DM, Verlicchi P (2021) Activated carbon coupled with advanced biological wastewater treatment: a review of the enhancement in micropollutant removal. Sci Total Environ 790:148050
Gutman J, Herzberg M, Walker SL (2014) Biofouling of reverse osmosis membranes: positively contributing factors of sphingomonas. Environ Sci Technol 48:13941–13950
Gwon E-m, Yu M-j, Oh H-k, Ylee Y-h (2003) Fouling characteristics of NF and RO operated for removal of dissolved matter from groundwater. Water Res 37:2989–2997

Hamida AB (1996) Controlling biological fouling in open sea intake RO plants without continuous chlorination. Desalin Water Reuse 6:40–45
Hamoda MF, Al-Ghusain I, Al-Mutairi NZ (2004) Sand filtration of wastewater for tertiary treatment and water reuse. Desalination 164:203–211
Harper DR, Parracho HMRT, Walker J, Sharp R, Hughes G, Werthén M, Lehman S, Morales S (2014) Bacteriophages and biofilms. Antibiotics 3:270–284
He Y, Hower J, Chen S, Bernards MT, Chang Y, Jiang S (2008) Molecular simulation studies of protein interactions with zwitterionic phosphorylcholine self-assembled monolayers in the presence of water. Langmuir 24:10358–10364
He M, Gao K, Zhou L, Jiao Z, Mengyuan W, Cao J, You X, Cai Z, Yanlei S, Jiang Z (2016a) Zwitterionic materials for antifouling membrane surface construction. Acta Biomater 40:142–152
He X, Wang J, Abdoli L, Li H (2016b) Mg2+/Ca2+ promotes the adhesion of marine bacteria and algae and enhances following biofilm formation in artificial seawater. Colloids Surf B: Biointerfaces 146:289–295
Herzberg M, Elimelech M (2007) Biofouling of reverse osmosis membranes: role of biofilm-enhanced osmotic pressure. J Membr Sci 295:11–20
Herzberg M, Kang S, Elimelech M (2009) Role of extracellular polymeric substances (EPS) in biofouling of reverse osmosis membranes. Environ Sci Technol 43:4393–4398
Hong S, Elimelech M (1997) Chemical and physical aspects of natural organic matter (NOM) fouling of nanofiltration membranes. J Membr Sci 132:159–181
Hoslett J, Massara TM, Malamis S, Ahmad D, van den Boogaert I, Katsou E, Ahmad B, Ghazal H, Simons S, Wrobel L, Jouhara H (2018) Surface water filtration using granular media and membranes: a review. Sci Total Environ 639:1268–1282
Hosseinzadeh M, Bidhendi GN, Torabian A, Mehrdadi N (2013) Evaluation of membrane bioreactor for advanced treatment of industrial wastewater and reverse osmosis pretreatment. J Environ Health Sci Eng 11:1–8
Huang S, Voutchkov N, Jiang S (2019) Balancing carbon, nitrogen and phosphorus concentration in seawater as a strategy to prevent accelerated membrane biofouling. Water Res 165:114978
Huang J, Luo J, Chen X, Feng S, Wan Y (2020) How do chemical cleaning agents act on polyamide nanofiltration membrane and fouling layer? Ind Eng Chem Res 59:17653–17670
Hurwitz G, Guillen GR, Hoek EMV (2010) Probing polyamide membrane surface charge, zeta potential, wettability, and hydrophilicity with contact angle measurements. J Membr Sci 349: 349–357
Jackson KD, Starkey M, Kremer S, Parsek MR, Wozniak DJ (2004) Identification of *psl*, a locus encoding a potential exopolysaccharide that is essential for *Pseudomonas aeruginosa* PAO1 biofilm formation. J Bacteriol 186:4466–4475
Jacobson JD, Kennedy MD, Amy G, Schippers JC (2009) Phosphate limitation in reverse osmosis: an option to control biofouling? Desalin Water Treat 5:198–206
Javier L, Farhat NM, Desmond P, Linares RV, Bucs S, Kruithof JC, Vrouwenvelder JS (2020) Biofouling control by phosphorus limitation strongly depends on the assimilable organic carbon concentration. Water Res 183:116051
Javier L, Farhat NM, Vrouwenvelder JS (2021) Enhanced hydraulic cleanability of biofilms developed under a low phosphorus concentration in reverse osmosis membrane systems. Water Res X 10:100085
Ji Y-L, An Q-F, Zhao Q, Sun W-D, Lee K-R, Chen H-L, Gao C-J (2012) Novel composite nanofiltration membranes containing zwitterions with high permeate flux and improved anti-fouling performance. J Membr Sci 390:243–253
Jun B-M, Kim SH, Kwak SK, Kwon Y-N (2018) Effect of acidic aqueous solution on chemical and physical properties of polyamide NF membranes. Appl Surf Sci 444:387–398
Kallioinen M, Sainio T, Lahti J, Pihlajamäki A, Koivikko H, Mattila J, Mänttäri M (2016) Effect of extended exposure to alkaline cleaning chemicals on performance of polyamide (PA) nanofiltration membranes. Sep Purif Technol 158:115–123

Kang G-d, Cao Y-m (2012) Development of antifouling reverse osmosis membranes for water treatment: a review. Water Res 46:584–600

Kang G-D, Gao C-J, Chen W-D, Jie X-M, Cao Y-M, Yuan Q (2007a) Study on hypochlorite degradation of aromatic polyamide reverse osmosis membrane. J Membr Sci 300:165–171

Kang G, Liu M, Lin B, Cao Y, Yuan Q (2007b) A novel method of surface modification on thin-film composite reverse osmosis membrane by grafting poly (ethylene glycol). Polymer 48: 1165–1170

Kang S, Herzberg M, Rodrigues DF, Elimelech M (2008) Antibacterial effects of carbon nanotubes: size does matter! Langmuir 24:6409–6413

Kappachery S, Paul D, Yoon J, Kweon JH (2010) Vanillin, a potential agent to prevent biofouling of reverse osmosis membrane. Biofouling 26:667–672

Katebian L, Hoffmann MR, Jiang SC (2018) Incorporation of quorum sensing inhibitors onto reverse osmosis membranes for biofouling prevention in seawater desalination. Environ Eng Sci 35:261–269

Kim T-S, Park H-D (2015) Tributyl tetradecyl phosphonium chloride for biofouling control in reverse osmosis processes. Desalination 372:39–46

Kim SH, Kwak S-Y, Sohn B-H, Park TH (2003) Design of TiO2 nanoparticle self-assembled aromatic polyamide thin-film-composite (TFC) membrane as an approach to solve biofouling problem. J Membr Sci 211:157–165

Kim J-H, Choi D-C, Yeon K-M, Kim S-R, Lee C-H (2011) Enzyme-immobilized nanofiltration membrane to mitigate biofouling based on quorum quenching. Environ Sci Technol 45:1601–1607

Kim E-S, Hwang G, El-Din MG, Liu Y (2012) Development of nanosilver and multi-walled carbon nanotubes thin-film nanocomposite membrane for enhanced water treatment. J Membr Sci 394-395:37–48

Kim HJ, Baek Y, Choi K, Kim D-G, Kang H, Choi Y-S, Yoon J, Lee J-C (2014) The improvement of antibiofouling properties of a reverse osmosis membrane by oxidized CNTs. RSC Adv 4: 32802–32810

Kim H-S, Lee S-H, Byun Y, Park H-D (2015) 6-Gingerol reduces Pseudomonas aeruginosa biofilm formation and virulence via quorum sensing inhibition. Sci Rep 5:1–11

Kim T-S, Park S-H, Park D, Lee J-H, Kang S (2019) Surface immobilization of chlorhexidine on a reverse osmosis membrane for in-situ biofouling control. J Membr Sci 576:17–25

Koc J, Simovich T, Schönemann E, Chilkoti A, Gardner H, Swain GW, Hunsucker K, Laschewsky A, Rosenhahn A (2019) Sediment challenge to promising ultra-low fouling hydrophilic surfaces in the marine environment. Biofouling 35:454–462

Kochkodan V, Tsarenko S, Potapchenko N, Kosinova V, Goncharuk V (2008) Adhesion of microorganisms to polymer membranes: a photobactericidal effect of surface treatment with TiO2. Desalination 220:380–385

Kwon B, Lee S, Cho J, Ahn H, Lee D, Shin HS (2005) Biodegradability, DBP formation, and membrane fouling potential of natural organic matter: characterization and controllability. Environ Sci Technol 39:732–739

Lee KP, Arnot TC, Mattia D (2011) A review of reverse osmosis membrane materials for desalination—development to date and future potential. J Membr Sci 370:1–22

Lee J, Won Y-J, Choi D-C, Lee S, Park P-K, Choo K-H, Hyun-Suk O, Lee C-H (2019) Micro-patterned membranes with enzymatic quorum quenching activity to control biofouling in an MBR for wastewater treatment. J Membr Sci 592:117365

Leenheer JA, Croué J-P, Benjamin M, Korshin GV, Hwang CJ, Bruchet A, Aiken GR (2000) Comprehensive isolation of natural organic matter from water for spectral characterizations and reactivity testing. In: Natural organic matter and disinfection by-products. American Chemical Society

Leroy C, Delbarre C, Ghillebaert F, Compere C, Combes D (2008) Effects of commercial enzymes on the adhesion of a marine biofilm-forming bacterium. Biofouling 24:11–22

Li Q, Elimelech M (2004) Organic fouling and chemical cleaning of nanofiltration membranes: measurements and mechanisms. Environ Sci Technol 38:4683–4693

Li Q, Xu Z, Pinnau I (2007) Fouling of reverse osmosis membranes by biopolymers in wastewater secondary effluent: role of membrane surface properties and initial permeate flux. J Membr Sci 290:173–181

Li Y, Yang Y, Feng Y, Jian P, Hou L-a (2021a) Combined effects of Pseudomonas quinolone signal-based quorum quenching and graphene oxide on the mitigation of biofouling and improvement of the application potential for the thin-film composite membrane. Sci Total Environ 760:143348

Li Y, Wang H, Wang S, Xiao K, Huang X (2021b) Enzymatic cleaning mitigates polysaccharide-induced refouling of RO membrane: evidence from foulant layer structure and microbial dynamics. Environ Sci Technol 55:5453–5462

Lin NH, Kim M-m, Lewis GT, Cohen Y (2010) Polymer surface nano-structuring of reverse osmosis membranes for fouling resistance and improved flux performance. J Mater Chem 20: 4642–4652

Lind ML, Ghosh AK, Jawor A, Huang X, Hou W, Yang Y, Hoek EMV (2009) Influence of zeolite crystal size on zeolite-polyamide thin film nanocomposite membranes. Langmuir 25:10139–10145

Liu G, Craig VSJ (2009) Improved cleaning of hydrophilic protein-coated surfaces using the combination of nanobubbles and SDS. ACS Appl Mater Interfaces 1:481–487

Liu C, Caothien S, Hayes J, Caothuy T, Otoyo T, Ogawa T (2001) Membrane chemical cleaning: from art to science. Pall Corporation, Port Washington, p 11050

Liu G, Li L, Qiu L, Shuili Y, Liu P, Zhu Y, Jun H, Liu Z, Zhao D, Yang H (2018) Chemical cleaning of ultrafiltration membranes for polymer-flooding wastewater treatment: efficiency and molecular mechanisms. J Membr Sci 545:348–357

Louie JS, Pinnau I, Ciobanu I, Ishida KP, Ng A, Reinhard M (2006) Effects of polyether–polyamide block copolymer coating on performance and fouling of reverse osmosis membranes. J Membr Sci 280:762–770

Madaeni SS, Samieirad S (2010) Chemical cleaning of reverse osmosis membrane fouled by wastewater. Desalination 257:80–86

Majamaa K, Johnson JE, Bertheas U (2012) Three steps to control biofouling in reverse osmosis systems. Desalin Water Treat 42:107–116

Matin A, Khan Z, Gleason KK, Mazen Khaled SMJ, Zaidi AK, Moni P, Yang R (2014) Surface-modified reverse osmosis membranes applying a copolymer film to reduce adhesion of bacteria as a strategy for biofouling control. Sep Purif Technol 124:117–123

Milho C, Silva MD, Sillankorva S, Harper DR (2021) Biofilm applications of bacteriophages. In: Bacteriophages: biology, technology, therapy, pp 789–822

Miqueleto AP, Dolosic CC, Pozzi E, Foresti E, Zaiat M (2010) Influence of carbon sources and C/N ratio on EPS production in anaerobic sequencing batch biofilm reactors for wastewater treatment. Bioresour Technol 101:1324–1330

Misdan N, Ismail AF, Hilal N (2016) Recent advances in the development of (bio)fouling resistant thin film composite membranes for desalination. Desalination 380:105–111

Mo H, Tay KG, Ng HY (2008) Fouling of reverse osmosis membrane by protein (BSA): effects of pH, calcium, magnesium, ionic strength and temperature. J Membr Sci 315:28–35

Monnot M, Laborie S, Cabassud C (2016) Granular activated carbon filtration plus ultrafiltration as a pretreatment to seawater desalination lines: impact on water quality and UF fouling. Desalination 383:1–11

Nguyen T, Roddick FA, Fan L (2012) Biofouling of water treatment membranes: a review of the underlying causes, monitoring techniques and control measures. Membranes 2:804–840

Oh H-S, Tan CH, Low JH, Rzechowicz M, Siddiqui MF, Winters H, Kjelleberg S, Fane AG, Rice SA (2017) Quorum quenching bacteria can be used to inhibit the biofouling of reverse osmosis membranes. Water Res 112:29–37

Ong CS, Goh PS, Lau WJ, Misdan N, Fauzi A, Ismail. (2016) Nanomaterials for biofouling and scaling mitigation of thin film composite membrane: a review. Desalination 393:2–15

Ostuni E, Chapman RG, Erik Holmlin R, Takayama S, Whitesides GM (2001) A survey of structure–property relationships of surfaces that resist the adsorption of protein. Langmuir 17:5605–5620

Pang CM, Liu W-T (2007) Community structure analysis of reverse osmosis membrane biofilms and the significance of Rhizobiales bacteria in biofouling. Environ Sci Technol 41:4728–4734

Park S-H, Hwang SO, Kim T-S, Cho A, Kwon SJ, Kim KT, Park H-D, Lee J-H (2018) Triclosan-immobilized polyamide thin film composite membranes with enhanced biofouling resistance. Appl Surf Sci 443:458–466

Paulus W (2005) Directory of microbicides for the protection of materials: a handbook. Springer

Perreault F, Tousley ME, Elimelech M (2014) Thin-film composite polyamide membranes functionalized with biocidal graphene oxide nanosheets. Environ Sci Technol Lett 1:71–76

Ponnusamy K, Paul D, Kweon JH (2009) Inhibition of quorum sensing mechanism and Aeromonas hydrophila biofilm formation by vanillin. Environ Eng Sci 26:1359–1363

Radu AI, Vrouwenvelder JS, van Loosdrecht MCM, Picioreanu C (2012) Effect of flow velocity, substrate concentration and hydraulic cleaning on biofouling of reverse osmosis feed channels. Chem Eng J 188:30–39

Rajala RL, Pulkkanen M, Pessi M, Heinonen-Tanski H (2003) Removal of microbes from municipal wastewater effluent by rapid sand filtration and subsequent UV irradiation. Water Sci Technol 47:157–162

Richardson SD (2003) Disinfection by-products and other emerging contaminants in drinking water. TrAC Trends Anal Chem 22:666–684

Ridgway HF, Kelly A, Justice C, Olson BH (1983) Microbial fouling of reverse-osmosis membranes used in advanced wastewater treatment technology: chemical, bacteriological, and ultrastructural analyses. Appl Environ Microbiol 45:1066

Ridgway HF, Rigby MG, Argo DG (1984) Adhesion of a Mycobacterium sp. to cellulose diacetate membranes used in reverse osmosis. Appl Environ Microbiol 47:61–67

Rodgers M, Mulqueen J, Healy MG (2004) Surface clogging in an intermittent stratified sand filter. Soil Sci Soc Am J 68:1827–1832

Saeki D, Nagao S, Sawada I, Ohmukai Y, Maruyama T, Matsuyama H (2013) Development of antibacterial polyamide reverse osmosis membrane modified with a covalently immobilized enzyme. J Membr Sci 428:403–409

Safarpour M, Vatanpour V, Khataee A, Zarrabi H, Gholami P, Yekavalangi ME (2017) High flux and fouling resistant reverse osmosis membrane modified with plasma treated natural zeolite. Desalination 411:89–100

Sagle AC, Van Wagner EM, Hao J, McCloskey BD, Freeman BD, Sharma MM (2009) PEG-coated reverse osmosis membranes: desalination properties and fouling resistance. J Membr Sci 340: 92–108

Salahi A, Gheshlaghi A, Mohammadi T, Madaeni SS (2010) Experimental performance evaluation of polymeric membranes for treatment of an industrial oily wastewater. Desalination 262:235–242

Schut F, Gottschal JC, Prins RA (1997) Isolation and characterisation of the marine ultramicrobacterium Sphingomonas sp. strain RB2256. FEMS Microbiol Rev 20:363–369

Shafi HZ, Zafarullah Khan R, Yang, and Karen K Gleason. (2015) Surface modification of reverse osmosis membranes with zwitterionic coating for improved resistance to fouling. Desalination 362:93–103

Shahalam AM, Al-Harthy A, Al-Zawhry A (2002) Feed water pretreatment in RO systems: unit processes in the Middle East. Desalination 150:235–245

Shang C, Pranantyo D, Zhang S (2020) Understanding the roughness–fouling relationship in reverse osmosis: mechanism and implications. Environ Sci Technol 54:5288–5296

Shannon MA, Bohn PW, Elimelech M, Georgiadis JG, Mariñas BJ, Mayes AM (2008) Science and technology for water purification in the coming decades. Nature 452:301–310

Shen L, Zhang X, Zuo J, Wang Y (2017) Performance enhancement of TFC FO membranes with polyethyleneimine modification and post-treatment. J Membr Sci 534:46–58

Siddiqui A, Pinel I, Prest EI, Bucs S, Van Loosdrecht MCM, Kruithof JC, Vrouwenvelder JS (2017) Application of DBNPA dosage for biofouling control in spiral wound membrane systems. Desalin Water Treat 68:12–22

Simon A, Price WE, Nghiem LD (2012) Effects of chemical cleaning on the nanofiltration of pharmaceutically active compounds (PhACs). Sep Purif Technol 88:208–215

Soroush A, Ma W, Myriam Cyr M, Rahaman S, Asadishad B, Tufenkji N (2016) In situ silver decoration on graphene oxide-treated thin film composite forward osmosis membranes: biocidal properties and regeneration potential. Environ Sci Technol Lett 3:13–18

Subramani A, Hoek EMV (2008) Direct observation of initial microbial deposition onto reverse osmosis and nanofiltration membranes. J Membr Sci 319:111–125

Świetlik J, Agata Dąbrowska U, Stanisławiak R, Nawrocki J (2004) Reactivity of natural organic matter fractions with chlorine dioxide and ozone. Water Res 38:547–558

Tan Y-J, Sun L-J, Li B-T, Zhao X-H, Tong Y, Ikuno N, Ishii K, Hong-Ying H (2017) Fouling characteristics and fouling control of reverse osmosis membranes for desalination of dyeing wastewater with high chemical oxygen demand. Desalination 419:1–7

Tang CY, Kwon Y-N, Leckie JO (2007a) Probing the nano-and micro-scales of reverse osmosis membranes—a comprehensive characterization of physiochemical properties of uncoated and coated membranes by XPS, TEM, ATR-FTIR, and streaming potential measurements. J Membr Sci 287:146–156

Tang CY, Kwon Y-N, Leckie JO (2007b) Fouling of reverse osmosis and nanofiltration membranes by humic acid—effects of solution composition and hydrodynamic conditions. J Membr Sci 290:86–94

Tang CY, Kwon Y-N, Leckie JO (2009) Effect of membrane chemistry and coating layer on physiochemical properties of thin film composite polyamide RO and NF membranes: II. Membrane physiochemical properties and their dependence on polyamide and coating layers. Desalination 242:168–182

Tchobanoglous G, Burton FL, David H, Stensel. (2003) Wastewater engineering treatment and reuse. McGraw-Hill Higher Education, Boston

Tiraferri A, Vecitis CD, Elimelech M (2011) Covalent binding of single-walled carbon nanotubes to polyamide membranes for antimicrobial surface properties. ACS Appl Mater Interfaces 3:2869–2877

Tong X, Zhao X-H, Yin-Hu W, Bai Y, Ikuno N, Ishii K, Hong-Ying H (2021) The molecular structures of polysaccharides affect their reverse osmosis membrane fouling behaviors. J Membr Sci 625:118984

Van der Hoek JP, Hofman JAMH, Bonné PAC, Nederlof MM, Vrouwenvelder HS (2000) RO treatment: selection of a pretreatment scheme based on fouling characteristics and operating conditions based on environmental impact. Desalination 127:89–101

van Oss C, Jan. (2008) Chapter three—The extended dlvo theory. In: van Oss CJ (ed) Interface science and technology. Elsevier

Van Wagner EM, Sagle AC, Sharma MM, La Y-H, Freeman BD (2011) Surface modification of commercial polyamide desalination membranes using poly(ethylene glycol) diglycidyl ether to enhance membrane fouling resistance. J Membr Sci 367:273–287

Vatanpour V, Sanadgol A (2020) Surface modification of reverse osmosis membranes by grafting of polyamidoamine dendrimer containing graphene oxide nanosheets for desalination improvement. Desalination 491:114442

Verma S, Daverey A, Sharma A (2017) Slow sand filtration for water and wastewater treatment–a review. Environ Technol Rev 6:47–58

Visscher JT (1987) Slow sand filtration for community water supply: planning, design, construction, operation and maintenance. Village Earth

Visvanathan C, Boonthanon N, Sathasivan A, Jegatheesan V (2003) Pretreatment of seawater for biodegradable organic content removal using membrane bioreactor. Desalination 153:133–140

Vrijenhoek EM, Hong S, Elimelech M (2001) Influence of membrane surface properties on initial rate of colloidal fouling of reverse osmosis and nanofiltration membranes. J Membr Sci 188: 115–128

Vrouwenvelder JS (2011) Biofouling of spiral wound membrane systems. Water Intelligence Online, p 10

Vrouwenvelder JS, Beyer F, Dahmani K, Hasan N, Galjaard G, Kruithof JC, Van Loosdrecht MCM (2010) Phosphate limitation to control biofouling. Water Res 44:3454–3466

Wang J, Yao Y, Yue Z, Economy J (2009) Preparation of polyelectrolyte multilayer films consisting of sulfonated poly (ether ether ketone) alternating with selected anionic layers. J Membr Sci 337:200–207

Wang J, Gao X, Yuan X, Wang Q, Zhang Y, Wang X, Gao C (2016) Ultrasonic-assisted acid cleaning of nanofiltration membranes fouled by inorganic scales in arsenic-rich brackish water. Desalination 377:172–177

Wang Y, Wang Z, Wang J, Wang S (2018) Triple antifouling strategies for reverse osmosis membrane biofouling control. J Membr Sci 549:495–506

Weber-Shirk ML, Dick RI (1997) Biological mechanisms in slow sand filters. Am Water Works Assoc J 89:72

Wei X, Wang Z, Chen J, Wang J, Wang S (2010) A novel method of surface modification on thin-film-composite reverse osmosis membrane by grafting hydantoin derivative. J Membr Sci 346: 152–162

Whittaker C, Ridgway H, Olson BH (1984) Evaluation of cleaning strategies for removal of biofilms from reverse-osmosis membranes. Appl Environ Microbiol 48:395–403

Wibisono Y, El Obied KE, Cornelissen ER, Kemperman AJB, Nijmeijer K (2015) Biofouling removal in spiral-wound nanofiltration elements using two-phase flow cleaning. J Membr Sci 475:131–146

Williams C, Wakeman R (2000) Membrane fouling and alternative techniques for its alleviation. Membr Technol 2000:4–10

Wolf PH, Siverns S, Monti S (2005) UF membranes for RO desalination pretreatment. Desalination 182:293–300

Wood TL, Guha R, Tang L, Geitner M, Kumar M, Wood TK (2016) Living biofouling-resistant membranes as a model for the beneficial use of engineered biofilms. Proc Natl Acad Sci 113: E2802–E2E11

Xu J, Wang Z, Wang J, Wang S (2015) Positively charged aromatic polyamide reverse osmosis membrane with high anti-fouling property prepared by polyethylenimine grafting. Desalination 365:398–406

Xue Y, Xiao H (2016) Characterization and antipathogenic evaluation of a novel quaternary phosphonium tripolyacrylamide and elucidation of the inactivation mechanisms. J Biomed Mater Res A 104:747–757

Yeon K-M, Cheong W-S, Hyun-Suk O, Lee W-N, Hwang B-K, Lee C-H, Beyenal H, Lewandowski Z (2009) Quorum sensing: a new biofouling control paradigm in a membrane bioreactor for advanced wastewater treatment. Environ Sci Technol 43:380–385

Yi M, Lau CH, Xiong S, Wei W, Liao R, Shen L, Ang L, Wang Y (2019) Zwitterion–Ag complexes that simultaneously enhance biofouling resistance and silver binding capability of thin film composite membranes. ACS Appl Mater Interfaces 11:15698–15708

Yin J, Yang Y, Hu Z, Deng B (2013) Attachment of silver nanoparticles (AgNPs) onto thin-film composite (TFC) membranes through covalent bonding to reduce membrane biofouling. J Membr Sci 441:73–82

Yin IX, Zhang J, Zhao IS, Mei ML, Li Q, Chu CH (2020) The antibacterial mechanism of silver nanoparticles and its application in dentistry. Int J Nanomedicine 15:2555–2562

Zhang M, Jiang S, Tanuwidjaja D, Voutchkov N, Hoek EMV, Cai B (2011) Composition and variability of biofouling organisms in seawater reverse osmosis desalination plants. Appl Environ Microbiol 77:4390

Zhang S, Qiu G, Ting YP, Chung T-S (2013) Silver–PEGylated dendrimer nanocomposite coating for anti-fouling thin film composite membranes for water treatment. Colloids Surf A Physicochem Eng Asp 436:207–214

Zhao Y, Anichina J, Xiufen L, Bull RJ, Krasner SW, Hrudey SE, Li X-F (2012) Occurrence and formation of chloro-and bromo-benzoquinones during drinking water disinfection. Water Res 46:4351–4360

Zips A, Schaule G, Flemming HC (1990) Ultrasound as a means of detaching biofilms. Biofouling 2:323–333

Zodrow KR, Tousley ME, Elimelech M (2014) Mitigating biofouling on thin-film composite polyamide membranes using a controlled-release platform. J Membr Sci 453:84–91

Zou L, Vidalis I, Steele D, Michelmore A, Low SP, Verberk JQJC (2011) Surface hydrophilic modification of RO membranes by plasma polymerization for low organic fouling. J Membr Sci 369:420–428

Biofilm in Equine and Other Veterinary Wounds

Elin Jørgensen and Stine Jacobsen

Abstract Horses frequently sustain wounds due to their "fight and flight" nature. Equine body wounds heal well whereas limb wounds often display impaired healing. This disparity in healing patterns has been attributed to factors like hypoxia and prolonged chronic inflammation in equine limb wounds. Recent research suggests biofilm infections to also be part of the pathogenesis in equine chronic limb wounds, as biofilms are present in equine limb wounds and further impair the healing, whereas biofilms are seldom and do not affect healing in body wounds. However, specific veterinary research, horses included, within wound biofilm infections is scarce and just emerging.

In human wounds, biofilms have been studied for a longer period and are considered one of the causes of the healing pattern in chronic wounds like diabetic foot ulcers and venous leg ulcers. As human chronic wounds and equine limb wounds share many similarities, such as chronicity, weak inflammation, hypoxia, biofilms, and increased pH, the horse might be a relevant animal model species to study interactions between host and biofilms and to investigate future treatments strategies.

1 Introduction

Horses are nervous animals by nature and they often become wounded due to their "fight and flight" behaviour. A large UK survey of horse owners has shown that every year approximately 20% of adult horses (under 15 years) will sustain a wound (Owen et al. 2012). A recent Nigerian study found a wound prevalence of 10.1% in a

E. Jørgensen (✉)
Department of Immunology and Microbiology, Costerton Biofilm Center, Faculty of Health and Medical Sciences, University of Copenhagen, Copenhagen, Denmark
e-mail: elin.jorgensen@sund.ku.dk

S. Jacobsen
Department of Veterinary Clinical Sciences, Faculty of Health and Medical Sciences, University of Copenhagen, Taastrup, Denmark

K. Richter, K. Nørskov Kragh (eds.), *Antibiofilm Strategies*, Springer Series on Biofilms 11, https://doi.org/10.1007/978-3-031-10992-8_21

sample of 207 horses (Agina and Ihedioha 2017). In a 2015 report, The National Animal Health Monitoring System from the United States Department of Agriculture found that 27.8% of euthanasia cases in horses less than 1 year old and 16.3% of cases of euthanasia in adult horses (1–20 years old) were related to wounds, injury, and/or trauma. In addition, wounds, injury, and/or trauma affected 8% of young horses (1–5 years old) and 4.3% of adult horses (5–20 years old) (Anonymous 2016). Across surveys, wounds most commonly occur on the distal limbs (Wilmink et al. 2002; Spaas et al. 2013; Owen et al. 2012). Equine limb wounds frequently need to be managed by second intention healing, due to often severe tissue loss, contamination, and tightness of the skin, all of which prevent suturing of the wound. Second intention healing is prone to be impaired for equine limb wounds. As described below (Knottenbelt 1997; Wilmink et al. 2002), this impaired limb wound healing pattern is remarkably different from equine body wounds that normally heal well (Jacobs et al. 1984; Wilmink et al. 1999a; Jørgensen et al. 2019). Thus, especially limb wounds pose a substantial economic burden on the equine industry besides the obvious welfare issues (Theoret et al. 2016; Anonymous 2016; Perkins et al. 2005).

In both humans and horses, chronic inflammation and hypoxia are considered main contributors to the impaired healing pattern seen in chronic limb wounds (Bundgaard et al. 2016; Celeste et al. 2011; Mustoe 2004; Sørensen et al. 2014; Uccioli et al. 2015; Wilmink et al. 1999b). Since the beginning of this millenium, a new research topic within chronic wounds has been biofilm infections (Bjarnsholt et al. 2008; James et al. 2008). Biofilms have been found ubiquitously within human chronic wounds and there seems to be a link between biofilm and wound chronicity in wounds in humans (Malone et al. 2017; Schultz et al. 2017). However, it is not known exactly how and when biofilms impair wound healing and further research is needed in this area. It seems plausible to believe that bacteria would behave similarly in animals as in humans, so it is highly likely that biofilms also play a role in many chronic infection in veterinary medicine. However, biofilm as a research topic is quite new in the veterinary medicine area.

In this chapter, the impaired healing of equine limb wounds is described, to detail its resemblance to chronic wounds in humans. Biofilm in human chronic wounds and the few biofilm investigations within equine and other veterinary wounds are reviewed, and as biofilms seems to be yet another similarity between equine and human chronic wounds, the possibility of using the horse as a model animal is proposed.

2 Impaired Healing in Horses

As mentioned above, equine limb wounds are among the most frequent injuries in horses, and these wounds often need to heal by second intention due to tissue loss, contamination, and/or break-down of primary closure (Theoret et al. 2016; Theoret and Wilmink 2013). Limb wounds healing by second intention in horses heal

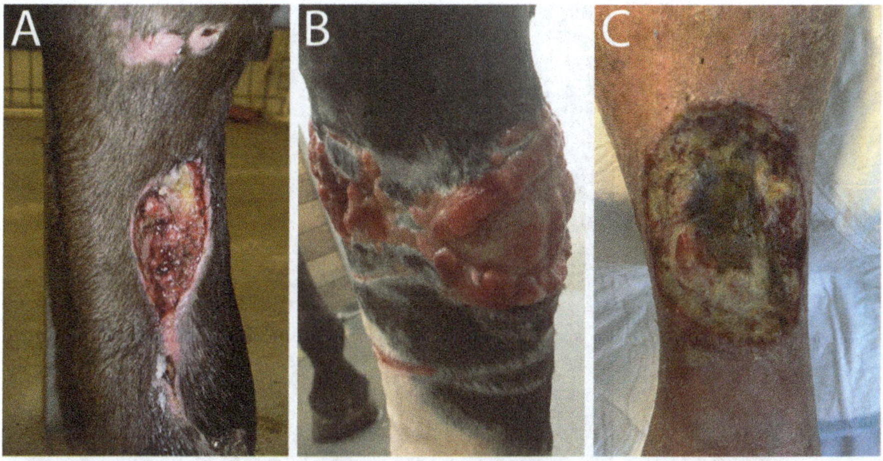

Fig. 1 Limb wounds healing by second intention. (**a**) Typical accidental equine limb wound with tissue loss hindering suturing and thus left to heal by second intention. (**b**) Unhealthy and partial exuberant granulation tissue in an equine accidental limb wound. (**c**) Venous leg ulcer in a human patient (courtesy of Klaus Kirketerp-Møller)

markedly slower and with more complications than body wounds (Jacobs et al. 1984; Wilmink et al. 1999a) and wounds in ponies (Bertone et al. 1985; Wilmink et al. 1999a, 2002). The equine contraction of limb wounds is inferior to body wounds, and a proposed reason could be the more chaotic arrangement of myofibroblasts in limb wounds (Wilmink et al. 1999b; Schwartz et al. 2002). The lack of contraction means that equine limb wounds rely mainly on epithelialisation for wound closure, which is also markedly slower in limb wounds than in body wounds. Reasons for the slower epithelialisation have been proposed, including exuberant granulation tissue (EGT) physically hindering epithelial cell migration, as well as inhibited keratinocyte mitosis (Wilmink et al. 1999b; Bertone 1989).

A frequent characteristic of impaired equine limb wound healing is the development of EGT, which very rarely occurs in body wounds (Jacobs et al. 1984; Sørensen et al. 2014; Theoret and Wilmink 2017; van den Boom et al. 2002). It is speculated that mitosis is not adequately inhibited in equine limb wounds when compared with body wounds, which may account for the often excessive formation of granulation tissue and non-healing wounds (Wilmink et al. 1999b) (Figs. 1b and 2a, b).

Two principal reasons for the significantly impaired healing of limb wounds with or without EGT formation have been proposed: (1) an insufficient acute inflammatory response and prolonged chronic inflammation (Bundgaard et al. 2016; Wilmink et al. 1999b) and (2) hypoxia, poor blood supply, and microvascular occlusion (Lepault et al. 2005; Sørensen et al. 2014; Celeste et al. 2011; Dubuc et al. 2006). These two aspects cause multiple imbalances in cytokines and growth factors (van

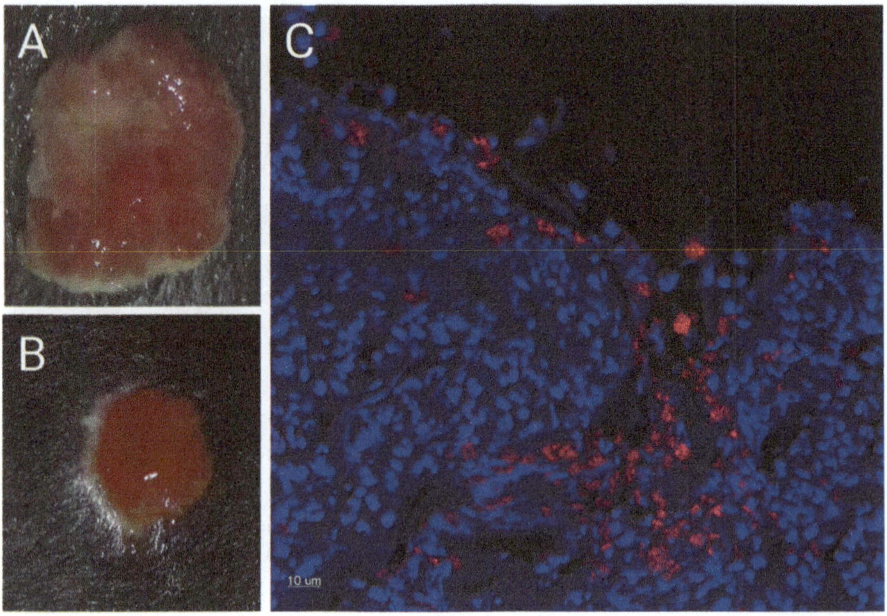

Fig. 2 Experimental equine wounds healing per second intention. (**a**) Limb wound 14 days after creation (10 days after inoculation), an unhealthy wound bed and exuberant granulation tissue protruding above the surface of the skin. (**b**) Body wound 14 days after creation (10 days after inoculation) displaying normal healing and healthy granulation tissue and the epithelial rim is just visible. Both wounds are to scale and started out the same size, thereby the marked difference in healing is evident. (**c**) *Staphylococcus aureus* biofilms near the surface of an equine experimental limb wound 14 days after creation (10 days after inoculation). A red fluorescent bacterial ribosomal peptide nuclei acid probe was used to identify *S. aureus*, equine cell nuclei appear blue (due to nuclear counterstain with DAPI stain)

den Boom et al. 2002; Theoret et al. 2001) as well as in collagen synthesis and degradation (Schwartz et al. 2002), thus driving a chronic wound healing trajectory.

How bioburden and potential presence of bacterial biofilms affect healing in equine traumatic wounds is unknown, though a negative influence is expected (Westgate et al. 2010). In our recent experimental equine wound study, biofilms seemed to impair limb wound healing further, while not affecting healing of body wounds (Jørgensen et al. 2019), as described below.

3 Human Chronic Wounds and Similarities to Equine Limb Wounds

Chronic wounds affect 1–2% of the human population in the Western world (Gottrup 2004; Frykberg and Banks 2015) and approximately 25% of residents in nursing homes (Rizzi et al. 2010). It is estimated that chronic wound care accounts for 2–5.5% of the healthcare budgets of developed countries (Olsson et al. 2019; Fife and Carter 2012; Phillips et al. 2016; Posnett and Franks 2008), and besides being a major economic burden on healthcare systems, losses related to productivity should also be considered (Augustin et al. 2014). Of even more importance is the severely reduced life quality of the patients due to pain, reduced mobility, social isolation, and both physical and emotional distress (Augustin et al. 2014; Herber et al. 2007; Olsson et al. 2019).

The prevalence of chronic wounds in humans is increasing. Obesity is an increasing public health threat that is leading to a global rise in the prevalence of diabetes and vascular diseases, as well as sequential diabetic foot ulcers (DFUs) and venous leg ulcers (VLUs) (Sen et al. 2009; Lipsky et al. 2016; Sen 2019). The ability to heal deteriorates with increasing age, and as life expectancy increases along with the proportion of elderly people, this further contributes to the increased prevalence of chronic wounds (Thomason and Hardman 2009; Mboi et al. 2018). Together with pressure ulcers, DFUs and VLUs comprise the majority of chronic wounds in humans, and these wound types mainly occur in elderly and/or diseased individuals (Mustoe 2004). Besides systemic and cellular ageing, common causative factors for chronic wound development are bacterial colonisation (including biofilm infection) and repeated ischaemia-reperfusion injury resulting in a chronic inflammatory phase failing to progress to healing (Mustoe 2004; Uccioli et al. 2015; Gurjala et al. 2011; Sen 2019).

DFU is one of the most common and serious complications of diabetes, affecting 15–25% of diabetic patients during their lifetime (Mendes and Neves 2012; Boulton et al. 2005; Singh et al. 2005). Approximately 2–3.6% of patients with a DFU will end up with a foot or limb amputation (Bondor et al. 2016; Margolis et al. 2011). In patients with diabetes, neuropathy causes insensible and deformed feet leading to ulcer development due to small intrinsic or extrinsic trauma. Often peripheral arterial diseases like atherosclerosis further contribute with ischaemia of the feet, thereby impairing healing (Mendes and Neves 2012). Treatment of DFU is centred on off-loading, correcting underlying metabolic imbalances (e.g. hyperglycaemia), revascularisation, and infection control (Uccioli et al. 2015).

In venous stasis syndrome, inactivity of calf muscle pumps causes chronic venous insufficiency and associated venous hypertension, which can eventually lead to a VLU (O'Brien et al. 2013). Between 1% and 2% of the population over 60 years of age will experience a VLU (Heit et al. 2001; Graham et al. 2003) (Fig. 1c). Treatment of VLUs includes compression and leg elevation (O'Meara et al. 2012; O'Brien et al. 2013). Exercise programmes focusing on the calf muscles can improve healing outcomes and prevent future incidences (O'Brien et al. 2013).

Pressure ulcers occur when sustained pressure and shearing forces over a bony prominence impair the flow through the tissue vasculature, thereby inducing hypoxia that will eventually injure the skin. Severely ill patients unable to reposition themselves are at particular risk, as well as patients with urinary and/or faecal incontinence. The prevalence of pressure ulcers in hospital patients is between 12% and 20% (Jenkins and O'Neal 2010; Gallagher et al. 2008).

The similarities between equine limb wounds and chronic wounds in humans have been highlighted and discussed over the last two decades (Cochrane 1997; Knottenbelt 1997; Textor et al. 2018; Lefebvre-Lavoie et al. 2005; Theoret and Wilmink 2013; Gordillo et al. 2013) (Fig. 1) and include but are not limited to:

I. Weak and prolonged inflammation. Chronic wounds are often restrained in a drawn-out and dysregulated inflammatory condition (Beidler et al. 2009; Bodaan et al. 2016; Diegelmann and Evans 2004; Theoret and Wilmink 2013; Wilmink et al. 1999b)
II. Local hypoxia, ischaemia, and decreased blood flow (Schreml et al. 2010; Sørensen et al. 2014; Jull et al. 2012)
III. Increased wound pH occur in chronic wounds (Percival et al. 2014; Schreml et al. 2014; Jørgensen et al. 2019)
IV. Lower extremities affected with impaired healing, whereas body wounds of both humans and horses normally heal well (Mustoe 2004; Knottenbelt 1997; Jacobs et al. 1984; Theoret 2017; Cochrane et al. 2003; Jørgensen et al. 2019)
V. Epithelialisation dominates the healing, while contraction only contributes to wound closure with approximately 30% in humans (Mogford and Mustoe 2001) and 15–20% in equids (Sardari et al. 2009; Wilmink et al. 1999a; Davidson et al. 2013)
VI. Protease activity is often increased (Cochrane 1997; Beidler et al. 2008; Jørgensen et al. 2020)
VII. *Staphylococcus aureus* and *Pseudomonas aeruginosa* are strong biofilm formers and common pathogens in wounds of both species (Freeman et al. 2009; Westgate et al. 2011; Hanson 2008; Serra et al. 2015; Kirketerp-Moller et al. 2008)

4 Biofilm in Human Chronic Wounds

A recent meta-analysis found 78% of chronic wounds in humans to contain biofilm, and the authors stated that biofilms must be considered ubiquitous in human chronic non-healing wounds (Malone et al. 2017; Johani et al. 2017). As discussed above, impaired wound healing in humans has different pathologies, yet biofilms also seem to play a role in the pathogeneses (Bjarnsholt et al. 2008; Malone et al. 2017; James et al. 2008). Additionally, in two recent consensus guidelines based on international panels of wound biofilm experts, biofilms were considered a major contributor to non-healing wounds in humans. The guidelines highlight that biofilm should always be suspected in wounds that do not follow the expected healing pattern (Schultz et al. 2017; Haesler et al. 2019). It is likely that biofilms have always been present in wounds and caused biofilm infections, but it is only over the past two decades that this interaction has been postulated and studied (Costerton et al. 1999; James et al. 2008; Bjarnsholt et al. 2008; Percival and Bowler 2004; Edwards and Harding 2004). The problem with biofilm in wounds is the biofilms' inherent tolerance towards antibiotics and the host responses, which complicate treatment markedly (Burmølle et al. 2010).

S. aureus and *P. aeruginosa* are common bacteria in chronic wounds in humans (Wolcott et al. 2016; Serra et al. 2015; Kirketerp-Moller et al. 2008; Smith et al. 2014); however, the majority of chronic wounds harbour between four and six bacterial species, and the bacterial composition may change over time (Gjødsbøl et al. 2006; Wolcott et al. 2016; Smith et al. 2014; Schultz et al. 2017). *S. aureus* is present at an early stage in most wounds, most likely due to its commensal position on the skin, whereas *P. aeruginosa* is often found in older wounds (Bjarnsholt et al. 2008; Mendes and Neves 2012). How bacteria settle in the wounds is uncertain, including whether they arrive as planktonic cells or as already aggregated small biofilms. Despite the presence of several bacterial species per wound, most biofilms detected in chronic wounds are mono-species (Burmølle et al. 2010; Kirketerp-Moller et al. 2008; Pastar et al. 2013), although multi-species biofilms have also been detected (Johani et al. 2017). When *S. aureus* and *P. aeruginosa* co-exist in wounds, they often settle in different areas; *P. aeruginosa* biofilms are often found deeper in the tissue (50–60 μm from the wound surface) and *S. aureus* closer (20–30 μm) to the wound surface (Fazli et al. 2009).

During biofilm infections, collateral damage is supposed to occur, as neutrophils try to kill the biofilms, but as often unsuccessful, their release of oxidative radicals and enzymes end up harming the wound/host tissue instead (Jensen et al. 2010). Thus, biofilm often induce an ineffective and unsuitable inflammation, not making the host capable of eradicating the infection. As described above for human chronic wounds, often hypoxia or low oxygen tension is involved, and this can be further attenuated by the activated leukocytes' and the biofilms' oxygen consumption (James et al. 2016a). Both hypoxia and weak/dysregulated inflammation most likely make it easier for bacteria to settle as biofilms in wounds; on the other hand, biofilms themselves can cause both conditions (James et al. 2016b; Alhede et al. 2009), thereby ending a vicious circle of wound chronicity.

5 Biofilm in Equine Wounds

Biofilm infections in relation to wounds have been studied to a very limited degree in veterinary medicine. The presence of biofilms has been reported in a number of traumatic wounds in three equine studies (Freeman et al. 2009; Van Hecke et al. 2017; Westgate et al. 2011). In 2009, Freeman and colleagues were the first to report biofilms in equine wounds. They examined biopsies from eight traumatic equine wounds using histology including Gram stain and reported biofilms to be found in the majority of the wounds (Freeman et al. 2009). Shortly after, Westgate and colleagues found biofilm in 6 out of 13 equine wounds (61.5%) using histology and Gram stain (Westgate et al. 2011). A third study found biofilm in 5 out of 50 equine wound samples (10%) using the same methods (Van Hecke et al. 2017). However, in these studies, suboptimal detection methods, i.e. histology and Gram stain, were used and not one of the gold standard methods from human biofilm research; confocal laser scanning microscopy (CLSM) or scanning electron microscopy is considered a gold standard technique for biofilm detection in human wounds (Hoiby et al. 2015; Burmølle et al. 2010). Often peptide nucleic acid (PNA) fluorescence in situ hybridisation (FISH) is combined with CLSM (Thomsen et al. 2010; Malic et al. 2009) to achieve great visualisation of the bacteria and biofilms in the tissue (Fig. 2c). It is further relevant to directly visualise the triggered immune cell response in the tissue to diagnose a biofilm, as a biofilm infection as opposed to contamination. Due to the suboptimal detection methods used in these equine studies, it is difficult to estimate the true prevalence of biofilm in equine traumatic limb wounds. With several similarities between equine and human impaired wound healing as described above (Cochrane 1997; Knottenbelt 1997; Lefebvre-Lavoie et al. 2005), it is most likely that biofilms also play a significant role in healing of chronic limb wounds in horses.

As biofilm is a relatively new topic in equine and veterinary wounds, reports also describe the in vitro biofilm forming capacity of bacterial isolates from equine wounds (Westgate et al. 2011; Nadzir et al. 2020). However, the same bacteria can behave markedly differently in vitro and in vivo (Cornforth et al. 2018; Roberts et al. 2015; Trevors 2011), thus the in vitro biofilm forming potential of bacteria does not reveal much information on whether they have been present in wounds as biofilms or not. This also emphasises the need for more in vivo investigations to better understand biofilm in (equine) wounds.

In our group, we have evaluated biofilm formation in equine experimental wounds in a series of studies. Our initial study (Jørgensen et al. 2017) investigated the presence of biofilms in 49 wound biopsies from equine body wounds and bandaged limb wounds created under clean surgical conditions and left open to heal by second intention. To detect biofilms, the gold standard method PNA FISH and subsequent CLSM was applied for the first time to equine wound samples. While no biofilms were detected in body wounds, biofilms occurred in all limb wounds within 1–2 weeks of wounding. From this study, it seems that the prevalence of biofilm in clean equine experimental limb wounds healing by second intention is 100% (Jørgensen et al. 2017).

Next, we established an equine biofilm-infected wound model (Jørgensen et al. 2019). Full-thickness wounds were created on the limbs and bodies of six horses and were bandaged and left to heal by second intention. Four days after wound creation, wounds either were inoculated with 10^4 *S. aureus* and 10^5 *P. aeruginosa* or were left non-inoculated (Fig. 2a, b). Wound healing was tracked until day 27 with clinical wound examinations that included wound scoring, pH measurements, and digital photography for area calculations.

Inoculated limb wounds healed even more slowly than non-inoculated limb wounds. Body wounds retracted markedly and thus increased in size immediately after creation, but healed more quickly than the limb wounds. Limb wounds had higher pH and poorer wound scores than body wounds. There were no significant differences in wound scores between inoculated and non-inoculated wounds, underlining the fact that biofilm infection is difficult to diagnose from clinical point of view (Jørgensen et al. 2019), as also well-known from human chronic wounds (Barker et al. 2017; Haesler et al. 2019; Johani et al. 2017; Percival et al. 2015; Schultz et al. 2017). In biopsies from body wounds, *P. aeruginosa* biofilms were detected immediately after inoculation and were quickly eradicated; non-inoculated body wounds did not contain biofilms at any time point. In biopsies from limb wounds, *S. aureus* biofilms were detected in some non-inoculated wounds and in the majority of inoculated wounds (Fig. 2c), where the biofilms were also larger. More seldom were *P. aeruginosa* biofilms detected in inoculated limb wounds. All detected biofilms were mono-species (Jørgensen et al. 2019).

As biofilms delayed healing in limb wounds, while body wound healing was not affected by the formed biofilms, the wound beds of equine limb and body wounds apparently influence the inoculating bacteria in different ways (Jørgensen et al. 2019). Similar findings are described for human wounds (Schultz et al. 2017; Ganesh et al. 2015). Prolonged inflammation as well as hypoxia in equine limb wounds might explain the settlement and persistence of biofilms in limb wounds. The elimination of biofilms in body wounds was quick and seemed uncomplicated, potentially due to the more competent inflammatory response in body wounds (Wilmink et al. 1999b; Bundgaard et al. 2016).

Biofilms seem to be widespread in equine experimental limb wounds and negatively affect healing. It is clear that diagnosing wound infections can be complicated, as wound scores and histology were affected to a very limited extent by the presence of biofilm (Jørgensen et al. 2019, 2020). It is likely that biofilms in equine limb wounds can be linked to wound chronicity similar to that in humans, as biofilms protracted the healing of equine limb wounds. Thus, veterinary clinicians must be aware of biofilms when treating equine wounds and suspect biofilm infections in wounds that despite appropriate treatment do not heal (Jacobsen 2017), as in human medicine (Woo and Sibbald 2009; Schultz et al. 2017; Haesler et al. 2019). For both human and equine wounds, no clinical detection/diagnostic methods are yet available to detect biofilm, i.e. PNA FISH and CLSM are only used in research so far. Hopefully, better and quicker methods to detect biofilm in wounds will be developed in the future. Further, no studies have investigated how to treat biofilm infection in equine wounds, and therefore at present the best strategy must be repeated surgical debridement as needed to keep the surface as "clean" as possible and applying topical antimicrobial strategies to eradicate biofilm/bacteria surviving the debridement before settlement as mature biofilms (Wolcott et al. 2010). In human medicine, no gold standard treatment for wound biofilm infections exists yet, so the treatment of both human and equine biofilm-infected wounds must rely on personal experience and best available evidence. It is beyond the scope of this chapter to list all the antimicrobial and anti-biofilm treatment options tested in human wound research. For equine wound treatment, topical treatment has been comprehensively reviewed (Jacobsen 2017).

As equine limb wounds share certain similarities with chronic wounds in humans, including the presence of biofilms, we suggest the horse as a relevant model animal for studying the interactions between wound biofilms and the host response more closely. The equine biofilm-infected wound model might further be useful to test biofilm-induced wound pathology, diagnostic methods, and improved anti-biofilm treatments.

6 Biofilm in Other Veterinary Wounds

Biofilms have also been detected in canine wounds, where König and colleagues have detected coccoid biofilms in relation to sutures in two chronic infected skin wounds and from an ovariohysterectomized uterus stump using histology (Konig et al. 2014b). They also examined 91 samples (either swab residues or tissue samples with intralesional suture material) from equine, canine, and feline wounds for the presence of biofilms in relation to the sutures. Similar to the equine studies described above suboptimal methodology (histology and Gram stain) was employed. Coccoid biofilm in relation to sutures in two canine wounds was detected (Konig et al. 2014a). A case report found coccoid biofilms in a dog with chronic non-healing pressure wounds over both elbows (Swanson et al. 2014). Further, a convenience sample of 140 *Staphylococcus pseudintermedius* isolates (including methicillin-resistant isolates) obtained from clinical infection or colonisation in dogs were examined for their biofilm forming ability, and not surprisingly the majority of these (96%) were reported as strong or moderate biofilm producers (Singh et al. 2013). However, as described above for the similar reports from equine samples, the biofilm forming abilities in vitro and the behaviour in vivo are not necessarily related.

7 Relevant Models of Biofilm Infection in Human Chronic Wounds: The Horse Has Its Place

Even though investigations in veterinary biofilm wounds infections are few, the clear similarity between equine and human chronic limb wounds might be useful to further understand and characterise the pathophysiology of biofilm wound infections. Wound biofilm research is a fairly young discipline and knowledge is still lacking on important topics like how and why biofilms settle in wounds, how and when biofilms affect healing, how biofilm infections can be diagnosed quickly in a clinical setting, and how biofilm infections in chronic wounds can be treated successfully (Snyder et al. 2017; Schultz et al. 2017). The overall aim of wound biofilm research is to improve healing and avoid amputations, thereby improving the quality of life for patients and reducing health costs. However, there seems to be a lack of good and relevant models to study biofilm and wound infections.

Biofilms have been studied intensively in standardised in vitro systems like flow cells and microtitre plate assays, which have provided important insights into biofilm metabolism, signalling pathways, and resistance and tolerance mechanisms (Haney et al. 2018; Roberts et al. 2015; Lebeaux et al. 2013). Biofilms studied in vitro are often attached to a surface and with a large and complex 3D structure ("mushroom structure"). As more information on in vivo wound biofilms is gathered, it is becoming clear that they do not resemble the in vitro biofilms, as in vivo biofilms are often fairly small with a simple structure and not attached to a surface (Roberts et al. 2015; Schultz et al. 2017). Furthermore, in vitro biofilms consist solely of bacteria-derived elements, whereas in vivo biofilms consist of both bacteria- and host-derived components (Bjarnsholt 2013). Wound biofilm assays have been developed to take into account the non-surface-attached nature of biofilms in wounds (Sun et al. 2008; Crone et al. 2015). However, one of the main concerns with in vitro biofilm models is naturally the absence of the host responses and wound healing. The transcriptomes of *P. aeruginosa* during in vitro conditions and in vivo infections are markedly different, indicating the clear differences in bacterial behaviour under laboratory conditions and clinical infections (Cornforth et al. 2018).

In vitro experiments are quick and can be useful, e.g., in screening potential therapeutics for anti-biofilm properties. While in vitro models not always align with in vivo conditions (Hengzhuang et al. 2014), they can be practical for the rapid evaluation of treatments before considering further studies in in vivo models (Schultz et al. 2017).

Information on how biofilms affect wound healing, interact with the immune system of the host, and respond to treatment in the complex setting of a clinical wound is essential and must be studied in vivo. The role of biofilms in wound healing is often studied in small laboratory animal models, but the validity of these models is questionable due to the healing patterns and skin anatomy being markedly different to that of humans (Trøstrup et al. 2016).

A direct negative effect of biofilms on wound healing has been proven in murine models, where staphylococcal biofilms delayed reepithelialisation (Schierle et al. 2009; Simonetti et al. 2008). Mainly, diabetic murine biofilm wound models (Watters et al. 2014; Zhao et al. 2012), rat wound models (Kanno et al. 2011), and rabbit ear wound models (Seth et al. 2012a; Leung et al. 2014; Gurjala et al. 2011) have been used to intensively study many aspects of wound healing, treatments options, and interactions between bacteria and immune defences. Some of the fundamental concerns regarding rodent and rabbit biofilm wound models include the relatively short duration of the studies, the young age and fast growth of the animals, and the more acute character of the wounds (Roy et al. 2014). Furthermore, the healing patterns and skin anatomy of these animals are strikingly different than those of humans; rodents and rabbits have loose skin that heal mainly by contraction due to their cutaneous trunci muscle, while humans have tight skin, especially on the lower limbs where most wounds occur, and human wounds heal mainly by epithelialisation (Seth et al. 2012b; Mustoe 2004; Lindblad 2008; Pastar et al. 2014). To inhibit contraction in rodents, many splinting strategies have been developed, but this does not change the fact that the healing is dominated by competent myofibroblasts (Davidson et al. 2013). The rabbit ear model uses the ear cartilage as a natural splint, but this forms an avascular wound base as the dermis is firmly attached to underlying cartilage, which is unlike human chronic wounds (Pastar et al.

2014). Rodent wound therapy studies have an estimated 53% concordance with studies in humans (Sullivan et al. 2001).

Porcine wound healing bear more resemblance to human wound healing than the small animal models as pigs' tight skin heal mainly by epithelialisation (Sullivan et al. 2001; Meyer et al. 1978; Klein et al. 2018). Porcine biofilm wound models (including diabetic models) have been developed and can give better insights into immune and treatment responses than the rodent and rabbit models (Roche et al. 2012; Pastar et al. 2013; Hirsch et al. 2008) and are therefore considered superior to other animal models (Klein et al. 2018; Gordillo et al. 2013; Ganesh et al. 2015). Disadvantages of porcine wound models are the animals relative young age and that they do not naturally suffer from a chronic healing pattern similar to humans (Klein et al. 2018). Porcine models are also more expensive, less accessible, and require other facilities and personnel training than for small animal models.

The translational value of any in vivo model is naturally important and choosing a model animal that to a larger extend resemble humans in terms of age, naturally occurring wound chronicity in extremities, and a similar healing pattern would seem clever. As listed above, human chronic wounds and equine limb wounds share many similarities, including biofilm formation. It might therefore be relevant to "think outside the box" regarding biofilm wound models and use the horse as a model animal to study biofilms in chronic wounds. Relevant research topics would be the pathophysiology of biofilm wound infections to improve our understanding of biofilm establishment, host responses, and possible treatment options.

Even though horses are large animals and thereby more expensive and equine facilities and experience are needed, several advantages exist for equine wound models. As several wounds can be created per horse and per limb due to their size, and as both normal (body wound) and impaired (limb wound) healing, in addition to inoculated and non-inoculated wounds can be studied directly in the same animal, stronger statistical analyses with repeated and paired measurements can be performed. As a result, fewer experimental animals are needed in accordance with the 3Rs (Replacement, Reduction, and Refinement) guideline on animal experimentation (Russell and Burch 1959). Previous studies in animal wound models have suggested that the number of wounds (rather than the number of animals) can be considered as the biological experimental unit, as no correlation exists between wounds in the same animal (Ansell et al. 2014). In our inoculated equine wound model, we included six horses with a total of 120 wounds (Jørgensen et al. 2019), so even with few animals more wounds could be investigated than in most biofilm wound model studies in pigs and rodents (Klein et al. 2018; Kanno et al. 2010; Zhao et al. 2010, 2012; Schierle et al. 2009; Pastar et al. 2013). Another advantage of an equine model in accordance with Refinement (3Rs) is working with an animal species that is used to handling and human interaction, including being bandaged, etc.

From a welfare point of view, the equine wound model with excisional wounds healing by second intention has its merits. As in a previous similar equine wound model without inoculation (Bundgaard et al. 2018), a systemic response to wounding was evident in our inoculated model, with increases in serum amyloid A (acute phase reactant), white blood cell count, and fibrinogen concentrations right after the wounding procedure. Inoculation did neither elicit a systemic response nor affect the well-being of the horses negatively (Jørgensen et al. 2019). Otherwise, the research horses experienced few and mild episodes of discomfort, pain, and lameness during the study, in line with previous non-inoculated equine wound studies (Bundgaard et al. 2018; Celeste et al. 2011; Theoret et al. 2001).

8 Conclusion and Future Perspectives

In horses, impaired healing is seen primarily in limb wounds whereas body wounds normally heal without complications. Chronic inflammation and hypoxia are the main characteristics of delayed wound healing in equine limbs, and biofilms seem to contribute to this healing pattern as well. However, biofilm formation in equine and veterinary wounds has been studied to a very limited degree and warrants further systematic investigation.

Horses might be relevant as a model animal for human chronic wounds since they: (1) naturally display impaired healing with chronic inflammation and hypoxia of limb wounds, (2) have thick and tight skin, and (3) heal mainly by epithelialisation in limb wounds, in the same way as humans. There are therefore many similarities between chronic wounds in horses and humans; and also for horses, biofilms are naturally occurring in these chronic limb wounds and seem to delay healing even further. Hopefully, equine wound models can be useful in future mechanistic and clinically relevant investigations to illuminate the intricacies of biofilm infections in wounds.

Compliance with Ethical Standards Parts of this chapter are based on a PhD project and thesis at the University of Copenhagen. The project was partly funded by a University of Copenhagen scholarship. All authors declare they have no conflict of interest. Informed consent from the patient whose wound is depicted in Fig. 1c was obtained.

References

Agina OA, Ihedioha JI (2017) Occurrence of wounds in Nigerian horses. J Appl Anim Welf Sci 20(4):372–380. https://doi.org/10.1080/10888705.2017.1343149

Alhede M, Bjarnsholt T, Jensen PO, Phipps RK, Moser C, Christophersen L, Christensen LD, van Gennip M, Parsek M, Hoiby N, Rasmussen TB, Givskov M (2009) Pseudomonas aeruginosa recognizes and responds aggressively to the presence of polymorphonuclear leukocytes. Microbiology 155(Pt 11):3500–3508. https://doi.org/10.1099/mic.0.031443-0

Anonymous (2016) Equine 2015 - baseline reference of Equine Health and Management in the United States, 2015. Accessed 27 July 2021

Ansell DM, Campbell L, Thomason HA, Brass A, Hardman MJ (2014) A statistical analysis of murine incisional and excisional acute wound models. Wound Repair Regen 22(2):281–287. https://doi.org/10.1111/wrr.12148

Augustin M, Brocatti LK, Rustenbach SJ, Schafer I, Herberger K (2014) Cost-of-illness of leg ulcers in the community. Int Wound J 11(3):283–292. https://doi.org/10.1111/j.1742-481X.2012.01089.x

Barker JC, Khansa I, Gordillo GM (2017) A formidable foe is sabotaging your results: what you should know about biofilms and wound healing. Plast Reconstr Surg 139(5):1184e–1194e. https://doi.org/10.1097/Prs.0000000000003325

Beidler SK, Douillet CD, Berndt DF, Keagy BA, Rich PB, Marston WA (2008) Multiplexed analysis of matrix metalloproteinases in leg ulcer tissue of patients with chronic venous insufficiency before and after compression therapy. Wound Repair Regen 16(5):642–648. https://doi.org/10.1111/j.1524-475X.2008.00415.x

Beidler SK, Douillet CD, Berndt DF, Keagy BA, Rich PB, Marston WA (2009) Inflammatory cytokine levels in chronic venous insufficiency ulcer tissue before and after compression therapy. J Vasc Surg 49(4):1013–1020. https://doi.org/10.1016/j.jvs.2008.11.049

Bertone AL (1989) Management of exuberant granulation-tissue. Vet Clin N Am Equine 5(3): 551–562. https://doi.org/10.1016/S0749-0739(17)30574-6

Bertone AL, Sullins KE, Stashak TS, Norrdin RW (1985) Effect of wound location and the use of topical collagen gel on exuberant granulation-tissue formation and wound-healing in the horse and pony. Am J Vet Res 46(7):1438–1444

Bjarnsholt T (2013) The role of bacterial biofilms in chronic infections. APMIS 121:1–58. https://doi.org/10.1111/apm.12099

Bjarnsholt T, Kirketerp-Møller K, Jensen PØ, Madsen KG, Phipps R, Krogfelt K, Høiby N, Givskov M (2008) Why chronic wounds will not heal: a novel hypothesis. Wound Repair Regen 16(1):2–10. https://doi.org/10.1111/j.1524-475X.2007.00283.x

Bodaan CJ, Wise LM, Wakelin KA, Stuart GS, Real NC, Mercer AA, Riley CB, Theoret C (2016) Short-term treatment of equine wounds with orf virus IL-10 and VEGF-E dampens inflammation and promotes repair processes without accelerating closure. Wound Repair Regen 24(6): 966–980. https://doi.org/10.1111/wrr.12488

Bondor CI, Veresiu IA, Florea B, Vinik EJ, Vinik AI, Gavan NA (2016) Epidemiology of diabetic foot ulcers and amputations in Romania: results of a cross-sectional quality of life questionnaire based survey. J Diabetes Res 2016:5439521. https://doi.org/10.1155/2016/5439521

Boulton AJ, Vileikyte L, Ragnarson-Tennvall G, Apelqvist J (2005) The global burden of diabetic foot disease. Lancet 366(9498):1719–1724. https://doi.org/10.1016/S0140-6736(05)67698-2

Bundgaard L, Bendixen E, Sorensen MA, Harman VM, Beynon RJ, Petersen LJ, Jacobsen S (2016) A selected reaction monitoring-based analysis of acute phase proteins in interstitial fluids from experimental equine wounds healing by secondary intention. Wound Repair Regen 24(3): 525–532. https://doi.org/10.1111/wrr.12425

Bundgaard L, Sorensen MA, Nilsson T, Salling E, Jacobsen S (2018) Evaluation of systemic and local inflammatory parameters and manifestations of Pain in an equine experimental wound model. J Equine Vet Sci 68:81–87. https://doi.org/10.1016/j.jevs.2018.05.219

Burmølle M, Thomsen TR, Fazli M, Dige I, Christensen L, Homøe P, Tvede M, Nyvad B, Tolker-Nielsen T, Givskov M, Moser C, Kirketerp-Møller K, Johansen HK, Høiby N, Jensen PØ, Sørensen SJ, Bjarnsholt T (2010) Biofilms in chronic infections – a matter of opportunity – monospecies biofilms in multispecies infections. FEMS Immunol Med Microbiol 59(3): 324–336. https://doi.org/10.1111/j.1574-695X.2010.00714.x

Celeste CJ, Deschene K, Riley CB, Theoret CL (2011) Regional differences in wound oxygenation during normal healing in an equine model of cutaneous fibroproliferative disorder. Wound Repair Regen 19(1):89–97. https://doi.org/10.1111/j.1524-475X.2010.00639.x

Cochrane CA (1997) Models in vivo of wound healing in the horse and the role of growth factors. Vet Dermatol 8(4):259–272. https://doi.org/10.1111/j.1365-3164.1997.tb00272.x

Cochrane CA, Pain R, Knottenbelt DC (2003) In-vitro wound contraction in the horse: differences between body and limb wounds. Wounds 15(6):175–181

Cornforth DM, Dees JL, Ibberson CB, Huse HK, Mathiesen IH, Kirketerp-Moller K, Wolcott RD, Rumbaugh KP, Bjarnsholt T, Whiteley M (2018) Pseudomonas aeruginosa transcriptome during human infection. Proc Natl Acad Sci U S A 115(22):E5125–E5134. https://doi.org/10.1073/pnas.1717525115

Costerton JW, Stewart PS, Greenberg EP (1999) Bacterial biofilms: a common cause of persistent infections. Science 284(5418):1318–1322

Crone S, Garde C, Bjarnsholt T, Alhede M (2015) A novel in vitro wound biofilm model used to evaluate low-frequency ultrasonic-assisted wound debridement. J Wound Care 24(2):64–72. https://doi.org/10.12968/jowc.2015.24.2.64

Davidson JM, Yu F, Opalenik SR (2013) Splinting strategies to overcome confounding wound contraction in experimental animal models. Adv Wound Care (New Rochelle) 2(4):142–148. https://doi.org/10.1089/wound.2012.0424

Diegelmann RF, Evans MC (2004) Wound healing: an overview of acute, fibrotic and delayed healing. Front Biosci 9:283–289

Dubuc V, Lepault E, Theoret CL (2006) Endothelial cell hypertrophy is associated with microvascular occlusion in horse wounds. Can J Vet Res 70(3):206–210

Edwards R, Harding KG (2004) Bacteria and wound healing. Curr Opin Infect Dis 17(2):91–96

Fazli M, Bjarnsholt T, Kirketerp-Moller K, Jorgensen B, Andersen AS, Krogfelt KA, Givskov M, Tolker-Nielsen T (2009) Nonrandom distribution of Pseudomonas aeruginosa and Staphylococcus aureus in chronic wounds. J Clin Microbiol 47(12):4084–4089. https://doi.org/10.1128/JCM.01395-09

Fife CE, Carter MJ (2012) Wound care outcomes and associated cost among patients treated in US outpatient wound centers: data from the US Wound Registry. Wounds 24(1):10–17

Freeman K, Woods E, Welsby S, Percival SL, Cochrane CA (2009) Biofilm evidence and the microbial diversity of horse wounds. Can J Microbiol 55(2):197–202. https://doi.org/10.1139/W08-115

Frykberg RG, Banks J (2015) Challenges in the treatment of chronic wounds. Adv Wound Care (New Rochelle) 4(9):560–582. https://doi.org/10.1089/wound.2015.0635

Gallagher P, Barry P, Hartigan I, McCluskey P, O'Connor K, O'Connor M (2008) Prevalence of pressure ulcers in three university teaching hospitals in Ireland. J Tissue Viability 17(4):103–109. https://doi.org/10.1016/j.jtv.2007.12.001

Ganesh K, Sinha M, Mathew-Steiner SS, Das A, Roy S, Sen CK (2015) Chronic wound biofilm model. Adv Wound Care 4(7):382–388. https://doi.org/10.1089/wound.2014.0587

Gjødsbøl K, Christensen JJ, Karlsmark T, Jørgensen B, Klein BM, Krogfelt KA (2006) Multiple bacterial species reside in chronic wounds: a longitudinal study. Int Wound J 3(3):225–231. https://doi.org/10.1111/j.1742-481X.2006.00159.x

Gordillo GM, Bernatchez SF, Diegelmann R, Di Pietro LA, Eriksson E, Hinz B, Hopf HW, Kirsner R, Liu P, Parnell LK, Sandusky GE, Sen CK, Tomic-Canic M, Volk SW, Baird A (2013) Preclinical models of wound healing: is man the model? Proceedings of the wound healing society symposium. Adv Wound Care 2(1):1–4. https://doi.org/10.1089/wound.2012.0367

Gottrup F (2004) A specialized wound-healing center concept: importance of a multidisciplinary department structure and surgical treatment facilities in the treatment of chronic wounds. Am J Surg 187:S38–S43. https://doi.org/10.1016/S0002-9610(03)00303-9

Graham ID, Harrison MB, Nelson EA, Lorimer K, Fisher A (2003) Prevalence of lower-limb ulceration: a systematic review of prevalence studies. Adv Skin Wound Care 16(6):305–316

Gurjala AN, Geringer MR, Seth AK, Hong SJ, Smeltzer MS, Galiano RD, Leung KP, Mustoe TA (2011) Development of a novel, highly quantitative in vivo model for the study of biofilm-impaired cutaneous wound healing. Wound Repair Regen 19(3):400–410. https://doi.org/10.1111/j.1524-475X.2011.00690.x

Haesler E, Swanson T, Ousey K, Carville K (2019) Clinical indicators of wound infection and biofilm: reaching international consensus. J Wound Care 28(Sup3b):s4–s12. https://doi.org/10.12968/jowc.2019.28.Sup3b.S4

Haney EF, Trimble MJ, Cheng JT, Valle Q, Hancock REW (2018) Critical assessment of methods to quantify biofilm growth and evaluate antibiofilm activity of host Defence peptides. Biomol Ther 8(2):biom8020029. https://doi.org/10.3390/biom8020029

Hanson RR (2008) Complications of equine wound management and dermatologic surgery. Vet Clin North Am Equine Pract 24(3):663–696. https://doi.org/10.1016/j.cveq.2008.10.005

Heit JA, Rooke TW, Silverstein MD, Mohr DN, Lohse CM, Petterson TM, O'Fallon WM, Melton LJ 3rd (2001) Trends in the incidence of venous stasis syndrome and venous ulcer: a 25-year population-based study. J Vasc Surg 33(5):1022–1027. https://doi.org/10.1067/mva.2001.113308

Hengzhuang W, Hoiby N, Ciofu O (2014) Pharmacokinetics and pharmacodynamics of antibiotics in biofilm infections of Pseudomonas aeruginosa in vitro and in vivo. Methods Mol Biol 1147: 239–254. https://doi.org/10.1007/978-1-4939-0467-9_17

Herber OR, Schnepp W, Rieger MA (2007) A systematic review on the impact of leg ulceration on patients' quality of life. Health Qual Life Outcomes 5(1):44. https://doi.org/10.1186/1477-7525-5-44

Hirsch T, Spielmann M, Zuhaili B, Koehler T, Fossum M, Steinau HU, Yao F, Steinstraesser L, Onderdonk AB, Eriksson E (2008) Enhanced susceptibility to infections in a diabetic wound healing model. BMC Surg 8:5. https://doi.org/10.1186/1471-2482-8-5

Hoiby N, Bjarnsholt T, Moser C, Bassi GL, Coenye T, Donelli G, Hall-Stoodley L, Hola V, Imbert C, Kirketerp-Moller K, Lebeaux D, Oliver A, Ullmann AJ, Williams C, Biofilms ESGf, Consulting External Expert Werner Zimmerli (2015) ESCMID guideline for the diagnosis and treatment of biofilm infections 2014. Clin Microbiol Infect 21(Suppl 1):S1–S25. https://doi.org/10.1016/j.cmi.2014.10.024

Jacobs KA, Leach DH, Fretz PB, Townsend GG (1984) Comparative aspects of the healing of excisional wounds on the leg and body of horses. Vet Surg 13(2):83–90

Jacobsen S (2017) Topical wound treatments and wound-care products. In: Theoret C, Schumacher J (eds) Equine wound management. Wiley Blackwell, New York, pp 75–103

James GA, Swogger E, Wolcott R, Ed P, Secor P, Sestrich J, Costerton JW, Stewart PS (2008) Biofilms in chronic wounds. Wound Repair Regen 16(1):37–44. https://doi.org/10.1111/j.1524-475X.2007.00321.x

James GA, Ge Zhao A, Usui M, Underwood RA, Nguyen H, Beyenal H, deLancey PE, Agostinho Hunt A, Bernstein HC, Fleckman P, Olerud J, Williamson KS, Franklin MJ, Stewart PS (2016a) Microsensor and transcriptomic signatures of oxygen depletion in biofilms associated with chronic wounds. Wound Repair Regen 24(2):373–383. https://doi.org/10.1111/wrr.12401

James GA, Zhao AG, Usui M, Underwood RA, Nguyen H, Beyenal H, Pulcini ED, Hunt AA, Bernstein HC, Fleckman P, Olerud J, Williamson KS, Franklin MJ, Stewart PS (2016b) Microsensor and transcriptomic signatures of oxygen depletion in biofilms associated with chronic wounds. Wound Repair Regen 24(2):373–383. https://doi.org/10.1111/wrr.12401

Jenkins ML, O''Neal E (2010) Pressure ulcer prevalence and incidence in acute care. Adv Skin Wound Care 23(12):556–559. https://doi.org/10.1097/01.ASW.0000391184.43845.c1

Jensen PO, Givskov M, Bjarnsholt T, Moser C (2010) The immune system vs. Pseudomonas aeruginosa biofilms. FEMS Immunol Med Microbiol 59(3):292–305. https://doi.org/10.1111/j.1574-695X.2010.00706.x

Johani K, Malone M, Jensen S, Gosbell I, Dickson H, Hu H, Vickery K (2017) Microscopy visualisation confirms multi-species biofilms are ubiquitous in diabetic foot ulcers. Int Wound J 14(6):1160–1169. https://doi.org/10.1111/iwj.12777

Jørgensen E, Bay L, Bjarnsholt T, Bundgaard L, Sorensen MA, Jacobsen S (2017) The occurrence of biofilm in an equine experimental wound model of healing by secondary intention. Vet Microbiol 204:90–95. https://doi.org/10.1016/j.vetmic.2017.03.011

Jørgensen E, Bay L, Skovgaard LT, Bjarnsholt T, Jacobsen S (2019) An equine wound model to study effects of bacterial aggregates on wound healing. Adv Wound Care (New Rochelle) 8(10): 487–498. https://doi.org/10.1089/wound.2018.0901

Jørgensen E, Hjerpe FB, Hougen HP, Bjarnsholt T, Berg LC, Jacobsen S (2020) Histologic changes and gene expression patterns in biopsy specimens from bacteria-inoculated and noninoculated excisional body and limb wounds in horses healing by second intention. Am J Vet Res 81(3): 276–284. https://doi.org/10.2460/ajvr.81.3.276

Jull AB, Arroll B, Parag V, Waters J (2012) Pentoxifylline for treating venous leg ulcers. Cochrane Database Syst Rev 12:CD001733. https://doi.org/10.1002/14651858.CD001733.pub3

Kanno E, Toriyabe S, Zhang L, Imai Y, Tachi M (2010) Biofilm formation on rat skin wounds by Pseudomonas aeruginosa carrying the green fluorescent protein gene. Exp Dermatol 19(2): 154–156. https://doi.org/10.1111/j.1600-0625.2009.00931.x

Kanno E, Kawakami K, Ritsu M, Ishii K, Tanno H, Toriyabe S, Imai Y, Maruyama R, Tachi M (2011) Wound healing in skin promoted by inoculation with Pseudomonas aeruginosa PAO1: the critical role of tumor necrosis factor-alpha secreted from infiltrating neutrophils. Wound Repair Regen 19(5):608–621. https://doi.org/10.1111/j.1524-475X.2011.00721.x

Kirketerp-Moller K, Jensen PO, Fazli M, Madsen KG, Pedersen J, Moser C, Tolker-Nielsen T, Hoiby N, Givskov M, Bjarnsholt T (2008) Distribution, organization, and ecology of bacteria in chronic wounds. J Clin Microbiol 46(8):2717–2722. https://doi.org/10.1128/JCM.00501-08

Klein P, Sojka M, Kucera J, Matonohova J, Pavlik V, Nemec J, Kubickova G, Slavkovsky R, Szuszkiewicz K, Danek P, Rozkot M, Velebny V (2018) A porcine model of skin wound infected with a polybacterial biofilm. Biofouling 34(2):226–236. https://doi.org/10.1080/08927014.2018.1425684

Knottenbelt DC (1997) Equine wound management: are there significant differences in healing at different sites on the body? Vet Dermatol 8(4):273–290. https://doi.org/10.1111/j.1365-3164.1997.tb00273.x

Konig L, Klopfleisch R, Kershaw O, Gruber AD (2014a) Prevalence of biofilms on surgical suture segments in wounds of dogs, cats, and horses. Vet Pathol. https://doi.org/10.1177/0300985814535609

Konig LM, Klopfleisch R, Hoper D, Gruber AD (2014b) Next generation sequencing analysis of biofilms from three dogs with postoperative surgical site infection. Int Sch Res Notices 2014: 282971. https://doi.org/10.1155/2014/282971

Lebeaux D, Chauhan A, Rendueles O, Beloin C (2013) From in vitro to in vivo models of bacterial biofilm-related infections. Pathogens 2(2):288–356. https://doi.org/10.3390/pathogens2020288

Lefebvre-Lavoie J, Lussier JG, Theoret CL (2005) Profiling of differentially expressed genes in wound margin biopsies of horses using suppression subtractive hybridization. Physiol Genomics 22(2):157–170. https://doi.org/10.1152/physiolgenomics.00018.2005

Lepault E, Celeste C, Dore M, Martineau D, Theoret CL (2005) Comparative study on microvascular occlusion and apoptosis in body and limb wounds in the horse. Wound Repair Regen 13(5):520–529. https://doi.org/10.1111/j.1067-1927.2005.00073.x

Leung KP, D'Arpa P, Seth AK, Geringer MR, Jett M, Xu W, Hong SJ, Galiano RD, Chen T, Mustoe TA (2014) Dermal wound transcriptomic responses to infection with Pseudomonas aeruginosa versus Klebsiella pneumoniae in a rabbit ear wound model. BMC Clin Pathol 14:20. https://doi.org/10.1186/1472-6890-14-20

Lindblad WJ (2008) Considerations for selecting the correct animal model for dermal wound-healing studies. J Biomater Sci Polym Ed 19(8):1087–1096. https://doi.org/10.1163/156856208784909390

Lipsky BA, Aragon-Sanchez J, Diggle M, Embil J, Kono S, Lavery L, Senneville E, Urbancic-Rovan V, Van Asten S, International Working Group on the Diabetic Foot, Peters EJ (2016) IWGDF guidance on the diagnosis and management of foot infections in persons with diabetes. Diabetes Metab Res Rev 32(Suppl 1):45–74. https://doi.org/10.1002/dmrr.2699

Malic S, Hill KE, Hayes A, Percival SL, Thomas DW, Williams DW (2009) Detection and identification of specific bacteria in wound biofilms using peptide nucleic acid fluorescent in situ hybridization (PNA FISH). Microbiology 155(Pt 8):2603–2611. https://doi.org/10.1099/mic.0.028712-0

Malone M, Bjarnsholt T, McBain AJ, James GA, Stoodley P, Leaper D, Tachi M, Schultz G, Swanson T, Wolcott RD (2017) The prevalence of biofilms in chronic wounds: a systematic review and meta-analysis of published data. J Wound Care 26(1):20–25. https://doi.org/10.12968/jowc.2017.26.1.20

Margolis DJ, Malay DS, Hoffstad OJ, Leonard CE, MaCurdy T, Lopez de Nava K, Tan Y, Molina T, Siegel KL (2011) Prevalence of diabetes, diabetic foot ulcer, and lower extremity amputation among medicare beneficiaries, 2006 to 2008. Data points #1. In: Data Points Publication Series. Agency for Healthcare Research and Quality (US), Rockville, MD

Mboi N, Murty Surbakti I, Trihandini I, Elyazar I, Houston Smith K, Bahjuri Ali P, Kosen S, Flemons K, Ray SE, Cao J, Glenn SD, Miller-Petrie MK, Mooney MD, Ried JL, Nur Anggraini Ningrum D, Idris F, Siregar KN, Harimurti P, Bernstein RS, Pangestu T, Sidharta Y, Naghavi M, Murray CJL, Hay SI (2018) On the road to universal health care in Indonesia, 1990-2016: a systematic analysis for the Global Burden of Disease Study 2016. Lancet 392(10147):581–591. https://doi.org/10.1016/S0140-6736(18)30595-6

Mendes JJ, Neves J (2012) Diabetic foot infections: current diagnosis and treatment. J Diabetes Foot Complications 4(2):26–45

Meyer W, Schwarz R, Neurand K (1978) The skin of domestic mammals as a model for human skin, with special reference to the domestic pig. Curr Probl Dermatol 7:39–52

Mogford JE, Mustoe TA (2001) Experimental models of wound healing. In: Falanga V (ed) Cutaneous wound healing, 1st edn. Martin Dunitz Ltd, London, pp 109–122

Mustoe T (2004) Understanding chronic wounds: a unifying hypothesis on their pathogenesis and implications for therapy. Am J Surg 187(5A):65S–70S. https://doi.org/10.1016/S0002-9610(03)00306-4

Nadzir NAA, Zakaria Z, Adzahan NM, Mayaki AM (2020) Antibiotic susceptibilities of biofilm producing bacteria isolated from horse wounds. Explor Anim Med Res 10(1):42–49

O'Meara S, Cullum N, Nelson EA, Dumville JC (2012) Compression for venous leg ulcers. Cochrane Database Syst Rev 11(11):CD000265. https://doi.org/10.1002/14651858.CD000265.pub3

O'Brien J, Edwards H, Stewart I, Gibbs H (2013) A home-based progressive resistance exercise programme for patients with venous leg ulcers: a feasibility study. Int Wound J 10(4):389–396. https://doi.org/10.1111/j.1742-481X.2012.00995.x

Olsson M, Jarbrink K, Divakar U, Bajpai R, Upton Z, Schmidtchen A, Car J (2019) The humanistic and economic burden of chronic wounds: a systematic review. Wound Repair Regen 27(1):114–125. https://doi.org/10.1111/wrr.12683

Owen KR, Singer ER, Clegg PD, Ireland JL, Pinchbeck GL (2012) Identification of risk factors for traumatic injury in the general horse population of north-west England, Midlands and north Wales. Equine Vet J 44(2):143–148. https://doi.org/10.1111/j.2042-3306.2011.00387.x

Pastar I, Nusbaum AG, Gil J, Patel SB, Chen J, Valdes J, Stojadinovic O, Plano LR, Tomic-Canic M, Davis SC (2013) Interactions of methicillin resistant Staphylococcus aureus USA300 and Pseudomonas aeruginosa in polymicrobial wound infection. PLoS One 8(2):e56846. https://doi.org/10.1371/journal.pone.0056846

Pastar I, Stojadinovic O, Yin NC, Ramirez H, Nusbaum AG, Sawaya A, Patel SB, Khalid L, Isseroff RR, Tomic-Canic M (2014) Epithelialization in wound healing: a comprehensive review. Adv Wound Care (New Rochelle) 3(7):445–464. https://doi.org/10.1089/wound.2013.0473

Percival SL, Bowler P (2004) Biofilms and their role in wound healing. Wounds 16(7):234–240

Percival SL, McCarty S, Hunt JA, Woods EJ (2014) The effects of pH on wound healing, biofilms, and antimicrobial efficacy. Wound Repair Regen 22(2):174–186. https://doi.org/10.1111/wrr.12125

Percival SL, Vuotto C, Donelli G, Lipsky BA (2015) Biofilms and wounds: an identification algorithm and potential treatment options. Adv Wound Care (New Rochelle) 4(7):389–397. https://doi.org/10.1089/wound.2014.0574

Perkins NR, Reid SW, Morris RS (2005) Risk factors for musculoskeletal injuries of the lower limbs in thoroughbred racehorses in New Zealand. N Z Vet J 53(3):171–183. https://doi.org/10.1080/00480169.2005.36502

Phillips CJ, Humphreys I, Fletcher J, Harding K, Chamberlain G, Macey S (2016) Estimating the costs associated with the management of patients with chronic wounds using linked routine data. Int Wound J 13(6):1193–1197. https://doi.org/10.1111/iwj.12443

Posnett J, Franks PJ (2008) The burden of chronic wounds in the UK. Nurs Times 104(3):44–45

Rizzi SC, Upton Z, Bott K, Dargaville TR (2010) Recent advances in dermal wound healing: biomedical device approaches. Expert Rev Med Devices 7(1):143–154. https://doi.org/10.1586/erd.09.57

Roberts AE, Kragh KN, Bjarnsholt T, Diggle SP (2015) The limitations of in vitro experimentation in understanding biofilms and chronic infection. J Mol Biol 427(23):3646–3661. https://doi.org/10.1016/j.jmb.2015.09.002

Roche ED, Renick PJ, Tetens SP, Ramsay SJ, Daniels EQ, Carson DL (2012) Increasing the presence of biofilm and healing delay in a porcine model of MRSA-infected wounds. Wound Repair Regen 20(4):537–543. https://doi.org/10.1111/j.1524-475X.2012.00808.x

Roy S, Elgharably H, Sinha M, Ganesh K, Chaney S, Mann E, Miller C, Khanna S, Bergdall VK, Powell HM, Cook CH, Gordillo GM, Wozniak DJ, Sen CK (2014) Mixed-species biofilm compromises wound healing by disrupting epidermal barrier function. J Pathol 233(4):331–343. https://doi.org/10.1002/path.4360

Russell WMS, Burch RL (1959) The sources, incidence, and removal of inhumanity. In: The principles of humane experimental technique, 1st edn. Methuen & Co Ltd, London

Sardari K, Kazemi H, Emami M, Movasaghi A, Goli A (2009) Role of collagen cross-linking on equine wound contraction and healing. Comp Clin Pathol 18(3):239–247. https://doi.org/10.1007/s00580-008-0789-0

Schierle CF, De la Garza M, Mustoe TA, Galiano RD (2009) Staphylococcal biofilms impair wound healing by delaying reepithelialization in a murine cutaneous wound model. Wound Repair Regen 17(3):354–359. https://doi.org/10.1111/j.1524-475X.2009.00489.x

Schreml S, Szeimies RM, Prantl L, Karrer S, Landthaler M, Babilas P (2010) Oxygen in acute and chronic wound healing. Br J Dermatol 163(2):257–268. https://doi.org/10.1111/j.1365-2133.2010.09804.x

Schreml S, Meier RJ, Kirschbaum M, Kong SC, Gehmert S, Felthaus O, Küchler S, Sharpe JR, Wöltje K, Weiß KT, Albert M, Seidl U, Schröder J, Morsczeck C, Prantl L, Duschl C, Pedersen SF, Gosau M, Berneburg M, Wolfbeis OS, Landthaler M, Babilas P (2014) Luminescent dual sensors reveal extracellular pH-gradients and hypoxia on chronic wounds that disrupt epidermal repair. Theranostics 4(7):721–735. https://doi.org/10.7150/thno.9052

Schultz G, Bjarnsholt T, James GA, Leaper DJ, McBain AJ, Malone M, Stoodley P, Swanson T, Tachi M, Wolcott RD, Global Wound Biofilm Expert Panel (2017) Consensus guidelines for the identification and treatment of biofilms in chronic nonhealing wounds. Wound Repair Regen 25(5):744–757. https://doi.org/10.1111/wrr.12590

Schwartz AJ, Wilson DA, Keegan KG, Ganjam VK, Sun Y, Weber KT, Zhang J (2002) Factors regulating collagen synthesis and degradation during second-intention healing of wounds in the thoracic region and the distal aspect of the forelimb of horses. Am J Vet Res 63(11):1564–1570

Sen CK (2019) Human wounds and its burden: an updated compendium of estimates. Adv Wound Care (New Rochelle) 8(2):39–48. https://doi.org/10.1089/wound.2019.0946

Sen CK, Gordillo GM, Roy S, Kirsner R, Lambert L, Hunt TK, Gottrup F, Gurtner GC, Longaker MT (2009) Human skin wounds: a major and snowballing threat to public health and the economy. Wound Repair Regen 17(6):763–771. https://doi.org/10.1111/j.1524-475X.2009.00543.x

Serra R, Grande R, Butrico L, Rossi A, Settimio UF, Caroleo B, Amato B, Gallelli L, de Franciscis S (2015) Chronic wound infections: the role of Pseudomonas aeruginosa and Staphylococcus aureus. Expert Rev Anti-Infect Ther 13(5):605–613. https://doi.org/10.1586/14787210.2015.1023291

Seth AK, Geringer MR, Gurjala AN, Abercrombie JA, Chen P, You T, Hong SJ, Galiano RD, Mustoe TA, Leung KP (2012a) Understanding the host inflammatory response to wound infection: an in vivo study of Klebsiella pneumoniae in a rabbit ear wound model. Wound Repair Regen 20(2):214–225. https://doi.org/10.1111/j.1524-475X.2012.00764.x

Seth AK, Geringer MR, Hong SJ, Leung KP, Mustoe TA, Galiano RD (2012b) In vivo modeling of biofilm-infected wounds: a review. J Surg Res 178(1):330–338. https://doi.org/10.1016/j.jss.2012.06.048

Simonetti O, Cirioni O, Ghiselli R, Goteri G, Scalise A, Orlando F, Silvestri C, Riva A, Saba V, Madanahally KD, Offidani A, Balaban N, Scalise G, Giacometti A (2008) RNAIII-inhibiting peptide enhances healing of wounds infected with methicillin-resistant Staphylococcus aureus. Antimicrob Agents Chemother 52(6):2205–2211. https://doi.org/10.1128/AAC.01340-07

Singh N, Armstrong DG, Lipsky BA (2005) Preventing foot ulcers in patients with diabetes. JAMA 293(2):217–228. https://doi.org/10.1001/jama.293.2.217

Singh A, Walker M, Rousseau J, Weese JS (2013) Characterization of the biofilm forming ability of Staphylococcus pseudintermedius from dogs. BMC Vet Res 9:93. https://doi.org/10.1186/1746-6148-9-93

Smith ME, Robinowitz N, Chaulk P, Johnson K (2014) Comparison of chronic wound culture techniques: swab versus curetted tissue for microbial recovery. Br J Community Nurs 19:S22–S26. https://doi.org/10.12968/bjcn.2014.19.Sup9.S22

Snyder RJ, Bohn G, Hanft J, Harkless L, Kim P, Lavery L, Schultz G, Wolcott R (2017) Wound biofilm: current perspectives and strategies on biofilm disruption and treatments. Wounds 29(6): S1–S17

Sørensen MA, Pedersen LJ, Bundgaard L, Toft N, Jacobsen S (2014) Regional disturbances in metabolism and blood flow in equine limb wounds healing with formation of exuberant granulation tissue. Wound Repair Regen 22(5):647–653

Spaas JH, Broeckx S, Van de Walle GR, Polettini M (2013) The effects of equine peripheral blood stem cells on cutaneous wound healing: a clinical evaluation in four horses. Clin Exp Dermatol 38(3):280–284. https://doi.org/10.1111/ced.12068

Sullivan TP, Eaglstein WH, Davis SC, Mertz P (2001) The pig as a model for human wound healing. Wound Repair Regen 9(2):66–76. https://doi.org/10.1046/j.1524-475x.2001.00066.x

Sun Y, Dowd SE, Smith E, Rhoads DD, Wolcott RD (2008) In vitro multispecies Lubbock chronic wound biofilm model. Wound Repair Regen 16(6):805–813. https://doi.org/10.1111/j.1524-475X.2008.00434.x

Swanson EA, Freeman LJ, Seleem MN, Snyder PW (2014) Biofilm-infected wounds in a dog. J Am Vet Med Assoc 244(6):699–707. https://doi.org/10.2460/javma.244.6.699

Textor JA, Clark KC, Walker NJ, Aristizobal FA, Kol A, LeJeune SS, Bledsoe A, Davidyan A, Gray SN, Bohannon-Worsley LK, Woolard KD, Borjesson DL (2018) Allogeneic stem cells Alter gene expression and improve healing of distal limb wounds in horses. Stem Cell Transl Med 7(1):98–108. https://doi.org/10.1002/sctm.17-0071

Theoret C (2017) Physiology of wound healing. In: Theoret C, Schumacher J (eds) Equine wound management, 3rd edn. Wiley, Ames, Iowa, pp 1–13

Theoret CL, Wilmink JM (2013) Aberrant wound healing in the horse: naturally occurring conditions reminiscent of those observed in man. Wound Repair Regen 21(3):365–371. https://doi.org/10.1111/wrr.12018

Theoret C, Wilmink J (2017) Exuberant granulation tissue. In: Theoret C, Schumacher J (eds) Equine wound management, 3rd edn. Wiley, Ames, Iowa, pp 369–384

Theoret CL, Barber SM, Moyana TN, Gordon JR (2001) Expression of transforming growth factor β1, β3, and basic fibroblast growth factor in full-thickness skin wounds of equine limbs and thorax. Vet Surg 30(3):269–277. https://doi.org/10.1053/jvet.2001.23341

Theoret CL, Bolwell CF, Riley CB (2016) A cross-sectional survey on wounds in horses in New Zealand. N Z Vet J 64(2):90–94. https://doi.org/10.1080/00480169.2015.1091396

Thomason HA, Hardman MJ (2009) Delayed wound healing in elderly people. Rev Clin Gerontol 19(03):171–184. https://doi.org/10.1017/S095925980999027X

Thomsen TR, Aasholm MS, Rudkjøbing VB, Saunders AM, Bjarnsholt T, Givskov M, Kirketerp-Møller K, Nielsen PH (2010) The bacteriology of chronic venous leg ulcer examined by culture-independent molecular methods. Wound Repair Regen 18(1):38–49. https://doi.org/10.1111/j.1524-475X.2009.00561.x

Trevors JT (2011) Viable but non-culturable (VBNC) bacteria: gene expression in planktonic and biofilm cells. J Microbiol Methods 86(2):266–273. https://doi.org/10.1016/j.mimet.2011.04.018

Trøstrup H, Thomsen K, Calum H, Høiby N, Moser C (2016) Animal models of chronic wound care: the application of biofilms in clinical research. Chronic Wound Care Manag Res 3:123–132. https://doi.org/10.2147/CWCMR.S84361

Uccioli L, Izzo V, Meloni M, Vainieri E, Ruotolo V, Giurato L (2015) Non-healing foot ulcers in diabetic patients: general and local interfering conditions and management options with advanced wound dressings. J Wound Care 24(4 Suppl):35–42. https://doi.org/10.12968/jowc.2015.24.Sup4b.35

van den Boom R, Wilmink JM, O'Kane S, Wood J, Ferguson MW (2002) Transforming growth factor-beta levels during second-intention healing are related to the different course of wound contraction in horses and ponies. Wound Repair Regen 10(3):188–194

Van Hecke LL, Hermans K, Haspeslagh M, Chiers K, Pint E, Boyen F, Martens AM (2017) A quantitative swab is a good non-invasive alternative to a quantitative biopsy for quantifying bacterial load in wounds healing by second intention in horses. Vet J 225:63–68. https://doi.org/10.1016/j.tvjl.2017.06.004

Watters C, Everett JA, Haley C, Clinton A, Rumbaugh KP (2014) Insulin treatment modulates the host immune system to enhance Pseudomonas aeruginosa wound biofilms. Infect Immun 82(1):92–100. https://doi.org/10.1128/IAI.00651-13

Westgate SJ, Percival SL, Knottenbelt DC, Clegg PD, Cochrane CA (2010) Chronic equine wounds: what is the role of infection and biofilms. Wounds 22(6):138–145

Westgate SJ, Percival SL, Knottenbelt DC, Clegg PD, Cochrane CA (2011) Microbiology of equine wounds and evidence of bacterial biofilms. Vet Microbiol 150(1–2, 152):–159. https://doi.org/10.1016/j.vetmic.2011.01.003

Wilmink JM, Stolk PWT, Van Weeren PR, Barneveld A (1999a) Differences in second-intention wound healing between horses and ponies: macroscopic aspects. Equine Vet J 31(1):53–60. https://doi.org/10.1111/j.2042-3306.1999.tb03791.x

Wilmink JM, Van Weeren PR, Stolk PWT, Van Mil FN, Barneveld A (1999b) Differences in second-intention wound healing between horses and ponies: histological aspects. Equine Vet J 31(1):61–67. https://doi.org/10.1111/j.2042-3306.1999.tb03792.x

Wilmink JM, van Herten J, van Weeren PR, Barneveld A (2002) Retrospective study of primary intention healing and sequestrum formation in horses compared to ponies under clinical circumstances. Equine Vet J 34(3):270–273. https://doi.org/10.2746/042516402776186047

Wolcott RD, Rumbaugh KP, James G, Schultz G, Phillips P, Yang Q, Watters C, Stewart PS, Dowd SE (2010) Biofilm maturity studies indicate sharp debridement opens a time-dependent therapeutic window. J Wound Care 19(8):320–328

Wolcott RD, Hanson JD, Rees EJ, Koenig LD, Phillips CD, Wolcott RA, Cox SB, White JS (2016) Analysis of the chronic wound microbiota of 2,963 patients by 16S rDNA pyrosequencing. Wound Repair Regen 24(1):163–174. https://doi.org/10.1111/wrr.12370

Woo KY, Sibbald RG (2009) A cross-sectional validation study of using NERDS and STONEES to assess bacterial burden. Ostomy Wound Manag 55(8):40–48

Zhao G, Hochwalt PC, Usui ML, Underwood RA, Singh PK, James GA, Stewart PS, Fleckman P, Olerud JE (2010) Delayed wound healing in diabetic (db/db) mice with Pseudomonas aeruginosa biofilm challenge: a model for the study of chronic wounds. Wound Repair Regen 18(5):467–477. https://doi.org/10.1111/j.1524-475X.2010.00608.x

Zhao G, Usui ML, Underwood RA, Singh PK, James GA, Stewart PS, Fleckman P, Olerud JE (2012) Time course study of delayed wound healing in a biofilm-challenged diabetic mouse model. Wound Repair Regen 20(3):342–352. https://doi.org/10.1111/j.1524-475X.2012.00793.x

The manufacturer's authorised representative in the EU is Springer Nature Customer Service Centre GmbH, Europaplatz 3, 69115 Heidelberg, Germany. If you have any concerns regarding our products, please contact ProductSafety@springernature.com

Printed and bound by CPI Group (UK) Ltd, Croydon, CR0 4YY

30/09/2026

02225242-0019